Handbook of Derivatives for Chromatography

Edited by

KARL BLAU and GRAHAM S. KING

Bernhard Baron Memorial Research Laboratories,
Queen Charlotte's Maternity Hospital, London, W6 0XG.

HEYDEN

London · Philadelphia, PA. · Rheine

Heyden & Son Ltd., Spectrum House, Hillview Gardens, London NW4 2JQ.
Heyden & Son Inc., 247 South 41st Street, Philadelphia, PA 19104, U.S.A.
Heyden & Son GmbH, Münsterstrasse 22, 4440 Rheine/Westf., Germany.

Reprinted with corrections, 1978

ISBN 0 85501 206 4

Printed in Great Britain by J. W. Arrowsmith Ltd, Bristol BS3 2NT

Contents

48.00

Foreword

Chromatography is a complicated business these days. When I first became the proud co-owner (with Evan Horning) of a gas chromatograph at the National Institutes of Health in the middle 1950s, however, there was not much a biochemist could do with the thing except fatty acid analyses. Historians of the field who inspect the literature from 1957 to 1960 will certainly be puzzled by the extraordinary amount of scientific manpower that went into that one subject of fatty acid analysis. Of course, it was not the perfection of the chromatographic separation alone that provided the driving force for this work. Much was done to delineate the theory of chromatography, many new liquid phases were invented, new detectors were described, support-coating procedures were developed, and several new firms were stocking their shelves with chromatographic supplies and announcing that they were ready to provide the chromatographer with everything he needed.

The explosive growth of gas–liquid chromatography since 1960 has depended in large measure on the development of lightly loaded columns and on the growing realization that 'almost anything' can be made volatile under the proper circumstances. One is impressed now if a compound (outside the macromolecular domain) is found that cannot be volatilized; I confess that I do not know of many such examples.

Unlike the efforts devoted to perfection of the chromatographic process with fatty acids, work in the amino acid field was involved with the problem of quantitative derivative formation with all of the amino acids, and with selection of derivative and partition phases that would maximize separation of the mixture in a reasonable period of time. Derivatives and the reactions by which they are formed thus became extremely important considerations in the chromatographic procedure. In this and many other examples, derivatization ranks equally with sample selection, preprocessing and chromatography as an important part of the overall analytical technique.

Derivatization in liquid chromatography has a different rationale and hence quite different rules regarding choice of reagents, etc. The derivative group(s) added to the molecule almost inevitably become the detectable residue in the derivative. Sensitivity is of primary concern, while the molecular weight of the product is of little or no consequence. Reactions are usually carried out *after* the chromatographic step, whereas derivatization precedes gas–liquid chromatography.

Here is a book about the chemistry of derivative reactions that will be consulted by experts and beginners alike. It contains detailed information about the chemical reactions of a wide variety of functional groups and compares strategies of derivatization of common classes of compounds such as amino acids, steroids and carbohydrates. The book should help overcome the initial problem in making choices among various derivatives and will certainly steer the reader to appropriate literature about yields, derivative purification, stability, etc. Processing of biological samples is an entirely separate subject and the editors have appropriately not considered this aspect of the overall chromatographic procedure. The chapter on ion-pair extraction and chromatography does address itself to this subject to some extent, but extraction and chromatography are treated together in a smooth and comprehensive review of a technique that should gain in popularity.

This is a thoroughly delightful book to me personally, as it contains so much of the information I and my students need to refer to when dealing with new compounds and different kinds of biological samples. The handbook will enjoy a wide and well-deserved popularity and should be a reference work in every laboratory where chromatography is a serious business.

East Lansing, Michigan
March, 1977

Charles C. Sweeley

List of Contributors

DUŠAN ANDERLE, Institute of Chemistry, Šlovak Academy of Sciences, 809 33 Bratislava, Czechoslovakia (p. 201)

KARL BLAU, Prenatal Biochemistry, Bernhard Baron Memorial Research Laboratories, Queen Charlotte's Maternity Hospital, London W6 0XG, United Kingdom (pp. 1, 104, 317)

K. O. BORG, Department of Analytical Pharmaceutical Chemistry, Biomedical Centre, University of Uppsala, Box 574, S-751 23 Uppsala, Sweden (p. 500)

HANS BRANDENBERGER, Gerichtlich-Medizinisches Institut der Universität Zürich, Chemistry Department, Zürichbergstrasse 8, Zürich, Switzerland (p. 234)

ROBERTA H. BRANDENBERGER, Gerichtlich-Medizinisches Institut der Universität Zürich, Chemistry Department, Zürichbergstrasse 8, Zürich, Switzerland (p. 234)

WILLIAM C. BUTTS, Clinical Chemistry Section Group Health Co-operative of Puget Sound, Group Health Hospital 200–15th Avenue East, Seattle, Washington 98112, U.S.A. (p. 411)

ANDRÉ DARBRE, Department of Biochemistry, University of London King's College, Strand, London WC2R 2LS, United Kingdom (pp. 39, 262)

LOTHAR DEMISCH, Max-Planck-Institut für Hirnforschung und Zentrum der Psychiatrie und Klinikum der Universität, Frankfurt am Main—Niederrad, West Germany (p. 346)

DAVID J. EDWARDS, University of Pittsburgh School of Medicine, Department of Psychiatry, Western Psychiatric Institute and Clinic, 3811 O'Hara Street, Pittsburgh, Pa. 15261, U.S.A. (p. 391)

BERT HALPERN, Department of Chemistry, University of Wollongong, P.O. Box 1144 Wollongong, N.S.W. 2500, Australia (p. 457)

MAY N. INSCOE, Organic Chemicals Synthesis Laboratory, Agricultural Environmental Quality Institute, U.S. Department of Agriculture, Research Center East, Beltsville, Maryland 20705, U.S.A. (p. 317)

GRAHAM S. KING, Mass Spectrometry Unit, Bernhard Baron Memorial Research Laboratories, Queen Charlotte's Maternity Hospital, London W6 0XG, United Kingdom (pp. i, 104, 317)

PAVOL KOVÁČ, Institute of Chemistry, Slovak Academy of Sciences, 809 33 Bratislava, Czechoslovakia (p.201)

R. MODIN, Department of Analytical Pharmaceutical Chemistry, Biomedical Centre, University of Uppsala, Box 574, S-751 23 Uppsala, Sweden (p. 500)

PAUL MUSHAK, Department of Pathology, School of Medicine, University of North Carolina at Chapel Hill, Chapel Hill, N.C. Z7514, U.S.A. (p. 433)

BENGT-ARNE PERSSON, Department of Analytical Pharmaceutical Chemistry, Biomedical Centre, University of Uppsala, Box 574, S-751 23 Uppsala, Sweden (p. 500)

COLIN F. POOLE, Department of Organic Chemistry, University of Ghent, Krijslaan 271 (S.4), B-9000 Ghent, Belgium (p. 152)

GÖRAN SCHILL, Department of Analytical Pharmaceutical Chemistry, Biomedical Centre, University of Uppsala, Box 574, S-751 23 Uppsala, Sweden (p. 500)

NIKOLAUS SEILER, Max-Planck-Institut für Hirnforschung und Zentrum der Psychiatrie und Klinikum der Universität, Frankfurt am Main—Niederrad, West Germany (p. 346)

CHARLES C. SWEELEY, Department of Biochemistry, Michigan State University, East Lansing, Michigan 48824, U.S.A. (p. xi)

List of Abbreviations

The following is a list of abbreviations in the *Handbook*. Other less common abbreviations are explained by the authors at their point of use. A number of generally accepted abbreviations such as those used for units and dimensions are not defined here

Bns	5-Di-*n*-butylamino-naphthalene-1-sulfonyl
BOC	Butyloxycarbonyl
b.p.	Boiling point
BSA	*N,O*-Bistrimethylsilylacetamide
BSTFA	*N,O*-Bistrimethylsilyltrifluoracetamide
Dis-Cl	2-*p*-Chlorosulfophenyl-3-phenylindene
DMAA	*N,N*-Dimethylacetamide
DMF	*N,N*-Dimethylformamide
DMSO	Dimethylsulfoxide
DNP	Dinitrophenyl
2,4-DNP	2,4-Dinitrophenylhydrazine
Dns	'Dansyl' i.e. 5-Dimethyl-aminonaphthalene-1-sulfonyl
EC	Electron capture
ECD	Electron capture detector or detection
ECD–GC	Electron capture gas chromatography
EDTA	Ethylenediaminetetraacetic acid
FID	Flame ionization detector or detection
GC	Gas chromatography
GC–MS	Combined gas chromatography–mass spectrometry
GLC	Gas liquid chromatography
GLC–EC	Electron capture gas chromatography
GSC	Gas solid chromatography
HFB	Heptafluorobutyryl
HFBA	Heptafluorobutyric anhydride
HMDS	Hexamethyldisilazane
HPLC	High-pressure liquid chromatography
i.r.	infrared
MBTFA	Methyl-bis-trifluoracetamide
Mns	6-Methylanilinonaphthalene-2-sulfonyl
MO–TMS	Combined methoxime-trimethylsilyl derivative
m.p.	Melting point
MS	Mass spectrometry
MTH	Methylthiohydantoin
n.m.r.	Nuclear magnetic resonance
Nbd-Cl	4-Chloro-7-nitrobenzo-(c)-1,2,5-oxadiazole
OD	Optical density
OPT	*o*-Phthalaldehyde
PFB	Pentafluorobenzoyl—Chapter 3 Pentafluorobenzyl—Chapter 2 Pentafluorobenzimidyl—Chapter 6
PFBCl	Pentafluorobenzoyl chloride
PFP	Pentafluoropropionyl
PFPA	Pentafluoropropionic anhydride
PTC	Phenylthiocarbamyl
PTFE	Polytetrafluorethylene (e.g. Teflon®)
PTH	Phenylthiohydantoin

RGC	Reaction gas chromatography
RLC	Reaction liquid chromatography
RTLC	Reaction thin layer chromatography
SIM	Selected ion monitoring (also known as mass fragmentography, multiple ion detection, specific ion monitoring, etc.)
TFA	Trifluoracetyl
TFAA	Trifluoracetic anhydride
TLC	Thin layer chromatography
TMCS	Trimethylchlorosilane
TMS	Trimethylsilyl
TMSIM	Trimethylsilylimidazole
TNP	Trinitrophenyl
u.v.	Ultraviolet
v/v	Volume for volume
w/v	Weight for volume

CHAPTER 1

Introduction to the Handbook

Graham S. King and Karl Blau

Bernhard Baron Memorial Research Laboratories,
Queen Charlotte's Maternity Hospital, London W6 OXG

1 INTRODUCTION

The enthusiastic adoption of chromatographic methods has had a great impact on chemical analysis, not only in chemistry and biochemistry but also in pharmacology, toxicology, clinical sciences, genetics, forensic science, environmental science and many other fields. Indeed chromatography has contributed greatly to recent advances in carbohydrate and lipid chemistry. The previously demanding discipline of amino acid, peptide and protein structural analysis has expanded and experienced unprecedented growth, transforming our understanding of large areas of biology. Unfortunately not all compounds are accessible to direct analysis by chromatography and much ingenuity has been expended into devising ways of making chemical derivatives that will render them easier to analyse. A whole new branch of chemistry has evolved around this problem. Much of the work is widely scattered, many derivatization schemes that were worked out for a limited objective are potentially much more widely applicable, and in spite of the extensive achievements that have already been described, a great deal of further development can still be anticipated.

For these reasons we feel that this is a good time to gather some of this experience in derivatization together into a more systematic arrangement. We hope that the underlying principles will, as a result, become more apparent, and that derivatization chemistry will become more readily appreciated, providing a solid foundation on which workers can base the development of new methods of analysis. There have already been a few useful reviews and monographs devoted to specific problems or techniques. The derivatization of amino acids for gas chromatography has received considerable attention[1–4] this is also true for steroids[5–7] and drugs and pesticides.[8] Review articles and a recent monograph have covered derivatization for liquid chromatography,[9,10] gas chromatography[11,12,21] and the analysis of pharmaceuticals by gas chromatography.[14]

The use of fluorescent derivatives[10,15] and of chiral resolving agents[16] has also received attention. The techniques of silylation have been thoroughly covered in Pierce's excellent *Silylation of Organic Compounds* which should be at the side of every practising gas chromatographer.[17] Several books have been published on the subject of functional group analysis by gas chromatography. The book by Leathard and Shurlock[18] deals briefly with some of the simpler methods of derivatization which were current in the late sixties and also describes many useful techniques of abstraction in GC and pyrolysis gas chromatography. Crippen[19] has published his original and valuable approach to functional group identification, which necessarily relies upon derivatization to a large extent. Recently Ma and Ladas[20] have covered the area of functional group analysis in a slightly different way, briefly reviewing some areas of derivative formation, providing selected practical details and giving a very good coverage to the subject of reaction gas chromatography and abstraction techniques. The comprehensive *Handbook of Chromatography*, edited by Zweig and Sherma,[21] contains a useful section on derivatization which is a first attempt, but necessarily a brief one, at a general coverage of this subject. Our plan is to cover derivatization chemistry along quite different lines, and to systematize the chemical reactions used on the basis of the chemical processes involved in derivative formation. We do not intend to duplicate or extend the publications on functional group analysis or reaction gas chromatography but we aim to provide a practical approach to derivative selection for the analyst. Our approach is two-fold: not only to present group reactions that have already been applied to specific problems and so reveal their underlying principles, common basis and applicability, but also to provide the information and ideas by which these principles may be extended to new applications and perhaps even to new derivatization methods.

Direct chromatographic analysis of certain classes of compounds is difficult. For example, the application of paper chromatography to lipophilic substances was hampered for years by their hydrophobic nature, and the expedient of 'reversed-phase' chromatography was used until the development of thin-layer chromatography on silica gel.[22] Difficulties were often experienced with the gas chromatography of very polar compounds such as free acids and amines, with thermolabile substances and with more complex polar compounds such as the amino sugars and carbohydrates. It was soon found that chemical modification of polar functional groups improved the compound's accessibility to gas chromatographic analysis. In amino acid analysis it was found that the coloured dinitrophenyl derivatives, and later the fluorescent 'dansyl' derivatives had distinct advantages for paper, thin-layer and liquid chromatography. During these early stages in the development of derivatization techniques, the chemical reactions used were mostly those currently available from preparative organic chemistry and the groups introduced were those commonly used in organic synthesis for 'blocking' or protecting groups. These were usually designed to be easily removed at the end of a synthetic sequence. It was soon obvious that for chromatographic analysis, the susceptibility of the protecting group to easy removal was of no importance and might even be a disadvantage. This in turn led to the expansion of a new area of derivatization chemistry: the preparation of chemical derivatives selected or specifically designed to improve chromatograph analysis. This might involve incorporating a chemical function which gives a high detector sensitivity, or a particularly volatile derivative, or good chromatographic separation or even a simple and convenient preparation method. One of the most important contributions was the development of silylation methods; Pierce's handbook on silylation techniques[17] was an important publication and helped to transform a relatively unfamiliar chemical reaction into one of the most widely used derivatization methods. Silylation effectively blossomed and scattered seeds in all directions; recent advances are covered in this volume and considerable progress has been made since the publication of Pierce's book. To some extent Pierce's book has prompted the presentation of our *Handbook* and we gladly acknowledge its influence.

There is now a very wide choice of reactions for the preparation of derivatives to assist in chromatographic analysis. The selection of suitable methods depends on many factors. We hope to convey an appreciation of these factors which, apart from those already mentioned, may also involve considerations such as chemical stability of the substrate and the analytical sensitivity required. It may not be very easy at first to choose the best derivative for a given application from the many methods presented here. It may not even be apparent that the compound might chromatograph quite well without any derivative formation. However, we hope to make it easier to reach a rational decision by presenting the information in a convenient and fairly unified form in a single volume. Much of the book is basic chemistry that has stood the test of time and describes methods that have become standard and so we do not anticipate that it will be out of date as soon as publications in some other areas of research.

The procedures and discussion given by the different authors are inevitably varied, both in presentation and in scope. We have encouraged the contributors to standardize only to the extent that each chapter includes enough details to make the book a truly practical handbook; we have not attempted to cover isolation methods (apart from the chapter on ion-pair extraction), because they are so diverse that a whole separate volume would hardly cover the field. It must be stressed that the isolation of a compound from the sample matrix in good yield and ready for derivatization and chromatography is of paramount importance, and it is this that is very often the most difficult part of an assay. Nevertheless, both isolation and analysis are essential links of the chain. Solutions to the problems of extraction, derivatization and separation are more easily tackled once the analytical procedure has been worked out with standard samples.

We have attempted to obtain some degree of uniformity between chapters and also a degree of balance, especially between different types of chromatography. The interests of the editors and of the contributors have obviously influenced the scope and selection of examples. We hope that readers, far from objecting to this, will find that it adds to the *Handbook's* variety and makes it more readable. We have been fortunate in obtaining contributions from world experts and naturally we have given them the widest scope as to how they cover their various subjects. We have certainly attempted to ensure that no startling omissions have occurred; however, this is a fairly new subject and, particularly for techniques such as high-pressure liquid chromatography, the coverage may be a little fragmented. We hope that these shortcomings may be remedied in time. On the other hand, derivatization for use with gas chromatography and biochemical analysis may seem to be especially prominent. There are two reasons for this: it reflects the research interests of the editors and their choice of authors, and derivatization techniques seem to have made a considerable impact in these areas—GLC and GC–MS are very powerful and sensitive instrumental methods. Derivatization for liquid chromatography is the subject of a recent monograph.[10]

The chapters vary in the extent to which they present a theoretical background to the subjects covered. Although this is designed to be a very practical book, we feel that a sound grasp of basic principles in some areas will help readers to use the available derivatization techniques more effectively.

2 THE ORGANIZATION OF THE BOOK

The approach we have adopted is an empirical one, because there is no *a priori* 'best solution' to any analytical problem: a number of solutions can be tried out and one will emerge as the most acceptable for a number of reasons. Different users will very likely prefer different methods, depending upon personal choice and experience, the instrumentation available, urgency, cost, reagent availability and so on. We have therefore accepted that there are innumerable methods in the literature, and have tried simply to make a selection of them accessible and to provide informed comment. In short we suggest that this book should be used as a source of ideas and of practical tested methods to get started with a problem, but *not necessarily* as an exhaustive and academic review, although we consider that the authors of each chapter have achieved very comprehensive coverage of the literature in the subjects with which they have dealt. In practice the *Handbook* only requires the user to know the chemical groupings present in the substances to be analysed and to have some knowledge of the general types of analysis which may be available. The introductory Tables will then show which of the processes covered by the book will be of interest and indicate the appropriate chapter or chapters to consult. A rapid scan of the Chapter or Chapter Tables should reveal whether the compound of interest has been derivatized or whether an analogous compound has been covered. Because of the nature of this *Handbook* it is quite likely that the specific compound is not mentioned but that a similar compound has been used to illustrate a particular analytical point. As an example, the user should not expect to find a precise retention index for the methyl ester of oleic acid in the chapter on esterification: for this a book of chromatography data such as that by Zweig and Sherma[21] should be consulted. That chapter does however give methods for ester formation of long chain fatty acids and comments on their individual advantages and disadvantages. Some chapters which deal with more specific areas such as those on optical resolution or inorganic anions contain a lot of detailed chromatographic information which we believe is essential for good coverage of these subjects. We do not guarantee that the *Handbook* will invariably lead to success: obviously we cannot foresee every eventuality. However, even when the answer to a problem may not be spelled out, we hope that on looking through the book, such a variety of analytical techniques will be found that enough ideas will emerge to lead towards a solution.

The last two chapters in the book are slightly outside the mainstream of the approach we have described, and provide descriptions and methods of two techniques which we feel will prove invaluable. Both chromatographic optical resolution, and ion-pair extraction have been applied with great effect, and we hope that these chapters will stimulate the reader into thinking about their further application. To a large extent these comments also apply to the chapters on inorganic cation and anion gas chromatography, but the approach underlying these techniques is basically the same as that in the rest of the book. The chapter on ion-pair extraction may describe somewhat unfamiliar ground, but will repay the effort spent in understanding the principles and their implications for analytical chemistry and biochemistry. The ideas in it are very simple and the potential applications in both isolation and analysis are many. The derivative formation is reversible and this gives the analyst flexibility and control over the whole process.

2.1 Using the Handbook

The format of this book is based upon a limited number of chemical reactions and processes which are described in the Contents List at the beginning of the book. The reference tables in this chapter are designed for the reader to determine the types of basic reaction which could be applied to the compound of interest. For example, by reference to Table 1 (p. 6) it may be seen that a compound with a carboxylic acid grouping could be derivatized by either alkyl or aryl esterification (Chapter 2), silyl ester formation (Chapter 4), reduction (Chapter 8) and then treatment as an alcohol and so on. If the acid is optically active, then Chapter 13 could well be of interest, its sub-section on acids describes ways of tackling resolution of the enantiomers. Similarly, if isolation of the compound proved difficult, or if liquid chromatographic

TABLE 1

Synoptic chart on derivative formation with oxygenated groupings

Functional group	Procedure	Examples of products	Chapter	Comments
—OH (1°, 2° and 3° Alcohols; phenols; carbohydrates)	Silylation	$—O—Si(CH_3)_3$	4	
	Acylation	$—O—CO—CH_3$; $—O—CO—CF_3$	3	
	Benzoylation	$—O—CO—C_6H_5$; $—O—CO—C_6F_5$	3	
	Alkylation	$—O—CH_3$; $—O—CH_2—C_6F_5$	5	
	Oxidation	—CHO; >C=O; —COOH	8	Oxidation products to be derivatized (see below)
	Dansylation	Ar—O—Dns	9	Fluorescent derivative (phenols)
	Reaction with Dis-Cl	—O—Dis	9	Fluorescent derivatives of phenols and alcohols. Consult Chapter 9 for details
	Reaction with FDNB	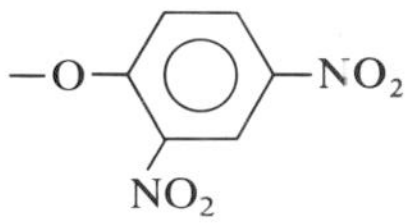	10	For GLC of phenols with ECD
	Reaction with NBD-Cl	7-nitrobenzofurazan	9	Fluorescent derivative (phenols)
	Ion-pair formation	$Ar—O^-M^+$	14	For phenols; 'M' can be a variety of counter-ions
>C=O (Aldehydes and ketones)	Oxime formation	>C=N—OH; $>C=N—O—CH_3$	6	May form *syn* and *anti* isomers
	Oxime formation and silylation	$>C=N—O—Si(CH_3)_3$	6	May form *syn* and *anti* isomers

TABLE 1—continued

Functional group	Procedure	Examples of products	Chapter	Comments
>C=O (Aldehydes and ketones) —*continued*	Ketal/acetal formation	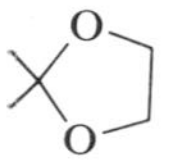	7	
	Hydrazone formation	>C=N—NH—C_6H_5	6, 9, 10	Fluorescent and electron-capturing derivatives available
	Schiff's base formation	>C=N—R	6	
	Silylation	=C(—O—$Si(CH_3)_3$)—CO—O—$Si(CH_3)_3$	4	Only when enol formation is favoured, e.g. pyruvate
	Oxidation	—COOH	8	For aldehydes and methyl ketones (iodoform reaction); derivatized as carboxylic acids
	Reduction	—OH	8	Derivatized as alcohols
	Aldononitrile formation	—CH=N—OH → —CN	8	Mainly for GLC of aldoses
—COOH (Carboxylic acids)	Esterification (alkyl)	—CO_2—CH_3; —CO_2—CH_2CF_3	2	
	Esterification (aryl)	—CO_2—$CH_2C_6H_5$; —CO_2—$CH_2C_6F_5$	2	
	Silylation	—CO_2—$Si(CH_3)_3$	4	
	Reduction	—CH_2—OH	8	Derivatized as alcohols

TABLE 1—continued

Functional group	Procedure	Examples of products	Chapter	Comments
—COOH (Carboxylic acids) —*continued*	Decarboxylation	R—COOH → R—H	8	Rarely used; may occur also in pyrolysis, or be done in reaction gas chromatography
	Cyclization	Depends on parent compound	7	Miscellaneous examples with amino acids, hydroxy acids, dibasic acids, etc.
	Ion-pair formation	$R—COO^{-}M^{+}$	14	'M' may be a variety of counter-ions
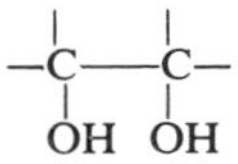 (Glycols)	As for —OH, but also: Cyclic boronate formation	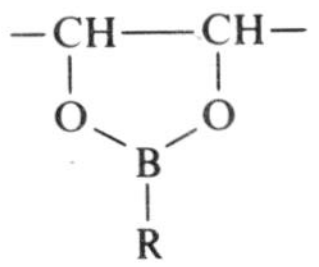	7	R = alkyl (most often butyl), or phenyl
	Acetal or ketal formation	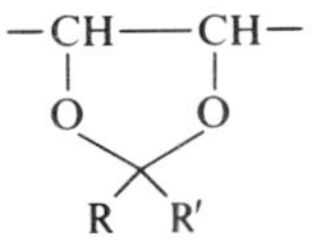	7	
	Oxidative cleavage	R—COOH + R′—COOH	8	Not applicable to catechol systems
—CH(OH)—COOH (α-Hydroxy acids)	As for the individual groupings, but also: Boronation	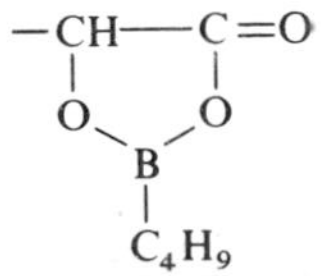	7	

TABLE 1—continued

Functional group	Procedure	Examples of products	Chapter	Comments
—CH—COOH (with —OH on CH) (α-Hydroxy acids) —*continued*	Reduction	—CH—CH_2OH (with —OH on CH)	8	Product derivatized as glycol
	Oxidative cleavage (periodate)	—COOH	8	Product derivatized as acid
	Oxidation (mild)	—CO—COOH	8	Product derivatized as pyruvate
	Simultaneous acylation and esterification	—CH—COO—C_2H_5 (with —O—CO—C_3F_7 on CH)	2, 3	A number of variations of this approach are available
—CO—COOH (α-Keto acids)	As for the individual groupings, but also: Cyclization with 1,2-diaminobenzene followed by silylation	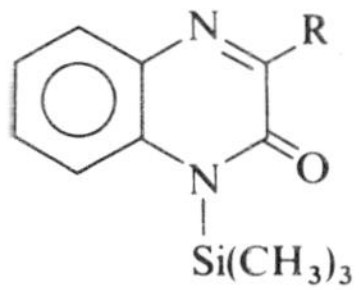	7	
	Reduction	—CH—CH_2—OH (with —OH on CH)	8	Product derivatized as glycol
	Oxidative cleavage	—COOH	8	Product derivatized as acid

TABLE 1—continued

Functional group	Procedure	Examples of products	Chapter	Comments
R—CO—OR′ (Esters)	Esters may be analysed chromatographically without derivatization, but where R′ is involatile:			
	Ester interchange (transesterification)	R—CO—OCH_3	2	
	Reduction	R—CH_2—OH + R′—OH	8	Then derivatize the alcohols
	Alkaline hydrolysis	R—COOH + R′—OH	8	Then derivatize the acid and alcohol
R—CO, R′—CO bridged by O (Acid anhydrides)	Direct chromatographic analysis			Direct GLC may be possible, especially with cyclic anhydrides e.g. succinic, phthalic, etc.
	Reduction	R—CH_2—OH + R′—CH_2—OH	8	Then derivatize as alcohols
	Hydrolysis	R—COOH + R′—COOH	8	Then derivatize as acids
	Esterification	R—CO—OR″ + R′—CO—OR″	2	Treat exactly the same as an acid, using e.g. EtOH/HCl
R—O—R′ (Ethers)	Ethers may be chromatographed without derivatization			
	Cleavage with hydriodic acid	R—I + R′—I		Classical Zeisel method for methyl ethers[20]

analysis was thought to be necessary, the section on ion-pair extraction would repay close attention.

The simplicity of this approach grew out of our own attempts to develop derivatization methods for a variety of compounds with the constraint of needing to use gas chromatography with specific detectors. Since we were obliged to use electron-capture detection because we needed the high sensitivity theoretically available, we found it necessary to devise ways of introducing electron-capturing functions into the molecule. A particularly difficult analysis may require the use of GC–MS for sensitivity in the face of extraction problems; by using selected-ion monitoring[23] (SIM, also known as multiple-ion detection, mass fragmentography, multiple-ion monitoring, etc.) a high degree of specificity is theoretically obtainable. However, if the molecule fragments easily under electron impact, by introducing derivatives which effect cyclization it may be possible to get a more abundant and diagnostic ion pattern which would improve the analysis. To some extent the approach is self-evident but a practical example together with some discussion of the possibilities may illustrate the general approach. We present the kind of scenario that could go through one's mind as one uses the *Handbook* to try to get a practical laboratory solution to a particular problem.

Let us assume that it is necessary to assay the phenolic amine dopamine (1). Firstly, because of the moderately amphoteric nature of the molecule and the sensitivity of the catechol function to oxidation in alkaline solutions, it would probably be worth considering whether an ion-pair extraction method (Chapter 14) is feasible for its isolation. In considering counter-ions and solvent systems it should be remembered that the molecule contains an amino group which would assist in the formation of an ion-pair with an acidic reagent and a phenolic-diol system which might form some type of bridged or metal chelation ion-pair.

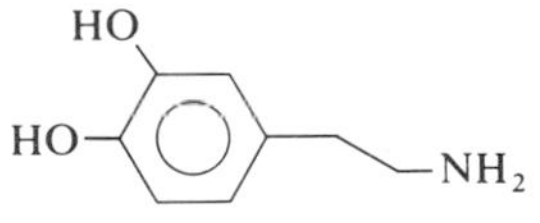

(1)

Assuming that a reasonable isolation scheme has been thought out by a modification of published procedures or even by evolving a completely new process, this will be developed and refined hand-in-hand with the actual analytical chromatography system, perhaps paper, thin-layer, liquid or gas chromatography, and the choice of derivative (if necessary) will rest largely upon the method used. Paper and thin-layer chromatography are the least sensitive, and detection of the underivatized material requires specific spray reactions or quenching of the fluorescence of activated plates. Such methods would not allow the detection of less than several micrograms. Liquid chromatography might give better sensitivity by the use of an ultra-violet spectrophotometric detector at the u.v. absorption maximum of the compound. Recent reports have indicated that electrochemical detection methods may allow detection of as little as several hundred picograms using high-pressure liquid chromatography on underivatized material.[24] Isolation methods and actual chromatography would demand conditions which avoid oxidative losses. Direct gas chromatography is not possible with such a polar compound.

Derivative formation is therefore almost essential to analyse this compound chromatographically. One choice to improve both separation and sensitivity using paper or thin-layer would be to form a stable fluorescent derivative, which may be visualized by ultra-violet irradiation and quantified by a scanning device. The chapter on fluorescent derivatives (Chapter 9) describes a number of suitable reagents, some general, such as dansyl chloride, and some specific methods of catecholamines. This approach would also improve detection with high-pressure liquid chromatography (HPLC) using either a fluorescence or absorption detector. Under favourable conditions, HPLC with fluorescence detection can extend limits of sensitivity well into the picogram region.[25] A considerable increase in sensitivity with u.v. absorption detectors can often be secured by using protecting groups with high extinction coefficients in the ultra-violet region.[26]

Gas chromatography is quite a different proposition: some form of volatile derivative formation is unavoidable. The separation potential of GLC is higher than any other chromatography system, especially with capillary columns or GC-MS. Consequently, we have much more freedom to tailor the derivative to suit our needs of volatility, ease of formation, or detector compatibility. Again, by using the tables in this chapter we can see that there is a variety of ways of derivatizing dopamine. The first step is to divide the molecule into its reactive functional constituents: these are primary amine, phenolic hydroxy or phenolic diol.

Acylation is an obvious choice for the primary amine and this is top of the list of available methods in Table 2. The most obvious form of acylation is acetylation and this may be carried out in a variety of ways (Chapter 3). Either the amino group alone or all three protic groups will be acetylated depending upon the reaction conditions. However the acetamido group is not always easy to chromatograph on the common silicone phases as it is still fairly polar; it tends to 'tail' and the volatility of the derivative is not always impressive. The acetyl function does not improve the electron-capture response of the molecule and for a low-concentration biogenic amine like dopamine we will need to harness the sensitivity of the electron-capture detector. Of course, if it is considered sufficient to detect sample sizes of around 10–50 ng using a flame-ionization detector, then the acetyl group might suffice. Combined GC–MS could well allow detection of a non-capturing derivative at very low sample sizes using SIM.

TABLE 2

Synoptic chart on the formation of derivatives from nitrogen-containing groupings

Functional group	Procedure	Examples of products	Chapter	Comments
$—NH_2$ (Primary amines; amino acids; amino sugars)	Acylation	$—NH—CO—CH_3$; $—NH—CO—CF_3$	3	
	Benzoylation	$—NH—CO—C_6H_5$; $—NH—CO—C_6F_5$	3	
	Silylation (mild)	$—NH—Si(CH_3)_3$	4	Mixtures may be obtained
	Silylation (vigorous)	$—N[Si(CH_3)_3]_2$	4	
	Treatment with CS_2	$—N=C=S$	6	Volatile, for use with GLC
	Thiourea formation	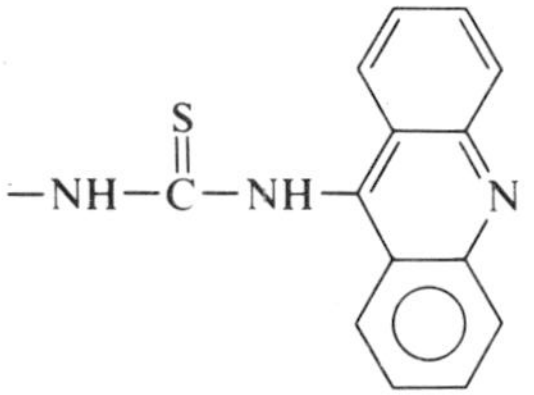	9	Fluorescent derivative
	Schiff's base formation	$—N=CH—C_6F_5$	9	
	2,4-Dinitrophenylation	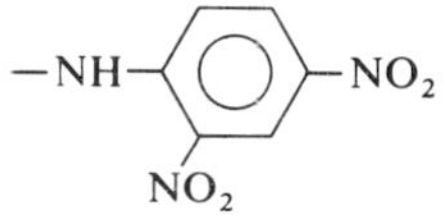	10	
	Sulfonamide formation	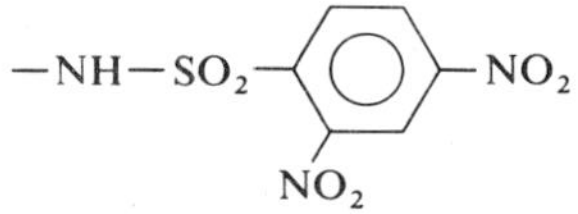	10	
		$—NH—Dns$	9	Fluorescent derivative (several variants available)

TABLE 2—continued

Functional group	Procedure	Examples of products	Chapter	Comments
$-NH_2$ (Primary amines; amino acids; amino sugars) —*continued*	Carbamate formation	$-NH-CO_2-CH_3$	3, 13	
	Carbylamine reaction	$-N\equiv C$	8	
	Treatment with nitrous acid	$-OH$		Derivatize as alcohol; nitriles may be formed in side-reaction; alcohol formation may not occur in some cases
	Treatment with fluorescamine	*N*-substituted 2-phenyl-pyrrolin-4-ones	9	Specific fluorigenic reagent for primary amino groups
	Treatment with pyridoxal	Pyridoxylidine derivative	9	Semi-specific fluorogenic reagent for primary amines
	Treatment with NBD—Cl	7-Nitrobenzo-furazan	9	Fluorescent product
	Alkylation	$-N(CH_3)_2$	5	
	Ion-pair formation	$R-NH_3^+X^-$	14	'X' may be a variety of counter-ions
$-NH-R$ (Secondary amines, imino acids, substituted amino sugars)	Acylation	$-N(R)-CO-CH_3$; $-N(R)-CO-CF_3$	3	As with primary amines
	Benzoylation	$-N(R)-CO-C_6H_5$; $-N(R)-CO-C_6F_5$	3	As with primary amines
	Silylation	$-N(R)-Si(CH_3)_3$	4	May need 'forcing' conditions

TABLE 2—continued

Functional group	Procedure	Examples of products	Chapter	Comments
—NH—R (Secondary amines, (imino acids, substituted amino sugars) —*continued*	2,4-Dinitro-phenylation	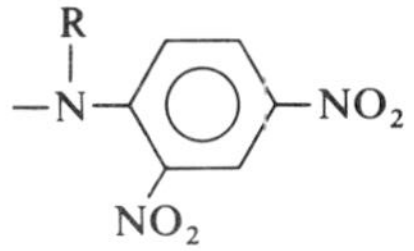	10	As with primary amines
	Sulfonamide formation		10, 9	As with primary amines (more probability of side-reactions)
	Treatment with NBD—Cl		9	As with primary amines
	Ion-pair formation	$RR'NH_2^+X^-$	14	'X' may be a variety of counter-ions
R—N(R″)—R′ (Tertiary amines)	Hofmann degradation	→ alkenes	8	
	Carbamate formation	$RR'N—CO—CH_2—C_6F_5$	3, 8	
	Ion-pair formation	$R_3NH^+X^-$	14	'X' may be a variety of counter-ions
Quaternary ammonium salts	Thermal decomposition	Tertiary amines	8	
	Dealkylation with sodium benzene thiolate	Tertiary amines		

TABLE 2—continued

Functional group	Procedure	Examples of products	Chapter	Comments
Quaternary ammonium salts—*continued*	Ion-pair formation	$R_4N^+X^-$	14	'X' may be a variety of counter-ions
$-CO-NH_2$ (Amides)	Silylation (vigorous)	$-C(-O-Si(CH_3)_3)=N-Si(CH_3)_3$	4	Silylamides are themselves powerful silylating reagents
	Acylation (vigorous)	$-CO-NH-CO-C_6F_5$	3	
	Alkaline hydrolysis	$-COOH$	8	Treat as carboxylic acid
	Reduction	$-CH_2-NH_2$	8	Treat as primary amine
	Dehydration	$-C\equiv N$	8	
	Alkylation	$-CO-N(CH_3)_2$	5	e.g. CH_3I/NaH/DMSO
$-CO-NH-R$ (Alkylamides)	Acylation (vigorous)	$-CO-N(R)-CO-C_6F_5$	3	
	Silylation	$-CO-N(Si(CH_3)_3)-R$	4	Silylamides are themselves powerful silylation reagents
	Alkylation	$-CO-N(CH_3)-R$	5	e.g. CH_3I/NaH/DMSO
	Reduction	$-CH_2-NH-R$	8	Treat as secondary amine
	Hydrolysis	$-COOH + R-NH_2$	8	Treat as individual functional groups

TABLE 2—continued

Functional group	Procedure	Examples of products	Chapter	Comments
—CO—N(R)(R′) (Dialkylamides)	Reduction	—CH_2—N(R)(R′)	8	May offer no great improvement in volatility
	Hydrolysis	—COOH + (R)(R′)NH	8	Derivatize individual functional groupings
R—NH—CO—NH—R′ (Substituted ureas; carbamides)	Acylation (vigorous)	R—N($CO—CF_3$)—CO—N(R′)—$CO—CF_3$	3	
	Alkaline hydrolysis	R—NH_2; R′—NH_2	8	Derivatize as amines
R—NH—C(=NH)—NH—R′ (Substituted) guanidines)	Acylation (vigorous)	R—N($CO—CF_3$)—C(=N—$CO—CF_3$)—NR′—$CO—CF_3$	3	Acylation is difficult, and stability of the products is poor
	Hydrolysis	Substituted ureas	8	
	Cyclic derivatives		7	
—CH(NH_2)—COOH (Amino acids)	See Table 4, p. 24 and 25		2–5, 7 and 10	

TABLE 2—continued

Functional group	Procedure	Examples of products	Chapter	Comments
$-CH(OH)-CH(NH_2)-$ (Amino alcohols)	As for individual groups, but see also:			
	Cyclic boronate formation	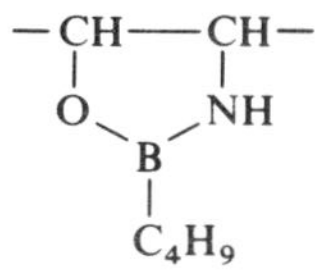	7	
	Simultaneous acylation and silylation	$-CH(O-Si(CH_3)_3)-CH(-)-NH-CO-CF_3$	3	Using *N*-methyl-*N*-trimethylsilyl-trifluoroacetamide for example
$-NO_2$ (Nitro compounds)	Chromatograph without derivatization		10	Electron-capturing
	Reduction	$-NH_2$	8	Derivative as primary amine; use catalytic, or $LiAlH_4$, or Sn/HCl reduction
$>C=N-OH$ (Oximes)	Silylation	$>C=N-O-Si(CH_3)_3$	4	
	Acylation	$>C=N-O-CO-CH_3$	3	
		$>C=N-O-CO-CF_3$		
	Dehydration	$-C\equiv N$	8	Aldoximes → aldonitriles

TABLE 2—continued

Functional group	Procedure	Examples of products	Chapter	Comments
>N—NO (Nitrosoamines)	Chromatograph without derivatization			Potent mutagens—CAUTION!
—C≡N (Nitriles)	May be chromatographed without derivatization, but also:			
	Hydrolysis	—COOH	8	Derivatize as carboxylic acid
	Reduction	$—CH_2—NH_2$	8	Derivatize as primary amine
—N≡C (Isocyanides)	May be chromatographed directly		8	
—CNO (Cyanates)	May be chromatographed directly			
—NCO (Isocyanates)	May be chromatographed directly		8	
—CNS (Thiocyanates)	May be chromatographed directly			
—NCS (Isothiocyantes)	May be chromatographed directly		9	Used in derivatization of primary amines

TABLE 3

Synoptic chart on the formation of derivatives with miscellaneous groupings

Functional group	Procedure	Examples of products	Chapter	Comments
—SH (Thiols, sulfydryls; sensitive to oxidation)	Acylation	$—S—CO—CH_3$; $—S—CO—CF_3$	3	
	Silylation	$—S—Si(CH_3)_3$	4	
	Alkylation	$—S—CH_3$	5	May be difficult to stop at this stage; product still somewhat sensitive to oxidation
	Benzoylation	$—S—CO—C_6H_5$; $—S—CO—C_6F_5$	3	
	Treatment of with NBD-Cl	7-Nitrobenzofurazan	9	Fluorescent derivative
	Various other fluorescent derivative formation schemes		9	
	Reaction with 2,4-DNFB	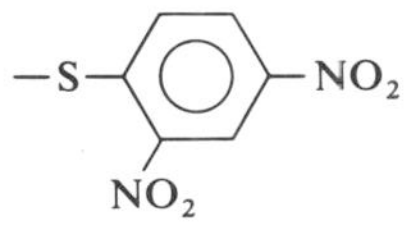	10	
	Treatment with iodoacetate	$—S—CH_2—COOH$		Derivatize as carboxylic acids
	Treatment with acrylonitrile and hydrolysis	$—S—CH_2—CH_2—COOH$		Derivatize as carboxylic acids
	Reduction with Raney nickel	Hydrocarbon	8	Chromatograph the hydrocarbon

TABLE 3—continued

Functional group	Procedure	Examples of products	Chapter	Comments
—SH (Thiols, sulfydryls; sensitive to oxidation)—*continued*				
	Oxidation	—S—S—	8	Products tend not to be very volatile
		$—SO_3H$	8	See following section
$—SO_3H$ (Sulfonic acids)	Ion-pair formation	$R—SO_3^-M^+$	14	'M' may be a variety of counter-ions
	Halogenation	$R—SO_2—Cl$	8	May be chromatographed direct, or converted to the ester; see following section
	Esterification	$R—SO_2—OR$	2	Best analysed by TLC or HPLC. These esters are not very volatile
—S—R (Thioethers)	May be chromatographed directly, but also:			
	Oxidation	See following two sections		
$>S \rightarrow O$ (Sulfones)	May be chromatographed directly			Less volatile than the corresponding thioether
$>S(\rightarrow O)_2$ (Sulfoxides)	May be chromatographed directly			Less volatile than the corresponding sulfone

TABLE 3—continued

Functional group	Procedure	Examples of products	Chapter	Comments
>C=C< (Alkenes, also applicable to alkynes)	Hydrogenation	>CH—CH<	8	
	Epoxide formation	—C(—O—)C— (epoxide: —C—C— bridged by O)	8	Double-bond location technique
	Epoxide formation and hydrolysis	—C(OH)—C(OH)—	8	Treat as diol; double-bond location technique
	Oxidation	—C(OH)—C(OH)—	8	Treat as diol; double-bond location technique
	Oxidative cleavage	>C=O O=C<	8	Treat as aldehydes, ketones or carboxylic acids depending on the conditions; double-bond location technique
	Ozonolysis	>C=O O=C<	8	As above

TABLE 3—continued

Functional group	Procedure	Examples of products	Chapter	Comments
—CH——CH— CH_2 (Cyclopropyl derivatives)	Hydrogenolysis	$—CH(CH_3)—CH_2—$ $—CH_2—CH(CH_3)—$ $—CH_2—CH_2—$	8	Yields three products
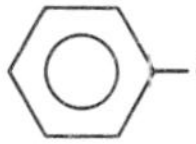$—CH_2—X$ (Benzyl derivatives)	Hydrogenolysis	Toluene; X—H	8	X may be *O*-alkyl, NH-alkyl, *O*-aryl, NH-aryl etc.
R—S—S—R′ (Dilsulfide)	May be chromatographed directly, but also			Use TLC: spray with iodoplatinate for detection. Disulfides are not very volatile
	Reduction	R—SH + R′—SH	8	Derivatize as thiols
R—X (Halogenated compounds)	May be chromatographed directly			Electron-capturing, see Chapter 3. Some halides, especially bromo- and iodo-compounds may be unstable to heating.

TABLE 4

Derivatives of amino acids that have been used for chromatographic separation[c]

Procedure	Product	Chapter	Comments
A. Treatment with 2,4-dinitrofluorobenzene	*N*-(2,4-dinitrophenyl) amino acids	10	For isolation by solvent extraction and analysis by liquid chromatography methyl esters may be analysed by GLC (see E)
B. Treatment with phenyl isothiocyanate	Amino acid phenylthiohydantoins	7	For peptide sequencing and analysis by TLC. These and the methylthiohydantoins may also be analysed by GLC
C. Treatment with dimethylaminonaphthalene-sulfonyl chloride (Dns-Cl)	*N*-Dns-amino acids	9	For high-sensitivity analysis by TLC using fluorescence methods
D. Treatment with fluorescamine	*N*-substituted 2-phenyl-pyrrolin-4-ones	9	Analysis by TLC and HPLC and fluorimetry
E. Volatile derivatives for analysis by GLC. Many methods have been devised, some for specific purposes, others for general application; the reader is referred to special reviews of the subject.[1,3] A selection of ten GLC methods is presented here; each of these has special advantages. Although new methods may be developed to supersede these, it is doubtful whether any of those already tried out and not listed here will have a renaissance.			
Esterification followed by acylation	1. *N*-TFA amino acid *n*-butyl esters	2, 3	The most widely used for GLC analysis of amino acids. Extensive MS data available
	2. *N*-TFA amino acid methyl esters	2, 3	One of the most volatile derivatives
	3. *N*-HFB amino acid *n*-propyl esters	2, 3	For GLC with ECD. Has been used with capillary columns[39–41]

TABLE 4—*continued*

Procedure	Derivative	Chapter	Comments
Esterification followed by acylation—*continued*			
	4. *N*-Acetyl amino acid *n*-propyl esters	2, 3	Highly stable derivatives used in automated derivatizers and with solvent-free injection [42]
Acylation in aqueous solution, solvent extraction and esterification	5. *N-iso*-Butyloxycarbonyl amino acid methyl esters	3, 2	Amino acids are isolated as well as derivatized [43]
Acylation followed by silylation	6. *N*-TFA amino acid TMS esters	3, 4	For mild derivatization, especially esterification [44]
Silylation	7. *N*-TMS amino acid TMS esters	4	Derivatization occurs in a single step
Treatment with 2,4-dinitrofluorobenzene in aqueous solution, followed by solvent extraction and esterification	8. *N*-DNP amino acid methyl esters	2, 10	Amino acids can be isolated as well as derivatized by this procedure, giving electron-capturing derivatives [45]
Alkylation, combined with esterification	9. *N*-Dimethylaminomethyl amino acid methyl esters	5	A one-stage reaction leading to derivatives with enhanced response to nitrogen-specific detectors [46,47]
Acylation and cyclization	10. Substituted oxazolin-5-ones or oxazolidin-5-ones	2, 5, 7	Derivatization to cyclic compounds (also applicable to α-hydroxy- and α-thiol acids) with electron-capturing properties [48–50]

[a] Amino acids are generally analysed without derivatization, using paper chromatography, TLC or liquid chromatography on ion-exchange resins. Chromatographic methods involving derivatization of amino acids prior to separation have been developed for special reasons, and are presented here.

The trifluoracetyl group instead of acetyl would lead to a marked improvement in volatility and chromatographic behaviour. Reaction with trifluoracetic anhydride (TFAA) would protect both the amino and hydroxy groups simultaneously. However, the sensitivity of the electron-capture detector to TFA derivatives is not necessarily high enough (see the discussion in Chapter 3) and for this reason it would be better to use the heptafluorobutyryl function, which might also produce a further drop in volatility. Trifluoracetyl derivatization could still be a better choice for GC–MS because the increase in molecular weight for the TFA group is 96 and for HFB, 196. High molecular weights can cause problems with SIM measurements.

Perfluoracylated catechols are sensitive to moisture, and it might be found that sample stability is a problem with our model compound and that a more stable alternative is required. Silyl derivatives tend to store well, because they are normally kept and injected in the silyl reagent itself, but silylation would not be a good choice because the N—Si bond is not very stable and most silylation techniques would give a mixture of one and two silyl groups on the nitrogen atom. Another problem with silylation is that we are again limited to flame ionization sensitivity unless using SIM. Injection of an HFB derivative in the anhydride reagent should be avoided: excess acid in the solution would considerably shorten the life of the GC column. This could be avoided by using the HFB-Imidazole reagent, but excess of the reagent would have to be removed when using electron-capture detection (see Chapter 3).

It might prove possible to get round some of these problems by using a two-stage derivative formation process. Mild silylation following reaction with pentafluorobenzoyl chloride would give a stable, strongly electron-capturing derivative (2), which could be stored and injected into the gas chromatograph in the silylation reagent. Alternatively the Schiff's base (3) could be prepared from pentafluorobenzaldehyde, or the two hydroxy groups could be bridged using a butyl boronate function (4).

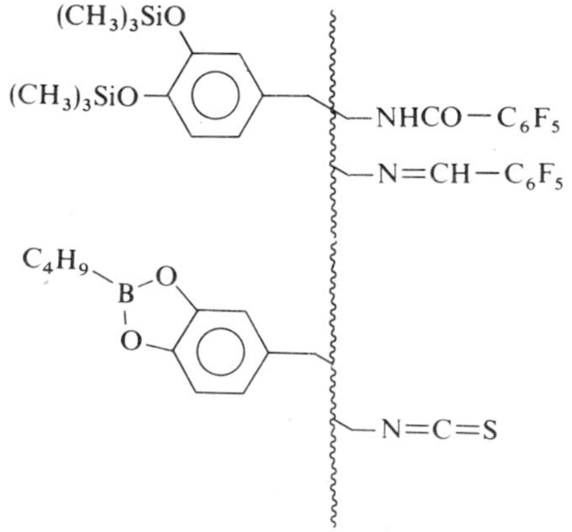

(2)

(3)

(4)

(5)

If it were decided to use GC–MS for the analysis, different considerations would apply. The most important requirement of a molecule for SIM is that it should give abundant and distinctive ions in its mass spectrum. There is, to some extent, more flexibility in the choice of a derivative because the compound's electron-capturing ability is not important. Very often it is wise to consider whether a particular derivative will have an abundant molecular ion and distinctive fragmentation pattern; in general, cyclic and aromatic compounds will stabilize a positive charge and so it is worth considering whether some form of cyclic derivative may be formed. The substituents in our example will affect the way in which the molecule breaks down under electron impact; for example considering the compound (2), most of the ionic charge would be expected to be stabilized on fragments including the pentafluorobenzoyl function. This benzamido system would also favour cleavage of the aliphatic chain. It might be worth considering whether to use an isothiocyanate derivative (5), because this would reduce the tendency to cleave in the side chain and give a more abundant molecular ion.

This does not exhaust the possibilities for dopamine derivatives but it should show the way in which the *Handbook* is designed to help with this type of analytical problem, when starting from the very beginning. A closer look will also reveal that ready-made methods covering dopamine are already presented in other chapters of the book, so the first thing to do might well be to choose one of those and to see how it works out in practice.

3 PRACTICAL CONSIDERATIONS

A great deal of the more detailed practical information needed to perform particular derivatization reactions is contained in the chapters describing the main topics. Some general comments on techniques and precautions seem advisable at this point for those workers new to the field. Most of these general comments centre around a few main principles.

1. In the main, derivatization is carried out on small quantities of material for reasons of economy, availability and convenience—the equipment used must be adapted or designed with this microchemistry in mind.
2. Because very often traces of material are analysed, great care must be taken to minimize contamination and losses of the sample.
3. Many of the reagents used are reactive, often unstable and often highly noxious; many of the derivatives formed have poor stability and precautions must be taken to preserve them. Working on a small scale helps to minimize the hazards, and makes it easier to take the necessary precautions.

3.1 The reagent

Having decided upon a possible method for derivatizing a compound the analyst's next problem is to find a source of the appropriate reagent. There are now several companies which specialize in the supply of suitable reagents for derivatization and these have agency arrangements throughout the world. However, we would like to offer a few words of caution. It is usually unwise to assume that a reagent is pure or that it is always exactly what it appears to be; healthy scepticism is always advisable—even the most rigorous manufacturer's quality control can go wrong sometimes. As an example, a few years ago a batch of pentafluoropropionic acid found its way on to the chemical market which was heavily contaminated with trifluoroacetic acid. As a result, a whole variety of PFP reagents became available with this contamination and some of it is still finding its way to users at the time of writing.

The availability of a reagent as a 'foolproof' kit for a specific purpose does not necessarily mean that it is perfect and that it is the best way to obtain the reagent; it may be expensive and a quick check of a local chemical manufacturer's catalogue may reveal the reagent at one tenth of the price. If the reagent will be in regular use, it might be better to buy from the cheaper source, redistil the chemical, seal it into small 'one-reaction' size ampoules or vials (possibly under dry nitrogen) and store these in a freezer. Presentation and packaging of the reagent can be very important. If for example, phthalate plasticizer impurities are a potential problem for the analysis and the bottle has a plastic cap with no lining, then this will not be satisfactory and may lead to problems. Similarly if, for example, a sample of BF_3-etherate has a white crust around the outside of the cap, or a sample of butyl boronic acid appears 'sticky' or heptaflurobutyryl-imidazole arrives with debris floating in the liquid, these may all cause problems. We have encouraged our authors to consider the purification of reagents, but if a particular reagent is not mentioned, then it would be advisable to take it through the last purification stage of its synthesis again. It is always *essential* to take a control sample ('reagent blank') through the derivatization process to identify any contaminants and to analyse this control *first*, to avoid cross-contamination with actual samples. Finally, many reagents are potent chemicals with harmful toxic properties; in the main they should be handled with gloves, safety spectacles should be worn and a fume cupboard or hood used. All of these comments also apply if the analyst is driven to synthesizing a particular reagent.

3.2 Storage of reagents, purified solvents and derivatives

It is always worth considering the division of a newly arrived reagent into smaller containers as soon as possible. Continual opening of the stock bottle will allow water vapour, oxygen and carbon dioxide (to name just a few constituents of 'breathable' air!) into the reagent. In any case, the new bottle should be dated, cooled and kept in the dark. When dividing reagents into smaller containers, it is necessary to consider the material which will be in direct contact with the reagent. All such materials should be clean and anything in its fabric should not leach out into the stored chemical. We have occasionally encountered difficulties with polytetrafluorethylene (PTFE) in contact with reactive reagents for electron-capture—capturing materials seem to leach out into the reagent liquid. Obviously glass ampoules are a very good idea but they must be sealed in a flame while cooling the contents in a dry-ice bath. This can be a tricky procedure and there must be no hazard of fire, explosion or degradation in the operation. With the extremely dangerous reagent diazomethane, it should be prepared as required and if it must be stored, this should be done in small quantities. Diazomethane solutions should be kept cold and allowance made for the escape or build-up of nitrogen gas; ground-glass joints and stoppers are not advisable. Many of the above comments also apply to purified solvents. These will probably be stored over dessicants or stabilizers as described in a solvent purification handbook.

Very often, derivatives are not stable. If they must be kept then a few precautions should help. A derivative's stability will vary greatly with its chemical nature: some will best be stored in a reagent; some might degrade and are best stored dry after removal of excess reagent and by-product. In general, it is better to store them in the absence of a solvent. A useful technique for storing small sealed tubes containing derivatives is to put them into a small, sealed jar (possibly under nitrogen) with some fresh self-indicating silica gel and then freeze them. On removal from the freezer (−40 °C) they should be allowed to warm to room temperature before opening; this will avoid condensation of atmospheric water. Sensitive reagents may also be stored in this way and should certainly be stored in a dark dessicator. Stretchable plastic film is invaluable for wrapping round the outside of stoppered bottles to ensure a good seal.

3.3 Solvents and common co-reagents (see also p. 227)

The best thing we can advise on this subject is to use a good solvent handbook[27] and check the solvent's purity by the method of analysis. Such a control sample should always be analysed first. Remember that published purification methods are often designed to remove specific impurities and these may not necessarily be the cause of an analytical problem; the purification method should be tailored to remove the specific impurities causing the interference problem. It should also be remembered that solid drying agents are not necessarily totally clean and that ground-glass joints should not be greased (although they will not give such a tight seal unlubricated). Oxygen, water and light are often detrimental to pure solvents. Unfortunately, manufacturers are compelled to add preservatives to certain solvents, such as pyrogallol to diethyl ether and ethanol to chloroform; if in doubt it is worth checking with the manufacturer, and a telephone call may save a lot of problems. Some manufacturers now supply extremely high-purity solvents in all-glass systems; for a tricky problem this could be the answer. The source of a particular solvent could also be important; for example ethyl acetate may be made from synthetic or fermented ethanol, and the synthetic material has fewer interfering impurities for electron-capture gas chromatography.[38]

3.4 The reaction vessel

The fundamental requirement for a reaction vessel to be used in microderivatization is that it should be small. Miniature thick-walled reaction vials (Fig. 1) with screw-on tops have become widely accepted. These vials are usually available in 0.3, 1.0, 5.0 and 10.0 ml sizes and one of their advantages is that they are tapered inside to make even the smallest trace of liquid accessible with a microlitre syringe. When sealed with a PTFE cap, they may be heated to quite high temperatures. The minor disadvantages are their cost, which is quite high if a number are required, and the fact that the reaction mixture is in contact with the fluoropolymer coating of the cap. The vials are designed to be re-used. The authors have found that for trace analysis using fluorocarbon reagents to prepare electron-capturing derivatives, an all-glass system should be checked against the PTFE-capped vial to ensure that there is only a low level of contamination (see Chapter 3). The great value of small reaction vessels of this type is that *all* manipulations may be performed in the one container.

Cheaper alternatives to thick-walled reaction vials are small plastic-capped 2 and 4 ml sample bottles with PTFE liners inside the caps (Fig. 1). Sample bottles are available from most equipment suppliers and a small disc of PTFE-coated material may be cut from a sheet using a cork-borer, or even purchased in bulk from a suitable manufacturer. These containers are not as leak-tight as the reaction vessels and certainly should not be heated very much when sealed because of the risk of breakage and of leaks. They are very suitable for room temperature reactions and at this temperature if an all-glass system is required it may be sufficient simply to rest a small glass ball or microscope cover-slip on the top of the bottle. Small tapered test-tubes or centrifuge tubes with ground-glass stoppers are also ideal (Fig. 1).

A simple and useful idea has been published by Darbre[28] using a Quickfit® Rotaflo® stopcock to seal a reaction vessel (Fig. 6). The stopper may be removed to introduce reagents with a pipette and air is removed by evacuation through the side arm; if necessary air may be swept out previously using a stream of nitrogen. The evacuated vessel may be heated up to 150 °C in an oven. For prolonged heating at temperatures above 150 °C, there is probably no better method than to seal the sample and reagents in a thick glass tube and heat it in a small oven inside a safety container.

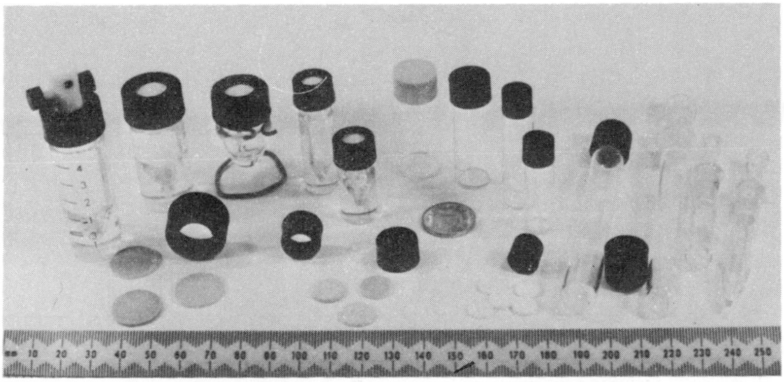

Fig. 1 Reaction vials, liners and caps (left); small plastic-capped tubes with liners (centre); small ground-glass stopper tubes (right).

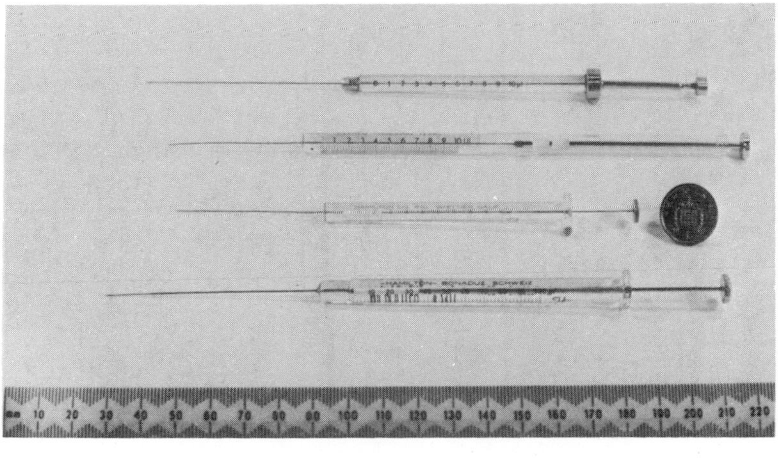

Fig. 2 Ten microlitre and hundred microlitre syringes.

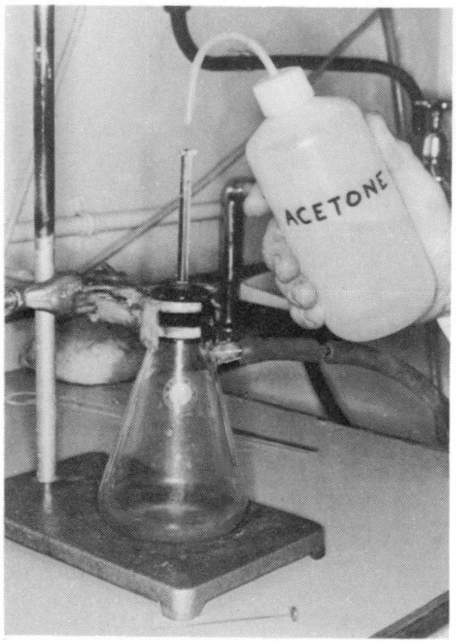

Fig. 3 Washing a syringe.

Fig. 4 A small thermostatically controlled heating block and reaction vials.

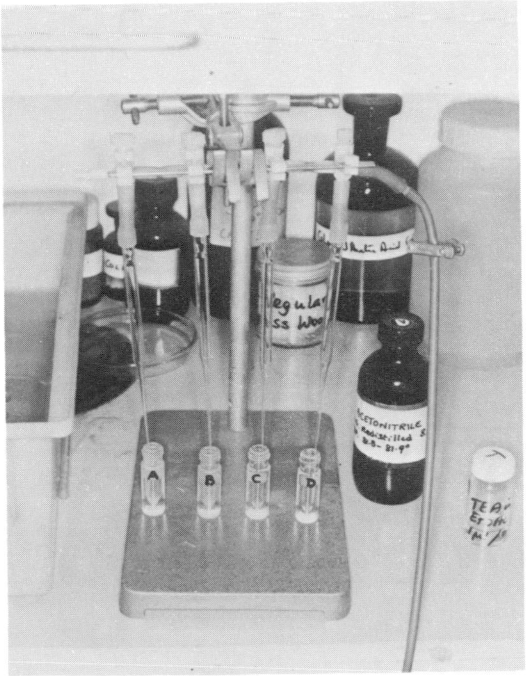

Fig. 5 Evaporation of excess of volatile reagent and solvent from four samples simultaneously, using glass pipettes and a stream of nitrogen under a fume-hood.

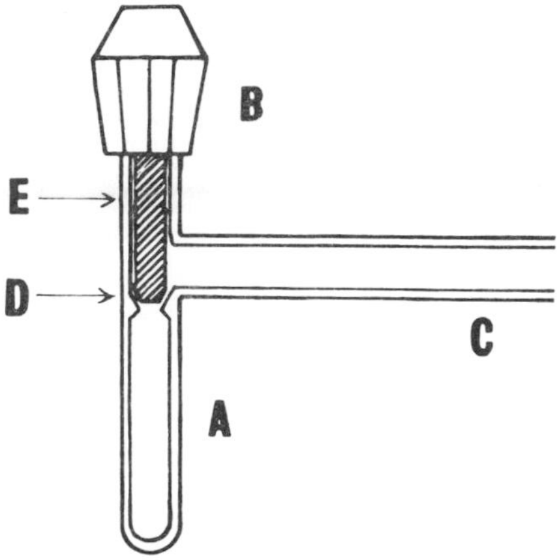

Fig. 6 Reaction vessel sealed with Quickfit® Rotaflo® stopcock. A, Lower limb sealed; B, Knurled plastic stopper; C, Open tube; D, Position of Teflon®–glass seal; E, Teflon® rod expandable by turning a screw inside nut B to give a gas-tight seal (A. Darbre, Ref. 28. Reproduced by kind permission of the author.)

3.5 Sample manipulation

Microlitre syringes are invaluable for manipulation of small volumes of non-viscous samples and reagents. It is useful to have at least one 10 μl and one 100 μl fixed-needle syringe (Fig. 2). It must be remembered that these syringes tend to retain traces of a liquid after a transfer operation. The best way to clean them is to insert the needle through a septum into a reservoir evacuated by a water pump, remove the plunger and wash the syringe through with a suitable volatile solvent from a wash bottle or pipette (Fig. 3). The plunger should be gently wiped with a tissue moistened with the solvent. The syringe should be left inserted until it is dried by the passage of air through the barrel. For solid materials the end of a small melting-point capillary is a useful micro-spatula. When performing manipulations on a sample it should be kept stoppered or sealed for as much of the time as possible. It should be remembered that no solvent is totally pure, and that it is sound practice in trace analysis to minimize the volumes of reagents and solvents used.

3.6 Heating the reaction vessel

There are a number of thermostatically controlled heating blocks commercially available for heating sample vials (Fig. 4) and these are ideal. A cheaper alternative is to use an ordinary laboratory oven, providing that the whole vial will withstand the temperature. A simple micro-reflux arrangement is described in Chapter 8. A steam-bath is a convenient method of heating a vial to precisely 100 °C, but precautions must be taken to exclude moisture.

3.7 Removing excess of reagents

Obviously this is largely dependent upon the precise method of preparation, but several general comments can be made:

1. It is important not to remove the more volatile components of the sample itself.
2. It is important to remove all traces of the reagents if they interfere with the chromatography system, e.g. perfluoracyl reagents and gas chromatography electron-capture detection.
3. If the reaction mixture is acceptable as a chromatography solvent then it may be used as such, e.g. silylation in gas chromatography.

One of the simplest ways of removing volatile solvents from a reaction mixture is in a stream of dry nitrogen from a small pipette (Fig. 5). It is often convenient when handling many samples to use a small manifold with a number of pipettes or small needles connected to a nitrogen supply (Figs 7 and 8). This method is quite satisfactory when a compound of low volatility is dissolved in solvents of high volatility. If the solvent is difficult to remove in this way it might be necessary to use a pumping system and cold trap; with vacuum methods

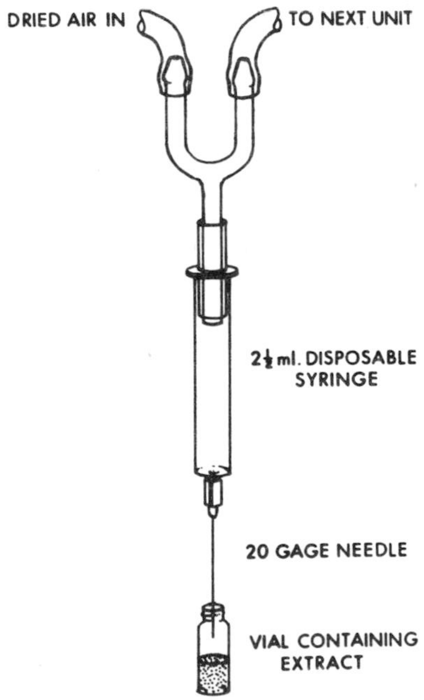

Fig. 7 One unit of a drying assembly. Units may be connected in series for simultaneous evaporation of several extracts (K. Blau, Ref. 35).

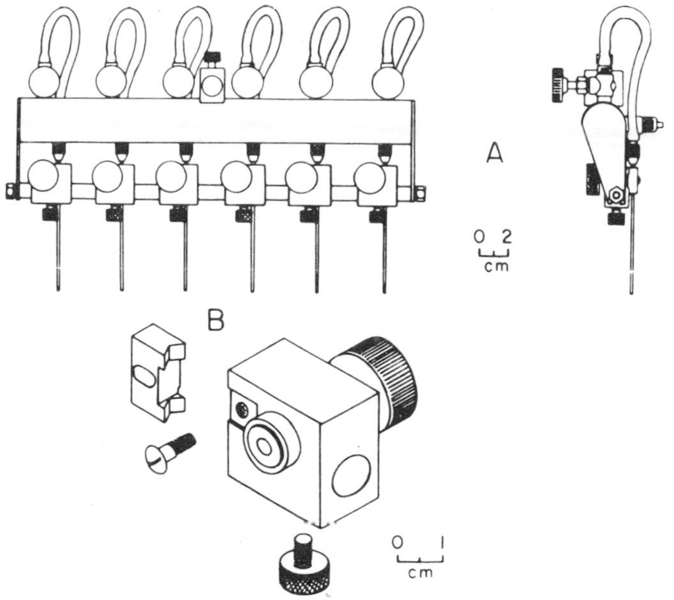

Fig. 8 Small-sample evaporator showing the front and side view (A) and a more detailed view of the nozzle-directing component (B) (R. J. Komarek, Ref. 36. Reproduced by kind permission of *Journal of Lipid Research.*)

great care must be taken to control the rate of evaporation to avoid 'bumping'. An advantage of evaporation *in vacuo* is that many samples can be handled simultaneously. With some reagent systems, such as methanol-HCl, a small freeze-drying assembly can be used. In any case with analytical methods, such as ECD–GLC, when using a strongly capturing reagent it is often wise to remove the last traces of reagent in a vacuum system; the same applies to solvents such as pyridine which can 'tail' badly on some GLC columns. A suitable small evaporation device for pyridine is shown in Fig. 9. A useful

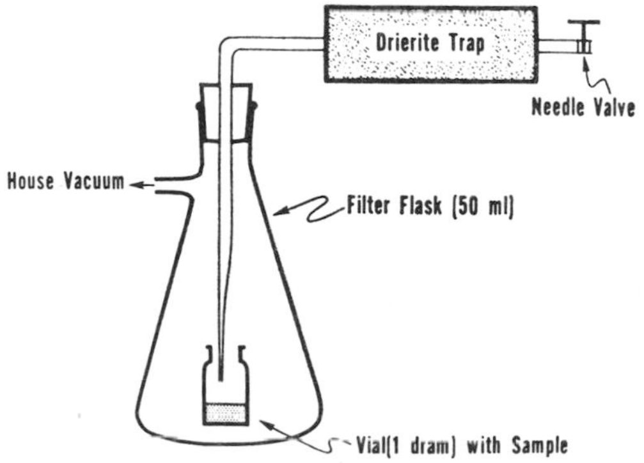

Fig. 9 Apparatus for the removal of pyridine from TMS sugar mixtures (J. Lehrfeld, Ref. 37. Reproduced by kind permission of *Journal of Chromatographic Science.*)

device for concentrating large volumes of solution to a small volume with the minimum of attention and contamination has been described by Beroza and Bowman.[29]

As an example of the type of problem which can arise, the mixture which results from esterification for gas chromatography with alcoholic boron trifluoride cannot be injected directly because the boron trifluoride causes column deterioration. Extraction procedures are often used, but recoveries may not be quantitative. The addition of an organic base such as triethylamine or pyridine to cause precipitation of the insoluble BF_3 complexes, followed by centrifugation and direct injection of the reaction mixture, is reported to improve yields of the esters.[30]

Similarly, excess 5-(dimethylamino)-1-naphthalenesulfonyl chloride (dansyl chloride) in dansylation reactions may cause difficulties in TLC analyses, and the presence of the hydrolysis product may cause fluorescent streaks on the developed plate. Addition of proline to the reaction mixture to remove excess reagent produces a better chromatogram (always assuming that proline analysis is not the object of the experiment!).

Evaporation may be necessary to concentrate extracts, to change solvents for improved chromatographic separations, or to remove excess reagents. Extraction is another means of changing solvents or removing excess reagents or side-products. When small volumes are involved, the transfer of a liquid layer by pipette is more satisfactory than the use of a separatory funnel. Filtration for the removal of drying agents or precipitates can often be avoided by using centrifugation. In a GLC procedure for the separation of vitamin B_6 compounds as their heptafluorobutyryl derivatives, prepared with heptafluorobutyrylimidazole, solids were not separated at all; instead the reaction mixture was allowed to stand ten minutes before injection on to the column, to give time for crystals in the mixture to grow to a size that would not deposit in the syringe.[31] Such simplification of manipulations reduce losses and improve recoveries in microprocedures.

With some volatile reagents, such as acetic anhydride, it is possible to prepare derivatives by exposing material deposited on a solid support, such as a stainless-steel gauze or ball of platinum wire, to the vapours;[32–34] the sample can then be introduced into the gas chromatograph by a solid-injection technique. Since no solvent is required, this procedure may have advantages for GC–MS. Other advantages are the relative ease with which several samples can be treated at one time, reduction of the need for handling, the economical use of reagents, and the fact that only the amount of sample actually used in the chromatographic step undergoes derivatization, which improves the overall sensitivity of the analysis.

A further simplification is possible when the derivatization reaction is performed as a part of the chromatographic process; this 'reaction-chromatography' finds applications in many fields. Such procedures are particularly valuable when limited amounts of material are available, because only that amount which is to be chromatographed is used in the reaction. Due to the ease and rapidity of these procedures, they are popular even when sample size is not a limitation.

4 CONCLUSION

Many schemes of analysis that have used derivatization followed by chromatography are needlessly complicated. This may be because they are scaled-down versions of earlier methods or because extra steps have been added to overcome imagined difficulties without checking whether the analysis might not work just as well if they were omitted. If an analysis is likely to be used extensively, the time and trouble involved in trying several of the published versions of a procedure and optimizing the conditions is often a sound investment. There is inevitably some duplication of procedures in various sections of this book and this is not necessarily a bad thing. If confronted with such a situation we would advise trying the simplest one first.

REFERENCES

1. K. Blau in *Biomedical Applications of Gas Chromatography*, Vol. 2, H. A. Szymanski (Ed.), Plenum Press, New York, 1967, p. 1.
2. C. W. Gehrke, *J. Assoc. Offic. Anal. Chem.* **55**, 449 (1972).
3. P. Hušek and K. Máček, *J. Chromatogr.* **113**, 139 (1975).
4. W. J. McBride and J. D. Klingman, in *Lectures on Gas Chromatography*, L. R. Mattick and H. A. Szymanski (Eds), Plenum Press, New York, 1967, p. 25.
5. J. K. Grant (Ed.) *The Gas Chromatography of Steroids*, Cambridge University Press, Cambridge, 1967.
6. R. Scholler and M. F. Jayle, *Gas Chromatography of Hormonal Steroids*, Gordon and Breach, London, 1968.
7. K. B. Eik-Nes and E. C. Horning, *Gas Phase Chromatography of Steroids: Monographs on Endocrinology*, Vol. 2, Springer-Verlag, Berlin, 1968.
8. B. J. Gudziewicz, *Gas Chromatographic Analysis of Drugs and Pesticides*, Marcel Dekker, New York, 1967.
9. (a). R. W. Frei and W. Santi, *Z. Anal. Chem.* **277**, 303 (1975).
   (b). J. F. Lawrence and R. W. Frei, *J. Chromatogr.* **98**, 253 (1974).
10. J. F. Lawrence and R. W. Frei, *Chemical Derivatization in Liquid Chromatography*, Elsevier, Amsterdam, 1976.
11. J. Drozd, *J. Chromatogr.* **113**, 303 (1975).
12. L. M. Cummins in *Recent Advances in Gas Chromatography*, I. I. Domsky and J. A. Perry (Eds), Marcel Dekker, New York, 1971, p. 313.
13. K. Hammerstrand and E. S. Bonelli, *Derivative Formation in Gas Chromatography*, Varian Instrument Division, Palo Alto, California, 1974.
14. S. Ahuja, *J. Pharm. Sci.* **65**, 163 (1976).
15. S. Udenfriend, *Fluorescence Assay in Biology and Medicine*, Academic Press, New York, 1962 and 1969.
16. C. H. Lochmüller and R. W. Souter, *J. Chromatogr.* **113**, 283 (1975).
17. A. E. Pierce, *Silylation of Organic Compounds*, Pierce Chemical Co., Rockford, Illinois, 1968.
18. D. A. Leathard and B. C. Shurlock, *Identification Techniques in Gas Chromatography*, Wiley–Interscience, New York, 1970.

19. R. C. Crippen, *Identification of Organic Compounds with the Aid of Gas Chromatography*, McGraw-Hill, New York, 1973.
20. T. S. Ma and A. S. Ladas, *Organic Functional Group Analysis by Gas Chromatography*, Academic Press, London, 1976.
21. G. Zweig and J. Sherma, *Handbook of Chromatography*, two volumes, Chemical Rubber Corporation Press, Cleveland, Ohio, 1972.
22. E. Stahl (Ed.), *Thinlayer Chromatography—A Laboratory Handbook*, Academic Press, New York, 1965.
23. W. H. McFadden, *Techniques of Combined Gas Chromatography-Mass Spectrometry*, Wiley–Interscience, London, 1973.
24. P. T. Kissinger, L. J. Felice, R. M. Riggin, L. A. Pachla and D. C. Wenke, *Clin. Chem.* **20**, 992 (1974).
25. E. Bayer, E. Grom, B. Kattenegger and R. Uhlmann, *Anal. Chem.* **48**, 1106 (1976).
26. N. E. Hoffman and J. C. Liao, *Anal. Chem.* **48**, 1104 (1976).
27. J. A. Riddick and W. B. Bunger, *Techniques of Chemistry* (*Vol. 2*)—*Organic Solvents*, 3rd Edition, Wiley–Interscience, New York, 1970.
28. A. Darbre, *Lab. Pract.* **9**, 726 (1971).
29. M. Beroza and M. C. Bowman, *Anal. Chem.* **39**, 1200 (1967).
30. P. A. Biondi and M. Cagnosso, *J. Chromatogr.* **109**, 389 (1975).
31. A. K. Williams, *J. Agr. Food Chem.* **22**, 107 (1974).
32. J. K. Norymberski and A. Riondel, *Experientia* **23**, 318 (1967).
33. B. Teuwissen and A. Darbre, *J. Chromatogr.* **49**, 298 (1970).
34. F. L. Rigby, H. J. Karvolas, D. W. Norgard and R. C. Wolf, *Steroids* **16**, 703 (1970).
35. K. Blau, *Clin. Chim. Acta* **27**, 5 (1970).
36. R. J. Komarek, *J. Lipid Res.* **8**, 287 (1967).
37. J. Lehrfeld, *J. Chromatogr.* **9**, 757 (1971).
38. B. L. Goodwin, F. Karoum, C. R. J. Ruthven and M. Sandler, *Clin. Chim. Acta* **40**, 269 (1972).
39. C. W. Moss, M. A. Lambert and F. J. Diaz, *J. Chromatogr.* **60**, 134 (1971).
40. J. Jönsson, J. Eyem and J. Sjöquist, *Anal. Biochem.* **51**, 204 (1973).
41. J. F. March, *Anal. Biochem.* **69**, 420 (1975).
42. R. F. Adams, *J. Chromatogr.* **95**, 189 (1974).
43. M. Makita, S. Yamamoto and M. Kono, *J. Chromatogr.* **120**, 129 (1976).
44. M. Schwarz and G. Michael, *J. Chromatogr.* **118**, 101 (1976).
45. N. Ikekawa, O. Hoshino, R. Watanuki, H. Orimo, T. Fujita and M. Yoshikawa, *Anal. Biochem.* **17**, 16 (1966).
46. J. P. Thenot and E. C. Horning, *Anal. Lett.* **5**, 519 (1972).
47. I. Horman and F. J. Hesford, *Biomed. Mass Spectrom.* **1**, 115 (1974).
48. O. Grahl-Nielsen and E. Solheim, *Anal. Chem.* **47**, 333 (1975).
49. O. Grahl-Nielsen and B. Møvik, *Biochem. Med.* **12**, 143 (1975).
50. P. Hušek, *J. Chromatogr.* **91**, 475 (1974).

CHAPTER 2

Esterification

André Darbre

Department of Biochemistry, University of London, King's College, Strand, London WC2R 2LS.

1 THEORETICAL BASIS

1.1 General introduction

Esterification is one of the most common chemical reactions. General organic chemistry textbooks, which give a good introduction to the subject,[1,2] and a comprehensive text on the chemistry of carboxylic acids and esters are available.[3] A useful list of esterification agents[4] and methods for preparing esters are given.[5,6] *Ortho*-esters (ester-acetal derivatives of the hydrates of carboxylic acids),

$$RC(OR')(OR'')(OR''')$$

where R, R′, R″ and R‴ are alkyl or aryl groups, are considered separately.[7,8] Practical details for preparing some carboxylic acid esters in gram quantities have been published.[2,9,10] Esters of many inorganic acids—boric, silicic, nitrous, nitric, phosphorous, phosphoric, sulfurous, sulfuric, hypochlorous and perchloric have been considered[11] (see Chapter 11).

An attempt has been made to give a general survey of the preparation of esters for analytical work in the hope that the reader may find useful ideas and references which may be of value. Some emphasis has been placed on various classes of compounds because of widespread interest in them. When compounds in biological samples are to be determined, it may be necessary to carry out preliminary purification steps before derivatization is possible. These are not considered here in detail, but reference may be made to the original papers and to specialist reviews on this subject.[12]

Sometimes esterification alone is insufficient to derivatize a compound for satisfactory analysis, and a second chemical reaction may be needed to derivatize other reactive groups which may be present. This is particularly true of amino acids.[13–16] As an example, the methyl ester of 3,4-dihydroxyphenylacetic acid is insufficiently volatile to be conveniently determined by gas-liquid chromatography (GLC). However, if the —OH groups are also converted to the *O*-TMS[17] or *O*-pentafluoropropionyl[18] derivatives, then GLC of the methyl[17] or ethyl esters[18] is no longer difficult. Where necessary, brief details are given for carrying out the second derivatization step. General reviews on derivatization for GLC,[19,20] and other chapters in this book, should also be consulted. The methods (mostly GLC) used for analysing the esters have been mentioned where these are useful.

It is usual to form the ester first, before carrying out a second chemical reaction to derivatize other functional groups which may be present. The reverse process is sometimes used and may be illustrated by the following example for the preparation of trifluoroacetyl homovanillic acid hexafluoroisopropyl ester for GLC-EC:[21]

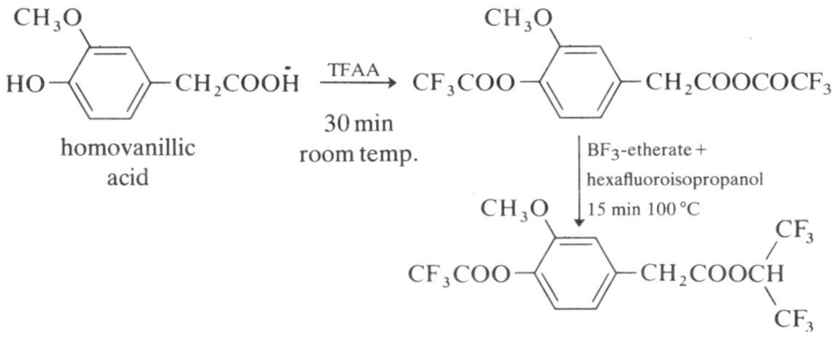

Homovanillic acid (from urine) was reacted with TFAA for 30 min at room temperature. Excess TFAA was removed and the 4-trifluoroacetyl mixed anhydride derivative of homovanillic acid was then reacted with BF_3-etherate and hexafluoroisopropanol. Another example is the preliminary *N*-trifluoroacetylation of amino acids followed by trimethylsilyl esterification.[22]

The derivatization of a compound by esterification (or any other method) for the purposes of analytical determination is only recommended when it is necessary to do so. Since it involves chemical reaction, time and manual manipulation are involved and errors may be introduced into quantitative determinations. Thus, it is necessary to take due precautions and to include suitable control experiments. Internal standards having similar chemical structures should be added to the unknown sample, e.g. norleucine is often added to a mixture of protein amino acids, and these are carried through the entire reaction sequence. It is less satisfactory to add a different type of standard compound just prior to the analysis of an unknown mixture, e.g. docosane added to a sample of fatty acid esters.

The practical details for the preparation of esters include old and well-tried methods, as well as recently published ones which appear to show considerable scope for future successful development. It should be noted that where a method of esterification is described for a particular compound or group of compounds, that method would be applicable to a wide variety of other carboxylic acids. Details are also given at the end of this chapter for the preparation of some esterification reagents.

1.2 Requirement for esterification

Many simple monocarboxylic acids are sufficiently volatile to be separated directly by GLC[23] or GSC.[24,25] However, adsorption on to the chromatographic support often leads to tailing peaks, and it is better to esterify the acids for GLC.[26–29] Non-esterified fatty acids ($C_{6:0}$—$C_{18:3}$)[30] and plasma fatty acids (C_{14}—C_{18})[31] were determined quantitatively, but an important source of error is ghosting.[32–35] It has been claimed that ester formation of fatty acids may be unnecessary with chemical ionization–mass spectrometry.[36]

For the purposes of gas chromatography, esterification may be used as a method of converting a relatively non-volatile and polar compound into one having better characteristics, i.e. one which may be gas-chromatographed at a reasonable temperature (usually below 250 °C) and also one which gives better separation of components in a mixture by giving sharp elution peaks without tailing. It is the acid, and not the alcohol, component which normally requires esterification. Thus succinic acid, $HOOCCH_2CH_2COOH$, does not distil—it decomposes at 235 °C; whereas succinic dimethyl ester distils at 196.4 °C. Simple straight-chain alcohols are usually gas-chromatographed unchanged, derivatization is only necessary for compounds having more than one alcoholic function, e.g. glycerol, $CH_2OHCHOHCH_2OH$, which decomposes at 290 °C, and these are better converted to ether-type derivatives (see Chapters 4 and 5).

1.3 Catalysis for esterification

Although some esters may be prepared without added catalyst, e.g. benzyl alcohol and formic acid,[2] in general carboxylic acids are unreactive to alcohols alone, but they can usually be esterified by electrophilic catalysis in the presence of a trace of concentrated sulfuric acid or hydrogen chloride (about 3% w/v). Other catalysts have been used, such as dichloroacetic acid and trifluoroacetic acid,[37] benzene sulfonic acid,[38] polyphosphoric acid,[39] *p*-toluene sulfonic acid,[38,40] sulfuryl chloride,[41] thionyl chloride,[42] $POCl_3$, PCl_3,[43] or strong acid cation exchange resins,[44] such as Zeo Carb[2] or Dowex-50.[45]

Graphite bisulfate ($C_{24}^{+}HSO_4^{-} \cdot 2H_2SO_4$, an electrolytic lamellar compound of graphite with intercalated sulfuric acid) was used to prepare some esters of primary alcohols in high yield by reaction in dry cyclohexane at room temperature. Tartaric dimethyl ester was prepared in 99% yield with retention of configuration. This method offers scope for development, because of the mild conditions used. No excess alcohol was required to obtain a maximum yield, and solubility of the reactants did not appear to be an important factor.[46]

In an unusual reaction, tin(II) oxalate was used to esterify 2,4-dicarboxybenzene sulfonate with ethylene glycol.[47]

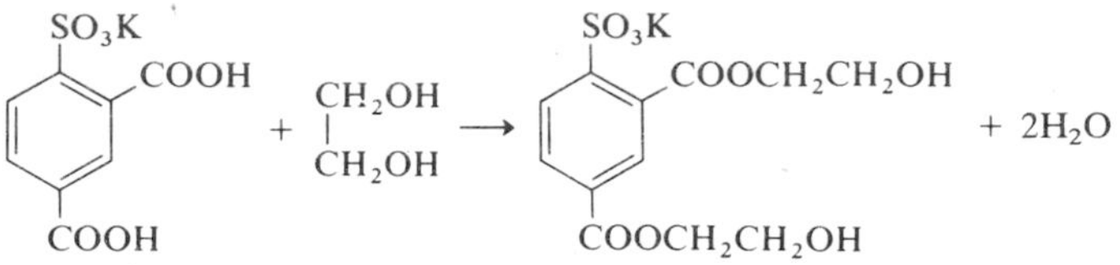

For analytical work esterification is best catalysed with a volatile catalyst, such as HCl or thionyl chloride, which can be removed along with excess reagent. Sulfuric acid was recommended,[48,49] but use of this reagent (as with some others mentioned here) nearly always requires the extraction of the esters into a non-polar solvent, which must subsequently be reduced in volume.

1.4 Reaction mechanisms of esterfication

In the esterification of a carboxylic acid with an alcohol, reaction occurs by acyl-oxygen (1) or alkyl-oxygen (2) heterolysis:

(1)

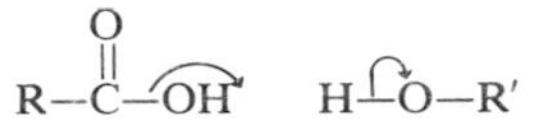

(2)

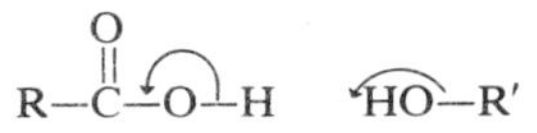

Ingold[50] showed that acid-catalysed esterification involves three possible mechanisms, denoted as $A_{AC}1$, $A_{AC}2$ and $A_{AL}1$, where A is the substrate (conjugate acid of the carboxylic acid), AC and AL denote acyl and alkyl bond heterolysis, and the numbers represent the molecularity of the rate-determining step.

1.4.1 $A_{AC}2$ MECHANISM

The esterification reaction is mainly bimolecular, and goes via acyl–oxygen fission, where the rate-limiting step is the attack of alcohol on the protonated carboxylic acid. Ingold proposed the following schemes:

(1)

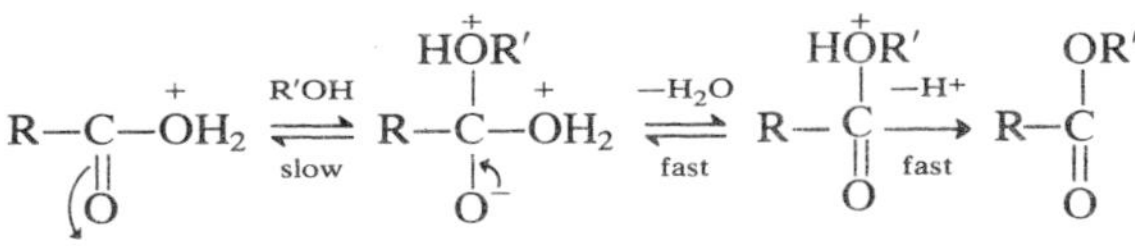

(2)

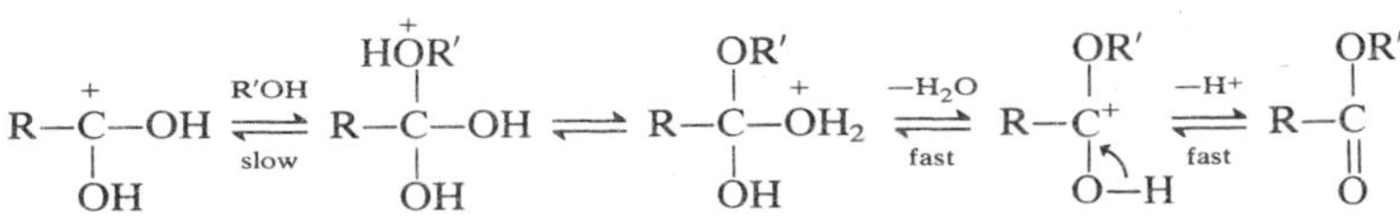

The carbon atom becomes tetrahedrally bonded, and for this reason retardation of the reaction rate may occur where —R and —R' are bulky groups which may interact sterically.

1.4.2 $A_{AL}1$ MECHANISM

This mechanism is common for tertiary alcohols and explains the formation of racemized esters when esterifying with an optically active alcohol. This reaction is slow for primary alcohols. The alcohol is first protonated and then loses water to form a carbonium ion which reacts rapidly with the acid.

$$R'—OH \underset{}{\overset{H^+}{\rightleftharpoons}} R'—OH_2^+ \underset{slow}{\overset{-H_2O}{\rightleftharpoons}} R'^+$$

$$R—\overset{O}{\overset{\|}{\underset{OH}{C}}} + R'^+ \underset{fast}{\rightleftharpoons} R—\overset{OR'}{\overset{|}{\underset{O—H}{C^+}}} \overset{-H^+}{\rightleftharpoons} R—\overset{OR'}{\overset{|}{\underset{O}{C}}}$$

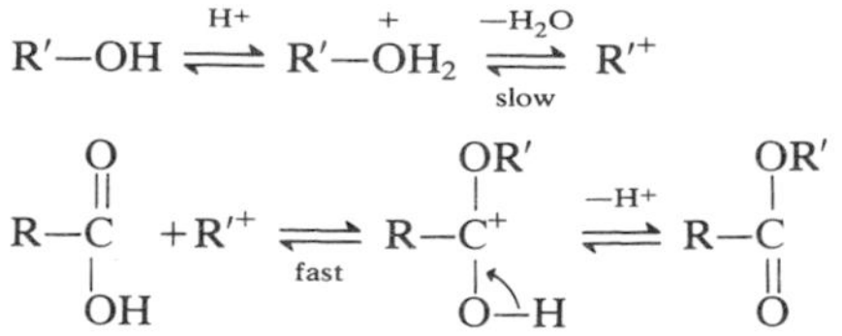

However, the carbonium ion can also attack the alcohol or anions to form by-products.

1.4.3 $A_{AC}1$ MECHANISM

This occurs with sterically hindered acids, e.g. 2,4,6-trimethylbenzoic acid, using sulfuric acid catalysis, where the acylium ion intermediate reacts with the alcohol.

$$R—\overset{O}{\overset{\|}{C}}—OH + H_2SO_4 \rightleftharpoons R—\overset{+}{C}\overset{OH}{\underset{OH}{\diagup}} + HSO_4^-$$

$$R—\overset{+}{C}\overset{OH}{\underset{OH}{\diagup}} + H_2SO_4 \rightleftharpoons R—\overset{+}{C}=O + HSO_4^- + H_3O^+$$

$$R—\overset{+}{C}=O + R'OH \rightleftharpoons R—\overset{}{\underset{O}{\overset{\|}{C}}}—OR' + H^+$$

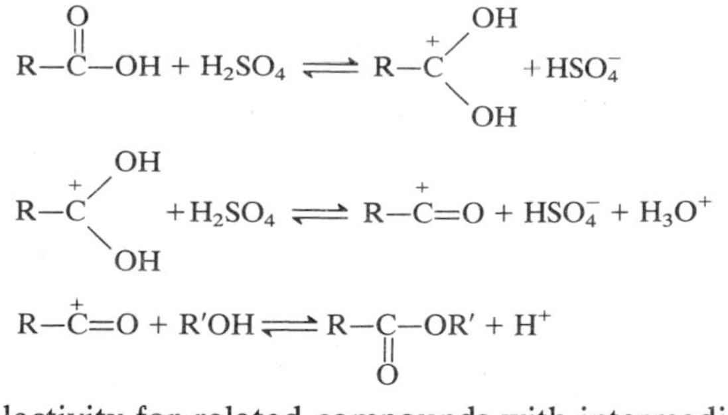

A reaction selectivity for related compounds with intermediate steric hindrance was shown by Sniegoski,[51] who studied by GLC the relative esterification rates of eight isomeric hexanoic acids, with methanolic HCl, and reported an esterification rate as follows: hexanoic > 4-methyl pentanoic > 3-methyl pentanoic > 2-methyl pentanoic > 3,3-dimethylbutanoic > 2-ethylbutanoic = 2,3-dimethylbutanoic > 2,2-dimethylbutanoic acid. It is of interest here that when the samples were injected directly on to the GLC column some esterification (about 5% for n-hexanoic acid) occurred due to the presence of the HCl catalyst. Previous workers had discounted the need to quench the HCl catalyst before injection. In a further report by Sniegoski[52] based on the esterification rates of ten α-methyl and β-methyl isomers of carboxylic acids possessing between four and eight carbon atoms, the following generalization was given: a β-methyl-substituted saturated aliphatic acid esterifies with methanol at a greater rate ($\approx$1.8 times as fast at 40 °C) than the corresponding α-methyl acid. This contradicted previously accepted ideas published in many textbooks that α-methyl aliphatic acids were more reactive.

An increase of temperature increases the speed of esterification. The reaction rate is proportional to the acid concentration and is reduced as the size of the alcohol increases.[53] The reaction mechanisms of esterification[50] and the kinetics of the formation of methyl esters of higher fatty acids[54] and the acid-catalysed esterification of aliphatic acids[55–57] have been reported. Removal of water (see below) may drive the reaction to give a quantitative yield of ester. However, when preparing *n*-butyl esters of amino acids, it was found that in less than 15 min, *n*-butanol containing 2.7 M HCl became more than 0.2 M with respect to water at 100 °C and more than 2.0 M at 150 °C.[58] The reactions involved were as follows:

$$CH_3CH_2CH_2CH_2OH + HCl \rightarrow CH_3CH_2CH_2CH_2Cl + H_2O \qquad (1)$$

$$2CH_3CH_2CH_2CH_2OH \rightarrow (CH_3CH_2CH_2CH_2)_2O + H_2O \qquad (2)$$

1.5 Removal of water

Water present in the reagents, or formed as a result of esterification, may be removed in a variety of ways such as using a Soxhlet thimble containing desiccant, a Dean-Stark trap or a chemical reagent.

When the reaction mixture boils above 100 °C, water may be removed as it forms. Thus, there is an advantage in preparing esters with *n*-butanol (b.p. 117 °C) and higher homologues. However, the problem arises of solubility of the acid in the esterifying alcohol medium, and for this reason methyl esters may be formed and transesterification used to convert these to higher alcohols, as carried out for amino acid *n*-butyl ester formation[59,60] (see p. 67). The order of reactivity of the three isomeric butyl alcohols is primary > secondary ≫ tertiary, and for this reason primary alcohols are normally used for preparing derivatives for chromatography.

Azeotropic distillation used to remove water from the reaction mixture promotes rapid esterification and high yields,[61] and may be useful when dry reagents are not available (see p. 60). The method is not easily applicable to very small quantities. Water has been removed from a condensing azeotropic mixture by allowing it to pass through an extraction thimble filled with calcium carbide,[62] molecular sieve Linde type 4A or 5A[63] or anhydrous magnesium sulfate[64,65] before allowing the organic phase to return to the reaction vessel. To avoid the use of large quantities of anhydrous magnesium sulfate, aliphatic and aromatic esters were prepared by refluxing sulfuric acid- or ethane sulfonic acid-catalysed esterification with methanol in the presence of a large excess of ethylene dichloride.[66] The high boiling point solvents (to raise the temperature of the reaction mixture), *o*-dichlorobenzene, anisole, phenetole and nitrobenzene with naphthalene β-sulfonic acid catalyst have been used to prepare gallic acid esters.[67] Other ways of removing water are to allow a volatile alcohol to distil away from the reaction mixture, entraining water with it,[68] or to dry it over roasted potassium carbonate, before continuously returning it to the reaction vessel.[69]

The use of a water scavenger may avoid the need for high temperature to remove water. Graphite bisulfate is both an acid catalyst and dehydrating agent.[46] A new and rapid method of esterification introduced for sterically hindered aromatic carboxylic acids was the reaction with 100% sulfuric acid. The acid–sulfuric mixture is then poured into excess alcohol and the ester subsequently recovered by extraction.[70] A rapid semimicro method using methanol–sulfuric acid was described for the preparation of methyl esters from triglycerides.[71] Lorette and Brown[72] first reported the use of 2,2-dimethoxypropane for the preparation of fatty acid *n*-butyl esters, using *n*-butanol–HCl esterification. It was also used for amino acid *n*-propyl[73] and *n*-butyl esters[74] and methyl esters from rat epidermal fat.[75] Different methanolic–HCl concentrations were studied for the preparation of methyl esters from fats by transesterification, and overnight reaction at 22 °C was recommended with 2,2-dimethoxypropane, methanolic–HCl (10% w/v) and benzene (1:5:14, by vol.).[76] Alcoholic–HCl solutions containing 2,2-dimethoxypropane darken with increasing severity of the conditions due to polymer formation, although the addition of dimethylsulfoxide to alcoholic HCl (10% w/v) prevented this when esterifying at room temperature.[77]

1.6 Transesterification

Esters may be solvolysed by alcohols,[78] and this process of transesterification is catalysed by acid or base. Consider the base-catalysed hydrolysis of an ethyl ester:

$$\mathrm{RCOOEt + OH^- \rightleftharpoons RCOOH + EtO^-}$$

The carboxylic acid is essentially all removed by ionization under the basic conditions, and the reaction goes to completion. But in transesterification, where the nucleophile is usually an alkoxide ion, the balance between reactants and products is fairly even.

$$\mathrm{RCOOEt + R'O^- \rightleftharpoons RCOOR' + EtO^-}$$

However, using a large excess of the alcohol R′OH, ester formation may be driven to completion because the alkoxide anion is removed, due to the following equilibrium:

$$\mathrm{EtO^- + R'OH\,(excess) \rightleftharpoons EtOH + R'O^-}$$

Transesterification of a methyl ester was driven to completion by adding molecular sieve Linde type 3A to the reaction mixture which preferentially absorbs methanol.[79]

Transesterification has been applied, particularly for the preparation of fatty acid and amino acid esters (see p. 68). Lamkin and Gehrke[59] found that direct esterification of amino acids was difficult because of the insolubility of cystine and basic amino acids in *n*-butanol. They tried direct esterification with HCl, HBr and *p*-toluenesulfonic acid as catalysts, direct esterification in a polar solvent and transesterification with HCl, sulfuric acid, Dowex-50, boron trichloride and boron trifluoride as catalysts, but finally adopted the method of transesterification with HCl catalysis.[59,60,80–82]

Boron tribromide has also been proposed as a convenient catalyst for transesterification.[83] Lipids and conjugated compounds may be saponified and the fatty acid salts subsequently esterified. Alternatively, transesterification may be used.[84] Thus, bilirubin conjugates were transesterified and characterized by TLC and MS as their methyl esters, and these were distinguished from free bilirubin which was directly converted to its dimethyl ester with diazomethane.[85] Plant galactolipids were treated with 1.5 M HCl in dry methanol at 80 °C for 4–16 h or with 10% (w/v) sulfuric acid in methanol at 100 °C for 2 h for the preparation of methyl esters.[86] Methyl esters from glycerides, phosphatides and cholesterol esters were obtained by transesterification with 5% (w/v) HCl[87] and with BF_3-methanol, which was reported to be better than BCl_3-methanol.[88] Transmethylation of lipids was carried out in methanol with 2%[89] and 6%[90] sulfuric acid and 10% (v/v) sulfuric acid in methanol with silica gel added.[91–92] Fatty acid methyl and ethyl esters prepared from glycerolipids by transesterification with sulfuric acid (2% w/v) in alcohol–light petroleum (4 : 1, v/v) at 70 °C for 4 h were brominated and analysed by GLC, TLC and MS.[93] A rapid semimicro method was described using sulfuric acid–methanol.[71] Instead of boron trifluoride–methanol for transesterification of long chain fatty acids, perchloric acid–methanol was recommended,[94] but a warning on the hazards has been published.[95]

Sodium[96–98] and potassium[97] methoxide in anhydrous methanol were used to transesterify fats. It was claimed that all lipid fractions were completely transesterified by boiling a benzene solution of the lipid with 0.5 M sodium methoxide for 2 min.[99] In addition, glycerides and glycerophosphatides on TLC plates were transesterified by spraying with 2 M sodium methoxide in methanol, and the methyl esters were subsequently eluted. These methods were further developed for micro quantities.[99,100] Transmethylation of lipid fractions from TLC plates was used to prepare methyl esters for GLC.[101,102] Lithium methoxide was used to transesterify polyester resins.[103]

Whereas the preparation of fatty acid methyl esters from fats required 30 min refluxing with sodium methoxide,[104] it was found that 5 min with potassium methoxide was sufficient for the transesterification of fats and steroid esters.[105] Sodium methoxide used for preparing amino acid methyl esters caused racemization.[106] In quantitative studies on triglycerides in serum, isopropanol solutions of sodium methoxide, sodium ethoxide and sodium hydroxide were equally effective at room temperature as transesterification reagents for the preparation of isopropyl esters. Sodium hydroxide (100 mM) in isopropanol gave quantitative hydrolysis and the efficiency of transesterification was not affected by the presence of water up to 1% (v/v).[107]

A procedure for the preparation of methyl and ethyl fatty acids from milk fat[108] was later modified with the formation of *n*-butyl esters by heating the fat on a steam bath for only 2 min with di-*n*-butyl carbonate and sodium butoxide.[109]

1.7 Thionyl chloride

Thionyl chloride has not gained wide acceptance as an esterification catalyst, although any excess is easily removed at the end of the reaction, together with the volatile end-products SO_2 and HCl. The general method of use involves adding a few drops of thionyl chloride to the alcohol solution of carboxylic acid in the cold and then refluxing for about 30 min. It was used for preparing methyl esters of hydroxy and keto acids,[110] amino acids,[42,106,111–113] non-volatile plant acids,[114] phenylalanine in serum,[115] 2-bromomethyl phenylacetic acid[116] and methylaminosuccinic acid,[117] benzyl esters of amino acids[118] and L-menthyl esters of isoprenoid fatty acids.[119]

Thionyl chloride (0.7–3.5 M) in methanol at 40 °C for 1 h in closed tubes (with air) was used for amino acid methyl ester preparation. After trifluoroacetylation, cysteine was the only amino acid which failed to give a GLC peak. Some of the cysteine (64%) was oxidized to cystine. The remainder was probably converted to cysteic acid.[120]

1.8 Boron halides

The conditions of esterification are mild, and boron trifluoride etherate may be used for a wider variety of different classes of carboxylic acids than any other single reagent.[121,122] If the acid possesses other functional groups, such as OH, NH_2 or C=O, additional reagent must be used on an equivalent basis to allow for complex formation. Esterification may be achieved by refluxing the acid for a few minutes with boron trifluoride or trichloride in methanol or other simple aliphatic alcohol, but many different conditions have been used, e.g. fatty acids evaporated to dryness, as their ammonium salts to avoid evaporation losses were treated with BF_3–methanol (12.5% w/v) at room temperature overnight.[123] Boron tribromide[83] (see Transesterification, p. 45) and alcohols other than methanol have not been widely used.

Boron trifluoride–etherate in alcohol was used to prepare eight methyl to *tert*-pentyl esters of *p*-aminobenzoic acid[124] and the methyl esters of aliphatic and aryl unsaturated carboxylic acids.[125] The dimethyl ester of *p,p*-biphenyldicarboxylic acid[126] (which is inert to diazomethane and proton-catalysed reaction conditions), and the methyl, ethyl and isopropyl esters of 1,4-dihydrobenzoic acid[126] (which can be isomerized to conjugate isomers, and which is oxidized by atmospheric oxygen to benzoic acid) were also prepared. This reagent was considered to be unique for the esterification of heterocyclic carboxylic acids.[127]

Boron trifluoride–methanol was used to prepare the methyl esters of 24 aromatic acids,[128] of triterpene acids from latex after hydrolysis,[129] of fatty acids,[130–133] octadecadienoate isomers,[134] unsaturated fatty (dimer) acids,[135] methylmalonic[136] and succinic acids[136,137] and of Krebs cycle acids, both in the free state[138] and trapped on alumina.[139]

Low recoveries of short chain fatty acid methyl esters with BF_3–methanol were reported,[140] but in a modified rapid procedure this reagent was found to be quicker for fatty acids with 12 or more carbons and the results were equivalent to those obtained with sulfuric acid–methanol.[141] In determinations of phospholipid fatty acids, BF_3–methanol gave equivalent results to those obtained with diazomethane.[142] The esterification catalysts HCl, BF_3 and thionyl chloride were found to be equally suitable for fatty acid esterification. In order to minimize losses, the esters were extracted into heptane before adding NaCl.[143]

BF_3 gas was passed into trifluoroethanol to prepare 2,2,2-trifluoroethyl esters of citric, fumaric, malic and succinic acids for GLC-EC.[144] Boron trifluoride was not satisfactory for amino acid methyl esters,[112] probably because of complex formation with nitrogen.[121] Methyl esters of γ-hydroxy butyric acid were produced in higher yield than the ethyl, *n*-propyl or *n*-butyl esters using BF_3–alcohol,[145] but the *n*-butyl esters were selected to determine α-, β- and γ-hydroxybutyric acids in nanomol quantities.[146] Methylmalonic,[147] 2,4-dichlorophenoxy acetic acid[148] and benzoic acid[149] butyl esters and fatty acid *n*-propyl esters[150] were also prepared. Normally, at the end of the reaction, excess BF_3 is decomposed by adding water or a salt solution and the esters extracted into an organic solvent such as chloroform. This procedure mediates against accurate determinations. A recent study with BF_3–methanol, –isopropanol and –*n*-butanol[151] avoided this problem. Excess BF_3 was neutralized with an organic base, triethylamine or pyridine, to form a complex[121] that was insoluble in a mixture of diethyl ether and *n*-pentane. After extraction and centrifugation 1 μl of the supernatant was injected on to a GLC column. The limited number of results obtained[151] showed higher recoveries than those obtained when using water decomposition of BF_3. This method merits further study.

Boron trichloride was first reported for the transesterification of lipids and found to be more active than BF_3, whilst for methyl esterification of fatty acids it showed equivalent reactivity.[152] Both reagents were used as etherates. Boron trichloride–methanol was better than BF_3–methanol for unsaturated fatty acids and triglycerides.[153] Because of its dehydrating action, samples need not be pre-dried when using BCl_3. This reagent was also reported to be equivalent to diazomethane as an esterifying agent for cyclopropane fatty acids, but BF_3 could not be used.[154,155] 2-Chloroethyl esters of short chain fatty acids were prepared with BCl_3 catalysis (see p. 62).

1.9 Diazoalkanes

Diazoalkanes enable esters of carboxylic acids to be prepared very elegantly, by reaction in neutral media at room temperature. Only volatile end-products are normally produced, although some polymeric material may be formed.[156,157] Any excess diazoalkane is easily volatilized or destroyed by the addition of acetic acid or HCl. Reviews of the chemistry of diazomethane have been published.[158,159]

The preparation of diazomethane requires care (see p. 92). It is reported to be both explosive and carcinogenic. In addition, starting materials used for its preparation such as *N*-methyl-*N*-nitro-*N*′-nitroso-guanidine are strongly mutagenic, and some persons develop skin sensitivity towards them. Special distillation kits featuring smooth clear-seal joints, for eliminating the explosion hazard occasioned by the use of ground-glass joints, have been designed for the production of diazomethane and may be purchased commercially. Diazomethane is fairly safe in dilute solution in inert solvents such as ether or dioxan.[159]

The esterification of carboxylic acids with diazomethane is normally carried out in ethereal solution. Werner[160] first showed that an alcoholic medium could be used. Diazomethane reacts under mild conditions. The gas is passed into the solution of the appropriate acid until excess is present, as shown by the formation of nitrogen gas evolution and by the persistence of the yellow colour. Reaction normally proceeds for about 5 min at room temperature, although some authors allow longer periods of time, even overnight in the case of bile acids.[161] It was shown that esterification of long chain fatty acids was instantaneous in ether containing 10% methanol, whereas in pure ether reaction was slow or incomplete.[156]

It is of interest to note the specific action of diazomethane which may be useful in certain cases.[162] The methylation of oestriol monoglucosiduronic acid (I) with diazomethane gave rise to the methoxy methyl ester (II), but in the presence of BF_3 all the hydroxyl groups were methylated (III).

The preparation of fatty acid methyl esters with the diazomethane and methanol–BF_3 methods were compared and found to be equivalent.[142] However, Vorbeck *et al.*[163] found that diazomethane gave higher yields of fatty acid methyl esters (butyric to stearic) with lower standard deviation than either the methanolic–HCl, ion exchanger or methanol–BF_3 methods. There was no evidence that diazomethane caused losses of unsaturated fatty acids due to the formation of pyrazolines which, on heating, decompose to cyclopropane

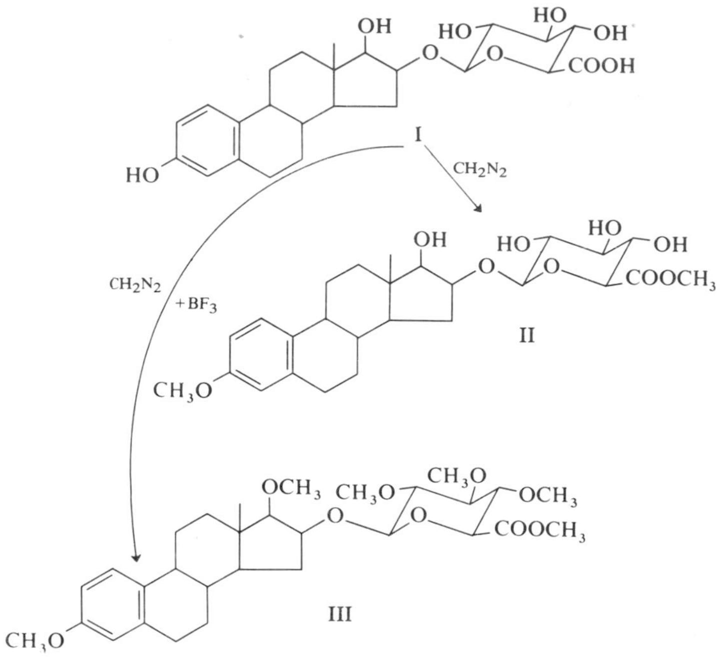

derivatives.[156,163] When preparing methyl esters of aliphatic and aromatic dibasic acids, diazomethane was reacted in methanol–ether solution at −60 °C to prevent side reactions, although methanol–dioxan at 22 °C was recommended to shorten the reaction time.[164] Six methods of fatty acid esterification were studied, comparing methanol with sulfuric acid, acetyl chloride and BF_3–etherate with diazomethane and silver salt-methyl iodide. Diazomethane was best.[165]

Fatty acid methyl esters were formed when lipid fractions, such as egg yolk lecithin or brain phosphatidyl inositol, were reacted with diazomethane under u.v. irradiation. It was suggested that by using the correct choice of conditions the fatty acid composition of glycerides and glycerophosphatides might be determined in the presence of other lipids.[166]

Krebs cycle acids were converted to their methyl esters with diazomethane,[167,168] but dimethyl fumarate decomposed above 90 °C during GLC.[167] Multiple reaction products were shown with some of these acids.[168] Diazomethane was first reported to react with the ammonium salts of benzoic and propionic acids[169] to give their methyl esters and those of several dibasic acids to give the corresponding dimethyl esters in high yields.[170] The esterification of octanoic acid with diazomethane by reaction gas chromatography was explored as a method of saving time and eliminating the hazards associated with this reagent.[171]

Esterification of amino acids with diazomethane leads to difficulties with simultaneous *N*-methylation. The methylation of hydroxyl groups normally only occurs with phenolic hydroxyl groups or those adjacent to the carboxyl group. Diazoethane, -*n*-propane and -*n*-butane were used to esterify phenolic acids, and it was reported that the carboxyl and hydroxyl groups were better derivatized in the presence of 0.007% BF_3 catalyst. Diazomethane was unsatisfactory.[172] Methoxymethyl esters were formed in small amounts with excess of diazomethane in ether. The methyl ether was formed only with hydroxy fatty acids having a primary hydroxyl group,[173] and this could be reduced by using diethyl ether containing 10% methanol.[156] When carbohydrate molecules possess carboxyl groups, as in glucuronides, the uronic acid carboxyl group may be esterified with diazomethane, leaving the hydroxyl groups to be derivatized by trimethylsilylation[174] or by other methods.[175,176]

Multiple peaks were obtained by GLC after preparing the methyl esters of phenylpyruvic, phenyllactic and *o*-hydroxyphenylacetic acids,[177] pyruvic and α-ketoglutaric acids[168] with diazomethane. For this reason keto-acids may be better transformed into trimethylsilyl (see Chapter 4) or quinoxalinol derivatives[178] (see Chapter 7).

Diazomethylation of 2,4-dichlorophenoxyacetic acid was about 98% efficient.[179] Diazoalkylation, although efficient, gave rise to impurities and comparative studies showed mineral acid catalysis to be time-consuming and inefficient, trimethylsilyl esters to be unstable, but BF_3–alcohol gave the methyl, *n*-butyl and *n*-octyl esters in 96%, 98% and 99% yield respectively.[180] Diazomethane was used to prepare methyl esters of glycollic,[181] 2-hydroxybutyric,[182] *p*-chlorophenoxyisobutyric,[183] 5-decynedioic,[184] ritalinic,[185] *m*-hydroxyphenylhydracrylic,[186] cyclohexanecarboxylic,[187] benzene tricarboxylic[188] and *N*-benzyloxyamino acids,[189] N^{α}-BOC-L-histidine,[190] and acids of cigarette smoke,[191] urine,[192–194] discharge waters from mills,[195] abscissic acid on TLC plates[196] and prostaglandins.[197]

Polyene antibiotics, e.g. nystatin, were converted to their methyl to *n*-butyl ester derivatives.[198] Phenyldiazomethane was used for the benzyl esters of formic acid,[199] aliphatic and aromatic acids,[200] 2-hydroxy- and 2-oxo-dicarboxylic acids[201] and short and medium chain fatty acids[202] and C_1–C_{20} fatty acids.[203]

1.10 Trimethylsilylation

Trimethylsilyl esters are conveniently prepared with a variety of reagents[204,205] (see Chapter 4). In their preparation, other reactive groups may be trimethylsilylated at the same time. Derivatives of 250 biochemically significant compounds, amino acids, organic acids, amines, purines, pyrimidines and glycine conjugates were gas-chromatographed on both OV-1 and OV-17 liquid phases.[206] Trimethylsilyl derivatives are sensitive to moisture, and they have only been separated by GLC. It was also reported that they break down on some stationary phases.[207] The trimethylsilyl esters were reported of a wide range of urinary organic acids,[208] *cis* and *trans* isomers of cinnamic acids,[209] chlorophenoxy acids,[210] Krebs cycle acids,[211] neuraminic acids,[212] acetylsalicylic and salicylic acids,[213] C_1 to C_5 fatty acids,[214] urinary indole-3-acetic acid[215] and tetronic acids.[216] Excellent separations were shown with trimethylsilyl amino acid derivatives[217–219] (see also Ref. 16 for extensive account). This method was used specifically for lysine and methionine in foods[220] and enabled asparagine and glutamine to be differentiated from aspartic and glutamic acids.[221]

1.11 Crown ethers

Pedersen, in 1967, first described the cyclic polyethers or crown ethers for complexing many cations, especially potassium,[222–225] and their properties and use have been reviewed.[226] Crown ethers hold the cation in the centre of the main ring, principally by electrostatic forces, and the complexing ability depends on the relative sizes of the cation and the diameter of the hole in the macrocycle. As a result, the anion becomes very reactive, especially carboxylate anions.[227,228] The cation complex aids the extraction of the 'naked' anion[228,229] into an aprotic solvent, where reaction is rapid under mild conditions. Using *α*-*p*-dibromoacetophenone as an alkylating agent, phenacyl esters were prepared, with esterification occurring according to the following scheme:[230]

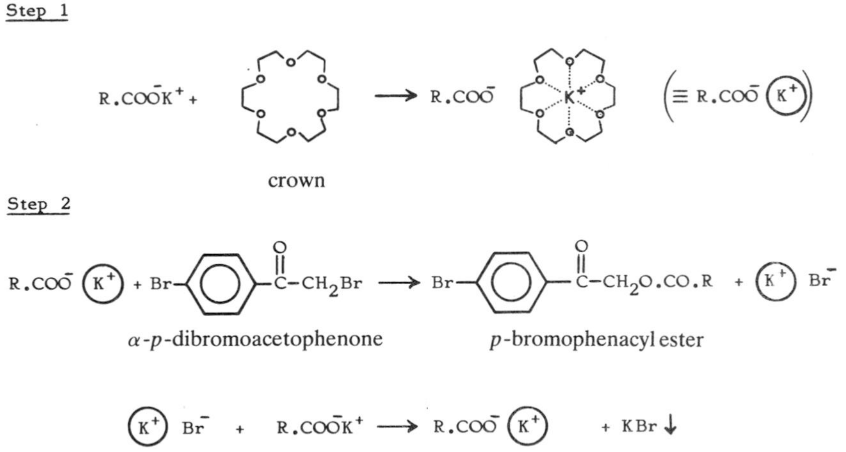

Solvents such as benzene, cyclohexane, methylene chloride and carbon tetrachloride may be used, but reaction rates varied and acetonitrile was preferred.[227–229] Rigorous anhydrous conditions were not necessary. A variety of esters were prepared[228,230] (see p. 78). Highly activated bromides, such as benzyl bromide, reacted rapidly at room temperature, but primary alkyl halides required heating.[227] Esters were obtained in quantitative yield

(>97%), free of by-products.[230] Sterically hindered acids could be successfully esterified by crown catalysis[227] (see p. 89). Crown ether catalysis is being developed for use in many chemical reactions. It should prove extremely useful in esterification, e.g. after concentrating acids in solution as their salts to avoid losses due to volatility.

1.12 Alkyl halides

The reaction between methyl iodide and the silver salt of an acid is a standard method for esterification and was used to derivatize fatty acids for GLC.[231] It is expensive and has not been used extensively for analytical purposes. Fatty acids trapped as their silver salts in a narrow tube were reacted with methyl iodide and the tube was then inserted into the inlet port of a gas chromatograph[232] (see p. 75). This method was developed for short chain C_1 to C_5 fatty acids which were reacted with benzyl chloride.[233] It should be noted that methyl iodide also converts hydroxyl groups to methyl ether derivatives and replaces active hydrogens in peptides. Fatty acids were refluxed for 10–60 min with methyl, allyl and benzyl iodides in the presence of potassium carbonate to prepare the corresponding esters.[234] In order to determine acetyl salicylic acid in the presence of salicylic acid, methyl iodide and potassium carbonate were used for the preparation of the methyl esters.[213,235,236] This method minimized hydrolysis of the acetyl group. Using the alkylating agent 2-bromopropane, amino acids were derivatized in a single-step reaction for GLC by the preparation of *N,O*-isopropyl amino acid isopropyl esters.[237] This method was extended to GC–MS studies of urinary amino acids and amines[238] (see p. 76).

A new approach was to react a steroid acid with methyl iodide and sodium bicarbonate in *N,N*-dimethylacetamide at room temperature for 48 h to prepare the methyl ester.[239] Greeley[240] developed this method with studies in which stearic acid was reacted in *N,N*-dimethylacetamide-methanol solution with methyl-, ethyl-, 1-butyl- or 2-butyl-iodide in the presence of phenyltrimethyl- or tetramethyl-ammonium hydroxide (see p. 76). The esters were separated by GLC. The basic reaction was the S_N2 attack of a soluble organic carboxylic anion on an alkyl iodide in a very polar solvent. The reaction was completed in 10 min, depending on the alkyl group, and virtually quantitative (98–100%) under mild conditions:

$$\text{RCOOH} \xrightarrow{\text{R}'_4\text{N}^+\text{OH}^-} \text{RCOO}^-\text{R}'_4\text{N}^+ \xrightarrow{\text{R}''\text{I}} \text{RCOOR}''$$

The 1-butyl esters of bile acids were better separated than the methyl esters from the large amounts of cholesterol present in many biological samples. This method was used with the more volatile solvent *N,N*-dimethylformamide for the determination of acids in polyester resins[241] (see p. 77).

The preparation of esters with large molecular structures avoids many problems associated with excessive volatility, and halide-containing derivatives may be analysed at the nanogram level by GLC–EC. Pentafluorobenzyl bromide was used to prepare the pentafluorobenzyl esters of prostaglandin $F_{2\alpha}$[242] and a range of organic acids.[243] Using the appropriate alkyl halides, C_1 to C_3 fatty acids were converted to *p*-bromo-, *p*-methyl- and *p*-nitro-benzyl esters[244] and C_2 to C_{10} acids were converted to *p*-bromo- and *p*-phenyl-benzyl esters[245] (see p. 79). Long chain fatty acid 2-naphthyl esters were prepared with 2-naphthacyl bromide and diisopropylethylamine by reaction for 20 h at room temperature.[246]

Iodomethyltetramethylmethyldisiloxane esters were prepared by a two-step reaction in which the bromo derivatives were first formed, and were then converted to the iodo derivatives for study by GC–MS[247] (see p. 77). Both acids and alcohols having a strong affinity for free electrons reacted by this method. Acids reacted with bromomethyldimethylchlorosilane (BMDCS) in the following manner:

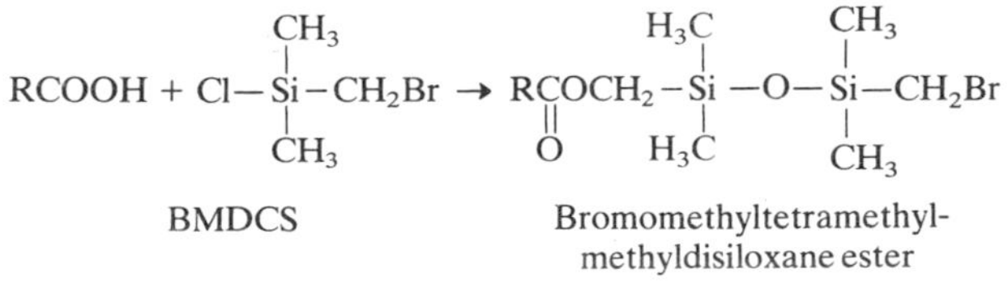

Alcohols reacted as follows:

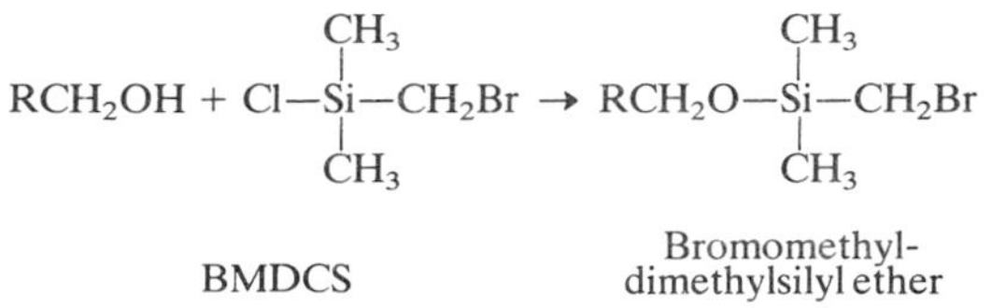

The second chemical step was carried out with sodium iodide in acetone.

1.13 Mixed anhydrides

Esters may be formed via the formation of mixed anhydrides. Trifluoroacetic anhydride was first shown to be useful as a catalyst,[248] particularly where the alcohol and the acid were sterically hindered[249] (see Ref. 250 for review). Trichloroethyl esters were prepared with trichloroethyl alcohol and trifluoroacetic[251] and heptafluorobutyric[252] anhydrides (see p. 72). The preparation of 4-trifluoroacetyl homovanillic hexafluoroisopropyl ester,[21] using the trifluoracetic mixed anhydride as intermediate, was previously discussed. Pyrrolidone carboxylate and γ-glutamyl amino acids were converted to their pentafluoro-1-propyl esters using pentafluoropropionic anhydride and

2,2,3,3,3-pentafluoro-1-propanol[253] (see p. 74). The simultaneous esterification of carboxyl and hydroxyl groups with alcohols and heptafluorobutyric anhydride was described[254] (see p. 73). In a different approach, triterpene alcohols were esterified with acetic acid by reaction in the presence of butyric anhydride.[255]

1.14 Coupling methods

Sheehan and Hess[256] introduced dicyclohexylcarbodiimide (DCCI) as a coupling agent for peptide bond formation between a free carboxyl and a free amino group. The chemistry of carbodiimides[257,258] and coupling reagents[259] has been reviewed. The activation of the carboxyl group may be carried out in the presence of moisture, unlike other methods which make use of mixed anhydride formation. When the active ester reacted with an alcohol virtually quantitative esterification occurred:[260]

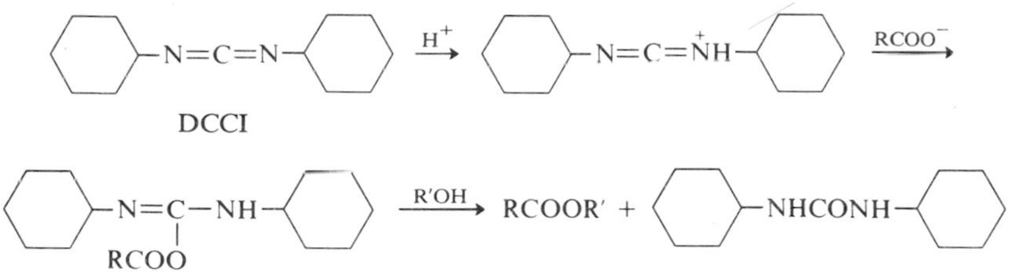

N,N'-dicyclohexylurea

The reaction may be catalysed with a tertiary amine such as tributylamine or pyridine and by raising the temperature (see p. 82). This coupling reagent was used to couple α-hydroxyphosphonic dialkyl esters with a carboxylic acid to prepare acylated α-hydroxyphosphonate esters[261] (see p. 82). Dicyclohexyl urea, which is an end-product of the reaction, is soluble to some extent in many organic solvents and is not always easily isolated from the ester formed. The products of the reaction depend on the carbodiimide, the acid, the solvent and the temperature.[257]

Phosphorus-containing derivatives may be detected at high sensitivity with the alkaline flame ionization detector, as used with steroid phosphinic esters,[262] *N*-diethylphosphate amino acid methyl esters[263] and 1-hydroxy-2,2,2-trichloroethylphosphonic acid esters (insecticides).[264]

When *N*-*N'*-carbonyldiimidazole is used for coupling, the end-product imidazole is soluble and easily removed[265] (see p. 83). This reagent, however, is sensitive to moisture and must be protected.

Sulfonates of strongly acidic *N*-hydroxy compounds were introduced as novel coupling reagents.[266,267] These offered fewer side-reactions than those given by the DCCI method, with easier removal of by-products of the reaction, no observed racemization and with no dehydration from ω-amide groups of *N*-protected asparagine or glutamine. The active esters were generally stable and crystalline, and the method is recommended for labile carboxylic acids and alcohols. Coupling with 6-chloro-1-*p*-chlorobenzenesulfonyloxybenzotriazole[267] is described on p. 84.

The silver ion-induced reaction of 2-pyridyl esters of thiocarboxylic-*S*-acids with alcohols at room temperature under neutral conditions was reported.[268] This is a new approach to esterification, as follows:

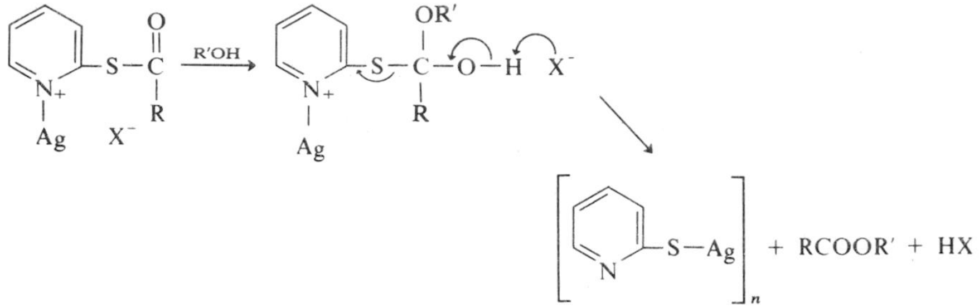

By attachment of the silver ion to the pyridine nitrogen, the acceptor activity of the 2-thiopyridyl residue was increased for rapid nucleophilic attack by the alcohol at room temperature. A neutral polynuclear complex consisting of equimolar amounts of silver ion and 2-pyridine thiolate anion was released. The 2-pyridyl ester of 3-phenylthiopropionic acid and 2-propanol were reacted in benzene solution with an equimolar quantity of silver tetrafluoroborate or silver perchlorate for 10 min at room temperature to give 3-phenylpropionic 2-propyl ester in 70–85% yield.[268]

Esters may also be formed from *N,N'*-dicyclohexyl-*O*-alkyl isoureas in high yield,[269] with the formation of dicyclohexylurea (see p. 85):

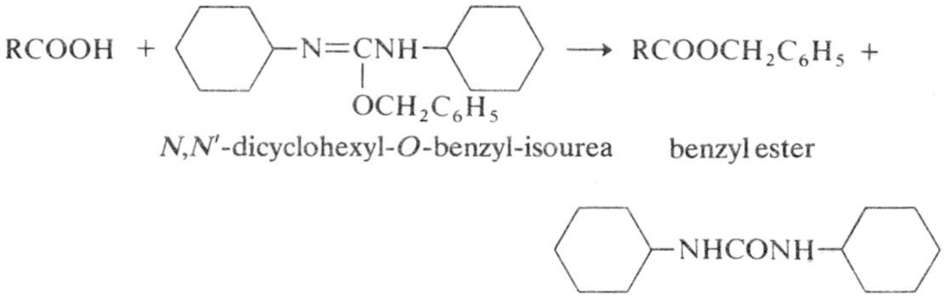

N,N'-dicyclohexyl-*O*-benzyl-isourea benzyl ester

N,N'-dicyclohexylurea

Methyl, ethyl, iso-propyl, *n*-butyl and benzyl esters were prepared in yields greater than 90%, but aromatic *t*-butyl esters less than 40%.[269]

Benzyl esters of hydroxymonocarboxylic and dicarboxylic acids[201] and C_1 to C_{12} fatty acids[202] were prepared by the isourea method, as an alternative method to the use of phenyldiazomethane. Oxaloacetic and oxoglutaric acids, gave no benzyl esters, either by this method or with phenyldiazomethane, but lactic and 3-hydroxybutyric acids were esterified and alkylated to give the benzoxy ether benzyl ester derivatives.[201]

1.15 *N,N*-dimethylformamide dialkyl acetals

The preparation of amide acetals (amino acetals) was first described by Meerwein.[270] By reaction with dialkylacetals of *N,N*-dimethylformamide, esters of both acids[271,272] and phenols[271] were prepared with yields of 64% to 92%. The reaction mechanism was of the S_N2 type with alkylation of the carboxylate oxygen atom:

$$RCOO^- + (CH_3)_2NCH(OR')_2 \rightarrow RCOOR' + R'OH + (CH_3)_2NCHO$$

dialkyl acetal of *N,N*-dimethylformamide; dimethylformamide

This reaction was developed by Thenot *et al.*[273] for use with GLC. Fatty acids were reacted with *N,N*-dimethylformamide dialkyl acetals under slightly alkaline conditions, with pyridine at 60 °C for 10–15 min in a reaction vial, and an aliquot was taken for analysis by GLC.[273] In an alternative procedure, the reagents were directly introduced on to a GLC column with a microsyringe using a sandwich injection technique. Reaction occurred in the heated inlet port, and the esters were swept through the column by the carrier gas. A limited number of results were published for the quantitative determination of long chain fatty acid methyl, ethyl, *n*-propyl, *n*-butyl and *t*-butyl esters[273] (see p. 74). The method is very promising, although a critical report of this work stated that the reaction, though rapid, was not necessarily quantitative.[274]

N-protected amino acids and peptides were converted to their benzyl and *p*-dodecylbenzyl esters[272] and *N*-protected amino acids to their methyl, ethyl, isobutyl, neopentyl and benzyl ester derivatives.[275] The method was extended to amino acids,[276,277] where *N*-dimethylaminomethylene alkyl esters,

$$\begin{array}{l} RCHN{=}CHN(CH_3)_2 \\ \;\; | \\ \;COOR' \end{array}$$

were prepared and analysed by GLC[276] and MS.[276,277] Reaction occurred with carboxyl, amino and guanidino groups but not with hydroxyl groups.

1.16 Reaction gas chromatography

Chemical reactions can be carried out in conjunction with gas chromatography, by allowing reaction to occur in a pre-column reactor, in the column or in a post-column reactor. Esterification can easily occur with alcohol catalysis by strong acid, or with diazomethane (see Chapter 8 and Ref. 278 for short review).

Methylation of aliphatic and aromatic carboxylic acids by thermal degradation (pyrolysis) of their quaternary ammonium salts in the heated inlet zone of a gas chromatograph, first shown by Robb and Westbrook[279] was further developed[280] (see p. 85). Tetramethylammonium salts in methanol solution were pyrolysed in the inlet port, and the corresponding methyl esters were formed. This method of liquid injection was criticized because it gave satisfactory yields only when high concentrations were used.[281] A method was developed for aqueous solutions of the tetramethyl ammonium salts. These were dried in a special probe device,[281] which was then heated in the gas chromatograph for pyrolysis of the dried salts.[281–283] Polyunsaturated fatty acids could be quantitatively determined only if the salt solution was adjusted to pH 7.5–8.0 prior to drying and pyrolysis.[283] This problem of destruction caused by excess alkalinity was solved in a different way by 'ester neutralization' (see below). Results obtained with fatty acid methyl esters by pyrolysis were comparable with those obtained using BF_3–methanol esterification.[281] Tetramethylammonium hydroxide was used to saponify fat, and the liberated salts were then directly pyrolysed to their methyl esters.[284]

Because of difficulties with the GLC determination of formic and lactic acids by pyrolysis of the tetramethylammonium salts, the tetrabutylammonium salts were prepared.[285] Single peaks for each of the *n*-butyl esters of formic, acetic, propionic, isobutyric, *n*-butyric, lactic, valeric and caproic acids were separated in 34 min on a column with 15% Silicone DC-550. As reported with other pyrolysis methods, malonic and oxalic acids were decomposed. Phenylalanine in serum was determined by GC–MS as its *N*-neopentylidene methyl ester by pyrolysis of the tetramethylammonium salt of phenylalanine in the presence of pivaldehyde.[286]

Trimethylanilinium salts of fatty acids were also pyrolysed.[287] Although *N,N*-dimethylaniline was known to be an end-product of the reaction, the formation of interfering amounts of anisole was confirmed by GC–MS.[288] It was recommended that injections with solutions of anilinium hydroxides should be made in dimethylformamide.[288] Urinary homovanillic, vanillylmandelic[289] and phenolic acids[290] were pyrolysed as their trimethylanilinium salts and converted to their *O*-methyl methyl ester derivatives. Chlorophenoxyisobutyric acid in plasma and urine was also pyrolysed as the trimethylanilinium salt.[291]

Trimethyl (α,α,α-trifluoro-*m*-tolyl)-ammonium hydroxide was used to prepare the salts of serum free fatty acids (14:0 to 20:4)[292] (see p. 86). The better the leaving group the lower the inlet temperature which is required, and this decreases the chance of side-reactions and degradation of the compounds under test.

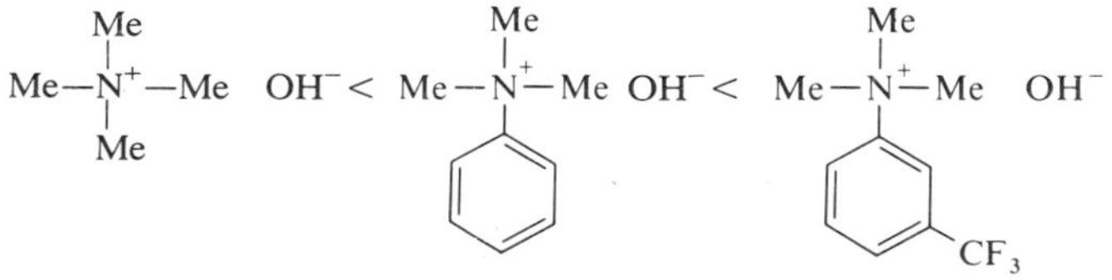

Ease of leaving group during methylation

The fluoro-reagent caused least destruction of polyunsaturated fatty acids, and when the sample was injected with a stoichiometric excess of methyl propionate the excess alkalinity of the reagent was neutralized by so-called 'ester neutralization' and no degradation of linoleate, linolenate or arachidonate was detected.[292] The most important limitation on the use of pyrolytic procedures is the availability of a suitable injection system, such as the probe described,[281] although liquid injection by microsyringe has been used.[279,280,288] Inlet port temperatures of less than 300 °C are normally used and are available on most modern GLC apparatus.

Sodium or potassium salts of fatty acids were reacted with potassium ethyl sulfate at 300 °C for about 10 s. The yield of ethyl esters by GLC was not quantitative, but the method was useful in the presence of large amounts of water.[293]

Trimethylsilyl derivatives of long chain fatty acids[294] and methyl esters (by the use of diazomethane) were formed by on-column reaction.[171] Boron trifluoride-methanol esterification is not satisfactory as an on-column technique, because BF_3 accumulates on the column and affects subsequent injections.

Amino acids were reacted in a reaction vial with trimethylanilinium hydroxide and pivaldehyde at 80 °C for 15 min, and a sample was then injected for pyrolysis in the gas chromatograph.[295] The *N*-neopentylidene amino acid methyl esters were studied by GC–MS. Arginine and cysteine gave several peaks, but histidine did not chromatograph. A variety of *N*-protected di- and tri-peptide trimethylanilinium salts were inserted into the probe for MS studies of pyrolytic methylation.[296]

Thermal decomposition of carboxylic acid benzyldimethylanilinium salts by refluxing in solution was used to prepare benzyl esters. This method was useful for sterically hindered carboxylic acids[297] (see p. 89), and could be applied to reaction-gas chromatography.

1.17 Methods for sterically hindered carboxylic acids

Problems may occur if other functional groups are sensitive to the acidic conditions often used for esterification. The equilibrium between the acid and the ester may be unfavourable or steric hindrance may cause a slow and incomplete reaction. Low yields are usually obtained with mineral acid catalysis, and sometimes better methyl esterification is obtained by the use of diazomethane, reaction of the acid chloride with methanol, or use of the silver salt-methyl iodide method. Crown ether catalysis should be explored further; for example, the *p*-bromophenacyl ester of the hindered 2,4,6-trimethylbenzoic (mesitoic) acid was prepared in 98% yield[227] (see p. 89).

The difficulty in esterifying hindered acids is probably due to an increased non-bonded interaction in the tetrahedral intermediate (see previous discussion on $A_{AC}2$ mechanism of esterification) and it was considered that the use of trialkyloxonium salts would prevent this difficulty by causing attack to occur at the more remote oxygen atom of the carboxyl group rather than at the acyl carbon atom:[298]

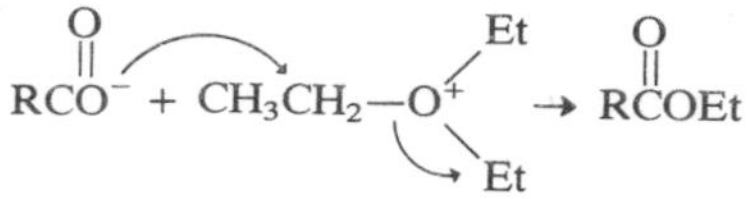

The carboxylate ion was generated by adding the bulky organic base, diisopropylethylamine[299] and ethyl esterification was achieved with triethyloxonium fluoroborate[300] (see p. 87).

Using the BF_3–methanol-etherate reagent, *p,p'*-diphenyldicarboxylic acid was converted under reflux to the corresponding dimethyl ester, and the very sensitive 1,4-dihydrobenzoic acid was similarly converted to its ethyl ester by refluxing with BF_3-etherate in ethanol.[126]

Dimethyl sulfate is more often used for phenols and alcohols, although it is also used for carboxylic acid esterification.[301] It reacts by hydrolysis of the dimethylsulfate by the carboxylate anion, as follows:

$$RCOO^{-}CH_3OSO_2OCH_3 \rightarrow RCOOCH_3 + {}^{-}OSO_2OCH_3$$

Mesitoic and triphenylacetic acids were esterified under basic conditions in high yield, >95%.[301]

In the presence of catalytic amounts of various proton donating agents, e.g. sulfuric acid or *p*-toluenesulfonic acid, carboxylic acids react with alkyl *t*-butyl ethers to form esters:[302]

$$R'COOH + ROC(CH_3)_3 \xrightarrow{H^+} R'COOR + CH_2{=}C(CH_3)_2 + H_2O$$

The reactants are heated under reflux for a few minutes without solvent, with the quantitative evolution of isobutylene.

The sodium salts of aliphatic, aromatic and sterically hindered acids were converted to their esters by pyrolysis of their primary alkyl chlorosulfites.[303] Highly hindered acids were converted to their tetramethylammonium salts, and these were pyrolysed to give the methyl esters in 63–90% yield.[304] Benzyl esters were prepared by thermal decomposition of their benzyldimethylanilinium salts[297] (see p. 88). Several hindered methylbenzoic acids were esterified in the presence of conc. sulfuric acid in 90% yield[70] and by using trifluoracetic anhydride.[249] Ethyl esters were prepared by heating ethylorthoformate with the hindered acid.[305] Hindered carboxylic acids were converted to their sodium salts, and these were reacted with alkyl halide in the presence of hexamethylphosphoramide[306,307] (see p. 89).

2 PRACTICAL METHODS

2.1 Azeotropic method[61]

A 500 ml round-bottomed two-necked flask is equipped with a magnetic stirring bar, a thermometer, a condenser with water separator and a calcium chloride drying tube. Into the flask is put 0.1 mol of an amino acid or its hydrochloride, 200 ml of 95% ethanol, 150 ml of benzene (N.B. carcinogenic) and 20–30 ml of conc. HCl. The reaction is heated on an oil bath at 80–85 °C with stirring. In about 30 min, 200 ml of the azeotropic mixture (alcohol–benzene–water, b.p. 66 °C) is removed and the temperature will rise to 72 °C. At this point, 50 ml of ethanol and 150 ml of benzene are added and distillation continued for another 30 min until the temperature again reaches 72–74 °C. After refluxing gently for a further 90 min, 50 ml of ethanol and 50 ml of benzene are added and in 15 min a further 100 ml of azeotropic mixture is removed. After refluxing gently for a further 60 min, the reaction mixture is concentrated at 50 °C with reduced pressure under N_2 atmosphere until a dry or oily residue is obtained. After cooling, the mixture is stirred with 100 ml of diethyl ether, filtered by suction and the amino acid ethyl ester hydrochloride dried at about 50 °C and 0.1 mm Hg (13.3 Nm^{-2}).

2.1.1 COMMENTS

The method is of general application. Other azeotrope-forming alcohols can also be used. The method permits the use of 95% alcohol and ordinary conc. HCl. Starting with a pure amino acid, the final product was claimed to be equal to the purified material described in the literature.

2.2 Direct esterification of an ammonium salt[308]

Method for *n*-butyl acetate: 1 mol of ammonium acetate is refluxed for 5–10 h with 2.5 mol of *n*-butanol using a Dean and Stark water separating trap. Ammonia is removed by passing non-condensing vapour into boric acid.

2.2.1 COMMENTS

This may be useful where the acid is sensitive to other reactants. Yields were 40–70%. No esters were obtained from ammonium citrate, 2-ethyl hexanoate, salicylate or phthalate.

2.3 Methyl or ethyl esters with sulfuric acid catalysis[93]

Free fatty acids are methylated with 2–3 ml of 2% sulfuric acid in a methanol–light petroleum (b.p. 40–60 °C) mixture (4 : 1, v/v) at 70 °C for 1 h. Glycerolipids are transesterified with the same reagent at 70 °C for 4 h. Ethanol may be substituted to give the ethyl esters.

2.3.1 COMMENTS

The method is simple. It is not claimed to be quantitative. Details are given for brominating unsaturated fatty acid esters and for assessing the relative proportions of the main fatty acid classes in a mixture, using GLC, TLC and MS.

2.4 Methods with BF_3 or BCl_3

1. To the acid (1–20 mg) in a test tube with a ground glass joint is added 1 ml of BF_3–methanol (14% w/v) or BCl_3–methanol (10% w/v) and the solution is boiled under reflux over a steam bath for 2 min. With short chain (up to 10 carbons) fatty acids, the esters are extracted with a hydrocarbon such as *n*-pentane or *n*-heptane. With long chain fatty acids petroleum ether (b.p. 40–60 °C) is used (see also Refs. 131 and 140). BF_3–butanol (14% w/v) is used to make *n*-butyl esters with a refluxing time of 10 min. A 50% aqueous solution of NaCl (1 ml) is added, shaken and centrifuged if necessary to separate out the butanol layer containing the butyl esters. (See Ref. 309 for comparison of methods using HCl, BF_3-*n*-butanol and di-*n*-butyl carbonate.)

2. Butylation of the herbicide 2,4-dichlorophenoxy acetic acid was about 97% efficient[310] using the following method (see also Refs. 148 and 180): the acid sample is refluxed for 20 min with 10 ml of *n*-butanol and 3 ml of BF_3-*n*-butanol reagent (0.2 g ml^{-1}).

3. The acid (25–100 μg) is put in a screw-capped 3 ml tube with BF_3 (14% w/v) in the selected alcohol (0.15 ml) and left to stand at 100 °C for 20 min,[151] and cooled in an ice-bath. After adding 0.2 ml of dry diethyl ether, 0.2 ml of *n*-pentane and 0.15 ml of a mixture of *n*-pentane and triethylamine or pyridine (1:1, v/v) with shaking after each addition (neutralization is exothermic), the mixture is centrifuged and the supernatant (1 μl) injected on to the GLC column. An internal standard, such as phenanthrene, may be added.

2.4.1 COMMENT

Using this method, the isopropyl and *n*-butyl esters of benzoic, fumaric and stearic acids were prepared.[151]

2.5 2-Chloroethyl esters with BCl_3[311]

About 10 μmol of short chain fatty acids are heated with 1–2 ml of BCl_3-2-chloroethanol mixture (1:9) at 100 °C for 30 min. The mixture is cooled in ice and the esters extracted into petroleum ether (b.p. 30–60 °C). The organic layer is passed through a filter paper containing anhydrous sodium sulfate and concentrated by evaporation with nitrogen for analysis by GLC–EC or TLC.

2.5.1 COMMENTS

These esters may also be prepared from the free acids or from the methyl esters by transesterification with 2-chloroethanol and sulfuric acid catalysis [98:2 (v/v) at 60 °C for 2 h]. They are then extracted into petroleum ether for GLC–FID.[312] Methods for preparing these esters were compared and

9–11% BF_3 required 10 min, whereas 5–7% HCl catalysis required 1 h at 100 °C.[313] Chloroethanol is best purified by extracting 10 times with an equal vol. of hexane to remove β,β'-dichlorodiethyl ether. Sufficient chloroethanol for a few days only should be purified.[312]

2.6 Methods with thionyl chloride

1. The acid is placed in a round-bottomed flask with 20 ml of dry methanol (or other alcohol) and cooled in a solid CO_2-acetone bath. Thionyl chloride (2–4 ml) is added slowly with caution and thorough agitation of the flask. The flask is closed with a $CaCl_2$ drying tube and allowed to stand at 40 °C for 2 h. The flask contents may be taken to dryness with a rotary evaporator.
2. The dry peptide is suspended in 1 ml of MeOH and cooled in a solid CO_2-acetone bath and 0.4 ml of thionyl chloride are slowly added with shaking. After a few minutes, the reaction is heated to 45 °C for 30 min, taken to dryness, and acylated with 200 μl of pentafluoropropionic anhydride. It is allowed to stand at room temperature for 15 min, taken to dryness and then dissolved in dry dioxan for GLC on a column of 1% Dexsil 300 GC.[314]
3. The peptide in methanol is refluxed for 30 min with the addition of 2 drops of thionyl chloride. The solution is concentrated on a rotary evaporator.[315]

2.6.1 COMMENTS

The first method is based on that of Brenner *et al.*,[42] which depends on the formation of methyl sulfite as intermediate. Thionyl chloride should be redistilled from linseed oil before use.[112]

2.7 Methods for trimethylsilyl esters[206]

1. The dried sample to be reacted (1 mg or less) is put into a reaction vial and 100 μl of dry pyridine, and 100 μl of bis-trimethylsilyltrifluoroacetamide containing 1% trimethylchlorosilane are added. After thorough mixing, the closed vial is left for 16 h at 60 °C. The time of incubation may be reduced but overnight is convenient. Samples may be withdrawn by microsyringe through the septum and injected onto the GLC column of 3% OV-1 or 3% OV-17.
2. In a small reaction vial 1–2 mg of sample is shaken with 100 μl of *N*-methyl-*N*-trimethylsilylacetamide for 5 min at room temperature. If the sample does not go into solution, heating to 60–100 °C may be needed.[317]

2.7.1 COMMENTS

The mixture of pyridine, hexamethyldisilazane and trimethylchlorosilane (1:0.2:0.1, by vol.) was introduced in 1963,[316] and is still widely used.[211,212]

2.8 Trimethylsilyl esters of amino acids[219]

This method enables all the reactive groups of amino acids to be trimethylsilylated for GLC in a one-step chemical reaction. The reagent *N,O*-bis-(trimethylsilyl)trifluoracetamide (BSTFA) yields more volatile by-products.

2.8.1 PROCEDURE

The amino acid sample in a reaction vial is carefully dried with a stream of N_2 gas at 60–70 °C, and 0.5 ml of CH_2Cl_2 is added and taken to dryness to remove water azeotropically. This azeotropic distillation may be repeated several times. An accurately known amount of internal standard (e.g. phenanthrene) dissolved in acetonitrile may be added. A total of 0.25 ml of acetonitrile should be added for each 1 mg of total amino acids. Then 0.25 ml of BSTFA (30 mol excess) is added for each 1 mg of total amino acids. The vial is securely closed and placed in an ultrasonic bath for 1 min, then heated for 2.5 h at 150 °C in an oil bath. After cooling a sample is injected onto a GLC column of 10% OV-11 avoiding exposure to atmospheric moisture.

2.8.2 COMMENTS

Silylation for 15 min at 150 °C is adequate for reproducible derivatization of 16 amino acids, but 2.5 h is required for glycine, arginine, lysine and glutamic acid. Asparagine and glutamine are differentiated from aspartic and glutamic acids:[221] this does not occur with other GLC methods for amino acids. Several other relevant papers have been published.[217,218,221] It should be noted that some amino acids give multiple peaks. An MS study of trimethylsilylated amino acids and amines has been published.[318]

2.9 Trimethylsilyl esters of C_1 to C_5 fatty acids in serum and urine[214]

2.9.1 PROCEDURE

A closed reaction vial with 50 μl of an ether solution containing the fatty acids and 500 μl of trimethylsilylimidazole is heated at 60 °C for 15 min. A GLC column of 3% OV-17 is used for analysis.

Preparation of urine sample: two ml of urine are adjusted to pH 13 with 5 M NaOH, then saturated with NaCl and extracted three times with diethyl ether. The ether extracts are discarded. The aqueous phase is acidified to pH 1–2 with 5 M HCl and extracted three times with 2 ml of ether. The combined extracts are dried over anhydrous sodium sulfate. An aliquot is taken for silylation. Preparation of serum sample: serum (1–2 ml) is deproteinized by adding an equal volume of ethanol. After centrifuging, the supernatant is analysed in the way described for urine.

2.9.2 COMMENT

The excess of the reagent trimethylsilylimidazole did not interfere with the GLC separation of the first 16 esters of short chain carboxylic acids. It obscured the GLC profile thereafter.

2.10 Methyl esters of amino acids with HCl[319,320]

The amino acid sample (2 mg or less) is dried in a test tube with a B14 ground-glass joint; about 2 ml of 4 M HCl in dry methanol is added and incubated, preferably after flushing with N_2 gas, at 70 °C for 90 min. The solution is taken to dryness on the rotary evaporator, or with a stream of N_2 gas at about 50 °C. To remove traces of water, 100 μl of dry CH_2Cl_2 is added and taken to dryness. The ester hydrochlorides in the stoppered tube must be kept under anhydrous conditions. An alternative method of esterification, giving essentially similar results, is the use of 1 M HCl in methanol at 70 °C for 16 h; 0.1 ml of trifluoroacetic anhydride is added, the stoppered tube left at room temperature for 20–30 min, and taken to dryness at 0 °C to prevent losses due to volatility. The residue is dissolved in 100 μl of methyl ethyl ketone or acetonitrile for GLC.

2.10.1 COMMENTS

These were compromise conditions giving the optimum overall yield for all the protein amino acids.[120] Valine and isoleucine were quantitatively esterified (±2%), but methionine was partially degraded (10%). Many of the amino acids, including methionine, were quantitatively esterified with 4 M HCl in methanol at 70 °C in 30 min or less.[321]

Figure 1 shows the separation achieved with a biological sample, using this method.

It should be noted that, with all methods for gas chromatography of protein amino acid derivatives, there are problems. These relate mostly to the acylation reaction, particularly with certain amino acids, such as histidine, arginine and cystine, and it is necessary to refer to the original publications for specific details.

2.11 Methyl esters of peptides with HCl

Dipeptides were reacted with 1 M HCl in methanol at 20 °C for 2 h.[322] Di- and tri-peptides have also been reacted with 1–2 M HCl in methanol overnight at room temperature.[323]

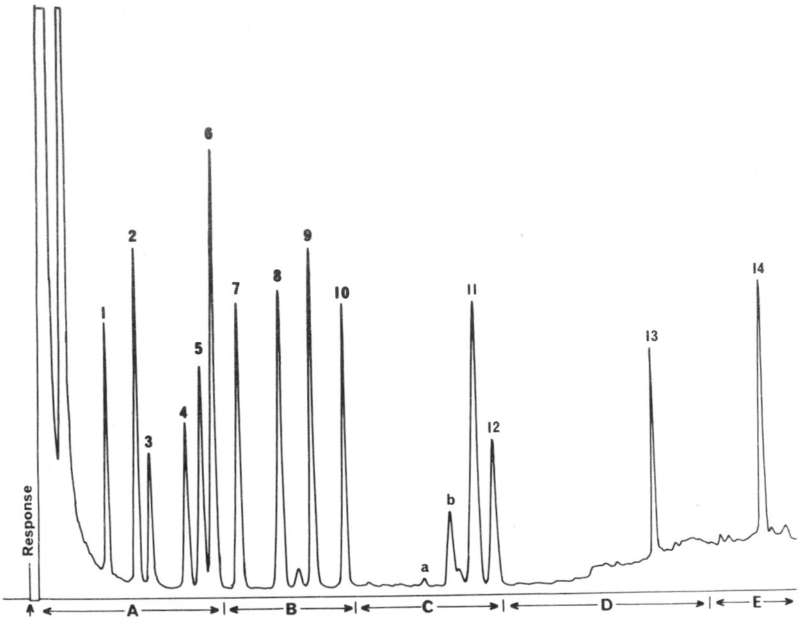

Fig. 1 GLC separation of trifluoroacetylated amino acid methyl esters from a hydrolysate of human finger-nail clippings. Hydrolysis with 6 M HCl at 110 °C for 22 h. Methyl esters prepared with 4 M HCl in methanol at 70 °C for 90 min and trifluoroacetylation at 150 °C for 10 min. Mixed silicone stationary phase.[319] Temperature programme: A—100 °C isothermal; B—100 °C × 1.5 °C min^{-1} for 11 min; C—116.5 °C × 4 °C min^{-1}; D—140 °C isothermal; E—140 °C × 6 °C min^{-1} to 210 °C. GLC peaks: 1—ala; 2—val; 3—gly; 4—ile; 5—thr; 6—leu; 7—nle, internal standard; 8—ser; 9—pro; 10—asp; 11—glu; 12—phe; 13—tyr; 14—lys. Breakdown product of (a) methionine and (b) glutamic acid. (Unpublished material, by kind permission of Miss Mavis S. Greaves, Sheffield Centre for the Investigation and Treatment of Rheumatic Diseases, Nether Edge Hospital, Sheffield, U.K.)

2.12 Tri-deuterated methyl esters

The fatty acid (1 mg) is allowed to react with 30 μl of tri-deuter-8 (Pierce) for 15–30 min at 50–60 °C. Perdeuterated fatty acid esters were analysed by GC–MS.[324]

2.13 *n*-Propyl esters of amino acids with HCl

2.13.1 SINGLE-STEP ESTERIFICATION

1. To the dried amino acid sample 2 ml of 3.5 M HCl in *n*-propanol is added, and shaken or sonicated until dissolved and heated at 110 °C for 30 min.[325]
2. The dried amino acids are treated with 1 ml of 8 M HCl in *n*-propanol and 50 μl of dimethoxypropane and heated at 110 °C for 20 min.[73]

3. The dried residue is treated with 10.0 ml of 7 M HCl in *n*-propanol and heated at 100 °C for 10 min.[326]

4. The dry amino acids are treated with 0.4 ml of 6 M HCl in *n*-propanol and heated at 150 °C for 3.5 min.[327]

2.13.2 TWO-STEP ESTERIFICATION

The dried amino acids are treated with 0.4 ml of 8 M HCl in *n*-propanol, heated at 100 °C for 10 min, taken to dryness, and the esterification procedure repeated.[328–330]

2.13.3 COMMENTS

The *n*-propyl esters were further derivatized for GLC. The heptafluorobutyryl derivatives [325,327,329,330] and the acetylated derivatives[73,326,328] were all shown to give excellent GLC separations on conventional packed columns, except for one separation on a short capillary column.[330]

2.14 *n*-Butyl esters of amino acids with HCl[331,332]

2.14.1 PROCEDURE

A solution of amino acids (5–100 μg of each) is placed in a micro-reaction vial. The vial is placed on a black surface and evaporated just to dryness under an infrared lamp. A clean watch glass is suspended over the vial to exclude air-borne contamination. After all visible moisture is removed, the vial is sealed, allowed to cool and 50 μl of dry CH_2Cl_2 added and evaporated under the lamp to remove any moisture azeotropically. Then 100 μl of 3 M HCl in *n*-butanol are added and the tightly-capped vial put into an ultrasonic bath for about 30 s and then transferred to a 100 °C sand-bath for 30 min. Only the lower part of the vial is submerged in the sand, thus allowing reflux to occur. The vial is removed from the sand-bath, allowed to cool, and evaporated to dryness under the lamp. After sealing, the vial is allowed to cool, and 50 μl of CH_2Cl_2 added and evaporated to dryness.

Acylation is done by adding 100 μl of trifluoroacetic anhydride—CH_2Cl_2 solution (1:10 v/v), capping, sonicating for about 30 s and heating at 100 °C for 20 min (150 °C for 5–10 min is better for arginine, tryptophan and cystine). Analysis is on a GLC column of 0.65% EGA.

2.14.2 COMMENTS

This method described[331] was developed for very small amounts of amino acids[332] and emphasizes the use of dry reaction conditions, pure reagents and clean glassware. See other papers for the separation of these esters,[333] for the preparation of biological samples,[334] and sources of contamination.[335] Earlier papers by Gehrke and co-workers described methods using transesterification[59,60,80–82] and its application to biological samples.[336,337] The use of a solvent-venting device was also described.[331] Other workers have used similar methods.[58,338–341] Histidine[342] and tryptophan[343] have been considered in detail separately.

2.15 Isobutyl esters with HCl catalysis and by transesterification

The amino acid sample (1.0–5.0 μmol total amino acids) is dried in a reaction vial either by freeze-drying or under a stream of N_2 gas. Any remaining traces of water are removed azeotropically with 200 μl of CH_2Cl_2.

2.15.1 TRANSESTERIFICATION METHOD[344]

A solution of 1.25 M HCl in methanol (200 μl) is placed in an ultrasonic bath for 30 s and left at room temperature for 30 min. The reagent is removed with a stream of N_2 gas. After addition of 200 μl of CH_2Cl_2 and evaporation, the residue is dissolved in 200 μl of 1.25 M HCl in isobutanol and the vial heated at 110 °C for 150 min. After cooling, the sample is taken to dryness.

Acylation is with 50 μl of ethyl acetate and 20 μl of HFBA at 150 °C for 10 min. After cooling, the solution is evaporated to dryness and the residue dissolved in ethyl acetate for GLC with 3% SE 30. The sample is injected onto the GLC column with acetic anhydride in the ratio 2:1 (v/v).

2.15.2 DIRECT ESTERIFICATION[345]

The dried sample is dissolved in 3 M HCl in isobutanol and heated at 120 °C for 20 min. After evaporating the excess reagent the sample is acylated.

2.15.3 COMMENTS

The optimum esterification temperature varied for different amino acids.[345] From the published relative molar response values obtained by GLC, there is little difference between the two methods. Direct esterification is quicker.

2.16 Isoamyl esters of amino acids by transesterification[346]

2.16.1 PROCEDURE

A protein hydrolysate or amino acid solution containing up to 50 μg of amino acids in a reaction vial is dried under a stream of N_2 gas at 70 °C. A freshly prepared solution of acetyl chloride in methanol (0.4 ml of 20% v/v) is added and the closed vessel sonicated for 15–20 s after cavitation occurs. After heating at 70 °C for 30 min, the mixture is evaporated to dryness at 50 °C under a stream of N_2 and 0.4 ml of isoamyl alcohol–acetyl chloride solution (prepared as before) added. The vial is flushed with N_2, closed *in vacuo*, and after sonication as before, heated at 100 °C for 2.5 h, then dried at 70 °C under a stream of N_2.

Further derivatization is carried out by adding 100 μl of ethyl acetate, and 20 μl of HFBA, flushing with N_2 and closing the vial *in vacuo*. The vial is sonicated and then heated at 150 °C for 5 min. The authors recommend special care when evaporating this final solution, for GLC on 3% SP 2100.

2.16.2 COMMENTS

It was claimed that 17 protein amino acids were quantitatively determined by this method, which was a modification of one previously published.[347]

Repeated evacuation and N_2 flushing of the vials was through a stainless steel needle piercing the Teflon® coated disc of the reaction vial. Transesterification of methionine must be done *in vacuo*.

2.17 Esters with diazoalkanes

Diazoalkanes in methanol, ether or other solvent may be prepared and stored in the cold (see p. 92). These may then be used as required. For regular use and for critical work, the generation of diazomethane for immediate use is recommended (see p. 70).

CAUTION: diazomethane presents a health hazard and an explosion risk. Use efficient ventilation and avoid the use of ground-glass joints.

2.17.1 PROCEDURE

The acids to be esterified are reacted with excess of a solution of diazoalkane, preferably at 0 °C, and the temperature may then be allowed to rise. The reaction is completed in a few minutes, as seen by cessation of N_2 evolution. If desired, excess diazoalkane is destroyed with a drop of dilute HCl or acetic acid. The ester solution may be concentrated with a stream of N_2 gas or *in vacuo*. Precautions against volatility losses must be taken.

2.18 *n*-Propyl and *n*-butyl esters with diazoalkanes

The acid (*N*-acetyl aspartic acid from a brain extract)[348] is dissolved in 2 ml of methanol and then stirred with 5 ml of an ethereal solution of diazopropane. After 5 min the sample is taken to dryness, and the residue taken up in ethylene chloride for GLC on a column of 10% Carbowax 20M.

Amino acid hydrochlorides (5 mg) are dissolved in 4 ml of methanol and reacted with excess ethereal solution of diazopropane or diazobutane at room temperature for about 5 min, for the preparation of *n*-propyl or *n*-butyl esters respectively.[349]

2.19 Benzyl esters with phenyldiazomethane[201–203]

Phenyldiazomethane in *n*-pentane is added to a solution of carboxylic acids in *n*-pentane and left at room temperature for about 15 min or until the evolution of N_2 gas ceases and the red colour of phenyldiazomethane persists. Hydroxy- and keto-acids are reacted at 0 °C for at least 30 min. Benzylation depends upon acidity, and long chain fatty acids take longer to react. The excess of phenyldiazomethane may be destroyed by adding 1% ethereal phosphoric acid solution and 1 μl of the decolourized solution may be injected onto the GLC column. It was reported that long chain fatty acids reacted more rapidly in chloroform : methanol solution (2 : 1, v/v).

2.19.1 COMMENTS

Excellent GLC separations of fatty acid benzyl esters were shown with 3% SE-30[201,202] and 3% EGSS–X.[202,203] Some small peaks of unknown origin occurred between the solvent peak and the peak of formic acid benzyl ester. A peak for dibenzyl ether was also detected. The presence of water in the reaction mixture (which reacted with phenyldiazomethane to a small extent only at 0 °C to form dibenzyl ether) was not a problem. Benzene cannot be used as a solvent for phenyldiazomethane because of the formation of phenylcycloheptatriene. An alternative method with *N,N'*-dicyclohexyl-*O*-benzyl-isourea for these esters gave no extraneous GLC peaks[202] (see p. 85).

2.20 Methyl esters with bubbling of gaseous diazomethane[350]

The apparatus shown in Fig. 2 enables about 20 successive samples to be reacted with diazomethane with one charge of reagents.

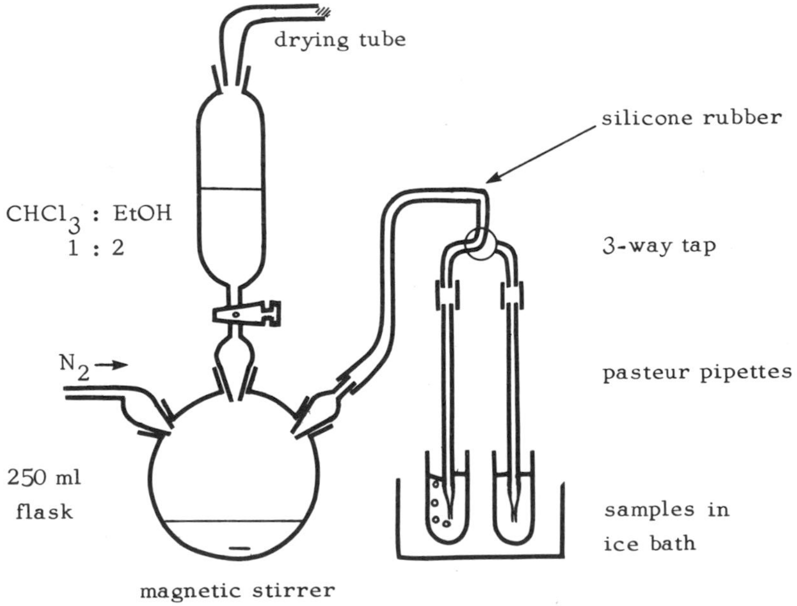

Fig. 2 Apparatus for generating gaseous diazomethane (adapted from original, Ref. 350).

2.20.1 REAGENTS REQUIRED

Hydrazine hydrate; saturated solution of KOH in dry ethanol; chloroform : ethanol (1 : 2, v/v).

2.20.2 PROCEDURE

The chloroform–ethanol mixture is placed in the dropping funnel fitted with a drying tube. The samples to be esterified are dissolved in methanol and/or ether and placed in position in the ice bath. After placing 1 ml of hydrazine hydrate and 14 ml of ethanolic KOH in the 250 ml three-necked round-bottomed flask, stirring is started gently and a slow stream of N_2 gas is passed. The chloroform–ethanol OH solution is added drop by drop. The production of diazomethane will be seen by its yellow colour. An excess of diazomethane in the sample is present when the sample solution remains yellow. The three-way tap allows successive samples to be reacted and the pasteur pipettes to be changed. All glassware and solutions must be kept dry.

2.20.3 COMMENTS

Work in a fume cupboard, away from bright light, and use a safety screen. Similar procedures have also been described for the rapid methylation of micro quantities of acids.[156,351] The use of freshly generated diazomethane at a low temperature, combined with a minimum reaction time, minimizes the likelihood of side reactions,[158,163] especially with biological samples. The preparation of ^{14}C-diazomethane has also been described.[156]

2.21 Methyl esters with gaseous diazomethane prepared *in situ*[352]

2.21.1 REAGENTS REQUIRED

Diethyl ether (peroxide free), redistilled from KOH pellets; KOH, aqueous 50% (w/v).

N-methyl-*N*-nitroso-*p*-toluenesulfonamide (MNTS), 1 g, is dissolved in 3 ml of ether and 2 ml of 2-(2-ethoxy)ethoxy ethanol. This solution is reported to be stable at room temperature in a closed vessel for at least 4 months.

2-(2-Ethoxy) ethoxy ethanol is purified by heating to 110 °C for 1 h with 5% KOH and then distilling under reduced pressure at about 90 °C at 12 mm Hg pressure. The pure solvent does not turn yellow in the presence of KOH.[156]

2.21.2 PROCEDURE

A capillary tube about 2.5 cm long is half-filled with Celite 545 to form a short compact column. The acid sample (0.5–5.0 μg) in ether solution (2 μl) is applied to the column with a microsyringe, and the sides of the capillary are washed down with 2×2 μl portions of ether. The capillary is inserted into a rubber septum, with the wetted end down. 10 μl each of MNTS solution and 50% aqueous KOH are placed in a 1 ml serum bottle with a small magnetic stirrer and the septum is pushed into position immediately. Diazomethane is generated with stirring for 2 min. The capillary is removed and the methyl esters are eluted by injecting 10–15 μl of diethyl ether (or other volatile solvent) into the end of the capillary. The first 5–6 μl of solvent emerging from the column contain all the esters, which are taken into a microsyringe for injection on to the GLC column.

The amount of MNTS taken gives approximately 7 μmol of diazomethane.

2.21.3 COMMENTS

This sensitive method for GLC eliminates many of the hazards associated with the use of diazomethane. There is scope for improvement. If a tube having suitable dimensions is used for diazomethane treatment, this could be inserted directly into the heated inlet zone of the GLC column,[353,354] in this way avoiding the need for solvent elution.

2.22 General method with trifluoroacetic anhydride (TFAA)

2.22.1 PROCEDURE

The carboxylic acid (1 mol) in TFAA (1.2–1.5 mol) is allowed to react with the alcohol (1 mol acid/hydroxy group). The reaction is rapid but gentle warming may be necessary.

Less reactive primary, secondary, tertiary and polyhydroxy alcohols, phenols and thiophenols give better yields than reactive primary alcohols (e.g. methanol), which tend to form TFA esters (see Ref. 249).

2.23 Trichloroethyl esters with trifluoroacetic anhydride[251]

The carboxylic acids (10 mg total) are refluxed on a steam bath for 10 min with 1 ml of 2,2,2-trichloroethanol-TFAA mixture (1:9, v/v). The reaction mixture is diluted with ethyl acetate or *n*-heptane for GLC with a flame ionization detector. When an ECD instrument is used, the reaction mixture is diluted with 100 ml of ethyl acetate and passed down a short silica gel column to remove residual trichloroethanol. The effluent may be further diluted for GLC on a column of 3% OV-17.

2.23.1 COMMENTS

The trichloro group is strongly electron-capturing for high sensitivity detection by GLC–EC. The method of preparation is simple. Several substituted benzoic acids were studied by this method. Yields greater than 98% were reported. For preparative synthesis (10 mmol) a different method with sulfuric acid catalysis was proposed. The silica gel method of clean-up caused losses of short chain fatty acids, and for these acids the following method should be used.[252]

2.24 Trichloroethyl esters with heptafluorobutyric anhydride[252]

A chloroform solution of C_2–C_8 aliphatic acids (50 μl) is treated with 50 μl of freshly prepared 2,2,2-trichloroethanol–chloroform solution (1:99, v/v) and 150 μl of HFBA in a closed tube at room temperature for 30 min. After addition of 50 μl of palmitic acid in chloroform (0.25 g/ml) and 50 μl of HFBA the solution is left to react for 15 min. (This removes unreacted trichloroethanol as the palmitic ester which does not interfere with the GLC of the more volatile short chain fatty acid esters.) Further purification is achieved by shaking with 100 μl of chloroform and 100 μl of 0.1 M HCl. The aqueous layer is discarded and the organic layer is washed with 100 μl of 0.1 M NaOH. The organic phase is evaporated to dryness, and the residue dissolved in 0.1 ml of diethyl ether for GLC–EC.

2.25 Heptafluorobutyryl esters of hydroxyacids with heptafluorobutyric anhydride[254]

An ether solution containing about 1–10 μmol of standard hydroxyacids is evaporated almost to dryness in a reaction vial and 100 μl of chloroform is added. An extract from a biological fluid is prepared and treated in the same way.

Both the sample and the standards are treated with 10 μl of an ethanol–chloroform mixture (1:4, v/v). After mixing, 10 μl of pyridine–chloroform mixture (1:3, v/v) are added. After mixing, 30 μl of HFBA are added. After mixing, the solution is concentrated with a stream of N_2 gas to a final volume of about 20 μl. The reaction vial is closed and heated at 100 °C for 4 min. After cooling, 140 μl of chloroform are added, followed by 90 μl of 0.1 M HCl. After thorough shaking and removal of the aqueous layer containing substances which interfere, subsequent GLC analysis may be carried out by withdrawing all of the solution into a Pasteur pipette and then allowing the lower chloroform layer to run back into the reaction vial. The washing procedure is repeated with 90 μl of 0.1 M NaOH. At this stage a microsyringe passing into the chloroform layer can be used to take a sample for GLC. Alternatively, the aqueous layer may be removed and the chloroform layer concentrated for GLC. An internal standard could also be added at this final stage.

2.25.1 COMMENTS

Hydroxyacids are usually fully derivatized for GLC by trimethylsilylation[355] or by esterification followed by acylation.[17,18,146,356] This method carries out both steps simultaneously with HFBA in the presence of an alcohol. Ethanol acts as both the esterifying agent and as a stabilizing agent, to prevent the formation of extraneous GLC peaks with HFBA. Quantitative results were not given and, with experience, the various steps described might be improved or even eliminated, depending on the sample under investigation. The method is sensitive, small volumes of reagents are used and acids, hydroxyacids and amines are derivatized. The derivatives are electron-capturing for GLC–EC and the method described was used for the analysis of traces of hydroxyacids present in synovial fluid. A procedure was also described for drying aqueous samples with magnesium sulfate prior to derivatization.

2.26 Pentafluoropropionyl esters[253]

The dried acids in a reaction vial are treated with 0.25 ml of a mixture of 2,2,3,3,3-pentafluoro-1-propanol in pentafluoropropionic anhydride (1:4, v/v) and heated at 75 °C for 15 min. After cooling and evaporation just to dryness with a stream of N_2, the reaction is completed by further heating with 0.1 ml of pentafluoropropionic anhydride at 75 °C for 5 min. After evaporation just to dryness with nitrogen gas, the residue is dissolved in 1 ml of ethyl acetate for GLC–EC on 3.0% OV-17 or 3.0% DC-200.

2.26.1 COMMENTS

Pyrrolidone carboxylate and γ-glutamyl amino acids gave single, sharp symmetrical GLC peaks, and this suggested that all functional groups were derivatized. In fact this scheme is generally applicable to amino acids.

2.27 Fatty acid esters for GLC with *N,N*-dimethylformamide dialkyl acetals[273]

2.27.1 REAGENTS

Amide acetals: *N,N*-dimethylformamide dialkyl acetals (methyl, ethyl, propyl, *n*-butyl, *t*-butyl), distilled in a moisture-free atmosphere.

2.27.2 PROCEDURE

Either: Sandwich injection technique: a microsyringe is charged in the following order with 1 μl of *N,N*-dimethylformamide dimethylacetal, 1 μl of pyridine containing 2 μg of docosane (or other internal standard) and 1 μl of pyridine containing 2 μg of stearic acid. No mixing is necessary. Inject on to a GLC column of 1% SE-30 at 200 °C, with inlet port at 190 °C, and nitrogen carrier gas flow rate at 40 ml min^{-1}.

Or: Allow the acid in benzene or other solvent to react with pyridine and with the reagent in a microreaction vial at 60 °C for 10–15 min. The time required for reaction depends on the solubility of the acid in the reaction mixture, i.e. when a clear solution is obtained, the reaction is complete.

2.27.3 COMMENTS

The method is rapid and simple, but it is in an early stage of investigation. Quantitative results were assumed (see Ref. 274). Few details were published.

2.28 Amino acid *N*-dimethylaminomethylene alkyl esters with *N,N*-dimethylformamide dialkyl acetals[276]

2.28.1 PROCEDURE

The amino acid in a reaction vial is treated with 100–200 μl of a reagent–acetonitrile mixture (1 : 1, v/v) at 100 °C for 20 min, or until a clear solution is obtained. Aspartic acid required 1 h for solution. Analysis is on a GLC column of 1% SE-30.

2.28.2 COMMENTS

Reaction is rapid once solution has been effected. Different solvents may be used. Hydroxyl groups did not react. Mass spectral studies confirmed the structures of the derivatives.[276,277]

2.29 Methyl esters with methyl iodide[236]

2.29.1 REAGENTS

Redistilled methyl iodide; redistilled acetone dried over Molecular Sieve (Linde 4A); K_2CO_3 dried over P_2O_5 at 170 °C.

2.29.2 PROCEDURE

To the dried sample 2 ml of acetone, 2 ml of methyl iodide and 30–50 mg of K_2CO_3 are added. After refluxing at 60–70 °C under a $CaCl_2$ drying tube for 20–30 min, the solution is made up to 5 ml with acetone and 1 μl injected on to a GLC column of 3% OV-17.

2.29.3 COMMENTS

This method was used for salicylic and acetylsalicylic acids. A similar reaction for barbituric and fatty acids[234] and a microdistillation apparatus[357] have been described.

2.30 Methyl esters with methyl iodide–silver salts[232]

Non-esterified fatty acids are gas chromatographed and trapped from the eluent gas in tubes packed with silver oxide. The trapped fatty acids are reacted in the sealed tube with methyl iodide at 100 °C for 2 min, and the tube is then inserted into the injection port of the gas chromatograph for GLC of the methyl esters. A mixture of fatty acids can be similarly applied to a tube packed with silver oxide. Hydroxyl groups on the fatty acids are converted at the same time to methyl ether derivatives. Benzyl chloride in ether (4% v/v) was used for short chain fatty acid benzyl esters.[233]

2.31 Isopropyl esters with 2-bromopropane[237]

The method offers a one-step reaction for the production of amino acid derivatives for GLC.

$$NH_2CHRCOOH + CH_3CHBrCH_3 \xrightarrow[DMSO]{NaH} \underset{CH_3}{\overset{CH_3}{CHNHCHRCOOCH}} \underset{CH_3}{\overset{CH_3}{}}$$

2.31.1 PROCEDURE

The amino acid (1–10 mg) in a reaction vial is suspended in 2 ml of dimethyl sulfoxide (dried over molecular sieve Linde type 4A). An excess weight of sodium hydride in a separate tube is washed four times with 5 ml portions of dry hexane to remove the oil. The washed sodium hydride is transferred to the reaction vial and excess 2-bromopropane is added. The vial is closed and allowed to stand overnight at room temperature. NaCl-saturated distilled water (3 ml) is added and the derivative is extracted into benzene or chloroform for GLC.

2.31.2 COMMENTS

Mass spectrometry studies confirmed that N-, S- and O-isopropyl derivatization occurred, but derivatization of arginine was not reported. The reaction is applicable to carboxylic acid groups in general. The authors suggested the use of heptafluoro-2-iodopropane for high sensitivity ECD analysis. This reaction was also studied with amines.[238]

2.32 Esters with tetra-alkyl ammonium hydroxide and alkyl halide[240]

2.32.1 REAGENTS

Tetramethylammonium hydroxide (24% (w/v) in methanol), or phenyltrimethylammonium hydroxide (0.1 M in methanol). Appropriate primary aliphatic iodides such as iodomethane, iodoethane or 1-iodobutane.

2.32.1 PROCEDURE

1. Stearic acid (25 mg, 88 μmol) is dissolved in 4 ml of N,N-dimethylacetamide and 0.95 ml of methanol. A 50 μl portion of tetramethylammonium hydroxide solution is added (104 μmol). After addition of iodomethane (880 μmol) and vigorous mixing with a mechanical shaker for 3 min, tetramethylammonium iodide precipitates as the reaction proceeds. A sample from the supernatant may be taken for GLC of the methyl ester (yield 100%).

Shaking for 10 min is necessary for quantitative esterification with ethyl iodide (ethyl ester) or 1-iodobutane (*n*-butyl ester).

2. Stearic acid (14.2 mg, 50 μmol) is dissolved in 4 ml of *N,N*-dimethylacetamide and 1.0 ml of 0.1 M phenyltrimethylammonium hydroxide in methanol (100 μmol). After addition of 1-iodobutane (880 μmol) and shaking for 10 min, the butyl ester is analysed by GLC.

2.32.2 COMMENTS

The method is very mild, proceeding at room temperature. Many primary aliphatic iodides may be used. The method has been applied successfully to bile acids, and promises to be of immense value when fully exploited. 1-Iodohexadecane was used to prepare the ester of benzoic acid.

2.33 Methyl esters from lipids etc. using tetramethylammonium hydroxide and methyl iodide[241]

The lipid material in a test tube (100–200 mg) is treated with tetramethylammonium hydroxide (12% w/v) in methanol (1.0 ml). The fat is saponified by heating on a steam bath until all fat globules disappear. This takes 5–20 min. *N,N*-dimethylformamide (5 ml) is added, shaken, and methyl iodide (0.5 ml) is added, shaken and then tetramethylammonium iodide allowed to precipitate out. Portions of the supernatant may be injected on to the GLC column. Alternatively, it is possible to extract the methyl esters into hexane for GLC.

2.33.1 COMMENTS

The method was applied to lipids, alkyd paint resins and ester plasticizers. *N,N*-dimethylformamide is more volatile than *N,N*-dimethylacetamide used in the previous method.[240]

2.34 Iodomethyltetramethylmethyldisiloxane (IMTMMDS) esters of short-chain acids and iodomethyldimethysilyl (IMDMS) ethers of hydroxyl groups[247]

The acids to be derivatized should be dried in ether solution over anhydrous magnesium sulfate. The ether solution is evaporated in a 0.3 ml reaction vial to small volume (25 μl or less). A pipette having a disposable tip is used to add 10 μl of the catalyst solution consisting of diethylamine and chloroform (1:11, v/v) and 10 μl of a solution of bromomethyldimethylchlorosilane in chloroform (1:11, v/v). After mixing and incubation at 80 °C for 1 h, the reaction is left to cool and the chloroform removed with a stream of N_2. Now the bromo- to iodo-derivative conversion is performed with 100 μl of a saturated (25 °C) solution of sodium iodide in acetone, and incubation at 35 °C for 30 min. After cooling and addition of 0.9 ml (or less) of chloroform, the solution is centrifuged, if necessary, and 1.0 μl of the clear solution injected on to a GLC column of 3% OV-1. The temperature programme is from 125° to 225 °C at 3 °C per min.

2.34.1 COMMENTS

The reacting solutions must be dry. The effects of using different-sized reaction tubes were reported, and this method requires further study to ensure quantitative results. The ester group is large, and short chain fatty acids may be reacted to give derivatives which minimize problems caused by excessive volatility. These iodine-containing compounds may be detected at the picomol level by GLC–EC with a ten-fold increase of sensitivity over the bromo-derivatives, which were previously described.[358] The number of interfering peaks are fewer.

2.35 Phenacyl and other esters using crown ether[230]

2.35.1 REAGENTS

85% KOH in dry methanol (w/v); alkylating agent, *α-p*-dibromoacetophenone; 18-Crown-6-(1,4,7,10,13,16-hexaoxacyclooctadecane) or dicyclohexyl-18-Crown-6-(2,3,11,12-dicyclohexyl-1,4,7,10,13,16-hexaoxacyclooctadecane). Prepare the alkylating agent–crown solution in acetonitrile with a molar ratio 10:1.

2.35.2 PROCEDURE

Either: The total acids (1.0–50 μmol) in a 50 ml pear-shaped flask are dissolved in methanol and neutralized to phenolphthalein end-point by careful addition of alkaline methanol. A rotary evaporator is used to take the solution to small volume, this is then transferred with methanol washings to a 3 ml reaction vial. This solution is taken to dryness, with a stream of nitrogen or *in vacuo*, leaving a residue of potassium salts.

Or: The total acids are placed in a reaction vial with added potassium bicarbonate × 3–5 mol excess and taken to dryness.

Then: To the residue in the reaction vial an excess of a solution of alkylating agent and crown in acetonitrile is added. The total volume is made up to between 0.5 ml and 1.5 ml by addition of acetonitrile. The vial is closed with a Teflon® liner and heated with stirring at 80 °C for 15 min. The vial is then cooled and the solution may be chromatographed directly, or alternatively taken to dryness and the *p*-bromophenacyl esters taken up in a known volume of solvent containing an internal standard.

2.35.3 COMMENTS

These esters may be separated by liquid chromatography because they absorb strongly at 254 nm and this enables the detection of nanogram quantities. Small amounts of crown ethers in molar ratios 1:20 to 1:100 may be used to catalyse the phase transfer of carboxylate salts. The crown ether solid–liquid phase transfer may also be used for the preparation of benzyl, *p*-nitrobenzyl, *p*-bromobenzyl, *p*-chlorophenacyl, *p*-phenylphenacyl and 2-naphthacyl esters.

2.36 *p*-Bromophenacyl and *p*-phenylphenacyl esters in the presence of water[245]

Many methods of esterification utilize non-aqueous conditions. This method was developed to overcome the difficulties of recovering acids from aqueous media by such methods as solvent extraction or distillation. The reaction is as follows:

$$\text{RCOOK} + p\text{-BrC}_6\text{H}_4\text{COCH}_2\text{Br} \rightleftharpoons p\text{-BrC}_6\text{H}_4\text{COCH}_2\text{OCOR} + \text{KBr}$$

p-bromophenacyl bromide — *p*-bromophenacyl ester

$$\text{RCOOK} + p\text{-C}_6\text{H}_5\text{C}_6\text{H}_4\text{COCH}_2\text{Br} \rightleftharpoons p\text{-C}_6\text{H}_5\text{C}_6\text{H}_4\text{COCH}_2\text{OCOR} + \text{KBr}$$

p-phenylphenacyl bromide — *p*-phenylphenacyl ester

2.36.1 PROCEDURE

0.1 M Ethanolic solutions of *n*-C_2–C_{10} carboxylic acids are prepared, singly or as a mixture.

The 0.1 M ethanolic solution of carboxylic acid (2.5 ml) and 2.5 ml of water are put into a 50 ml stoppered flask. After neutralization to phenolphthalein with 5 M KOH, the solution is just made acidic by adding 3 drops of 1 M HCl. Addition of 2.3 mmol of *p*-bromophenacylbromide or *p*-phenacylphenacyl bromide (depending on the ester to be prepared) is followed by refluxing for 10 min. Any solid reagent which remains after bringing to the boil is dissolved by adding a few ml of ethanol. The reaction mixture is cooled and extracted with 25 ml of ethyl acetate. Ice-cold water may be run into the flask until the ethyl acetate layer separates clearly. A centifuge may be used to aid the separation. A portion of the ethyl acetate phase is dried over anhydrous sodium sulfate and samples taken for GLC on 2.5% sodium dodecylbenzene sulfonate or 1% Apiezon L.

2.36.2 COMMENTS

The columns used did not satisfactorily separate formic acid from the reagents used or from acetic acid. The esters gave good GLC peaks. Problems of excessive volatility were minimal. It was claimed that the *n*-C_2–C_{10} carboxylic acids were determined quantitatively as free acids or as salts. The method requires further study for adapting to small quantities. The *p*-bromophenacyl esters are strongly electron capturing and are ideal for GLC–EC.

2.37 *p*-Nitrophenacyl esters[359]

Prostaglandin PGE_2 (4.55 mg, 12.9 μmol) is dissolved in 1 ml of acetonitrile. *p*-Nitrophenacyl bromide (9.38 mg, 38.4 μmol, 200% mol excess) is added, followed by *N,N*-diisopropylethylamine (2.91 μl, 16.9 μmol, 30% mol excess). Esterification is completed after 6 min at room temperature. Analysis is with HPLC and detection by u.v. (λ_{max} 263 nm).

2.37.1 COMMENTS

Other prostaglandins may require slightly different conditions. Heating to 45 °C may reduce the time for esterification but dehydration of the prostaglandins may occur. Using nanomol quantities, particularly for GLC–EC analysis, it is necessary to reduce the amount of *p*-nitrophenacyl bromide to avoid overloading the detector.

2.38 *p*-Bromo-, *p*-methyl- and *p*-nitro-benzyl esters[244]

2.38.1 REAGENTS REQUIRED

KOH in ethanol (3 mg ml^{-1}); *p*-bromobenzyl, *p*-methylbenzyl or *p*-nitrobenzyl iodide dissolved in ethanol (3 mg ml^{-1}); carboxylic acids in ethanol (1 mg ml^{-1}).

2.38.2 PROCEDURE

A microsyringe is used to introduce 10 μl of carboxylic acid solution, 3 μl of KOH and 5 μl of the appropriate benzyl halide solution into a capillary melting point tube. The tube is sealed by flame and incubated at 110 °C for 1 h. After cooling, a sample is injected onto a GLC column of 5% OV-17.

2.38.3 COMMENT

This micromethod was developed for C_1 to C_3 monocarboxylic acids

2.39 Phenacyl esters with phenacyl bromide[360]

2.39.1 REAGENTS REQUIRED

Phenacyl bromide, recrystallized from pentane, 12 mg ml^{-1} in acetone; triethylamine, distilled before use, 10 mg ml^{-1} in acetone. These solutions should be stored at 0 °C.

2.39.2 PROCEDURE

To a reaction vial containing up to 100 μg of fatty acids, 10 μl of phenacyl bromide and 10 μl of triethylamine solution are added, and heated at 50 °C for 2 h, or allowed to stand at room temperature for 8 h or longer.

2.39.3 COMMENT

A mixture of long-chain saturated and unsaturated fatty acid phenacyl esters, from C12:0 to C24:1 were separated by high-pressure liquid chromatography, and detected by u.v. absorption at 254 nm.

2.40 Pentafluorobenzyl esters

2.40.1 MICROSCALE PREPARATION[243]

The acids (5 μmol of each) are dissolved in 100 ml of acetone and refluxed for 3 h with 25-fold mol excess of pentafluorobenzyl bromide (N.B. strong lachrymator) and 10-fold mol excess of alkaline reagent (potassium carbonate or alcoholic KOH). After cooling, 500 ml of diethyl ether and 20 ml of ethyl acetate are added to the reaction mixture. After a brief wash with 10 ml of water saturated with ether, the ether solution is dried over anhydrous sodium sulfate. Solvent is removed with a rotary evaporator at 50 mm Hg (6650 N m^{-2}). The residue is dissolved in 100 ml of hexane containing 1% (v/v) each of acetone and diethyl ether, and 1 ml of this solution is made up to 100 ml with the same solvent for GLC analysis with an ECD on a column of 1:1 (w/w) of 5% FS-1265 and 3% DC-200.

2.40.2 COMMENTS

Pentafluorobenzyl esters of 17 organic acids were prepared in high yield and were detected at the nanogram level by GLC–EC.[243] Losses were possible during the ether washing process and also when taking solutions to dryness *in vacuo*. Phenols and mercaptans reacted under these conditions to give the corresponding ethers and thioethers, respectively.

2.40.3 SUB-MICROGRAM METHOD

This method was developed for prostaglandin $F_{2\alpha}$.[242]

2.40.4 REAGENTS REQUIRED

Pentafluorobenzyl bromide (58 mg, 0.22 mmol) in 100 μl acetonitrile; diisopropylethylamine (8.86 mg, 0.07 mmol) in 100 μl acetonitrile.

2.40.5 PROCEDURE

The dried residue of acid in a reaction vial is treated with 10 μl of each of the two reagents at 40 °C for 5 min, and taken to dryness with N_2 taking precautions against volatility losses. This process is repeated. Prostaglandin is trimethylsilylated with 10 μl of *N,O*-bis-(trimethylsilyl)acetamide at room temperature for 5 min for analysis by GLC–EC on 3% OV-1.

2.41 Coupling with *N,N'*-dicyclohexylcarbodiimide[260]

2.41.1 PROCEDURE

About 10 mmol of carboxylic acid in 25 ml of alcohol and 4 ml of pyridine is treated with 12 mmol (slight excess) of *N,N'*-dicyclohexylcarbodiimide (DCCI) and stirred gently for 30–120 min at room temperature. *N,N'*-dicyclohexylurea is left to settle out, or centrifuged. The clear supernatant (1 μl) is injected onto the GLC column.

2.41.2 COMMENTS

Some aromatic and fatty acids were esterified quantitatively with methanol at room temperature, but for some acids or with other alcohols it was necessary to heat the mixture at 40–80 °C for 30–120 min. If excess DCCI at the end of the reaction interferes with the analysis, it can be decomposed by adding glacial acetic acid. In certain cases less pyridine may be used, and this reduces the size of the tailing GLC peak of pyridine. On the other hand, a solid alcohol such as 1-menthol may need an increased pyridine concentration.

CAUTION: DCCI must be handled with great care. Do not expose the eyes. Use rubber gloves, protective goggles and a fume cupboard.

2.42 Acylated α-hydroxyphosphonate esters using *N,N'*-dicyclohexylcarbodiimide[261]

The analytical procedure enables traces of carboxylic acids to be esterified by α-hydroxyphosphonic acid esters in the presence of dicyclohexylcarbodiimide (DCCI) under mild conditions. The FID, or the more sensitive and specific thermionic (alkaline flame ionization) and flame photometric detectors may be used for GLC detection in the picogram range.

$$RCOOH + (CH_3O)_2\underset{\underset{O}{\|}}{P}CH_2OH \xrightarrow{DCCI} RCOOCH_2\underset{\underset{O}{\|}}{P}(CH_3O)_2$$

2.42.1 REAGENTS REQUIRED

1. DCCI (25.0 g) and pyridine (12.5 g, 12.8 ml) made up to 100 ml with dry benzene. DCCI dissolves with shaking.
2. α-Hydroxyphosphonic dimethyl ester.

2.42.2 PROCEDURE

A benzene solution of the fatty acid sample in a reaction vial (100 μl) is treated with 150 μl of reagent 1 followed by 50 μl of reagent 2. The vial is shaken vigorously to mix and allowed to stand for 1 h at 50 °C. The reaction is allowed to cool and samples are injected for GLC on 15% Apiezon L, 10% Carbowax 20 M, 5% NPGS or 7.5% OV-7.

2.42.3 COMMENTS

Reagent excess was vented by a multivalve switching system before it entered the main GLC column. Conditions for optimizing the yields were studied and retention indexes given for a variety of esters of C_1–C_{10} fatty acids.

Various derivatives typified by Reagent 2 were prepared by condensation of dialkyl phosphites (dimethyl-, diethyl- and methyl-ethyl) with formaldehyde, acetaldehyde, acetone and butan-2-one.[361,362]

2.43 Coupling with *N,N'*-carbonyldiimidazole[265]

The general reaction scheme is as follows:

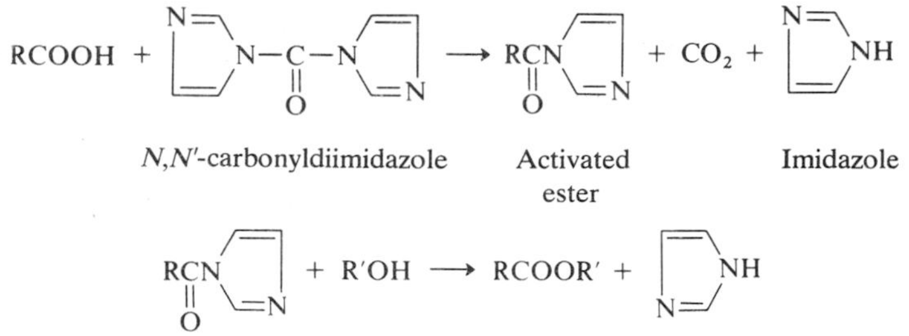

2.43.1 REAGENTS REQUIRED

0.033 M H_2SO_4; 1.0 M aqueous NaOH, saturated with carbon tetrachloride; 0.325 M *N,N'*-carbonyldiimidazole in freshly-distilled ethanol or hydrocarbon-stabilized chloroform, prepared daily (the reagent is sensitive to water and must be stored under anhydrous conditions); triethylamine 10% (v/v) in dry methanol; heptane : isopropanol mixture, 3 : 7 (v/v) containing pentadecanoic acid internal standard; silicic acid washed with heptane activated at 110 °C overnight.

2.43.2 PROCEDURE FOR ESTERIFICATION

The carbonyldiimidazole solution (0.1 ml) is added to the dried residue of fatty acids in a reaction vial and the vial is agitated for about 1 min until the residue is completely dissolved. After addition of 0.1 ml of triethylamine-methanol solution and mixing, the reaction is left for one minute (or more), before addition of 3 ml of the sodium hydroxide solution and mixing on a Vortex mixer for 2 min. The septum of the vial is pierced to release excess pressure, then centrifuged. A microsyringe is used to withdraw a sample from the bottom chloroform layer for GLC on a column of 6% LAC-28.

2.43.3 PROCEDURE FOR EXTRACTING FATTY ACIDS FROM PLASMA

In a glass-stoppered 15 ml centrifuge tube are placed 0.1–1.0 ml of fresh plasma or serum and 0.017 M H_2SO_4 to bring to a total volume of 3 ml. After addition of 4 ml of heptane-isopropanol extraction mixture and vigorous mixing for 10 s with a Vortex mixer, the phases are allowed to separate, or centrifuged briefly if necessary. About 1 ml of the upper heptane phase is put into a small test-tube or vial containing 20–50 mg of activated silicic acid and mixed by rotation for about 10 min. The clear heptane phase is put into a clean vial and taken just to dryness with a stream of N_2. (Excessive drying may lead to losses by volatility.)

2.43.4 COMMENTS

The method was developed for the GLC determination of plasma free fatty acids ($C_{14:0}$–$C_{18:2}$), and, because the reaction conditions are exceptionally mild, triglycerides and cholesteryl esters are not transesterified and phospholipids are less than 20% transesterified. The concentration of carbonyldiimidazole used gave maximum esterification. The reaction was complete in about 0.5 min and even after 5 h, 95% of the maximal value was obtained. Methanolysis of the imidazolide was essentially complete in 1 min with 10% triethylamine in methanol. Standard solutions of fatty acids should be taken through the procedure.

2.44 Coupling with 6-chloro-1-*p*-chlorobenzenesulfonyl-oxybenzotriazole[267]

The general reaction scheme is as follows:

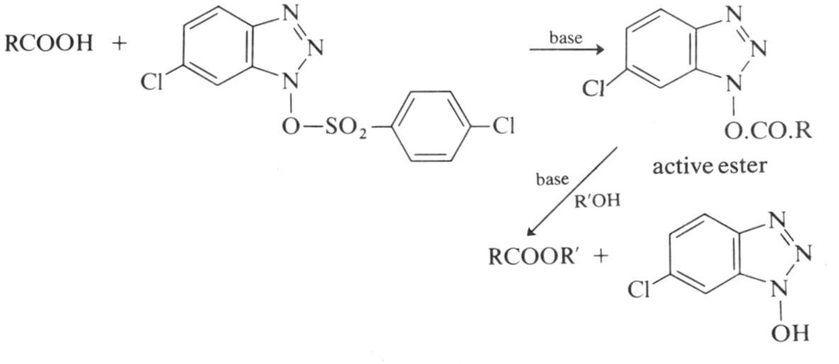

2.44.1 PROCEDURE

Benzoic acid (10 mmol) and triethylamine (10 mmol) in dry chloroform (15 ml) are stirred in an ice bath with addition of 6-chloro-1-*p*-chlorobenzene-sulfonyloxybenzotriazole (10 mmol). The temperature rises, and stirring is continued for 1.5 h at room temperature. After cooling to 5 °C, methanol (40 mmol) and triethylamine (10 mmol) are added drop by drop. Stirring is continued for 1.5 h at room temperature, followed by evaporation. Take to dryness. After ether extraction, the extract is washed successively with sodium bicarbonate solution, water, dilute HCl and water, followed by drying over anhydrous magnesium sulfate, and evaporation to dryness. Yield 93%.

2.44.2 COMMENTS

The reagent used here is one of a series of novel coupling agents.[266] Acetonitrile is a better solvent for stronger carboxylic acids, such as *p*-nitrobenzoic acid.

2.45 Benzyl esters with *N,N'*-dicyclohexyl-*O*-benzyl-isourea[202]

2.45.1 PROCEDURE

A solution of the carboxylic acids in benzene, dioxan, tetrahydrofuran or carbon tetrachloride is refluxed for 1 h with a slight molar excess of *N,N'*-dicyclohexyl-*O*-benzylisourea. After cooling, and centrifuging if necessary to remove dicyclohexylurea, 1.0 μl of the clear superantant solution is injected onto the GLC column. No interfering by-products of the reaction were detected during GLC, over the entire range of the separation (C_1–C_{12}).

2.46 Methyl esters for GLC by pyrolysis of tetramethylammonium salts[280]

2.46.1 PROCEDURE

Preparation of tetramethylammonium salts: the acid is dissolved in methanol and titrated with tetramethylammonium hydroxide in methanol solution (24% w/v) to phenolphthalein end-point. The titrated solution is diluted to give between 20 and 25 mg of acid per ml. An ion-exchange method of preparation is also described.[279]

GLC conditions: glass column (1.9 m × 0.6 cm o.d.) packed with 30 60 mesh Chromosorb W coated with 20% Carbowax 20 M (w/w); column temperature programmed from 75 °C by 6° min^{-1} to 250 °C; inlet temperature 360–380 °C; helium carrier gas flow rate 100 ml min^{-1}; sample injected, 1–5 μl of a solution containing 20–25 mg ml^{-1} of tetramethylammonium salt. A thermal conductivity detector was used.[280]

2.46.2 COMMENTS

The method is a modification of one previously published.[279] Neutral components in a mixture do not have to be separated from acids. High reproducibility was claimed with peak heights equal to that of standard methyl esters injected directly.[280] The syringe needle must penetrate through the septum a minimum distance into the injection port, which is loosely packed with glass wool. Up to 50 μl of solution may be injected. The geometry of the vaporizer heating elements affects conversion.[279] The effects of the presence of water, $CHCl_3$ and CCl_4 on the production of di- and mono-methyl esters of dibasic acids were reported.[280]

High yields (about 90%) of methyl esters of 13 acids were reported, including succinic, lactic, benzoic and fatty acids. Several dicarboxylic acid salts, e.g. oxalic acid, decomposed instead of being converted to their methyl esters.[279] Non-volatile tetramethylammonium salts may be concentrated without losses due to volatility.

2.47 Methyl esters for GLC by pyrolysis with trimethyl (α,α,α-trifluoro-*m*-tolyl) ammonium hydroxide (TMTFTH)[292]

2.47.1 REAGENTS

Preparation of trimethyl (α,α,α-trifluoro-*m*-tolyl) iodide (TMTFTI): 10 ml of α,α,α-trifluoro-*m*-toluidine and 20 ml of methyl iodide in a 250 ml Erlenmeyer flask are left in the dark for 24 h at room temperature. Methanol (50 ml) and pieces of bumping stone are added and the suspension of crystals is heated in a fume cupboard until the vapours in the neck of the flask reach a temperature between 60 and 65 °C. After cooling the flask in an ice-bath for 10 min, the crystals are collected on a Buchner funnel and washed with a small volume of ice-cold methanol. The damp product is recrystallized from 25 ml of methanol. The solution is chilled, and the crystals are washed in the cold as before. Dry *in vacuo* at 50 °C. Yield 1.3–1.8 g.

Preparation of aqueous TMTFTH: 0.66 g of TMTFTI, 0.35 g of silver oxide and 4.0 ml of distilled water in a test tube are shaken until all the TMTFTI has dissolved, and centrifuged at 600 *g* for 1 min. The supernatant is tested for iodide by diluting a portion ten-fold with distilled water, acidifying with 6 M nitric acid and adding 0.1 M silver nitrate. (If iodide is present, the mixing, centrifugation and testing procedures are repeated.) The reagent solution (0.5 M) is stored in a refrigerator.

Preparation of methyl propionate-methanol: 1 volume of methyl propionate, pretreated with anhydrous Na_2CO_3, is mixed with 2 volumes of methanol.

2.47.2 PROCEDURE FOR TOTAL SAPONIFIABLE FATTY ACIDS FROM 100 μl OF SERUM

Serum (100 μl) is mixed with 0.5 ml of methanolic KOH (2.7 M) in a 15 ml glass-stoppered centrifuge tube and heated for 30 min at 65 °C. After cooling and addition of 0.7 ml of 1 M phosphoric acid, the reaction is briefly Vortex mixed and 5 ml of hexane is added, mixed vigorously for 1 min and then centrifuged at 600 *g* for 1 min. The hexane layer containing the free fatty acids is transferred into a 5 ml glass-stoppered tapering centrifuge tube. Care must be taken to avoid transferring any of the aqueous phase. After adding 10 μl of aqueous TMTFTH reagent, the tube is shaken and centrifuged. The free fatty acids should all be extracted into the TMTFTH extract (lower phase).

2.47.3 GLC CONDITIONS

A glass column with removable glass inserts in the injection port is used, packed with Gas Chrom P coated with 10% EGSS–X. The injection port temperature was 240 °C, the oven temperature 180 °C. Nitrogen carrier gas flow rate was 38 ml min^{-1}.

Injection of sample: A 10 μl syringe is wetted with methyl propionate–methanol (1:2, v/v) and the needle is wiped. The TMTFTH extract (1 μl) of fatty acids is drawn into the syringe and the needle is again wiped. The methyl propionate–methanol solution (0.5 μl) is drawn into the syringe. The plunger of the syringe is pulled back to the 5 μl mark and then returned to the 3 μl mark in order to mix the contents of the syringe, and the sample is injected onto the GLC column.

2.47.4 COMMENTS

The procedure takes less than 45 min and was developed for the analysis of small biological samples, fatty acids in serum (5–100 μl) and needle biopsy tissue specimens (3–5 mg).

The method is applicable to extracted lipids, cholesteryl, phospholipid and triglyceride esters after preliminary saponification. It is important to use sufficient TMTFTH for extraction of all the fatty acids in a sample. Free fatty acids inadvertently injected on to the GLC column give rise to 'ghost' peaks and these become more serious when determining small quantities of fatty acids.

2.48 Methods for sterically hindered carboxylic acids

2.48.1 TRIALKYLOXONIUM FLUOROBORATE[298]

Triethyloxonium fluoroborate (ethyl esters) and trimethyloxonium fluoroborate (methyl esters) should be stored under diethylether at about −80 °C.

One mmol of diisopropylethylamine[299] is added to a solution of hindered (or non-hindered) carboxylic acid (1.0 mmol) and triethyloxonium fluoroborate (1.1 mmol) in 13 ml of methylene chloride and left to stand at room temperature in a stoppered flask for 24 h. The reaction mixture is extracted 3 times each with 9 ml of 1 M HCl and 1 M $KHCO_3$ and finally once with 9 ml of saturated aqueous NaCl. The methylene chloride solution is then dried over sodium sulfate. Bulb-to-bulb distillation (1 mm Hg (133 N m^{-2}) at about 90 °C) yielded ethyl 2,4,6-trimethylbenzoate with purity greater than 99% by n.m.r. and GLC. Ethyl esters of benzoic, 2-phenylisobutyric and triethylacetic acids were prepared in high yields (90% or greater).

Trimethyloxonium fluoroborate is insoluble in methylene chloride and is used as a suspension in this solvent for the preparation of methyl esters.

2.48.2 DIMETHYLSULFATE[301]

The hindered acid (1 mol) and dimethylsulfate (1.1 mol) in dioxane solution are stirred efficiently at room temperature and aqueous 6 M NaOH (1.1 mol) is added, followed by refluxing for 30 min to complete the reaction.

Alternatively an acetone solution may be used with anhydrous K_2CO_3 (1.1 mol), followed by refluxing for 3 h.

2.48.3 HEXAMETHYLPHOSPHORAMIDE (HMPA) AND ALKYL HALIDE[306,307]

Either: A solution of the acid (5 mmol) in 10 ml of ethanol : HMPA (1 : 1, v/v) with added powdered KOH (5.3 mol) is heated at 50 °C with stirring until all the solid has dissolved. The alkyl halide (e.g. *n*-butyl bromide for *n*-butyl esters), (10 mmol) is added and heated for 30 min at 50 °C. The reaction mixture is poured into water acidified with HCl and the ester extracted with hexane. Potassium carboxylates react more quickly than sodium carboxylates. The sterically hindered mesitoic acid was methylated in 96% yield after 20 min, but slower rates were observed for pelargonic acid, $CH_3(CH_2)_7COOH$.[306]

Or: To a solution of 20 mmol of the carboxylic acid in 50 ml of HMPA is added 30 mmol of NaOH (added as a 25% aqueous solution) or sodium hydride (50% dispersion in mineral oil). This is stirred at room temperature for 1 h, followed by addition of 80 mmol of alkyl chloride or bromide and stirring 20–24 h at room temperature. The solution is poured into 100 ml of 1 M HCl and extracted with two 75 ml portions of ether. The combined ether extracts are washed with water, dried over sodium sulfate and evaporated.[307]

2.48.4 COMMENTS

Yields of 95% to 100% were reported for eleven methyl, ethyl, isopropyl and secondary butyl esters. It was claimed that sodium salts in HMPA reacted faster at room temperature than did potassium or sodium salts in 50% ethanol : HMPA at 50 °C.

2.48.5 ALKYL-*t*-BUTYL ETHERS[302]

The alkyl-*t*-butyl ether (0.1 mol) is mixed with carboxylic acid (0.12–0.15 mol) and 20–50 μl of concentrated sulfuric acid and heated under reflux or on a steam bath for 2–5 min until the evolution of isobutylene ceases. The reaction mixture is cooled, diluted with 50–100 ml of diethyl ether, washed with aqueous $NaHCO_3$ and then water and finally dried over anhydrous sodium sulfate. After removal of diethyl ether, the residue is distilled to give the corresponding ester.

2.48.6 COMMENTS

Yields varied from 53% for benzyl benzoate to 95% for *n*-butyl acetate. The reaction mixtures were monitored by GLC but only excess carboxylic acid and the ester were detected. No alcohol that might be derived from alkyl-*t*-butyl ether by acid hydrolysis was found.

2.48.7 THERMAL DECOMPOSITION OF BENZYLDIMETHYLANILINIUM SALTS[297]

The reaction is as follows:

$$RCOO^{-}C_6H_5CH_2\overset{\displaystyle CH_3}{\underset{\displaystyle CH_3}{N^{+}}}C_6H_5 \xrightarrow{\text{heat}} RCOOCH_2C_6H_5 + C_6H_5N(CH_3)_2$$

Benzyldimethylanilinium salt — Benzyl ester — Dimethyl aniline

2.48.8 PREPARATION OF BENZYLDIMETHYLANILINIUM HYDROXIDE IN METHANOL

A mixture of freshly distilled *N,N*-dimethylaniline (40 mmol) and benzyl chloride (40 mmol) is allowed to stand at room temperature for several days. The crystals are filtered off and washed with dry diethyl ether. The benzyldimethylanilinium chloride (1.8 g, 7.3 mmol) and silver oxide (1.8 g, 7.8 mmol) are stirred in dry methanol (50 ml) for 2 h. The solution is filtered and the filtrate is stored in the cold over molecular sieve (Linde type 3A).

2.48.9 PROCEDURE

The acid (6.5 mmol) is dissolved in methanolic benzyldimethylanilinium hydroxide (6.5 mmol, 46.5 ml) in a small round-bottomed flask and the solvent is removed *in vacuo* with the rotary evaporator. The oily residue of the benzyldimethylanilinium salt is shaken vigorously with benzene (N.B. carcinogenic) and the benzene is then removed with the rotary evaporator under reduced pressure. The residue of salt is refluxed with 20 ml of toluene for 1 h. Unreacted acid and dimethylaniline are removed by extracting the toluene solution successively with 20 ml portions of water, 1 M HCl and finally water, and the aqueous extracts are discarded. The final product of benzyl ester is purified by distillation under reduced pressure.

2.48.10 COMMENTS

The procedure is mild and offers great potential for carboxylic acids and phenols. The sterically hindered 2,4,6-trimethylbenzoic acid was esterified in 77% yield. Yields for seven aromatic and dicarboxylic acids varied from 65 to 86%. Development of this method on the microscale would be useful.

2.48.11 CROWN ETHERS[227]

Potassium acetate (100 mg, 1.02 mmol), *α-p*-dibromoacetophenone (278 mg, 1.00 mmol) and dicyclohexyl-18-crown-6 (or 18-crown-6)-(15 mg, 0.04 mmol) are suspended in 10 ml of acetonitrile, refluxed for 15 min and acetonitrile is removed under reduced pressure. The residue may be washed through a short column of dry silica gel with benzene to remove catalyst.

2.48.12 COMMENTS

The reaction time is short, the conditions mild and steric problems were not shown.

3 DETERMINATIONS AND REAGENTS

3.1 Determination of carboxylic esters[363]

If it is necessary to establish the yield of an esterification reaction, the ester may be determined in a portion of the derivatization product. In this method, the esters are converted to the corresponding hydroxamate and development with ferric ion gives a red-violet colour.

3.1.1 REAGENTS

Hydroxylamine hydrochloride in methanol (12.5% w/v) (1); sodium hydroxide in methanol (12.5% w/v) (2); mix equal volumes of 1 and 2 and filter. This reagent (3) is stable for 4 h; (stock solution (4))—dissolve 5.0 g of ferric perchlorate in 20 ml of 35% perchloric acid and add 80 ml of ethanol with cooling (store at 0–4 °C); to 4 ml of stock solution (4) add 1.2 ml of 70% perchloric acid, and make up to 100 ml with ethanol, with cooling (reagent 5).

3.1.2 PROCEDURE

To 0.5 ml of ester sample in ethanol, 0.3 ml of reagent 3 is added and heated at 70 °C for 5 min, cooled and diluted to 5 ml with reagent 5 and read at 530 nm. The wavelength of maximum absorption varies with the ester, e.g. ethyl malonate 520 nm, benzyl benzoate 550 nm.

3.1.3 COMMENTS

The factors affecting the stability of the coloured complex,[364] the hydrogen ion concentrations of the final solution[363] and the kinetics of acetohydroxamic acid formation from ethyl acetate[365] have been evaluated. Hydroxamate formation was also used for ester determination.[366,367] Alcohols and phenols may be determined by quantitative esterification of these compounds with an acetic anhydride–pyridine mixture.[368] The analysis of carboxylic acids and esters was reviewed.[369]

3.2 Preparation of alcoholic solutions of HCl

The alcohol should be thoroughly dried by refluxing for 2 h over magnesium turnings. A trace of iodine may be used to catalyse the reaction. The alcohol should be distilled into closed vessels with protection using drying tubes. Hydrogen chloride gas may be bubbled into known amounts of alcohol in a tared vessel to the desired weight or alternatively the molarity may be adjusted after titration with standardized alkali solution to phenolphthalein indicator.

Hydrogen chloride may be produced by the action of concentrated sulfuric acid on NaCl or fused ammonium chloride, but a smoother reaction is obtained with concentrated HCl. It is recommended that concentrated HCl in a separating funnel should be run drop by drop into concentrated H_2SO_4 contained in a two-necked flask, preferably with magnetic stirring to avoid the danger of layering. The gas evolved should be dried by passing it through two sulfuric acid drying towers or bubbling chambers. This method requires apparatus to be set up permanently in a fume cupboard (see Ref. 10).

If a cylinder of HCl is available, the gas should be passed through a similar drying procedure. In addition, this removes any volatile iron compounds which may be present, because these will cause lowered yields of methionine esters (personal observation; see also Ref. 337).

It is convenient to use acetyl chloride as a source of HCl when alcoholic HCl solutions are required only occasionally. Thus, to prepare 4 M HCl in methanol, redistilled acetyl chloride (0.04 mol, 3.14 g) is added cautiously with several additions to the alcohol, e.g. methanol (about 8 ml), in a 10 ml measuring flask which is kept in an ice bath. When addition is complete, the flask is made up to volume and stored in the cold. Or, the required amount of acetyl chloride may slowly be added to 1 ml of alcohol at 0 °C and used immediately.[346]

Alcoholic HCl solutions give rise to the corresponding ethers and alkyl halides and water on standing; thus for quantitative work it is better to avoid storage for periods longer than 2 weeks; however, alkyl halides were added up to 50% (v/v) to alcoholic HCl solutions without affecting esterification yields of amino acids (Blau and Darbre, unpublished).

3.3 Preparation of boron trifluoride and boron trichloride reagents

Boron trifluoride and boron trichloride are very reactive and should be handled with care in a fume cupboard.

Commercial BF_3-etherate darkens owing to air oxidation and should be redistilled[370] by adding 50 ml of ether to 500 ml of etherate and distilling from 2 g of calcium hydride at 46 °C and 10 mm Hg. A water-clear product is obtained. It should be stored at 0 °C in the dark.

Alcohol–BF_3 reagent is prepared by bubbling BF_3 gas into the alcohol at 0 °C until a concentration of 2.0 M is attained. The BF_3 content may be determined by adding a known excess of pyridine and titrating the base with 0.1 M HCl to pH 3.0. Boron trifluoride in ether and methanol is stable for 4 months and does not interfere with the GLC of fatty acid methyl esters.[163]

Boron trichloride reagent may be prepared in a similar manner.

3.4 Determination of diazomethane or other diazoalkanes

A standard 1.0 mol solution of benzoic acid in methanol is made up, and an accurately known amount of benzoic acid (1 mmol) is added to dry diethyl ether (10 ml). A known volume of diazomethane solution is now added and allowed to react until the solution is decolourized. The benzoic acid must be in excess. Water is added and the excess benzoic acid is titrated with 0.1 M alkali to a phenolphthalein end-point (see Ref. 10).

Reviews on the chemistry of diazomethane should be consulted.[158,159]

CAUTION: Diazomethane is carcinogenic. Avoid breathing any of the gaseous reactant, work in a well-ventilated fume cupboard, do not overheat when preparing or reacting diazomethane (danger of explosion). Workers wishing to prepare large quantities should refer to the original papers.

3.5 Preparation of diazomethane in ether and alcohol solution[371]

The following procedure for preparing small quantities of diazomethane is simple and has been used for many years by the author, and takes less than 30 min.

Potassium hydroxide (0.4 g) is dissolved in 10 ml of 90% aqueous ethanol. A distillation apparatus with a receiving vessel of about 25 ml capacity (preferably a 25 ml stoppered measuring flask), in an ice-water bath is set up in an efficient fume cupboard. *N*-methyl-*N*-nitroso-*p*-toluene sulfonamide (2.14 g, 0.01 mol) is dissolved in 30 ml of diethyl ether in a 250 ml round-bottomed flask and cooled in an ice bath. To the cold ethereal solution 10 ml of KOH–ethanol is added and mixed. The flask is connected to the distillation apparatus, and the reaction vessel is left to stand at room temperature for a few minutes. Distillation is initiated by heating on an oil or water bath at about 35–40 °C. The distillate is collected (about 12 ml) until no colour is apparent in the distilling fraction. The yellow ethereal solution (containing some alcohol) is conveniently stored in a measuring flask fitted with a drying tube, at −20 °C or lower temperature. (If the flask is stoppered, any pressure caused by liberated N_2 gas should be released daily.) It will keep for at least one week. Yield claimed 70–78%.

The original paper also gives details for preparing gaseous diazomethane.

3.6 Preparation of diazomethane or diazoethane without co-distillation[372]

The apparatus shown in Fig. 3 is made from standard glassware. It is a modification of one which is commercially available. Chemical reaction occurs in the small tube and diazomethane which is liberated is dissolved in the desired solvent in the outer tube.

About 3 ml of ether (or other solvent) is put into the large tube which stands in an ice bath. Into the small tube is placed 1 mmol (133 mg) or less of *N*-

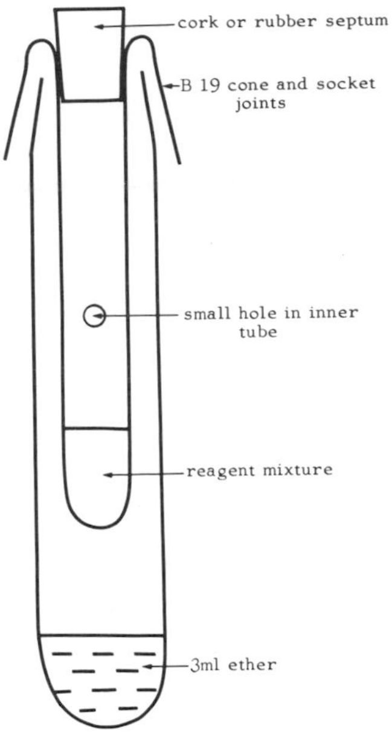

Fig. 3 Apparatus for preparing diazomethane or diazoethane (adapted from original, Ref. 372).

methyl-*N*-nitroso-*N*′-nitroguanidine (CAUTION: This is a potent mutagen); 1.0 ml of 5 M NaOH is added and the top of the tube is quickly stoppered; alternatively, the alkali is injected through the rubber septum with a syringe. The amount of diazomethane trapped in the solvent is time-dependent and in 45 min a maximum of about 60% theoretical yield is obtained.[372] A small bar magnet in the lower tube may be used to create turbulence and thus increase gaseous diffusion. *N*-Methyl-*N*-nitrosourea also reacts rapidly at room temperature but *N*-methyl-*N*-nitroso-*p*-toluene-sulfonamide reacts too slowly for use in this apparatus. The diazoalkane may be collected in solvents other than ether. Because the hydrogens are easily exchanged in alkali,[373] tritiated diazomethane may be prepared from tritiated alkali.[372]

Diazoethane is obtained by using *N*-ethyl-*N*-nitroso-*N*′-nitroguanidine.

3.7 Preparation of diazoalkanes without distillation[374]

About 4 g of the appropriate *N*-alkyl-*N*-nitrosourea is added to 10 ml of 60% aqueous KOH covered with 80 ml of dry diethyl ether in a large boiling tube at −10 °C and the temperature is allowed to rise to 0 °C. The reaction is completed when the ether becomes deeply coloured and the solid matter in the aqueous layer loses its yellow colour.

The upper ether layer is decanted onto KOH pellets and after one day into a closed vessel for storage at −10 °C. Solutions will keep for one week, and may be used without distillation. Diazo-ethane, -*n*-propane, -isobutane and -*n*-butane were prepared.

Diazomethane may be prepared in the same way with 4 g of *N*-methyl-*N*-nitroso-*N'*-nitroguanidine and 20 ml of 30% KOH.

3.8 Preparation of diazopropane without distillation[348]

50% aqueous KOH is added to *N*-propyl-*N'*-nitrosoguanidine dissolved in diethyl ether to prepare diazopropane.

3.9 Preparation of phenyldiazomethane (diazotoluene) for benzyl esters

3.9.1 FROM AZIBENZIL[375]

A solution of 8 g NaOH in 15 ml of H_2O and 100 ml of methanol is added to azibenzil (5.56 g, 25 mmol) dissolved in 125 ml of diethyl ether. If a white precipitate (sodium benzoate?) appears on standing, it will dissolve on addition of traces of water and/or methanol. The reaction is loosely stoppered and left to stand at room temperature for 8 h. It is filtered if necessary. The clear red solution is treated with 100 ml of 10% aqueous NaOH. The ethereal layer is washed successively four times with 25 ml portions of 10% aqueous NaOH. The ether solution is dried over anhydrous sodium sulfate. Yield approx. 70%. (Store at −20 °C.)

Azibenzil may be prepared from benzil monohydrazone.

3.9.2 FROM *N*-NITROSO-*N*-BENZYL-*p*-TOLUENESULFONAMIDE (I)[376]

Over a period of 1 h, *N*-nitroso-*N*-benzyl-*p*-toluenesulfonamide (14.5 g, 0.05 mol) is added in portions to 100 ml of a well-stirred solution of methanol–water (4:1, v/v) saturated with NaOH. When the addition is complete, it is refluxed for 15–20 min. The reaction is cooled and extracted with petroleum spirit (b.p. 30–40 °C), and the extract is washed with water, dried over anhydrous sodium sulfate, and then concentrated to about 5 ml *in vacuo* at 0 °C.

Further purification may be carried out by vacuum distillation (phenyldiazomethane, b.p. 28 °C at 1 mm Hg, 133 Nm^{-2}). Store as a 50% solution in petroleum spirit (v/v) at −20 °C.

3.9.3 FROM *N*-NITROSO-*N*-BENZYL-*p*-TOLUENESULFONAMIDE (II)[202]

Into a 500 ml two-necked round-bottomed flask fitted with a paddle stirrer in an ice-bath is added 90 ml of diethyl ether, 15 ml of methanol and 3 g of sodium methoxide, and whilst stirring strongly over a period of 1 h, 14.5 g of *N*-nitroso-*N*-benzyl-*p*-toluenesulfonamide. When addition is complete, the reaction is refluxed for 20 min in an oil bath under a calcium chloride drying tube and evaporated *in vacuo* at 25 °C in a rotary evaporator. The residue is dissolved in 100 ml of *n*-pentane (benzene is unsuitable, as phenylcycloheptatriene might be formed) and filtered into a 250 ml round-bottomed flask. On cooling the filtrate to −20 °C, a liquid phase separates out and solidifies. The supernatant *n*-pentane solution is decanted and reduced in volume to about 20 ml. This concentrated solution is stirred and distilled *in vacuo* (0.1 mm Hg) at room temperature into a receiving vessel kept at −30 °C or below. The solution of phenyldiazomethane keeps well for several weeks in a closed vessel at −20 °C. The product is claimed to be purer than one previously described.[377]

3.10 Preparation of diazobenzene for phenyl esters[376]

N-Nitroso-*N*-phenylurea (7.7 g, 0.05 mol) is added in portions over a period of 1 h to 100 ml of a well-stirred solution of methanol–water (4:1, v/v) saturated with NaOH. When the addition is complete, the mixture is extracted with petroleum spirit (b.p. 30–40 °C), and the extract washed with water, and then dried over anhydrous sodium sulfate and concentrated to about 5 ml *in vacuo* at 0 °C. It is stored at −20 °C.

ACKNOWLEDGEMENTS

This work was commenced during a period as Visiting Scientist, Wellcome Research Laboratories, Beckenham, Kent, and I am grateful for the help received, particularly from the Library and Photographic Staff.

REFERENCES

1. L. F. Fieser, *Experiments in Organic Chemistry*, D. C. Heath Ltd., Boston, Mass., 1957, p. 77.
2. C. A. Buehler and D. E. Pearson, *Survey of Organic Synthesis*, Wiley, New York, 1970, p. 801.
3. S. Patai (Ed.), *The Chemistry of Carboxylic Acids and Esters*, Interscience, London, 1969.
4. G. Zweig and J. Sherma (Eds), *Handbook of Chromatography*, Vol. II, CRC Press, The Chemical Rubber Co., Cleveland, Ohio, 1972, p. 226.
5. I. T. Harrison and S. Harrison, *Compendium of Organic Synthetic Methods*, Vol. 1, Wiley, New York, 1971, p. 271.
6. I. T. Harrison and S. Harrison, *Compendium of Organic Synthetic Methods*, Vol. 2, Wiley, New York, 1974, p. 108.
7. S. R. Sandler and W. Karo, *Organic Functional Group Preparations*, Vol. II, Academic Press, New York, 1971, p. 41.
8. E. H. Cordes in S. Patai (Ed.) *The Chemistry of Carboxylic Acids and Esters*, Interscience, London, 1969, p. 623.
9. S. R. Sandler and W. Karo, *Organic Functional Group Preparations*, Vol. I, Academic Press, New York, 1968, p. 245.
10. A. I. Vogel, *Practical Organic Chemistry*, Longman, London, 1972, p. 379.
11. J. G. Buchanan, N. A. Hughes and G. A. Swan, in S. Coffey (Ed.) *Rodd's Chemistry of Carbon Compounds*, Vol. 1, Elsevier, Amsterdam, 1965, p. 54.

12. E. Reid, *Analyst (London)* **101**, 1 (1976).
13. B. Weinstein in D. Glick (Ed.) *Methods of Biochemical Analysis*, Vol. XIV, Interscience, Wiley, New York, 1966, p. 203.
14. K. Blau in H. A. Szymanski (Ed.) *Biomedical Applications of Gas Chromatography*, Vol. 2, Plenum Press, New York, 1967, p. 1.
15. V. Marek, *Chem. Listy*, **68**, 250 (1974).
16. P. Hušek and K. Macek, *J. Chromatogr.* **113**, 139 (1975).
17. F. Karoum, C. R. J. Ruthven and M. Sandler, *Clin. Chim. Acta* **20**, 427 (1968).
18. M. W. Weg, C. R. J. Ruthven, B. L. Goodwin and M. Sandler, *Clin. Chim. Acta* **59**, 249 (1975).
19. J. Drozd, *J. Chromatogr.* **113**, 303 (1975).
20. S. Ahuja, *J. Pharm. Sci.* **65**, 163 (1976).
21. S. W. Dziedzic, L. M. Bertani, D. D. Clarke and S. E. Gitlow, *Anal. Biochem.* **47**, 592 (1972).
22. M. Schwarz and G. Michael, *J. Chromatogr.* **118**, 101 (1976).
23. G. C. Cochrane, *J. Chromatogr. Sci.* **13**, 440 (1975).
24. D. P. Collins, P. G. McCormick and M. G. Schmitt Jr, *Clin. Chem.* **20**, 1235 (1974).
25. W. R. White and J. A. Leenheer, *J. Chromatogr. Sci.* **13**, 386 (1975).
26. W. R. Supina in H. A. Szymanski (Ed.) *Biomedical Applications of Gas Chromatography*, Plenum Press, New York, 1964, p. 271.
27. R. G. Ackman and L. D. Metcalfe (Eds.), *J. Chromatogr. Sci.* **13**, 397 (1975).
28. R. G. Ackman and L. D. Metcalfe (Eds.), *J. Chromatogr. Sci.* **13**, 453 (1975).
29. R. G. Ackman and L. D. Metcalfe (Eds.), Analysis of Fatty Acids and Fatty Acid Esters, Part III, *J. Chromatogr. Sci.* **14**, 1 (1976).
30. G. H. DeVries, P. Mamunes, C. D. Miller and D. M. Hayward, *Anal. Biochem.* **70**, 156 (1976).
31. D. Sampson and W. J. Hensley, *Clin. Chem.* **61**, 1 (1975).
32. C. Van Eenaeme, J. M. Bienfait, O. Lambot and A. Pondant, *J. Chromatogr. Sci.* **12**, 398 (1974).
33. C. Van Eenaeme, J. M. Bienfait, O. Lambot and A. Pondant, *J. Chromatogr. Sci.* **12**, 404 (1974).
34. J.-J. Van Huyssteen, *Water Res.* **4**, 645 (1970).
35. R. G. Ackman in R. T. Holman (Ed.) *Progress in the Chemistry of Fats and other Lipids*, Vol. 12, Pergamon Press, Oxford, 1972, p. 165.
36. T. Murata, S. Takahashi and T. Takeda, *Anal. Chem.* **47**, 573 (1975).
37. T. Mitchell, Jr., *Anal. Chem.* **36**, 2050 (1964).
38. J. D. Cipera and R. V. V. Nicholls, *Chem. Ind. (London)*, 16 (1955).
39. B. F. Erlanger and R. M. Hall, *J. Am. Chem. Soc.* **76**, 5781 (1954).
40. T. Perlstein, A. Eisner and I. Schmeltz, *J. Am. Oil Chem. Soc.* **51**, 335 (1974).
41. E. Taschner and C. Wasielewski, *Liebigs Ann. Chem.* **640**, 139 (1961).
42. M. Brenner, H. R. Müller and R. W. Pfister, *Helv. Chim. Acta* **33**, 568 (1950).
43. J. Wolinski, *Acta Polon. Pharm.* **32**, 303 (1975).
44. M. L. Bender, *Chem. Rev.* **60**, 53 (1960).
45. H. A. Saroff and A. Karmen, *Anal. Biochem.* **1**, 344 (1960).
46. J. Bertin, H. B. Kagan, J.-L. Luche and R. Setton, *J. Am. Chem. Soc.* **96**, 8113 (1974).
47. J. Vejrosta and J. Málek, *J. Chromatogr.* **109**, 101 (1975).
48. M. Rogozinski, *J. Gas Chromatogr.* **2**, 136 (1964).
49. M. Rogozinski, *J. Gas Chromatogr.* **2**, 328 (1964).
50. C. K. Ingold, *Structure and Mechanism in Organic Chemistry*, Cornell University Press, Ithaca, N.Y., 1969, p. 1128.
51. P. J. Sniegoski, *J. Chromatogr. Sci.* **10**, 644 (1972).
52. P. J. Sniegoski, *J. Org. Chem.* **39**, 3141 (1974).
53. E. K. Euranto in S. Patai (Ed.) *The Chemistry of Carboxylic Acids and Esters*, Interscience, London, 1969, p. 505.

54. V. Peterka and M. Zbirovsky, *Fette, Seifen Anstrichm.* **76**, 397 (1974).
55. H. A. Smith, *J. Am. Chem. Soc.* **61**, 254 (1939).
56. H. A. Smith, *J. Am. Chem. Soc.* **61**, 1176 (1939).
57. H. A. Smith, *J. Am. Chem. Soc.* **62**, 1136 (1940).
58. J. P. Hardy and S. L. Kerrin, *Anal. Chem.* **44**, 1497 (1972).
59. W. M. Lamkin and C. W. Gehrke, *Anal. Chem.* **37**, 383 (1965).
60. C. W. Gehrke and D. L. Stalling, *Separation Sci.* **2**, 101 (1967).
61. M. Dymicky, E. F. Mellon and J. Naghski, *Anal. Biochem.* **41**, 487 (1971).
62. E. Thielepape, *Ber. Deut. Chem. Ges.* **66**, 1454 (1933).
63. R. L. Stern and E. N. Bolan, *Chem. Ind. (London)*, 825 (1967).
64. R. R. Baker, *J. Am. Chem. Soc.* **65**, 1572 (1943).
65. B. R. Baker, M. V. Querry, S. R. Safir and S. Bernstein, *J. Org. Chem.* **12**, 138 (1947).
66. R. O. Clinton and S. C. Laskowski, *J. Am. Chem. Soc.* **70**, 3135 (1948).
67. W. C. Ault, J. K. Weil, G. C. Nutting and J. C. Cowan, *J. Am. Chem. Soc.* **69**, 2003 (1947).
68. D. E. Johnson, S. J. Scott and A. Meister, *Anal. Chem.* **33**, 669 (1961).
69. J. Kenyon, *Org. Synth. Coll.* **1**, 258 (1932).
70. M. S. Newman, *J. Am. Chem. Soc.* **63**, 2431 (1941).
71. K. V. Peisker, *J. Am. Oil. Chem. Soc.* **41**, 87 (1964).
72. N. B. Lorette and J. H. Brown Jr, *J. Org. Chem.* **24**, 261 (1959).
73. R. F. Adams, *J. Chromatogr.* **95**, 189 (1974).
74. C. Zomzely, G. Marco and E. Emery, *Anal. Chem.* **34**, 1414 (1962).
75. S. B. Tove, *J. Nutrition*, **75**, 361 (1961).
76. M. E. Mason and G. R. Waller, *Anal. Chem.* **36**, 583 (1964).
77. P. G. Simmonds and A. Zlatkis, *Anal. Chem.* **37**, 302 (1965).
78. R. S. Juvet Jr and F. M. Wachi, *J. Am. Chem. Soc.* **81**, 6110 (1959).
79. D. P. Roelofsen, J. A. Hagendoorn and H. van Bekkum, *Chem. Ind. (London)* 1622 (1966).
80. C. W. Gehrke, W. M. Lamkin, D. L. Stalling and F. Shahrokhi, *Biochem. Biophys. Res. Commun.* **19**, 328 (1965).
81. D. L. Stalling, G. Gille and C. W. Gehrke, *Anal. Biochem.* **18**, 118 (1967).
82. C. W. Gehrke, D. Roach, R. W. Zumwalt, D. L. Stalling and L. L. Wall, *Quantitative Gas–Liquid Chromatography of Amino Acids in Proteins and Biological Substances*, Analytical Biochemistry Laboratories, Columbia (1968).
83. H. Yazawa, K. Tanaka and K. Kariyone, *Tetrahedron Lett.* 3995 (1974).
84. C. Litchfield, *Analysis of Triglycerides*, Academic Press, New York (1972).
85. M. Salmon, C. Fenselau, J. O. Cukier and G. B. Odell, *Life Sci.* **15**, 2069 (1974).
86. J. P. Williams, G. R. Watson, M. Khan, S. Leung, A. Kuksis, O. Stachnyk and J. J. Myher, *Anal. Biochem.* **66**, 110 (1975).
87. W. Stoffel, F. Chu and E. H. Ahrens Jr, *Anal. Chem.* **31**, 307 (1959).
88. W. R. Morrison and L. M. Smith, *J. Lipid Res.* **5**, 600 (1964).
89. C. Litchfield, M. Farquhar and R. Reiser, *J. Am. Oil Chem. Soc.* **41**, 588 (1964).
90. J. J. Myer, L. Marai and A. Kuksis, *Anal. Biochem.* **62**, 188 (1974).
91. A. Kuksis, L. Marai, W. C. Breckenridge, D. A. Gornall and O. Stachnyk, *Canad. J. Physiol. Pharmacol.* **46**, 511 (1968).
92. A. Kuksis, J. J. Myher, L. Marai, S. K. F. Yeung, I. Steiman and S. Mookerjea, *Can. J. Biochem.* **53**, 519 (1975).
93. L. Gosselin and J. de Graeve, *J. Chromatogr.* **110**, 117 (1975).
94. P. J. Mavrikos and G. Eliopoulos, *J. Am. Oil Chem. Soc.* **50**, 174 (1973).
95. H. W. Wharton, *J. Am. Oil Chem. Soc.* **51**, 35 (1974).
96. V. L. Davison and M. J. D. Dutton, *J. Lipid Res.* **8**, 147 (1968).
97. S. W. Christopherson and R. L. Glass, *J. Dairy Sci.* **52**, 1289 (1969).
98. C. R. Scholfield, *Anal. Chem.* **47**, 1417 (1975).
99. K. Oette and M. Doss, *J. Chromatogr.* **32**, 439 (1968).
100. K. Oette, M. Doss and M. Winterfeld, *Z. Klin. Chem. Klin. Biochem.* **8**, 525 (1970).

101. B. J. Holub, *Biochim. Biophys. Acta* **369**, 111 (1974).
102. S.-N. Lin and E. C. Horning, *J. Chromatogr.* **112**, 483 (1975).
103. G. G. Esposito and M. H. Swann, *Anal. Chem.* **34**, 1048 (1962).
104. B. M. Craig and N. L. Murty, *J. Am. Oil Chem. Soc.* **36**, 549 (1959).
105. F. E. Luddy, R. A. Barford and R. W. Riemenschneider, *J. Am. Oil Chem. Soc.* **37**, 447 (1960).
106. M. Brenner and W. Huber, *Helv. Chim. Acta* **36**, 1109 (1953).
107. J. L. Giegel, A. B. Ham and W. Clema, *Clin. Chem.* **21**, 1575 (1975).
108. R. L. Glass, R. Jenness and H. A. Troolin, *J. Dairy Sci.* **48**, 1106 (1965).
109. J. Sampugna, R. E. Pitas and R. G. Jensen, *J. Dairy Sci.* **49**, 1462 (1966).
110. C. A. Atkins and D. T. Canvin, *Can. J. Biochem.* **49**, 949 (1971).
111. G. Kupryszewski and T. Sokotowska, *Acta Biochem. Pol.* **4**, 85 (1957).
112. P. B. Hagen and W. Black, *Can. J. Biochem.* **43**, 309 (1965).
113. S. Makisumi and H. A. Saroff, *J. Gas Chromatogr.* **3**, 21 (1965).
114. M. Gee, *Anal. Chem.* **39**, 1677 (1967).
115. E. Jellum, V. A. Close, W. Patton, W. Pereira and B. Halpern, *Anal. Biochem.* **31**, 227 (1969).
116. G. Cignarella, F. Savelli and P. Sanna, *Synthesis* 252 (1975).
117. F. C. Uhle, *J. Org. Chem.* **27**, 4081 (1962).
118. R. P. Patel and S. Price, *J. Org. Chem.* **30**, 3575 (1965).
119. R. G. Ackman, S. N. Hooper, M. Kates, A. K. Sen Gupta, G. Eglington and I. Maclean, *J. Chromatogr.* **44**, 256 (1969).
120. A. Islam, University of London Ph.D. Thesis (1970).
121. A. V. Topchiev, S. V. Zavgorodnii and Ya. M. Paushkin, *Boron Fluoride and its Compounds as Catalysts in Organic Chemistry*, Pergamon Press, London, 1959.
122. P. K. Kadaba, *J. Pharm. Sci.* **63**, 1333 (1974).
123. E. Hautala and M. L. Weaver, *Anal. Biochem.* **30**, 32 (1969).
124. P. K. Kadaba, M. Carr, M. Tribo, J. Triplett and A. C. Glasser, *J. Pharm. Sci.* **58**, 1422 (1969).
125. P. K. Kadaba, *Synthesis*, 316 (1971).
126. J. L. Marshall, K. C. Erickson and T. K. Folsom, *Tetrahedron Lett.* 4011 (1970).
127. P. K. Kadaba, *Synthesis* 628 (1972).
128. G. Hallas, *J. Chem. Soc. (London)* 5770 (1965).
129. F. Warnaar, *Anal. Biochem.* **71**, 533 (1976).
130. L. D. Metcalfe, A. A. Schmitz and J. R. Pelka, *Anal. Chem.* **38**, 514 (1966).
131. F. D. Gunstone and I. Ismail, *Chem. Phys. Lipids*, **1**, 209 (1967).
132. F. E. Luddy, R. A. Barford, S. F. Herb and P. Magidman, *J. Am. Oil Chem. Soc.* **45**, 549 (1968).
133. J. L. Harwood and A. T. James, *Eur. J. Biochem.* **50**, 325 (1975).
134. M. S. F. Lie Ken Jie, *J. Chromatogr.* **109**, 81 (1975).
135. J. P. Nelson and A. J. Milun, *J. Am. Oil Chem. Soc.* **52**, 81 (1975).
136. C. V. Warner and G. V. Vahouny, *Anal. Biochem.* **67**, 122 (1975).
137. A. J. Giorgio, E. Malloy and T. Black, *Anal. Lett.* **5**, 13 (1972).
138. M. A. Harmon and H. W. Doelle, *J. Chromatogr.* **42**, 157 (1969).
139. E. M. De Silva, *Anal. Chem.* **43**, 1031 (1971).
140. L. D. Metcalfe and A. A. Schmitz, *Anal. Chem.* **33**, 363 (1961).
141. D. Van Wijngaarden, *Anal. Chem.* **39**, 848 (1967).
142. H. Goldfine and E. Panos, *J. Lipid Res.* **12**, 214 (1971).
143. S. J. Kubacki, W. Kasprowicz, M. Jamiolkowska and T. Poskorz, *Pr. Inst. Lab. Badaw. Przem. Spozyw.* **23**, 535 (1973).
144. G. Aguggini and P. A. Biondi, *Arch. Vet. Ital.* **24**, 213 (1973).
145. J. D. Doherty, O. C. Snead and R. H. Roth, *Anal. Biochem.* **69**, 268 (1975).
146. J. B. Brooks and C. C. Alley, *Anal. Chem.* **46**, 145 (1974).

147. K. R. Millar and P. P. Lorentz, *J. Chromatogr.* **101**, 177 (1974).
148. B. Henshaw, S. S. Que Hee, R. G. Sutherland and C. C. Lee, *J. Chromatogr.* **106**, 33 (1975).
149. R. L. Geison, B. O. N. Rowley and T. Gerritsen, *Clin. Chim. Acta* **60**, 137 (1975).
150. A. J. Appleby and J. E. O. Mayne, *J. Gas Chromatogr.* **5**, 266 (1967).
151. P. A. Biondi and M. Cagnasso, *J. Chromatogr.* **109**, 389 (1975).
152. J. I. Peterson, H. De Schmertzing and K. Abel, *J. Gas Chromatogr.* **3**, 126 (1965).
153. W. E. Klopfenstein, *J. Lipid Res.* **12**, 773 (1971).
154. B. L. Brian, R. W. Gracy and V. E. Scholes, *J. Chromatogr.* **66**, 138 (1972).
155. D. E. Minnikin and N. Polgar, *Chem. Commun.* 312 (1967).
156. H. Schlenk and J. L. Gellerman, *Anal. Chem.* **32**, 1412 (1960).
157. W. R. Morrison, T. D. V. Laurie and J. Blades, *Chem. Ind.* (*London*) 1534 (1961).
158. R. Roper and T. S. Ma, *Microchem. J.* **1**, 245 (1957).
159. J. S. Pizey, *Synthetic Reagents*, Vol. 2, Ellis Harwood Ltd., Chichester and Wiley, New York, 1974, p. 65.
160. E. A. Werner, *J. Chem. Soc.* (*London*) **15**, 1093 (1919).
161. R. Spears, D. Vukusich, S. Mangat and B. S. Reddy, *J. Chromatogr.* **116**, 184 (1976).
162. M. Neeman and Y. Hashimoto, *J. Am. Chem. Soc.* **84**, 2972 (1962).
163. M. L. Vorbeck, L. R. Mattick, F. A. Lee and C. S. Pederson, *Anal. Chem.* **33**, 1512 (1961).
164. O. Mlejnek, *J. Chromatogr.* **70**, 59 (1972).
165. M. Jankovsky, P. Bobak and J. Hubacek, *Prum. Potravin*, **25**, 127 (1974).
166. W. O. Ord and P. C. Bamford, *Chem. Ind.* (*London*) 2115 (1967).
167. F. L. Estes and R. C. Bachmann, *Anal. Chem.* **38**, 1178 (1966).
168. P. G. Simmonds, B. C. Pettitt and A. Zlatkis, *Anal. Chem.* **39**, 163 (1967).
169. M. Frankel and E. Katchalski, *J. Am. Chem. Soc.* **65**, 1670 (1943).
170. M. Frankel and E. Katchalski, *J. Am. Chem. Soc.* **66**, 763 (1944).
171. D. B. De Oliveira and W. E. Harris, *Anal. Lett.* **6**, 1107 (1973).
172. M. Wilcox, *Anal. Biochem.* **32**, 191 (1969).
173. P. J. Holloway and A. H. B. Deas, *Chem. Ind.* (*London*) 1140 (1971).
174. F. Marcucci, R. Bianchi, L. Airoldi, M. Salmona, R. Fanelli, C. Chiabrando, A. Frigerio, E. Mussini and S. Garattini, *J. Chromatogr.* **107**, 285 (1975).
175. G. G. S. Dutton in R. S. Tipson and D. Horton (Eds) *Advances in Carbohydrate Chemistry and Biochemistry*, Vol. 28, Academic Press, New York, 1973, p. 11.
176. G. G. S. Dutton in R. S. Tipson and D. Horton (Eds) *Advances in Carbohydrate Chemistry and Biochemistry*, Vol. 30, Academic Press, New York, 1974, p. 9.
177. J. M. Ruhlmann and A. J. C. Stahl, *Bull. Soc. Pharm. Strasburg*, **16**, 131 (1973).
178. U. Langenbeck, H.-U. Möhring and K.-P. Dieckmann, *J. Chromatogr.* **115**, 165 (1975).
179. L. C. Erickson and H. Z. Hield, *J. Agr. Food Chem.* **10**, 204 (1962).
180. J. Horner, S. S. Que Hee and R. G. Sutherland, *Anal. Chem.* **46**, 110 (1974).
181. J. D. Mahon, K. Egle and H. Fock, *Can. J. Biochem.* **53**, 609 (1975).
182. S. Landaas, *Clin. Chim. Acta* **58**, 23 (1975).
183. T. C. Cuong and A. Tuong, *J. Chromatogr.* **106**, 97 (1975).
184. S. Lindstedt and G. Steen, *Clin. Chem.* **21**, 1964 (1975).
185. R. M. Milberg, K. L. Rinehart Jr, R. L. Sprague and E. K. Sleator, *Biomed. Mass Spectrom.* **2**, 2 (1975).
186. J. H. Duncan, M. W. Couch, G. Gotthelf and K. N. Scott, *Biomed. Mass Spectrom.* **1**, 40 (1974).
187. E. M. Rho and W. C. Evans, *Biochem. J.* **148**, 11 (1975).
188. M. Schnitzer and J. G. Desjardins, *J. Gas Chromatogr.* **2**, 270 (1964).
189. T. Kolasa, A. Chimiak and A. Kitowska, *J. Prakt. Chem.* **317**, 252 (1975).
190. S. M. Kalbag and R. W. Roeske, *J. Am. Chem. Soc.* **97**, 440 (1975).
191. L. D. Quin and M. E. Hobbs, *Anal. Chem.* **30**, 1400 (1958).
192. K. B. Hammond and S. I. Goodman, *Clin. Chem.* **16**, 212 (1970).
193. D. Gompertz and G. H. Draffan, *Clin. Chem. Acta* **40**, 5 (1972).

194. T. Kitagawa, *Clin. Chim.* **20**, 1543 (1974).
195. H. W. Mahood and I. H. Rogers, *J. Chromatogr.* **109**, 281 (1975).
196. J. D. Mann, N. G. Porter and J. E. Lancaster, *J. Chromatogr.* **92**, 177 (1974).
197. S. M. M. Karim and K. Hillier in S. M. M. Karim (Ed.), *The Prostaglandins*, Medical and Technical Publishing, Oxford, 1972, p. 1.
198. T. Bruzzese, M. Cambieri and F. Recusani, *J. Pharm. Sci.* **64**, 462 (1975).
199. E. K. Doms, *J. Chromatogr.* **105**, 79 (1975).
200. D. L. Corina, *J. Chromatogr.* **87**, 254 (1973).
201. J. Oehlenschläger, U. Hintze and G. Gercken, *J. Chromatogr.* **110**, 53 (1975).
202. H. P. Klemm, U. Hintze and G. Gercken, *J. Chromatogr.* **75**, 19 (1973).
203. U. Hintze, H. Röper and G. Gercken, *J. Chromatogr.* **87**, 481 (1973).
204. A. E. Pierce, *Silylation of Organic Compounds*, Pierce Chemical Co., Rockford, Ill., 1968.
205. E. D. Smith and K. L. Shewbart, *J. Chromatogr. Sci.* **7**, 704 (1969).
206. W. C. Butts, *Anal. Biochem.* **46**, 187 (1972).
207. E. D. Smith, J. M. Oathout and G. T. Cook, *J. Chromatogr. Sci.* **8**, 291 (1970).
208. J. A. Thompson and S. P. Markey, *Anal. Chem.* **47**, 1313 (1975).
209. R. D. Hartley and E. C. Jones, *J. Chromatogr.* **107**, 213 (1975).
210. R. H. Collier and G. S. Grimes, *J. Assoc. Off. Anal. Chem.* **57**, 781 (1974).
211. A. Pinelli and A. Colombo, *J. Chromatogr.* **118**, 236 (1976).
212. J. Casals-Stenzel, H.-P. Buscher and R. Schauer, *Anal. Biochem.* **65**, 507 (1975).
213. S. L. Ali, *Chromatographia* **8**, 33 (1975).
214. O. A. Mamer and B. F. Gibbs, *Clin. Chem.* **19**, 1006 (1973).
215. J. A. Hoskins and R. J. Pollitt, *J. Chromatogr.* **109**, 436 (1975).
216. J. A. Thompson, S. P. Markey and P. V. Fennessey, *Clin. Chem.* **21**, 1892 (1975).
217. C. W. Gehrke, H. Nakamoto and R. W. Zumwalt, *J. Chromatogr.* **45**, 24 (1969).
218. C. W. Gehrke and K. Leimer, *J. Chromatogr.* **53**, 201 (1970).
219. C. W. Gehrke and K. Leimer, *J. Chromatogr.* **57**, 219 (1971).
220. R. Gerstl and K. Ranfft, *Mitt. Lebensmittelunters. Hyg.* **65**, 399 (1974).
221. D. L. Stalling, C. W. Gehrke and R. W. Zumwalt, *Biochem. Biophys. Res. Commun.* **31**, 616 (1968).
222. C. J. Pedersen, *J. Am. Chem. Soc.* **89**, 7017 (1967).
223. C. J. Pedersen, *J. Am. Chem. Soc.* **92**, 386 (1970).
224. C. J. Pedersen, *J. Am. Chem. Soc.* **92**, 391 (1970).
225. R. N. Greene, *Tetrahedron Lett.* 1793 (1972).
226. C. J. Pedersen and H. K. Frensdorff, *Angew. Chem. Internat. Edit.* **11**, 16 (1972).
227. H. D. Durst, *Tetrahedron Lett.* 2421 (1974).
228. C. L. Liotta, H. P. Harris, M. McDermott, T. Gonzalez and K. Smith, *Tetrahedron Lett.* 2417 (1974).
229. C. L. Liotta and H. P. Harris, *J. Am. Chem. Soc.* **96**, 2250 (1974).
230. H. D. Durst, M. Milano, E. J. Kikta, Jr., S. A. Connelly and E. Grushka, *Anal. Chem.* **47**, 1797 (1975).
231. C. W. Gehrke and D. F. Goerlitz, *Anal. Chem.* **35**, 76 (1963).
232. C. B. Johnson and E. Wong, *J. Chromatogr.* **109**, 403 (1975).
233. C. B. Johnson, *Anal. Biochem.* **71**, 594 (1976).
234. W. Dünges, *Chromatographia* **6**, 196 (1973).
235. S. L. Ali, *Chromatographia* **6**, 478 (1973).
236. S. L. Ali, *Chromatographia* **7**, 655 (1974).
237. B. C. Pettitt and J. E. Stouffer, *J. Chromatogr. Sci.* **8**, 735 (1970).
238. B. Blessington and N. I. Y. Fiagbe, *J. Chromatogr.* **78**, 343 (1973).
239. F. S. Alvarez and A. N. Watt, *J. Org. Chem.* **33**, 2143 (1968).
240. R. H. Greeley, *J. Chromatogr.* **88**, 229 (1974).
241. J. C. West, *Anal. Chem.* **47**, 1708 (1975).
242. A. J. F. Wickramasinghe and R. S. Shaw, *Biochem. J.* **141**, 179 (1974).

243. F. K. Kawahara, *Anal. Chem.* **40**, 2073 (1968).
244. J. R. Watson and P. Crescuolo, *J. Chromatogr.* **52**, 63 (1970).
245. E. O. Umeh, *J. Chromatogr.* **56**, 29 (1971).
246. M. J. Cooper and M. W. Anders, *Anal. Chem.* **46**, 1849 (1974).
247. J. B. Brooks, J. A. Liddle and C. C. Alley, *Anal. Chem.* **47**, 1960 (1975).
248. E. J. Bourne, M. Stacey, J. C. Tatlow and J. M. Tedder, *J. Chem. Soc.* 2976 (1949).
249. R. C. Parish and L. M. Stock, *J. Org. Chem.* **30**, 927 (1965).
250. J. M. Tedder, *Chem. Rev.* **55**, 787 (1955).
251. R. V. Smith and S. L. Tsai, *J. Chromatogr.* **61**, 29 (1971).
252. C. C. Alley, J. B. Brooks and G. Choudhary, *Anal. Chem.* **48**, 387 (1976).
253. S. Wilk and M. Orlowski, *Anal. Biochem.* **69**, 100 (1975).
254. J. B. Brooks, C. C. Alley and J. A. Liddle, *Anal. Chem.* **46**, 1930 (1974).
255. B. Wilkomirski and Z. Kasprzyk, *J. Chromatogr.* **103**, 376 (1975).
256. J. C. Sheehan and G. P. Hess, *J. Am. Chem. Soc.* **77**, 1067 (1955).
257. H. G. Khorana, *Chem. Rev.* **53**, 145 (1953).
258. F. Kurzer and K. Douraghi-Zadeh, *Chem. Rev.* **67**, 107 (1967).
259. Y. S. Klausner and M. Bodansky, *Synthesis*, 453 (1972).
260. E. Felder, U. Tiepolo and A. Mengassini, *J. Chromatogr.* **82**, 291 (1973).
261. P. Schulz and R. Vîlceanu, *J. Chromatogr.* **111**, 105 (1975).
262. W. Vogt, K. Jacob and M. Knedel, *J. Chromatogr. Sci.* **12**, 658 (1974).
263. G. Ertingshausen, C. W. Gehrke and W. A. Aue, *Separation Sci.* **2**, 681 (1967).
264. P. Vîlceanu, P. Schulz, R. Drăghici and P. Soimu, *J. Chromatogr.* **82**, 285 (1973).
265. H. Ko and M. E. Royer, *J. Chromatogr.* **88**, 253 (1974).
266. M. Itoh, H. Nojima, J. Notani, D. Hagiwara and K. Takai, *Tetrahedron Lett.* 3089 (1974).
267. M. Itoh, D. Hagiwara and J. Notani, *Synthesis*, 456 (1975).
268. H. Gerlach and A. Thalmann, *Helv. Chim. Acta* **57**, 2661 (1974).
269. E. Vowinkel, *Chem. Ber.* **100**, 16 (1967).
270. H. Meerwein, P. Borner, O. Fuchs, H. J. Sasse, H. Schrodt and J. Spille, *Chem. Ber.* **89**, 2060 (1956).
271. H. Vorbrüggen, *Angew. Chem. Int. Edit.* **2**, 211 (1963).
272. H. Brechbühler, H. Büchi, E. Hatz, J. Schreiber and A. Eschenmoser, *Angew. Chem. Int. Edit.* **2**, 212 (1963).
273. J. P. Thenot, E. C. Horning, M. Stafford and M. G. Horning, *Anal. Lett.* **5**, 217 (1972).
274. Anon., *Chromatography Lipids* **7** (4), 2 (1973). Supelco Inc., Bellefonte, Pa.
275. H. Brechbühler, H. Büchi, E. Hatz, J. Schreiber and A. Eschenmoser, *Helv. Chim. Acta* **48**, 1746 (1965).
276. J. P. Thenot and E. C. Horning, *Anal. Lett.* **5**, 519 (1972).
277. I. Horman and F. J. Hesford, *Biomed. Mass. Spectrom.* **1**, 115 (1974).
278. W. E. Harris, *J. Chromatogr. Sci.* **13**, 514 (1975).
279. E. W. Robb and J. J. Westbrook III, *Anal. Chem.* **35**, 1644 (1963).
280. J. J. Bailey, *Anal. Chem.* **39**, 1485 (1967).
281. D. T. Downing, *Anal. Chem.* **39**, 218 (1967).
282. D. T. Downing and R. S. Greene, *Lipids* **3**, 96 (1968).
283. D. T. Downing and R. S. Greene, *Anal. Chem.* **40**, 827 (1968).
284. J. B. F. Lloyd and B. R. G. Roberts, *J. Chromatogr.* **77**, 228 (1973).
285. J. W. Schwarze and M. N. Gilmour, *Anal. Chem.* **41**, 1686 (1969).
286. B. Halpern, W. E. Pereira Jr, M. D. Solomon and E. Steed, *Anal. Biochem.* **39**, 156 (1971).
287. B. S. Middleditch and D. M. Desiderio, *Anal. Lett.* **5**, 605 (1972).
288. K. M. Williams and B. Halpern, *J. Chromatogr.* **97**, 267 (1974).
289. M. S. Roginsky, R. S. Gordon and M. J. Bennett, *Clin. Chim. Acta* **56**, 261 (1974).
290. I. Gan, J. Korth and B. Halpern, *J. Chromatogr.* **92**, 435 (1974).
291. R. Gugler and C. Jensen, *J. Chromatogr.* **117**, 175 (1976).
292. J. MacGee and K. G. Allen, *J. Chromatogr.* **100**, 35 (1974).

293. J. W. Ralls, *Anal. Chem.* **32**, 332 (1960).
294. G. G. Esposito, *Anal. Chem.* **40**, 1902 (1968).
295. K. M. Williams and B. Halpern, *Anal. Lett.* **6**, 839 (1973).
296. G. M. Schier and B. Halpern, *Aust. J. Chem.* **27**, 2455 (1974).
297. K. Williams and B. Halpern, *Synthesis* 727 (1974).
298. D. J. Raber and P. Gariano, *Tetrahedron Lett.* 4741 (1971).
299. S. Hünig and M. Kiessel, *Chem. Ber.* **91**, 380 (1958).
300. H. Meerwein, *Org. Synth.* **46**, 120 (1966).
301. J. Grundy, B. G. James and G. Pattenden, *Tetrahedron Lett.* 757 (1972).
302. V. A. Derevitskaya, E. M. Klimov and N. K. Kochetkov, *Tetrahedron Lett.* 4269 (1970).
303. M. S. Newman and W. S. Fones, *J. Am. Chem. Soc.* **69**, 1046 (1947).
304. R. C. Fuson, J. Corse and E. C. Horning, *J. Am. Chem. Soc.* **61**, 1290 (1939).
305. H. Cohen and J. D. Mier, *Chem. Ind. (London)* 349 (1965).
306. P. E. Pfeffer, T. A. Foglia, P. A. Barr, I. Schmeltz and L. S. Silbert, *Tetrahedron Lett.* 4063 (1972).
307. J. E. Shaw, D. C. Kunerth and J. J. Sherry, *Tetrahedron Lett.* 689 (1973).
308. E. M. Filachione, E. J. Costello and C. H. Fisher, *J. Am. Chem. Soc.* **73**, 5265 (1951).
309. R. W. Parodi, *Aust. J. Dairy Technol.* **22**, 144 (1967).
310. J. Horner, S. S. Que Hee and R. G. Sutherland, *J. Agr. Food. Chem.* **22**, 726 (1974).
311. Anon., Gas–Chrom Newsletter, Applied Science Laboratories Inc., State College, PA., **11**(4), 1 (1970).
312. A. Karmen, *J. Lipid Res.* **8**, 234 (1967).
313. K. Oette and E. H. Ahrens Jr, *Anal. Chem.* **33**, 1847 (1961).
314. R. M. Caprioli, W. E. Seifert Jr and D. E. Sutherland, *Biochem. Biophys. Res. Commun.* **55**, 67 (1973).
315. H. Falter, K. Jajasimhulu and R. A. Day, *Anal. Biochem.* **67**, 359 (1975).
316. C. C. Sweeley, R. Bentley, M. Makita and W. W. Wells, *J. Am. Chem. Soc.* **85**, 2497 (1963).
317. L. Birkofer and M. Donike, *J. Chromatogr.* **26**, 270 (1967).
318. F. P. Abramson, M. W. McCaman and R. E. McCaman, *Anal. Biochem.* **57**, 482 (1974).
319. A. Darbre and A. Islam, *Biochem. J.* **106**, 923 (1968).
320. A. Islam and A. Darbre, *J. Chromatogr.* **71**, 223 (1972).
321. A. J. Cliffe, N. J. Berridge and D. R. Westgarth, *J. Chromatogr.* **78**, 333 (1973).
322. H. Lindley and P. C. Davis, *J. Chromatogr.* **100**, 117 (1974).
323. D. H. Calam, *J. Chromatogr.* **70**, 146 (1972).
324. J. A. McCloskey in J. M. Lowenstein (Ed.) *Methods in Enzymology*, Academic Press, New York, **35**, 340 (1975).
325. M. A. Kirkman, *J. Chromatogr.* **97**, 175 (1974).
326. R. F. McGregor, G. M. Brittin and M. S. Sharon, *Clin. Chim. Acta.* **48**, 65 (1973).
327. J. F. March, *Anal. Biochem.* **69**, 420 (1975).
328. J. R. Coulter and C. S. Hann, *J. Chromatogr.* **36**, 42 (1968).
329. C. W. Moss, M. A. Lambert and F. J. Diaz, *J. Chromatogr.* **60**, 134 (1971).
330. J. Jönsson, J. Eyem and J. Sjöquist, *Anal. Biochem.* **51**, 204 (1973).
331. D. Roach and C. W. Gehrke, *J. Chromatogr.* **44**, 269 (1969).
332. R. W. Zumwalt, K. Kuo and C. W. Gehrke, *J. Chromatogr.* **57**, 193 (1971).
333. C. W. Gehrke and H. Takeda, *J. Chromatogr.* **76**, 63 (1973).
334. F. E. Kaiser, C. W. Gehrke, R. W. Zumwalt and K. C. Kuo, *J. Chromatogr.* **94**, 113 (1974).
335. J. J. Rash, C. W. Gehrke, R. W. Zumwalt, K. C. Kuo, K. A. Kvenvolden and D. L. Stalling, *J. Chromatogr. Sci.* **10**, 444 (1972).
336. R. W. Zumwalt, K. Kuo and C. W. Gehrke, *J. Chromatogr.* **55**, 267 (1971).
337. R. W. Zumwalt, D. Roach and C. W. Gehrke, *J. Chromatogr.* **53**, 171 (1970).
338. W. J. McBride and J. D. Klingman, *Anal. Biochem.* **25**, 109 (1968).
339. J. Metz, W. Ebert and H. Weicker, *Chromatographia* **4**, 259 (1971).
340. P. Cancalon and J. D. Klingman, *J. Chromatogr. Sci.* **12**, 349 (1974).

341. M. Sakamoto, K. Kajiyama and H. Tonami, *J. Chromatogr.* **94**, 189 (1974).
342. I. M. Moodie, *J. Chromatogr.* **99**, 495 (1974).
343. C. W. Gehrke and H. Takeda, *J. Chromatogr.* **76**, 77 (1973).
344. S. L. MacKenzie and D. Tenaschuk, *J. Chromatogr.* **97**, 19 (1974).
345. S. L. Mackenzie and D. Tenaschuk, *J. Chromatogr.* **111**, 413 (1975).
346. P. Felker and R. S. Bandurski, *Anal. Biochem.* **67**, 245 (1975).
347. J. P. Zanetta and G. Vincendon, *J. Chromatogr.* **76**, 91 (1973).
348. F. Marcucci and E. Mussini, *J. Chromatogr.* **25**, 11 (1966).
349. E. Mussini and F. Marcucci, *J. Chromatogr.* **26**, 481 (1967).
350. C. Crotte, A. Mulé and N. E. Planche, *Bull. Soc. Chim. Biol.* **52**, 108 (1970).
351. M. J. Levitt, *Anal. Chem.* **45**, 618 (1973).
352. D. P. Schwartz and R. S. Bright, *Anal. Biochem.* **61**, 271 (1974).
353. A. Darbre and A. Islam, *J. Chromatogr.* **49**, 293 (1970).
354. K. W. M. Davy and C. J. O. R. Morris, *J. Chromatogr.* **116**, 305 (1976).
355. J. B. Brooks, D. S. Kellogg, L. Thacker and E. M. Turner, *Can. J. Microbiol.* **18**, 157 (1972).
356. L. Jansén and O. Samuelson, *J. Chromatogr.* **57**, 353 (1971).
357. W. Dünges, *Anal. Chem.* **45**, 963 (1975).
358. J. B. Brooks, C. C. Alley, J. W. Weaver, V. E. Green and A. M. Harkness, *Anal. Chem.* **45**, 2083 (1973).
359. W. Morozowich and S. L. Douglas, *Prostaglandins* **10**, 19 (1975).
360. R. F. Borch, *Anal. Chem.* **47**, 2437 (1975).
361. V. S. Abramov, *Dokl. Akad. Nauk SSSR* **73**, 487 (1950).
362. E. K. Fields, U.S. Pat. 2 579 810 (1951).
363. R. F. Goddu, N. F. Leblanc and C. M. Wright, *Anal. Chem.* **27**, 1251 (1955).
364. R. E. Notari and J. W. Munsen, *J. Pharm. Sci.* **58**, 1060 (1969).
365. R. E. Notari, *J. Pharm. Sci.* **58**, 1069 (1969).
366. F. Snyder and N. Stephens, *Biochim. Biophys. Acta* **34**, 244 (1959).
367. Y. Kasai, T. Tanimura and Z. Tamura, *Anal. Chem.* **47**, 34 (1975).
368. A. Verley and F. R. Bölsing in G. H. Schenk (Ed.) *Organic Functional Group Analysis*, Pergamon Press, Oxford, 1968, p. 157.
369. T. S. Ma in S. Patai (Ed.) *The Chemistry of Carboxylic Acids and Esters*, Interscience, London, 1969, p. 871.
370. G. Zweifel and H. C. Brown, *Organic Reactions* **13**, 28 (1963).
371. Th. J. De Boer and H. J. Backer, *Recl. Trav. Chim. Pays Bas* **73**, 229 (1954).
372. H. M. Fales, T. M. Jaouni and J. F. Babashek, *Anal. Chem.* **45**, 2302 (1973).
373. W. B. Denmore, H. D. Pritchard and N. Davidson, *J. Am. Chem. Soc.* **81**, 5874 (1959).
374. M. Wilcox, *Anal. Biochem.* **16**, 253 (1966).
375. P. Yates and B. L. Shapiro, *J. Org. Chem.* **23**, 759 (1958).
376. D. L. Corina and P. M. Dunstan, *Anal. Biochem.* **53**, 571 (1973).
377. C. G. Overberger and J.-P. Anselme, *J. Org. Chem.* **28**, 592 (1963).

CHAPTER 3

Acylation

Karl Blau and Graham S. King

Bernhard Baron Memorial Research Laboratories,
Queen Charlotte's Maternity Hospital,
London W6 OXG, United Kingdom.

1 INTRODUCTION

Acylation has been widely used in derivatization for chromatographic analyses for several important reasons. First, acylation reduces the polarity of amino, hydroxy and thiol groups, which in the underivatized form may not chromatograph very well because of non-specific absorption effects such as tailing and ghost-peaks. Secondly, acylation may confer volatility on substances such as carbohydrates and amino acids which have so many polar groupings that they are involatile, and so these substances become accessible to gas chromatographic analysis. Here acylation is an alternative to silylation, giving more stable derivatives with amines for example. Thirdly, acylation may help to separate closely related substances which in the underivatized state may be difficult to resolve. Finally, it may be used to introduce electron-capturing groupings or other detector-oriented substituents into molecules which are thus detectable with very high sensitivity. Other advantages may also come from acylation, for example catecholamines which in the free state are very easily oxidized, are to some extent stablized by acylation. All of these reasons have led to a proliferation of applications, and this may at first seem rather bewildering. However, the underlying principles and methods are in fact very straightforward, and once they are mastered, it will not be difficult to select one or two of the more promising acylation procedures to apply to any particular problem with a good chance of a successful analysis.

2 REACTION MECHANISMS

2.1 General

The chemistry of acylation involves the introduction of an acyl group $R{-}\overset{\overset{\displaystyle O}{\|}}{C}{-}$ into a molecule by substitution of one of the replaceable hydrogen atoms (see also Table 1).

$$R-\overset{O}{\overset{\|}{C}}-X + RY'-H \rightarrow R-\overset{O}{\overset{\|}{C}}-RY' + HX \qquad (1)$$

Less commonly, an acyl group may be added across a double bond.

$$R-\overset{O}{\overset{\|}{C}}-X + \overset{\diagdown}{\diagup}C=C\overset{\diagup}{\diagdown} \rightarrow R-\overset{O}{\overset{\|}{C}}-\overset{|}{C}-\overset{|}{C}-X \qquad (2)$$

The acylating agent $R-\overset{O}{\overset{\|}{C}}-X$ can lose the group $-X$ by (a) electrophilic, (b) nucleophilic or (c) free radical mechanisms, represented in Scheme (3).

$$R-\overset{O}{\overset{\|}{C}}{}^{+} + X^{-} \qquad (3a)$$

$$R-\overset{O}{\overset{\|}{C}}-X \longrightarrow R-\overset{O}{\overset{\|}{C}}{}^{-} + X^{+} \qquad (3b)$$

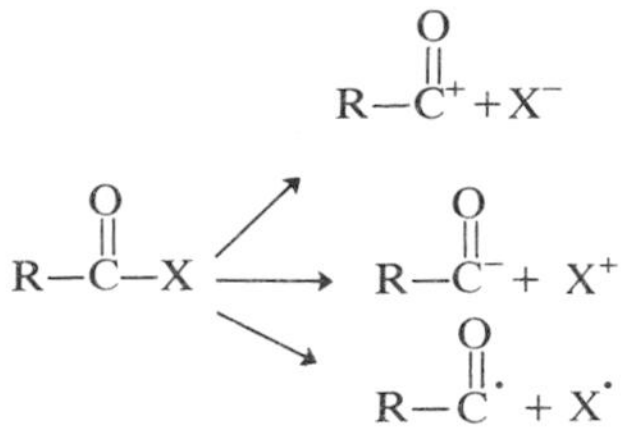

$$R-\overset{O}{\overset{\|}{C}}{}^{\bullet} + X^{\bullet} \qquad (3c)$$

Direct electrophilic acylations of type (3a) are by far the most common mode of acylation.[1] It is probably unrealistic to propose cleavages as shown in Scheme 3, but the mechanism may be viewed as the formation of a carbonyl addition intermediate:

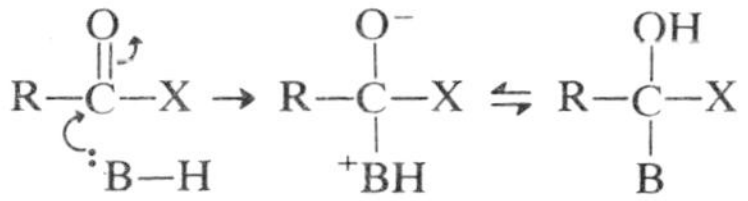

$$R-\underset{\overset{|}{\underset{B-H}{\overset{\cdot\cdot}{C}}}}{\overset{O}{\overset{\|}{C}}}-X \rightarrow R-\underset{\overset{|}{\overset{+}{BH}}}{\overset{O^-}{\overset{|}{C}}}-X \leftrightharpoons R-\underset{\overset{|}{B}}{\overset{OH}{\overset{|}{C}}}-X \qquad (4a)$$

followed by elimination of HX (4b).

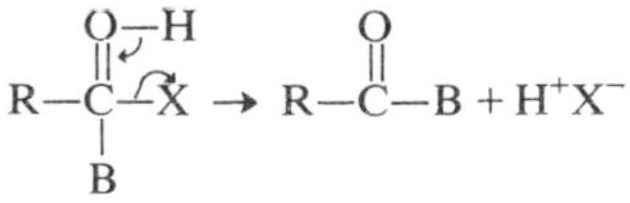

$$R-\underset{\overset{|}{B}}{\overset{O-H}{\overset{\|}{C}}}X \rightarrow R-\overset{O}{\overset{\|}{C}}-B + H^+X^- \qquad (4b)$$

However, synchronous displacement (5) is also a possibility:

$$R-\overset{O}{\overset{\|}{C}}-X + H-B \rightarrow R-\overset{O}{\overset{\|}{C}}-B + HX \qquad (5)$$

In any particular acylation reaction the precise mechanism by which it proceeds probably lies somewhere in between these extreme formulations.[2] The mechanistic considerations of acylation are of course essentially the same as those of esterification, so for a more detailed consideration of these points Chapter 2 may be consulted.

2.2   The reactivity of acyl imidazoles

Normal amides have a low reactivity in nucleophilic reactions because the carbon atom of the amide carbonyl group is more negatively charged than in normal esters and ketones (6),

$$R-\overset{\displaystyle O}{\overset{\|}{C}}-O-R' \qquad (6)$$

$$R-\overset{\displaystyle O}{\overset{\|}{C}}-NH-R' \leftrightarrow R-\overset{\displaystyle O^-}{\overset{|}{C}}=\overset{+}{N}H-R' \qquad (7)$$

due to the delocalization of the lone pairs of electrons on the nitrogen into the carbonyl group (7). At first sight the high reactivity of imidazolides is unusual; they are acylating agents of comparable strength to acid anhydrides and chlorides. This can be seen in a comparison of the susceptibilities to hydrolysis of *N*-acetylpyrrole and *N*-acetylimidazole (8).

(8)

Halflife in water at 25 °C and pH 7: ∞ (N-acetylpyrrole); 41 min (N-acetylimidazole)

The high reactivity of the acyl imidazole is due to the fact that the nitrogen electrons are delocalized into the heterocyclic ring. This delocalization is encouraged by increasing the number of nitrogen atoms in the ring, because this tends to increase its aromaticity. As an example, the hydrogen atom in imidazole is quite acidic, because of bond polarization.

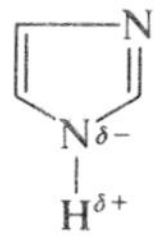

(9)

Considerable variation in stability can be seen with different kinds of acyl groups:

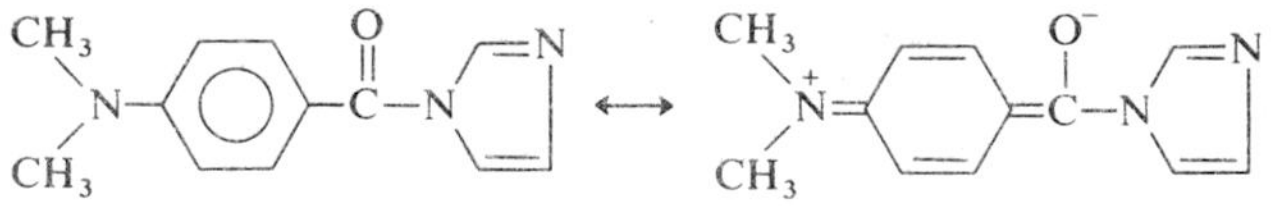

halflife in excess of one week (10)

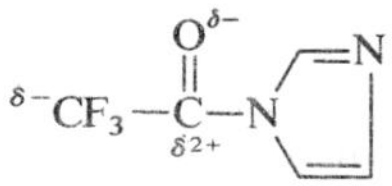

almost instantaneous hydrolysis

Perfluoroacyl-imidazoles, acetyl-imidazole and pentafluorobenzoyl-imidazole are commercially available from several sources; others may be prepared from the acid chloride and imidazole in tetrahydrofuran, or from the acid and diimidazolecarbodiimide.[3]

Derivatives of benzimidazole (11) are more stable than imidazole itself:

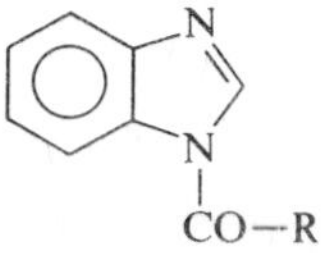

(11)

This may be a useful way of obtaining a more stable reagent for perfluoroacylation or formylation, but the by-products could be a problem. The imidazole moiety of imidazole reagents, which is the product of acylation with these reagents, is volatile enough not to interfere with most GLC analyses.

3 TYPES OF ACYLATING AGENTS

In general, acylation reactions for chromatographically useful derivatizations make use of three main types of acylating agents: acid anhydrides, acid halides, and reactive acyl derivatives such as acylated imidazoles, acylated amides or acylated phenols. There are reasons for choosing each particular type. Acyl halides are highly reactive and this may be important with compounds that are difficult to acylate, such as amides, or where acylation may be unfavourable because of steric factors. A drawback to using acyl halides is that a basic acceptor is usually needed for the halogen acid produced in the reaction:

$$R{-}NH_2 + R'COCl + B \rightarrow R{-}NH{-}CO{-}R' + B^{+}HCl^{-} \qquad (12)$$

The removal of excess of the acid halide, and of the halide salt formed with the acceptor, may be a troublesome necessity. Anhydrides are usually easier to remove at the end of the reaction, yielding a clean product:

$$R{-}NH_2 + R'{-}CO{-}O{-}CO{-}R' \rightarrow R'{-}CO{-}NH{-}R + R'{-}COOH \qquad (13)$$

However, anhydrides are less volatile than the corresponding acid halides, and may therefore interfere with chromatography of the more volatile products. Also, the reaction medium is acidic, which may make this type of acylation reagent unsuitable with acid-sensitive compounds. For such compounds an acylation reagent which results in a basic leaving group on reaction, such as an acyl-imidazole, may be preferable:

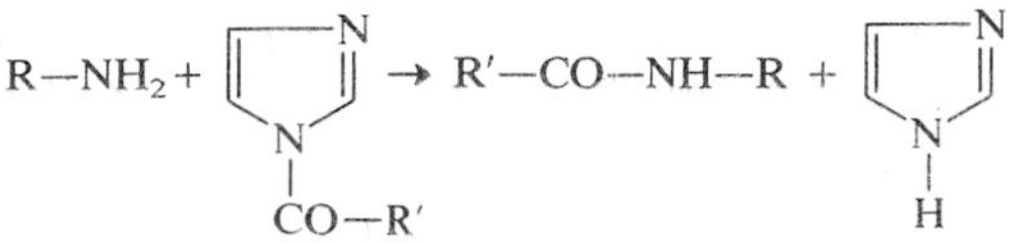

(14)

A drawback is that excess of the reagent may have to be removed by some kind of extraction step.

Small-scale acylations, like other derivatization reactions for chromatography, are usually done with a large molar excess of acylation reagent, to drive the reaction to completion. The nature of the solvent used may have a considerable influence on the yield and even on the course of the reaction, as will be seen from the specific methods that are described later. It has been found that in most cases it is preferable to get rid of excess of the acylation reagents, because their great chemical reactivity may have adverse effects on subsequent analyses. Problems that have arisen include irreversible alteration of the chromatographic column, corrosion or other damage within the chromatographic or GC—MS system, and re-derivatization of non-volatile residues at the top of the column to give 'ghost' peaks. For these reasons the acid anhydrides, which are often easier to remove, and often do not require a basic acceptor molecule (which may itself pose chromatographic problems), have been more popular than the acyl halides.

4 DETECTOR-ORIENTED ACYLATIONS

The electron-capture detector (ECD) is a gas chromatographic detector which responds to those molecules best able to stabilize an attached electron. These include halogen-containing molecules, and to a lesser extent those containing nitro and keto groups. The detector may respond to such compounds with a degree of sensitivity one or more orders of magnitude greater than that of the flame-ionization detector (FID), while having a diminished response to compounds lacking these structural features (usually including the solvent) by a similar factor.[4] The ECD thus discriminates greatly in favour of electron-capturing substances, and is the most sensitive means of analysing such compounds. This has made gas chromatography with electron-capture detection the method of choice for the analysis of pesticides which are usually halogen-containing compounds that do not require derivatization.[5] However, the great sensitivity obtainable with the ECD has attracted those working with compounds which are present in solution at very low concentrations to use carefully selected reagents that will specifically introduce halogen-containing groupings to give electron-capturing derivatives for use with the ECD. Examples of such compounds are biogenic monoamines; drugs and their metabolites in blood, urine and tissues; and phenolic compounds. As shown in Table 1, acylation reactions are particularly suitable for making derivatives from compounds such as these, and many electron-capturing groups have been selected for conferring electron-capturing properties to such substances and are described in this chapter (Table 3).

Although these derivatives all have enhanced responses in the ECD, more than one mechanism of electron-capture has been described,[6] and a knowledge of these may help in the choice of the most suitable derivative to use in any given application. These factors are described more fully later in this chapter, but as a rough guide the sensitivity of detection increases in the order:

$$F < NO_2 \approx Cl < Br < I$$

TABLE 1

The main types of groupings capable of being acylated

Chemical structure	Description
$-NH_2$, $-NH-$, $-\overset{\vert}{N}-$	Primary, secondary and tertiary amines
$-NH-CO-R$	Amides
$-CH_2-OH$, $-\overset{OH}{\overset{\vert}{C}H}-$, $-\overset{OH}{\overset{\vert}{\underset{\vert}{C}}}-$	Primary, secondary and tertiary alcohols
$-SH$	Thiols
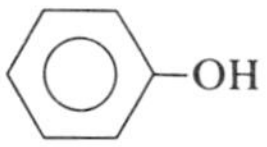	Phenols
$-C=\overset{OH}{\overset{\vert}{C}}-$	Enols
$-\overset{HO}{\overset{\vert}{C}}-\overset{OH}{\overset{\vert}{C}}-$	Glycols
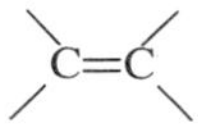	Unsaturated compounds
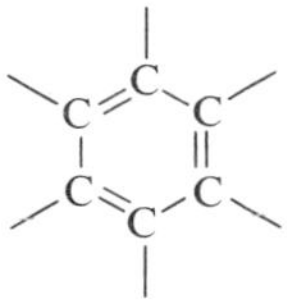	Aromatic rings

It also increases as the number of electron-capturing groupings in the molecule increases. However, there are exceptions to this, and the response of electron-capturing derivatives is compared later in this chapter (Table 4). The resulting increased sensitivity of detection is, however, often accompanied by an increased retention time, although in the perfluoroalkanoyl series the increase in the retention times as one ascends from trifluoroacetyl (TFA) to heptafluorobutyryl (HFB) is not too great. There may, for some compounds, even be a minimum in the retention time for the pentafluoropropionyl (PFP) derivatives.[7] In general it has been found preferable to introduce five fluorine atoms rather than two chlorines, because the fluorinated derivatives are likely to be more volatile, and can therefore be separated at lower temperatures. The volatility of the products is an important consideration in choosing which derivative to make. For substances that are not very volatile it is advantageous to make a derivative that is more volatile, while very volatile substances should be converted to less volatile derivatives as early in the analytical scheme as possible, to prevent losses caused by unintentional volatilization. These considerations must be reconciled with the temperature limits of the chromatographic column to be used.

Derivatives containing nitro groups are generally much less volatile, and tend to 'tail' on chromatography, so that they have been used rather less widely for GLC (see Chapter 10). Of the compounds that have been evaluated, the halogenated acyl derivatives in Table 3 have shown the most useful chromatographic properties.

A detector-oriented derivatization scheme using phosphorus instead of halogens has been described: this made use of diethylphosphoryl derivatives with the flame thermionic detector, again giving greatly enhanced sensitivity of analysis, and discrimination against non-phosphorus compounds.[8] If the use of satisfactory phosphorus-specific detectors (outside the pesticide analysis field) should gain ground, then this approach offers a great deal of promise.

5 PRACTICAL APPLICATIONS

5.1 Introduction

As with esterification (Chapter 2), there are innumerable acylation procedures, and only a selection of these can be presented here, to give a wide but by no means exhaustive coverage of the available approaches. It soon becomes evident on reading through the 'recipes' that follow, that the mechanistic unity described earlier for acylation reactions taken as a whole, is reflected in the general resemblance found among procedures for acylation. This has the practical consequence that it is not difficult to devise an acylation procedure for a new compound, which accounts for the popularity of this type of derivatization.

Since analyses require only a few microlitres of the final solution of the derivatives, it is not necessary to perform acylations (or any other derivatization reaction) on anything like as large as the preparative scale, unless it is necessary to characterize the derivative by rigorous classical chemical methods. This is especially true of electron-capturing derivatives, where sensitivity of detection may be so high that analyses at the nanogram or even picogram scale are involved. In these cases it is satisfactory to work with quite small amounts of reagents in small vessels. Of great value have been the various kinds of special reaction vessels now widely available, and described in Chapter 1. However, the caution mentioned there about the use of Teflon® for lining the caps of reaction vials particularly applies with acylation reactions. These Teflon® liners are still able to extract acylation reagents and products by physical solution into the polymer itself, where physical losses and hydrolysis may occur. Teflon® is particularly likely to absorb fluorine-containing materials, so the performance of a reaction in a Teflon®-lined screw-capped vessell should be compared with the same reaction done in an all-glass system such as a ground-glass stoppered tube, to make sure that it is safe to use Teflon® as part of the closure (see Fig. 1).

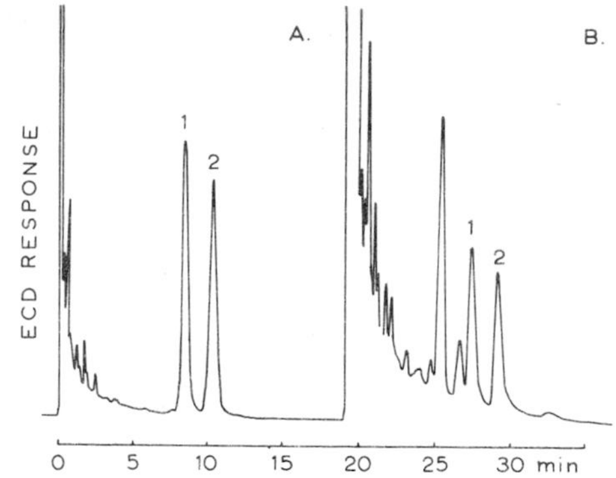

Fig. 1 Illustration of the effect of Teflon® sealing a reaction vessel in the derivatization of small amounts of substances. Pentafluorobenzoyl derivatives of 10 ng each of β-phenylethylamine (peak 1) and β-(*p*-tolyl)ethylamine (peak 2) were made as described in Section 5.13.2 (p. 133) in identical 2 ml vials. One was sealed with a glass microscope cover-slip (chromatogram A), the other with a Teflon®-lined screw-cap (chromatogram B). Note the reduced peak heights, by-product peaks and slower fall of the solvent peak in B. Analysis was on a 180×0.3 cm column of 2.5% OV-225 at 215 °C.

As with esters (Chapter 2), acyl derivatives usually have good chromatographic properties. Some amides retain a residual polar character which may reveal itself in a tendency to 'tail' on GLC, but this can usually be overcome by using a moderately polar stationary phase. The stability of the various acylated derivatives will be discussed wherever appropriate in the practical procedures that follow.

5.2 Acetylation methods

5.2.1 ACETYLATION WITH ACETIC ANHYDRIDE AND ACETIC ACID

This is the classical way to perform the acetylation reaction. The substance to be acetylated (20 mg) is dissolved in chloroform (20 ml), and acetic anhydride (0.5–1 ml) in acetic acid (1–5 ml) is added. The reaction proceeds at 50 °C for 2–16 h, and excess of the reagents is removed *in vacuo*, before dissolving the residue in chloroform to a final volume of 20 ml for GLC analysis.[9]

5.2.2 ACETYLATION OF URINARY CARBOHYDRATES WITH ACETIC ANHYDRIDE AND SODIUM ACETATE

This is another classical acetylation procedure, here used on carbohydrates extracted from human urine. The dried residue from the extraction procedure is converted to the oximes in the presence of sodium acetate (12.5 mg), and acetic anhydride (0.3 ml) is subsequently added. The reaction is done at 100 °C for 1 h. The same results are also obtained by using pyridine as base and catalyst (see Section 5.2.4 below).[10]

5.2.3 ACETYLATION OF ALCOHOLS WITH NEAT ACETIC ANHYDRIDE

The anhydrous alcohol, or mixture of alcohols obtained by prior isolation (20 mg or less) is refluxed with freshly distilled acetic anhydride (1 ml) or heated with the anhydride in a sealed reaction vial at 100 °C for 15 min. The reaction mixture is poured into water (5 ml) with intermittent shaking, and alkali is added to destroy excess of the anhydride, to a final pH of 10. The esters are extracted with ether (3 × 1 ml) or other suitable solvent, and the extract dried over anhydrous magnesium sulfate. The solution is concentrated in a stream of N_2 for analysis by GLC.[11]

5.2.4 ACETYLATIONS WITH ACETIC ANHYDRIDE AND PYRIDINE

Acetic anhydride is often used with pyridine, which is not only an excellent solvent, especially for carbohydrates and steroids, but also acts as a basic catalyst with the ability to promote smooth reactions and as an acceptor for the acid formed in the reaction. The proportions may be varied to achieve the best results, and depend on the compounds to be acetylated. It should be pointed out that acetic anhydride can form a derivative with pyridine[12] identified as *N*-acetyl-1,2-dihydro-2-pyridylacetic acid, and this may interfere in some chromatographic analyses. Three typical acetylations of this type are presented.

Acetylation of alditols: the alditols used as such, or resulting from the sodium borohydride reduction of sugars, are dissolved in a 1:1 acetic anhydride:pyridine mixture and heated at 100 °C for 20 min. The reaction mixture can be injected directly for GLC analysis, or it may be evaporated to dryness under vacuum and the residue dissolved in a suitable solvent such as ethyl acetate, for analysis. A column of 3% ECNSS—M gives the best separation. The advantage of this procedure is that sugars reduced to the alditols and acetylated give a single peak for each sugar, as distinct from the multiple peaks obtained with silylation.[13]

Acetylation of paracetamol: the drug, or a dried extract from a physiological fluid (obtained by extraction with ethyl acetate) to a total weight of up to 50 μg, is treated with pyridine (5 μl) and acetic anhydride (15 μl), mixed and heated at 45 °C for 20 min. The mixture (1–3 μl) is injected directly for GLC on 3% Hi-Eff 8BP or OV-17 at 200–220 °C.[14]

Acetylation of amino acid n-*propyl esters*: amino acids (mg scale or less) are esterified with dry 7–8 M propanolic HCl at 100 °C for 10 min. After removal of the alcohol, the residue is treated with acetic anhydride:pyridine 1:4 (0.4 ml) at room temperature for 10 min. Excess reagents are removed either *in vacuo* or in a stream of N_2. The residue is dissolved in ethyl acetate for GLC analysis.[15,16]

A more basic acetylation reagent was used by Adams,[17] with the advantage of rapid acetylation and quantitative yield. The amino acids are first isolated on a microcolumn of Dowex 50-X8 (100–120 mesh, 50 mg). The dried amino acid isolate (approx. 1 μmol) is converted to the propyl esters with 8 M propanolic HCl (1 ml) and dimethoxypropane (50 μl) as a water scavenger at 100 °C for 20 min, and evaporated to dryness in a stream of N_2, taking care to remove all traces of acid. The dried residue is acetylated with a freshly prepared 5:2:1 mixture of acetone-triethylamine-acetic anhydride (1 ml) at 60 °C for 30 s. The excess reagents are blown off cautiously to avoid volatilization losses, and the dry residue dissolved in ethyl acetate (100 μl) for GLC on a special mixed stationary phase.[17]

5.2.5 ACETYLATION OF GLUCURONIDES WITH ACETIC ANHYDRIDE AND METHANESULFONIC ACID

The sample is dissolved in acetic anhydride (1 ml) and methanesulfonic acid (50 μl) and is left at room temperature for 24 h. The reaction mixture is poured into ice-water and extracted into chloroform. The chloroform extract is dried over anhydrous sodium sulfate and concentrated *in vacuo*. The dried residue is methylated with ethereal diazomethane, evaporated to dryness *in vacuo*, and dissolved in chloroform for GLC analysis on 5% SE-30 at 280 °C.[18] These derivatives have been studied by mass spectrometry.[19] (See also methods 5.2.9, 5.2.10, and 5.2.12.)

5.2.6 ACETYLATION OF PHENOLIC AMINES OR ALCOHOLS IN AQUEOUS SOLUTION WITH ACETIC ANHYDRIDE (SCHOTTEN–BAUMANN CONDITIONS)[20,21]

The aqueous solution of amines (or amine extract from tissues or biological fluids), totalling about 4 ml, is treated with acetic anhydride (0.3 ml). A slight excess of solid sodium bicarbonate is added in small portions with mixing, allowing effervescence to subside each time. The solution is extracted with dichloromethane (2 × 10 ml), and the combined extracts evaporated to dryness in a stream of N_2. Under these conditions phenolic hydroxyl groups (but not alcoholic hydroxyls) are also acetylated.[22,23] This is also essentially the method used by Röder and Merzhäuser to acetylate amines for HPLC.[23] The amino group is sufficiently reactive so that its acetylation in an aqueous solution is more rapid than hydrolysis of the acetylation reagent. This is the basis of the Schotten–Baumann type acylation reactions.

5.2.7 ACETYLATION OF AMINO ACIDS OR AMINES IN AQUEOUS SOLUTION WITH KETENE

Ketene is produced by the pyrolysis of acetone:

$$(CH_3)_2C{=}O \rightarrow CH_2{=}C{=}O + CH_4 \qquad (15)$$

It is an effective and reactive acetylating agent.[24] If it is bubbled into an aqueous solution of an amine or amino acid it smoothly acetylates all the amino groups,[25] but none of the hydroxyls, and the products can be isolated as described in the previous section. It has not been widely used in derivatization, because a special generator must be made,[26] and because it is a respiratory tract irritant.

CAUTION: an efficient fume hood is necessary if ketene is to be used.

5.2.8 ACETYLATION OF HEXOSAMINES WITH SIMULTANEOUS SILYLATION[27]

Hexosamines (5 mg) are mixed with dimethylformamide (50 μl) containing *N*-acetylglucosaminitol (1 mg ml^{-1}) as internal standard, pyridine (40 μl), acetic anhydride (10 μl) to acetylate the amino groups, and hexamethyldisilazane (30 μl) to silylate the hydroxyl groups. After vigorous mixing for 1 min, the reaction is left at room temperature for 30 min, and a few microlitres injected directly for analysis. Yields are 92–97%.

5.2.9 ACETOLYSIS OF POLYSACCHARIDES

It is often possible to save a step during an analytical scheme, by getting two reactions to occur simultaneously, as in the previous method. Another example is acetolysis, where cleavage and acetylation proceed at the same time. The polysaccharide (0.1 g) is stirred with a 1:1:1 mixture of acetic anhydride: acetic acid: conc. sulfuric acid at 40 °C overnight (or longer if necessary) and the reaction mixture is then neutralized with pyridine and ice-cooling. The mixture is diluted with water and the acetylated saccharides extracted with chloroform. The chloroform solution is washed with water and evaporated to dryness for chromatography.[28]

5.2.10 ACETOLYSIS OF SULFATE ESTER CONJUGATES

Another useful acetolysis of a similar kind is the replacement of a sulfate ester group by an acetyl group: this is a relatively mild procedure, which can be used to characterize sulfate esters which are too unstable for conventional conditions of hydrolysis.

The sulfate ester (10 mg) is mixed with an acetylation mixture of 40:1 acetic anhydride: methanesulfonic acid (2 ml) at 0 °C. The mixture is heated at 100 °C for 30 min, and then cooled to 0 °C. Ice (2 g) is added, and the mixture extracted with benzene (3×4 ml). The combined extracts are evaporated to a small volume for analysis. Yields vary with different compounds, and may be improved by longer heating and by increasing the proportion of acetylation mixture to sulfate ester.[29]

5.2.11 ACETYLATION OF HYDROXYANTHRANILIC ACID WITH ACETYL CHLORIDE

Hydroxyanthranilic acid (approx. 1 mg) is first esterified with diazomethane overnight. This also methylates the phenolic hydroxyl group. After evaporating in a stream of N_2, the residue is left in the dark with benzene (0.5 ml) and acetyl chloride (0.5 ml) at room temperature for 30 min. The mixture is evaporated again with N_2 at 37 °C, and the dry residue is dissolved in methanol for GLC analysis.[30]

Acetyl chloride is not much used for acetylation, and a base should preferably be used with it to act as acceptor for the acidic products of the reaction.

5.2.12 RAPID ACETYLATION OF ALCOHOLS ON A CELITE–ACETYL METHANESULFONATE COLUMN

This procedure is unusual in using a microcolumn (1.5 cm in length in a melting-point capillary tube) filled with celite onto which the mixed anhydride of acetic acid and methanesulfonic acid has been absorbed. The alcohols (1–3 mg) in a suitable solvent (carbon tetrachloride, hexane, benzene, carbon disulfide etc., 1–10 μl) are injected into one end of the column with a microlitre syringe, and the column is eluted with one bed-volume of solvent, using gentle air or N_2 pressure. The eluates are retrieved with the syringe at the other end, and can be injected directly, or the acidic products may first be removed on an adjacent microcolumn in the same tube, of K_2HPO_4-impregnated celite (1.5 cm length). For further details it is recommended that the original paper be consulted. Most primary and secondary alcohols are acetylated smoothly, rapidly and quantitatively. Even hindered tertiary alcohols (with a few exceptions) react, but less than quantitatively.[31]

5.3 Methods for making propionyl derivatives

5.3.1 DIRECT METHOD FOR AMINES WITH PROPIONIC ANHYDRIDE

This procedure is like acetylation with neat anhydride (see Section 5.2.3). The amine or amine hydrochloride (5 mg) is heated with freshly distilled propionic anhydride (0.5 ml) in a sealed reaction vial at 100 °C for 30 min. The product is poured onto water (5 ml), and cautiously basified with alkali; the amides are extracted with methylene chloride (3 × 1 ml).

5.3.2 *N*-PROPIONYL AMINO ACID ESTERS

This procedure, due to Youngs, has been used to characterize changes in the amino acid pools of growing micro-organisms.[32] Amino acids (5 mg) are esterified with isopentanol saturated with HCl (10 ml) at 100 °C for 30 min. The alcohol is pumped off *in vacuo*, and the residue taken up in carbon tetrachloride (5 ml). Anhydrous sodium carbonate (100 mg) is added, followed by propionyl chloride (0.2 ml). After refluxing for 30 min, the carbonate is filtered off, the solvent concentrated, and the solution analysed by GLC on 10% QF-1 at 150 °C and 3° min^{-1} up to 250 °C.

5.4 Methods for making formyl derivatives

5.4.1 DIRECT PREPARATION OF FORMAMIDO DERIVATIVES

The base or its salt (1–5 mg) is treated with formic acid (20 μl), anhydrous sodium formate (10 mg) and acetic anhydride (200 μl) at 25 °C for 30 min in a reaction vial sealed with a Teflon®-lined cap. Excess reagents are removed *in vacuo*, and the residue is extracted with a suitable solvent e.g. ethyl acetate. Yields usually exceed 90%.[33–35]

5.4.2 FORMYLATION OF AMINES USING DICYCLOHEXYLCARBODIIMIDE

The free base dissolved in chloroform, or liberated from its salts with 1 molar equivalent of triethylamine, in chloroform, is treated at 0 °C with 1 molar equivalent of formic acid in the presence of 1 molar equivalent of dicyclohexy-carbodiimide. After addition of excess 1 M potassium bicarbonate, the product is extracted into a suitable solvent. Yields of 60–90% have been quoted.[36] Neither of these methods is supposed to cause racemization with optically active bases. Formylation has been used in amino acid and peptide studies,[37,38] and sensitive groupings such as hydroxyls should first be protected by a suitable blocking group (e.g. *t*-butyloxycarbonyl). Formamido acids must be esterified for GLC.[37] Formyl derivatives have been little used for analytical purposes. The formyl group is the smallest protecting group that can be put onto an amino function. Other derivatives of amines and amino acid esters such as the TFA group are, however, more volatile than the corresponding formyl derivatives.

5.5 Benzoylation methods

5.5.1 SCHOTTEN–BAUMANN PROCEDURE FOR AMINES IN AQUEOUS SOLUTION

The amine solution (3 ml) is made alkaline with an equal volume of 7.5 M NaOH, and benzoyl chloride (50 μl) is added. The reaction mixture is shaken vigorously until the benzoyl chloride has all gone. The benzamides are extracted with diisopropyl ether (2 × 1 ml), and the ether solution may be dried and blown down with a stream of N_2.[39]

5.5.2 BENZOYLATION WITH BENZOYL CHLORIDE AND PYRIDINE

Benzoyl derivatives can be made from volatile alcohols and thiols as well as from amines, by treating the volatiles concentrate, isolated by benzene extraction, with pyridine (1 ml) and benzoyl chloride (0.5 ml), and shaking intermittently at room temperature over several hours. The pyridine is extracted with 2 M HCl, and excess chloride destroyed by shaking with water for 12 h. The benzene solution is washed with 2 M sodium carbonate, dried over anhydrous sodium sulfate and concentrated to approx. 0.5 ml for analysis by GLC on 5% SE-30 between 125° and 185 °C.[40]

Benzoyl derivatives have not been widely used for analytical purposes.

5.6 Phthaloyl and succinyl derivatives of amino acids and peptides

These derivatives, popular in synthetic work for their facile removal under mild conditions, have been little used except for mass spectrometry.[41,42] They were not readily separated by TLC and had unfavourable GLC properties.[7]

5.7 Carbobenzoxy, *t*-butyloxycarbonyl and related derivatives of amino acids

These derivatives, generally made by Schotten–Baumann type procedures, are familiar in peptide chemistry, and were very useful for synthetic work because they could easily be split off at the end of the synthesis, but have been little used in the analysis of amino acids. 'Carbobenzoxy' (benzyloxycarbonyl) amino acid methyl and ethyl esters were found to be rather involatile for GLC.[43,44] The *t*-butyloxycarbonyl derivatives of L-amino acid-(+)-4-methyl-2-pentylamides were used to resolve amino acids to study racemization in peptide bond formation.[45] *N*-Isobutyloxycarbonyl amino acid methyl esters have however been successfully used for analysis of amino acids by GLC.[46] Acylation was performed first, and the procedure looks simple and convenient enough to be worth quoting.

Amino acids (2 mg total) are dissolved in 0.5 M sodium carbonate (4 ml) and shaken with isobutyl chloroformate (0.1 ml) at room temperature for 10 min. Excess reagent is extracted with ether (2×3 ml), and after acidification to pH 1–2 the acylated amino acids are extracted with ether (3×3 ml). After esterification with diazomethane and evaporation to dryness, the derivatives are dissolved in ethyl acetate (0.1–0.2 ml), dried over anhydrous sodium sulfate and analysed by GLC on lightly loaded columns containing Poly-A-101A and FFAP. The application of this acylation procedure to the GLC analysis of amines and phenols is also indicated.[46]

Carbethoxy derivatives (urethanes) can also be made, either with ethyl chloroformate as above, or by reaction of amines with diethyl pyrocarbonate. The chromatographic behaviour of these derivatives was not impressive, but the column used (SE-30) is probably not ideal for these derivatives.[47]

5.8 Pivaloyl derivatives

Pivaloyl derivatives are easy to prepare and stable to air, moisture, heat and acid hydrolysis. They have good chromatographic properties. Amino and hydroxy groups are both derivatized simultaneously. Pivalic anhydride and pivaloyl chloride are stable for several months if kept cool and dry. Although not at first sight an obvious choice, this combination of properties has led to several applications of pivaloyl derivatives. The non-polar nature of the *tert*-butyl portion of the pivaloyl group, with its screen of methyl groups reminiscent of the trimethylsilyl group has also found application for blocking primary amines in a facile base-carbonyl group condensation with pivaldehyde (Chapter 6).

5.8.1 PIVALIC ANHYDRIDE AND BASE TO ACYLATE AMINO ACID ESTERS

$$R{-}NH_2 + \begin{matrix}(CH_3)_3C{-}CO \\ (CH_3)_3C{-}CO\end{matrix}\Big\rangle O \rightarrow R{-}NH{-}CO{-}C(CH_3)_3 + (CH_3)_3C{-}COOH \qquad (16)$$

The base, amino acid ester or its hydrochloride (approx. 1 mg) with methanol or tetrahydrofuran (10–20 μl) to dissolve it, is heated with pivalic anhydride (200 μl) and triethylamine (10–15 μl) at 70–110 °C for 30 min. Alternatively, a mixture of pivalic anhydride : triethylamine : methanol (20 : 1 : 1, v/v; 250 μl) may be added. The reaction mixture is centrifuged if necessary, to remove any precipitated triethylammonium salts, and the supernatant is collected and evaporated to dryness in a stream of nitrogen or *in vacuo*. The product may be extracted into a suitable solvent (hexane, benzene, chloroform etc.).[48–50]

5.8.2 PIVALOYL CHLORIDE FOR THE ACYLATION OF AMINES

$$R{-}NH_2 + (CH_3)_3C{-}CO{-}Cl \rightarrow R{-}NH{-}CO{-}C(CH_3)_3 + HCl \qquad (17)$$

The amine or its hydrochloride (up to 5 mg) is suspended in chloroform (200 μl) and 'light' magnesium carbonate (50 mg) is added. To the stirred suspension is added a 110% molar equivalent of pivaloyl chloride via a microlitre syringe. After stirring for a further 30 min at room temperature, the suspension is centrifuged and the supernatant is collected and evaporated to dryness in a stream of N_2 or *in vacuo*. The product, which is usually quite pure, may be dissolved in a suitable solvent for analysis by TLC or GLC.[51,52]

5.9 Preparation of 3,5-dinitrobenzoates

The dinitrobenzoyl group is useful for making coloured derivatives of alcohols and amines for TLC and also to make derivatives for GLC which are considerably less volatile than the parent compounds, so minimizing evaporation losses. The acid chloride is used, with an alkali (Schotten–Baumann conditions) or a tertiary amine as acceptor for the HCl produced in the reaction.

5.9.1 DINITROBENZOYL CHLORIDE AND PYRIDINE ON ALCOHOLS AND AMINES

The alcohol or amine (up to 10 μl) is dissolved in benzene (0.1 ml) and heated on a boiling water-bath with a solution of 3,5-dinitrobenzoyl chloride : pyridine : benzene (1 : 2 : 8; 0.4 ml) for 30 min. After cooling, the solution is washed with 2 M KOH (2×5 ml), water (2×1 ml), 5 M HCl (2×1 ml) and water (2×1 ml). The benzene solution is dried over anhydrous sodium sulfate and applied to the TLC plate for development.[53]

5.9.2 DINITROBENZOYL CHLORIDE IN BENZENE

The alcohol or amine hydrochloride (up to 10 mg) in 1:1 aqueous ethanol (2 ml) is shaken up with a 1:9 mixture of dinitrobenzoyl chloride:benzene (2 ml). The benzene solution may be directly injected for GLC with a flame-ionization detector. For use with an ECD the excess of reagent must first be removed by a washing procedure like that in the previous method (Section 5.9.1).[54]

5.10 Preparation of sulfonyl phenolic acids

A more polar volatile derivative of phenolic acids was developed in order to allow their GLC peaks to be moved away from contaminant peaks by manipulation of the polarity of the stationary phase.[55] The acids (5 mg) are dissolved in ethyl acetate (2 ml) and derivatized with chlorosulfonic acid (0.3 ml) at 60 °C for 5 min. After cooling, water (2 ml) is slowly added to destroy excess of the reagent, followed by ethyl acetate (3 ml). The aqueous layer is re-extracted with ethyl acetate (2×3 ml) and the combined ethyl acetate solution is extracted with 1 M sodium carbonate (2×2 ml). The basic solution is acidified with 5 M HCl, and extracted with ether (3×4 ml) which is then dried over anhydrous sodium sulfate. The ether solution is evaporated to dryness and the residue converted to trimethylsilyl esters with bis(trimethylsilyl)acetamide for GLC analysis.

5.11 Perfluoroacyl derivatives: anhydride procedures

The trifluoroacetyl (TFA) group was first used in carbohydrate research[56] and introduced for peptides and amino acids by Weygand and his group.[57] When the ECD was developed, and halogenated derivatives were sought, TFA derivatives were examined, and the analogues pentafluoropropionyl (PFP) and heptafluorobutyryl (HFB) were also developed. In general it was found that the more fluoro-groups, the greater the sensitivity of detection, although there are specific differences with different types of halogen-containing groupings, as will be discussed later. It was also found that *O*-perfluoroacyl derivatives are generally far more electron-capturing than the corresponding amine derivatives.[58]

In general, in the derivatization schemes that follow, the reagents TFA-anhydride, PFP-anhydride and HFB-anhydride may be used interchangeably, although they differ somewhat in reactivity and volatility. Thus TFA-anhydride is the most reactive and the most volatile. On the other hand, for most compounds the HFB derivative gives the greatest sensitivity with electron-capture detection, while the derivatives that give the lowest temperature of analysis may be the PFP derivatives.[7] The choice of which to use therefore depends on the analytical priorities, and for these reasons no preference is given in many of the following schemes. It will be seen that many of these procedures show similarities and only vary in detail, because of the different acylation requirements of amines, steroids, carbohydrates, phenols, amino acid esters and other compounds.

It must be added that these electron-capturing derivatives, with their good chromatographic properties and ease of formation, are also entirely suitable for analysis with other detectors such as the FID or the thermal conductivity detector, although, of course, not at such low levels, and therefore these derivatives should not be ruled out when choosing the best ones to use, merely because an ECD is not available.

5.11.1 TRIFLUOROACETYL DERIVATIVES OF PHENOLS

The phenol is treated with 10% excess molar equivalents of neat TFA (or PFP or HFB) anhydride containing a trace of 10 M NaOH, and refluxed until solution is complete and obvious reaction ceases. Excess of the reagent is evaporated in a stream of N_2 or *in vacuo*, and the residue is dissolved in ethyl acetate for analysis.[59]

Neat TFA (or PFP or HFB) anhydride can also be used with many other compounds, including amines and amino acid esters, (see method described in Section 5.11.10) and prostaglandins,[60,61] and removed quite simply at the end of the reaction, but it is more common to use a solvent and also a catalyst. Solvents such as acetonitrile are popular because they act to some extent in both capacities. One reason for avoiding neat anhydride is that it may act as a dehydrating agent, because the acid activity in the reagent may be high, causing protonation of hydroxy groups followed by loss of water.

5.11.2 ACYLATION OF AMINES WITH ANHYDRIDE IN ETHYL ACETATE

The phenolic amine mixture or an amine extract from a biological system, is evaporated to dryness in a 1 ml reaction vessel, and derivatized with PFP anhydride (100 μl) in ethyl acetate (20 μl). The vessel is tightly capped and heated at 65 °C for 1 h. After cooling the mixture is evaporated with N_2 and the residue taken up in ethyl acetate (100 or 200 μl) containing 0.25 $\mu g\ ml^{-1}$ of lindane as reference standard. Analysis is by ECD–GLC on 10% SE-54 at 200 °C.[62] See also the method described in Section 5.11.20.

5.11.3 ANHYDRIDE AND ACETONITRILE ON AMINES AND PHENOLIC ALCOHOLS

The amines or amine extract are obtained dry as in the previous method (Section 5.11.2), but derivatization with acetonitrile (0.5 ml) and TFA anhydride (0.05 ml) gives acylation at room temperature in 5 min, and the reaction mixture, because of the low concentration of anhydride, can be injected directly. This is apparently useful with 5-hydroxyindole-3-acetic acid, whose derivative decomposes on evaporation to dryness; it cannot be recommended for use with an ECD.[63] For indolic alcohols, PFP anhydride (100 μl) and acetonitrile (100 μl) are used, at 60 °C for 20 min, but excess of the reagents are removed before analysis using an ECD and 10% SE-54 at 200 °C.[64] The acylation of phenolic amines with solvents ethyl acetate and acetonitrile has been investigated, and the optimal proportions of ethyl acetate, acetonitrile and anhydride were found to vary with different amines. These reactions were done at 60 °C for 2 h. Best results were obtained for all the amines studied using 100 μl acetonitrile, without any ethyl acetate.[65]

5.11.4 PREPARATION OF AMINE DERIVATIVES WITH THE PERFLUOROACYL ANHYDRIDE IN ETHER

The amines are dissolved in ether and the perfluoroacyl anhydride is added drop by drop with ice-cooling. After dilution with more ether, the ethereal solution is washed with sodium bicarbonate solution and water, dried, and then evaporated to dryness.[7,66]

5.11.5 COMBINED ISOLATION AND ACYLATION OF ALIPHATIC AMINES

Amines (in ether) are isolated on a column of dried Amberlyst 15 resin, and washed on-column to remove non-basic contaminants. Derivatization is also on-column, with 1:1 anhydride-ether, followed by ether elution of the products. The ethereal solution is washed as in the previous method (Section 5.11.4), and dried over anhydrous sodium sulfate, before direct injection for the analysis of the derivatives of primary and secondary amines. Tertiary amines are then eluted from the column with 3 M methanolic HCl; the eluate is basified and the amines extracted into ether. They are analysed by GLC as the free bases.[67] Derivatization of tertiary amines is, however, possible (see Section 5.25).

A number of other solvents have been successfully used, such as pentane,[68] hexane,[69,70] cyclohexane[71] and dichloromethane,[72] indicating that the precise nature of the solvent is not very important.

5.11.6 ACYLATION WITH ANHYDRIDES AND VARIOUS BASES

Although amines do not have to be in the free base form for acylation by perfluoroacyl anhydrides, and react smoothly with excess anhydride even as their salts, some workers have included bases such as pyridine or triethylamine partly as catalysts and partly also to act as acceptors for the acids formed in the acylations.[73,74]

The amine is dissolved in benzene (0.5 ml) in a 5 ml centrifuge tube with a gound glass stopper, and a 0.05 M solution of triethylamine in benzene (0.1 ml) is added, followed by the anhydride (10 μl). The tube is tightly stoppered and heated at 50 °C for 15 min. The solution is shaken with water (1 ml) and then with 3 M ammonium hydroxide (1 ml), and then analysed by GLC with ECD.[75,76] In an alternative procedure, benzene (4 ml), pyridine (0.1 ml) and the anhydride (0.1 ml) are used, and the reaction is left at room temperature for at least 4 h, and preferably overnight.[77]

5.11.7 CHLORODIFLUOROACETYLATION OF 1-(2-PHENYLADAMANTYL)-2-METHYLAMINOPROPANE

An essentially similar procedure is described for the rather uncommon chlorodifluoroacetic anhydride. The reaction conditions are that the base, extracted from blood or urine, is dissolved in chloroform (20 μl) and derivatized with the anhydride (3 μl) and triethylamine (1.5 μl) at 50 °C for 30 min. Excess of the anhydride is neutralized with alkali, and the derivative is extracted into chloroform and concentrated to dryness. It is dissolved in toluene for GLC on 5% XE-60 at 235 °C.[78]

5.11.8 BIS-PERFLUOROACYLATION OF PRIMARY AMINES

The amine is dissolved at 10 mM in benzene and the solution (0.2 ml) is mixed with 0.3 M trimethylamine in benzene (0.2 ml). TFA or HFB anhydride (25 μl) is added and the reaction is left at room temperature for 30 min. The mixture is washed with pH6 phosphate buffer (2 ml) for just 15 s and centrifuged. The benzene solution is diluted to a suitable concentration with more benzene and injected for GLC on 3% OV-1 or OV-17. The derivatives are reasonably stable, give good peaks, and are more volatile than the monoacyl derivatives. They can be determined with great sensitivity by ECD.[79]

Since this is essentially the perfluoroacylation of first the amine and subsequently of the resulting amide, it follows that this method is also applicable to amides (see Section 5.11.22)[80] and to sulfonamides.[81] The procedure has been optimized to yield the products quantitatively. The concentration of trimethylamine is particularly critical. It is difficult to reconcile the known ease of hydrolysis of these derivatives with the buffer wash, but the authors claim that the brief contact is safe.

5.11.9 TRIFLUOROACETYLATION IN THE PRESENCE OF A MERCURY CATALYST

Amphetamine (isolated from urine, for example, by solvent extraction) after evaporation to dryness, is transferred to a small reaction vial in 3:1 benzene:ether (100 μl) and treated with 10 μl of a well-shaken reagent made by suspending mercuric acetate (50 mg) in TFA anhydride (20 ml) and leaving it at room temperature for 1 h. Acylation is completed by standing the sealed vial on an 80 °C hotplate for 30 s. The solution is injected directly for GLC.[82] The use of mercury was prompted by the probably erroneous belief that acylation calls for the amine to be in the free base form. The possibility of volatile mercury derivatives entering the ECD or a GC-MS system cannot be contemplated with equanimity.

5.11.10 ACYLATION OF AMINO ACID ESTERS WITH TFA-ANHYDRIDE

As can be seen from Table 4 in Chapter 1, there have been many different approaches to the acylation of the amino groups of amino acids to yield volatile derivatives for GLC.[83] Indeed, the impetus towards the use of perfluoroacyl groups generally, came from the introduction of the TFA group.[57] This group has remained the most widely used for amino acid work. Acetyl propionyl derivatives have also been used, and the PFP and HFB derivatives have been explored more recently. With very few exceptions,[84,85] the amino acids have been esterified first: if the acylation were done first, the acyl group, and particularly the TFA group, might come off again under acidic esterification conditions, although acylated amino acids can be smoothly esterified with diazomethane and other diazoalkanes. (See Chapter 2 for more details on the esterification of amino acids.)[85–87] The two most widely used approaches are the direct acylation of the amino acid with neat anhydride, and the acylation with the anhydride dissolved in methylene dichloride. Other methods are briefly described. It is surprising that solvents such as acetonitrile have not been more widely applied to the acylation of amino acid esters, as it appears that they promote rapid and clean reactions;[51] this is another example of how different areas of research may be isolated from one another, and of how this handbook may help to unify fragmented fields of study.

The amino acids (2 mg total) in a ground-glass stoppered tube are esterified with the appropriate alcohol containing about 3 M HCl. Excess of the alcoholic HCl is removed *in vacuo* (for further details of the esterification procedures for amino acids, see Chapter 2). Neat TFA-anhydride (or other perfluoroacyl anhydride; 0.2 ml) is added to the dried residue, and the reaction is left to proceed at room temperature for 30 min, or in the case of arginine, at 140 °C for 10 min. The derivatives are evaporated to dryness *in vacuo* for just 4 min. This avoids evaporative losses of the more volatile components of the mixture (derivatives of alanine, valine and glycine) which may occur if excess of the reagent is blown down with N_2.[86] The dried residue is dissolved in dry methylene dichloride for GLC analysis.[88] A neat variation of this method is to acylate the amino acid esters on the microgram scale in a sealed vessel supported on a glass wool or platinum wire plug.[89,90]

5.11.11 TRIFLUOROACETYLATION OF AMINO ACID ESTERS IN METHYLENE DICHLORIDE

The esters (usually the *n*-butyl esters) made from 1–20 mg of total amino acids, are treated with a 1 : 3 mixture of TFA anhydride in methylene dichloride (1 ml for each 10 mg of total amino acids) and placed into a reaction vial or tube sealed with a Teflon®-lined reaction vial. Acylation is at 150 °C for 5 min. This procedure can be scaled down to the microgram level, using only 60 μl of the acylating mixture for each 100 μg of amino acids (minimum final volume 80 μl). Excess of the reagents is blown down to a small volume (but not to dryness) with a stream of N_2: the reason for choosing the *n*-butyl esters is to minimize evaporation losses at this stage. The residual sample solution is then directly analysed by GLC.[91]

These procedures are widely used for the GLC analysis of amino acids, and have been carefully optimized, so that it is important to follow the procedure closely. For fuller details it is best to consult the original papers. The procedures are also directly transferable to the PFP and HFB derivatives.[92–98] The electron-capturing properties of these higher perfluoroacyl amino acid esters have not been so well established, whereas this has been the main thrust in using these derivatives for amines, another example of isolation in research topics. As surmised by Pollock,[92] the *N*-TFA derivatives do not respond well to electron-capture detection and neither do the PFP derivatives, but as found by Clark *et al.*,[58] the PFP derivatives of the hydroxylated amino acids can be analysed by GLC with high sensitivity using electron-capture detection.[51] It seems that the electron-capturing ability of *N*-perfluoroacylated derivatives only begins to become really appreciable with the *N*-HFB derivatives.

This procedure has also been used to prepare derivatives of nucleosides and of DNA hydrolysates for mass spectrometry,[99] and with the addition of catalytic amounts (3%) of dimethylformamide, for chlorpromazine metabolites.[100]

5.11.12 TRIFLUOROACETYLATION OF ALDITOLS WITH TFA-ANHYDRIDE AND SODIUM TRIFLUOROACETATE

As already mentioned, the advantage of reducing sugars to the alditols is that on GLC of the acylated alditols each sugar gives a single peak. A number of workers have made alditol trifluoroacetyl derivatives, which are more volatile than the acetates.

The mixture of sugars after either methylation to the glycosides or reduction to the alditols (1–3 mg) is treated with 0.8 M sodium trifluoroacetate in acetonitrile (0.1 ml) and TFA-anhydride (0.1 ml) and heated in a sealed tube at 60 °C for 10–15 min with occasional shaking. Excess of the anhydride is pumped off *in vacuo*, and the residue is dissolved in methylene dichloride. The supernatant is examined by GLC.[101] These derivatives are sensitive to moisture. An apparent improvement is to substitute 0.3 mg of sodium acetate in 0.05 ml of formamide, which leads to much milder reaction conditions: room temperature for 10 min.[102] These derivatives have also been made by the method described in Section 5.11.6,[103] and their mass spectra have been determined.[104]

5.11.13 TRIFLUOROACETYLATION OF CARBOHYDRATES WITH TFA-ANHYDRIDE AND PYRIDINE

In order to obtain derivatives of carbohydrates that accurately reflect their isomer composition without alterations during the derivatization, König *et al.* used mild acylation conditions on the microgram scale.

The sugars (100 μg) are acylated in methylene dichloride (65 μl) with TFA-anhydride (30 μl) and pyridine (5 μl) at 20 °C for 3 h. The reaction mixture is injected directly for GC–MS. As mentioned earlier for acetylation with pyridine, occasional interference is observed from a reaction product of pyridine with TFA-anhydride.[105] Tables of GLC properties of these and other sugar derivatives on several different stationary phases have been published,[106] where the TFA derivatives were in fact made by the method described in 5.11.11.

5.11.14 ACYLATION OF SUGARS WITH TFA-ANHYDRIDE IN TETRAHYDROFURAN

The sugars (1–5 mg) are treated with tetrahydrofuran (0.2 ml) and TFA-anhydride (0.05 ml) at 40–50 °C for 10 min with occasional shaking. The solution is evaporated to dryness in a stream of N_2, and dissolved in acetonitrile (1 ml) for GLC. The tetrahydrofuran was found to be better for derivatization because each sugar gave a single peak, while with acetonitrile multiple peaks were obtained; however, the derivatives, once made, were more stable dissolved in acetonitrile.[107] They were found to be of comparable volatility to the TMS derivatives, had good chromatographic properties, and were electron-capturing.

5.11.15 METHYL-TRIFLUOROACETYL DERIVATIVES OF ETHER GLUCURONIDES

These derivatives (see also Section 5.2.4) are prepared by methylation with methanolic diazomethane, evaporated to dryness under N_2 at 50 °C, and acylated by procedure 5.11.2.[108]

5.11.16 STEROID HEPTAFLUOROBUTYRATES: ANHYDRIDE AND ACETONE METHOD

The dry steroid (1 μg or less) is treated with 1:4 HFB-anhydride in acetone (0.25 ml) at 25 °C for 1 h. The solution is blown down with a stream of nitrogen, and the residue dissolved in dry benzene for GLC on 1% OV-1 or OV-17 at 200 °C. These conditions lead to acylation of the 17-hydroxyl and the enol of 3-ketosteroids.[109] These derivatives were stable in benzene at 10 °C, and are strongly electron-capturing. They are also considerably more volatile than the parent steroid.

Care must be taken to avoid the use of too much acetone: the quoted concentration is optimal.[110]

5.11.17 ANALYSIS OF PLASMA TESTEROSTERONE: HFB-ANHYDRIDE IN TETRAHYDROFURAN

The extract of steroids from blood is purified by TLC, and the testosterone area scraped up and extracted with ethanol, which is then evaporated. The residue is treated with HFB-anhydride (20 μl), *n*-hexane (1 ml) and tetrahydrofuran (40 μl) at 50 °C for 30 min. The reaction mixture is pumped dry *in vacuo*, and purified by TLC on silica gel with 19:1 benzene : ethyl acetate. The derivative is extracted into acetone for GLC.[111]

5.11.18 MONOACYLATION OF TESTEROSTERONE AT THE 17-POSITION

The steroid (100 mg) is dissolved in benzene (2 ml) and tetrahydrofuran (0.5 ml) and heated with anhydride (HFB-dichlorofluoroacetic, pentafluorobenzoic or chloroacetic) at 60 °C overnight. This procedure leaves the 3-keto group intact, and of course it may be derivatized as in Section 5.11.16,[112] or by one of the derivatization methods used for carbonyl groups, such as methoxime formation.

The steroid HFB derivatives have been widely used because they are more volatile and have better chromatographic properties than the parent steroids, and because they are often strongly electron-capturing and therefore lead to a sensitive analysis, which is valuable where their concentration is low. Not all of these derivatives are equally stable, and some tend to decompose on TLC or GLC. Dehydration has been observed when cholesterol is acylated with HFB-anhydride. These possibilities should be borne in mind when using these derivatives.[113]

5.11.19 HFB DERIVATIVES OF PHENOLS

This method is similar to that in Section 5.11.8, but the proportions are different. The phenol (up to 1 mg) in benzene (0.5 ml) is mixed with 0.1 M trimethylamine in benzene (0.1 ml) and acylated with HFB-anhydride (10 μl) at room temperature for 10 min. After shaking with pH 6 phosphate buffer for 30 s, the phases are separated by centrifugation and about 2 μl of the benzene layer is taken for GLC analysis on 3% OV-1 or OV-17 at 190 °C. The products are relatively stable, and show good sensitivity with the ECD. PFP and HFB derivatives show comparable sensitivities, which are greater than those found with the corresponding amine or alcohol derivatives.[114]

5.11.20 PERFLUOROACYLATION OF PHENOLIC GLYCOLS AND ALCOHOLS IN ETHYL ACETATE

These compounds, mainly 3-methoxy-4-hydroxy-phenethylene glycol (MHPG) and 3-methoxy-4-hydroxyphenylethanol (MHPE) are catecholamine metabolites, and have been derivatized mainly as TFA and PFP derivatives which are electron-capturing. All the methods are variants of the one in Section 5.11.2 with varying proportions of ethyl acetate and anhydride, and varying times and temperatures: all of these are summarized in Table 2.

The use of pyridine is not recommended, because it causes interference from extra peaks[115] (see also Section 5.11.13) although when acetyl derivatives of these compounds were made by the method in Section 5.2.4, pyridine was found to be essential for good yields and a clean product.[122]

At the end of the reaction the excess of reagent and solvent is removed by a stream of N_2, and the residue is dissolved in ethyl acetate. These derivatives are subject to hydrolysis by atmospheric moisture, and up to 2% of the appropriate anhydride is sometimes added at this stage to keep the ethyl acetate solution

TABLE 2
Derivatization of MHPG and MHPE

Volume of anhydride (TFA or PFP)	Volume of ethyl acetate	Temp.	Time	Ref.
0.25 ml	2 ml	room	1 h	115
50 μl	400 μl	room	1 h	116
0.25 ml	2 ml	70 °C	15 min	117
100 μl	500 μl	room	30 min	118
0.25 ml	1 ml	room	30 min	119
0.05 ml	0.1 ml	60 °C	10 min	120
100 μl	20 μl	65 °C	40 min	121

protected against hydrolysis. An 'injection standard' such as α-lindane may be included in the ethyl acetate used to dissolve the products of the reaction for GLC analysis. This acts as a check on derivatization, injection volume and the response of the ECD.

5.11.21 TRIFLUOROACETYLATION OF COMPOUNDS IN THE VITAMIN B_6 GROUP WITH TETRAHYDROFURAN

Pyridoxine, pyridoxamine and pyridoxic acid lactone were acylated directly. Pyridoxal is first converted to the methoxime or pyridoxylidine-*n*-propylamine derivative. The vitamers (0.01–0.5 mg) are dissolved in tetrahydrofuran (0.2 ml) and TFA-anhydride (0.1 ml) is added. The reaction is left at room temperature for 10 min with occasional shaking, and injected directly for GLC on 2% XF-1105 or 5% DC-550 at 160 °C. For electron-capture detection the samples would need to be blown dry with N_2 or evaporated *in vacuo*, as TFA anhydride should not be allowed to enter the detector. The derivatives capture well at sub-nanogram levels. PFP or HFB derivatives might be even better.[123]

5.11.22 TRIFLUOROACETYLATION OF UREA IN BENZENE

Urea is first extracted and purified with $Zn(OH)_2$ precipitation and ether fractionation. Aliquots are evaporated to dryness in a screw-capped reaction tube and treated with benzene (0.2 ml) and TFA-anhydride (0.05 ml) at room temperature for 10 min. A useful way of removing excess of the acidic reagent and by-products is to use a small column of celite filter-aid mixed with a little 3.5 M sodium acetate, through which the benzene solution is passed. This can only be done where the reaction product is not sensitive to moisture. After elution with 1 : 1 ether : benzene, the eluate is made up to a known volume and analysed by electron-capture GLC.[124]

5.11.23 TRIFLUOROACETYLATION OF AMIDES IN CARBON DISULFIDE

Amides, being 'already derivatized', are not easy to derivatize further without using forcing conditions, but we have already seen that amines can be bis-perfluoroacetylated (Section 5.11.8). The present method is illustrated with the drug phenacetin. The amide (0.04–0.4 mg) is dissolved in carbon disulfide (0.5 ml) and TFA anhydride (20 μl) is added. The reaction is left at room temperature for 10 min, and aliquots are injected directly for GLC with an FID.[80] These derivatives are not very stable, and neither are the TMS derivatives of amides. The best derivative for amides appears to be the pentafluorobenzoyl (PFB) derivative, and preparation of these is described in Section 5.13.[80]

5.11.24 PERFLUOROACYLATION OF PHENOLIC ACIDS FOR ELECTRON-CAPTURE GLC

Phenolic acids such as homovanillic acid and 5-hydroxyindole acetic acid, need to be derivatized on the phenolic (and other) hydroxy groups as well as esterified, and this provides the opportunity of introducing electron-capturing perfluoroacyl groups for detecting these acids at the very low concentrations at which they may be found in biological fluids. Not surprisingly, the hydroxyls are generally derivatized by the same procedures used to derivatize the same groupings on the biochemical precursor molecules, the phenolic amines (see, for example, Sections 5.11.1, 5.11.2, 5.11.3 and 5.11.20[125,126] and 5.11.10).[127] The esterification may be done first [125,126] or second.[127] However, there are distinct advantages to derivatizing both hydroxyl and carboxyl functions at the same time, and some of these are presented in the next section.

5.11.25 SIMULTANEOUS PERFLUOROACYLATION AND ESTERIFICATION

The basis for this approach seems to have been the use by Smith and Tsai[128] of TFA-anhydride as an esterification catalyst. Wilk and Orlowski used this idea to prepare the *N*-PFP derivative and pentafluoropropyl ester of pyrrolidone carboxylic acid.[129] The method has also been used for double derivatization of phenolic acids[130] and γ-glutamyl amino acids.[131] It has also proved potentially useful for amino acids.[132]

The compounds to be derivatized (up to 1 mg) in a 3 ml ground-glass stoppered centrifuge tube, are treated with a 1:4 mixture of 2,2,3,3,3-pentafluoro-1-propanol in PFP-anhydride (0.25 ml) at 75 °C for 15 min. After evaporation to dryness with N_2, the acylation is completed with neat PFP-anhydride (0.1 ml) at 75 °C for 5 min. The solution is again evaporated to dryness and the residue is dissolved in ethyl acetate for injection into the GLC column. A similar procedure was developed by Brooks *et al.*, using HFB-anhydride with pyridine as catalyst and chloroform as solvent[133] (see also Chapter 2).

5.11.26 RING ACYLATION OF TERTIARY AROMATIC AMINES

Tertiary amines are frequently accessible to GLC analysis without derivatization, and columns lightly 'doped' with alkali or coated with very polar stationary phases are used, but with an FID the sensitivity of detection may be insufficient. Walle[134] developed a procedure of introducing a TFA group into the ring of aromatic tertiary amines, and the resulting trifluoroacetophenone derivatives could be determined at very low levels with the ECD. Best results are obtained in the presence of dimethylformamide or trimethylamine. The response of the ECD to the products decreased appreciably with increasing detector temperature.

The aromatic tertiary amine (1 mg) is dissolved in benzene (200 μl) and a 1 M solution of trimethylamine in benzene (100 μl) is added. The mixture is heated at 50 °C for 1 h, cooled and washed with 0.5 M NaOH (500 μl) with shaking for 30 s. After centrifugation the supernatant is analysed directly by electron-capture GLC on 3% OV-17 at 120 °C.

Acylation of a carbon chain was also reported with TFA-anhydride and ethyl acetate on the tricyclic antidepressive drug, desipramine. Method 5.11.18 was used (50 μl ethyl acetate, 100 μl TFA-anhydride, 35 °C, 60 min).[135] Introduction of the second TFA group was on the carbon atom β to the methylamino group of the side chain, and was accompanied by dehydrogenation (18).

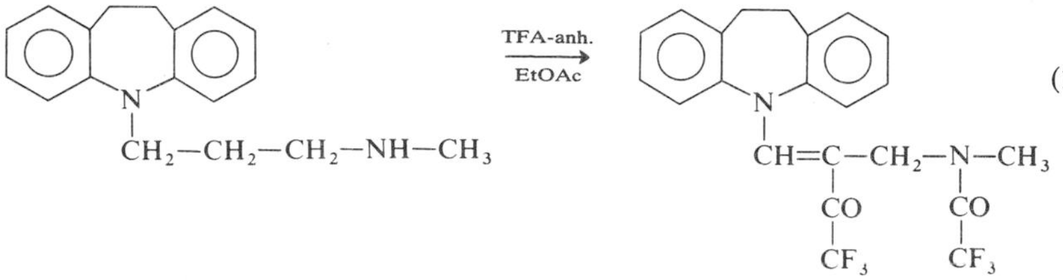

(18)

5.11.27 CONCLUSIONS

The great variety of conditions that have been used in these acylation reactions with perfluoroacyl anhydrides, and the fact that many of these compounds have been acylated to the same derivatives via different methods, strongly suggests that the precise conditions used in acylation are probably interchangeable, and initially a particular analytical problem involving acylation can be solved with any likely looking procedure, although for quantitative analyses, further work may be needed to develop conditions that will give the highest yield of derivative. It seems safe to conclude that any suitable solvent may be used for the reaction, provided it does not react with the anhydride and so reduce its effective concentration, and that catalysis by a base or basic solvent promotes the reaction and permits reduction of reaction times and temperatures.

5.12 Alternative perfluoroacylation procedures

In addition to the many anhydride procedures already described, other methods of introducing perfluoroacyl groups have been devised for various reasons, and are presented here. The most significant are the methods using perfluoracyl-imidazoles, for which the rationale and reaction mechanisms have already been described. The great advantage of these reagents is that they are used in a non-acidic environment, and are therefore suitable for compounds such as indoles which are sensitive to acid conditions. Acid conditons tend to cause dehydration and this can occur with anhydrides,[136] another advantage in using imidazole reagents. These reagents derivatize alcohols, primary and secondary amines. The reagents themselves are sensitive to moisture and are easily hydrolysed to the imidazole and acid.

5.12.1 HFB-IMIDAZOLE IN THE ACYLATION OF INDOLEAMINES AND INDOLE ALCOHOLS

The indole derivative (1–2 mg) is dissolved in HFB-imidazole (0.1–0.2 ml) and heated in a Teflon®-lined screw-capped vial at 80 °C for 2–3 h. After cooling the product is extracted with hexane (3 × 5 ml). The hexane extracts are chilled to − 14 °C to precipitate traces of reagent, and aspirated from the precipitate, which is washed with hexane (2 × 5 ml). The combined hexane solutions are concentrated to 0.5 ml and 1–3 μl aliquots of the concentrate analysed on 5% SE-30 at 200 °C.[137–138]

An alternative treatment is to decompose excess of the reagent with water (1 ml) and toluene (2 ml). After thorough vortex mixing a further 5 × 2 ml toluene extractions are done, the toluene layers are washed with water (3 × 2 ml) and the toluene layer filtered through dry filter paper and made up to 100 ml with toluene. If necessary, this is further diluted before electron-capture GLC analysis.[139] A small-scale version of this method suitable for cerebrospinal fluid is also described.[140] In another similar procedure, the HFB derivative of protryptyline was prepared and the toluene solution washed with 1 M NaOH (0.5 ml) and 0.1 M sulfuric acid (0.5 ml) indicating the stability of the *N*-perfluoroacyl bond.[141] A novel way of removing the aqueous layer is to freeze the two layers in a dry-ice/acetone bath, and to withdraw the upper layer with a Pasteur pipette from the ice.[142]

5.12.2 ACYLATION OF PHENOLIC ACIDS WITH PFP-IMIDAZOLE IN ETHYL ACETATE

The acids, isolated from the biological matrix by extraction at low pH with ethyl acetate, are first esterified with methanolic HCl, blown dry with N_2, and heated in a sealed reaction vial with 1 : 9 PFP-imidazole : ethyl acetate (50 μl) at 70 °C for 10 min. This solution was used for SIM.[143]

5.12.3 HFB- OR TFA-IMIDAZOLES IN A COMBINED SILYLATION AND ACYLATION PROCEDURE

This is a 'single pot' reaction in which the sample, e.g. a phenolic amine, is treated with TMS-imidazole to silylate the hydroxyls and with HFB- or TFA-imidazole to acylate the amino groups, all in a non-acidic medium. The sample (0.1 mg) is put into a reaction vial and dissolved in acetonitrile (0.1 ml). TMS-imidazole (0.2 ml) is added, and the vial is sealed with a Teflon®-lined screw-cap, and heated at 60 °C for 3 h. After cooling, HFB- or TFA-imidazole

(0.1 ml) is added, the vial is resealed, and heated at 60 °C for a further 30 min. The reaction mixture may be injected directly for FID or extracted into hexane (3 × 0.4 ml) and concentrated for ECD.[144]

5.12.4 REPLACEMENT OF TMS- BY AN HFB-GROUP

In this interesting variant of the previous procedure, the action of HFB-imidazole in the presence of catalytic amounts of HFB-acid, displaces some of the TMS-groups. These experiments, on zooecdysones, show more rapid reaction of the equatorial -OTMS groups than the axial ones.[145] The steroid (0.5 mg) in a Teflon®-lined capped reaction vial is dissolved in TMS-imidazole (20 μl) and heated at 100 °C for 1 h. After cooling, HFB-imidazole (20 μl) and HFB-acid (2 μl) are added, and the mixture heated at 50 °C for 2 h. Benzene (0.2 ml) is added, and the solution washed with ice-cold 0.6 M sodium bicarbonate (0.2 ml) and water (0.2 ml). The solution is dried over anhydrous sodium sulfate. This reaction may be scaled up to permit a further stage of purification using TLC.

5.12.5 CONCLUSIONS

The use of perfluoroacylated imidazoles is almost as convenient as that of the anhydrides, and has some advantages in specific applications. It is probably best with primary and secondary amines, whose perfluoroacyl derivatives are stable enough to permit washing out excess of the reagents with aqueous solutions. The derivatives of alcohols are likely to be more rapidly hydrolysed by moisture, but for FID work portions of the reaction mixture may be injected directly without washing, as the reagent and the by-products (acid and imidazole itself) are volatile enough to emerge at the start of the run together with the solvent peak or close to it. However, this does not apply with ECD, and excess of the reagent must first be removed, because it would probably depress the standing current and the detector response for long enough to interfere with reliable quantitation of the earlier peaks of interest.

5.12.6 THE USE OF *N*-METHYL-BIS(TRIFLUOROACETAMIDE)

An interesting acylation reagent is *N*-methyl-bis(trifluoroacetamide) or MBTFA:

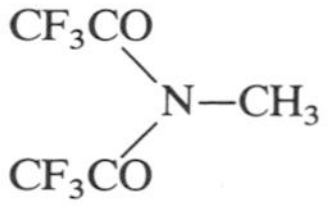

This reagent was developed by Donike,[146] together with the unmethylated analogue, for acylations under mild conditions. Acylation of primary and secondary amines is rapid and quantitative, of hydroxyls rather less so.

The compound to be analysed (2 mg) is dissolved in MBTFA (0.5 ml) and left at room temperature for 30 min. For compounds that do not react readily, warming or heating up to 120 °C is recommended. With compounds that fail to dissolve, the reagent may be mixed 1:4 with acetonitrile, pyridine, dimethylsulfoxide, tetrahydrofuran, etc. The reaction mixture is injected directly, and the excess of reagent together with the by-products of the reaction (19) elute early in the chromatogram.

$$R{-}NH_2 + (CF_3{-}CO)_2N{-}CH_3 \rightarrow R{-}NH{-}CO{-}CF_3 + CF_3CO{-}NH{-}CH_3 \quad (19)$$

5.12.7 COMBINED SILYLATION AND ACYLATION WITH MBTFA

If compounds contain both hydroxyl and amino groups, it is possible to silylate the hydroxyls first, and acylate the amino groups afterwards without isolation of the intermediates, as was described in Section 5.12.3. The silylation may be done with *N*-methyl-*N*-TMS-trifluoroacetamide (MSTFA) or bis-TMS-trifluoroacetamide (BSTFA).[147,148]

The hydroxylated amino compound, e.g. phenolic amine or amino alcohol (20 μg), is dissolved in silylation reagent, e.g. BSTFA (50 μl), and left either at room temperature overnight or at 80 °C for 3 h. Following addition of MBTFA (5 μl or more) the reaction is heated to 80 °C for 5 min to effect trifluoroacetylation. Any *N*-TMS groups are replaced by TFA groups[146,148] (compare with the method in Section 5.12.4).

5.12.8 PERFLUOROOCTANOYL, NONANOYL AND UNDECANOYL DERIVATIVES MADE WITH THE ACID CHLORIDES AND PYRIDINE

The reason for using these perfluoroacyl derivatives is the hope that the increased number of fluorine atoms will give greater detection-sensitivity to the derivatives with ECD, and to increase the retention times of rather too volatile substances or to move GLC peaks into positions on the chart that are free of interfering peaks.

The substance to be acylated, e.g. a steroid, (up to 1 mg) is treated with a 1:19 solution of the perfluoroacyl chloride in pyridine (0.1 ml) at 55 °C for 30 min. After evaporation to dryness, unreacted steroid and reagents are removed by a single partition between hexane:methanol:water 50:7:3 (1 ml). The hexane layer is either concentrated and the product isolated by TLC, or injected directly for GLC.[149]

5.13 Pentafluorobenzoyl derivatives

The pentafluorobenzoyl (PFB) group appears to be the one that confers the greatest sensitivity for ECD of amines[150] and phenols.[151] These derivatives are easy to make, are stable, and have good chromatographic properties. Pentafluorobenzoyl chloride (PFBCl) must be stored dry at a temperature below 0 °C to minimize hydrolysis and intramolecular halogen-exchange reactions. If the strongly electron-capturing pentafluorobenzoic acid, formed by hydrolysis of PFBCl, is found to interfere, it may safely be removed by extracting the ethyl acetate solution of an amine-PFB derivative with about one fourth its volume of 1 M NaOH, followed by a quick centrifugation. The supernatant is then used for GLC.[51]

5.13.1 PFB DERIVATIVES OF AMINES MADE WITH PFBCl AND TRIMETHYLAMINE

The amine (up to 1 mg) and an equimolar amount of trimethylamine are dissolved in ethyl acetate (1 ml) and PFBCl (50% molar excess) is added dropwise with shaking. The reaction is left at room temperature for 2 h, evaporated to dryness and dissolved in alcohol for analysis on 3% OV-17 at 190 °C.[152] For ECD it might be preferable to dilute to a suitable volume with ethyl acetate, as PFB derivatives of alcohols can be formed, and might interfere.

5.13.2 PFB-PHENYLETHYLAMINE MADE WITH PFBCl IN ETHER

The amine or amine hydrochloride (0.1 mg) is treated with 1:19 PFBCl: ether (0.1 ml) and left at room temperature for 20 min. The solution is evaporated to dryness *in vacuo*, and the residue dissolved in ethyl acetate (0.1 ml) for direct analysis by GLC with ECD on 2.5% OV-225 at 210 °C.[51]

5.13.3 PFB-DERIVATIVES OF PHENOLS: THE USE OF PFBCl AND SODIUM HYDRIDE

The phenol (25 mg) is dissolved in ether (10 ml) and an excess of solid sodium hydride is added in small portions until no more effervescence occurs. PFBCl: ether 1:9 (10% molar excess) is added, and the mixture is stirred at room temperature for 4 h. The solution is filtered and evaporated to dryness. The product (90% yield) is dissolved in a suitable solvent for analysis by TLC or GLC with FID or ECD on 1% SE-52 at 150 °C.[151]

5.13.4 *N*-PFB-DERIVATIVES OF AMIDES

The amide (0.2 mg) in hexane (0.5 ml) is mixed with 1.4 M trimethylamine in hexane (0.2 ml) and PFBCl (25 μl) and left at room temperature for 2 h. For some amides addition of acetone (0.1 ml) may be needed to obtain a clear solution. Some amides, e.g. benzamide, may be dehydrated to the nitrile. Although in the original procedure the reaction is diluted with hexane at this stage, for direct analysis by GLC with electron-capture,[80] it is preferable to remove excess of the reagents *in vacuo* and to take up the residue in ethyl acetate for injection. PFBCl is itself extremely electron-capturing, and will therefore produce a large peak at the start of the chromatogram, and may also depress the ECD response for some time. The PFB derivatives of amides are stable to moisture, and appear to be the only stable derivatives of amides that have been reported for the sensitive analysis of amides. Although amides form TMS and perfluoroacyl derivatives, these are much less stable,[80] and are not readily made in good yield for routine analysis.

5.13.5 PFB-DERIVATIVES MADE WITH PFB-IMIDAZOLE IN BENZENE

The sample of polar compound (0.1–2 mg) is dissolved in benzene (0.2–0.5 ml) and PFB-imidazole (5–10 μl) is added. The mixture is heated at 60 °C for 15–30 min. Sterically hindered compounds may need 2–6 h. For FID the mixture may be injected directly. For ECD the mixture is evaporated to dryness *in vacuo* and the residue extracted with hexane (0.4–0.8 ml), which may be used directly or concentrated first.[153]

5.14 Miscellaneous chloroacetylations

Chloroacetyl derivatives have, of course, been available for a long time, and turn out to be good electron-capturing derivatives for hydroxylated compounds, e.g. steroids, giving approximately the same sensitivity as the TFA group.[112] However, they are not so volatile. On the other hand chloroacetyl derivatives of amines, again like the TFA derivatives, do not capture to any significant extent.[154]

5.14.1 CHLOROACETYL DERIVATIVES OF STEROIDS MADE BY THE ANHYDRIDE-PYRIDINE METHOD

The steroid (100 mg) in tetrahydrofuran (5 ml) is treated with chloroacetic anhydride (1 g) and pyridine (0.2 ml) and left in the dark at room temperature for 18 h. Water (5 ml) is added and the solution is extracted with ether (3 × 5 ml). The ether extracts are combined and washed with 6 M HCl (1 × 5 ml) and water (2 × 5 ml), dried over anhydrous sodium sulfate and evaporated to dryness. The residue may be purified further by TLC with 4 : 1 benzene : ethyl acetate, and dissolved in ethyl acetate for GLC with ECD on 1% XE-60 at 210 °C.[155,156]

Alternatively, the steriods, either as an extract from biological sources or after isolation by TLC, may be dried down *in vacuo*, and treated with 1 : 99 chloroacetic anhydride in tetrahydrofuran (0.5 ml) and pyridine (0.1 ml). The reaction takes place in the dark at room temperature overnight, and is stopped by the addition of water (1 ml). The solution is extracted with ethyl acetate (3 × 1 ml) and the combined extracts washed with 6 M HCl (1 ml) and water (2 × 1 ml). It is then dried and concentrated for chromatography as before.[156]

5.14.2 CHLOROACETYLATION OF PHENOLS (SCHOTTEN–BAUMANN CONDITIONS)

The phenols (10 μl of each) in benzene (1 ml for each mg of total phenols) are mixed with 0.25 M NaOH (15 ml) and shaken with chloroacetic anhydride (1 g for every 200 ml of benzene) at room temperature for 2 min. After separation of the layers, the benzene solution (5 μl) is directly injected for GLC with ECD on 1% XE-60 at 170 °C.[157]

5.14.3 TRICHLOROACETYLATION OF AMPHETAMINE WITH TRICHLOROACETYL CHLORIDE

The trichloroacetyl derivatives of amines are amongst the most electron-capturing, second only to the PFB-derivatives.[154] To derivatize amphetamine, the amine is extracted twice from its biological matrix (blood, urine, tissues) after alkalinization to pH 12 with 10 M NaOH, using an equal volume of hexane, or 3 : 1 hexane-isooctane. These extractions lead to an emulsion which is broken by centrifugation. The hexane layer sometimes forms a stable gel with blood or urinary constituents, which can be dispersed by gentle stirring of the hexane layer with a glass rod, and re-centrifuging if necessary. Trichloroacetyl chloride (50 μl) is added to the clear hexane extract (4 ml), the solution is mixed, and then left at room temperature for 20 min. The hexane solution is washed with 1 M NaOH (2 ml) and centrifuged. The upper layer is injected for electron-capture GLC on 3% JXR at 150 °C.[158]

5.15 Derivative formation from tertiary amines with chloroformates

Although tertiary amines can be analysed chromatographically without making derivatives, their basic nature is often associated with poor chromatographic properties such as non-specific absorption, poor response and tailing. It is not easy to derivatize tertiary amines, but it has been found that treatment with chloroformates can displace one of the groups (particularly a methyl group, if one is present) from the nitrogen, to form a carbamate (20).

$$(R_1)(R_2)N{-}CH_3 + C_6F_5{-}CH_2{-}O{-}CO{-}Cl \rightarrow (R_1)(R_2)N{-}CO{-}O{-}CH_2{-}C_6F_5 \qquad (20)$$

The amine (microgram scale) is dissolved in heptane (0.2 ml) and treated with pentafluorobenzyl chloroformate (50 μl) and powdered anhydrous sodium carbonate (approx. 10 mg) in a reaction vial closed with a Teflon®-lined screw-cap. The reaction is heated at 100 °C for 30 min, cooled, and the mixture shaken with 1 M NaOH (1 ml). The upper layer can be used directly for GLC with an FID. For ECD the reaction mixture is pumped down to dryness *in vacuo*, and the product taken up in heptane (2 ml), washed with 1 M NaOH (1 ml) and then with water (2 × 1 ml) and dried over anhydrous sodium sulfate, before analysis by GLC.[159,160] A similar procedure has been described using trichloroethyl chloroformate for the analysis of pethidine.[161]

6 ELECTRON-CAPTURING ABILITY AND CHEMICAL STRUCTURE

6.1 Introduction

The electron capture detector (Fig. 2) is a flow-through cell containing a source of β-radiation (electrons) and a collector electrode which has a voltage

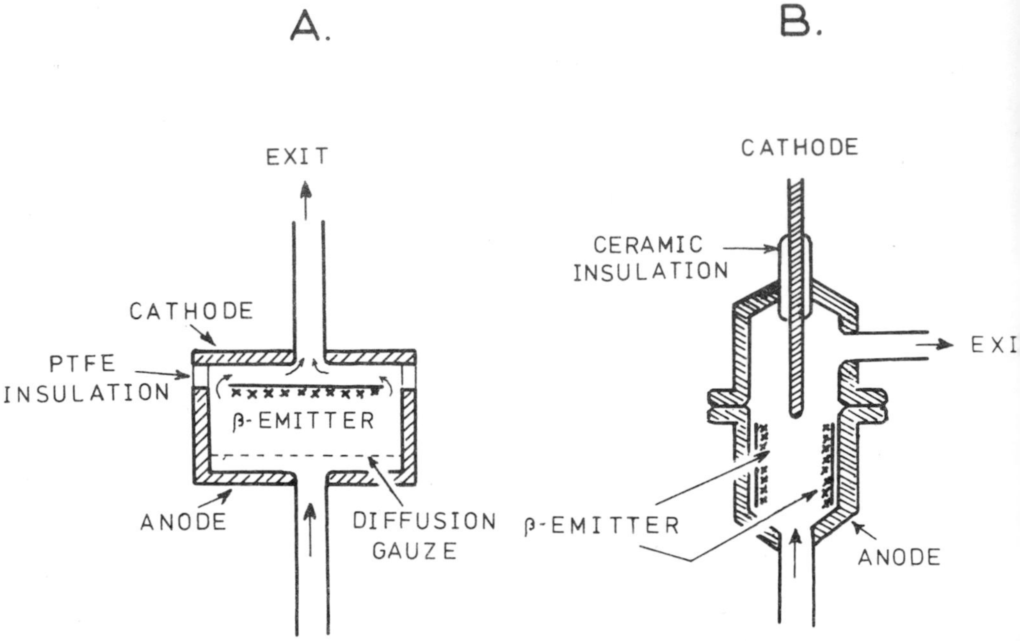

Fig. 2 Electron-capture detectors: diagrammatic drawings of two different configurations. A, parallel-plate type; B, concentric pin-cup type.

applied to it. The continuous electron-flux through the cell is collected, either continuously, or with a series of voltage pulses at intervals, giving rise to a 'standing current' even when only carrier gas is passing through. When an electron-capturing compound elutes as a peak from the chromatographic column, it causes a reduction in the electron-flux which results in a decrease in the standing current (Fig. 3).[162] This is converted electrically into the familiar display of peaks rising above a base-line. It may be helpful to compare the e.c.d. with the cuvette of a spectrophotometer, in which the light which falls onto the photomultiplier is decreased by the presence of an absorbing species, but the output is presented in the form of absorbance or optical density which increases with the concentration of the absorbing substance.

The capturing mechanism can involve either electron attachment to give a stable negative molecular ion (attachment process Eqn. 21)

$$A\text{—}B + e^- \rightarrow A\text{—}B^- \tag{21}$$

or dissociation to give fragment ions (dissociative process, Eqn. 22).

$$A\text{—}B + e^- \rightarrow A \quad + B^- \tag{22}$$

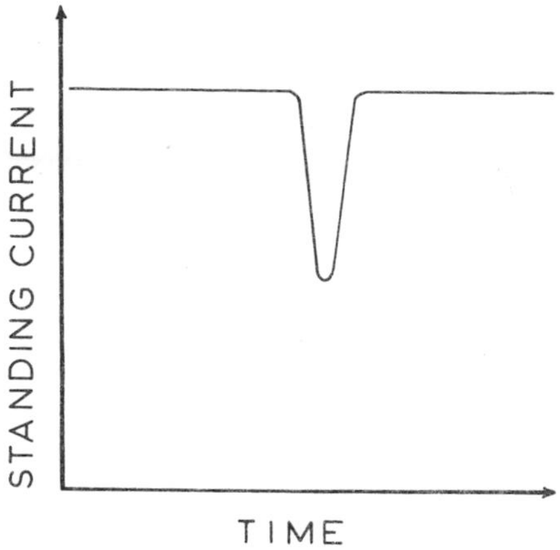

Fig. 3 The standing current given by an ECD with carrier gas alone passing through, and the effect on the standing current of the emergence of an electron-capturing compound from the chromatographic column.

Low detector temperatures apparently encourage the attachment process, and high temperatures the dissociative process.[163] Not all organic molecules are capable of reducing the thermal electron population: only those which are 'electron-capturing' do so. Certain structural features confer this property and when molecules are derivatized for detection by the ECD, electron-capturing functions or 'electrophores' are introduced to produce a derivative capable of decreasing the electron-flux even at a very low vapour pressure. Table 3 gives a summary of the main electrophores that are available for this purpose.

Although a great variety of chemical functions are quoted in Table 3, and only some of them are in fact acyl groups, the subject is dealt with in this chapter, because acylation reagents are by far the most widely used. A more detailed discussion of the practical applications of the other types of reagents that confer electron-capturing ability will be found in the relevant chapters as indicated in Table 3.

6.2 Halogenated electrophores: factors affecting sensitivity

A great variety of halogen-substituted reagents have been used to prepare electron-capturing derivatives, as we have seen in the previous section. The response of the ECD to various electrophores is however not predictable. Landowne and Lipsky[155] showed the advantages of the ECD for the analysis of steroids by protecting the hydroxyl of cholesterol with a variety of haloacetates. They established that the order of sensitivity is:

$$ClCH_2{-}CO{-} < CHCl_2{-}CO{-} < BrCH_2{-}CO{-} < CCl_3{-}CO{-} < CF_3{-}CO{-} \quad (24)$$

TABLE 3

Electron-capturing substituent groupings together with the classes of compounds into which they may be introduced [a]

Electrophore	Functions to be derivatized	Chapter
CF_3CO—, C_2F_5CO—, C_3F_7CO— . . . $C_9F_{19}CO$— . . . etc.	—OH, —NH_2, —NHR, —SH	
CCl_3CO—, $CHCl_2CO$—, CH_2ClCO—	—OH, —NH_2, —NHR, —SH	
C_6F_5CO—, $C_6F_5CH_2CO$—	—OH, —NH_2, —NHR, —SH, —NHCOR	3
CH_2FCO—, $CClF_2CO$—, CH_2BrCO—	—OH, —NH_2, —NHR, —SH	
p-Trifluoromethylbenzoyl-	—OH, —NH_2, —NHR, —SH	
Pentafluorobenzylidine-	primary amines	6
2,4-Dinitrophenyl-	primary and secondary amines	
2,6-Dinitro-4-trifluoromethylphenyl-	primary and secondary amines	10
2-Nitro-4-trifluoromethylphenyl-	primary and secondary amines	
2,4-Dinitrobenzenesulphonyl-	primary and secondary amines	
Pentafluorobenzyl-	—OH, —COOH, —SH, —SO_2NH—	5
2-Chloroethyl-	—COOH	
Trichloroethyl-	—COOH	
Pentafluoropropyl-	—COOH	
p-Chlorobenzyl-	—COOH	2
p-Bromobenzyl-	—COOH	
p-Nitrobenzyl-	—COOH	
p-Bromophenacyl-	—OH	
Chloromethyldimethylsilyl $ClCH_2$—$Si(CH_3)_2$—	—OH, —COOH	
Bromomethyl dimethylsilyl $BrCH_2$—$Si(CH_3)_2$—	—OH	4
Iodomethyldimethylsilyl ICH_2—$Si(CH_3)_2$—	—OH	
Pentafluorophenyldimethylsilyl (flophemesyl) C_6F_5—$Si(CH_3)_2$—	—OH	
C_6F_5—NH—N=	C=O	6
C_6F_5—O—N=	C=O	
$(CF_2Cl)_2C$=O forms oxazolidinones of	α-amino, —OH, and —SH acids	7
Pentafluorobenzyloxycarbonyl-	tertiary amines	3, 8
Trichloroethyloxycarbonyl-	tertiary amines	3

[a] This list also includes electrophores additional to those used for acylation, with cross references to the chapters in which they are described in more detail.

These results were surprising at the time, because it was, and still is generally accepted that the order of the halogens is:

$$F < Cl < Br < I \qquad (25)$$

and up to a point it seems that the more halogen atoms there are in the molecule, the better.[165] These generalizations should be interpreted with caution, because with some acylations dehydration is a common side-reaction,[113] resulting in a less than quantitative yield, and because where there is a marked temperature dependence, the sensitivity of detection may not have been optimized, so that we are not able to make strictly valid comparisons.[163] A number of comparative studies of various electrophores for a single substance or a limited range of substances have appeared,[58,75,151,154,165,166] and the results are shown in Table 4. It is plain from this table that ECD responses are not exactly predictable, although certain trends are apparent. Part of the problem is that we have not really reached a detailed understanding of the mechanisms of electron-capture: it is not precisely known what happens in the dissociative process, which bonds break, where the electron is stabilized. Perhaps the fragmentation processes of negative chemical or atmospheric pressure ionization mass spectrometry will clarify the subject. Two useful and detailed discussions of the various aspects of the ECD response have appeared recently.[167,168]

6.3 The choice of electron-capturing derivatives

Looking at the subject from a practical viewpoint, what can we conclude about the selection of derivatives to give high sensitivity with the ECD? If one classifies compounds into three broad categories, those that are very easy to volatilize, those that are moderately easy to volatilize, and those that are difficult to volatilize, then the first category contains compounds that probably should not be derivatized with any of the small electrophores, since the products will be too volatile for good quantitation because of unavoidable evaporative losses. This permits the use of the PFB-, pentafluorobenzyl, pentafluorobenzylidine, flophemesyl and similar large electron-capturing groups, which not only help to reduce the volatility problem, but also confer very great sensitivity to ECD on the derivatives. Response to them is invariably very high, and they probably approach the theoretical maximum for electron-capture detection fairly closely (see below). In the second class of compounds the choice is much wider, and to achieve the best response one might choose derivatives containing C_6F_5- or CCl_3- functions. Problems of volatility would only arise where more than one of these groups in the molecule might push the temperature of analysis up beyond the limit of the columns available. In such cases one can solve the problem by using mixed derivatives, such as one PFB and one TMS group. For the last category of compounds one is in practice restricted to HFB- or PFP- derivatives in acylation reactions, and to trichloroethyl- or pentafluoropropyl- in esterification. With the higher homologues the small increase in sensitivity that is obtained with, e.g. the perfluorooctanoyl group scarcely compensates for the greater preparative and volatility problems.[169] HFB- and PFP- groups in general confer more volatility than TMS groups. The *O*-PFP and *O*-HFB derivatives are much less stable than the *O*-TMS derivatives. On the other hand, the *N*-PFP and *N*-HFB

TABLE 4

Comparisons of electron-capture responses of different derivatives of thymol, some amines and testosterone

Compound	Derivative	Relative response
Thymol[151]	Monofluoroacetyl	0.007
	Monochloroacetyl	0.3
	Pentafluoropropionyl	1.3
	Heptafluorobutyryl	1.0
	Pentafluorobenzyl	5.9
	Pentafluorobenzoyl	6.9
	2,4-Dinitrophenyl	0.3
Benzylamine[58]	Acetyl	0.05
	Monochloroacetyl	37.5
	Trifluoroacetyl	1.0
	Pentafluoropropionyl	286
	Heptafluorobutyryl	894
Amphetamine[154]	Trifluoroacetyl	<0.1
	Pentafluoropropionyl	4
	Heptafluorobutyryl	9
	Perfluorooctanoyl	23
	Monochloroacetyl	0.1
	Trichloroacetyl	54
	Isothiocyanate	0.2
	Maleamide	13
	p-Trifluoromethylbenzoyl	12
	Pentafluorobenzoyl	77
	(α-Lindane	100)
Amines:[165]		
Cyclopropylamine	Heptafluorobutyryl	15
Cyclohexylamine	Heptafluorobutyryl	1
Aniline	Heptafluorobutyryl	670
Benzylamine	Heptafluorobutyryl	25
Dibenzylamine	Heptafluorobutyryl	83
Heptylamine	Heptafluorobutyryl	5
Cyclopropylamine	Pentafluorobenzoyl	1000
Cyclohexylamine	Pentafluorobenzoyl	500
Aniline	Pentafluorobenzoyl	133
Benzylamine	Pentafluorobenzoyl	153
Dibenzylamine	Pentafluorobenzoyl	23
Heptylamine	Pentafluorobenzoyl	670
Testosterone[112]	Acetyl	0.1
	Trifluoroacetyl	0.4
	Monochloroacetyl	4
	Pentafluoropropionyl	5
	Heptafluorobutyryl	19
	Chlorodifluoroacetyl	34
	Pentafluorobenzoyl	50
	Perfluorooctanoyl	60

derivatives are much more stable than the *N*-TMS derivatives, many of which are in fact silylating agents.

It is clear that the main factors in the choice of derivatives are sensitivity of detection, ease of preparation and volatility of the resulting product. These factors are not always entirely independent of one another. The decision about which to use is therefore bound to be a compromise which depends to a large extent on the analytical priorities of the user. In the past, for no very compelling reasons, fashions in research have led to amino acids being acylated mainly with the TFA group, phenolic amines with the PFP group and steroids with the HFB group. It is doubtful whether these choices were made by a rational assessment as outlined above, and they may not have been the best choices in all cases. There are hopeful signs that empirical solutions are about to give way to more purposefully designed choices.

5.4 Achieving the optimum response

Once the choice of derivative has been made, there are a number of ways of assuring the maximum response with the ECD. The first is obviously to work out the optimum reaction conditions for derivatization. The factors that must be looked at are proportion of reagent to compound to be derivatized, choice of solvents and catalysts, reaction temperature and duration, and the working-up procedure. A little systematic work at this stage is well worthwhile, because it will savc time later, as well as giving confidence when all these factors are understood. For example, if a reaction is completed in five minutes there is little point in doing it for an hour, yet it is clear from many of the methods quoted in this chapter that unnecessarily prolonged reaction times are often used, perhaps 'to be on the safe side'. This may be of doubtful value, as yields can go down on prolonged exposure of derivatives to powerful reagents. This applies to derivatization generally, and is not limited to those described in this chapter.

Next, it is valuable to determine the temperature dependence of the detector response to the chosen derivative. The response of the detector, as peak area for injection of a fixed amount of derivative, is determined for a range of detector temperatures. Care must be taken not to use detector temperatures below those of the column.[6,162,163,170] If A is the peak area, then ln $AT^{3/2}$ as ordinate is plotted against the abscissa $1/T$. The resulting graph will have a positive slope for a non-dissociative detection mechanism, and a negative slope for a dissociative process. The result will permit extrapolation to obtain the optimum conditions of detection.

Sullivan[171] has described the determination of how closely a derivative approximates to the 'theoretical limit of detection' by using some simple relationships derived from collision theory and applying them to a constant-current variable pulse-frequency ECD. Collision theory makes it possible to calculate the theoretical electron-capture rate constant K_T, while practical measurement gives experimental values K_E. If we compare some unknown compound with the pesticide standard α-lindane, using the same instrumental conditions for both, then with a constant-current detector and Sullivan's calculations for lindane:

$$K_E = 0.0124 \times \frac{W_L A_U M_U}{W_U A_L} \quad (25)$$

W_L and W_U are the injected weights of lindane and of the unknown, respectively; A_L and A_U are the peak areas of lindane and unknown, respectively; and M_U is the molecular weight of the unknown.

The molar refraction of a compound can be calculated from published tables[172] for various bond groupings and elements, and the various individual coefficients are additive, to make up a value R for the whole molecule. The theoretical value of the electron-capturing rate constant is given by Eqn. (26):

$$K_T = 6.29 \times 10^{-8} \mathrm{R} \quad (26)$$

A comparison of K_E and K_T will indicate just how close to the theoretical detection limit the compound approaches (Table 5). It can be seen that

TABLE 5

Electron-capture rate constants of some capturing compounds. The experimental rate constants, K_E, were determined on a gas chromatograph with an ECD containing 15 mCi ^{63}Ni operated in the pulse-modulated mode. Detector temperature was 300 °C

Compound	Molecular weight	Molar refraction ml	K_E	K_T
			(ml molec^{-1} s^{-1} $\times 10^{-7}$)	
Lindane	291	56.91	3.61	4.74
Dieldrin	381	76.99	4.49	5.52
p,p-DDT	354	83.99	3.10	5.76
Phenylethylamine trichloroacetyl derivative	280	66.84	2.86	5.14
Testosterone HFB derivative	484	102.10	0.72	6.39

polyhalogenated pesticides are not far from their theoretical limits, but that the use of an HFB derivative of testosterone still leaves room for considerable improvement. Sullivan suggests that temperature dependence is least important in those compounds which closely approach their theoretical maximum rate constant. This kind of part theoretical, part practical way of trying to achieve the highest response should lead to a closer approach to the best sensitivity available for the analysis of compounds by electron-capture GLC. A somewhat different way of evaluating the theoretical limit of detection is given by Pellizzari.[168]

6.5 Other electrophores

A number of chemical groupings not yet referred to in Table 3 are also electrophores, but to a rather lesser extent. These groupings include keto and ester groups, and the detector response to these is comparable to that of the FID. In a number of instances compounds which are not easily derivatized may be converted into capturing products by microchemical reactions (see Chapter 8). For example, phthalate esters are almost universally used as plasticizers,

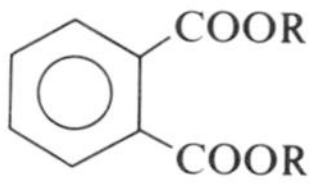

and have become almost ubiquitous as environmental pollutants and as trace impurities in commercial solvents. The ECD responds to these esters and they are best detected at as low a detector temperature as possible.[173] The drug terodiline is oxidized using chromic acid, to give benzophenone, which is detected at low levels with the ECD (Eqn. 27).[174]

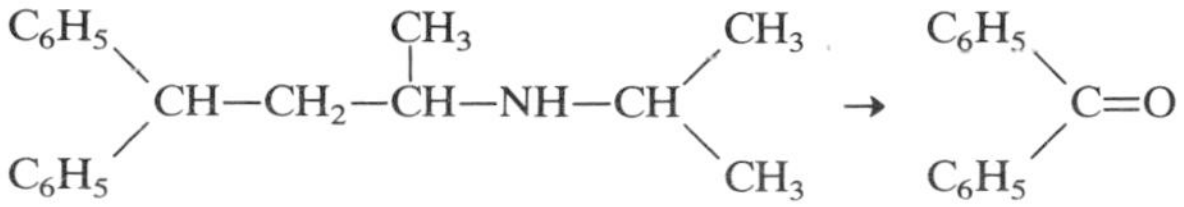

(27)

Terodiline

As a final example, the insect moulting hormones, the ecdysones, have also been determined as their TMS ethers by ECD, the electrophore in this case being the α,β-unsaturated keto system (28).[167]

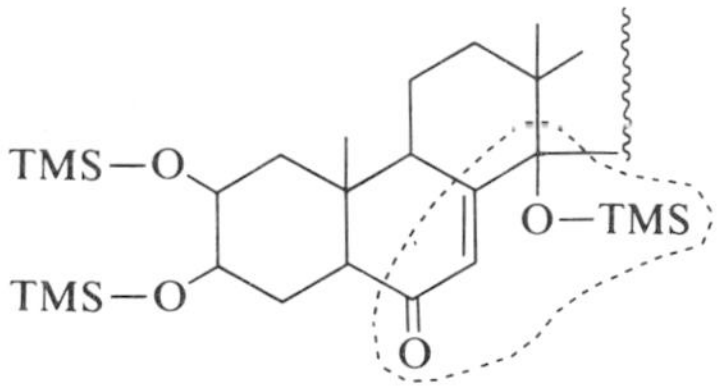

(28)

In general, these groupings involve oxygen atoms, usually as part of a carbonyl group and conjugated π-electron systems.

7 MASS SPECTROMETRY OF PERFLUORACYL DERIVATIVES

Considerable work has been reported on the mass spectra of these derivatives, much of it in verification of the proposed structures resulting from analytical derivatization schemes which have been described in this chapter. In most cases the $C_nF^+_{2n+1}$ ions (*m/e* 69 for TFA groups, *m/e* 119 for PFP groups and *m/e* 169 for HFB groups) are abundant. These are unfortunately of little use in structural assignment or for SIM. The $C_nF_{2n+1}CO^+$ ions (*m/e* 97, 147, 197 respectively) are also usually quite strong. However, perfluoroacyl derivatives usually give more useful and easily identifiable ions in the high mass region of the mass spectrum than ordinary acetates. This holds for carbohydrates such as alditols,[104] and for phenols[175,176] and phenolic amines.[177] For TFA derivatives of simple phenols the $M^{+\cdot}$ is usually dominant and major fragments correspond to loss of CO, $CF_3^{\cdot}$ and $COCF_3^{\cdot}$. These derivatives would be a good choice for SIM because of their sensitivity and specificity. There is still a quite pronounced tendency for higher-mass ions to be less abundant in the trifluoroacetates of the carbohydrates. The $M^{+\cdot}$ fragments with the loss of a TFA-radical (TFA˙), followed by molecules of TFA-acid (TFAc), i.e. an 'unzipping' of the molecule:[104]

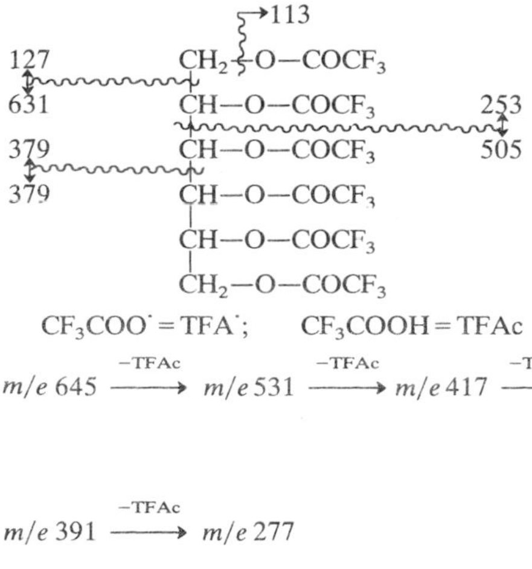

(29)

$M^{+\cdot} = 758$; $CF_3COO^{\cdot} = TFA^{\cdot}$; $CF_3COOH = TFAc$

C_6: $M^{+\cdot} = m/e\ 758 \xrightarrow{-TFA^{\cdot}} m/e\ 645 \xrightarrow{-TFAc} m/e\ 531 \xrightarrow{-TFAc} m/e\ 417 \xrightarrow{-TFAc} m/e\ 303$ (29a)

C_5: *m/e* 631 only (29b)

C_4: $m/e\ 505 \xrightarrow{-TFAc} m/e\ 391 \xrightarrow{-TFAc} m/e\ 277$ (29c)

The loss of $C_nF_{2n+1}COO^{\cdot}$ is in general a favourable process, and a fragmentation process equivalent to the loss of 42 a.m.u. ($CH_2{=}C{=}O$) seen with acetates is never observed. Fragmentation of MHPG-PFP is typical of this type of compound:

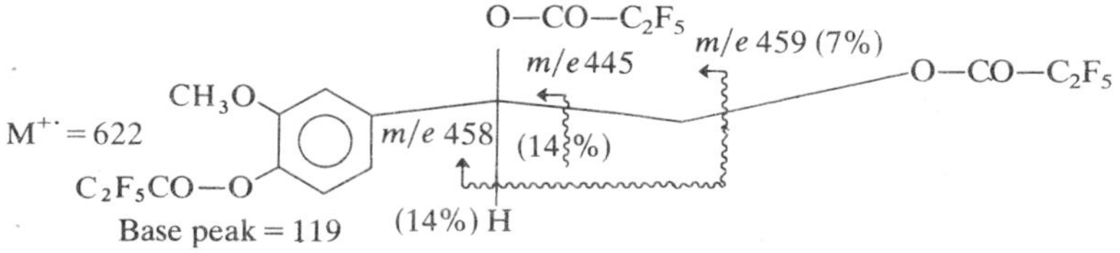

(30)

$m/e\ 445 \xrightarrow{-CO} m/e\ 417$ (8%)

$m/e\ 458 \xrightarrow{-C_2F_5CO} m/e\ 311$ (8%)

Most fragmentations are simple cleavages which tend to give charge-retention on the aromatic system. Perfluoroacyl derivatives can undergo McLafferty rearrangements such as that leading to m/e 458, or eliminations of CO from m/e 445 to give m/e 417—a fairly common process.[179]

Aliphatic amines tend to give abundant ions corresponding to $C_nF_{2n+1}CONH{-}CH_2^+$, e.g. m/e 126 due to $CF_3{-}CO{-}NH{-}CH_2^+$ in the mass spectrum of *N*-ethyltrifluoroacetamide,[175] and m/e 190 in the mass spectrum of the PFP derivative of adrenaline, which is due to $C_2F_5CO{-}N(CH_3)CH_2^+$. Phenylethylamines show a McLafferty rearrangement corresponding to a loss of $C_nF_{2n+1}CO{-}NH_2$ from $M^{+\cdot}$, but aliphatic amines retain the charge on the amide portion of the molecule. Several investigations of the mass spectra of TFA-amines have been published.[175,180–183] A GC–MS study of TFA-amino acid *n*-butyl esters was undertaken by Gelpi *et al.*[178] The dominant fragment is generally loss of butyloxycarbonyl $(M{-}CO_2C_4H_9)^+$. The mass spectra of some low molecular weight trifluoroacetic-non-protein amino acid butyl esters were also reported.[184] At the moment the *N*-TFA *n*-butyl esters of amino acids are the derivatives of choice, for identification by GC–MS, because of the larger amount of mass spectral data available for these derivatives than any of the others.

GC–MS properties of trifluoroacetylated nucleotides have been reported by Koenig *et al.* in considerable detail.[185]

8 FUTURE PROSPECTS OF ACYLATION FOR CHROMATOGRAPHY

8.1 Introduction

Although a great variety of acylation methods have been presented, a careful reading of this chapter soon makes it clear that these are variations on a limited number of themes. These methods in fact seem to satisfy almost any requirements for acylation, no matter for what analytical purposes, be they qualitative, quantitative or mass spectrometric. Nevertheless, numerous other acylation reactions, methods and reagents are available which have not been described here or which have not been applied to chromatography, usually for the good reason that they do not lend themselves to such applications. A few of these are, however, presented now, partly for the sake of completeness, partly also because they may give readers new ideas or because they hold the key to solutions of problems which we have not yet even thought of.

8.2 Derivatives of phosphorus oxyacids

The use of the diethylphosphoryl group in combination with an alkali FID for phosphorus detection has already been mentioned.[8] Dimethylphosphinic derivatives of steroids have been made for similar reasons.[186,187] The wider use of phosphorus-sensitive detectors is a prerequisite for the further exploitation of this approach.

8.3 Miscellaneous kinds of acylation

An interesting idea of using a mixed carbonic-carboxylic anhydride resin for acylation is foreshadowed by some work which describes such a resin used to acylate amines to their *p*-chlorobenzoyl, benzoyl and cinnamoyl derivatives. Alcohols did not react.[188] Although not developed to derivatize amines for chromatography, this might be a useful application for such a resin, because not having to remove excess of reagent would give a simpler work-up and probably a clean product in good yield.

Another kind of mixed anhydride that might be used for derivatization are the mixed sulfonic-carboxylic anhydrides described by Karger and Mazur.[189] These were shown to acylate aromatic ethers and hydrocarbons in the ring, and although only ring-acetylations were described, this procedure could probably also be applied for the introduction of electron-capturing groups into such aromatic systems for GLC analyses.

Mention should also be made of acylation with pyromellitic anhydride.[190] This reagent reacts smoothly and quantitatively with amines and alcohols (31).[191]

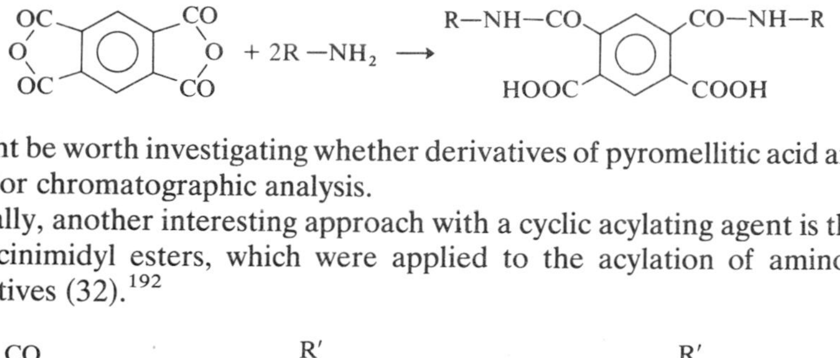

(31)

It might be worth investigating whether derivatives of pyromellitic acid are any good for chromatographic analysis.

Finally, another interesting approach with a cyclic acylating agent is the use of succinimidyl esters, which were applied to the acylation of amino acid derivatives (32).[192]

(32)

This reagent may be used analogously to the imidazole derivatives, but a wider range of acyl groups seems to be potentially available for introduction by this procedure.

9 CONCLUSION

The recent period of great activity in the development of acyl derivatives for chromatography seems likely to be followed by a consolidation phase, in which the many methods that have been presented in this chapter will continue to be applied to an increasing range of analytical problems in many different fields. We hope that the collection of these methods into one place will contribute to this widespread application, and that the comments we have made at appropriate points in this chapter will make it easier for others to choose and use the methods which are best suited to their needs.

REFERENCES

1. D. P. N. Satchell, *Rev. Chem. Soc.* **17**, 160 (1963).
2. D. P. N. Satchell, *Chem. Ind.* (*London*) 683 (1974).
3. H. A. Staab and W. Rohr, in W. Foerst (Ed.) *Newer Methods of Preparative Organic Chemistry*, Academic Press, New York, 1968.
4. J. E. Lovelock and S. R. Lipsky, *J. Am. Chem. Soc.* **82**, 431 (1960).
5. G. Zweig and J. Sherma, *Analytical Methods for Pesticides and Plant Growth Regulators*, Vol. 6. *Gas Chromatographic Analysis*, Academic Press, New York, 1972.
6. W. E. Wentworth and E. Chen, *J. Gas Chromatogr.* **5**, 170 (1967).
7. M. Pailer and W. J. Hübsch, *Mh. Chem.* **97**, 1541 (1966).
8. G. Ertinghausen, C. W. Gehrke and W. A. Aue, *Separation Sci.* **2**, 681 (1967).
9. D. G. Saunders and L. E. Vanatta, *Anal. Chem.* **46**, 1319 (1974).
10. C. D. Pfaffenberger, J. Safranek, M. G. Horning and E. C. Horning, *Anal. Biochem.* **63**, 501 (1975).
11. G. S. King, unpublished observation.
12. I. Fleming and J. B. Mason, *J. Chem. Soc.* 2509 (1969).
13. J. S. Sawardeker, J. H. Sloneker and A. Jeanes, *Anal. Chem.* **39**, 121 (1967).
14. L. F. Prescott, *J. Pharm. Pharmacol.* **23**, 807 (1971).
15. J. R. Coulter and C. S. Hann, *J. Chromatogr.* **36**, 42 (1968).
16. R. F. McGregor, G. M. Brittin and M. S. Sharon, *Clin. Chim. Acta* **48**, 65 (1973).
17. R. F. Adams, *J. Chromatogr.* **95**, 189 (1974).
18. J. B. Knaak, J. M. Eldridge and L. J. Sullivan, *J. Agr. Food Chem.* **15**, 605 (1967).
19. G. D. Paulson, R. G. Zaylskie and M. M. Dockter, *Anal. Chem.* **45**, 21 (1973).
20. F. C. Chattaway, *J. Chem. Soc.* 2495 (1931).
21. M. Hagopian, R. I. Dorfman and M. Gut, *Anal. Biochem.* **2**, 387 (1961).
22. R. Laverty and D. F. Sharma, *Brit. J. Pharmacol.* **24**, 538 (1965).
23. E. Röder and J. Merzhäuser, *Anal. Chem.* **34**, 272 (1974).
24. G. Quadbeck, *Angew. Chem.* **68**, 361 (1956).
25. M. Bergmann and F. Stern, *Ber. Dtsch. Chem. Ges.* **63B**, 437 (1930).
26. K. Blau, *Chem. Ind.* (*London*) 33 (1963).
27. S. Hara and Y. Matsushima, *J. Biochem.* (*Japan*) **71**, 907 (1972).
28. T. S. Stewart and C. E. Ballou, *Biochemistry* **7**, 1855 (1968).
29. G. D. Paulson and C. E. Portnoy, *J. Agr. Food Chem.* **18**, 180 (1970).
30. D. P. Rose and P. A. Toseland, *Clin. Chim. Acta* **17**, 235 (1967).
31. D. P. Schwartz, *Anal. Biochem.* **71**, 24 (1967).
32. P. S. S. Dawson, *Biochim. Biophys. Acta* **111**, 51 (1965).
33. J. C. Sheehan and D.-D. H. Yang, *J. Am. Chem. Soc.* **80**, 1154 (1958).
34. V. du Vigneaud, R. Dorfmann and H. S. Loring, *J. Biol. Chem.* **98**, 577 (1932).
35. K. Heyns and H.-F. Grützmacher, *Z. Naturforsch.* **16b**, 293 (1961).
36. J. O. Thomas, *Tetrahedron Lett.* 335 (1967).
37. G. Losse, A. Losse and J. Stöck, *Z. Naturforsch. Teil B* **17**, 785 (1962).
38. S. V. Shlyapnikov and M. Ya. Karpeisky, *Biokhimiya* **29**, 1067 (1964).
39. G. A. R. Decroix, J. G. Gobert and R. de Deurwaerder, *Anal. Biochem.* **25**, 523 (1968).

40. L. Gasco and R. Barrera, *Anal. Chim. Acta* **61**, 253 (1972).
41. R. T. Aplin and J. H. Jones, *J. Chem. Soc.* (*C*) 1770 (1968).
42. L.-A. Svensson, *Acta Chem. Scand.* **26**, 2663 (1972).
43. H. M. Fales and J. J. Pisano in *Biomedical Applications of Gas Chromatography*, H. A. Szymanski (Ed.), Plenum Press, New York, 1964, p. 39.
44. K. Morita, F. Irreverre, F. Sakiyama and B. Witkop, *J. Am. Chem. Soc.* **85**, 2832 (1963).
45. B. Halpern, L. F. Chew and J. W. Westley, *Anal. Chem.* **39**, 399 (1967).
46a. M. Makita, S. Yamamoto, M. Kono, K. Sakia and M. Shiraishi, *Chem. Ind.* (*London*) 355 (1975).
46b. M. Makita, S. Yamamoto and M. Kono, *J. Chromatogr.* **120**, 129 (1976).
47. T. Gejvall, *J. Chromatogr.* **90**, 157 (1974).
48. J. E. Stouffer, *J. Chromatogr. Sci.* **7**, 124 (1969).
49. E. M. Volpert, N. Kundu and J. B. Dawidzik, *J. Chromatogr.* **50**, 507 (1970).
50. N. N. Nihei, M. C. Gershengorn, T. Mitsuma, L. R. Stringham, A. Cordy, B. Kuchmy and C. S. Hollander, *Anal. Biochem.* **43**, 433 (1971).
51. K. Blau, unpublished observations.
52. J. C. Cavadore, G. Nota, G. Prota and A. Previero, *Anal. Biochem.* **60**, 608 (1974).
53. I. M. Hais and K. Macek (Eds) *Paper Chromatography*, Academic Press, New York, 1963, pp. 832–833.
54. W. G. Galetto, R. E. Kepner and A. D. Webb, *Anal. Chem.* **38**, 34 (1966).
55. P. H. Scott, *J. Chromatogr.* **70**, 67 (1972).
56. E. J. Bourne, C. E. M. Tatlow and J. C. Tatlow, *J. Chem. Soc.* 1367 (1950).
57a. F. Weygand and E. Csendes, *Angew. Chem.* **64**, 136 (1952).
57b. F. Weygand and R. Geiger, *Chem. Ber.* **89**, 647 (1956).
58. D. D. Clarke, S. Wilk and S. E. Gitlow, *J. Gas Chromatogr.* **4**, 310 (1966).
59. A. T. Shulgin, *Anal. Chem.* **36**, 920 (1964).
60. M. J. Levitt, J. B. Josimovitch and K. D. Broskin, *Prostaglandins* **1**, 121 (1972).
61. B. S. Middleditch and D. M. Desiderio, *Prostaglandins* **2**, 195 (1972).
62. K. P. Wong, C. R. J. Ruthven and M. Sandler, *Clin. Chim. Acta* **47**, 215 (1973).
63. I. L. Martin and G. B. Ansell, *Biochem. Pharmacol.* **22**, 521 (1973).
64. H. Ch. Curtius, M. Wolfensberger, U. Redweik, W. Leimbacher, R. A. Maibach and W. Isler, *J. Chromatogr.* **112**, 523 (1975).
65. J. Segura, F. Artigas, E. Martinez and E. Gelpi, *Biomed. Mass Spectrom.* **3**, 91 (1976).
66. W. J. Irvine and M. J. Saxby, *J. Chromatogr.* **43**, 129 (1969).
67. M. Pailer and W. Hübsch, *Mikrochim. Acta* 912 (1967).
68. R. B. Bruce and W. R. Maynard Jr, *Anal. Chem.* **41**, 977 (1969).
69. A. K. Cho, B. Lindeke, B. J. Hodshon and D. J. Jenden, *Anal. Chem.* **45**, 570 (1973).
70. M. Ervik, *Acta Pharm. Suecica* **6**, 393 (1969).
71. J. W. Blake, R. S. Ray, J. S. Noonan and P. W. Murdick, *Anal. Chem.* **46**, 288 (1974).
72. D. E. Coffin, *J.A.O.A.C.* **52**, 1044 (1969).
73. R. A. Morrissette and W. E. Link, *J. Gas Chrom.* **3**, 67 (1965).
74. A. Zeman and I. P. G. Wirotama, *Z. Anal. Chem.* **247**, 158 (1969).
75. T. Walle and H. Ehrsson, *Acta Pharm. Suec.* **7**, 389 (1970).
76. D. A. Garteiz and T. Walle, *J. Pharm. Sci.* **61**, 1728 (1972).
77. L. M. Cummins and M. J. Fourier, *Anal. Lett.* **2**, 403 (1969).
78. A. F. Cockerill, D. N. B. Mallen, D. J. Osborne and D. M. Price, *J. Chromatogr.* **114**, 151 (1975).
79. H. Ehrsson and H. Brötell, *Acta Pharm. Suec.* **8**, 591 (1971).
80. H. Ehrsson and B. Mellström, *Acta Pharm. Suec.* **9**, 107 (1972).
81. O. Gyllenhaal and H. Ehrsson, *J. Chromatogr.* **107**, 327 (1975).
82. A. Wu, *Clin. Toxicol.* **8**, 225 (1975).
83. P. Hušek and K. Macek, *J. Chromatogr.* **113**, 139 (1975).
84. J. Wagner and G. Winkler, *Z. Anal. Chem.* **183**, 1 (1961).

85. M. L. Rueppel, L. A. Suba and J. T. Marvel, *Biomed. Mass Spectrom.* **3**, 28 (1976).
86. A. Darbre and K. Blau, *J. Chromatogr.* **17**, 31 (1963).
87. E. Mussini and F. Marcucci, *J. Chromatogr.* **26**, 48 (1967).
88. A. Darbre and A. Islam, *Biochem. J.* **106**, 923 (1968).
89. H. A. Saroff, *4th Internat. Symp. Gas Chromatogr.*, École Polytechnique, Paris, 1965.
90. B. Teuwissen and A. Darbre, *J. Chromatogr.* **49**, 298 (1970).
91. D. Roach and C. W. Gehrke, *J. Chromatogr.* **44**, 269 (1969).
92. G. E. Pollock, *Anal. Chem.* **39**, 1194 (1967).
93. C. W. Moss, M. A. Lambert and F. J. Diaz, *J. Chromatogr.* **60**, 134 (1971).
94. J. P. Zanetta and G. Vincendon, *J. Chromatogr.* **76**, 91 (1973).
95. J. Jönsson, J. Eyem and J. Sjöquist, *Anal. Biochem.* **51**, 204 (1973).
96. S. L. MacKenzie and D. Tenaschuk, *J. Chromatogr.* **97**, 19 (1974).
97. P. Felker and R. S. Bandurski, *Anal. Biochem.* **67**, 245 (1975).
98. J. F. March, *Anal. Biochem.* **69**, 420 (1975).
99. W. A. Koenig, L. C. Smith, P. F. Crain and J. A. M. McCloskey, *Biochemistry* **10**, 3968 (1971).
100. C.-G. Hammar and B. Holmstedt, *Experientia* **24**, 98 (1968).
101. M. Vilkas, Hiu-I-Jan, G. Boussac and M.-C. Bonnard, *Tetrahedron Lett.* **14**, 1441 (1966).
102. T. Ueno, N. Kurihara and M. Nakajima, *Agr. Biol. Chem.* **31**, 1189 (1967).
103. J. Shapira, *Nature (London)* **222**, 792 (1969).
104. O. S. Chizlov, B. A. Dmitriev, B. M. Zolotarev, A. Ya. Cherniak and N. K. Kochetkov, *Org. Mass Spectrom.* **2**, 947 (1969).
105. W. A. König, H. Bauer, W. Voelter and E. Bayer, *Chem. Ber.* **106**, 1905 (1973).
106. K. Yoshida, N. Honda, N. Iino and K. Kato, *Carbohydr. Res.* **10**, 333 (1969).
107. Z. Tamura and T. Imanari, *Chem. Pharm. Bull* **15**, 246 (1967).
108. H. Ehrsson, T. Walle and S. Wikström, *J. Chromatogr.* **101**, 206 (1974).
109. L. A. Dehennin and R. Scholler, *Steroids* **13**, 739 (1969).
110. M. Hiroi and S. Kushinski, *Mikrochim. Acta* 1160 (1969).
111. W. P. Collins, J. M. Sisterson, E. N. Koullapis, M. D. Mansfield and I. F. Sommerville, *J. Chromatogr.* **37**, 33 (1968).
112. L. Dehennin, A. Reiffstock and R. Scholler, *J. Chromatogr. Sci.* **10**, 224 (1972).
113. C. F. Poole and E. D. Morgan, *J. Chromatogr.* **90**, 380 (1974).
114. H. Ehrsson, T. Walle and H. Brötell, *Acta Pharm. Suecica* **8**, 319 (1971).
115. H. Dekirmenjian and J. W. Mass, *Anal. Biochem.* **35**, 113 (1970).
116. E. K. Gordon and J. Oliver, *Clin. Chim. Acta* **35**, 145 (1971).
117. P. A. Bond, *Biochem. Med.* **6**, 36 (1972).
118. C. Braestrup, *J. Neurochem.* **20**, 519 (1973).
119. S. Wilk, K. L. Davis and S. B. Thacker, *Anal. Biochem.* **39**, 498 (1971).
120. F. Karoum, C. R. J. Ruthven and M. Sandler, *Biochem. Med.* **5**, 505 (1971).
121. L. Fellows, P. Riederer and M. Sandler, *Clin. Chim. Acta* **59**, 255 (1975).
122. Z. Kahane, W. Ebbinghausen and P. Vestergaard, *Clin. Chim. Acta* **38**, 413 (1972).
123. T. Imanari and Z. Tamura, *Chem. Pharm. Bull.* **15**, 896 (1967).
124. P. W. Miller, *J. Agr. Food Chem.* **19**, 941 (1971).
125. E. Änggård and G. Sedvall, *Anal. Chem.* **41**, 1250 (1969).
126. B. L. Goodwin, C. R. J. Ruthven and M. Sandler, *Clin. Chim. Acta* **62**, 439 (1975).
127. S. W. Dziedzic, L. M. Bertani, D. D. Clarke and S. E. Gitlow, *Anal. Biochem.* **47**, 592 (1972).
128. R. V. Smith and S. L. Tsai, *J. Chromatogr.* **61**, 29 (1971).
129. S. Wilk and M. Orlowski, *FEBS Lett.* **33**, 157 (1973).
130. C.-G. Fri, F.-A. Wiesel and G. Sedvall, *Life Sci.* **14**, 2469 (1974).
131. S. Wilk and M. Orlowski, *Anal. Biochem.* **69**, 100 (1975).
132. L. Fellows, G. S. King and K. Blau, unpublished observation.
133. J. B. Brooks, C. C. Alley and J. A. Liddle, *Anal. Chem.* **46**, 1930 (1974).
134. T. Walle, *J. Chromatogr.* **111**, 133 (1975).

135. M. Ervik, T. Walle and H. Ehrsson, *Acta Pharm. Suecica* **7**, 625 (1970).
136. M. Sugiura and K. Hirano, *J. Chromatogr.* **90**, 169 (1974).
137. J. Vessman, A. M. Moss, M. G. Horning and E. C. Horning, *Anal. Lett.* **2**, 81 (1969).
138. G. Zweig and J. Sherma (Eds), *Handbook of Chromatography*, Vol. 2, C. R. C. Press, Cleveland, 1972, p. 220.
139. F. Benington, S. T. Christian and R. D. Morin, *J. Chromatogr.* **106**, 435 (1975).
140. S. T. Christian, F. Benington, R. D. Morin and L. Corbett, *Biochem. Med.* **14**, 191 (1975).
141. S. F. Sisenwine, J. A. Knowles and H. W. Ruelins, *Anal. Lett.* **2**, 315 (1969).
142. P. H. Degen, J. R. Do Amaral and J. D. Barchas, *Anal. Biochem.* **45**, 634 (1972).
143. F. Karoum, J. C. Gillin, R. J. Wyatt and E. Costa, *Biomed. Mass Spectrom.* **2**, 183 (1975).
144. M. G. Horning, A. M. Moss, E. A. Boucher and E. C. Horning, *Anal. Lett.* **1**, 311 (1968).
145. H. Miyazaki, M. Ishibashi, C. Mori and N. Ikekawa, *Anal. Chem.* **45**, 1164 (1973).
146. M. Donike, *J. Chromatogr.* **78**, 273 (1973).
147. M. Donike, *J. Chromatogr.* **103**, 91 (1975).
148. G. Schwedt and H. H. Bussemas, *J. Chromatogr.* **106**, 440 (1975).
149. M. Kirschner and J. P. Taylor, *Anal. Chem.* **30**, 346 (1969).
150. A. C. Moffat, E. C. Horning, S. B. Matin and M. Rowland, *J. Chromatogr.* **66**, 255 (1972).
151. N. K. McCallum and R. J. Armstrong, *J. Chromatogr.* **78**, 303 (1973).
152. G. R. Wilkinson, *Anal. Lett.* **3**, 289 (1970).
153. Pierce Chemical Co., Rockford, Ill., General Catalog 1976–77, p. 254.
154. E. Änggård and A. Hankey, *Acta Chem. Scand.* **23**, 3110 (1969).
155. R. A. Landowne and S. R. Lipsky, *Anal. Chem.* **35**, 532 (1963).
156. A. C. Brownie, H. J. van der Molen, E. E. Nishizawa and K. B. Eik-Nes, *J. Clin. Endocr.* **24**, 1091 (1964).
157. R. J. Argauer, *Anal. Chem.* **40**, 122 (1968).
158. J. S. Noonan, P. W. Murdick and R. S. Ray, *J. Parmacol. Exp. Ther* **168**, 205 (1969).
159. P. Hartvig and J. Vessman, *Anal. Lett.* **7**, 223 (1974).
160. P. Hartvig and J. Vessman, *Acta Pharm. Suecica* **11**, 115 (1974).
161. P. Hartvig, K.-E. Karlsson, L. Johansson and C. Lindberg, *J. Chromagor.* **121**, 235 (1976).
162. C. F. Poole, *Lab. Pract.* **25**, 309 (1976).
163. C. F. Poole, *J. Chromatogr.* **118**, 280 (1976).
164. A. Zlatkis and J. E. Lovelock, *Clin. Chem.* **11**, 259 (1965).
165. L. M. Cummins, in I. I. Domsky and J. A. Perry (Eds), *Recent Advances in Gas Chromatography*, Marcel Dekker, New York, 1971, p. 313.
166. A. C. Moffat and E. C. Horning, *Anal. Lett.* **3**, 205 (1970).
167. C. F. Poole, *Chem. Ind.* (*London*) 479, (1976).
168. E. D. Pellizzari, *J. Chromatogr.* **98**, 323 (1974).
169. K. Nagawa, N. L. McNiven, E. Forchielli, A. Vermeulen and R. T. Dorfman, *Steriods* **7**, 329 (1966).
170. B. C. Pettit, P. G. Simmonds and A. Zlatkis, *J. Chromatogr. Sci.* **7**, 645 (1969).
171. J. J. Sullivan, *J. Chromatogr.* **87**, 9 (1973).
172. R. R. Dreisbach, *Physical Properties of Chemical Compounds*, American Chemical Society, Washington, 1955.
173. D. E. Durbin, W. E. Wentworth and A. Zlatkis, *J. Am. Chem. Soc.* **92**, 5131 (1970).
174. J. Vessman and S. Stromberg, *Acta Pharm. Suecica* **6**, 505 (1969).
175. M. J. Saxby, *Org. Mass Spectrom.* **2**, 33 (1969).
176. M. J. Saxby, *Org. Mass Spectrom.* **4**, 133 (1970).
177. M. J. Saxby, *Org. Mass Spectrom.* **2**, 835 (1969).
178. E. Gelpi, W. A. Koenig, J. Gilbert and J. Oró, *J. Chromatogr. Sci.* **7**, 604 (1969).
179. G. S. King, *Org. Mass Spectrom.* **9**, 1239 (1974).
180. M. J. Saxby, *Chem. Ind.* (*London*), 1316 (1968).
181. A. Zeman and I. P. G. Wirotama, *Z. Anal. Chem.* **247**, 158 (1969).
182. A. Prox and J. Schmid, *Org. Mass Spectrom.* **2**, 121 (1969).

183. A. Prox and J. Schmid, *Org. Mass Spectrom.* **2**, 105 (1969).
184. J. G. Lawless and M. S. Chadha, *Anal. Biochem.* **44**, 473 (1971).
185. W. A. Koenig, L. C. Smith, P. F. Crain and J. A. McCloskey, *Biochemistry* **10**, 3968 (1971).
186. W. Vogt, K. Jacob and M. Knedel, *J. Chromatogr. Sci.* **12**, 658 (1974).
187. K. Jacob, W. Vogt and M. Knedel, *Biomed. Mass Spectrom.* **3**, 64 (1976).
188. M. B. Shambhu and G. A. Digenis, *Chem. Commun.* 619 (1974).
189. M. H. Karger and Y. Mazur, *J. Org. Chem.* **36**, 540 (1971).
190. S. Siggia, J. C. Hanna and R. Culmo, *Anal. Chem.* **33**, 900 (1961).
191. W. Selig, *Microchem. J.* **21**, 92 (1976).
192. A. Paquet, *Can. J. Chem.* **54**, 733 (1976).

CHAPTER 4

Recent Advances in the Silylation of Organic Compounds for Gas Chromatography

Colin F. Poole

Department of Organic Chemistry, University of Ghent, Krijgslaan 271, (S.4), B-9000 Ghent, Belgium

1 INTRODUCTION

Early developments in the formation of silyl derivatives of organic compounds have been summarized in the book by Pierce.[1] It is not intended to cover the same ground here, but to concentrate on the more recent developments and trends in this rapidly expanding field. The term silylation will be treated in its broadest sense to include all reactions of the type shown below.

$$\mathrm{R{-}\underset{R}{\overset{R}{Si}}{-}X + HY \rightarrow R{-}\underset{R}{\overset{R}{Si}}{-}Y + HX} \qquad \text{R = alkyl or halocarbon} \tag{1}$$

A specific example is the formation of trimethylsilyl (TMS) ethers in which R = methyl. For convenience, the different silylating reagents will be referred to by their abbreviated form. These are summarized in Table 1 along with their chemical structure. References are given to a convenient synthesis of the reagents for those workers wishing to prepare their own for reasons of economy.

Of all the derivatives described for improving the chromatographic properties of polar molecules, none are as versatile or as popular as the trimethylsilyl ethers and esters. The general technique is applicable to all compounds containing active hydrogen functions (2). Other functional groups such as ketones can be converted to enol–TMS ethers, but in general, these are better handled by alternative techniques.

On silylation, the character of a molecule is changed from polar and active to non-polar and inert. This is generally coupled with an increase in volatility and thermal stability which are the properties most desired for gas chromatography. Interactions between the column support and the derivatized form of the molecule are reduced to a minimum and peak shape is usually Gaussian.

TABLE 1

Reagents for the formation of silyl derivatives

Reagent	Structure	Abbreviation	Approximate silyl donor power	Reference to synthesis
Trimethylsilyl ethers				
Trimethylchlorosilane	$(CH_3)_3SiCl$	TMCS	**	
Hexamethyldisilazane	$(CH_3)_3SiNHSi(CH_3)_3$	HMDS	*	3
N-methyl-*N*-(trimethylsily)acetamide	$CH_3-C(=O)-N(CH_3)-Si(CH_3)_3$	MSTA	***	4
N-methyl-*N*-(trimethylsilyl)trifluoroacetamide	$CF_3-C(=O)-N(CH_3)-Si(CH_3)_3$	MSTFA	****	5
N,*O*-bis-(trimethylsilyl)acetamide	$CH_3-C(O-Si(CH_3)_3)=N-Si(CH_3)_3$	BSA	*****	6
N,*O*-bis-(trimethylsilyl)-trifluoroacetamide	$CF_3-C(O-Si(CH_3)_3)=N-Si(CH_3)_3$	BSTFA	*****	7
N-trimethylsilylimidazole	$(CH_3)_3Si-N$ (ring: $C=N$, $C=C$)	TMSIM	*****	8
N-trimethylsilyldiethylamine	$(CH_3)_3Si-N(C_2H_5)_2$	TMSDEA	***	9
N-trimethylsilyldimethylamine	$(CH_3)_3Si-N(CH_3)_2$	TMSDMA	***	
N,*N*,*N*′,*N*′-tetrakis-(trimethylsilyl)-1,2-diaminoethane	$((CH_3)_3Si)_2N-CH_2CH_2-N(Si(CH_3)_3)_2$	TTDE	*	10

TABLE 1—continued

Reagent	Structure	Abbreviation	Approximate silyl donor power	Reference to synthesis
Trimethylsilyl ethers—continued				
N-trimethylsilylpiperidine	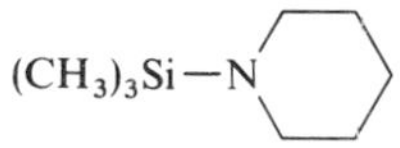	TMSPI	*****	11
N-trimethylsilylpyrrolidine	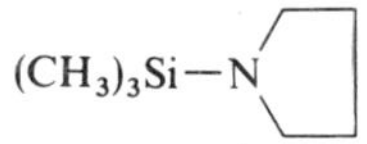	TMSPY	*****	11
N-trimethylsilylmorpholine	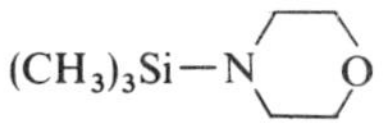	TMSM	***	11
N-trimethylsilylacetanilide	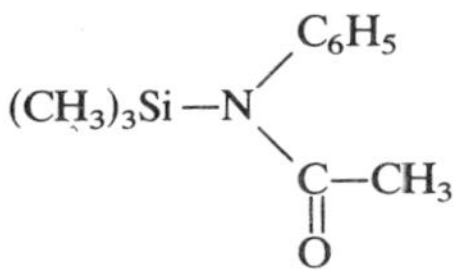	TMSA	****	12
N-trimethylsilyl-*p*-ethoxyacetanilide		TMSEA	*****	12
Trialkylsilyl ethers				
R = ethyl, *n*-propyl, *n*-butyl or *n*-hexyl	R_3SiCl	—	—	13
t-Butyldimethylchlorosilane	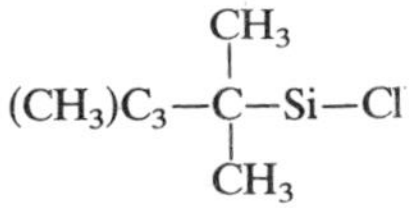	—	—	14
Dimethylsilyl ethers (*DMS*)			**	
Dimethylchlorosilane	$H(CH_3)_2SiCl$		*	
1,1,3,3,-Tetramethyldisilazane	$H(CH_3)_2SiNHSi(CH_3)_2H$			

TABLE 1—continued

Reagent	Structure	Abbreviation	Approximate silyl donor power	Reference to synthesis
Dimethylsilyl ethers (DMS)—continued				
N,O-bis-(dimethylsilyl)-acetamide	$CH_3-C(O-Si(CH_3)_2H)=N-Si(CH_3)_2H$	BDSA	*****	
Halocarbondimethylsilyl ethers				
Chloromethyldimethylchlorosilane	$ClCH_2Si(CH_3)_2Cl$	CMDMCS		
1,3-Bis(chloromethyl)-1,1,3,3-tetramethyldisilazane[a]	$ClCH_2-Si(CH_3)_2-NH-Si(CH_3)_2-CH_2Cl$	CMTMDS		
Pentafluorophenyldimethylchlorosilane	$C_6F_5Si(CH_3)_2Cl$	flophemesyl chloride		15
Pentafluorophenyldimethylsilyldiethylamine	$C_6F_5Si(CH_3)_2N(C_2H_5)_2$	flophemesyldi-ethylamine		15
Pentafluorophenyldimethylsilylamine	$C_6F_5Si(CH_3)_2NH_2$	flophemesylamine		16

[a] The bromomethyl reagents are also available. The iodomethylsilyl ethers are prepared by halide ion-exchange (see p. 194).

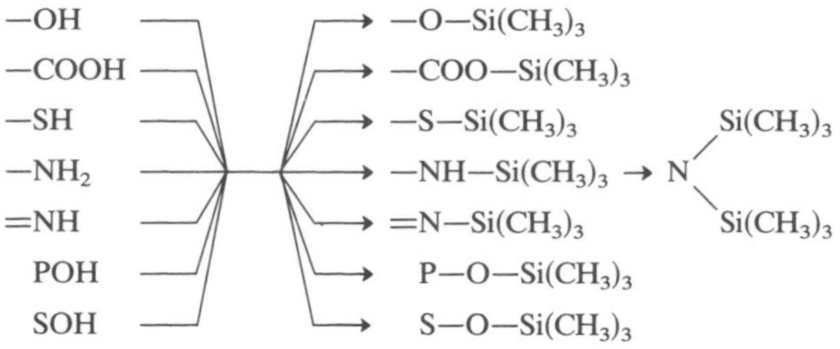

(2)

Mass spectrometry has also gained from the introduction of silyl ethers. Again the main requirement was to convert high molecular weight polyfunctional substances into volatile derivatives that were thermally stable and sufficiently volatile to be vaporized without decomposition at moderate temperatures.[2] The coupling of the mass spectrometer to the gas chromatograph gave the analyst, for the first time, a method of separating and identifying the components of a complex organic mixture in a simple yet highly diagnostic manner. The present role of the gas chromatograph–mass spectrometer (GC–MS) combination as a powerful tool for biomedical research owes much to the ability of silyl ethers to stabilize thermally labile substances to both gas chromatography and mass spectrometry. The commercial availability of perdeuteriated trimethylsilyl reagents has been an important development in aiding the interpretation of electron-impact induced mass spectra of trimethylsilyl ethers.

Other important developments have been the introduction of silyl ethers containing higher alkyl homologues attached to silicon. These reagents are less volatile than the trimethylsilyl ethers and are useful in the separation of complex mixtures containing components with different numbers of functional groups which are incompletely resolved as their trimethylsilyl ethers. Replacement of one of the methyl groups in the trimethylsilyl ethers with a halocarbon group enables derivatives to be prepared which can be selectively detected at trace levels with an electron-capture detector. The electron-capture detector is one of the most sensitive gas chromatography detectors available at present but is limited in application to those molecules which strongly capture thermal electrons. The formation of halocarbondimethylsilyl ethers provides a convenient technique for the introduction of an electron capturing group into otherwise unresponsive compounds for their determinations down to the picogram (10^{-12} g) level.

2 THE FORMATION OF SILYL ETHERS

An important advantage of the silyl ethers is their ease of formation with a wide variety of functional groups. Typically, the compound and reagent, perhaps in the presence of solvent, are mixed together and the rate of the ensuing reaction depends on a series of factors which will be discussed below. The reactions are often instantaneous and are conveniently carried out in screw-capped vials containing a Teflon®-faced septum through which additions and subtractions can be made by syringe without exposure to the atmosphere. The success, simplicity and wide applicability to all types of functional groups, of the solution technique has not encouraged the search for alternative reaction methods. The use of a pre-column reactor,[17] pre-column trap[18] as well as on-column derivatization[19–23] have been evaluated. On-column silylation by direct injection of a solution of the sample and the silyl reagent either sequentially or as a 'sandwich' injection is a useful method for the formation of silyl ethers of compounds which are good silyl acceptors (e.g. unhindered alcohols, phenols, carboxylic acids) with powerful silyl donors (e.g. BSA, BSTFA, TMSIM). This technique lessens the possibility of hydrolysis of unstable TMS ethers due to accidental exposure to moisture.

The silylation mechanism and therefore the conditions which favour the reaction in solution are incompletely understood. The available information has been summarized by Pierce.[1] However, certain practical observations can be made which aid the selection of the best conditions for a particular sample. Individual functional groups show different orders of reactivity towards silyl donors. For any particular silyl reagent, the ease of reaction follows the order:

alcohols > phenols > carboxylic acids > amines > amides.

Alcohols react in the order:

primary > secondary > tertiary

and amines:

primary > secondary.

With primary amines the introduction of one silyl group hinders access of a second which can be introduced only with difficulty. Steric factors can have an overriding influence on the rate of reaction. The trimethylsilyl group has similar geometry to the *t*-butyl group but is larger, so that unimpeded access to the functional group is important.

The silyl donor ability of the trimethylsilyl reagents in Table 1 differs between individual reagents in a way which cannot be adequately explained at present. They are all characterized by having good leaving groups as the active part of the reagent. As a working hypothesis, the silylation reaction is considered to involve the formation of a transition state as indicated in (3) below.[1]

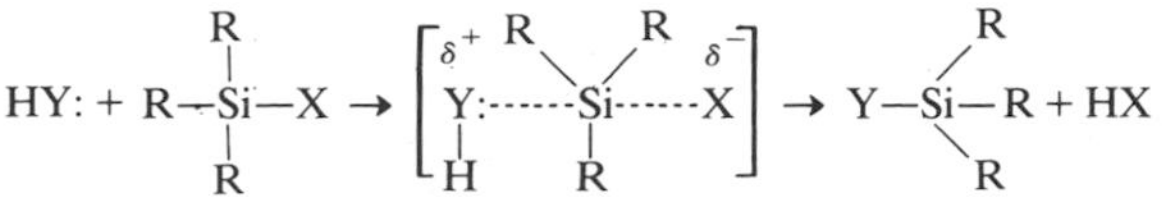

(3)

The properties most desired of X, the leaving group, are low basicity, the ability to stabilize a negative charge in the transition state during silyl ether formation and little or no (p → d) π back-bonding between X and silicon. The ability to stabilize a negative charge in the transition state is thought to be very important and explains why TMSIM and BSA are potent silylating reagents as the leaving group (imidazole, acetamide) can stabilize a partial negative charge through resonance. Poor silylating reagents such as TMCS or HMDS cannot stabilize a developing negative charge in this way. Back-bonding between the leaving group and silicon leads to a strengthening of the Si—X bond and obviously does not favour the forward reaction. The basicity of the leaving group should be less than that of Y as the formation of the transition state is a reversible process which is favoured by X as a weak base. For a new series of *N*-trimethylsilylheterocyclic reagents, the silyl donor ability was found to mirror the basicity of the leaving group.[11] Leaving groups of low basicity such as piperidine made more powerful silyl donor reagents than the more basic morpholine. The trimethylsilyl reagents can be placed in the following approximate order of silyl donor ability:

TMSIM > BSTFA > BSA > MSTFA > TMSDMA
> TMSDEA > MSTA > TMCS (with base) > HMDS.

There are exceptions to this order and in some cases the difference in silyl donor ability between neighbouring entries is small (e.g. BSA, BSTFA). A notable anomally is TMSIM which reacts easily with hydroxyl functional groups but not at all with aliphatic primary amines.[24] The new reagents TMSPI, TMSPY, TMSA and TMSEA have been shown to have similar reactivity to TMSIM and BSA.[11,12]

The rate of reaction leading to silyl ether formation can be increased by the addition of an acid or base catalyst. Many substances have been tried for this purpose, but none is as successful as TMCS.[1,25] It has become conventional to consider TMCS as an 'acid catalyst' although it has been clearly demonstrated that in reagent mixtures it performs as a silyl donor in its own right.[1,26] TMCS is often used as a catalyst with HMDS, but a by-product of this reaction is ammonium chloride which appears as a fine precipitate and may be undesirable for some applications. The strongest silylating reagent mixture for all purposes is a mixture of BSA : TMSIM : TMCS (1 : 1 : 1).

The correct choice of solvent can influence the rate of reaction. The chief requirement of any solvent is that it should dissolve both reagent and sample without reacting with either. When a choice of solvent is available, polar solvents are preferred as they tend to promote a faster reaction. The silylating reagents themselves have good solubilizing properties for many compounds and can be used without additional solvent. The most frequently used solvents are pyridine, dimethyl sulfoxide, dimethylformamide, acetonitrile, dioxan, tetrahydrofuran and chloroform. When TMCS is used as a reagent, then the reaction is much faster in the presence of a base such as pyridine or an alkylamine. Dimethyl sulfoxide is immiscible with silyl reagents and so when this solvent is used, the mixture must be constantly agitated or a solvent of intermediate polarity such as dioxane added to promote homogeneity.[27] Dimethylformamide shows a tendency to condense with primary amines,[28] reacts partially with HMDS to produce an addition product,[29] and is unsuitable for the storage of TMS ether derivatives.[25] Pyridine often tails badly on the chromatogram and can interfere with the analysis of volatile substances.[30–32] For these derivatives a different solvent should be used for the injection. The pyridine can be removed with a stream of nitrogen or by the addition of hexane followed by washing with water or dilute acid.[32,33]

An increase in temperature of the reaction mixture increases the rate of reaction and enables some difficult-to-derivatize functional groups to be quantitatively converted to the silyl ethers in a convenient time. Temperatures as high as 200 °C have been recorded in the literature, but normal reaction temperatures are room temperature, 60 °C, 100 °C and 150 °C. Plastic-capped vials can usually be used up to 150 °C before the plastic softens and the septum seal leaks. Higher temperatures require the use of sealed glass tubes.

The optimum conditions for a particular reaction are achieved by appropriate changes in the parameters discussed above. It is obvious that several sets of conditions will give the same result and the governing feature is the need for a quantitative reaction in a convenient time period. For example, to increase the rate of reaction, one could choose to increase the temperature of the reaction or add catalyst to the medium or both. Many silylation reactions are instantaneous with a powerful silyl donor.

2.1 Analysis of silyl ethers by gas chromatography

The silylation reaction produces a silyl ether in quantitative yield and this is the only product which the analyst wishes to see on the chromatogram. The conditions of the reaction are such that interference from other components can influence the results obtained. Usually the reagents and solvent are used in a large molar excess and impurities in either of these can eclipse the peak due to the derivative itself. It is difficult to avoid contact with moisture, so the disiloxane produced may appear on the chromatogram as well as the protonated leaving group which is a product of the formation of the silyl derivative proper. For silyl reagents with two donor centres such as BSA, the monosub-

$$CH_3{-}\underset{CH_3}{\overset{CH_3}{\underset{|}{\overset{|}{Si}}}}{-}X + H{-}R \rightarrow CH_3{-}\underset{CH_3}{\overset{CH_3}{\underset{|}{\overset{|}{Si}}}}{-}R + HX \qquad (4)$$

Silyl reagent — Reactive compound — Silyl product — Protonated leaving group

$$2\,CH_3{-}\underset{CH_3}{\overset{CH_3}{\underset{|}{\overset{|}{Si}}}}{-}X + H_2O \rightarrow CH_3{-}\underset{CH_3}{\overset{CH_3}{\underset{|}{\overset{|}{Si}}}}{-}O{-}\underset{CH_3}{\overset{CH_3}{\underset{|}{\overset{|}{Si}}}}{-}CH_3 + 2HX \qquad (5)$$

Silyl reagent — Hexamethyldisiloxane

stituted silyl donor (*N*-trimethylsilylacetamide) is a further product. A recent analysis of commercially available samples of BSA indicated that the reagent contained an average 92 mol% of BSA with the principal impurities being hexamethyldisiloxane and *N*-trimethylsilylacetamide.[34] Problems such as these may become important when analysing low molecular weight compounds, or higher molecular weight substances at high sensitivity with the flame ionization detector by direct injection of the reaction mixture. In the former case the reagents and by-products may have retention times similar to the derivative, making identification difficult. Fluorinated silyl reagents such as BSTFA are often used with advantage in this situation as the fluorinated reagents and by-products are more volatile than their hydrocarbon analogues and have greatly reduced retention times. As a bonus, the fluorinated reagents are useful in keeping the flame ionization detector clean, as fewer deposits of silicone dioxide are formed due to removal of silicon as the volatile silicon tetrafluoride. With less volatile derivatives at high sensitivity, tailing of the solvent front into the region of the chromatogram of interest may be a problem. This can often be avoided by removal of excess reagents *in vacuo* or with a stream of nitrogen and injection of the sample in a different solvent.

The formation of a silyl derivative changes the character of the molecule from polar to non-polar, and consequently those stationary phases suitable for the separation of hydrocarbon mixtures are more appropriate for silyl ethers. By far the most popular liquid phases are the silicone oils, as these can provide a moderate range of selectivities with high-temperature thermal stability. The use of OV-1 or OV-101 (polymethylsilicone gum or oil similar to SE-30) allows the separation of the silyl ethers by boiling-point which is the preferred method of separation. The substitution of phenyl or cyanide groups for methyl into the polymer structure increases the polarity. The silicone oils OV-17 or OV-225 (similar to XE-60 or XE-1150 but superior to both) are often used as a more polar second column and a foil to OV-101 for the identification of derivatives by retention time parameters. OV-210 (Similar to QF-1) shows useful selectivity for free ketone groups and can be used to advantage for the separation of silyl ethers containing different numbers of ketone groups. More polar columns of polyester phases can also be used but provide little extra selectivity unless the derivative contains unreacted functional groups. Fully silylated derivatives elute faster on polyester phases than on the silicone oils but tend to bunch together giving poorer separations. Very polar liquid phases containing active proton groups (e.g. Carbowax, etc.) are unsuitable as they react with the silyl reagents.

2.2 Problems with the silylation of compounds containing ketone groups

2.2.1 ENOL FORMATION

Secondary products can be formed under the normal conditions for silyl ether formation with compounds containing unprotected ketone groups. These products arise from the formation of enol–TMS ethers (6):

$$RCH_2-\underset{\underset{O}{\|}}{C}-R \rightleftharpoons R-CH=\underset{\underset{OH}{|}}{C}-R \dashrightarrow R-CH=\underset{\underset{OTMS}{|}}{C}-R \qquad (6)$$

Keto–enol equilibrium enol–TMS ether

The amount of enol–TMS ether increases with reaction time suggesting a forward reaction as shown above rather than a shift in the position of the keto–enol equilibrium. The presence of acid catalyst such as TMCS results in the formation of high yields of the enol–TMS ether depending on the reactivity of the ketone group and the silyl donor power of reagents used. Weak silylating acid-catalysed reagent combinations such as HMDS : TMCS (10 : 11) produce a moderate yield and strong silyl reagents with acid catalyst such as BSA : TMSIM : TMCS (3 : 2 : 2) produce amounts of the derivative from moderate to a quantitative yield.[35]

Trimethylsilylbromide is an even more effective catalyst than TMCS in promoting the formation of enol–TMS ethers.[36] Most silyl reagents in the absence of catalyst give a negligible yield of the enol–TMS ether. MSTFA is an exception in being able to yield considerable amounts of the enol–TMS ether without the addition of acid catalyst.[18,37] When testosterone was treated under vigorous conditions of heating for 72 h at 60 °C with a mixture of BSA : TMCS (10 : 1) in addition to the expected isomeric enol–TMS ethers a minimum of four other products was indicated by gas chromatography (Fig. 1). These secondary products identified by mass spectrometry were formed by the addition of a trimethylsiloxy group (presumably as a free radical) to the enol–ether double bond with subsequent elimination of the enol trimethylsilyl group.[35]

In general, enol–TMS ethers are very hydrolytically and thermally unstable as well as difficult to form in quantitative yield. One of the few successful applications of this type of derivative is found in the analysis of steroids containing a dihydroxyacetone side chain (corticosteroids).[38,39] The formation of silyl ethers by BSTFA in the presence of solid potassium acetate at room temperature overnight gives a quantitative yield of the TMS–enol–TMS derivative (7).

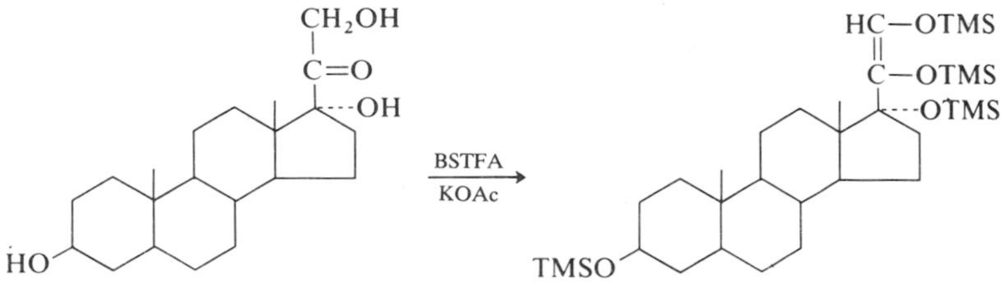

(7)

Substitution at C-21 as given by the 21-OH group in hormonal steroids appears to be necessary for quantitative formation of the enol–TMS form of the 20-ketone. Under the conditions employed 4-ene-3-one groups give multiple products but these are unstable and can be decomposed by removing the solvent with nitrogen and redissolving in another solvent for analysis. In

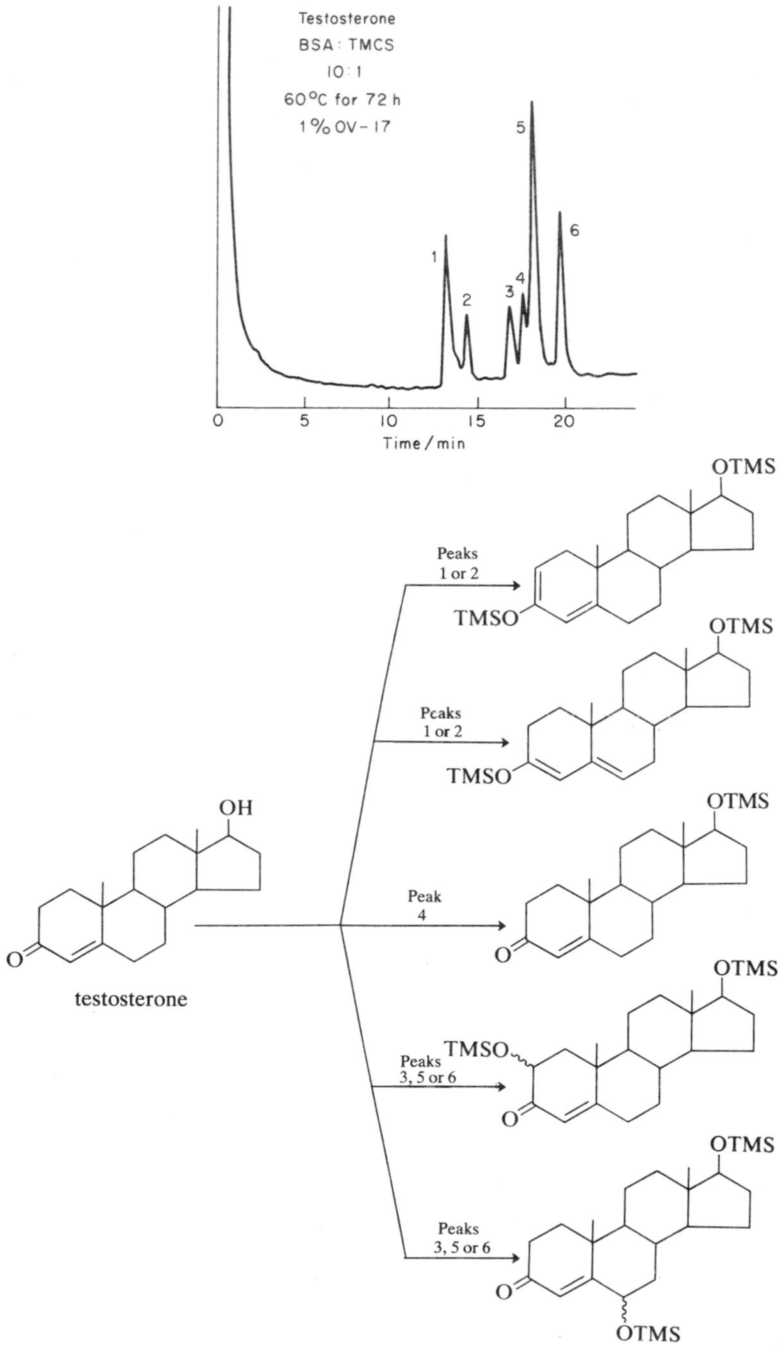

Fig. 1 Complex chromatogram obtained in testosterone with BSA–TMCS.

another study, it was shown that the 7-ene-6-one group in ecdysones reacts with TMSIM in the presence of potassium acetate to produce predominantly the expected enol ether, depending on the substituents present on neighbouring carbon atoms.[40]

Compounds containing unsaturated lactone rings can form enol–TMS ethers and in the presence of TMCS with a powerful silyl donor the reaction can be made quantitative.[41–43]

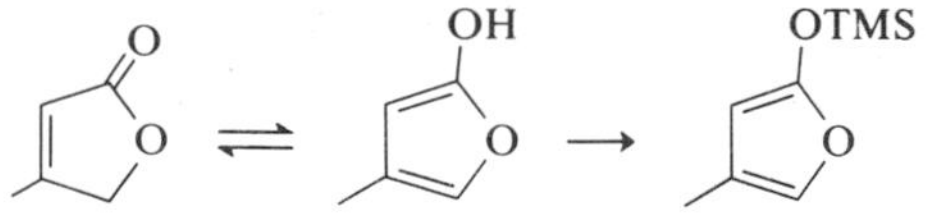

(8)

Perhaps the most studied example of multiple products obtained on silylation is the analysis of sugars in solution. Reducing sugars at equilibrium in solution can exist in more than one isomeric form known as anomers.[44,45] The ring forms of sugars are hemiacetals formed by reaction of the aldehyde with the hydroxyl group in the same molecule. Each of the ring forms for D-glucose, for example (Fig. 2), generates two isomers, diastereomers, differing only in the

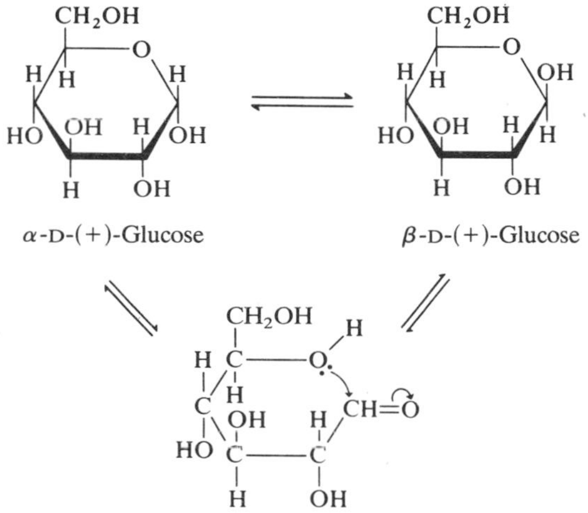

Fig. 2 An equilibrium mixture of α- and β-glucose anomers in solution, formed via an open-chain intermediate.

configuration of the hemiacetal group. These isomers are mutually interconvertible through an open-chain aldehyde intermediate, giving rise to the phenomenon known as mutarotation. The two anomers have different physical properties and their TMS ethers are usually easily separable by GLC.[46] By this means if pure β-(+)-D-glucose is dissolved in a solvent and then silylated, there is the possibility of generating two peaks on the chromatogram, as in solution two anomeric forms can exist. To silylate pure reducing sugars the conditions employed should be both mild and rapid, but for sugars isolated from natural sources they can be expected to be a mixture reflecting the average composition of the anomers at equilibrium in the extract.[46–49]

Heating galactose in pyridine prior to silyl ether formation produced four anomeric derivatives which were separated by GLC (Fig. 3).[50] Two were due to

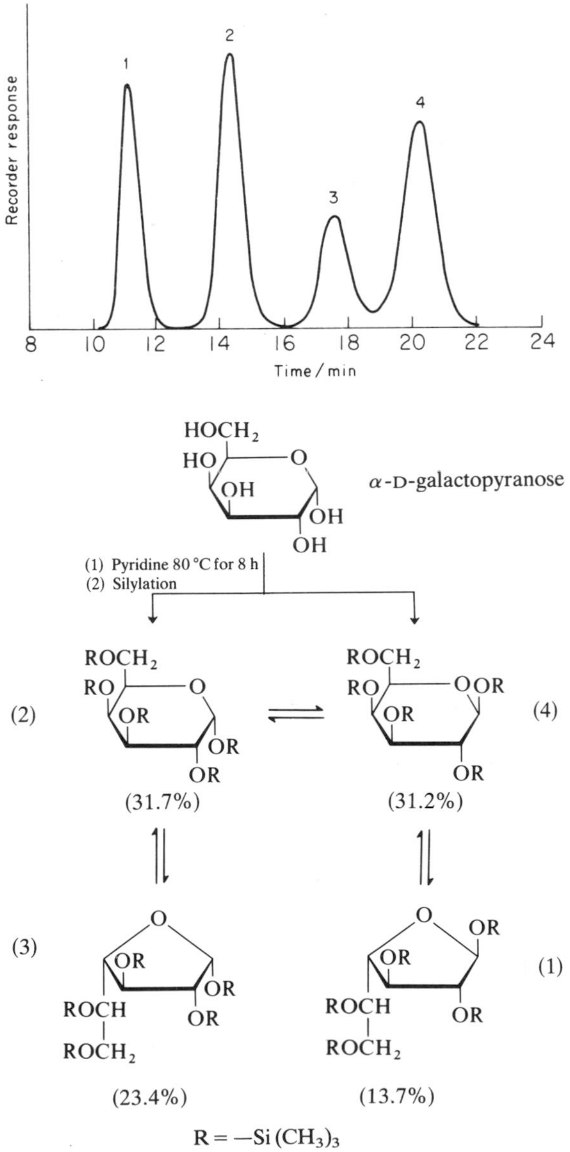

Fig. 3 An anomeric mixture formed by heating galactose in pyridine. (From T. E. Acree, R. S. Shallenberger and L. R. Mattick, *Carbohydr. Res.* **6**, 498 (1968)).

the expected $\alpha \rightleftharpoons \beta$ equilibrium and the other two due to pyranose$\rightleftharpoons$furanose conversion occurring through an enol intermediate.[45,50] The acyclic aldonic acids (general formula $HOCH_2(CHOH)_nCOOH$) and aldoric acids (general formula $HOOC(CHOH)_nCOOH$) represent a further complex problem, as in solution the acyclic acids with hydroxyl groups at C-4 and C-5 exist also as the 1,4- and 1,5-lactones (9).[51] Uronic acids, which are cyclic sugars containing both carboxylic acid and aldehyde groups, can exist in as many as six different forms at equilibrium in solution due to a combination of lactone and anomer formation.

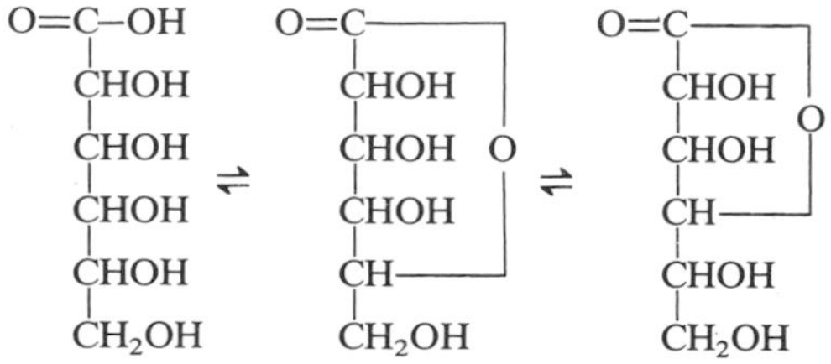

(9)

For the successful formation of silyl ethers of compounds containing ketone groups that can form enol–TMS ethers there are four alternatives:

1. Quantitatively form the enol–TMS ether. This is the least favoured alternative as the enol–TMS ethers are often difficult to form quantitatively, and once formed are generally unstable.
2. Use a strong silyl donor such as TMSIM or BSA which can react quantitatively and rapidly with the target group without the addition of acid catalyst. TMSIM, whose by-product (imidazole) is weakly amphoteric, does not tend to promote enol–TMS ether formation.[40]
3. Protect the ketone group from enolization by the formation of methoxime derivatives.[52] The methoxime derivatives are easy to form, stable and well explored for this purpose (see Chapter 6). On selective liquid phases, the methoximes may produce two peaks due to the separation of *syn*- and *anti*-geometric isomers.[53,54] In this case single peaks can often be obtained by analysis on non-selective silicone oil phases.
4. Reduction of the ketone group to an alcohol which can be silylated in the usual way for hydroxyl groups (see section on carbohydrates, p. 169).

The method selected will depend on the class of compound being analysed and further discussion is deferred to the section dealing with the systematic formation of silyl ethers.

2.3 Addition and cleavage reactions involving trimethylsilyl reagents

The formation of trimethylsilyl ethers is remarkably free from unwanted secondary reaction products caused by further reaction of the TMS reagent with the silyl derivative. The formation of enol–TMS ethers and the possibility of the addition of a trimethylsiloxy group to the enol double bond with unprotected ketone groups was discussed in the previous section. It was also shown that appropriate methods are available to eliminate enol–ether formation.

Compounds containing highly activated electron-deficient sites such as mesoxalic acid can produce secondary addition products with BSTFA and BSA:[55]

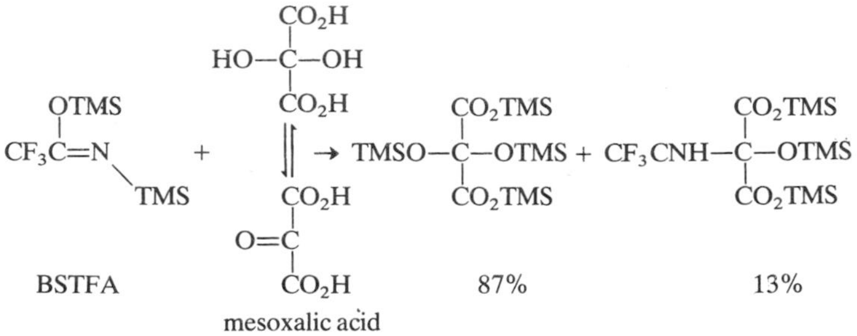

(10)

The addition of TMCS to epoxides with formation of the TMS ether of the chlorohydrin aduct is a useful reaction for the characterization of epoxides. Heating the epoxide in a mixture of TMCS–pyridine (1 : 2) at 60 °C for 30 min gives a quantitative yield of the derivative.[56,57]

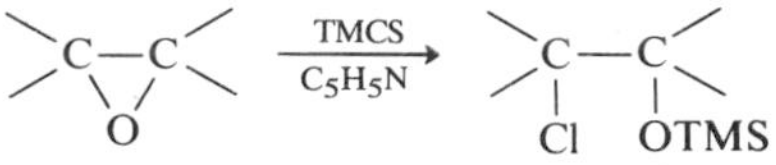

(11)

Any additional functional groups in the molecule are silylated by adding BSA to the reaction mixture after the formation of the trimethylsilyl ether chlorohydrin aduct. An investigation of the reaction of anthocyanins (colour pigments, such as delphinidin, in plants and fruit) with HMDS–TMCS by GC–MS indicated that as well as the silylation of all of the phenol groups in the molecule, the introduction of a $(CH_3)_3SiNH$ group had also occurred (12). This product was stable at normal temperatures, but on GLC it eliminated a molecule of trimethylsilanol to form a quinoline-like derivative.[58] Heating an anthocyanin at 80 °C with a mixture of HMDS–TMCS (2 : 1) until dissolved produces a single peak on GLC. Reaction of anthocyanins with BSA or TMSIM gave multiple products under a variety of conditions. The attempted silylation of norethynodrel with TMSIM gave two products on GLC, the expected TMS ether and a second product due to aromatization of ring A. The aromatic product was obtained quantitatively by heating the reaction mixture at 100 °C (13).[59]

The TMS reagents are generally compatible with other reagents for the formation of mixed derivatives of polyfunctional molecules. The TMS ethers exhibit moderate hydrolytic stability and are cleaved under the conditions normally used for the formation of alkyl esters, acetates, oximes, acetamides,

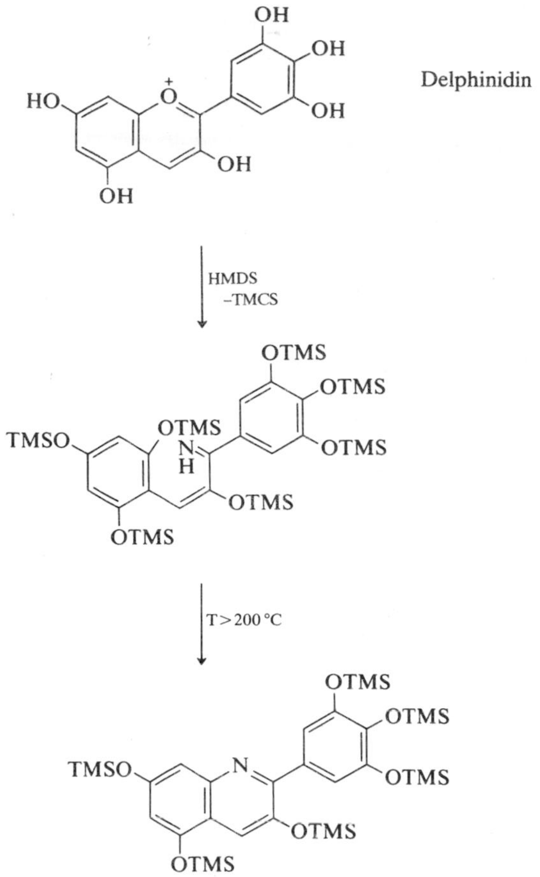

(12)

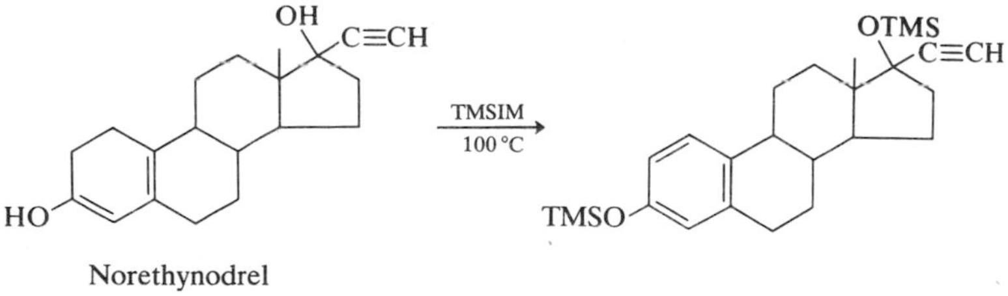

(13)

cyclic boronic esters, acetonides and most reactions employing acid catalysis. As a consequence for mixed derivative formation, the TMS ethers are formed as the last stage of a reaction sequence. The TMS reagents do not normally cleave any of the common protecting groups used for gas chromatography. In our laboratory, it has been found that TMSIM will slowly cleave acetonide and boronic ester derivatives of diols at room temperature. BSA and BSTFA with or without TMCS catalyst did not produce a similar reaction, even upon heating. We also find that TMSIM cleaves flophemesyl ethers (see later) of steroids. The principal product is the TMS ether with a lesser amount of an unidentified secondary product.

2.4 The effect of water on silyl ether formation

The preparation of silyl ethers is usually carried out in anhydrous conditions. However, the addition of trace quantities of water to silyl reagents has a pronounced catalytic action with amines and steroids.[24,60,61] For example, with BSA–TMCS–acetonitrile (2 : 1 : 1) norepinephrine required heating at 60 °C for 5 h for complete reaction (50 h when pyridine was used as solvent). With the addition of water (1% of the volume of BSA) complete reaction was obtained in 2 h in acetonitrile (5 h in pyridine) at 60 °C.[24] Samples of TMSIM prepared by the author under anhydrous conditions have been found to be slow to react with hindered steroid hydroxyl groups. Brief exposure to the atmosphere restored the expected reactivity to the reagent. The reason for this phenomena is not known at present, but may partially explain the difficulty some workers have experienced in following the reaction conditions of others. In an investigation of some of the artifacts which affect the formation of silyl ethers, Lau found that the addition of small quantities of water did not generate secondary products.[25] Breakdown or interference with complete conversion to the silyl ethers was observed only when the reaction was heated after the addition of water, the combined effect presumably providing the impetus for the rehydrolysis of the TMS ethers.

It is often difficult to obtain samples from natural sources which are completely free of water. A typical example is natural sugars, which after extraction and evaporation are obtained as syrups often containing more than 50% by weight of water. Further drying by lyophilization can result in the loss of volatile material, and solvent extraction may not be appropriate if no suitable solvent exists which is immiscible with water and has a high partition coefficient for the material of interest. In this case it would be desirable to form the silyl ethers in the aqueous environment. Brobst and Lott have silylated sugar syrups with excess HMDS, pyridine and trifluoroacetic acid as catalyst.[62,63] For success, strict adherence to the experimental conditions is necessary and further experimental difficulties are introduced by the large amount of heat and ammonia generated by the reaction. Weiss and Tambawala have succeeded in forming TMS ethers of aqueous solutions of carbohydrates and polyols (25% w/w) using an excess of pyridine–HMDS–TMCS (4 : 2 : 1) and heating for up to 3 h at 35–40 °C.[64] TMSIM has been shown to be a particularly effective reagent for the analysis of polyols in aqueous solution when used in excess. Thus a 10% solution of raffinose was completely derivatized in 15 min at 60 °C, the time required for complete dissolution.[65] TMSIM is the reagent of choice for the formation of TMS ethers in aqueous solution.

3 THE FORMATION OF TMS ETHERS OF SOME BIOLOGICALLY IMPORTANT COMPOUNDS

Detailed information on the formation of trimethylsilyl ethers of a wide variety of compounds has been given by Pierce[1] and Drozd.[66] More specialist sources include Fishbein's summaries on the chromatography of environmental hazards,[67] a review of the chromatography of amino acids by Husek and Macek,[68] a review of chemical derivatization of pesticides by Cochrane[69] and a review of the chemical derivatization of drugs and pharmaceutical products by Ahuja.[70] Since the comprehensive survey by Pierce, there has been a trend away from the use of poor silyl donors such as HMDS, TMSDEA etc. in favour of the more powerful reagents BSA, BSTFA and TMSIM for nearly all purposes. The products and the extent of the silylation reaction are now better understood due to the widespread application of GC–MS for monitoring the reaction products.

3.1 Carbohydrates and polyols

This class of compounds is the easiest of all to silylate and most silyl donors react to completion. For rapid reaction the powerful silyl donors BSA, BSTFA and TMSIM are preferred and may be essential for hindered hydroxyl groups. For ketone-containing polyols TMSIM is preferred to BSA as it tends to promote fewer products due to enolization, particularly if samples are to be stored for a period of time. TMSIM has important advantages for the reaction of aqueous syrups of carbohydrates. Multiple products obtained from the silylation of natural sugars due to anomer formation and pyranose $\rightleftharpoons$ furanose interconversion have been discussed in a previous section.

Direct reaction of polyols with BSA, BSTFA or TMSIM with or without a TMCS catalyst leads to a quantitative reaction of all hydroxyl groups in a short time at room temperature or at 60 °C.[71–74] Sweeley *et al.* introduced the reagent combination of pyridine–HMDS–TMCS (10 : 2 : 1) for the silylation of carbohydrates and other hydroxyl compounds and this continues to be widely used, but probably offers no advantages over the more potent silyl donors described above.[29,46,75–81]

Several methods have been developed for the formation of single peaks on GLC from equilibrium mixtures of sugar anomers. Reduction of the carbonyl group of the cyclic sugar with sodium borohydride gives the appropriate acyclic sugar alcohol which after silyl ether formation gives a single peak on GLC (14).[46,82] The disadvantages of this method are that quantitative results are difficult to obtain as sugar alcohols may already be present in the extract and the very similar chromatographic properties of the isomeric sugar alcohol silyl ethers make them difficult to separate. Neutral aldohexoses can be oxidized with hypohalite to the corresponding aldonic acids, which on subsequent treatment with acid form their 1,4-lactones. Formation of the silyl ether

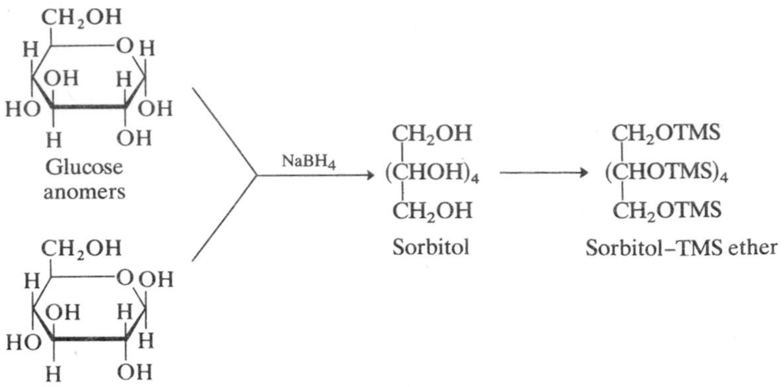

(14)

enables each aldose to be chromatographed as a single peak.[83–86] The exception is D-glucose which gives two peaks due to the formation of a mixture of the 1,4 and 1,5-lactones (15). The preferred method is the formation of the acyclic sugar methoxime followed by silyl ether formation in a two-step

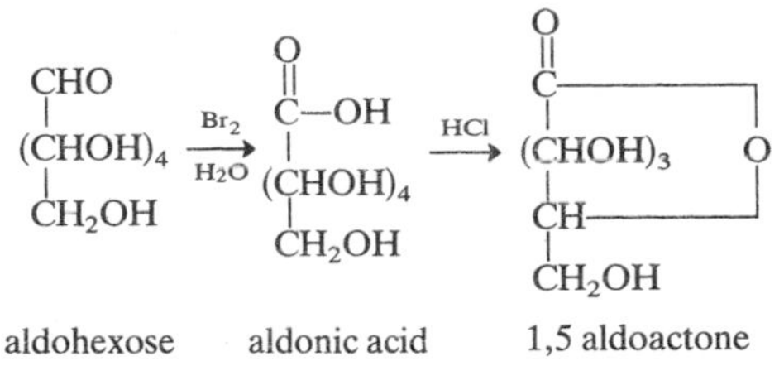

(15)

reaction.[23,51,87–91] This method is applicable to both ketoses and aldoses. Double peaks occur occasionally due to the separation of the *syn*- and *anti*-methoxime isomers on GLC.

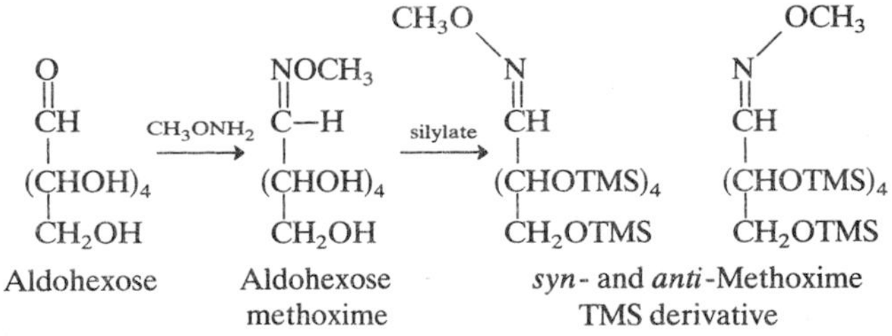

(16)

Carbohydrates found in nature are often combined with or contain carboxylic acid, phosphate ester, amine or nucleic acid bases as additional polar substituents. All these sugars and modified sugars can be analysed as their silyl derivatives if the precautions mentioned above are taken to avoid the formation of multiple products. A further constraint is that these additional polar functional groups are all more difficult to silylate than the sugar hydroxyl groups. The choice of silyl donor is thus very important. The popular HMDS–TMCS reagent combination lacks sufficient silyl donor power to form silyl derivatives with phosphate- or amine-containing sugars.[80,92–94] In the case of amine sugars, BSA is added to the above reagent mixture and silylation of the amine group is complete within 30 min at room temperature.[95] Wells *et al.* have analysed sugar phosphates by first forming the methyl ester of the phosphate group with diazomethane followed by conversion of the hydroxyl groups to silyl ethers with HMDS–TMCS–pyridine.[96] Sherman *et al.* found that a mixture of BSA–TMCS–pyridine (5 : 2 : 3) could rapidly silylate aldoses, ketoses and their terminal phosphate esters.[97] With this mixture the alcohol, carboxyl and phosphate groups were all silylated in 0.5–3 h at room temperature but two peaks were observed on chromatography due to the α and β anomers. Peterson silylated neutral and acidic sugars in their natural state or as their methoxime derivatives with a mixture of BSTFA–TMCS (4 : 1) in 2 h at room temperature,[51] in agreement with the results of Laine *et al.*[87] who prepared quantitative derivatives of neutral sugars in 15 min at 80 °C with BSTFA. Aldose-4-, -5- and -6-phosphates and ketose diphosphates were converted to their silyl derivatives with a mixture of BSTFA–TMCS–acetonitrile (2 : 1 : 2) at 80 °C for 10 min.[88] This method was unsuccessful for aldose-1-phosphates which were silylated with TMSIM–acetonitrile (1 : 1) at room temperature in approximately 5 min.[88]

3.1.1 ALDOSES AND KETOSES AS METHOXIME–TMS ETHER DERIVATIVES[87]

The ketone or aldehyde group was converted to the methoxime by heating a mixture of 1 mg of the sugar and 1 mg of methoxylamine hydrochloride in 50 μl of pyridine for 2 h at 80 °C. The hydroxyl groups were then converted to the TMS ethers by adding 50 μl of BSTFA and heating the mixture for a further 15 min at 80 °C.

3.1.2 ACYCLIC ALDONIC AND ALDORIC ACIDS AS TMS-ETHERS-TMS-ESTERS[51]

Aldonic acids and aldoric acids can exist in solution as an equilibrium mixture of the acyclic acid with the 1,4- and 1,5-lactones. Prior to silyl-ether formation the mixture was saponified to convert the sample completely to the sodium salt of the acyclic acid.

To the sample (0.1–10 mg) in water (2–5 ml) was added sufficient 0.05 M sodium hydroxide with an automatic titrator to maintain a pH or 8.5 for 4 h at room temperature. The 2-deoxyaldolactones are particularly stable and were saponified at pH 10 for 4 h. Water was removed *in vacuo* at 35 °C and the sample dried further by a double azeotropic evaporation with dichloromethane (1–2 ml) or by storing the residue for one day over P_2O_5 *in vacuo*. Several non-lactone-forming acids (e.g. glycolic, lactic and hydracrylic acids) were unstable to evaporation of the alkaline saponification medium. For these substances, adjustment of the pH to 7 with 0.01 M HCl, before evaporation was required. The sodium salts were then converted to the TMS derivative by the addition of a mixture of pyridine–BSTFA–TMCS (8 : 2 : 1) in excess. Reaction was complete in 2 h at room temperature.

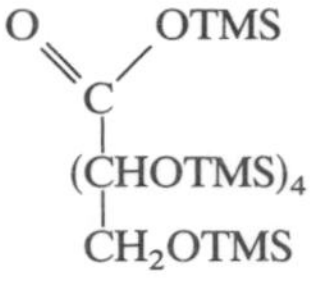

Aldonic–TMS–ether–TMS–ester (17)

3.1.3 URONIC ACIDS AS METHOXIME–TMS–ETHER–TMS–ESTERS[51]

The uronic acids are saponified and converted to their dried sodium salts as indicated for aldonic acids above. Glucuronic, mannuronic, guluronic and iduronic acids, which form furanurono-6,3-lactones, required 4 h at pH 8.5 for complete saponification, whereas shorter times could be used for other hexuronic acids and for penturonic and hexulsonic acids (these do not usually form lactones.)

To a solution or suspension of the sodium salt of the uronic acid (0.1–10 mg) in pyridine (1–2 ml) maintained in an ultrasonic water-bath was added a similar weight of methoxylamine hydrochloride and the mixture reacted for 2 h at 30 °C. The TMS derivative was prepared by the addition of BSTFA and TMCS to the above reaction mixture to give a final reagent combination of pyridine–BSTFA–TMCS (8 : 2 : 1). The reaction was complete in 2 h at room temperature:

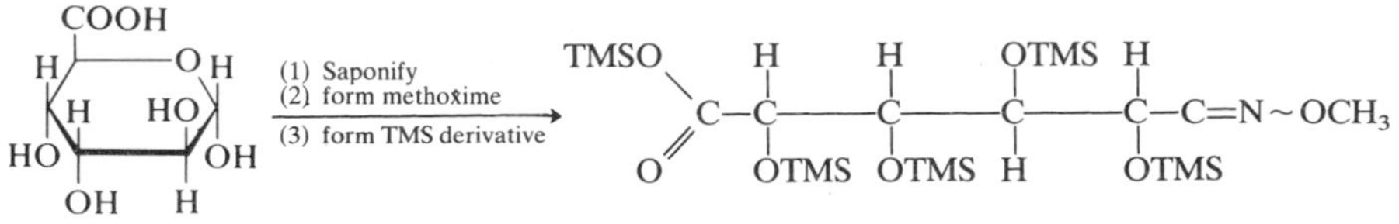

α-D-Glucuronic acid (18)

3.1.4 SUGAR AMINES AS TMS DERIVATIVES[95]

To the sugar amine was added an excess of BSA–TMCS–HMDS–pyridine (1 : 0.5 : 1 : 10). The reaction was complete in 30 min at room temperature:

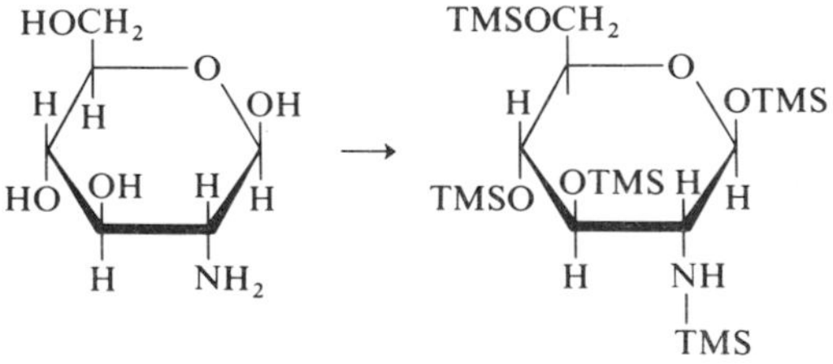

(19)

2-Amino-2-deoxy-D-glucose
(glucosamine)

3.1.5 SUGAR PHOSPHATES AS METHOXIME–TMS DERIVATIVES[88]

To the sugar phosphate (1.0 mg) in pyridine (0.2 ml) was added methoxylamine hydrochloride (1.0 mg) and the mixture was allowed to stand for 1 h at room temperature. The TMS derivatives were prepared by the addition of either BSTFA or TMSIM (0.1 ml) with heating at 60 °C for 5 min:

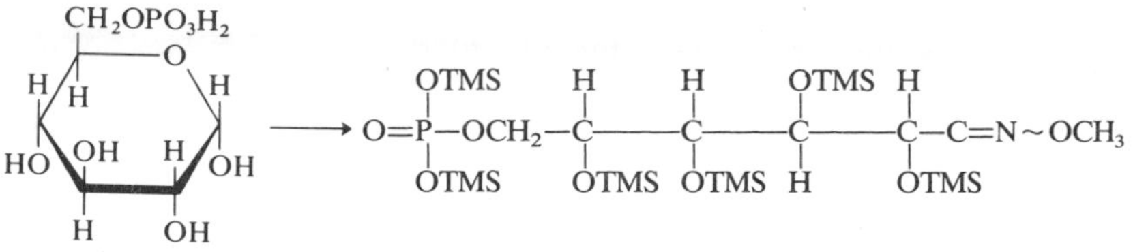

(20)

α-D-glucose-6-phosphate

3.2 Steroids and bile acids

Steroid hydroxyl groups have markedly different rates of reaction towards silyl reagents, firstly due to their nature (primary, secondary or tertiary) and secondly due to their steric environment. This latter point is very important and is often the rate-determining feature. The preferred reagent for the silylation of all hydroxyl groups is TMSIM. The use of less powerful silyl donors allows some hydroxyl groups to be reacted selectively. Steroid ketone groups are prone to enol–TMS ether formation in the presence of TMCS catalyst. The order of reactivity is 4-ene-3-one > -3-one > -17-one > -20-one ⋙ -11-one. For the universal analysis of steroid metabolic profiles methoxime formation followed by silylation with TMSIM is the recommended procedure.[98–101]

HMDS alone or in basic solution finds few applications in steroid chemistry. In pyridine at room temperature it forms silyl ethers of unhindered hydroxyl groups.[61] TMSDEA reacts very slowly with secondary hydroxyl groups and will react with equatorial hydroxyl groups in 4–10 h at room temperature—conditions under which axial groups do not react.[102] TMCS in neutral solvents does not react quantitatively with any steroid hydroxyl group but when used as a catalyst, in combination with HMDS and pyridine (HMDS–TMCS–pyridine; 10 : 1 : 10), the reagent mixture will silylate hydroxyl groups at C-3, 7, 16, 17 (*sec*), 20 and 21 readily and the hindered 11β-hydroxyl group in about 8 h at room temperature.[61,103] The secondary 11β-hydroxyl group reacts slower than other secondary hydroxyl groups as access to it is hindered by the methyl groups at C-18 and C-19. At room temperature BSA reacts with all secondary hydroxyl groups less hindered than the 11β-hydroxyl and is the reagent of choice for these groups.[61,104] BSTFA is a slightly more potent silylating reagent than BSA but has the disadvantage that it reacts slowly and incompletely with the 11β-hydroxyl group.[61] MSTFA shows a similar donor ability to BSA but tends to promote enol–TMS ether formation with unprotected ketone groups.[37] Mixtures of BSA or BSTFA with TMCS (5 : 1) react completely with the 11β-hydroxyl group in about 5 h at room temperature.[61] With the tertiary

17α-OH group in cortisol as its methoxime derivative (11β,17α,21-trihydroxy-pregn-4-en-3,20-dimethoxime) a mixture of BSA–TMCS (4:1) reacts very slowly and requires 30 h at 60 °C for complete reaction.[61,103] With a C-20 hydroxyl group as a near neighbour, silyl ether formation occurs rapidly at this position and hinders access to the 17α-OH group which does not react quantitatively with BSA–TMCS mixtures. The tertiary 17α-OH group when shielded by a 20-TMS ether is very difficult to react, the 20α-TMS ether providing a greater degree of steric hindrance than the 20β-TMS ether.[98] The potent silyl donor mixture TMSIM–BSA–TMCS (1:1:1) at 60 °C requires 20 h for complete reaction of the 17α-OH in 5β-pregnan-3α,11β,17α,20β,21-pentol and 72 h in 5β-pregnan-3α,11β,17α,20α,21-pentol. Under these conditions all other steroid hydroxyl groups react and unprotected ketone groups will be converted to a varying extent to enol–TMS ethers. With this reagent combination the formation of methoxime derivatives of ketones is essential. At high temperatures TMSIM without catalyst performs as well as the TMSIM–BSA–TMCS mixture.[105] Above 100 °C there is little difference in the rate of reaction between TMSIM and the reagent mixture. With TMSIM the 17α-OH group in 5β-pregnan-3α,11β,17α,20α,21-pentol requires approximately 1 h at 200 °C, 8 h at 150 °C or 48 h at 100 °C for complete reaction.[105] TMSIM is the only reagent that will react quantitatively with all steroid hydroxyl groups without the addition of TMCS catalyst.

Pregnanes such as cortisol (Fig. 4), with a dihydroxy-acetone side chain at C-17 have been a particular problem in steroid analysis by GLC due to their thermal lability. Direct injection of the unprotected cortisol results in the formation of the 17-oxo product. The formation of silyl ethers of the hydroxyl groups also results in a thermally labile product unless the C-20 ketone group is also protected as its methoxime. Several methods are available for the formation of thermally stable products of the corticosteroids and these are summarized in Fig. 4. Kelley formed a cyclic dimethylsiliconide of the side chain using dimethyldiacetoxysilane.[106] This bifunctional reagent forms unstable derivatives with isolated hydroxyl groups and is of limited application. The cyclic boronic esters[107–109] or the TMS–enol–TMS derivative[39] have not found such wide application as the methoxime–TMS derivative which is the method of choice for this class of steroid.[98–100]

The formation of the TMS ethers of ecdysones (insect moulting hormones) presents several difficulties (21). The use of BSA overnight allows the 2β, 3β, 22- and 25-hydroxyl groups to be protected.[50,54,110] The tertiary 14α-OH group is very unreactive and the C-20 hydroxyl is made less reactive by prior silyl ether formation at the neighbouring C-22 hydroxyl group. The addition of TMCS catalyst gives multiple products due to enol–TMS ether formation. The methoxime derivative gives two peaks on GLC due to separation of the *syn*- and *anti*-methoxime isomers.[54] TMSIM is the reagent of choice for these steroids as it has the necessary silyl donor power to react with all the hydroxyl groups without prior protection of the ketone function. The C-20 hydroxyl

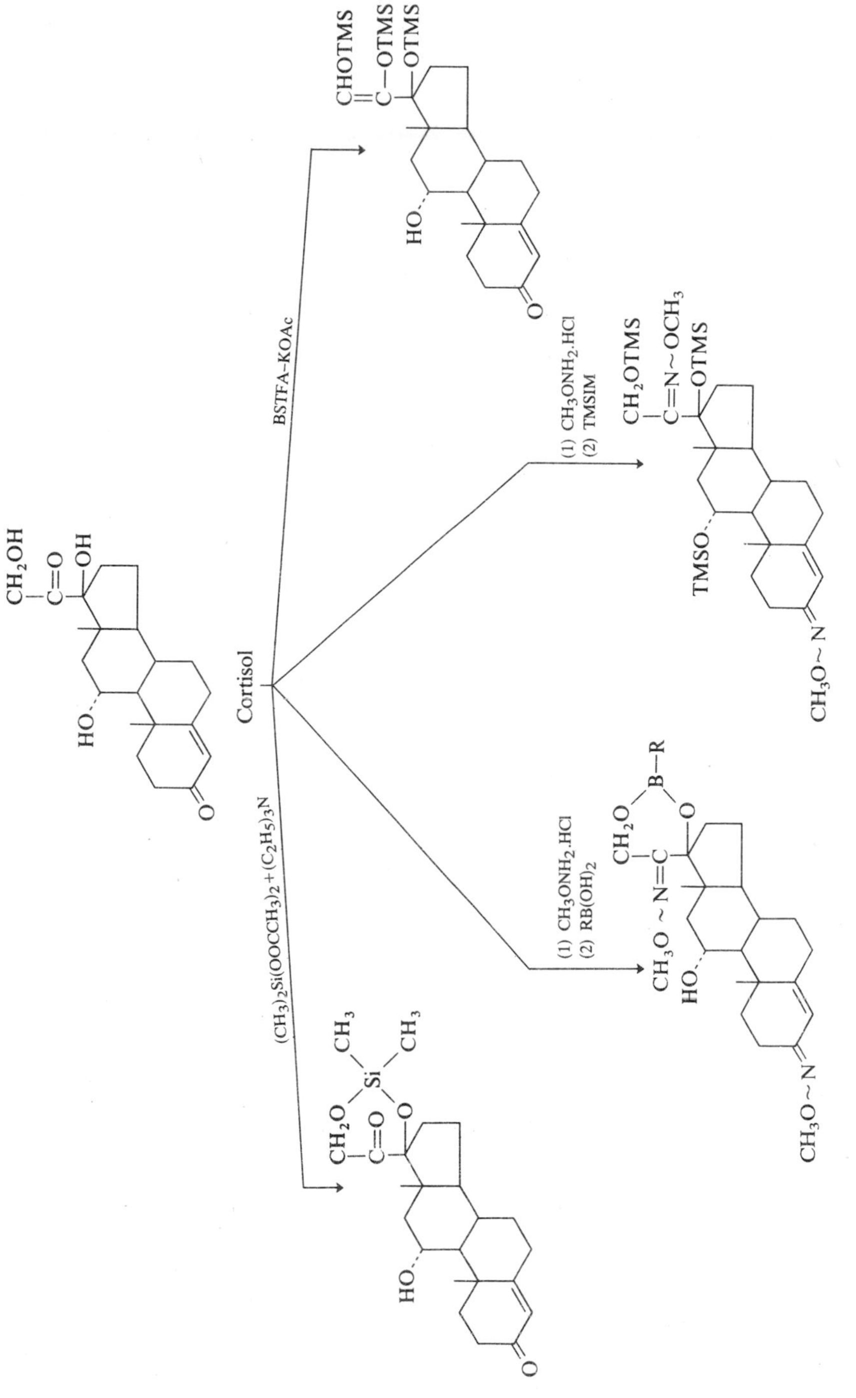

Fig. 4 A selection of derivatives for the analysis of cortisol by gas chromatography.

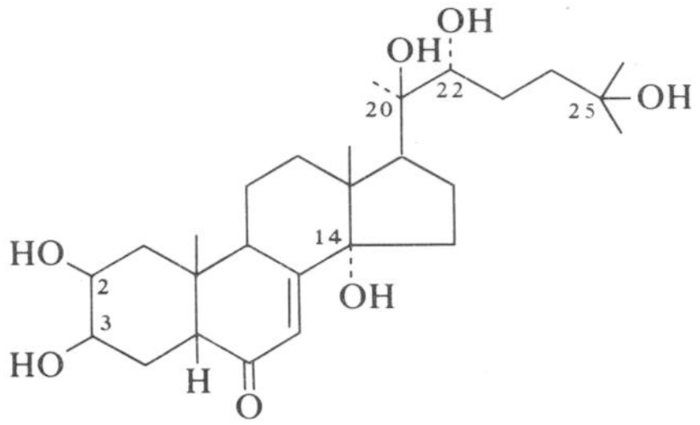

(21)

Ecdysterone

group can be silylated by heating for 4 h at 100 °C with TMSIM, under which conditions the 14α-OH group reacts slowly. The 14α-OH group requires heating at 140 °C for a minimum of 12 h for complete reaction.[40,111,112] Under comparable conditions the tertiary 17α-OH in 3α,17α,20α-trihydroxy-5β-pregnane formed a quantitative derivative after heating for 15 h at 140 °C. The 14β-OH group in cardiac aglycones (cardenolides) required heating for three days at 60 °C with TMSIM–BSA–TMCS (1 : 1 : 1) for complete reaction.[42] The rate of silyl ether formation of the 14α-OH group is slower in pyridine and using a mixture of TMSIM–pyridine (7 : 20), heating for approximately 20 h at 140 °C was required for complete reaction.[111]

The conditions indicated for insect and mammalian steroids are adequate for the formation of TMS ethers with steroids of microbiological and plant origin.[113] Bile acids are usually extracted as their methyl esters and can be silylated under conditions directly comparable to those used for steroids.[114]

Methods for either the selective reaction of target hydroxyl groups or the total formation of TMS derivatives are summarized in Table 2.

3.2.1 CORTICOSTEROIDS AS THEIR METHOXIME–TMS DERIVATIVES[100,101]

To the corticosteroid (1.0 mg) was added a solution of methoxylamine hydrochloride in pyridine (50 mg ml^{-1}, 200 μl). The mixture was heated at 60 °C for 15 min, the solvent removed with a stream of nitrogen, and the silyl ether formed by the addition of TMSIM (100 μl) followed by heating at 100 °C for 2 h.

3.3 Carboxylic acids and phenols

After alcohols, phenols and carboxylic acids are the easiest groups with which to form silyl derivatives. Consequently nearly all silyl donors will react with this class of compound, although a mixture of HMDS–TMCS is required for a quantitative yield with the more hindered groups.[115–118] The silyl derivatives of these compounds are often unstable when removed from the silyl medium, and their hydrolytic stability should be tested rather than assumed. BSA, BSTFA and TMSIM have been demonstrated to be the silyl donors of choice, promoting a smooth easy reaction in a few minutes at room tempera-

TABLE 2

Conditions for the selective or total formation of TMS ethers of steroid hydroxyl groups

Hydroxyl group environment	Quantitative reaction with TMSIM	Selective reaction BSA	Selective reaction BSA–TMCS (4 : 1)[a]
Mammalian Steroids			
All primary and normal secondary OH groups (less hindered than 11β-OH) phenolic groups and unhindred tertiary OH	Less than 1 h at room temp. Often instantaneous	1–4 h at room temp. 1 h at 60 °C	As for TMSIM
11β-OH	Less than 1 h at 100 °C	Does not react	5 h room temp.
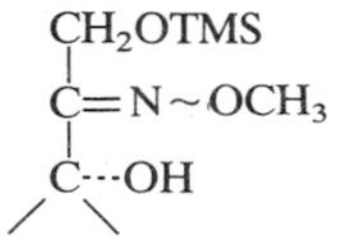	2 h at 100 °C	—	30 h at 60 °C
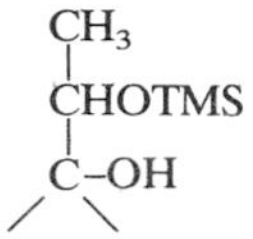	8 h at 150 °C	—	Non-quantitative reaction
Insect Steroids			
All primary and normal secondary OH groups and some unhindered tertiary OH groups.	Less than 1 h at room temp.	Overnight at room temp.	—
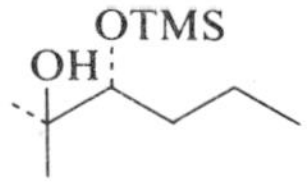	4 h at 100 °C		
hindered tertiary (e.g. 14α-OH)	12 h at 140 °C	—	—

[a] Ketone groups must be protected as their methoximes.

ture or if heated briefly at 100 °C.[119–123] For acids of low molecular weight BSTFA is superior to BSA and TMSIM as the reagent and its by-products are more volatile on GLC and do not interfere with the analysis of the early-eluting compounds. Phenols with bulky substituents (e.g. iodine, *t*-butyl group) on adjacent carbon atoms react slowly, or not at all.

Urinary acids can be silylated with BSTFA usually at an elevated temperature (e.g. 60 °C for 2 h, 130 °C for 10 min).[124–128] A mixture of BSTFA–TMCS

(99 : 1) has been used to allow the reaction to be performed at room temperature[129] or at 70 °C for 20 min for phenolic carboxylic acids.[127] For hydroxy-acids, the conditions required to silylate the carboxylic acid function are adequate for the hydroxyl group as well.[130–133] In the analysis of acidic substances of the Krebs cycle BSTFA at room temperature for 1 h was used to silylate the metabolites[134] and HMDS was successfully used with citric acid cycle intermediates.[135] A frequent problem with the analysis of acids of this type is the presence of a ketone group in the molecule which is readily enolized or reduces the thermal stability of the derivative, producing multiple peaks on analysis. For these acids (particularly α-keto acids), the formation of the methoxime derivative of the ketone group prior to silylation produces a thermally stable product.[136–140] As an alternative, for the analysis of α-keto acids the formation of silyl-quinoxalinol derivatives have been recommended[141–143] (see Chapter 7).

Phenols have been analysed as their sulfonyl trimethylsilyl esters which are useful in the separation of complex mixtures by GLC.[144] Other phenolic and acidic substances which have been analysed as their silyl derivatives include hydroxy anthraquinones and quinolines[145,146]; gossypol,[147] aminochromes,[148] flavonoids,[149] morphine,[150] salicyclic acid,[151] cannabinols[152] and biotin.[153]

3.3.1 α-KETO ACIDS AS METHOXIME–TMS ESTERS[139,140]

Methoxylamine hydrochloride (10–20 mg, a minimum of a five-fold molar excess), was added to the acid in pyridine (0.4 ml), thoroughly mixed and allowed to stand at room temperature for 2 h. BSA (0.5 ml) was then added and the mixture allowed to stand for at least 2 h at room temperature before GLC (22):

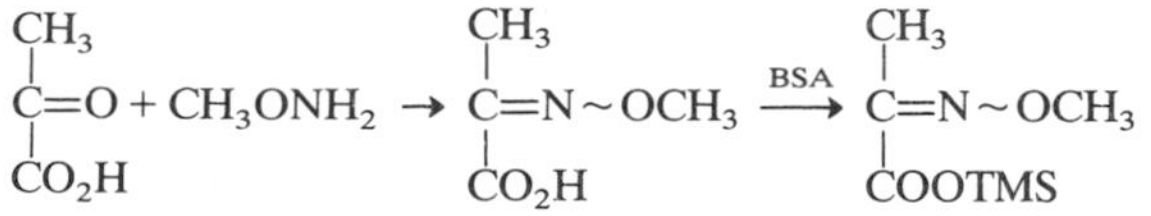

(22)

3.3.2 α-KETO AS SILYLQUINOXALINOL DERIVATIVES[143,154]

A mixture of 10 mmol of the α-keto acid dissolved in 96% ethanol (20 ml) and *o*-phenylenediamine (1.2 g) dissolved in 50% acetic acid (20 ml) in a stoppered test tube was placed in a boiling water-bath for one hour. The reaction mixture was extracted three times with an appropriate volume of dichloromethane, evaporated *in vacuo* and the residue dissolved in pyridine–BSTFA (1 : 1) and heated at 70 °C for 30 min (23). 2-Hydroxyphenylpyruvic acid and 2,5-dihydroxyphenylpyruvic acid in their lactone forms do not react by this procedure. Opening of the lactone ring was accomplished by boiling the ethanol solution with 14 mmol of NaOH for 5 min under N_2, followed by neutralization with 15 mmol of HCl.

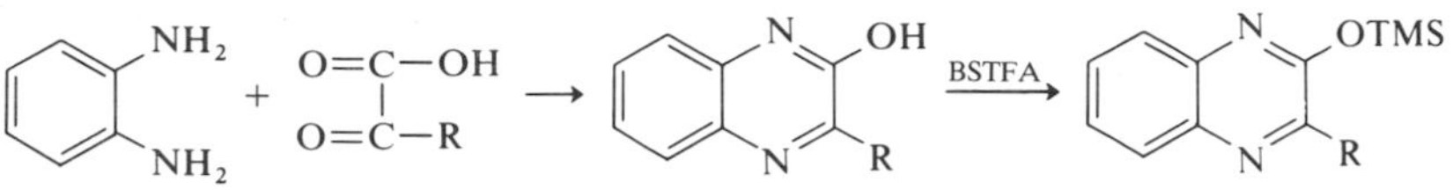

(23)

3.4 Amines, amides and amino acids

The amine group is the most difficult to silylate and consequently strong silyl donors are preferred. With primary amines the possibility of replacing both protons with the silyl group can lead to two products appearing on the chromatogram (mono-TMS and di-TMS). The reactions are generally markedly solvent-dependent and both the rate of reaction and the product formed varies from one solvent to another. Polar solvents (e.g. acetonitrile) tend to favour the formation of the di-TMS derivative but in neutral solvents (e.g. dichloromethane) the mono-TMS derivative is favoured. The *N*-TMS bond is very hydrolytically unstable, and scrupulous avoidance of moisture in solvents and syringes is important. The recommended silyl donors are BSA and BSTFA. These reagents have superseded the TMS-amines for use with this group of compounds, largely due to their greater silyl donor power. TMSIM is unreactive with primary and secondary aliphatic amines, but it has been reported to react quantitatively, if slowly, with aniline.[11,12]

HMDS in dimethylformamide will silylate primary amines slowly but does not react with secondary amines.[28] The reaction is much faster in dimethylformamide than pyridine, but problems can arise with this solvent due to condensation with the primary amine group. Mixtures of dimethyl sulfoxide–dioxane have been recommended as promoting a faster reaction without the formation of condensation products.[155] BSA or BSA–TMCS will silylate the primary amine function of catecholamines twice (*N*-di-TMS derivative) and secondary amine groups slowly or not at all.[24] BSA reacts with alkylamines in 5 min at room temperature to produce stable products, but the degree of substitution was not determined.[156] The disubstituted amine derivatives are normally prepared by heating with mixtures of BSA–TMCS.[157–160] Butts derivatized some biologically important amines by heating them with BSTFA–TMCS (99:1) overnight at 60 °C without the formation of secondary products.[161]

TMSIM reacts smoothly with the hydroxyl groups of catecholamines but not at all with the amine groups.[162,163] Fully silylated derivatives were prepared by reacting the amine groups with BSA–TMCS (2:1) for 2 h at 60 °C.[24] Under these conditions no reaction occurred for *N*-methyl amines. In a useful modification of the above reaction, TMSIM was used to silylate all the hydroxyl groups in the catecholamines (60 °C, 2–3 h) and the primary amines were converted to the amide derivative by reaction with *N*-heptafluorobutyrylimidazole.[162] The primary amine group in catecholamines has also been converted to a pentafluorobenzylimine derivative and the hydroxyl groups to the silyl ethers with BSA (15 min, room temp).[164–166] Both these latter methods allow the determination of very low levels of catecholamines by electron capture detection or by mass fragmentography (SIM).

Amino acids have been analysed as their N-TMS–TMS–ester derivatives or as N-TMS–alkyl ester derivatives. Hardy and Kerrin formed N-TMS–butyl–ester derivatives of the 20 major mammalian amino acids by esterifying the carboxyl group with *n*-butanol 3 M in HCl at 150 °C for 15 min followed by silylation of the amine group with BSTFA at the same temperature for 90 min.[167] Using BSA in acetonitrile, 22 protein amino acids were converted to their silyl derivatives by heating the medium just below its boiling point for 30 min.[6] On analysis by gas chromatography arginine decomposed on the column and the peaks for alanine and glycine were obscured by N-TMS–acetamide produced during the reaction. The more volatile by-products of BSTFA make this the reagent of choice for all the amino acids and it is essential for the determination of alanine and glycine. Gehrke and co-workers have carried out an extensive study of the conditions required for the analysis of the major protein amino acids.[168–170] All 20 protein amino acids were silylated by heating in a sealed tube with BSTFA–acetonitrile (1:1) for 2.5 h at 150 °C. With the exception of glycine, arginine and glutamic acid, the other 17 amino acids reacted completely in 15 min at 150 °C. The selection of the solvent is important in controlling the degree of substitution on nitrogen and the number of peaks appearing in the chromatogram. Essentially the recommended conditions of Gehrke have been followed by others for the derivatization of isotopically labelled amino acids,[171,172] amino acids extracted from the waters of the North Atlantic Ocean,[173] amino acids in commercial tablet formulations[174,175] and for amino acids of plant origin.[176] A number of sulfo- and selenoamino acids were silylated with BSA-pyridine in 15 min at 90–100 °C.[177]

Shahrokhi and Gehrke recommend heating iodoamino acids with BSA–acetonitrile (3:1) at 150 °C for 30 min to form their silyl derivatives.[178] Under these conditions the phenolic group of diiodotyrosine and diiodothyronine is hindered by the adjacent iodine atoms and is assumed not to react. Silylation takes place at the carboxyl group and at one of the amine protons. Other workers have used BSA or BSA–TMCS (99:1) with solvent such as pyridine[179] or tetrahydrofuran[180] for 5–10 min at 50 °C. When picomole quantities are to be derivatized, the use of BSA alone is preferred.[181] The stability of small quantities of the derivatives in solution and on exposure to light is in doubt. The silylation procedure has been adapted to the routine assay of thyroid hormone preparations and drugs.[179,180]

3.4.1 CATECHOLAMINES AS $N(TMS)_2$–TMS ETHERS[24]

To the catecholamine (1.0 mg) in acetonitrile (0.1 ml) was added BSA (0.2 ml) and TMCS (0.1 ml) and the mixture heated at 60 °C for 5 h. The addition of water (2 μl) at the start of the reaction gave a quantitative yield of derivative in 15 min:

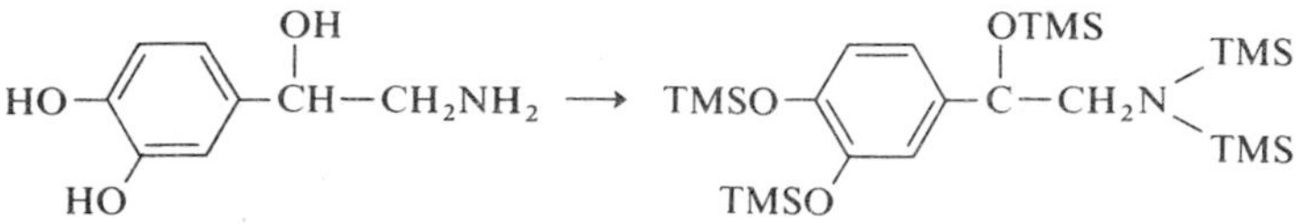

(24)

Norepinephrine

3.4.2 AMINO ACIDS AS *N*-TMS-*O*-BUTYL ESTER DERIVATIVES

The butyl esters were prepared by heating the amino acid for 15 min at 150 °C in butanol which was 3 M in HCl. Excess reagent was removed with a stream of nitrogen at 50 °C and the silyl ethers formed by the addition of an excess of BSTFA–acetonitrile (1 : 1) and heating at 150 °C for 90 min:

$$NH_2\text{—}\underset{\displaystyle R}{\underset{|}{CH}}\text{—}COOH \xrightarrow[\text{2) silylate}]{\text{1) esterify}} TMS\text{—}NH\text{—}\underset{\displaystyle R}{\underset{|}{CH}}\text{—}COOC_4H_9 \qquad (25)$$

3.4.3 AMINO ACIDS AS *N*-TMS–TMS–ESTERS

An aqueous aliquot containing the amino acid (0.5–6.0 mg) in a glass tube with cap was evaporated to dryness on a sand-bath at 70 °C with the aid of a stream of nitrogen. Dichloromethane (0.5 ml) was added and evaporated to remove the last traces of water as its azeotrope (repeated at least once more). To the residue was added acetonitrile (0.25 ml) and BSTFA (0.25 ml) for each 1.0 mg of amino acids, and the sample agitated in an ultrasonic bath for one minute prior to heating in an oil bath for 2.5 h at 150 °C:

$$NH_2\text{—}\underset{\displaystyle R}{\underset{|}{CH}}\text{—}COOH \xrightarrow{\text{BSTFA}} TMS\text{—}NH\text{—}\underset{\displaystyle R}{\underset{|}{CH}}\text{—}COOTMS \qquad (26)$$

3.5 Nucleic acid bases and nucleosides

Early reports indicated the possibility of analysing intact the nucleosides from natural sources by methods based on silylation with HMDS–TMCS–pyridine.[181–185] The sugar hydroxyl groups reacted quantitatively, but the position with the pyrimidine and purine bases was not clear. A comparison of various silylating reagents with guanine and hypoxanthine indicated that a variety of TMS–amine and TMS-acetamide reagents (but not TMSIM) were useful for their silylation.[186] The introduction of BSA[187] and later BSTFA[188–192] by Gehrke and co-workers allowed the major purine and pyrimidine bases to be analysed by GLC as single peaks, with the exception of cytosine and 5-methylcytosine which gave two peaks due to incomplete silylation of the amine group at C-4. The derivatives were formed by heating the base with BSTFA-acetonitrile (1 : 1) at 150 °C for 15 min. It has been shown that the problem with the derivatization of cytosine is the poor solubility of the base in BSTFA.[193] If the sample of cytosine in BSTFA is stirred, then a single peak is obtained after reaction at 150 °C for 15 min. Similarly, nucleosides were found to be poorly soluble in BSTFA but with the addition of pyridine and reaction at 150 °C single peaks were obtained with adenine after 15 min and with adenosine after 30 min.[193]

3.5.1 TMS DERIVATIVES OF NUCLEIC ACID BASES AND NUCLEOSIDES[193]

To the sample (1.0 mg) in pyridine (0.2 ml) was added BSTFA (0.5 ml) and a magnetic stirring bar. The sample tube was capped and heated at 150 °C for 30 min with stirring (27 and 28).

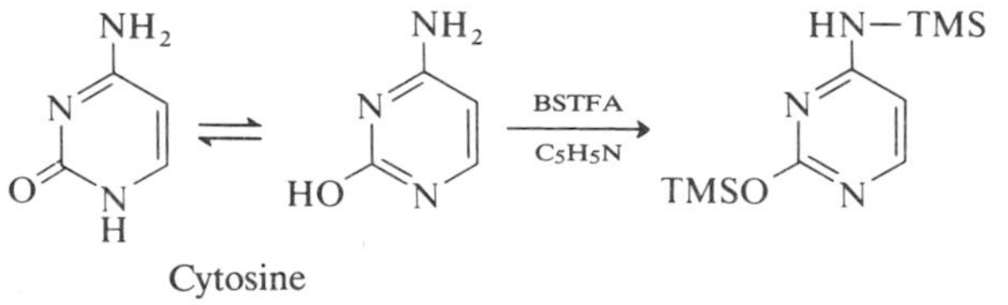

(27)

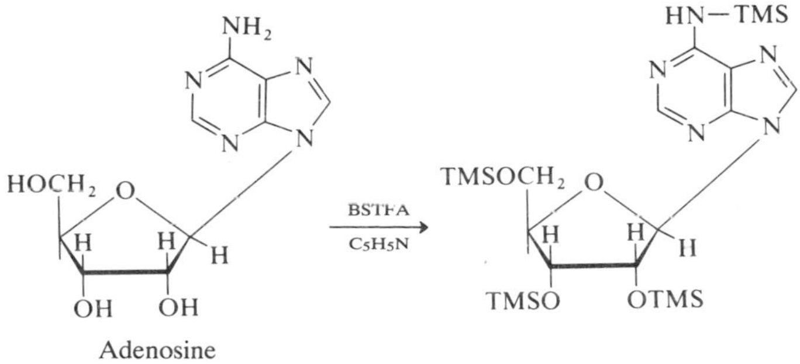

(28)

4 THE MASS SPECTRA OF TMS ETHERS

Combined gas chromatography–mass spectrometry (GC–MS) has enjoyed a spectacular growth in the last decade due to its unmatched ability both to separate and to identify the constitutents of complex organic mixtures. For many polyfunctional molecules, the formation of derivatives is essential to provide the necessary stability and volatility for their analysis by this technique. The TMS ethers are the most used derivatives for this purpose and a wealth of electron-impact induced mass spectra has now been recorded. In more recent times there has been a definite trend towards the use of alternative ionization sources such as chemical ionization or field emission to obtain complementary information to the electron-impact spectra as a further aid to compound identification. These methods have not been extensively applied to TMS ethers to date, but a considerable growth can be envisaged in this area.[194,195]

It is not intended to summarize the available information on the electron-impact mass spectra of TMS ethers here, but rather to indicate some of the very general features exhibited by these spectra.

Probably the most important piece of information available from the mass spectrum is the molecular ion as this indicates the molecular weight of the compound. Unfortunately, the molecular ions of TMS ethers are often weak or absent with the $[M-15]^+$ ion obtained by cleavage of a methyl to silicon bond (in part) being more prominent. This ion can be used to define the molecular weight, provided that it is not mistaken for the molecular ion itself. Recording the mass spectrum at lower electron energies (10–20 eV) compared to the usual conditions (70–80 eV) often enhances the relative intensity of the molecular ion region of the spectrum.

Secondary fragments formed by dissociation of the molecular ion often results in prominent fragment ions containing the ionized dimethylsiloxy group attached to a hydrocarbon part of the molecule. As for ethers in general, α-cleavage of the bond next to oxygen is favoured. Characteristic ions of hydroxy TMS ethers are:

$$[(CH_3)Si]^+ \quad m/e\ 73 \qquad [HO{=}Si(CH_3)_2]^+ \quad m/e\ 75$$

$$[(CH_3)_2Si{=}O{-}Si(CH_3)_3]^+ \quad m/e\ 147 \tag{29}$$

The ion m/e 73 is prominent in virtually all TMS spectra and is often the base peak.[196] The ion m/e 147 is common in polyhydroxy TMS compounds containing two or more TMS groups either on adjacent carbon atoms or brought near each other through expulsion of the central portion of the molecule.[197–199] The TMS group undergoes a prolific number of intramolecular migrations and rearrangements (including McLafferty rearrangements) to give prominent silicon containing ions.[199–202] In many cases the abundance of certain rearrangement ions has been found to be markedly dependent on structural, and in some cases stereochemical, factors. The presence of $[M+73]^+$ ions can arise from intermolecular transfer of a TMS group under electron-impact at high source pressures.[203]

A very important development as a diagnostic aid for the elucidation of mass spectral fragmentation pathways has been the commerical availability of perdeuterated silyl reagents. The formation of a d_9-TMS ether results in an increase of 9 a.m.u. compared to the TMS ethers and is invaluable for elucidating specific fragmentations and rearrangement processes based on a comparison of both spectra.[204,205] The observation of the shift in the molecular ion for the two spectra is a useful method of indicating the number of silyl groups introduced into the molecule. The use of the perdeuterosilyl reagents either alone or in conjunction with further heavy atoms (e.g. ^{18}O, ^{15}N, ^{13}C) substituted into the molecule is one of the principal tools used to probe the fragmentation of TMS derivatives.

5 TRIALKYLSILYL ETHERS OTHER THAN TRIMETHYLSILYL

An effective method of resolving the components of a complex biological mixture by GLC is by the preparation of a derivative such that a given group of components is selected and shifted relative to other components of the mixture. In some cases the formation of trimethylsilyl ethers may be inadequate for this purpose as the molecular weight increment produces an insufficient shift of the peaks in the chromatogram. Also, in a complex biological mixture the chromatogram will contain many peaks so that the components selected for separation should be shifted to an uncrowded region. By choosing alkylsilyl ether derivatives higher than trimethylsilyl, the difference in retention times between compounds containing different numbers of hydroxyl groups can be magnified. Those components which do not contain hydroxyl groups are not moved on the chromatogram by this technique. Harvey has described the use of triethylsilyl (TES), tri-*n*-propysilyl (TnPS), tri-*n*-butylsily (TnBS) and tri-*n*-hexylsilyl (TnHS) ethers for the separation of overlapping peaks in an extract of cannabis[13] and nutmeg.[206] The improvement in resolution is illustrated for the temperature programmed separation of an ethyl acetate extract of cannabis tincture analysed as a TMS derivative (Fig. 5) and TnBS derivative (Fig. 6). A

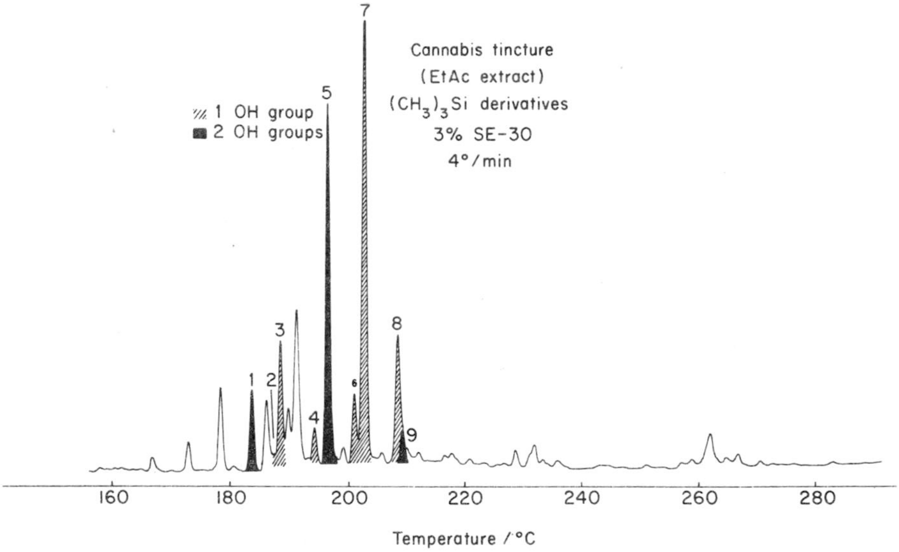

Fig. 5 Gas chromatogram of the TMS derivatives of an ethyl acetate extract of cannabis tincture obtained on a 6-ft 3% SE-30 column, temperature-programmed at 4 °C/min. The peaks were identified by mass spectrometry as: (1) propylcannabidiol; (2) propylcannabichromene; (3) propyl-1^1-tetrahydrocannabinol; (4) propylcannabinol; (5) cannabidiol; (6) cannabichromene; (7) *trans*-1^1-tetrahydrocannabinol; (8) cannabinol; (9) cannabigerol. (From D. J. Harvey and W. D. M. Paton, *J. Chromatogr.* **109**, 73 (1975)).

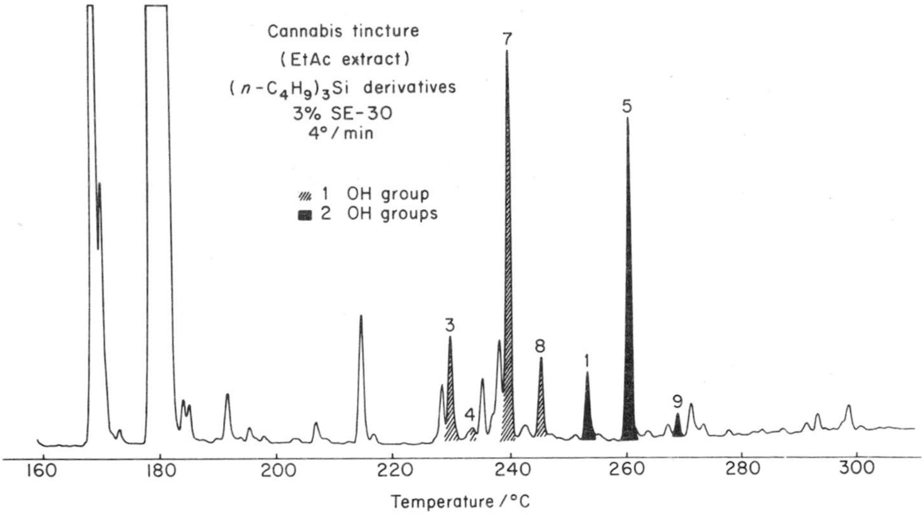

Fig. 6 Gas chromatogram of the tri-*n*-butylsilyl derivatives of an extract of cannabis tincture. The conditions and peak identification are the same as for Fig. 5. (Source as for Fig. 5.)

relatively large molecular weight increment (TMS = 72, TES = 114, TnPS = 156, TnBS = 196 and TnHS = 282) is introduced into the molecule which limits the technique to volatile components containing only a few hydroxyl groups, particularly with the higher molecular weight trialkylsilyl homologues.

The *t*-butyldimethylsilyl ethers are proving useful both as a selective protecting group for hydroxyls in prostaglandin[207–209] and nucleoside synthesis[210–212] and as derivatives for gas chromatography. The *t*-butyldimethylsilyl (*t*-BDMS) ethers are approximately 10^4 times less susceptible to solvolysis than the TMS ethers and are stable to acid and base under the normal conditions for acetate saponification, hydrogenolysis (H_2-Pd), mild chemical reduction (Zn-CH_3OH) and to phosphorylation.[208,210] Selective cleavage back to the alcohol is possible by treatment with 2–3 equivalents of tetra-*n*-butylammonium fluoride in tetrahydrofuran or with aqueous acetic acid.[208,210] The same features which are of value to the synthetic chemist might recommend this reagent to the analyst for further consideration.

The *t*-BDMS ethers have been used in the gas chromatography of nucleosides,[213,214] prostaglandins,[14,215] and steroids.[14] Derivatives are prepared by reaction with *t*-butyldimethylchlorosilane* and imidazole in dimethylformide. The *t*-butyldimethylchlorosilane can be synthesized from *t*-butyllithium and dimethyldichlorosilane according to the method of Sommer and Taylor.[216] For the analysis of those compounds which are unstable in the presence of acid catalyst, *t*-butyldimethylsilylimidazole can be prepared according to the method of Quilliam *et al.*[213] Reactions of *t*-butyldimethylchlorosilane in the

* Available from Willowbrook Laboratories, Waukesha, Wis. U.S.A. in 1976.

absence of imidazole are very slow and rarely quantitative. The reaction is also selective for hydroxyl groups in the presence of primary amines.[210] The introduction of a *t*-BDMS ether into a compound produces a substantial increase in retention time, approximately 2–3 times that of the TMS ether.[14,213] Kelley has introduced a scheme for the isolation of small quantities of *t*-BDMS ethers using a short column of Sephadex LH-20 which has advantages over techniques employing an aqueous work-up.[14]

The general features of the mass spectra of the alkylsilyl ethers resemble those of the TMS-ethers with some important characteristic differences. The distinguishing feature in the mass spectrum of *t*-BDMS ethers is the strong $[M-57]^+$ ion due to the ready loss of the *t*-butyl radical. In the absence of any competing fragmentation processes, the $[M-57]^+$ ion dominates the mass spectrum, is usually the base peak and is useful for quantitative analysis by single-ion monitoring.[14,215] The normal trialkylsilyl ethers of alcohols have weak molecular ions which fragment by elimination of alkenes from the silyl group and silicon-containing ions.[13] The elimination of ethylene from triethylsilyl derivatives has been well documented.[197,217,218]

The silyl donor power of the trialkylsilyl and *t*-BDMS reagents has not yet been determined with a wide range of functional groups. The two following methods are adequate for the analysis of hydroxyl-containing compounds.

5.1.1 THE FORMATION OF TRIALKYLSILYL ETHERS[13]

Into a stoppered centrifuge tube was placed pyridine (2 ml), diethylamine (0.5 ml) and the appropriate trialkylchlorosilane (1 ml). The mixture was cooled in ice water and centrifuged for one minute. The derivatives were formed by adding a portion of the clear supernatant (0.1 ml) to the sample (0.1 mg). The reaction was complete in 30 min at room temperature.

5.1.2 THE FORMATION OF *t*-BDMS ETHERS[14]

To the sample (0.1 mg) in the bottom of a 2×60 mm glass tube was added imidazole (50 μl of a 2 M solution in dimethylformamide) and *t*-butyldimethylchlorosilane (50 μl of a 2 M solution in dimethylformamide). The tube was sealed and heated at 100 °C for 1 h. The mixture was analysed by direct injection of the derivatives isolated by chromatography on Sephadex LH-20 (column 0.5×2.0 cm) swollen in heptane-ethyl acetate (3 : 1). The column was eluted with the same solvent (4 ml) and the eluate evaporated to dryness.

6 DIMETHYLSILYL ETHERS

The dimethylsilyl ethers are more volatile than the TMS ethers, allowing high molecular weight polyols to be analysed more rapidly or at a lower temperature. The relative retention times of the dimethylsilyl derivatives

(DMS) of carbohydrates may be half that of the TMS ethers although in the case of a monohydroxy steroid the retention time difference is much smaller and may be only 5% less than that of the TMS compound.[219] The DMS ethers are prepared from either a mixture of dimethylchlorosilane–tetramethyldisilazane–pyridine (1 : 3 : 9) or from bis-(dimethylsilyl)-acetamide under conditions directly comparable with the anlogous TMS reagents. The DMS-ethers have been used in the analysis of steroids,[26,219] carbohydrates,[219,220] phenols,[219] alcohols,[219,221–223] prostaglandins[215] and nucleosides.[224,225] The derivatives are not as stable as the TMS ethers, are very susceptible to hydrolysis and often develop spurious peaks when stored in the presence of the silyl reagent for more than a few hours.[219] The Si—H bond is considerably more reactive than the Si—CH_3 bond of the TMS ethers and could possibly function as a reducing agent or add to unsaturated systems.

The DMS-ethers are receiving more attention as reagents for mass spectrometry than for gas chromatography at present. In part this is due to the use of DMS-ether as a less expensive substitute for d_9-TMS derivatives in unravelling mass spectral rearrangements. The general features of the mass spectra of DMS derivatives resemble those of the TMS ethers,[222–224] but two important deviations from this general rule are known.[221,224] Unlike their TMS analogues, the DMS ethers of C_5—C_{10} primary aliphatic alcohols exhibit pronounced rupture of the carbon-to-carbon bond adjacent to the oxygen atom within the alkyl moiety (loss of an alkyl radical R) in marked preference to cleavage within the silyl substituent (loss of CH_3). The complementary nature of the two mass spectra (TMS, DMS) allows the nature of the hydroxyl group to be determined, as secondary and tertiary alcohols do not exhibit a preference for α-cleavage within the alkyl residue. The TMS and DMS mass spectra of secondary alcohols are similar, allowing for the appropriate mass shift difference between the TMS and DMS ethers.[221] The mass spectra of TMS ethers are characterized by ions of m/e 73 and m/e 75, but the analogous ions m/e 59 and m/e 61 are rarely seen in the DMS spectra.[221–224] The DMS-ethers have prominent molecular ions in comparison to the TMS-ethers of prostaglandins, and are potentially useful for single-ion monitoring.[215]

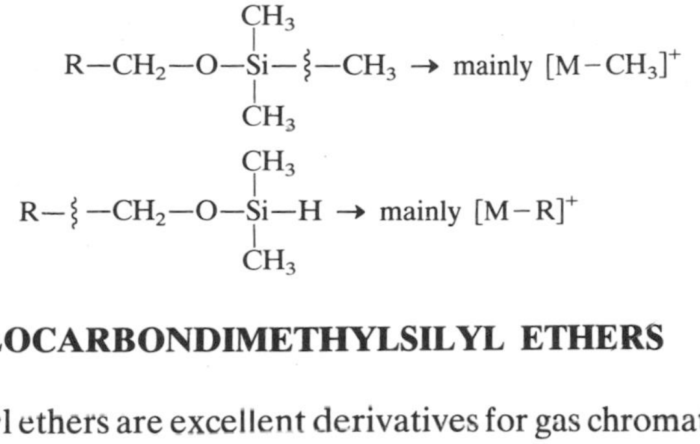

(30)

7 THE HALOCARBONDIMETHYLSILYL ETHERS

The alkylsilyl ethers are excellent derivatives for gas chromatography but are not amenable to determination at low levels with selective detectors. Detection limits were set by the sensitivity of the flame ionization detector which was inadequate for the analysis of many substances of physiological and environmental importance. Selectivity was achieved by coupling a gas chromatograph to an atomic absorption spectrometer and monitoring the silicon emission or absorption, but detection limits were inferior to those of the flame ionization detector.[226] In an attempt to lower the detection limit at which the silyl ethers could be determined, it was desirable to devise reagents that could be used with the selective and sensitive electron-capture detector. The alkylsilyl ethers are not naturally significant electron-capturing groups, so that new reagents have been synthesized in which one of the alkyl groups has been replaced by a halocarbon group. The halocarbon substituent provides the electrophore to which the electron capture detector responds. An electrophore can be defined, by analogy to a chromophore in optical spectroscopy, as that part of the molecule which is responsible for the initial capture of a thermal electron. For details of the selectivity and response of the electron-capture detector to different organic groupings, the reader is referred to Refs. 227–229.

Eaborn, Thomas and Walton introduced the halomethyldimethylsilyl ethers for use with the electron capture detector in which the electrophore was the CH_2X group (X = Cl, Br or I).[230,231] The sensitivity of the CH_2X group towards the electron capture detector mimics that of the halogens themselves (i.e. I > Br > Cl > F). As an approximate guide, the chloromethyldimethylsilyl (CMDMS) ethers have similar sensitivities to the electron-capture detector as to the flame-ionization detector; the bromomethyldimethylsilyl (BMDMS) derivatives are several-fold more sensitive and allow detection limits below the 1.0 ng level to be achieved; the iodomethyldimethylsilyl (IMDMS) derivatives can be determined at the picogram level with detection limits similar to heptafluorobutyrates in many cases. The volatility of the halomethyldimethylsilyl derivatives is the reverse order of their sensitivity to detection CMDMS > BMDMS > IMDMS. By comparison with the TMS derivatives the introduction of a CMDMS group increased the retention time approximately 2–3 times, a BMDMS group 5 times and a IMDMS group 7 times per silyl group introduced.[230,231]

The halomethyldimethylsilyl ethers are easy to prepare from either the appropriate 1,3-bis-(halomethyl)-1,1,3,3-tetramethyldisilazane and the halomethyldimethylchlorosilane or by addition of an aliquot from the reaction of excess halomethyldimethylchlorosilane with diethylamine in hexane.[230–232] The IMDMS ethers are prepared by halide ion-exchange with either the CMDMS or the BMDMS derivative and a saturated solution of sodium iodide in acetone.[231–234]

The halomethyldimethylsilyl ethers are usually prepared under mild reaction conditions, preferably at room temperature. The use of high reaction temperatures (particularly with acids and amines) can result in displacement of the halogen atom or expulsion of the halomethyl group to give a non-electron-capturing derivative or dimer (31).[229] A mass spectral study of the reaction of

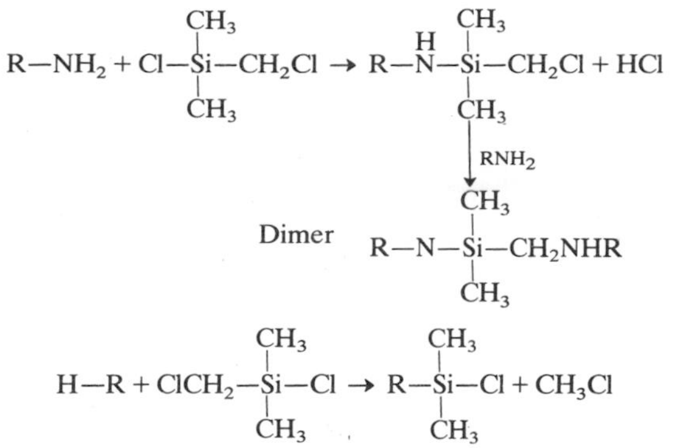

$$R-NH_2 + Cl-\underset{\underset{CH_3}{|}}{\overset{\overset{CH_3}{|}}{Si}}-CH_2Cl \rightarrow R-\underset{\underset{CH_3}{|}}{\overset{\overset{H}{}\overset{CH_3}{|}}{N-Si}}-CH_2Cl + HCl$$

$$\downarrow RNH_2 \qquad\qquad (31)$$

$$\text{Dimer} \quad R-\underset{\underset{CH_3}{|}}{\overset{\overset{CH_3}{|}}{N-Si}}-CH_2NHR$$

$$H-R + ClCH_2-\underset{\underset{CH_3}{|}}{\overset{\overset{CH_3}{|}}{Si}}-Cl \rightarrow R-\underset{\underset{CH_3}{|}}{\overset{\overset{CH_3}{|}}{Si}}-Cl + CH_3Cl$$

short chain aliphatic acids with bromomethyldimethylchlorosilane and di-ethylamine indicated that the products formed were bromomethyltetramethyl-methyldisiloxane (BMTMMDS) esters and not the expected derivatives (32).[235] The BMTMMDS esters were found to have good mass spectral and

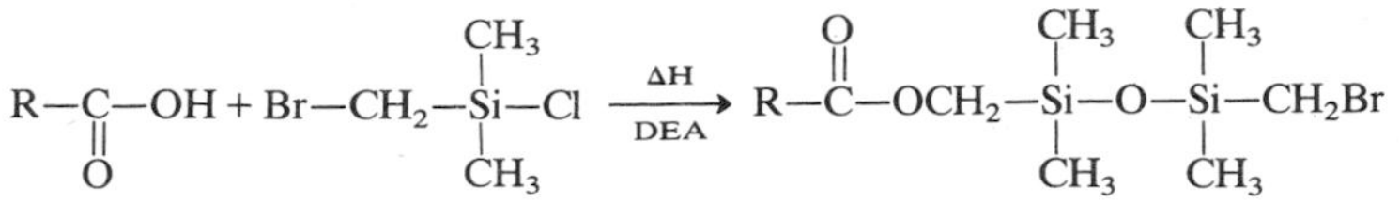

$$R-\underset{\underset{O}{\|}}{C}-OH + Br-CH_2-\underset{\underset{CH_3}{|}}{\overset{\overset{CH_3}{|}}{Si}}-Cl \xrightarrow[\text{DEA}]{\Delta H} R-\underset{\underset{O}{\|}}{C}-OCH_2-\underset{\underset{CH_3}{|}}{\overset{\overset{CH_3}{|}}{Si}}-O-\underset{\underset{CH_3}{|}}{\overset{\overset{CH_3}{|}}{Si}}-CH_2Br \qquad (32)$$

electron-capture properties. The additional molecular weight increment of the substituted disiloxane was no disadvantage for the gas chromatography of the short chain aliphatic acids.[235] For greater sensitivity towards the electron-capture detector the iodomethyltetramethylmethyldisiloxane (IMTMMDS) ester can be formed in the usual way by halide ion-exchange.

Under carefully controlled conditions, quantitative yields of derivatives can be obtained from steroids,[26,230–243] bile acids,[244] prostaglandins,[245] phenolic and acidic insecticides,[246] hydroxystilbenes[247] and carbohydrates.[248] The response of the electron-capture detector would be destroyed by injection of a large excess of silylating reagent into the gas chromatograph over a period of time. Consequently a method is required to separate the active silylating reagents from the derivative. Alumina column chromatography or vacuum evaporation are the methods of choice for this purpose.[231–233] The stability of the halomethyldimethylsilyl ethers to silica gel thin-layer chromatography is in doubt.[238]

The halomethyldimethylsilyl ethers of steroids have useful mass spectral properties.[238–241] A common feature is the elimination of halomethyldimethyl-silanol ($XCH_2Si(CH_3)_2OH$) and halomethylene (XCH_2) with a minimization of further fragmentation. Differences in stereochemistry of the original steroids are better represented with these derivatives than with the TMS ethers.[241] As the atom X decreases in electronegativity, so the bond from the silicon atom to the carbon bearing X becomes stronger and the loss of CH_2X less likely. The CMDMS ethers have relatively intense $[M-CH_2Cl]^+$ ions and are suitable for the determination of steroids at the picogram level by single-ion monitoring.[238–241]

Morgan and Poole have investigated the use of fluorocarbon groups as a method of introducing an electrophore into dimethylsilyl ethers.[15,16] Closely bound fluorine atoms in alkyl or aryl compounds are remarkable in that they show very little increase in boiling-point compared to hydrocarbons of a similar number of carbon atoms in spite of the increase in molecular weight, this increase being offset by a decrease in intermolecular bonding forces in the fluorocarbons.[249] Fluoroalkyldimethylsilanes with fluorine atoms bound to an α or β carbon atom are thermally unstable at the GLC temperatures necessary for sterols, giving alkenes by fluorine migration and elimination:

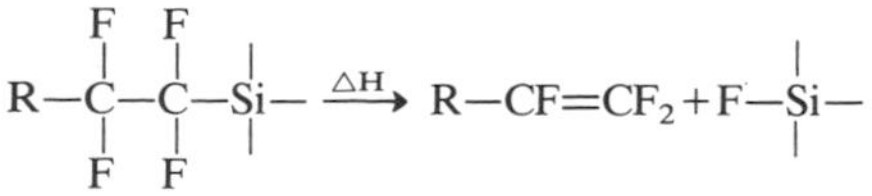

(34)

The relative volatility on gas chromatography and sensitivity towards the electron-capture detector of a series of fluorocarbondimethylsilyl ethers of cholesterol are given in Table 3. The fluoroalkyldimethylsilyl ethers have excellent volatility on gas chromatography but respond poorly to the electron-capture detector. The pentafluorophenyldimethylsilyl ether was nearly twenty times more sensitive to the electron-capture detector than the chloromethyldimethylsilyl ether and only slightly less volatile. This reagent was selected for further evaluation. For convenience, the abbreviated name FLOPHEMESYL, has been introduced for the more cumbersome expression pentafluorophenyldimethylsilyl as the generic stem for naming reagents and derivatives.[16]

TABLE 3

The relative volatility and sensitivity to ECD of a series of R—Si(CH$_3$)$_2$—cholesteryl ethers

R	Relative retention time[a]	Least detectable amount (ng)[b]
CH_3—	1.17	—
$CF_3(CH_2)_2$—	1.47	1500
$CF_3(CF_2)_2(CH_2)_2$—	1.60	115
$ClCH_2$—	2.54	75
C_6F_5—	3.68	4
$C_6F_5CH_2CH_2$—	—	200

[a] R_t cholesterol = 1.58 min^{-1}, 3 ft, 1% OV-101 on Gas Chrom Q, 250 °C, N_2 75 ml. min^{-1}.
[b] Detector oven temperature: 320 °C.

In the gas phase, electron capture by a molecule results in the formation of a negative ion with internal energy equal to the electron affinity of the compound formed. The excess energy of the ion may be dissipated by a dissociative process. In the case of perfluoroalkanes, this is unlikely from thermodynamic considerations due to the high energy of the carbon to fluorine bond. It is more likely that the ion will loose an electron again unless it is stabilized in some way. In small molecules such as CF_4, C_2F_6 and C_3F_8 this auto-ionization is apparently very rapid (less than 1 μs).[250] The auto-ionization mechanism probably explains the poor electron capture response of the heptafluoropentyl and the trifluoropropyl group. Also of note is the fact that the pentafluorophenyl group, when separated from silicon by an alkyl chain as in the 2′-pentafluorophenylethyldimethylsilyl ether, is fifty times less sensitive than the flophemesyl derivative. This can be explained if the captured electron buries itself in the π orbitals of the phenyl ring which is further stabilized by $(p \rightarrow d)\pi$ bonding with low energy orbitals of silicon. In the 2′-pentafluorophenylethyldimethylsilyl ether this type of bonding is absent.[251] For the determination of steroids by electron-capture detection, the introduction of one flophemesyl group allows detection at the nanogram level, whereas the introduction of a further flophemesyl group extends the range to the picogram level.[16]

In pyridine solution, the order of silyl donor power of the flophemesyl reagents* is:

flophemesylamine > flophemesyl chloride > flophemesyldiethylamine
> flophemesyldisilazane ≫ flophemesylimidazole (35)

Flophemesylamine is a particularly useful reagent being selective for unhindered primary and secondary hydroxyl groups in the presence of unprotected ketone groups. Mixtures of flophemesyldiethylamine and flophemesyl chloride provide the most potent silyl donor medium. The known properties of these new reagents and the extent of their reactivity with steroid hydroxyl groups is summarized in Tables 4 and 5.

The flophemesyl ethers have good mass spectral properties showing some marked differences to the TMS ethers of steroids.[252] They have prominent molecular ions with a greater percentage of the total ion current associated with steroid hydrocarbon fragments in comparison with the TMS ethers. Silicon-containing ions of diagnostic value are still easily distinguished but are less prominent. The mass spectra are not complicated by fragmentation of the pentafluorophenyl ring, nor is cleavage between silicon and the pentafluorophenyl ring a dominant process. The flophesmesyl derivatives are characterized by ions of m/e 58 and m/e 77. The relative intensity of the ions of m/e 58 and m/e 77 is variable with m/e 77 generally stronger, but rarely the base peak of the mass spectrum. Elimination of flophemesylanol ($C_6F_5Si(CH_3)_2OH$) is common from hydroxy steroid derivatives.[252] GC–MS of the reaction product from 5β-pregnan-3α,17α,20α-triol and flophemsylamine

$$\underset{m/e\ 58}{[(CH_3)_2Si]^+} \qquad \underset{m/e\ 77}{[(CH_3)_2SiF]^+} \tag{36}$$

* Commercially available in 1976 from Lancaster Synthesis, St. Leonard Gate, Lancaster, U.K.

TABLE 4

Reactivity of flophemesyl reagents

Reagent	Hydroxyl group environment
Flophemesyl chloride in pyridine	Unhindered secondary hydroxyl groups[a] where ketones are first protected
Flophemesylamine	Selectively reacts with unhindered secondary hydroxyl groups[b] in the presence of ketone groups. Does not react with tertiary or hindered secondary hydroxyl groups
Flophemesyldiethylamine and flophemesyl chloride (10:1)	Unhindered secondary hydroxyl groups and exposed tertiary hydroxyl groups[c] in the presence of ketones
Flophemesyldiethylamine and flophemesyl chloride (1:1)	Unhindered and hindered secondary, and exposed tertiary hydroxyl groups. Very hindered tertiary hydroxyl groups[d] do not react completely. Ketone groups must be protected.

[a] The 3β-OH of cholesterol is taken to be a typical unhindered secondary hydroxyl group.
[b] The 11β-OH of 11β-hydroxyandrost-4-en-3,17-dione is taken to be a typical hindered secondary hydroxyl group.
[c] The 17β-OH of 17β-hydroxy-17α-methylandrost-4-en-3-one is taken to be a typical exposed tertiary hydroxyl group.
[d] The 17α-OH of 17α,11β,21-trihydroxypregn-4-en-3,20-dione is taken to be a typical very hindered teriary hydroxyl group.

gave a single product identified as a cyclic 17,20-dimethylsilyl ether, probably arising by nucleophilic attack of the oxygen bonded to C-17 on silicon with expulsion of pentafluorobenzene either in solution or in the gas chromatograph (37):

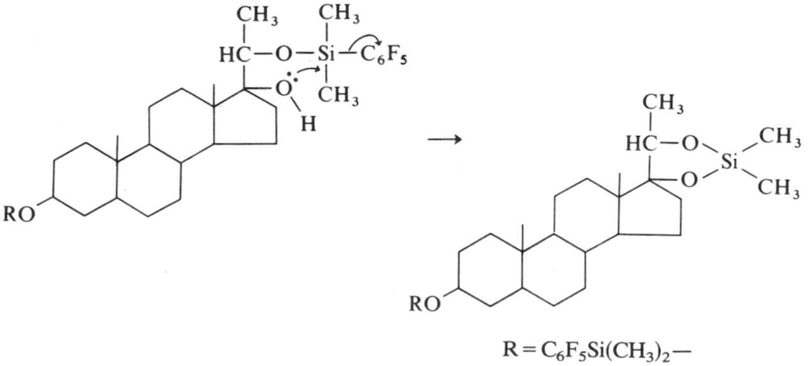

(37)

$R = C_6F_5Si(CH_3)_2—$

The flophemesyl ethers have only been evaluated for the analysis of steroids to date. A combination of their good volatility for gas chromatography, ability

TABLE 5

Conditions for the formation of flophemesyl steroid ethers

Steroid	Flophemesyl chloride in pyridine	Flophemesyl-amine	Flophemesyl-diethylamine: flophemesyl chloride 10:1	Flophemesyl-diethylamine flophemesyl chloride 1:1
Cholesterol	A	B	B	B
Ergosterol	A	B	B	B
Cholestanol	A	B	B	B
2β,3β-dihydroxy-5α-cholestane	A	B	B	B
2β,5α,6β-trihydroxycholestane	2β, 6β A	2β, 6β B	2β, 6β B	2β, 6β B
2β,5α-dihydroxycholestan-6-one	2β A[a]	2β B	2β B	NQ[a]
2β,3β,14α-trihydroxycholest-7-en-6-one	2β, 3β A[a]	2β, 3β B	2β, 3β B	2β, 3β B[a]
3α,20α-dihyroxy-5β-pregnane	A	B	B	B
17α-methyl-17β-hydroxy-androst-4-en-3-one	NQ[a]	NR	B	B[a]
11β-hydroxy-androst-4-en-3,17-dione	NR	NR	NR	A[a]
17α-hydroxypregn-4-en-3,20-dione	NR	NR	NR	C[a]
3α,17α,20α-trihydroxy-5β-pregnane	CP	CP	CP	CP
17α,21-dihydroxypregn-4-en-3,11,20-trione	IR[a]	IR	IR	—
17α,11β,21-trihydroxypregn-4-en-3,20-dione	IR[a]	IR	IR	NQ[a]

A = 3 h at 60 °C NR = no reaction
B = 0.25 h at room temp. NQ = none quantitative
C = 6 h at 85 °C CP = cyclic product (see text)
IR = not all hydroxyl groups react (incomplete reaction)

[a] Ketone protected as its methoxime.

to respond to the electron-capture detector and their useful mass spectral properties should find uses in many other areas of chemistry.

7.1.1 STEROLS AS THEIR CMDMS OR BMDMS ETHERS[232,234]

Solutions of diethylamine (0.5 ml) in hexane (5 ml) and halomethyldimethylchlorosilane (1.0 ml) in hexane (5 ml) were combined in a stoppered centrifuge tube and thoroughly mixed with a vortex mixer. The mixture was centrifuged for 15 min at 1500 r.p.m. and a portion of the supernatant (0.5 ml) added to the steroid (1.0 mg) in a stoppered tube. Polar samples may require the addition of tetrahydrofuran to aid solubility. Unhindered secondary hydroxyl groups were complete silylated in one hour at room temperature.

7.1.2 GENERAL PROCEDURE FOR THE FORMATION OF IODOMETHYLDIMETHYLSILYL ETHERS[234]

The iodomethyldimethylsilyl ethers were prepared from the lower halomethyldimethylsilyl ether by the process of halide exchange. The reagents used to form the halomethyldimethylsilyl ethers, as indicated in thc previous method, were evaporated to dryness with nitrogen at 37 °C, and 3–4 drops of acetone saturated with sodium iodide added. The tube was stoppered and heated at 37 °C for a further 30 min. Hexane (1 ml) was added to precipitate out the excess of sodium iodide, centrifuged and the supernatant and washings collected. This solution was evaporated to a residue with N_2 and redissolved in hexane for analysis.

6 CONCLUSION

If the foregoing discussion on silylation represents the recent past, then what does the future hold? No other protecting groups are as versatile or of such wide applicability as the silyl derivatives, and this position is unlikely to be challenged in the near future. An attack on the mechanism of the silylation reaction may produce a better understanding of those properties most desirable in a reagent and perhaps even greater selectivity in terms of reactivity with different functional groups. The appearance of new silyl reagents with optimum properties for reaction with different acceptor groups is both desirable and a possibility for the future. The use of silyl reagents with electron-capturing properties and the synthesis of new reagents is an area likely to show great expansion in the near future with the impetus coming from a continuing need to determine organic substances at ever lower concentrations. A specific and sensitive detector for silicon which is inexpensive and can be interfaced directly to a gas chromatograph would be a very useful innovation in the light of the ease with which silicon can be selectively introduced into organic molecules and of its natural low abundance in biological substances found in nature.

REFERENCES

1. A. E. Pierce, *Silylation of Organic Compounds*, Pierce Chemical Company, Illinois (1968).
2. A. G. Sharkey, A. Friedel and S. H. Langer, *Anal. Chem.* **29**, 770 (1957).
3. R. C. Osthoff and S. W. Kantor, *Inorg. Syn.* **5**, 55 (1957).
4. P. L. De Benneville and M. J. Hurwitz, *Chem. Abst.* **53**, 12321 (1959).
5. M. Donike, *J. Chromatogr.* **42**, 103 (1969).
6. J. F. Klebe, H. Finkbeiner and D. M. White, *J. Am. Chem. Soc.* **88**, 3390 (1966).
7. D. L. Stalling, C. W. Gehrke and R. W. Zumwalt, *Biochem. Biophys. Res. Commun.* **31**, 616 (1968).
8. R. Kuhn, L. Birkofer, P. Richter and A. Ritter, *Chem. Ber.* **93**, 2804 (1960).
9. R. Ruhlmann, *Chem. Ber.* **94**, 1876 (1961).
10. M. Donike, *J. Chromatogr.* **74**, 121 (1972).
11. R. Piekos, K. Osmialowski, K. Kobylczyk and J. Grzybowski, *J. Chromatogr.* **116**, 315 (1976).
12. R. Piekos, J. Teodorczyk, J. Grzybowski, K. Kobylczyk and K. Osmialowski, *J. Chromatogr.* **117**, 431 (1976).
13. D. J. Harvey and W. D. M. Paton, *J. Chromatogr.* **109**, 73 (1975).
14. R. W. Kelley and P. L. Taylor *Anal. Chem.* **48**, 465 (1976).
15. E. D. Morgan and C. F. Poole, *J. Chromatogr.* **89**, 225 (1974).
16. E. D. Morgan and C. F. Poole, *J. Chromatogr.* **104**, 351 (1975).
17. P. M. Wiesse and R. H. Hanson, *Anal. Chem.* **44**, 2393 (1972).
18. W. A. McGugan and S. G. Howsam, *J. Chromatogr.* **82**, 370 (1973).
19. H. V. Street, *J. Chromatogr.* **41**, 358 (1969).
20. E. C. Horning, M. G. Horning, N. Ikekawa, E. M. Chambaz, P. I. Jaakonmaki and C. J. W. Brooks, *J. Gas Chromatogr.* **5**, 283 (1967).
21. G. G. Esposito, *Anal. Chem.* **40**, 1902 (1968).
22. K. E. Rasmussen, *J. Chromatogr.* **114**, 250 (1975).
23. K. E. Rasmussen, *J. Chromatogr.* **120**, 491 (1976).
24. M. G. Horning, A. M. Moss and E. C. Horning, *Biochem. Biophys. Acta* **148**, 597 (1967).
25. H. L. Lau, *J. Gas Chromatogr.* **4**, 136 (1966).
26. W. J. A. Vandenheuvel, *J. Chromatogr.* **27**, 85 (1967).
27. S. Kawal and Z. Tamura, *J. Chromatogr.* **25**, 471 (1966).
28. P. Capella and E. C. Horning, *Anal. Chem.* **38**, 316 (1966).
29. S. Friedman and M. L. Kaufman, *Anal. Chem.* **38**, 144 (1966).
30. S. Sato and E. von. Rudolf, *Can. J. Chem.* **42**, 635 (1964).
31. R. D. Wood, P. K. Raju and R. Reiser, *J. Am. Oil Chem. Soc.* **42**, 161 (1965).
32. R. D. Partridge and A. H. Weiss, *J. Chromatogr. Sci.* **8**, 553 (1970).
33. J. Lehrfed, *J. Chromatogr. Sci.* **9**, 757 (1971).
34. E. D. Smith, *J. Chromatogr. Sci.* **10**, 34 (1972).
35. E. M. Chambaz, G. Maume, B. Maume and E. C. Horning, *Anal. Lett.* **1**, 749 (1968).
36. L. Aringer, P. Eneroth and J. Gustafsson, *Steroids* **17**, 377 (1971).
37. H. Gleispach, *J. Chromatogr.* **91**, 413 (1974).
38. E. M. Chambaz, C. Madani and A. Ross, *J. Steroid Biochem.* **3**, 741 (1972).
39. E. M. Chambaz, G. Defaye and C. Madani, *Anal. Chem.* **45**, 1090 (1973).
40. E. D. Morgan and C. F. Poole, *J. Chromatogr.* **116**, 333 (1976).
41. W. E. Wilson, S. A. Johnson, W. H. Perkins and J. E. Ripley, *Anal. Chem.* **39**, 40 (1967).
42. B. Maume, W. E. Wilson and E. C. Horning, *Anal. Lett.* **1**, 401 (1968).
43. F. C. Falkner, J. Frolich and J. T. Watson, *Org. Mass Spectrom.* **7**, 141 (1973).
44. I. M. Campbell and R. Bently, in *Carbohydrates in Solution*, G. F. Gould (Ed.), Advances in Chemistry Series 117, American Chemical Society, Washington, U.S.A. 1973, pp. 1–20.
45. W. Pigman and F. L. J. Anet, in *The Carbohydrates*, W. Pigman and D. Horton (Eds), Academic Press, New York, Vol. 1A, 1972, p. 165.

46. C. C. Sweeley, R. Bentley, M. Makita and W. W. Wells, *J. Am. Chem. Soc.* **85**, 2497 (1968).
47. R. Bentley and N. Botlock, *Anal. Biochem.* **20**, 312 (1967).
48. I. M. Campbell and R. Bentley, *Adv. Chem. Ser.* **117**, 1 (1973).
49. J. G. Buchanan and D. M. Clode, *J. Chem. Soc. Perkin Trans. I*, 388 (1974).
50. T. E. Acree, R. S. Shallenberger and L. R. Mattick, *Carbohydr. Res.* **6**, 498 (1968).
51. G. Petersson, *Carbohydr. Res.* **33**, 47 (1974).
52. H. M. Fales and T. Luukainen, *Anal. Chem.* **37**, 955, (1965).
53. F. Vane and M. G. Horning, *Anal. Lett.* **2**, 357 (1969).
54. E. D. Morgan and A. P. Woodbridge, *Chem. Commun.* 475 (1971).
55. S. P. Markey, *J. Chromatogr. Sci.* **11**, 417 (1973).
56. D. J. Harvey, L. Glaszener, C. Straton, D. B. Johnson, R. M. Hill, E. C. Horning and M. G. Horning, *Res. Commun. Chem. Path. Pharmacol.* **4**, 247 (1972).
57. D. J. Harvey, D. B. Johnson and M. G. Horning, *Anal. Lett.* **5**, 745 (1972).
58. E. Bombardelli, A. Bonati, B. Gabetta, E. M. Martinelli, G. Mustich and B. Daniel, *J. Chromatogr.* **120**, 115 (1976).
59. R. M. Thompson and E. C. Horning, *Steroids Lipids Res.* **4**, 135 (1973).
60. U. Henke and R. Tschesche, *J. Chromatogr.* **120**, 477 (1976).
61. E. M. Chambaz and E. C. Horning, *Anal. Biochem.* **30**, 7 (1969).
62. K. M. Brobst and C. E. Lott, *Cereal Chem.* **43**, 35 (1966).
63. C. E. Lott and K. M. Brobst., *Anal. Chem.* **38**, 1767 (1966).
64. A. H. Weiss and H. Tambawala, *J. Chromatogr. Sci.* **10**, 120 (1972).
65. G. D. Britain, J. E. Sullivan and L. R. Schewe, in *Recent Advances in Gas Chromatography*, I. I. Domsky and J. A. Perry (Eds), Marcel Dekker, New York, 1971, p. 223.
66. J. Drozd, *Chromatogr. Rev.* **113**, 303 (1975).
67. L. Fishbein, *The Chromatography of Environmental Hazards*, Vols 1–3, Elsevier, Amsterdam.
68. P. Husek and K, Macek, *J. Chromatogr.* **113**, 139 (1975).
69. W. P. Cochrane, *J. Chromatogr. Sci.* **13**, 246 (1975).
70. S. Ahuja, *J. Pharm. Sci.* **65**, 163 (1976).
71. G. Casparini, M. G. Horning and E. C. Horning, *Anal. Lett.* **1**, 481 (1968).
72. G. G. Esposito and H. M. Swann, *Anal. Chem.* **41**, 1118 (1969).
73. E. R. Atkinson and J. I. Calouche, *Anal. Chem.* **43**, 460 (1971).
74. R. B. Watts and R. G. O. Kekwick, *J. Chromatogr.* **88**, 15 (1974).
75. M. K. Withers, *J. Gas Chromatogr.* **6**, 242 (1968).
76. A. Rajiah, M. R. Subbaram and K. T. Achaya, *J. Chromatogr.* **38**, 35 (1968).
77. K. C. Liebman and E. Ortiz, *J. Chromatogr.* **32**, 757 (1968).
78. S. Inque and M. Miyawaki, *Anal. Biochem.* **65**, 164 (1975).
79. D. A. Kline and E. Fernandez-Flores and A. R. Johnson, *J. Off. Anal. Chem.* **53**, 1198 (1970).
80. J. K. Huttenen and T. A. Mrettin, *Anal. Biochem.* **29**, 441 (1969).
81. F. Eisenberg and A. Bolden, *Anal. Biochem.* **29**, 284 (1969).
82. J. S. Sawardeker, J. H. Sloueker and A. Jeans, *Anal. Chem.* **37**, 1602 (1965).
83. L. M. Morrison and M. B. Berry, *Can. J. Biochem.* **44**, 1115 (1966).
84. L. A. Th. Verhaar and H. G. J. De Wilt, *J. Chromatogr.* **41**, 168 (1968).
85. L. Janson and O. Samuelson, *J. Chromatogr.* **51**, 259 (1971).
86. S. D. Gangolli, R. C. Longland and W. H. Shilling, *Clin. Chim. Acta* **50**, 237 (1974).
87. A. Laine and C. C. Sweeley, *Anal. Biochem.* **43**, 533 (1971).
88. D. J. Harvey and M. G. Horning, *J. Chromatogr.* **76**, 51 (1973).
89. B. S. Mason and H. T. Slover, *J. Agric. Food Chem.* **19**, 551 (1971).
90. J. Havlicek, G. Petersson and O. Samuelson, *Acta Chem. Scand.* **26**, 2205 (1972).
91. R. A. Laine and C. C. Sweeley, *Carbohydr. Res.* **27**, 199 (1973).
92. J. R. Clamp, G. B. Dawson and L. Hough, *Biochim. Biophys. Acta* **148**, 342 (1967).
93. J. Karkkainen, A. Lehtonen and T. Mikkari, *J. Chromatogr.* **20**, 457 (1965).
94. R. A. Larson, G. R. Honold and W. G. Hobbs, *J. Chromatogr.* **90**, 345 (1974).

95. J. Karkkianen and R. Vihko, *Carbohydr. Res.* **10**, 113 (1969).
96. W. W. Wells, T. Katagi, R. Bentley and C. C. Sweeley, *Biochim. Biophys. Res. Commun.* **82**, 408 (1964).
97. W. R. Sherman, S. L. Goodwin and M. Zimbo, *J. Chromatogr. Sci.* **9**, 363 (1971).
98. J.-P. Thenot and E. C. Horning, *Anal. Lett.* **5**, 21 (1972).
99. E. C. Horning, M. G. Horning, J. Szafradek, P. Can Hout, A. L. German, J.-P. Thenot and C. D. Pfaffenberger, *J. Chromatogr.* **91**, 367 (1974).
100. W. G. Stillwell, A. Hung, M. Stafford and M. G. Horning, *Anal. Lett.* **6**, 407 (1973).
101. J. A. Luyten and G. A. F. M. Rutter, *J. Chromatogr.* **91**, 393 (1974).
102. *Organomet. Chem. Rev.* **7 (1)**, 102 (1971).
103. E. M. Chambaz and E. C. Horning, *Anal. Lett.* **1**, 201 (1967).
104. B. Maume, W. E. Wilson and E. C. Horning, *Anal. Lett.* **1**, 401 (1968).
105. N. Sakauchi and E. C. Horning, *Anal. Lett.* **4**, 1 (1971).
106. R. W. Kelley, *J. Chromatogr.* **43**, 229 (1969).
107. G. M. Anthony, C. J. W. Brooks, I. Maclean and I. Sangster, *J. Chromatogr. Sci.* **7**, 623 (1969).
108. C. J. W. Brooks and D. J. Harvey, *J. Chromatogr.* **54**, 193 (1971).
109. T. A. Baillie, C. J. W. Brooks and B. S. Middleditch, *Anal. Chem.* **44**, 30 (1972).
110. E. D. Morgan and A. P. Woodbridge, *Org. Mass Spectrom.* **9**, 475 (1974).
111. E. D. Morgan and C. F. Poole, *Adv. Insect Physiol.* **12**, 17 (1976).
112. E. D. Morgan and C. F. Poole, *J. Insect Physiol.* **22**, 885 (1976).
113. B. A. Knights *J. Gas Chromatogr.* **5**, 273 (1967).
114. W. H. Elliott, L. B. Walsh, M. M. Mui, M. A. Thorne and C. M. Siegfried, *J. Chromatogr.* **44**, 452 (1969).
115. J. Fitelson and G. L. Bowden, *J. Off. Anal. Chem.* **51**, 1224 (1968).
116. J. A. Vollmin, H. R. Bosshard, M. Muller, S. Rampini and H. Curtius, *Z. Klin. Chem. Klin. Biochem.* **9**, 402 (1971).
117. J. L. Wilson, W. J. Dunlap and S. H. Wender *J. Chromatogr.* **35**, 329 (1968).
118. L. Tullberg, I. B. Peetre and B. E. F. Smith, *J. Chromatogr.* **120**, 103 (1976).
119. R. F. Coward and P. Smith, *J. Chromatogr.* **45**, 230 (1969).
120. F. C. Dallos and K. G. Koeppl, *J. Chromatogr. Sci.* **7**, 565 (1969).
121. H. Moreta, *J. Chromatogr.* **71**, 149 (1972).
122. C. Grunwald and R. G. Lockwood, *J. Chromatogr.* **52**, 491 (1970).
123. M. G. Horning, G. Gasparrini and E. C. Horning, *J. Chromatogr. Sci.* **1**, 269 (1969).
124. J. A. Thompson and S. P. Markey, *Anal. Chem.* **47**, 1313 (1975).
125. B. A. Knights, M. Legendre, J. O. King and S. P. Markey, *Clin. Chem.* **19**, 586 (1973).
126. O. A. Mamer, J. C. Crawhall and S. S. Tjoa, *Clin. Chem.* **17**, 802 (1971).
127. T. A. Witter, S. P. Levine, J. O. King and S. P. Markey, *Clin. Chem.* **19**, 586 (1973).
128. N. E. Hoffman, A. Milling and D. Parmelee, *Anal. Biochem.* **32**, 386 (1969).
129. C. M. Schiller and G. K. Summer, *Clin. Chem.* **20**, 444 (1974).
130. T. J. Sprinkle, A. H. Porter, M. Green and C. M. Williams, *Clin. Chim. Acta* **25**, 409 (1969).
131. W. J. Esselman and C. O. Clagett, *J. Lipid Res.* **10**, 234 (1969).
132. R. Kannan, A. Rajiah, M. Subbaram and K. T. Achaya, *J. Chromatogr.* **55**, 402 (1971).
133. R. W. Pero, D. Harvan, R. G. Owens and J. P. Snow, *J. Chromatogr.* **65**, 501 (1972).
134. A. Pinelli and A. Colombo, *J. Chromatogr.* **118**, 236 (1976).
135. H. Rosenquist, H. Kallio and V. Nurmikko, *Anal. Biochem.* **46**, 224 (1972).
136. M. G. Horning, E. A. Boucher, A. Moss and E. C. Horning, *Anal. Lett.* **1**, 713 (1968).
137. U. Langenbeck and J. Seegmiller, *J. Chromatogr.* **78**, 420 (1973).
138. I. Anderson, B. Norkrans and G. Adham, *Anal. Biochem.* **53**, 629 (1973).
139. R. A. Chalmers and R. W. E. Watts, *Analyst* **97**, 951 (1972).
140. R. A. Chalmers and R. W. E. Watts, *Analyst* **97**, 958 (1972).
141. N. E. Hoffman, K. M. Gooding, K. Sheeham and C. A. Tylenda, *Res. Commun. Chem. Path. Pharmacol.* **2**, 87 (1971).
142. N. E. Hoffman and S. J. Houstein, *Anal. Lett.* **5**, 1 (1972).

143. U. Langenbeck, H.-U. Mohring and K. P. Dieckman, *J. Chromatogr.* **115**, 65 (1975).
144. P. H. Scott, *J. Chromatogr.* **70**, 67 (1972).
145. F. Furiuya, S. Shibata and H. Iizuba, *J. Chromatogr.* **21**, 116 (1966).
146. M. P. Gruber, R. W. Klein, M. E. Fox and J. Campisi, *J. Pharm. Sci.* **61**, 1147 (1972).
147. M. A. McClure, *J. Chromatogr.* **54**, 25 (1971).
148. R. A. Heacock and J. E. Forrest, *J. Chromatogr.* **81**, 57 (1973).
149. T. Katagi, A. Horrii, Y. Oomura, H. Miyakawa, T. Kyu, Y. Ikeda and K. Isoi, *J. Chromatogr.* **79**, 45 (1973).
150. G. R. Wilkinson and E. L. Way, *Biochem. Pharmacol.* **18**, 1435 (1969).
151. S. Patel, J. H. Perrin and J. J. Windheuser, *J. Pharm. Sci.* **61**, 1794 (1972).
152. N. E. Hoffman and K. A. Peteranetz, *Anal. Lett.* **5**, 589 (1972).
153. V. Viswanathan, F. P. Mahn, V. S. Venturella and B. S. Senkowski, *J. Pharm. Sci.* **59**, 400 (1970).
154. K. H. Nielsen, *J. Chromatogr.* **10**, 463 (1963).
155. S. Kawai and Z. Tamura, *Chem. Pharm. Bull Jpn* **15**, 1493 (1967).
156. L. D. Metcalfe and R. J. Martin, *Anal. Chem.* **44**, 403 (1972).
157. W. J. A. Vanden Heuvel, *J. Chromatogr.* **36**, 354 (1968).
158. P. W. Albro and L. Fishbein, *J. Chromatgr.* **55**, 297 (1971).
159. R. Piekos, K. Kobylczyk and J. Grzybowski, *Anal. Chem.* **47**, 1157 (1975).
160. P. W. Atkinson, W. V. Brown and A. R. Gilby, *Anal. Biochem.* **40**, 236 (1971).
161. W. C. Butts, *Anal. Biochem.* **46**, 187 (1972).
162. M. G. Horning, A. M. Moss, E. A. Boucher and E. C. Horning, *Anal. Lett.* **1**, 311 (1968).
163. Y. Maruyama and A. E. Takemori, *Anal. Biochem.* **49**, 240 (1972).
164. A. C. Moffat and E. C. Horning, *Anal. Lett.* **3**, 248 (1970).
165. B. F. Maume, P. Bournot, J.-C. L'Huguenot, C. Baron, F. Barbier, G. Maume, M. Prost and P. Padieu, *Anal. Chem.* **45**, 1073 (1973).
166. J.-C. L'Huguenot and B. F. Maume, *J. Chromatogr. Sci.* **12**, 411 (1974).
167. J. P. Hardy and S. L. Kerrin, *Anal. Chem.* **44**, 1497 (1972).
168. C. W. Gehrke and K. Leimer, *J. Chromatogr.* **57**, 219 (1971).
169. C. W. Gehrke and K. Leimer, *J. Chromatogr.* **53**, 201 (1970).
170. C. W. Gehrke, H. Nakamoto and R. W. Zumwalt, *J. Chromatogr.* **45**, 24 (1969).
171. W. J. A. Vanden Heuvel, J. L. Smith and J. S. Cohen, *J. Chromatogr. Sci.* **8**, 567 (1970).
172. W. J. A. Vanden Heuvel, J. L. Smith, I. Putter and J. S. Cohen, *J. Chromatogr.* **50**, 405 (1970).
173. R. Pocklington, *Anal. Biochem.* **45**, 403 (1972).
174. J. R. Watson and R. C. Lawrence, *J. Chromatogr.* **103**, 63 (1975).
175. J. R. Watson, *J. Pharm. Sci.* **63**, 96 (1974).
176. J. L. Laseter, J. D. Weete, A. Albert and C. H. Walkinshaw, *Anal. Lett.* **4**, 671 (1971).
177. K. A. Caldwell and A. L. Tappel, *J. Chromatogr.* **32**, 635 (1968).
178. F. Shahrokhi and C. W. Gehrke, *Anal. Biochem.* **24**, 291 (1968).
179. B. M. R. Heml, H. M. Ortner and H. Spitzy, *J. Chromatogr.* **60**, 51 (1971).
180. R. Bilous and J. J. Windheuser, *J. Pharm. Sci.* **62**, 274 (1973).
181. L. B. Hansen, *Anal. Chem.* **40**, 1587 (1968).
182. K. Funakoshi and H. J. Cahnmann, *Anal. Biochem.* **27**, 150 (1969).
183. R. L. Hancock, *J. Gas Chromatogr.* **4**, 363 (1966).
184. R. L. Hancock, *J. Chromatogr. Sci.* **7**, 366 (1969).
185. T. Hashizume and Y. Sasaki, *Anal. Biochem.* **24**, 232 (1968).
186. H. Iwase, T. Kimura, T. Sugiyama and A. Murai, *J. Chromatogr.* **106**, 213 (1975).
187. C. W. Gehrke, D. L. Stalling and C. D. Ruyle, *Biochem. Biophys. Res. Commun.* **28**, 869 (1967).
188. C. W. Gehrke and D. B. Lakings, *J. Chromatogr.* **61**, 45 (1971).
189. D. B. Lakings and C. W. Gehrke, *J. Chromatogr.* **62**, 347 (1971).
190. D. B. Lakings, C. W. Gehrke and T. P. Waalkes, *J. Chromatogr.* **116**, 69 (1976).
191. D. B. Lakings and C. W. Gehrke, *Clin. Chem.* **18**, 81 (1972).

192. S. Y. Chang, D. B. Lakings, R. W. Zumwalt and C. W. Gehrke, *J. Lab. Clin. Med.* **83**, 816 (1974).
193. V. Miller, V. Pacakova and E. Smolkova, *J. Chromatogr.* **119**, 355 (1976).
194. W. Blum and W. J. Richter, *Tetrahedron Lett.* 835 (1973).
195. P. Vouros, D. J. Harvey and T. J. Odiorne, *Spectrosc. Lett.* **6**, 603 (1973).
196. G. R. Waller (Ed.), *Biomedial Applications of Mass Spectrometry*, Wiley–Intersceince, New York, 1972.
197. J. Diekman, J. B. Thomson and C. Djerassi, *J. Org. Chem.* **33**, 2271 (1968).
198. S. C. Hovlicek, M. R. Brennan and P. Scheuer, *Org. Mass Spectrom.* **5**, 1273 (1971).
199. G. H. Draffan, R. N. Stillwell and J. A. McClosky, *Org. Mass Spectrom.* **1**, 669 (1968).
200. E. White and J. A. McClosky, *J. Org. Chem.* **35**, 4241 (1970).
201. G. Petersson, *Org. Mass Spectrom.* **6**, 565 (1972).
202. G. Petersson, *Org. Mass Spectrom.* **6**, 577 (1972).
203. D. J. Harvey, M. G. Horning and P. Vouros, *Anal. Lett.* **3**, 489 (1970).
204. J. A. McCloskey, R. N. Stillwell and A. M. Lawson, *Anal. Chem.* **40**, 233 (1968).
205. P. Vouros and D. J. Harvey, *Anal. Chem.* **45**, 7 (1973).
206. D. J. Harvey, *J. Chromatogr.* **110**, 91 (1975).
207. E. J. Corey and R. K. Varma, *J. Am. Chem. Soc.* **93**, 7319 (1971).
208. E. J. Corey and A. Venkateswarlu, *J. Am. Chem. Soc.* **94**, 6190 (1972).
209. G. Stork and P. F. Huorlik, *J. Am. Chem. Soc.* **90**, 4462 (1968).
210. K. K. Ogilvie, *Can. J. Chem.* **51**, 3799 (1973).
211. K. K. Ogilvie, K. L. Sadana, E. A. Thompson, M. A. Quilliam and J. B. Westmore, *Tetrahedron Lett.* 2861 (1974).
212. K. K. Ogilvie, E. A. Thompson, M. A. Quilliam and J. B. Westmore, *Tetrahedron Lett.* 2865 (1974).
213. M. A. Quilliam, K. K. Ogilvie and J. B. Westmore, *J. Chromatogr.* **105**, 297 (1975).
214. M. A. Quilliam, K. K. Ogilvic and J. B. Westmore, *Biomed. Mass Spectrom.* **1**, 39 (1974).
215. J. T. Watson and B. J. Sweetman, *Org. Mass Spectrom.* **9**, 39 (1974).
216. L. H. Sommer and L. J. Taylor, *J. Am. Chem. Soc.* **76**, 1030 (1954).
217. J. Diekman, J. B. Thompson and C. Djerassi, *J. Org. Chem.* **32**, 3904 (1967).
218. J. Diekman and C. Djerassi, *J. Org. Chem.* **32**, 1005 (1967).
219. W. R. Supina, R. F. Kruppa and R. S. Henly, *J. Am. Oil Chem. Soc.* **44**, 74 (1966).
220. W. W. Wells, C. C. Sweeley and R. Bentley in *Biomedical Applications of Gas Chromatography*, H. Szymanski (Ed.), Plenum Press, New York, 1964, p. 199.
221. W. J. Richter and D. Hunnemann, *Helv. Chim. Acta* **57**, 1131 (1974).
222. D. H. Hunnemann and W. J. Richter, *Org. Mass Spectrom.* **6**, 909 (1972).
223. D. H. Hunnemann, in *Mass Spectrometry in Biochemistry*, A. Frigerio and A. Castagnoli (Eds), Raven Press, New York 1974, p. 131.
224. J. B. Westmore, D. C. Lin, K. K. Ogilvie, H. Wayborn and J. Berestransky, *Org. Mass Spectrom.* **6**, 1243 (1972).
225. R. L. Hancock, *J. Gas Chromatogr.* **6**, 431 (1968).
226. R. W. Morrow, J. D. Dean, W. D. Shutts and M. R. Geurin, *J. Chromatogr. Sci.* **7**, 572 (1969).
227. C. F. Poole, *Chem. Ind. (London)*, 479 (1976).
228. E. D. Pellizzari, *Chromatogr. Rev.* **98**, 323 (1974).
229. L. M. Cummins, in I. I. Domsky and J. A. Perry (Eds), *Recent Advances in Gas Chromatography*, Marcel Dekker, New York, 1971, p. 313.
230. C. Eaborn, D. R. M. Walton and B. S. Thomas, *Chem. Ind. (London)*, 827 (1967).
231. C. Eaborn, C. A. Holder, D. R. M. Walton and B. S. Thomas, *J. Chem. Soc.* C 2502 (1969).
232. B. S. Thomas, *J. Chromatogr.* **56**, 37 (1971).
233. D. Exley and A. Dutton, *Steroids* **14**, 575 (1969).
234. E. Symes and B. S. Thomas, *J. Chromatogr.* **116**, 163 (1976).
235. J. B. Brooks, J. A. Liddle and C. C. Alley, *Anal. Chem.* **47**, 1960 (1975).
236. B. S. Thomas and D. R. M. Walton, *J. Endocrinol.* **41**, 203 (1968).

237. D. Y. Wang, R. D. Bulbrook, B. S. Thomas and M. Friedman, *J. Endocrinol.* **42**, 567 (1968).
238. C. J. W. Brooks and B. S. Middleditch, *Clin. Chim. Acta* **34**, 145 (1971).
239. C. J. W. Brooks and B. S. Middleditch, *Anal. Lett.* **5**, 611 (1972).
240. J. R. Chapman and E. Bailey, *J. Chromatogr.* **89**, 215 (1974).
241. J. R. Chapman and E. Bailey, *Anal. Chem.* **45**, 1636 (1973).
242. F. Berthou, L. Bardou and H. H. Flock, *J. Chromatogr.* **93**, 149 (1974).
243. D. B. Gower and B. S. Thomas, *J. Chromatogr.* **36**, 338 (1968).
244. W. J. A. VandenHeuvel and K. K. Brady, *J. Chromatogr.* **31**, 9 (1967).
245. G. H. Jouvenaz, D. H. Nugteren and D. A. Van Drop, *Prostaglandins* **3**, 175 (1973).
246. C. A. Bache, L. E. St. John and D. J. Lisk, *Anal. Chem.* **40**, 1241 (1968).
247. J. E. Sinsheimer and R. V. Smith, *J. Pharm. Sci.* **56**, 1280 (1967).
248. W. R. Supina, R. F. Kruppa and R. S. Henly, *J. Am. Oil Chem. Soc.* **44**, 965 (1967).
249. R. D. Chambers, *Fluorine in Organic Chemistry*, Wiley, New York, 1973, pp. 140 and 274.
250. L. A. Rajbenback, *J. Am. Chem. Soc.* **88**, 4275 (1966).
251. C. F. Poole, E. D. Morgan and N. Pacey, *Org. Mass Spectrom.* **10**, 1164 (1975).
252. C. F. Poole and E. D. Morgan, *Org. Mass Spectrom.* **10**, 537 (1975).

CHAPTER 5

Protective Alkylation

Pavol Kováč and Dušan Anderle

*Institute of Chemistry, Slovak Academy of Sciences,
809 33 Bratislava, Czechoslovakia.*

1 INTRODUCTION

In this chapter the term 'protective alkylation' refers to the small-scale reaction of organic substances that contain a reactive hydrogen, e.g. R—COOH, R—OH, R—SH, R—NH—R′, R—NH_2, R—$CONH_2$, R—CONH—R′, R—CO—CH_2—CO—R′, with a derivatizing agent. Replacement of such a hydrogen with an alkyl group is important in chromatographic analysis, because of the decreased polarity of the derivative when compared with the parent substance, facilitating analysis by chromatographic techniques. The decrease in polarity and intermolecular association is particularly important for analysis by gas chromatography and mass spectrometry because it permits their application to the analysis of compounds not amenable to these techniques because of low volatility. The proper choice of a particular type of derivative may be of special importance when dealing with heat-sensitive substances.

No chapter on alkylation methodology can hope to cover all facets of the topic, let alone make reference to all papers which bear on the subject. Here attention has been focused on the most commonly used types of derivatives and procedures which, due to their reliability, have found wide application. Since esterification with alcohols in the presence of acid catalysts is dealt with in Chapter 2, the procedure is considered here only with respect to simple glycosidation of sugars and the formation of benzyl ethers from substances related to catechol. Silyl ether derivatives are dealt with in Chapter 4.

One of the most important areas of chromatography where alkylation has been applied concerns carbohydrates, where methyl derivatives particularly, have been of the utmost value as a means of 'labelling' free hydroxyl groups in structural determination. Most methods which have been used for alkylation of carbohydrates can often be applied successfully in other areas, such as glycolipid, lipid, peptide, glycoprotein and protein chemistry. As substances submitted to the derivatization procedure often contain more than one type of reactive functional group it is important to consider the specificity of the reagent and the possible side reactions which may occur.

2 ALKYLATION PROCEDURES

Haworth originally devised the procedure of methylation in aqueous solution with dimethyl sulfate and sodium hydroxide;[1] similarly, the method of benzylation[2] with benzyl chloride and powdered alkali hydroxide with or without an additional solvent, and the old methylation procedure of Purdie and Irvine[3] are slow reactions: several treatments with the alkylating agent are often required with substrates bearing more than one active site. These methods are, however, still of utility where more powerful procedures cannot be applied because of substrate instability or side reactions, and can equally well be carried out with other reactive halides to introduce other alkyl groups. Purdie's technique was considerably improved by Kuhn and co-workers[4] who used *N,N*-dimethylformamide as a polar but aprotic solvent. Still more powerful alkylation techniques have been devised: when a substrate containing a replaceable hydrogen atom is treated with sodium hydride in a suitable solvent molecular hydrogen is evolved slowly and the resultant sodium salt reacts rapidly with alkyl halides (R′—X). Probably the most remarkable advance in protective alkylation was achieved by Hakomori[5] who developed this approach and used the strongly basic methylsulfinyl carbanion in dimethylsulfoxide for the methylation of glycolipids and polysaccharides. Hydroxyl functions are ionized under these conditions and they react readily with an alkyl halide. Rapid alkylations of other functional groups by this method probably proceed analogously.

Diazoalkanes are well established as alkylation agents. They preferentially alkylate moderately acidic functional groups. It would be beyond the scope of this chapter to discuss all the reactions of diazoalkanes with the whole spectrum of organic substances. The range of possible reactions is extensive, and the possibility of side reactions during alkylation of new compounds with diazoalkanes should therefore always be considered, and the data so obtained should be interpreted with care. An important extension to the use of diazoalkanes was the finding that in the presence of Lewis acids such as boron trifluoride etherate,[6] hydrogen tetrafluoroborate,[7] aluminium chloride,[6] stannous chloride,[8] and several other substances,[9] aliphatic alcohols can also be alkylated with these substances, and the procedure has been extensively used for the derivatization of organic substances.

The widespread use of gas chromatography and mass spectrometry necessitated fast, reliable and simple new derivatization techniques. Apart from silylation, most procedures for the formation of ethers from alcohols do not fulfil these requirements, but several useful reagents, most of them commercially available, have been introduced for rapid alkylation of the more acidic functions.

Many of the procedures given in the following sections are generally applicable. Comments are made on all the methods described and it is hoped that with a little chemical intuition, the reader will be able to choose the reaction conditions best suited for his particular application.

2.1 Alkylation with alkyl halides and silver oxide

$$\text{R—COOH} \rightarrow \text{R—COOR}'$$
$$\text{R—OH} \rightarrow \text{R—OR}'$$
$$\text{R—SH} \rightarrow \text{R—SR}'$$

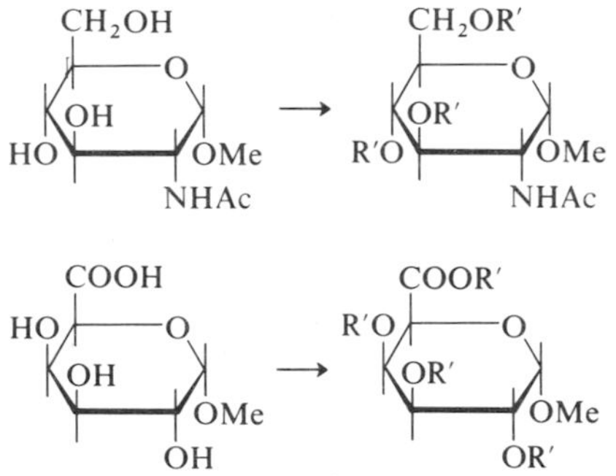

Silver oxide is added to a solution of the substrate in an excess of alkyl halide. The mixture is shaken in the dark until the reaction is complete; the progress of the reaction should be monitored by, e.g. TLC. Fresh portions of silver oxide are added as necessary at two to three-hourly intervals. The reaction mixture is filtered, the solids are washed with a suitable non-hydroxylic solvent and, after concentration, the procedure may, if necessary, be repeated using this partially alkylated product.

Notes: Alkyl iodides and bromides are far more reactive than chlorides and are the reagents of choice. When the substance is insoluble in the halide, a small amount of a suitable additional solvent may be used, but clearly hydroxylic solvents must only be used as a last resort. If hydroxylic solvents have to be used, the extra amount of alkylating agent that will be consumed in alkylating the solvent has to be allowed for.

The procedure will convert any non-hindered carboxylic function (or its salt) to the corresponding alkyl ester in minutes, and phenolic or thiol groups will also be alkylated rapidly. Alcoholic groups are alkylated more slowly. The process may be accelerated by stirring the reaction under reflux and in anhydrous conditions. The addition of dimethyl sulfide to the reaction mixture of methylation of hydroxylic compounds has been shown[28] to affect the rate of *O*-alkylation markedly, and to yield[29] permethylated products in cases where the standard Purdie procedure gives only partially methylated products. Evidence has been presented[28] against the trimethylsulfonium iodide being the alkylating agent in this case, and it has been suggested that the enhancement of

TABLE 1

A summary of derivatizing agents and procedures for their application to the alkylation of specific types of compounds

Reagent	Procedure No.	Recommended for	Not recommended for	Comments
Alkyl halide, silver oxide (no solvent)	2.1	C(1)-protected esterified uronic acids; substances which tend to undergo base-promoted degradation	Free sugars and other easily oxidizable substances	*O*-acyl groups may migrate
Alkyl halide, silver oxide in DMF	2.2	General; certain oligopeptides, amino acids, free sugars,[10] esterified uronic acids[11]	Peptides containing glutamic acid or tryptophane residues	*O*-acyl groups migrate; modification of sulfur-containing residues in peptides may occur[12,13]
Alkyl halide, barium oxide and/or barium hydroxide in DMF	2.3	General	Substances sensitive to base-catalysed degradation	*O*-acyl groups are replaced with *O*-alkyl groups
Alkyl halide, barium oxide in DMF	2.3	General		*O*-acyl migration is promoted and faster than with Ag_2O
Alkyl halide, sodium hydride in DMF	2.4	General	Esterified uronic acids and other substances likely to be modified under strongly basic conditions	*C*-alkylation instead of *O*-alkylation may occur;[14] *O*-acyl groups are replaced with *O*-alkyl groups
Alkyl halide, sodium hydride in DMSO	2.5	General, substances insoluble in common organic solvents; polysaccharides, lipopolysaccharides, peptides, sterically hindered functions, amides, amino acids	Esters and as stated for procedure 2.4; peptides containing sulfur-amino acids, histidine and arginine[15]	*O*-acyl groups are replaced with *O*-alkyl groups. For *N*-isopropyl amino acid, isopropyl ester formation (see Chapter 2)
Alkyl halide, sodium hydride in ether-type solvents	2.6	General; preferred over procedures 2.4 and 2.5 when substrates are soluble in ether-type solvents	As for procedure 2.4	For the derivatization of highly reactive functions sodium hydride may be replaced with potassium carbonate (see procedure 2.16.3 and Ref. 16)

TABLE 1—continued

Reagent	Procedure No.	Recommended for	Not recommended for	Comments
Alkyl halide, sodium hydride in DMAA	2.7	Peptides[17]	As for procedure 2.4	
Diazoalkanes	2.8	Selective alkylation of acidic functions in the presence of aliphatic hydroxyls; esterification of acidic polysaccharides[18,19]		The reaction should be 'alkyl' homogeneous (see *Notes*, procedure 2.8); de-*O*-acetylation may occur in the presence of lower alcohols[20]
Diazoalkanes, Lewis-acid	2.9	Less reactive functions; alkylation in the presence of base-sensitive substituents	Extremely acid-labile substances	*O*-acyl migration does not occur; polymethylene is the usual by-product of methylation
DMF-dialkyl acetals	2.10	Sterically hindered carboxylic acids, aldehydes, phenols, amines		*Cis*-diols will be simultaneously acetalized,[21] acetals may undergo *trans*-acetalization
3-Alkyl-*p*-tolyl-triazenes	2.11	Certain carboxylic acids and phenols		
Alkyl fluoro sulfonates	2.12	General		Ether cleavage and non-specific alkylation may occur
Trialkylanilinium hydroxide	2.13	*N*-alkylation of barbiturates,[22] other sedatives, phenolic alkaloids and related substances		Convenient flash-heater (or on-column) methylation for GLC analysis; the injector temp. should be set between 250–300 °C
Trialkyloxonium fluoroborate	2.14	Exocyclic *O*-alkylation in mesoionic ring systems[23]	*N*-acyl derivatives when these groups are to be preserved	*N*-deacylation occurs[24,25]
Alcohols, acid catalyst	2.15.1	Catechol and related substances		Two electron donating functions in the aromatic ring are essential for the alkylation to occur; possible risk of racemization of optically active substances

TABLE 1—continued

Reagent	Procedure No.	Recommended for	Not recommended for	Comments
Alcohols, acid catalyst	2.15.2	Free sugars		Losses due to degradations commencing at the reducing end of the sugars do not occur when these are first converted to glycosides; acid labile substituents are alcoholyzed and those at the reducing end are replaced with the alkyl group to give glycosides; oligosaccharides and polymeric glycosidically linked substances are alcoholyzed to variable extent giving lower glycosides
Pentafluorobenzyl chloroformate	2.16.1	Derivatization of tertiary amines for ECD-GLC analysis and mass spectrometry		Convenient introduction of fluorine when perfluoro acid chlorides or anhydrides cannot be used
O-(2,3,4,5,6-pentafluorobenzyl) hydroxylamine hydrochloride	2.16.2	Derivatization of steroids[26] and other ketones for ECD-GLC analysis and mass spectrometry		Convenient fluorine-introducing alkylation for ketones
Pentafluorobenzyl bromide	2.16.3	Preparation of pentafluorobenzyl ethers and esters for ECD-GLC and MS analysis; *N*-alkylation of barbiturates[16] and sulfonamides[27]		
Aldehydes, Pd/C catalyst	2.17	Amino acids		*N*-alkyl formation should be combined with esterification before GC and/or MS; aromatic nitroacids are reduced and *N*-alkylated in one operation

the rate of *O*-alkylation in the presence of dimethyl sulfide is due to a modification of silver oxide, e.g. by complex formation, in such a manner as to convert it to a more efficient base. This is supported by the fact, that degradation by β-elimination of uronic acid derivatives, similar to a strong base-induced analogous reaction, was observed to have taken place when substances of this class were methylated with Purdie's reagent in the presence of dimethyl sulfide.[30,31]

The conditions of procedure 2.1 are probably the mildest available for the alkylation of aliphatic hydroxyl groups. When the substrate contains *O*-acetyl groups these survive but may migrate; *N*-acetyl groups are also stable to the process, so that when *N*-acetamidodeoxy sugars or related substances are alkylated in this way, *O*-alkylated-*N*-acetamido derivatives are produced, the *N*-acetamido group remaining intact.[32] The procedure is not recommended for the alkylation of free sugars since oxidative degradation by silver oxide may occur.

2.2 Alkylation with alkyl halides and silver oxide in *N,N*-dimethylformamide

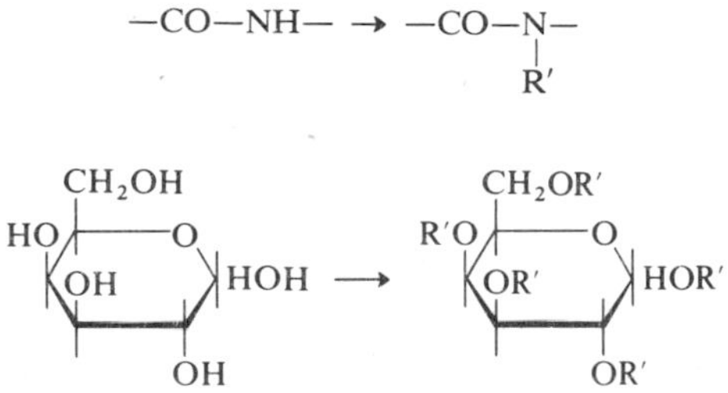

For further reactions see Section 2.1.

Silver oxide and the alkyl halide (1 : 20, w/v) are added to a solution of the substrate in *N,N*-dimethylformamide and the suspension is stirred at room temperature until the reaction is complete. The mixture is filtered, the solids washed with *N,N*-dimethylformamide, and chloroform (4 volumes) is added to the filtrate. The organic phase is washed with 5% KCN solution to remove silver salts which may have dissolved, then with water, and is then dried over anhydrous sodium sulfate to give, after concentration, the peralkylated product. Any residual *N,N*-dimethylformamide, if present, does not normally interfere with chromatographic analysis.

Notes: Complete substitution of some less reactive substrates may need several days' reaction; repeated treatment with the reagent is commonly required. The procedure has been applied[33] to permethylation of *N*-acyl oligopeptides, and was used in a number of cases for amino acid sequence determination. The presence of any existing *N*-methylamino acid residues in the peptides could be detected by mass spectrometry of the permethylated product when the above procedure was carried out with trideuteriomethyl iodide.[34,35] However, partial chain cleavage of certain peptides and modification of sulfur-containing residues have been observed during this alkylation process[12,13] and, consequently, the method is no longer employed for the permethylation of peptides. On the other hand, it has been established[10] that the methylation of unprotected aldoses, ketoses and uronic acids proceeds smoothly, since the substitution of the hemiacetal hydroxyl group occurs before the oxidative degradation, due to the presence of silver oxide, can take place. The methylation of free sugars may be carried out reliably by this method. The same procedure has been applied successfully[36] to the methylation of *N*-protected amino acids: the corresponding *N*-protected-*N*-methylamino acid methyl esters are obtained in excellent yield.

2.3 Alkylation with alkyl halides and barium oxide in *N,N*-dimethylformamide

For the chemistry see Section 2.2.

An excess of alkyl halide is added to a solution of the substrate in *N,N*-dimethylformamide (1:10–20, w/v) followed by finely divided barium oxide, and the mixture is stirred vigorously in a flask equipped with an efficient condenser (not necessary for high-boiling halides, in which case the flask is closed with a drying tube, or for milligram-scale reactions, where the reaction is carried out in a screw-capped vial). The temperature of the reaction mixture rises slowly, depending on the size of the reaction, and one hour after the temperature returns to room temperature the reaction is checked for completion. For isolation the mixture is diluted with a suitable water-immiscible solvent, filtered, the solids are washed, and the combined filtrates are washed with water (for water-soluble substances it may be necessary to backwash the aqueous washings). The desired product is obtained by concentration of the organic phase.

Notes: Compared to procedures 2.1 and 2.2 alkylation is faster under these more strongly basic conditions. When working under scrupulously dry conditions the *O*-acyl groups present in the substrate survive, but *O*-acyl migration is promoted. In the modified version[37] of the procedure barium oxide is replaced with barium hydroxide or barium oxide–barium hydroxide 1:1 as the base, and here *O*-acyl groups are completely replaced by *O*-alkyl groups. Variable results have been reported as to the completeness of *N*-alkylation using this procedure and prolonged or repeated treatment with the reagent may be necessary.

2.4 Alkylation with alkyl halides and sodium hydride in *N,N*-dimethylformamide

$$R{-}CONH_2 \rightarrow R{-}CON(R')_2$$

For further reactions see Section 2.5.

Powdered sodium hydride (~3 equiv./H to be replaced) is added to a solution of the substrate in *N,N*-dimethylformamide and the suspension is swirled for 15 min, with the exclusion of atmospheric moisture and carbon dioxide (KOH drying tube). The mixture is then cooled in ice and the halide (~2 equiv/H to be replaced) is introduced dropwise. The reaction mixture is stirred, again with the exclusion of moisture and CO_2, for 1–2 h (the content of the reaction vessel thickens sometimes, so that stirring becomes impossible, but later a clear solution is obtained) after which time the reaction is complete. Methanol is then added cautiously to destroy the excess of the alkylating reagents and, when effervescence stops, the mixture is partitioned between chloroform and water. The product is isolated by concentration of the organic phase.

Notes: Powdered sodium hydride has been used in the authors' laboratory without difficulty, and is preferred particularly when working on the milligram scale. When obtained commercially as a dispersion in mineral oil, the oil must be removed before use by washing the dispersion with dry ether under nitrogen on a sintered-glass funnel. When dimethylsulfoxide is used instead of *N,N*-dimethylformamide procedure 2.5 should be used.

This procedure is a more powerful method of alkylation than those described above. Sterically hindered groups will, as a rule, be quantitatively alkylated. However, the strongly basic reaction conditions of the procedure should be taken into account whenever sodium hydride is used as the base. Substances known to undergo base-catalysed transformations will do so under these conditions: thus, esterified uronic acid derivatives treated with sodium hydride will be more or less converted to 4,5-unsaturated-4-deoxy-hexuronates.[11] As these unsaturated sugars are not[38] the final products of the base-promoted β-elimination degradation of esterified hexuronic acids, one cannot expect a single product when such compounds are alkylated by this procedure. The same holds for the very efficient alkylation procedure described next (Section 2.5). Under the conditions of procedure 2.4, degradation by β-elimination of uronic acid derivatives was reported[39] not to occur as much as when dimethylsulfoxide was used as the solvent (Section 2.5), but it is still safer to alkylate this class of substances using the procedure described in Section 2.2 (see Ref. 11).

Some substances may undergo unwanted *C*-alkylation instead of, or in addition to, *O*-alkylation as for example when derivatives of L-ascorbic acid are treated with an alkyl halide in the form of their sodium salt.[14] *C*-Methylation as well as other unwanted side-reactions are more likely to be encountered when working with sub-milligram quantities of the substrate, as was the case for instance with peptides,[40] probably because in such cases the use of too large an excess of the base is difficult to avoid.

2.5 Alkylation with alkyl halides and methylsulfinyl anion in dimethylsulfoxide

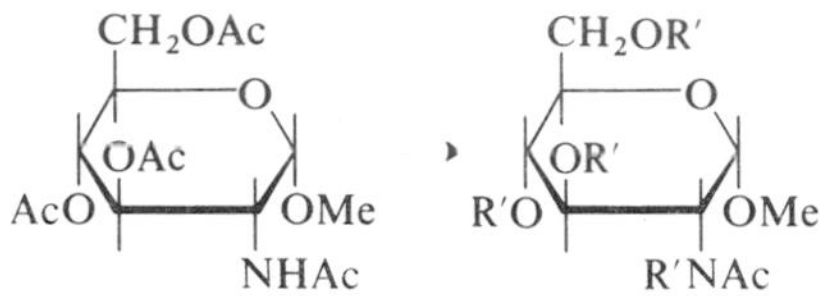

$$R{-}CO{-}NH_2 \longrightarrow R{-}CO{-}N(R')_2$$

$$R{-}CO{-}NH{-}CH_3 \xrightarrow{CD_3I} R{-}CO{-}N(CD_3){-}CH_3$$

The base is prepared by heating sodium hydride in dry dimethylsulfoxide at 50 °C in a flask protected from atmospheric moisture and CO_2 with a KOH drying tube. The solution becomes green and evolution of hydrogen gas ceases after approximately 1 h, after which time a solution of the substrate in dry dimethylsulfoxide is added. The mixture is stirred for 20 min at 50 °C and, after cooling to room temperature, the halide is introduced drop by drop with stirring, which is continued for 1 h. (The reaction is exothermic and, therefore, for alkylations with low-boiling halides the reaction flask is held in an ice-bath during the addition of the halide.) When the reaction is complete, as indicated by the formation of a clear solution, water is added cautiously and the solution is extracted with chloroform. The chloroform solution is backwashed with water, the organic phase dried with anhydrous sodium sulfate and concentrated. Polymeric material is most conveniently isolated by dialysis and freeze-drying.

Notes: CAUTION: Although no difficulties have been experienced in the authors' laboratory, care must be taken when heating sodium hydride in dry dimethylsulfoxide as there have been reports of violent explosions when the anion was prepared on a large scale.[40,41] The amount of the base and of the halide should be greater than 50% excess over the number of equivalents of replaceable hydrogen atoms.

The important difference between the alkylation procedures involving sodium hydride in polar aprotic solvents and those involving alkyl halides and metal salts, is that *O*-acyl groups are completely replaced by *O*-alkyl groups.[42] *N*-Acyl groups survive and *N*-alkyl-*N*-acylamido derivatives are produced. This alkylation can be performed on an extremely small amount of material and, thus, the procedure provides a rapid means for methylation-linkage analysis in the polysaccharide,[42] lipopolysaccharide,[43] and peptide[13,15] fields.

Although it has been claimed[44] in the earlier application of this procedure to acidic carbohydrates that elimination reactions do not occur under these conditions, convincing evidence has been presented[11,39,45] showing that base-sensitive substances will undergo side-reactions during treatment with this reagent, particularly on repeated[46] alkylation, i.e. when complete substitution is not achieved in one step. Consequently, substrates where base-promoted unwanted transformations are likely to occur should not be alkylated in this way, or substances from which on alkylation such structures are formed should not be realkylated using ths procedure. Another complication which may be encountered when other functional groups of esters are alkylated in the presence of the methylsulfinyl anion is the formation of methylsulfinyl ketones.[47] These difficulties may be largely overcome by saponification of the esters before treatment with the base. The ester formed will be exposed to the strong base only for a short time which will minimize unwanted side-reactions.

2.6 Alkylation with alkyl halides and sodium hydride in ether-type solvents

For typical reactions see Section 2.5.

Sodium hydride (3 equiv./H to be replaced) is added to a solution of the substrate (substrate–solvent ratio ~1:20, w/v) contained in a small flask which is then protected from atmospheric moisture and CO_2 with a KOH drying tube. The flask is shaken occasionally and when effervescence ceases the halide (2 equiv/H to be replaced) is introduced with stirring which is then continued for 15–60 min. When the reaction is complete the excess of the etherification reagents is destroyed by the addition of methanol, and the mixture is ready for chromatographic analysis. Alternatively, the mixture is diluted with water and the organic solvents are removed with a rotary evaporator, whereupon the product often separates. It is filtered off, dissolved in chloroform and the chloroform solution is washed in a separatory funnel with water until neutral. The product is isolated by concentration of the organic phase (dried over anhydrous calcium chloride or sodium sulfate). For water-soluble substances the aqueous phase obtained after evaporation of organic solvents is extracted with chloroform and treated as above.

Notes: This is probably the simplest of the very efficient methods of alkylation, possessing the advantage over the procedures involving *N,N*-dimethylformamide or dimethylsulfoxide in that the isolation of substances can be performed on a very small scale without the polar aprotic solvents interfering. Sterically hindered functional groups will be quantitatively converted to the fully substituted products although in this case, or when alkylation is carried out with less reactive halides, it may be necessary to prolong the reaction time or to apply gentle heating. Thus, for instance, tritylated carbohydrate derivatives were successfully converted[48] to per-*O*-methyl and per-*O*-benzyl derivatives in excellent yield. The procedure is a matter of choice when the substrate is soluble in any suitable anhydrous ether-type solvent. For possible side reactions see the notes to procedures 2.4 and 2.5.

2.7 Alkylation with alkyl halides and sodium hydride in *N,N*-dimethylacetamide

For typical reactions see Section 2.5.

Sodium hydride is added to *N,N*-dimethylacetamide and the solution is heated under a KOH drying tube at 120–125 °C until gas evolution ceases. The solution is cooled to room temperature and added to a solution of the substrate in *N,N*-dimethylacetamide, followed immediately by the halide. The mixture, protected from contact with air, is stirred for 1 h. For the substrate : solvent : reagent ratios and work-up see Section 2.4.

Notes: The procedure is closely related to the one using *N,N*-dimethylformamide as the solvent. Methylation of a peptide under these conditions was reported[17] to be cleaner than when methylsulfinyl carbanion was used as the base (procedure 2.5), although this may be due to the difference in the reaction time or the actual amount of the base.

2.8 Alkylation with diazoalkanes

$$\text{R—COOH} \xrightarrow{CH_2N_2} \text{R—COOMe}$$

$$\text{Ph—OH} \xrightarrow{CH_2N_2} \text{Ph—OMe}$$

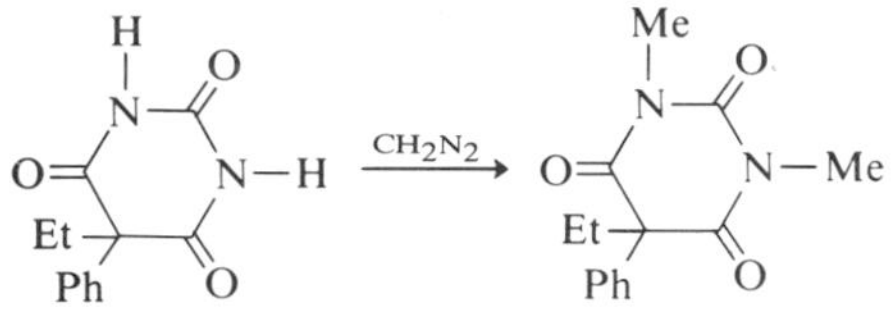

To a solution of the substrate a solution of diazoalkane is added slowly until a faint yellow colour persists and the evolution of nitrogen gas ceases. Concentration of the solution affords virtually pure alkylated product.

Notes: CAUTION: As diazomethane and related substances are both toxic and explosive, all work with then should be carried out behind a safety shield in an efficient hood. For further details concerning the preparation and handling of diazoalkanes see Chapter 2.

Except when used for etherification of aliphatic alcohols, in which case dichloromethane is most frequently used as the solvent (Section 2.9), ethereal and ether-alcoholic solutions are the most commonly used forms of diazoalkanes in alkylation reactions. The substrate may be dissolved in ether (preferred), alcohol, alcohol–water[49] or dimethylsulfoxide, a solvent which has been successfully applied for acidic polysaccharides.[18,19]

Methanol is reported [50,51] to catalyse the methylation of certain hydroxy compounds which are stable towards ethereal diazomethane. It has been suggested that diazomethane reacts with lower alcohols to give a new powerful alkylating agent in order to explain the observed[50] *n*-propylation with ethereal diazomethane in the presence of *n*-propylalcohol. Therefore, when diazoalkylation is carried out in the presence of alcohols the reaction mixture should be 'alkyl homogeneous'.

Diazoalkanes alkylate acidic and enolic groups rapidly and other groups with replaceable hydrogen slowly. Carboxylic and sulfonic acids, as well as phenols and enols will be almost instantaneously converted to the corresponding alkyl derivatives when treated with the reagent. In this way *N*-alkyl (methyl and ethyl) derivatives of several barbiturates possessing acidic NH groups have been prepared[52,53] by treatment with diazomethane and diazoethane, respectively. Catalysts are normally necessary with aliphatic alcohols and other substances bearing low reactivity hydrogen (Section 2.9), so that many carboxylic acids and phenols can be selectively alkylated by means of plain diazoalkanes in alcoholic solution. Occasionally, however, as in the case of tartaric acids and their esters, the hydroxyl groups were smoothly methylated[54,55] by diazomethane in ether. Further, aliphatic alcohols were also reported to be smoothly methylated by diazomethane when hexane or heptane replaced ether as the solvent.[56]

Diazoalkanes are among the most versatile reagents available to the organic chemist. Their high reactivity towards a number of types of organic substances has to be taken into account when they are used for alkylation, because in the presence of other groups which combine with this class of substances the resulting derivative may be far from the one expected. One most unusual example may be cited:[57] when adenosine was treated with diazoethane in aqueous 1,2-dimethoxyethane, mixed methylation and ethylation occurred as a result of solvent participation in the diazoalkylation reaction.

When *O*-acylated substances are alkylated with diazoalkanes in the presence of lower alcohols, the product should not be left for long in the resulting solution containing the excess of the reagent since de-*O*-acylation may occur.[20]

In view of commercial availability of deuteriodiazomethane precursor, diazoalkylation is a powerful means for the specific labelling of various classes of organic substances.

2.9 Lewis-acid-catalysed alkylation with diazoalkanes

$$\text{R—OH} \xrightarrow[\text{BF}_3\text{.Et}_2\text{O}]{\text{CH}_2\text{N}_2} \text{R—OMe}$$

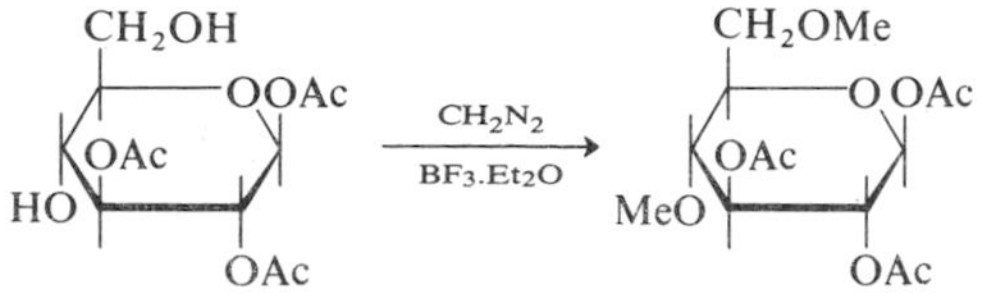

The substrate is dissolved in dichloromethane (1:10–50, w/v, depending upon the solubility at low temperature) and the solution is cooled to −10 °C. The solution is magnetically stirred while boron trifluoride etherate (1–3 drops, or its solution in dichloromethane) is added, followed by a solution of diazoalkane in dichloromethane until a faint yellow colour persists for at least 5 min. The solution is stirred at below 0 °C, protected from atmospheric moisture, and further portions of the reagents are periodically added until a suitable test shows satisfactory conversion of the starting material into the alkylated product. The solution is filtered to remove any polymeric material formed as a by-product, the filtrate is washed successively with a solution of sodium bicarbonate and water, dried over anhydrous sodium sulfate, and concentrated.

Notes: Alkylation of partially *O*-acylated substances, such as carbohydrates, with an alkyl halide and metal salts (Sections 2.1–2.3) frequently give rise to acyl migration. The more powerful alkylations, on the other hand, cause partial or complete replacement of the acyl groups with the alkyl groups. However, the use of diazomethane–boron trifluoride etherate reagent to methylate *O*-acylated sugars bearing free hydroxyl groups has been repeatedly shown[58–66] to give the corresponding methyl ethers without the migration of base-labile substituents, and it appears that this is the only safe methylation agent when such groups are present in the substrate.

Apart from complications due to the side-reactions already mentioned, the Lewis-acid may cause unwanted acid-catalysed transformations. When the substrate is likely to show acid-lability, to avoid high local concentration of the catalyst, the addition of a dilute solution of the catalyst in dichloromethane (1 : 100–1000, v/v) rather than of the neat catalyst is recommended. Furthermore, when the stereochemistry of the substrate is favourable, addition of diazomethane instead of, or in addition to, simple alkylation may occur. In this way when derivatives of methyl α-D-galacturonate having the hydroxyl group in the 4-position unsubstituted were methylated under the conditions of procedure 2.9, in addition to the formation of the C(4)-*O*-methyl ether (major product), addition of diazomethane to the methoxycarbonyl group followed by cyclization occurred, finally giving bicyclic tetrahydrofurone dimethyl acetal derivatives as by-products.[67]

Variable amounts of the catalyst with respect to the amount of the substrate, solvent and alkylating agent have been used, and the reaction when carried out with carbohydrates bearing more than one free hydroxyl group has rarely been complete (70–90% conversion of the starting material to the fully substituted product is normally achieved). Although any Lewis-acid would probably catalyse diazoalkylations of groups which, without a promoter, are resistant towards this type of reagent, boron trifluoride etherate is the one most frequently used. A successful use of $SnCl_2 2H_2O$ (10^{-3} mol g^{-1}) in the benzylation of nucleosides with phenyldiazomethane in 1,2-dimethoxyethane has been reported.[68] The same catalyst was effective in cases where boron trifluoride etherate failed to catalyse methylation with diazomethane of nucleosides because of the presence of the basic heterocycle which may have effectively complexed the boron trifluoride reagent. In most Lewis-acid-catalysed diazoalkylations dichloromethane has been used as the solvent, and the reaction has been conducted at temperatures ranging from −20 to 0 °C. Large excess of the catalyst and/or temperatures above 0 °C should be avoided as polymerization of diazoalkanes, i.e. the loss of the alkylating agent, will then be more pronounced. For side-reactions see *Notes*, procedure 2.8.

2.10 Alkylation with *N,N*-dimethylformamide dialkyl acetals

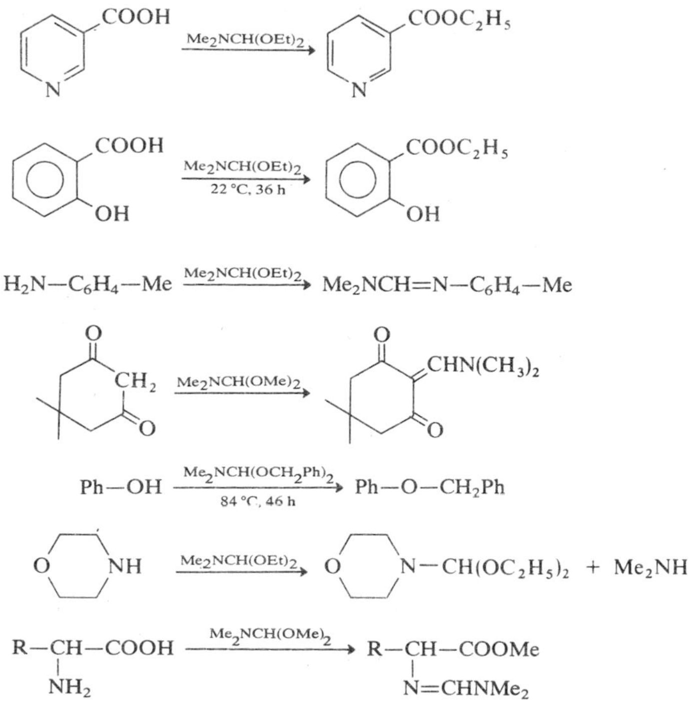

A concentrated solution of the substrate in a suitable solvent is treated with an excess of the reagent (1:4–6, w/v) and the mixture is heated at 50–60 °C for ~15 min in a flask protected from contact with air. The excess of the reagent is removed under reduced pressure and the solution of the residue in a water-immiscible solvent is washed with water, dried over anhydrous sodium sulfate and concentrated.

Notes: *N,N*-dimethylformamide dimethyl acetals are moisture-sensitive. They are hydrolysed to give *N,N*-dimethylformamide and the corresponding alcohol. The derivatization should therefore be carried out under scrupulously dry conditions. The reaction is, as a rule, complete immediately upon dissolution of the substrate and, therefore, heating is necessary only with samples which do not dissolve at room temperature. The final simple purification should not be omitted as extraneous peaks on gas chromatograms have been reported to appear when the reaction mixtures were directly injected.[69] Pyridine, benzene, alcohols, halogenated hydrocarbons, *N,N*-dimethylformamide, acetonitrile, and tetrahydrofuran have been used as the solvent.

Carboxylic acids,[70] phenols[70] and thiols[71] quickly react to give the corresponding alkyl derivatives. Thus, simple amino acids[72] are fully derivatized, *N*-dimethylamino methylamino acid alkyl esters being produced. Free amines and substances containing the —CO—NH— groups give,[21] respectively, *N*-dialkyl amino and *N*-alkyl derivatives. Although hydroxyl groups on hydroxy substituted amino acids were reported[72] not to react under the conditions of derivatization with *N,N*-dimethylformamide dialkyl acetals, *O*-alkylation in other systems has been observed (see Ref. 21) as was the simultaneous acetalization of *cis*-diol groups.[21]

Sterically hindered acids, e.g. trimethylbenzoic acid, react almost as rapidly with this type of reagent as unhindered acids, e.g. benzoic acid, and similarly hindered functions can be expected to react in the same way. This type of reagent will condense with active methylene compounds and is capable of exchange reactions to generate new acetals, and this fact has to be taken into account when substrates amenable to such reactions are being derivatized. A wide spectrum of *N,N*-dimethylformamide dialkyl acetals is now commercially available.

2.11 Alkylation with 3-alkyl-1-*p*-tolyltriazenes

$$\text{R—COOH} \xrightarrow{\text{R}'\text{—NH—N=N—C}_6\text{H}_4\text{—CH}_3} \text{R—COOR}'$$

$$\text{Ph—OH} \xrightarrow{\text{R}'\text{—NH—N=N—C}_6\text{H}_4\text{—CH}_3} \text{Ph—OR}'$$

The solution of the substrate (preferably in an ether-type solvent) is treated with a solution of the triazene (1.1 equiv./H to be replaced, or a larger excess for groups assumed to be less reactive) and the reaction mixture is stirred for 1 h while nitrogen is evolved. When the conversion is complete, as often indicated by ceased effervescence, excess of the reagent is destroyed by careful addition of 10% hydrochloric acid. The mixture is quantitatively transferred into a separating funnel and washed successively with 10% hydrochloric acid, then a dilute solution of sodium bicarbonate, and finally water. The organic phase is dried over a suitable dessicant and concentrated.

Notes: Gentle heating accelerates the reaction. For isolation of pure substances it may be necessary to decolourize the final solution with a little charcoal or silicic acid.

Carboxylic acid esters can be conveniently prepared in this manner using many commercially available 3-alkyl-1-*p*-tolyltriazenes. Alkylation of phenols, mercaptans and other less reactive functional groups requires more vigorous reaction conditions.[73]

CAUTION: The compounds of this class have been reported to exhibit carcinogenic activity and should be handled accordingly.

2.12   Alkylation with alkyl fluorosulfonates

$$R\text{—}OH \xrightarrow{CF_3SO_3Et} R\text{—}OEt$$

$$R_2\text{—}NH \xrightarrow{CF_3SO_3Et} EtR_2\overset{+}{\text{—}}NH \cdot CF_3SO_3^-$$

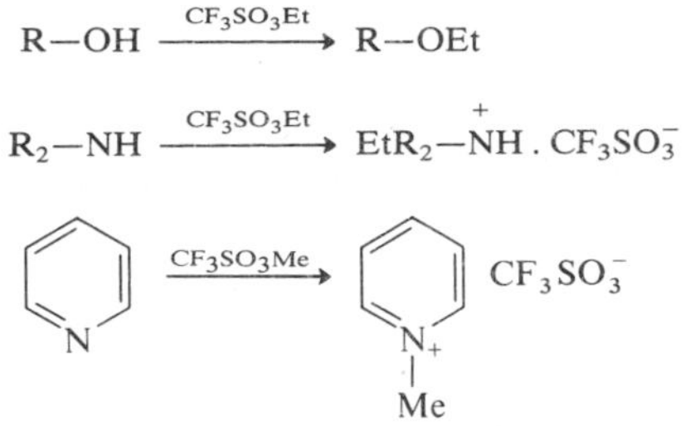

A mixture of the substrate and an alkyl fluorosulfonate (1 equiv./H to be replaced), protected from atmospheric moisture, is kept at room temperature for 5–16 h or, alternatively it is heated in a sealed tube at 100 °C for 2–8 h.

Notes: Alkyl fluorosulfonates are extremely powerful alkylating agents. Their high reactivity may therefore result in nonspecific alkylation. Among the possible side-reactions, alkylation of aromatic substances and *O*-alkylation by ether-cleavage may be cited.[74] Although numerous data are available in the literature[74–80] the full scope of alkylation with this type of reagent and, particularly, their value in the preparation of derivatives for analysis by GC and mass spectrometry has yet to be explored.

2.13   Alkylation with trimethylanilinium hydroxide

$$\underset{/}{\overset{\backslash}{N}}\text{—}H \xrightarrow{TMAH} \underset{/}{\overset{\backslash}{N}}\text{—}Me$$

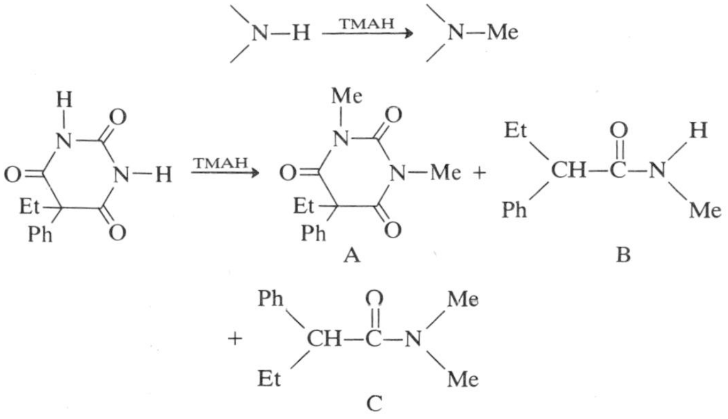

The substance is dissolved in 0.2 M alcoholic solution of trimethylanilinium hydroxide (1 : 5 w/v, for the effect of concentration of the derivatizing agent upon the quantitation of barbiturates see *Notes* and Ref. 22) and after 2 min the solution is injected into the gas chromatograph for on-column reaction and GC analysis.

Notes: Up to one thousand-fold molar excess of the reagent has been used.[81,82] Trimethylanilinium hydroxide will methylate substances bearing replaceable protons attached to nitrogen. It is currently accepted as the preferred *N*-alkylation reagent for barbiturates and related substances. Recent changes in the approach to the analysis of barbiturates has been evaluated.[83] It has been found by chemical ionization studies[22] that in the GC analysis of phenobarbital two additional products (B, C) were formed besides the expected *N,N'*-dimethyl derivative (A). While the amount of the second phenobarbital decomposition product (C) is insignificant, the amount of B is large and has to be taken into account in the quantitative analysis. The extent of the formation of B is affected by the concentration of the derivatizing agent and the time the drug is in contact with trimethylanilinium hydroxide prior to GC injection. Following a detailed study a procedure has been developed[22] under the conditions of which the amount of B formed is reproducible and can be used for more precise quantization of phenobarbital. Another application is the detection of xanthine bases[81] and phenolic alkaloids.[81]

2.14 Alkylation with trialkyloxonium fluoroborates

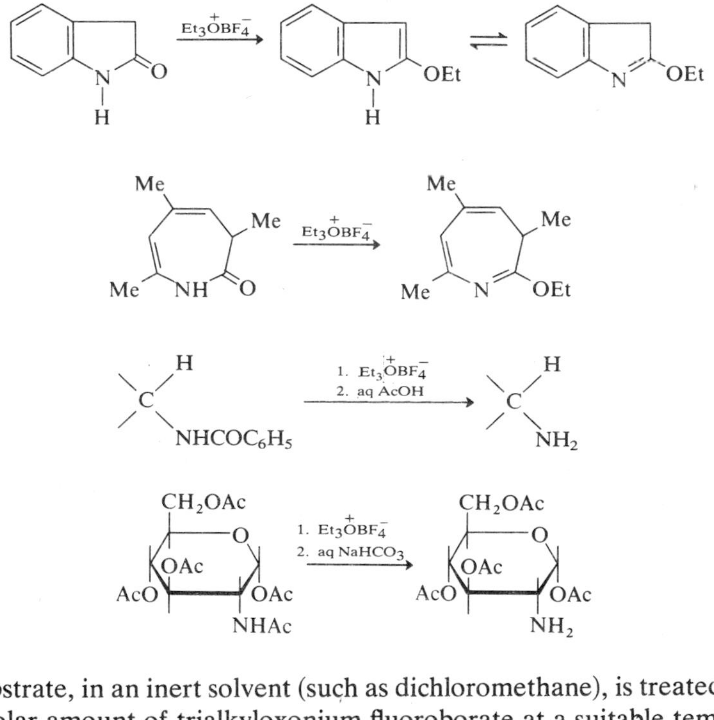

The substrate, in an inert solvent (such as dichloromethane), is treated with an equimolar amount of trialkyloxonium fluoroborate at a suitable temperature and with the exclusion of moisture. After concentration the product often crystallizes in the form of a fluoroborate salt which can be decomposed with a solution of sodium bicarbonate. Alternatively, when they do not affect other functions present, alkoxides in alcohols can be used. For the conversion of liquid substances, solid trialkyloxonium fluoroborate is added, and the mixture is stirred until the reagent dissolves. Warming to a suitable temperature may be necessary at this step.

Notes: Triethyloxonium fluoroborate and its methyl analogue can readily be prepared in the laboratory, according to Meerwein.[84,85] Alkylation using this method is limited to the lower members of the series because of the difficulty of preparation of the higher trialkyloxonium fluoroborates. This class of reagent alkylates[86] alcohols, phenols and carboxylic acids to give the corresponding *O*-alkyl derivatives, and ethers, sulfides, nitriles, ketones, esters, and amides are alkylated on oxygen, nitrogen or sulfur to give -onium fluoroborates. *O*-Alkylation on an exocyclic oxygen occurs readily with this reagent even in mesoionic ring systems[23] where, most of the time, alkylation cannot be achieved with alkyl halides. Temperatures ranging from −80 °C to +100 °C have been applied in the alkylations with this type of reagent. Triethyloxonium fluoroborate was proved to be useful not only for alkylations but also for mild *N*-deacylation, a reaction which otherwise requires drastic conditions. Thus, de-*N*-benzoylation[24] and de-*N*-acetylation[25] has been achieved using this reagent under conditions which left ester, acetal and glycosidic linkages unaffected. Hence, *N*-deacetylation can be selectively carried out with *O*-acetyl groups present, which provides an important means for *N*-acyl group interchange. Acylamino acids and peptides were esterified with an excess of triethyloxonium fluoroborate.[87]

2.15 Alkylation with alcohols in the presence of acid catalysts

2.15.1 FORMATION OF BENZYL ETHERS FROM SUBSTANCES RELATED TO CATECHOL

A solution of the substrate in alcoholic hydrochloric acid (1–3 M, substrate : reagent ratio 1 : 20–100, w/v) is heated in a sealed vial at 90–100 °C for 2 h, after which time the solution is evaporated to dryness under reduced pressure with more alcohol for complete dehydration.

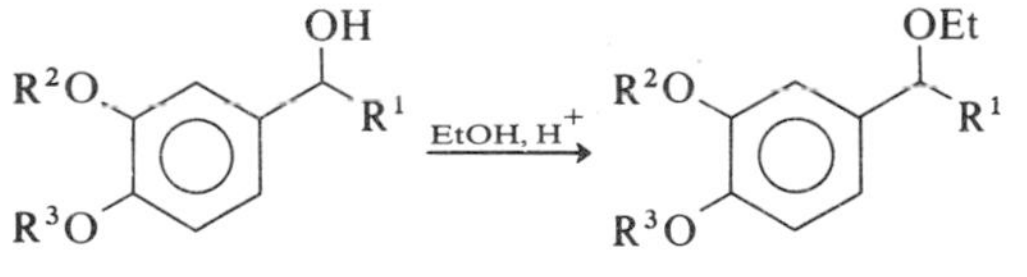

$R^1 = CH_2—NH_2, CH_2—OH$
$R^2 = H, CH_3$
$R^3 = H$
when R^1 is COOH, ethyl ester is simultaneously formed

2.15.2 CONDENSATION OF SIMPLE ALCOHOLS WITH FREE SUGARS (FISCHER'S METHOD OF GLYCOSIDATION)

A solution of the sugar in the respective alcohol containing the acid catalyst (for the substrate : alcohol : catalyst ratios see *Notes* below) is heated at the boiling point for a suitable period of time and, after removal of the acid, the solution is concentrated.

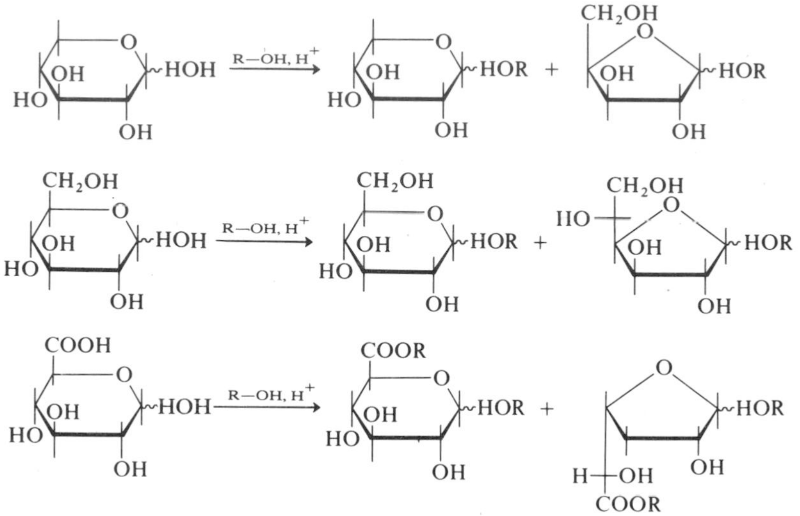

Notes: Substances related to catechol when treated with acidified alcohols undergo *O*-alkylation at the benzylic hydroxyl to give the corresponding benzyl alkyl ethers. Several substances of this class have been successfully analysed by GC–MS as *O*-ethyl ethers following this alkylation.[88] The reaction which occurs with[89] methanol, ethanol, *n*-butanol, *n*- and isopropanol forms derivatives which protects the sensitive benzyl hydroxyl group of this class of substances before further derivative formation and, thus, losses due to polymerization, enamine formation and other degradations can be avoided. The ease with which deuterio analogues are prepared when the reaction is carried out with deuterioalcohols is useful for specific-ion monitoring GC–MS assay. The reaction may proceed via a carbonium ion intermediate; if this is so, racemization of optically active compounds could occur during this process.

Treatment of higher sugars with anhydrous alcohols containing an acid catalyst, unlike that of the simplest members of the sugar series, glycoaldehyde and glyceraldehyde which in this way give dialkyl acetals, results in the formation of alkyl hemiacetals termed glycosides. This represents the simplest but least selective method for the preparation of glycosides and, hence it is a convenient means for the protection of the potential aldehyde or keto function of free sugars. The method is applicable to lower aliphatic alcohols but not to phenols. Hydrogen chloride (0.01–5%, w/v) is by far the most commonly used acid catalyst for this reaction although other acids, including Lewis-type acids,[90] and cation-exchange resins in their H^+ form[91] have also been used. When substances sensitive to mineral acids, such as ketoses, are to be converted to glycosides, losses due to decomposition may be minimized by catalysing the reaction with acetic acid[92] which has the additional advantage in that it can be readily removed from the reaction mixture by evaporation under reduced pressure. For the isolation of the products the acid is neutralized, most conveniently with an anion-exchange resin, and the solutions are concentrated under reduced pressure. Although occasionally[90,93] products were isolated without the removal of the acid catalyst, this step should not be omitted when the products are to be directly chromatographed or analysed by mass spectrometry.

Variable concentration of the sugars in the acidified alcohol (0.5–15%), temperature (3–100 °C) and the duration of the reaction (1 h–5 days) have been employed. While the first parameter appears to have little effect on the outcome of the reaction, the latter two significantly affect the rate at which the equilibrium between the α- and β-form of pyranosides and furanosides is established. A mixture of anomers is always obtained whenever the structure of the sugar molecule is appropriate. Bishop with his co-workers[94–97] thoroughly studied glycosidation of sugars and did complete GC analyses of equilibrium mixtures of glycosides formed from common sugars. It follows from their conclusions that pentoses tend to form furanosides more extensively than do hexoses and that substitution of free hydroxyl groups with methyl groups makes this tendency even more pronounced. Although exceptional cases exist, pyranosides are generally the principal constituents of the glycosidation mixtures when the reaction is allowed to go to completion. The course of the reaction is such that the decrease in the free sugar concentration is accompanied by a rapid but transient build-up of furanosides, which then isomerize slowly to the pyranosides. Thus, when Fischer's glycosidation reactions are prevented from reaching their equilibrium positions, thermodynamically less stable products may be isolated. It follows that for the preparation of glycoside mixtures rich in furanosides, mild reaction conditions should be used (concentration of the acid below 0.1%[98] and 40–60 °C or 0.5–1% and ambient or subambient temperature).[99,100]

Insignificant amounts of dimethyl acetals of higher sugars are formed, although their presence in the reaction mixtures of Fischer's glycosidation has been demonstrated.[101–103] This is particularly so at equilibrium and it appears that the true acetals are not the primary products from which glycosides are subsequently formed.

Glycosidation of uronic acids has not been studied in such detail as that of neutral sugars, but sufficient information has accumulated[103–107] to indicate that the reaction follows the same general pattern as in the case of aldoses. The carboxylic function is esterified before the glycosides are formed and furanoside formation is perhaps more pronounced.

Examination of the products of Fischer's glycosidation by GC–MS reveals the significant fact that a single product is not formed when the anomeric centre of sugars is derivatized in this way. GC is a powerful means of separating the various forms of sugar glycosides. Numerous data concerning the chromatographic properties of these substances are summarized in the excellent review by

Dutton[108,109] and for further references see also Tables 2 to 4. Anomeric isomers are normally not distinguished by mass spectrometry, but pyranoid and furanoid forms follow different fragmentation pattern so that their presence can be clearly demonstrated.[29,110–112]

2.16 Preparation of pentafluorobenzyl derivatives

2.16.1 FORMATION OF CARBAMATE DERIVATIVES USING PENTAFLUOROBENZYL CHLOROFORMATE

$$(R)_3N \xrightarrow{C_6F_5CH_2OCOCl} C_6F_5CH_2OCO{-}N(R)_2$$

The solution of a tertiary amine in *n*-heptane (1:200, w/v) is treated with the chloroformate (50 μl/mg of the substrate) and a pinch of anhydrous sodium carbonate. The stoppered tube is then heated for 1 h at 100 °C, and after cooling, a little 1 M NaOH solution is added. The mixture is shaken and the phases are allowed to separate. The *n*- heptane phase contains the derivatized product.

2.16.2 FORMATION OF OXIMES USING *O*-(2,3,4,5,6-PENTAFLUOROBENZYL) HYDROXYLAMINE HYDROCHLORIDE

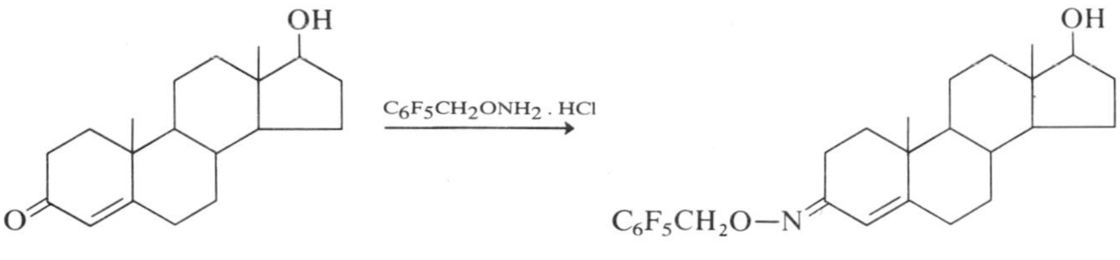

The ketone is dissolved in a stock solution of the hydrochloride in pyridine, and the stoppered tube is swirled at 65 °C for a half hour. The pyridine is removed by a stream of nitrogen and a little cyclohexane is added. After shaking, and addition of a little water, the phases are allowed to separate. The top cyclohexane layer is either chromatographed immediately, or transferred with the aid of a pipette to another tube to be dried over a few grains of anhydrous sodium sulfate. These conditions are discussed fully in Chapter 6.

2.16.3 FORMATION OF PENTAFLUOROBENZYL DERIVATIVES USING PENTAFLUOROBENZYL BROMIDE

$$R{-}OH \xrightarrow{C_6F_5CH_2Br} R{-}OCH_2C_6F_5$$

$$R{-}COOH \xrightarrow{C_6F_5CH_2Br} R{-}COOCH_2C_6F_5$$

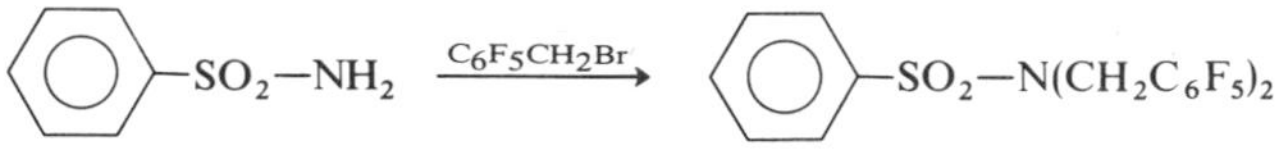

Pentafluorobenzyl bromide (~1.1 equiv./H to be replaced) is added to the solution of the substrate (~50 μg) in a suitable solvent (0.5 ml) followed by anhydrous potassium carbonate (slight excess based on the amount of the bromide) and the mixture is heated in a water bath near the boiling point for 1 h. A small amount of iso-octane is added and the mixture is evaporated to ~1–2 ml. When the solids settle the remaining solution is transferred onto a column of 1 g silica gel poured from a slurry in hexane. The solids are mixed with ~1 ml of hexane and the washings are added to the column followed by 5–10 ml of 5% benzene in hexane to remove the excess of pentafluorobenzyl bromide reagent. The derivatized substance, held on the column during these operations, is then eluted with a benzene–hexane mixture of appropriate polarity.

Notes: This class of derivatives is commonly used for electron-capture gas chromatographic analysis. The first two reagents are recommended for derivatizing substances which cannot be conveniently derivatized using perfluoro acid chlorides or anhydrides.

Pentafluorobenzyl chloroformate is a relatively new reagent and its full value has yet to be established but its use for the direct GC determination of tertiary amines[113,114] seems to be an improvement to the original[115] time-consuming method for the derivatization of these substances.

The above-described oxime formation has been successfully used for steroid chemistry[26] in the nanogram range. For the formation of oximes it is convenient to have handy a stock solution of the hydrochloride in pyridine containing 50 mg ml^{-1} of the reagent. For derivatization 1 mg of the sample is dissolved in 0.2 ml of the stock solution. Suitable dilutions of the stock solution with pyridine are made for the reactions at microgram and nanogram levels.

The method involving pentafluorobenzyl bromide is a convenient procedure for making esters and ethers and has accordingly been used in trace analysis. Sulfonamides have also been derivatized with the reagent and determined as *N*-pentafluorobenzyl derivatives.[27] The mildly basic conditions involving potassium carbonate as the base may not be sufficient for the conversion of sterically hindered groups, in which case stronger basic conditions (procedure 2.6) are recommended. The convenient clean-up procedure[116] which removes the excess of the reagent and other interfering substances is an improvement on the original directions where this was achieved by distillation.[117,118] The proper ratio of the benzene–hexane mixture for the elution of the products from the silica gel microcolumn should first be established by TLC.

CAUTION: Pentafluorobenzyl bromide is a strong lachrymator and should be handled with appropriate precautions; for routine work a stock solution of the reagent in acetone (1:50, w/v) may be prepared.[116]

It can be expected that new derivatizing agents forming fluorine-containing derivatives suitable for GC with ECD trace analysis will appear in future (see Chapter 3). A reagent of potential value with respect to this is certainly hexafluorobenzene which under the conditions of procedure 2.6 readily reacted with hydroxyl groups giving pentafluorophenyl ethers.[119]

Any glassware, including the pipettes and the syringes used in the preparation of derivatives at the nanogram level, must be most thoroughly cleaned.

2.17 Alkylation by means of reductive condensation of amines with aldehydes

$$\mathrm{NH_2{-}CH_2{-}CO{-}NH{-}} \xrightarrow{\mathrm{R'{-}CH{=}O,\ H_2}} \mathrm{(R'{-}CH_2)_2N{-}CH_2{-}CO{-}NH{-}}$$

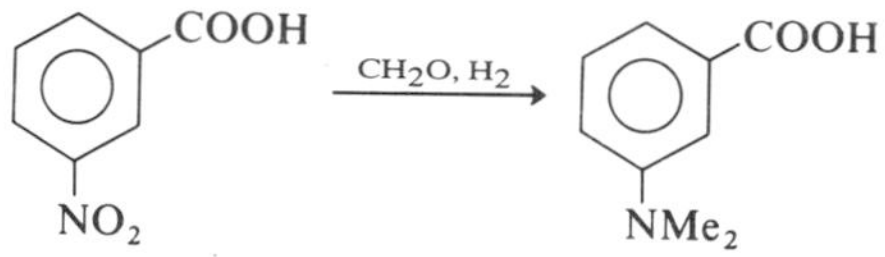

A mixture of the substrate and 10% palladium-on-charcoal catalyst (1:1, w/w) in ethanol or aqueous ethanol is stirred in the presence of the requisite aldehyde (300–400% excess) under hydrogen at room temperature and ordinary pressure until either reduction ceases or slightly more than the theoretical amount of hydrogen is absorbed (3–16 h). The mixture is then heated to boiling and filtered, the catalyst is washed with hot ethanol and the combined filtrates are evaporated to dryness to give the dialkylamino compound with some residual aldehyde.

Notes: The value of this procedure lies mainly in the fact, that when formaldehyde is used for the condensation *N,N*-dimethyl-amino acids can be conveniently prepared in this way,[120] whereas methylation of amino acids with methyl iodide in the presence of alkali leads to betaines, occasionally accompanied by monoalkylation. In the case of methylation, commercial 40% formaldehyde in water serves as both the reagent and the solvent and its amount should be at least twice the theoretical amount. Combined with suitable esterification[121,122] volatile derivatives of many amino acids, suitable for analysis by GC, are produced. These reactions are not applicable to all amino acids, and as tertiary amines the gas chromatographic properties of these derivatives are not always ideal.

With higher straight chain aldehydes, glycine and alanine furnish the corresponding *N,N*-dialkylamino acids, but other amino acids, when the reaction is carried out at ambient or subambient temperature, undergo monoalkylation owing to steric hindrance. When dialkylation is the aim this can often be achieved by conducting the condensation at a higher temperature and with prolonged reaction time. With aldehydes branched in the α-position, *N*-monoalkyl derivatives are produced.[123] The method has been successfully extended to the methylation of oligopeptides which undergo alkylation only at the end amino acid unit, leading to the identification of the *N*-terminal amino acid in polypeptides.[124]

TABLE 2

Relative retention times of *O*-acetyl-*O*-methyl-D-glucononitriles[125]

O-Methyl	*O*-Acetyl	RRT	
		3% QF-1 180 °C	3% SP-2340 200 °C
2,3,4,6	5	1.00 (8.03 min)	1.00 (6.53 min)
2,3,4	5,6	2.23	2.27
2,3,6	4,5	2.06	2.50
2,4,6	3,5	1.44	1.40
3,4,6	2,5	2.23	2.13
2,3	4,5,6	4.04	5.11
2,4	3,5,6	3.34	3.28
2,6	3,4,5	2.67	2.77
3,4	2,5,6	5.13	5.31
3,6	2,4,5	3.56	4.07
4,6	2,3,5	2.93	2.89
2	3,4,5,6	4.79	5.26
3	2,4,5,6	7.16	8.85
4	2,3,5,6	6.70	6.81
6	2,3,4,5	4.62	4.63

TABLE 3

TLC separation of selected carbohydrate *O*-methyl derivatives

Compound	R_S[a]	R_F[b]
D-arabinose		
4-*O*-methyl	0.11	
3-*O*-methyl	0.15	
2-*O*-methyl	0.24	
3,4-di-*O*-methyl	0.26	
5-*O*-methyl	0.40	
2,4-di-*O*-methyl	0.43	
2,3-di-*O*-methyl	0.56	
2,3,4-tri-*O*-methyl	0.63	
2,5-di-*O*-methyl	0.81	
3,5-di-*O*-methyl	0.81	
2,3,5-tri-*O*-methyl	1.00	
methyl tetra-O-*methyl*-		
β-D-glucopyranoside		0.46
α-D-glucopyranoside		0.29
β-D-glucofuranoside		0.39
α-D-glucofuranoside		0.31
β-D-galactopyranoside		0.28
α-D-galactopyranoside		0.21
β-D-galactofuranoside		0.44
α-D-galactofuranoside		0.29

[a] Silica gel F-254, single development in 2-butanone saturated with 3% ammonia.[126]

[b] Silica gel G, ether : toluene ratio 2 : 1.[127]

TABLE 4

Sources of GLC and TLC data of some carbohydrate alkyl derivatives[a]

	Reference
GLC data of	
methylated sugar derivatives, a review	128
permethylated alditols and aldonic acids	129
partially methylated alditols as TMS ethers	130
partially ethylated and methylated alditol acetates	131
methyl *O*-methyl-β-D-arabinofuranosides and methyl *O*-methyl-β-D-arabinopyranosides	126
methyl tri-*O*-acetyl-*O*-methyl-α-D-mannosides	132
O-methyl-mannoses	133
2-deoxy-*O*-methyl-2-methylamino-D-glucoses and D-galactoses	134
methyl (methyl *O*-methyl-α-D-glucopyranosid)uronates	135
methyl(methyl 4-deoxy-*O*-methyl-β-L-*threo*-hex-4-enopyranosid)uronates	136
TLC data of	
various carbohydrate derivatives, qualitative and quantitative, a review	137, 138
methyl *O*-methyl-D-arabinosides	139
substituted methyl-α-D-glucopyranosides	140–143
O-methyl-D-glucoses	141
O-methyl-D-galactoses	144
mono-*O*-methyl-D-fructoses	145
substituted methyl β-D-maltosides	146
aryl tetra-*O*-acetyl-D-glucopyranosides	147
halogenated aryl β-D-glucosides and thioglucosides	148
O-methyl-D-glucosamine	149
O-acetyl-*O*-methyl alditol acetates	150
branched chain sugars and their methyl glycosides	151

[a] For further data see Refs. 108 and 109.

TABLE 5

Examples of protective alkylation in GLC of various classes of organic substances

Class of substances	Reference
steroids	26, 152
amino acids	72
catecholamines	88
sulfonamides	27, 153
barbiturates	16, 22, 52, 83, 161
phenols	156–159
organic acids	162, 163

3 APPENDIX: PURIFICATION OF CHEMICALS

The procedures given below may not remove all the impurities but were sufficient to ensure the smooth course of the alkylations described.

Acetone, b.p. 56.2 °C

Successive small portions of potassium permanganate are added to a mixture of acetone and anhydrous potassium carbonate (20:1, v/w) at reflux until the violet colour persists. The solvent is then distilled off.

Acetonitrile, b.p. 81.6 °C

Reagent grade material is stirred with calcium hydride until no further hydrogen is evolved. The acetonitrile is then fractionally distilled under anhydrous conditions.

Boron trifluoride etherate, b.p. 67 °C at 43 mm Hg, 45 °C at 15 mm Hg

The commercial product is treated with a little ethyl ether to ensure an excess of this component and then distilled under reduced pressure from calcium hydride. The dry substance fumes in moist air. It should be stored refrigerated in a tightly closed bottle in the dark.

Diazoalkanes

For the preparation and handling of this class of substances see Chapter 2.

1,2-Dimethoxyethane, b.p. 84 °C

The commercial product should be refluxed with potassium and distilled from calcium hydride. Pure solvent should be stored over sodium hydride.

N,N-dimethylacetamide, b.p. 58 °C at 11 mm Hg

The procedure given for the purification of *N,N*-dimethylformamide should be used.

N,N-dimethylformamide, b.p. 153 °C, b.p. 76 °C at 39 mm Hg

Adequate drying is achieved by stirring with powdered barium oxide followed by decanting, before distillation, preferably at reduced pressure. When a high proportion of water is present, most of it may be eliminated by initial azeotropic distillation with benzene. *N,N*-Dimethylformamide forms an azeotrope with xylene (18.7% DMF, b.p. 135.1 °C) which is quite useful to know when pure substances are to be isolated from its solution. Alkylations are frequently carried out in this solvent in spite of its well-known drawbacks: it tends to dissolve silver salts and in several instances side reactions due to this solvent have been observed. The chemistry and applications of *N,N*-dimethylformamide have been dealt with comprehensively.[160]

Dimethyl sulfoxide, m.p. 18–18.5 °C, b.p. 190 °C, b.p. 75–76 °C at 12 mm Hg

Solvent suitable for most reactions is obtained by treatment with calcium hydride (2–5 g l^{-1}) for 2 h. The mixture is shaken occasionally and then distilled. It is then warmed to 80–90° at atmospheric pressure and distilled with the aid of a water pump.

Ethyl ether, b.p. 34.6 °C

Although tinned, commercially available, reagent grade dry ether is preferred, reagent grade solvent may be used without further purification. The amount of impurities and/or peroxides is not significant when working on a small scale. If necessary it may be passed through a column of activated alumina (80 g/700 ml).

***n*-Heptane, b.p. 98.4 °C (other aliphatic hydrocarbons can be purified in the same way)**

For extensive purification the commercial product is shaken with successive small portions of concentrated sulfuric acid until the lower (acid) layer remains colourless. The hydrocarbon is then washed successively with water, aqueous 10% sodium carbonate, water (twice), and dried with calcium sulfate or other similar dessicant. It is then distilled from sodium or calcium hydride.

Methyl iodide, b.p. 42.8 °C, and other alkyl halides

The commercial product often contains free iodine. Material suitable for methylation is obtained by shaking (under an efficient hood) with dilute sodium thiosulfate until colourless, then washing with water, dilute sodium carbonate, and more water, drying with anhydrous calcium chloride and distillation. The pure reagent should be kept refrigerated in the dark, in contact with anhydrous calcium chloride and a small amount of metallic mercury.

Alkyl halides are likely to be contaminated with the starting materials from which they were prepared, and should therefore be of the best grade commercially available. Common purification is by shaking with concentrated hydrochloric acid, followed by washing successively with water, 5% aqueous sodium bicarbonate and water. After drying with calcium chloride the halide is fractionally distilled through an efficient column. For solid halides the above purification is done in solution in a suitable solvent such as benzene, and finally by recrystallization.

CAUTION: Care should be taken in handling organic halogen compounds because of their toxicity. They should be kept in dark, tightly closed bottles to prevent oxidation.

Pyridine, b.p. 115 °C

Reagent grade pyridine is refluxed over solid potassium hydroxide, and decanted onto fresh potassium hydroxide pellets. The process is repeated twice. The pyridine is then fractionally distilled from fresh potassium hydroxide and stored over 3A-type molecular sieve. Care is required to exclude moisture from this very hygroscopic solvent. It forms a hydrate, b.p. 94.5 °C.

Silver oxide

Freshly prepared material is recommended. It is prepared by pouring slowly and with stirring a hot, filtered solution of barium hydroxide octahydrate in water (1 : 10, w/v) into a hot solution of silver nitrate (1 : 5, w/v) and filtering the precipitated silver oxide. After thorough washing with hot water the product is dried in a vacuum oven at 60 °C and stored in a tightly-closed dark container.

Tetrahydrofuran, b.p. 65.4 °C

Refluxing with solid potassium hydroxide and distilling from lithium aluminium hydride removes all impurities which may interfere with the procedures in which this solvent is used.

REFERENCES

1. W. N. Haworth, *J. Chem. Soc.* **107**, 8 (1915).
2. G. Zemplén, Z. Csürös and S. Angyal, *Ber. Deutsch. Chem. Ges.* **70**, 1848 (1937).
3. T. Purdie and J. C. Irvine, *J. Chem. Soc.* **83**, 1021 (1903).
4. R. Kuhn, H. Trischmann and I. Löw, *Angew. Chem.* **67**, 32 (1955).
5. S. Hakomori, *J. Biochem. (Tokyo)* **55**, 205 (1964).
6. E. Müller and W. Rundel, *Angew. Chem.* **70**, 105 (1958).
7. M. Neeman, M. C. Caseiro, J. D. Roberts and W. S. Johnson, *Tetrahedron* **6**, 36 (1959).
8. M. J. Robins, R. S. Naik, *Biochim. Biophys. Acta* **246**, 341 (1971).
9. M. J. Robins, A. S. K. Lee and F. Norris, *Carbohydr. Res.* **41**, 304 (1975).
10. H. G. Walker Jr, M. Gee and R. M. McCready, *J. Org. Chem.* **27**, 2100 (1962).
11. G. O. Aspinall and P. E. Barron, *Can. J. Chem.* **50**, 2203 (1973).
12. D. W. Thomas, B. C. Das, S. D. Gero and E. Lederer, *Biochem. Biophys. Res. Commun.* **32**, 519 (1968).

13. D. W. Thomas, *Biochem. Biophys. Res. Commun.* **32**, 483 (1968).
14. K. J. A. Jackson and J. K. N. Jones, *Can. J. Chem.* **43**, 450 (1965).
15. H. R. Morris, D. H. Williams and R. P. Ambler, *Biochem. J.* **125**, 189 (1971).
16. W. Dünges, H. Heinmann and K. J. Netter, in *Gas Chromatography* (S. G. Perry, Ed.), Applied Science, England, (1972).
17. K. L. Agarwal, G. W. Kenner and R. C. Sheppard, *J. Am. Chem. Soc.* **91**, 3096 (1969).
18. V. Zitko and C. T. Bishop, *Can. J. Chem.* **44**, 1275 (1966).
19. B. A. Dmitriev, L. V. Backinowsky, Yu. A. Knirel, V. L. Lvov and N. K. Kochetkov, *Izv. Akad. Nauk, Ser. Khim.* 2235 (1974).
20. H. Bredereck, R. Sieber and L. Kamphenkel, *Ber.* **89**, 1169 (1956).
21. J. Žemlička, *Collect. Czech. Chem. Commun.* **28**, 1060 (1963).
22. R. Osiewicz, V. Aggarwal, R. M. Young and I. Sunshine, *J. Chromatogr.* **88**, 157 (1974).
23. K. T. Potts, E. Houghton and S. Husain, *Chem. Commun.* 1025 (1970).
24. H. Muxfelt and W. Rogalski, *J. Am. Chem. Soc.* **87**, 933 (1965).
25. S. Hanessian, *Tetrahedron Lett.* 1549 (1967).
26. K. T. Koshy, D. G. Kaiser and A. K. VanDer Slik, *J. Chromatogr. Sci.* **13**, 97 (1975).
27. O. Gyllenhaal and H. Ehrsson, *J. Chromatogr.* **107**, 327 (1975).
28. B. Bannister and P. Kováč, *VIIth International Symposium on Carbohydrate Chemistry*, Bratislava, August 1974, Abstracts, Organic Chemistry Section, p. 8.
29. V. Kováčik and P. Kováč, *Chem. Zvesti* **27**, 662 (1973).
30. P. Kováč, *Carbohydr. Res.* **22**, 464 (1972).
31. P. Kováč, J. Hirsch, R. Palovčík, I. Tvaroška and S. Bystrický, *Collect. Czech. Chem. Commun.* **41**, 3119 (1976).
32. R. W. Jeanloz, *Adv. Carbohydr. Chem.* **13**, 189 (1958).
33. B. C. Das, S. D. Gero and E. Lederer, *Biochem. Biophys. Res. Commun.* **29**, 211 (1967).
34. B. C. Das, S. D. Gero and E. Lederer, *Nature (London)* **217**, 547 (1968).
35. D. W. Thomas, E. Lederer, M. Bodzanszky, J. Izdebski and I. Muramatsu, *Nature (London)* **220**, 580 (1968).
36. R. K. Olsen, *J. Org. Chem.* **35**, 1912 (1970).
37. R. Kuhn, H. H. Baer and A. Seeliger, *Liebigs Ann. Chem.* **611**, 236 (1958).
38. J. Hirsch, P. Kováč and V. Kováčik, *J. Carbohydr. Nucl. Nucl.* **1**, 431 (1974).
39. Z. Tamura and T. Imanari, *Chem. Pharm. Bull.* **12**, 1386 (1964).
40. F. A. French, *Chem. Eng. News*, April 11, 1966, p. 48.
41. G. L. Olson, *Chem. Eng. News*, June 13, 1966, p.7.
42. H. Björndal, C. G. Hellerqvist, B. Lindberg and S. Svensson, *Angew. Chem. Int. Ed.* **9**, 610 (1970).
43. H. Nikaido, *Eur. J. Biochem.* **15**, 57 (1970).
44. D. M. W. Anderson and G. M. Cree, *Carbohydr. Res.* **2**, 162 (1966).
45. R. Toman, Š. Karácsonyi and M. Kubačková, *Carbohydr. Res.* **43**, 111 (1975).
46. D. M. W. Anderson, I. C. M. Dea, P. A. Maggs and A. C. Munro, *Carbohydr. Res.* **5**, 489 (1967).
47. E. J. Corey and M. Chaykovsky, *J. Am. Chem. Soc.* **87**, 1345 (1965).
48. P. Kováč, *Carbohydr. Res.* **31**, 323 (1973).
49. R. Kuhn and H. A. Baer, *Chem. Ber.* **86**, 724 (1953).
50. A. Schönberg and A. Mustafa, *J. Chem. Soc.* 746 (1946).
51. C. M. Williams and C. C. Sweeley, in *Biomedical Application of Gas Chromatography*, Vol. 1, (H. A. Szymanski, Ed.), Plenum Press, New York, 1964, p. 231.
52. M. G. Horning, K. Lertratanangkoon, J. Nowlin, W. G. Stillwell, R. N. Stillwell, T. E. Zion, P. Kellaway and R. M. Hill, *J. Chrom. Sci.* **12**, 630 (1974).
53. J. G. Cook, C. Riley, R. F. Nunn and D. E. Budger, **6**, 182 (1961).
54. G. Hesse, F. Exner and H. Hertel, *Ann. Chem.* **609**, 60 (1957).
55. O. T. Schmidt and H. Kraft, *Ber.* **74**, 33 (1941).

56. H. Meerwein, T. Bersin and W. Burneleit, *Ber.* **62**, 1006 (1929), c.f. H. Meerwein and G. Hinz, *Ann. Chem.* **484**, 1 (1930).
57. L. M. Pike, M. K. A. Khan and F. Rottman, *J. Org. Chem.* **39**, 3674 (1974).
58. I. O. Mastronardi, S. M. Flematti, J. O. Deferrari and E. G. Gros, *Carbohydr. Res.* **3**, 177 (1966).
59. J. O. Deferrari, E. G. Gros and I. O. Mastronardi, *Carbohydr. Res.* **4**, 432 (1967).
60. G. J. F. Chittenden, *Carbohydr. Res.* **31**, 127 (1973).
61. E. J. Bourne, I. R. McKinley and H. Weigel, *Carbohydr. Res.* **25**, 516 (1972).
62. A. Lipták, *Acta Chim. Acad. Sci. Hung.* **66**, 315 (1970).
63. P. Kováč, *Carbohydr. Res.* **20**, 418 (1971).
64. P. Kováč and Ž. Longauerová, *Chem. Zvesti* **26**, 71 (1972).
65. J. Hirsch, P. Kováč and V. Kováčik, *Chem. Zvesti* **28**, 833 (1974).
66. P. Kováč, R. Palovčík, *Carbohydr. Res.* **36**, 379 (1974).
67. P. Kováč, R. Palovčík, V. Kováčik and J. Hirsch, *Carbohydr. Res.* **44**, 205 (1975).
68. L. F. Christensen and A. D. Broom, *J. Org. Chem.* **37**, 3398 (1972).
69. *Chromatography Lipids*, a publicity series available from Supelco Corp., Bellefonte, PA. U.S.A., Vol. VII, No. 4, p. 2 (1973).
70. H. Vorbrüggen, *Angew. Chem. Int. Ed.* **2**, 211 (1963).
71. A. Holý, *Tetrahedron Lett.* **7**, 585 (1972).
72. J. P. Thenot and E. C. Horning, *Anal. Lett.* **5**, 519 (1972).
73. V. Ya. Pochinok and A. P. Limarenko, *Ukr. Khim. Zh.* **21**, 628 (1955); *Chem. Abstr.* **50**, 11270 e (1956).
74. T. Gramstad and R. N. Haszeldine, *J. Chem. Soc.* 4069 (1957).
75. M. G. Ahmed, R. W. Adler, G. H. James, M. L. Sinnot and M. C. Whiting, *Chem. Commun.* 1533 (1968).
76. M. G. Ahmed and R. W. Adler, *Chem. Commun.* 1389 (1969).
77. R. F. Borch, *Chem. Commun.* 442 (1968).
78. T. Kametani, K. Takahashi and K. Ogasawara, *Synthesis*, 473 (1972).
79. R. L. Hansen, *J. Org. Chem.* **30**, 4322 (1965).
80. J. Burdon and V. C. R. McLoughlin, *Tetrahedron* **21**, 1 (1965).
81. E. Brochmann-Hansen and T. O. Oke, *J. Pharm. Sci.* **58**, 370 (1969).
82. J. MacGee, *Clin. Chem.* **17**, 357 (1971).
83. G. Kananen, R. Osiewicz and I. Sunshine, *J. Cromatogr. Sci.* **10**, 283 (1972).
84. H. Meerwein, *Org. Synth.* **46**, 113 (1966).
85. H. Meerwein, *Org. Synth.* **46**, 120 (1966).
86. L. F. Fieser and M. Fieser, *Reagents for Organic Syntheses*, Vol. 1 (c.f. Vol. II and Vol. III), J. Wiley and Sons, Inc., 1967.
87. O. Yonemitsu, T. Hamada and Y. Kanaoka, *Tetrahedron Lett.* 1819 (1969).
88. N. Narasimhachari, *J. Chromatogr.* **90**, 163 (1974).
89. W. Weg and G. S. King, personal communication.
90. W. Rhoads and P. G. Gross, *Z. Naturforsch. Teil B* **28**, 647 (1973).
91. J. E. Cadotte, F. Smith and D. Spriesterbach, *J. Am. Chem. Soc.* **74**, 1501 (1952).
92. Ľ. Stankovič, K. Linek and M. Fedoroňko, *Carbohydr. Res.* **35**, 242 (1973).
93. M. N. Oldham and H. Honeyman, *J. Chem. Soc.* 986 (1946).
94. C. T. Bishop and F. P. Cooper, *Can. J. Chem.* **40**, 224 (1962).
95. C. T. Bishop and F. P. Cooper, *Can. J. Chem.* **41**, 2743 (1963).
96. V. Smirnyagin, C. T. Bishop and F. P. Cooper, *Can. J. Chem.* **43**, 3109 (1965).
97. V. Smirnyagin and C. T. Bishop, *Can. J. Chem.* **46**, 3085 (1968).
98. I. Augestad and E. Berner, *Acta. Chem. Scand.* **8**, 251 (1954).
99. S. Baker and W. W. Haworth, J. Chem. Soc. 365 (1925).
100. E. E. Percival and R. Zobrist, *J. Chem. Soc.* 4307 (1952).
101. B. D. Heard and R. Barker, *J. Org. Chem.* **33**, 740 (1968).
102. R. J. Ferrier and L. R. Hatton, *Carbohydr. Res.* **6**, 75 (1968).

103. K. Larsson and G. Pettersen, *Carbohydr. Res.* **34**, 323 (1974).
104. H. W. H. Schmidt and H. Neukom, *Helv. Chim. Acta* **47**, 865 (1965).
105. H. W. H. Schmidt and H. Neukom, *Helv. Chim. Acta* **49**, 510 (1966).
106. L. N. Owen, S. Peat and W. J. G. Jones, *J. Chem. Soc.* 339 (1941).
107. E. F. Jansen and R. Jang, *J. Am. Chem. Soc.* **68**, 1475 (1946).
108. G. G. S. Dutton, *Adv. Carbohydr. Chem. Biochem.* **28**, 11 (1973).
109. G. G. S. Dutton, *Adv. Carbohydr. Chem. Biochem.* **30**, 9 (1974).
110. N. K. Kochetkov and O. S. Chizhov, *Methods Carbohydr. Chem.* **6**, 540 (1972).
111. K. Heyns, K. R. Sperling and H. F. Grützmacher, *Carbohydr. Res.* **9**, 79 (1969).
112. V. Kováčik and P. Kováč, *Carbohydr. Res.* **24**, 23 (1972).
113. P. Hartvig, J. Vessman and C. Svahn, *Anal. Lett.* **7**, 223 (1974).
114. P. Hartvig and J. Vessman, *Acta Pharm. Suecica* **11**, 115 (1974).
115. J. Vessman, P. Hartvig and M. Molander, *Anal. Lett.* **6**, 699 (1973).
116. L. G. Johnson, *J. AOAC* **56**, 1503 (1973).
117. F. K. Kawahara, *Anal. Chem.* **40**, 1009 (1968).
118. F. K. Kawahara, *Anal. Chem.* **40**, 2073 (1968).
119. A. H. Haines and K. C. Symes, *J. Chem. Soc. Perkin Trans. I* 53, (1973).
120. R. E. Bowman and H. H. Stroud, *J. Chem. Soc.* 1342 (1950).
121. K. Blau and A. Darbre, *Biochem. J.* **88**, 8P (1963).
122. E. K. Doms, *J. Chromatogr.* **105**, 79 (1975).
123. R. E. Bowman, *J. Chem. Soc.* 1346 (1950).
124. R. E. Bowman, *J. Chem. Soc.* 1349 (1950).
125. D. Anderle and P. Kováč, *Chem. Zvesti* in press.
126. P. A. Mied and Y. C. Lee, *Anal. Biochem.*, **49**, 534 (1972).
127. M. Gee, *Anal. Chem.* **35**, 350 (1963).
128. H. G. Jones, *Methods Carbohydr. Chem.* **6**, 25 (1972).
129. J. N. C. Whyte, *J. Chromatogr.* **87**, 163 (1973).
130. B. H. Freeman, A. M. Stephen and P. Van Der Bijl, *J. Chromatogr.* **73**, 29 (1972).
131. D. P. Sweet, P. Albersheim and R. H. Shapiro, *Carbohydr. Res.* **40**, 199 (1975).
132. B. Fournet, Y. Leroy and J. Montreuil, *J. Chromatogr.* **92**, 185 (1974).
133. B. Fournet and J. Montreuil, *J. Chromatogr.* **75**, 29 (1973).
134. P. A. J. Gorin and R. J. Magus, *Can. J. Chem.* **49**, 2583 (1971).
135. D. Anderle and P. Kováč, *J. Chromatogr.* **91**, 463 (1974).
136. D. Anderle, P. Kováč and J. Hirsch, *J. Chromatogr.* **105**, 206 (1975).
137. R. E. Wing and J. N. BeMiller, *Methods Carbohydr. Chem.* **6**, 42 (1972).
138. H. Scherz, G. Stehlik, E. Bancher and K. Kaindl, *Chromatogr. Rev.* **10**, 1 (1968).
139. S. C. Williams and J. K. N. Jones, *Can. J. Chem.* **45**, 275 (1972).
140. H. B. Sinclair, *J. Chromatogr.* **64**, 117 (1972).
141. G. W. Hay, B. A. Lewis and F. Smith, *J. Chromatogr.* **11**, 479 (1963).
142. A. Lipták and I. N. Jodál, *Acta Chim. (Budapest)*, **69**, 103 (1971).
143. P. J. Brennan, *J. Chromatogr.* **59**, 231 (1971).
144. M. L. Wolfrom, D. L. Palin and R. M. deLederkremer, *J. Chromatogr.* **17**, 488 (1965).
145. V. Prey, H. Berbalk and M. Kausz, *Microchim. Acta*, 449 (1962).
146. R. T. Sleeter and H. B. Sinclair, *J. Chromatogr.* **49**, 543 (1970).
147. T. D. Audichya, *J. Chromatogr.* **57**, 255 (1971).
148. J. L. Garraway and S. E. Cook, *J. Chromatogr.* **46**, 134 (1970).
149. Y. C. Lee and J. Scocca, *Anal. Biochem.* **39** 24 (1971).
150. Y. M. Choy, G. G. S. Dutton, K. B. Gibney, S. Kabir and J. N. C. Whyte, *J. Chromatogr.* **72**, 13 (1972).
151. R. J. Ferrier, W. G. Overend, G. A. Rafferty, H. M. Wall and N. R. Williams, *J. Chem. Soc.* C 1092 (1968).
152. R. B. Clayton, *Biochemistry* **1**, 357 (1962).
153. M. Ervik and K. Gustavii, *Anal. Chem.* **47**, 39 (1975).

154. W. Dünges, *Anal. Chem.* **45**, 963 (1973).
155. W. Dünges, *Chromatographia* **6**, 196 (1973).
156. C. Landault and G. Guiochon, *Anal. Chem.* **39**, 713 (1967).
157. A. C. Bhattacharyya, A. Bhattacherjee, O. K. Guha and A. N. Basu, *Anal. Chem.* **40**, 1873 (1968).
158. L. V. Semenchenko and V. T. Kaplin, *Zav. Labor.* **29**, 801 (1967).
159. H. G. Henkeı, *J. Chromatogr.* **20**, 596 (1965).
160. R, S. Kittila, *Dimethylformamide Chemical Uses*, E. I. Du Pont de Nemours, Wilmington, Delaware, U.S.A., 1968.
161. J. Pecci and T. J. Giovanniello, *J. Chromatogr.* **109**, 163 (1975).
162. P. A. Boindi and M. Cagnasso, *J. Chromatogr.* **109**, 389 (1975).
163. C. B. Johnson and E. Wong, *J. Chromatogr.* **109**, 403 (1975).

CHAPTER 6

Derivatives by Ketone-base Condensation

R. H. Brandenberger and H. Brandenberger

Chemical Division,
Institute of Forensic Medicine, University of Zürich,
Zürichbergstrasse 8, Zürich, Switzerland

The Schiff base or eneamine reaction has been employed to modify aldehydes and ketones by condensation with a primary amine (reaction I) and to condense primary amines with a carbonyl compound (reaction II). A similar reaction between primary amines and carbon disulfide yields isothiocyanates or mustard oils (reaction III). These three possibilities for derivative formation will be discussed in this chapter.

$$\text{I} \qquad \begin{matrix} R' \\ R'' \end{matrix}\!\!\!\!>C{=}O + H_2N{-}R \rightarrow \begin{matrix} R' \\ R'' \end{matrix}\!\!\!\!>C{=}N{-}R + H_2O$$

$$\text{II} \qquad R{-}NH_2 + O{=}C\!<\!\!\!\!\begin{matrix} R' \\ R'' \end{matrix} \rightarrow R{-}N{=}C\!<\!\!\!\!\begin{matrix} R' \\ R'' \end{matrix} + H_2O$$

$$\text{III} \qquad R{-}NH_2 + S{=}C{=}S \rightarrow R{-}N{=}C{=}S + H_2S$$

Of course, eneamine formation by I and II involves the same reaction scheme. However, the purposes and requirements for its application to the isolation, identification and quantitative determination of carbonyl compounds or amines differ. Therefore, I and II will be treated separately.

1 ANALYSIS OF CARBONYL COMPOUNDS AS SCHIFF BASES

The classical reagents for preparing derivatives from aldehydes and ketones include hydroxylamine (formation of oximes), semicarbazide, 4-phenylsemicarbazide and thiosemicarbazide (yielding the respective semicarbazones), as well as phenylhydrazine, *p*-nitrophenylhydrazine and 2,4-dinitrophenylhydrazine (formation of the corresponding hydrazones). Well known procedures for synthesizing these derivatives are found in many publications and textbooks.[34,60]

Before the age of chromatography, the characterization of unknown compounds necessitated the preparation of crystalline derivatives with definite melting points. For this purpose, hydroxylamine and phenylhydrazine are not always ideal reagents, particularly for low molecular weight carbonyl compounds. These derivatives possess rather low melting points or may even be oils at room temperature. The oximes are often very soluble and therefore difficult to purify, and mixtures of stereoisomers may also be formed. With phenylhydrazine, variations in the reaction procedure can lead to different derivatives: α-diketones yield mono- and dihydrazones and osazones; β-diketones form pyrazoles; α-hydroxy- aldehydes and ketones are oxidized in the presence of excess phenylhydrazine and condense with a second molecule of reagent to give phenylosazones, useful in the study of sugars. For characterization of derivatives by melting point, the other derivatives are usually to be preferred. Thiosemicarbazones are of special interest, since they form insoluble compounds with monovalent metals such as silver, copper and mercury, which may be an aid to their separation from mixtures and to their quantitative determination. The colour of the nitrated phenylhydrazine derivatives can be characteristic of the type of carbonyl group. *p*-Nitrophenylhydrazones of aliphatic compounds are yellow or brown, those of aromatic compounds red. 2.4-Dinitrophenylhydrazones of saturated aliphatic aldehydes and ketones are yellow or orange, of α,β-unsaturated carbonyls red, and of simple aromatic compounds yellow, orange or red. They are usually formed in excellent yields.

With the development of chromatographic separation and characterization methods, the classical carbonyl reagents have not been abandoned. However, the reasons for preparing these derivatives have changed to a large degree, and that, of course, influences the choice of reagent. Today, the main purposes of derivative formation are:

1. to isolate the carbonyl compounds from complex mixtures or from dilute solutions
2. to eliminate aldehydes and ketones from a mixture undergoing chromatographic investigation, i.e. by using a pre-column reaction
3. to convert carbonyls to compounds visible on paper, thin-layer plates or columns, or to compounds which can easily be revealed by spraying with a colour- or fluorescence-producing substance
4. to change the volatility and/or increase the differences between chemically related compounds for easier chromatographic separation
5. to improve the sensitivity of the chromatographic detection procedure, i.e. by conversion to compounds with high electron affinity
6. to convert the aldehydes and ketones to derivatives which can more readily be identified by combined gas chromatography–mass spectrometry.

A few examples will serve to illustrate these points:

1. The carbonyls from tobacco smoke,[11,58,64] the keto-acids from blood and urine,[14] the carbonyls from aroma concentrates,[29] and the dicarbonyls from oxidized whole milk powder[15,16] have been precipitated as 2,4-dinitrophenylhydrazones, removed by filtration or extraction and separated and identified by paper chromatography, liquid chromatography on columns, or gas chromatography. By mixing the 2,4-dinitrophenylhydrazones with α-ketoglutaric acid and heating to 240–260 °C, the aldehydes and ketones can be freed (flash exchange), swept into a column and separated by gas chromatography.[71,80] Steroids were isolated from tissues, blood plasma and urine as stable *O*-methyloxime-trimethylsilyl derivatives in good yields prior to their identification by gas chromatography–mass spectrometry[28,85] and determination by gas chromatography with specific ion monitoring (mass fragmentography).[59]

2. Semicarbazide has been used to remove carbonyls from mixtures in a pre-column reaction. The reagent was deposited on diatomaceous earth, placed in a short plug in the injection block of the gas chromatograph and heated to 115°. This resulted in a marked reduction or complete disappearance of the peaks due to carbonyl compounds.[17]

3. The coloured *p*-nitro- and 2,4-dinitrophenylhydrazones are ideal derivatives for the separation and characterization of carbonyls by paper, thin-layer and column chromatography. The oximes can easily be revealed on thin-layer plates by spraying with solutions of copper-II-chloride, copper-II-acetate (alcoholic) or iron-III-chloride.[44,69]

4. Good results with gas chromatography of keto-acids have been obtained by changing volatility and polarity by means of conversion to silylated oximes[81] and methyloximes.[42] If the separation of the derivatives is to be effected by gas chromatography with flame ionization or another general purpose detector, the non-substituted phenylhydrazones might be preferred over the derivatives with nitro groups on account of their greater volatility and lower retention characteristics.[39]

5. On the other hand, the detection limits can be improved by a factor of 500, if phenylhydrazine is replaced by 2,4-dinitrophenylhydrazine and the flame ionization by an electron capture detector.[46] Halogen-substituted phenylhydrazines permit even lower detection limits. Estrone in blood plasma has been determined in picogram amounts after conversion to the 3-methyl ether 17-pentafluorophenylhydrazone.[1]

6. The preparation of Schiff bases has been used to augment the differences between steroids prior to their separation and characterization. Earlier work was carried out with *N,N*-dimethylhydrazones[89] and pentafluorophenylhydrazones.[1,88] Recent investigations involved *O*-benzyl-[19,40] and *O*-methyloxime-trimethylsilyl derivatives.[24,59,85] The conditions for their preparation from steroids in biological materials and their gas chromatographic and mass spectrometric properties were studied,[35,40,85] and the formation of geometrical isomers was investigated.[43] The *O*-methyloxime-trimethylsilyl derivatives

have been used in a computer-based method for steroid identification by combined gas chromatography–mass spectrometry,[72] as well as for the trace analysis of steroids in blood and tissues by gas chromatography with mass specific detection.[59]

2 ANALYSIS OF PRIMARY AMINES AS SCHIFF BASES

The number of reagents that have been proposed for primary amines is probably larger than for any other class of compounds. But most of the classical derivatives lie beyond the scope of this chapter; they are covered in Chapters 3, 5, 10 and 12. Fluorescent derivatives (i.e. by condensation with fluorescamine) are discussed in Chapter 9.

In the past, Schiff base condensations of primary amines have been carried out mainly with benzaldehyde and *p*-nitrobenzaldehyde, less often with 2,4-dinitrobenzaldehyde. Derivative formation occurs rapidly on warming for 10–30 min, without or with a solvent (usually alcohol or acetic acid).[34] Benzylimines from aliphatic amines are stable oils which can be distilled without decomposition under atmospheric pressure. Aromatic benzylimines as well as aliphatic and aromatic *p*-nitro- and 2,4-dinitrobenzylimines are crystalline solids. The nitro-compounds are yellow.

These Schiff bases are obtainable in good yields. They are easy to purify by recrystallization and possess sharp melting points. But with the development of gas chromatographic techniques, reasons for derivative formation other than characterization by melting point became predominant. At present the main objectives are:

1. the isolation of primary amines from complex mixtures or from dilute solutions as a preconcentration step
2. their transformation to less polar and more stable compounds better suited for chromatographic separations
3. their conversion to coloured compounds for easier visibility upon separation by paper or thin-layer chromatography
4. their conversion to derivatives possessing high electron affinity for improved sensitivity upon gas chromatographic analysis with electron capture detection
5. their transformation to derivatives better suited for identification by mass spectrometry and quantification by mass specific detection.

Again, we would like to illustrate these points with a few examples. We will see that—quite unlike the situation existing for the analysis of carbonyl compounds—the classical reagents have been largely replaced.

1. In most cases, reagents for the separation of primary amines from mixtures are now chosen with regard for the chromatographic properties of the derivatives. Somewhat earlier work centred on the use of acetone, butanone, cyclobutanone and similar ketones to prepare Schiff bases from long chain

amines, diamines, aromatic amines and catecholamines,[9,10,12,36,88,91] prior to their gas chromatographic separation using general purpose detectors. More recently, pentafluorobenzaldehyde was chosen as a reagent, since electron capture gas chromatography was introduced as a final analytical step. This isolation procedure has been applied to biological extracts,[52,53,59] to pyrolysis products, to hydrazine in tobacco smoke and to technical maleic hydrazide.[55]

2. The gas chromatographic separation and estimation of free primary amines is hampered by their extreme polarity which causes peak tailing and favours irreversible adsorption on most of the common column packings.[62,88] The formation of nonpolar and more stable derivatives has therefore been the subject of many investigations. The eneamines fulfil these requirements to a large extent. SE-30 and OV-17 are satisfactory phases for their gas chromatographic separation. Even quite complex compounds such as pentafluorobenzaldehyde derivatives of catecholamines and related substances of biological importance, with their hydroxyl groups blocked by silylation, are formed quantitatively and can be separated by gas chromatography without decomposition. Very dilute solutions of the corresponding derivatives of dopamine and norepinephrine could be kept at room temperature for two weeks. Chromatographic checks failed to indicate signs of decomposition,[52] except for the Schiff base of 3,4-dimethoxyphenylethylamine.[63]

3. It is somewhat surprising that not much work has been done so far with paper or thin-layer chromatographic analysis of the yellow nitro- and 2,4-dinitrobenzylimines. This might be due to their insolubility in cold solvents.

4. The main attention is presently concentrated on the halogenated derivatives of primary amines. This development started with the introduction of electron-capture detection systems. They can be over 1000 times more sensitive to pentafluoro-derivatives than flame ionization methods. If hydroxyl groups are present, they are usually trimethylsilylated, thus augmenting the electron affinity of the compounds. Picogram quantities of catecholamines can be separated by methylsilicone phases on inactivated supports and detected by electron capture after derivative formation with pentafluorobenzaldehyde and silylation. This has been a great aid in the ultra-trace analysis of primary amines of biological interest. The mass spectra of underivatized primary amines are often not ideal for their characterization, especially if these compounds are only present in low concentration. Eneamine formation is helpful in this respect. These derivatives show the molecular ion as well as some intensive fragments permitting identification of the initial amine. Analogous to the work with electron-capture detection, pentafluorobenzylimines have usually been chosen for this purpose.[52,53,59] This does not seem justified: other eneamines might prove to be more suitable for mass spectrometric identification. Their introduction might lead to lower detection limits in the gas chromatographic trace analysis with mass specific detection.

An interesting application of the eneamine reaction for the dosage of non-volatile primary amines was described recently.[87] The amine solution was injected into the column of the gas chromatograph, followed by repeated injections of a constant amount of benzaldehyde in heptane. The resulting chromatograms showed only the elution of heptane and water until the total quantity of injected benzaldehyde exceeded that required to complete the Schiff base formation with the amine. In this titration, the amount of amine was calculated from the difference between the total amount of injected benzaldehyde and the excess determined from its peak area in the last chromatogram.

3 ANALYSIS OF PRIMARY AMINES AS MUSTARD OILS

Free primary and secondary aliphatic amines are almost instantly converted to dithiocarbamates by the action of carbon disulfide even at room temperature (reactions IV and VI), while tertiary amines do not react.[33] The dithiocarbamic acids of primary aliphatic amines decompose under the influence of metal salts (i.e. mercury or silver salts) and heat to the so-called mustard oils or isothiocyanates and metal sulfide (reaction V). This reaction is one of the most sensitive and characteristic for the detection of primary aliphatic amines.[47] Under less mild conditions, aromatic primary amines react with carbon disulfide (reaction VII) to form dialkylated thioureas with liberation of hydrogen sulfide.[26,47]

IV $$R{-}NH_2 + CS_2 \rightarrow R{-}NH{-}CS{-}SH \rightarrow R{-}N{=}C{=}S + H_2S$$ V

VI $$R^1{-}NH{-}R^2 + CS_2 \rightarrow (R^1)(R^2)N{-}CS{-}SH$$

VII $$2R{-}NH_2 + CS_2 \rightarrow R{-}NH{-}CS{-}NH{-}R + H_2S$$

The reaction of amines with carbon disulfide has been employed to improve the visibility of the separated compounds on thin-layer plates when viewed in u.v. light (usually 254 nm). Primary and secondary alkyl and heterocyclic amines, which ordinarily possess poor u.v. absorption, can be detected by a large increase in u.v. absorbance due to the formation of dithiocarbamates *in situ*. Since the plates should be weakly alkaline or neutral, they are sprayed with dilute ammonia before exposure to carbon disulfide.[82]

It was recently found[3,4] that β-aryl-substituted primary alkyl amines react with excess carbon disulfide even under mild conditions to yield isothiocyanates directly. There is evidence that this reaction proceeds according to a ketone-base condensation (reaction III). Since it is especially suitable for the biologically important series of phenylethylamines and phenylisopropylamines, it has attracted considerable attention, and its potential for the analytical characterization and the determination of primary amines by gas chromatography and combined gas chromatography–mass spectrometry has been extensively investigated.[3–7,54,66–68]

On the other hand, no reports on the use of this type of ketone-base condensation in conjunction with liquid chromatography on columns have appeared so far, although the shift from primary amines to isothiocyanates might also prove advantageous with this technique. The use of the mustard oil reaction prior to thin-layer chromatography is mentioned in two publications.[67] They deal with the detection of 3,4-dimethoxyphenylethylamine in urine samples. The isothiocyanate is separated on silica gel G and the plate sprayed with a mixture of equal volumes of sulfuric acid and methanol, then irradiated with u.v. light. An intense fluorescence develops which permits detection of nanogram quantities of 3,4-dimethoxyphenylethylamine. The derivative can be extracted with methanol for quantitative determination by spectrofluorimetry (fluorescence at 465 nm, excitation at 365 nm). The only other isothiocyanates which yielded a similar fluorescence were those of 3-methoxytyramine[67] and 4-*O*-methyldopamine.[54]

Up to now, all the primary amines reported to undergo direct isothiocyanate formation on addition of carbon disulfide are ethylamine or isopropylamine derivatives substituted in β-position with a benzene or an indole nucleus. They include a large range of biologically important β-phenylethylamines, β-phenylisopropylamines, β-indolethylamines and β-indolisopropylamines. Only in the case of amphetamine (and the 3,4-dimethoxyphenylethylamine mentioned above) has the mustard oil been isolated *in vitro* and characterized by u.v., i.r. and mass spectrometry.[3] For other isothiocyanates, only gas chromatographic retention data and mass spectra (obtained by GC–MS) have been reported.

The following points summarize the advantages of this type of derivative formation for gas chromatographic separation and detection:

1. The isothiocyanates are much less polar than their corresponding amines. Their separation does not require special column fillings. Less tailing and less irreversible adsorption occur, and lower detection limits are therefore possible.

2. Phenylethylamines of relatively low molecular weight such as the parent compound or amphetamine are quite volatile; losses occur during concentration of extracts by evaporation. Since the vapour pressure of the corresponding isothiocyanates is considerably lower, solutions of these derivatives can be evaporated to dryness without loss.

3. Quite often, both the free amine and the corresponding mustard oil can be separated by gas chromatography. The peak shift technique can therefore be used as an additional means of identification, as has been described for amphetamine.[3] It will also permit differentiation between primary and secondary amines with similar retention times, such as amphetamine and methylamphetamine.

4. For observing the sulfur-containing isothiocyanates, specific detection systems can be employed, alone or parallel to a flame ionization detector. The use of an electron-capture system has been described.[3] A sulfur-specific flame

photometric detector might be even more useful in order to improve chromatographic sensitivity and selectivity.

5. The degradation of primary amines by electron-impact mass spectrometry is often complex. The molecular ion, if visible, is of low intensity. A large number of fragment ions are formed; none of them are good tracer ions for following gas chromatography by mass specific detection, since they may have low intensity and are not always very diagnostic. On the other hand the different charge distribution in the isothiocyanates stabilizes the molecular ions, and favours the formation of usually only two high-intensity fragments resulting from the cleavage of the ethyl side chain between the α- and β-carbons (Fig. 1). One, two or all of these ions can be recorded during a chromatographic run with a pre-focused mass spectrometer as detection system. Since it is possible to work with much higher electrical amplification and larger ion slits if the mass spectrometer is used as a mass specific detector and not in the scanning mode, and since only high intensity ions have to be recorded, detection limits in the lowest picogram range are obtainable, as long as column adsorption does not interfere.[4–7]

Many of the biologically important phenylethylamines and indolethylamines contain alcoholic or phenolic hydroxyl groups or both. In order to make them suitable for gas chromatography, these polar groups must be blocked. This can be effected by silylation, simultaneously with or following the mustard oil formation. Brandenberger and Schnyder[5,6] used a one-step procedure. They added carbon disulfide and trimethylsilylimidazole to a solution of the primary amine in a polar solvent such as dimethylformamide or pyridine. Narasimhachari and Vouros[66,68] proceeded in two steps. They treated the free bases in ethyl acetate with carbon disulfide, evaporated to dryness, dissolved the mustard oil in pyridine and allowed it to react with a mixture of 99% bis(trimethylsilyl)trifluoroacetamide and 1% trimethylchlorosilane at 100 °C. This procedure has the advantage of also silylating the indole nitrogen, providing the reaction time is sufficiently long.[68]

Trace analyses were described of amphetamine,[4,7] the hallucinogens mescaline (3,4,5-trimethoxyphenylethylamine) and STP (2,5-dimethoxy-4-methylphenylisopropylamine),[5,6] as well as catecholamines such as dopamine (3,4-dihydroxyphenylethylamine) and norepinephrine (noradrenaline)[5,6] by gas chromatography of their mustard oils (the hydroxyls silylated if present) with mass-specific detection of the molecular ion or the base ion $[M-72]^+$ (for

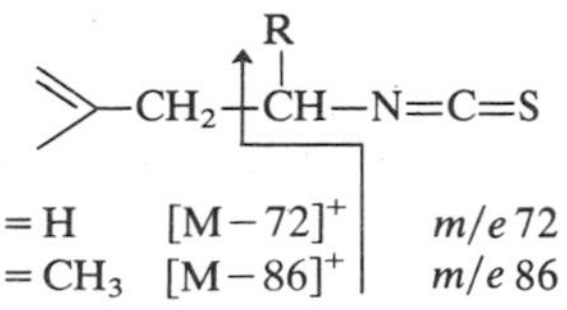

Fig. 1 Cleavage of β-phenyl and α-methyl-β-phenylisothiocyanates by electron impact mass spectrometry.

phenylethylamines) respectively $[M-86]^+$ (for phenylisopropylamines) (Fig. 1). In order to concentrate the mass spectrometric information in the few ions to be recorded, ionization with low electron voltage is recommended. The picogram detection limits obtained (sub-picogram limits would be possible without column interference) are good examples of the potential of mass-specific detection in gas chromatography. The wide application range of the mustard oil reaction with respect to biologically important phenylethylamines and tryptamine derivatives,[68] combined with this potential of the mass-specific detection method (high sensitivity without much loss of specificity) will probably lead to many other useful analytical applications in the fields of biochemistry and medical diagnostics.

4 SELECTED PROCEDURES

4.1 Schiff bases from carbonyl compounds

4.1.1 2,4-DINITROPHENYLHYDRAZONES FROM CARBONYL COMPOUNDS[46]

The carbonyl compounds (100 μl of each) were shaken with 100 ml saturated, 2,4-dinitrophenylhydrazine in aqueous 2M HCL. The mixture was allowed to stand overnight at room temperature. The precipitate was removed by filtration, washed with 2M HCl and H_2O and dried over silica gel in a vacuum desiccator. The derivatives were dissolved in EtOAc for GC with FID or in benzene for GC with ECD: 2% SE-30 or 12% F-60.

4.1.2 2,4-DINITROPHENYLHYDRAZONES OF VOLATILE CARBONYLS IN AROMA CONCENTRATES[29]

Apple aroma (3.5 ml from 1.5 l juice and 10 ml saturated solution of 2,4-dinitrophenylhydrazine in 2 N HCl) was shaken. 10 ml $CHCl_3$ was added, shaken and allowed to settle for 3 h at room temperature. The $CHCl_3$ layer was decanted and the aqueous layer was extracted with $CHCl_3$. The $CHCl_3$ extracts were washed with 2 M HCl until the washings were colourless, then with H_2O, then dried over anhydrous Na_2SO_4. The $CHCl_3$ was removed under vacuum. The derivatives were dissolved in 0.5 ml $CHCl_3$ and analysed by GC: SE-30 on Chromosorb G with FID.

4.1.3 *N,N*-DIMETHYLHYDRAZONES OF STEROIDS[89]

The steroid (0.1–1.0 mg) was dissolved in 0.1–0.2 ml anhydrous *N,N*-dimethylhydrazine and about 0.05 ml glacial HOAc was added, if necessary, as catalyst. After 1–2 h at room temperature, the excess reagent was removed by a stream of N_2. The residue was dissolved in tetrahydrofuran and the solution used directly for GC: 1% SE-30 on Gas Chrom P.

The position of the carbonyl group and the reaction conditions determined the extent of the condensation reaction. A quantitative reaction in 1–2 h at room temperature occurred for 3-ones. A catalytic amount of HOAc was necessary for 20-, 16-, and 17-ones. The 11-keto group failed to react after 12 h.

4.1.4 OXIMES OF CARBONYL COMPOUNDS[13]

To 10–100 mg carbonyl compound in 10 ml H_2O, 3.13 g hydroxylamine-HCl and 2.39 g anhydrous Na_2CO_3 were added. The mixture was extracted with ether continuously for 1 h. The ether extract was dried over Na_2SO_4 and the solvent evaporated until 1 ml solution remained. The residual solution was transferred to a 2 ml volumetric flask. Ether used to wash the distillation flask was employed to bring the volume to 2 ml. This solution was used for GC: 3% di-2-ethylhexylphthalate on Celite.

4.1.5 O-BENZYLOXIMES OF STEROIDS[19,40]

The steroid (1 mg) was dissolved in 0.5 ml dry pyridine solution of *O*-benzylhydroxylamine-HCl (20 mg ml^{-1}). The solution was heated to 70 °C overnight in a screw-capped tube fitted with a Teflon® gasket. The solvent was evaporated under N_2, the residue dissolved in 0.5 ml bis(trimethylsilyl)-trifluoroacetamide and the solution heated at 100 °C for 3 h. GC: 1% Dexsil 300 or 1% SE-30 on Gas Chrom P.

Alternatively, the concentration of *O*-benzylhydroxylamine-HCl was increased to 50 mg ml^{-1} dry pyridine and the *O*-benzyloxime was silylated with *N*-trimethylsilylimidazole at 150–160 °C for 3 h. GC–MS showed that the major metabolites of the adrenocortical steroid hormones were converted into *O*-benzyloxime and trimethylsilyl derivatives. All OH-groups reacted to yield TMS-derivatives.

4.1.6 *O*-METHYLOXIME-TRIMETHYLSILYL DERIVATIVES OF STEROIDS FROM URINE[76]

Urine (25 ml) was adjusted to pH 5.2 with HOAc or NaOH. 0.5 ml phosphate buffer pH 5.2 was added. To 10 ml of this solution, 0.25 ml Helix Pomatia enzyme preparation and 4 drops $CHCl_3$ were added. The mixture was incubated at 37 °C for 24 h and extracted twice with 20 ml EtOAc. The combined EtOAc layers were extracted twice with 20 ml NaOH (0.1 mol l^{-1}) and twice with 20 ml H_2O. The EtOAc layer was dried on $MgSO_4$. The solvent was removed until 0.25–0.5 ml remained. The residue was transferred quantitatively, using a minimum amount of EtOAc, to a small test tube with a Teflon® stopper and a pear-shaped bottom. Internal standards and reference compounds were added. EtOAc was removed under a stream of dry N_2, heating to 40 °C. To the evaporated residue was added 200 μl 10% methoxyamine in pyridine and the mixture allowed to react for 15 min at 60°. The pyridine was removed in an N_2 stream and 100 μl of bis(trimethylsilyl)acetamide/trimethylchlorosilane (4 : 1) was added. The mixture was kept for 1 h at 70 °C: GC on SE-30 or OV-1.

4.2 Schiff bases from amines

4.2.1 ENEAMINE FROM AMINES AND ACETONE (OR OTHER KETONES)[12]

Amine-HCl (0.5–1 mg) was dissolved in 0.05 ml dimethylformamide (DMF). Solid K_2CO_3 or $KHCO_3$ (2 mg) was added followed by 0.5 ml acetone (or other ketone). The mixture was shaken at room temperature for 3 h. Samples were injected directly into the GC: 10% F-60 or 7% F-60+1% EGSP-Z on Gas Chrom P.

Eneamine from amines containing OH-groups: to 0.5–1 mg amine in 0.05 ml DMF, 0.15 ml hexamethyldisilazane (HMDS) was added and allowed to stand 30 min at room temperature.

A ketone–HMDS mixture was prepared by adding 1 ml HMDS to 10 ml acetone (or other ketone), heating to boiling and cooling. A 0.4 ml portion of the ketone–HMDS mixture was added to the DMF solution and the reaction mixture was allowed to stand for 12 h. Precipitates were separated by centrifugation. The reaction products were stable for several days at −5 °C. Analyses by GC or GC–MS.

4.2.2 SCHIFF BASE FROM AMPHETAMINE AND ACETONE[78]

Urine (10 ml) was acidified with 0.4 ml 6 M HCl and extracted with 5.0 ml diethyl ether. This extract was discarded. The aqueous phase was made alkaline with 1.0 ml 20% KOH and extracted with 5.0 ml diethyl ether on a rotary mixer. The ether layer was transferred to a small narrow-necked vial and evaporated to dryness on a water bath at 60 °C. The residue was dissolved in 50 μl diethyl ether. Acetone (0.5 ml) was added and the mixture evaporated at 60 °C to about 50 μl. Both the ketone and some unchanged amphetamine were present. A 5 μl aliquot of this solution was chromatographed: GC on 5% SE-30 on Suprasorb with FID.

4.2.3 PENTAFLUOROBENZALDEHYDE-TRIMETHYLSILYL-DERIVATIVES OF CATECHOLAMINES[63]

The amine and pentafluorobenzaldehyde were heated in acetonitrile solution 1 h at 60 °C to form a Schiff base. A further hour at 60 °C after the addition of bis(trimethylsilyl)acetamide converted all OH-groups to *O*-TMS groups. The reaction mixture was injected directly for GC with FID or diluted with hexane for GC with ECD on 5% SE-30 or 5% OV-17. Epinephrine was converted to two isomeric substituted tetrahydrosioquinolines which interfered with the norepinephrine derivative peak on SE-30. OV-17 separated the norepinephrine and normetanephrine derivatives from each other and from the epinephrine derivative. The 3,4-dimethoxyphenethylamine–PFB derivative showed signs of losses during separation on SE-30.

4.2.4 PENTAFLUOROBENZALDEHYDE-TRIMETHYLSILYL-DERIVATIVES OF CATECHOLAMINES[52]

Microgram level: a sample of catecholamines in redistilled DMF (1 μg/50 μl) was added to 100 μl of an acetonitrile solution of repurified pentafluorobenzaldehyde (100 μg ml^{-1}). The mixture was heated at 85 °C for 5 min. Bis(trimethylsilyl)acetamide (BSA, 2 μl) was then added. Complete reaction was obtained in 15 min at room temperature. Derivatives were extracted with 850 μl of hexane, and injected into the gas chromatograph: 1% SE-30 or 1% OV-17 on Gas Chrom P with ECD and mass specific detection.

Nanogram level: Samples of 1–10 ng of catecholamines were introduced in silanized glass tubes, 10 μl of an acetonitrile solution of repurified pentafluorobenzaldehyde (4 μg ml^{-1}) were added. The tubes were heated to 85 °C for 15 min. After the addition of 1 μl BSA and 9 μl hexane, the tube was sealed again and heated to 60 °C for 5 min. The hexane phase was injected into the GC.

4.3 Isothiocyanates from primary amines

4.3.1 ISOTHIOCYANATES OF BIOLOGICALLY ACTIVE AMINES[5]

Amphetamine was extracted with ether from alkaline aqueous solutions (body fluids). If interfering compounds were present, the solution was first steam distilled, then the distillate made alkaline and extracted. The ether extract was mixed with an equal volume of CS_2, allowed to stand for 1 h and evaporated to a small volume.

Mescaline (3,4,5-trimethoxyphenylethylamine) and STP (2,5-dimethoxy-4-methylphenylisopropylamine) were extracted with ether from the alkaline aqueous solution. The ether extracts were mixed with CS_2 and allowed to stand for several hours or heated to 50 °C for 2 h. Then they were evaporated to a small volume.

The reactions with dopamine, norephedrine and noradrenaline were carried out in DMF. The solutions were mixed with equal volumes of CS_2 and trimethylsilylimidazole, then warmed under reflux for 3 h and evaporated to small volumes. Pyridine, formamide and tetrahydrofuran were also satisfactory solvents for converting dopamine.

4.3.2 ISOTHIOCYANATES OF BIOLOGICALLY ACTIVE AMINES[66,68]

A solution containing 1 mg free base in 5 ml EtOAc was shaken with 0.5 ml CS_2 for 30 min, evaporated to dryness under reduced pressure and redissolved in EtOAc (1 ml). Aliquots (1 μl) were used for GC: 1% SE-30 or 2.5% OV-225 on Gas Chrom Q.

Phenolic and indole amines: A 100 μg aliquot of the NCS-derivative was reacted with bis(trimethylsilyl)trifluoroacetamide/trimethylchlorosilane

(99 : 1) at 90 °C for 15 min. For 2- or 7-methyl substituted tryptamines, it was necessary to extend the reaction time to 1 h at 100 °C to ensure complete silylation of the indole nitrogen, presumably because of steric hindrance from the alkyl substituent.

Perdeutero–TMS derivatives were prepared by reacting the NCS-derivatives with a mixture of d_{18}-bistrimethylsilylacetamide and d_9-trimethylchlorosilane (10 : 1).

Trifluoroacetyl-derivatives were prepared by treating the NCS-compounds in EtOAc with a few drops of trifluoroacetic anhydride and allowing the solution to stand at room temperature for 30 min.[67]

5 TABULAR SUMMARY OF KETONE–BASE DERIVATIVES

Tables 1, 2 and 3 (pp. 247–258) are a summary of ketone-base derivatives for chromatography. The abbreviations appearing in them are as follows:

SB	Schiff base	BSTFA	Bis(trimethylsilyl)trifluoro-acetamide
PH	Phenylhydrazone		
DNPH	2,4-Dinitrophenylhydrazone	TMCS	Trimethylchlorosilane
MO	*O*-Methyloxime	TMCS-d_9	Trimethylchlorosilane-d_9
BO	*O*-Benzyloxime	TSIM	Trimethylsilylimidazole
TMS	Trimethylsilyl ether	TLC	Thin-layer chromatography
PFB	Pentafluorobenzylimine	GC	Gas chromatography
NCS	Isothiocyanate	MS	Mass spectrometry
DNP	2,4-Dinitrophenylhydrazine	SIM	Selected ion monitoring (in mass fragmentography or GC–MS)
DMF	*N,N*-Dimethylformamide		
HMDS	Hexamethyldisilazane		
BSA	Bis(trimethylsilyl)acetamide	FID	Flame ionization detection
BSA-d_{18}	*N,O*-Bis(trimethylsilyl-d_9)-acetamide	ECD	Electron-capture detection

TABLE 1

Analysis of carbonyl compounds as Schiff bases

Ketone	Reagent	Derivative	Separation	Detection	Reference
α-Keto acids	DNP	DNPH	Paper chromatography Butanol/aqueous NH_3 Butanol/aqueous EtOH	Aqueous NaOH Spectrometry	14
α-Ketoglutaric acid Pyruvic acid (blood, urine, tissues)	DNP	DNPH	Paper chromatography Butanol/EtOH	Aqueous NaOH Spectrometry	14
Aliphatic and aromatic aldehydes Aliphatic ketones	DNP	DNPH	Chromatography on paper impregnated with silicic acid Ether, acetone, tetrahydrofuran in petroleum ether	Aqueous NaOH	73
Aliphatic carbonyls	DNP	DNPH	Column chromatography Silicic acid/Celite Gradient of ethyl ether in petroleum ether	u.v. Spectrometry 356 nm	30
Aliphatic carbonyls	DNP	DNPH	Paper chromatography MeOH/heptane	Aqueous NaOH	61
Aliphatic carbonyls Benzaldehyde Anisaldehyde Veratraldehyde	DNP	DNPH	Chromatography on paper impregnated with phenoxyethanol/acetone Heptane saturated with phenoxyethanol	Spectrometry	57
Aliphatic carbonyls (tobacco smoke)	DNP	DNPH	Chromatography on paper impregnated with DMF *n*-Hexane saturated with DMF	Spectrometry	11

TABLE 1—continued

Ketone	Reagent	Derivative	Separation	Detection	Reference
Aliphatic carbonyls (cigarette smoke)	DNP	DNPH	Paper chromatography Isooctane saturated with DMF	Spectrometry 345–90 nm	64
Aliphatic aldehydes to C_{14}	DNP	DNPH	Chromatography on paper impregnated with propylene glycol/MeOH or Skellysolve C/vaseline Skellysolve C/aq. MeOH	Colour	22
2,3-Diones Glyoxal (tobacco smoke)	DNP	DNPH	Paper chromatography Petroleum ether/diethyl ether/MeOH	Aqueous KOH	58
Aliphatic aldehydes (irradiated meat fat)	DNP	DNPH	Column chromatography Silicic acid/Celite Hexane, benzene, diethyl ether, MeOH	Alcoholic NaOH Spectrometry 520–35 nm	93
Aliphatic and aromatic carbonyls	DNP	DNPH	Chromatography on paper impregnated with DMF/acetone Cyclohexane/cyclohexene	Aqueous NaOH Spectrometry	83
Aliphatic and aromatic carbonyls	DNP	DNPH	TLC on aluminium oxide Ether, benzene/hexane Centrifugal chromatography on paper impregnated with DMF/EtOH Cyclohexane	Spectrometry	74
Aromatic carbonyls Acetaldehyde	DNP	DNPH	TLC on silica gel G Benzene/petr. ether Benzene/EtOAc	Spectrometry	20

TABLE 1—continued

Ketone	Reagent	Derivative	Separation	Detection	Reference
Saturated and unsaturated aliphatic aldehydes	DNP	DNPH	Column chromatography Magnesia/Celite $CHCl_3$/hexane	Spectrometry 345–73 nm	77
Saturated and unsaturated aliphatic carbonyls (oxidized milk powder)	DNP	DNPH	Partition chromatography Celite with methylcyclohexane/acetonitrile/H_2O or, Celite/alumina with EtOAc/methylcyclohexane/acetonitrile/H_2O	Spectrometry 335–80 nm	15, 16
C_1–C_5 *n*-Aldehydes (oxidation or acid decomposition of organic compounds)	DNP	DNPH	TLC on Kieselgel GF_{254} Ether/petr. ether	$SnCl_2$ and *p*-dimethylaminobenzaldehyde Spectrometry 254 nm	8
Aliphatic aldehydes (hydrolysis of acetals)	DNP	DNPH	TLC on silica gel G Benzene/petr. ether Benzene/EtOAc GC: SE-30 on Chrom. W	FID Thermal conductivity	27
Aliphatic carbonyls Benzaldehyde	DNP	DNPH	Flash exchange chromatography Mix DNPH with α-ketoglutaric acid Exchange at 240–60° regenerates carbonyls GC on Carbowax		71

TABLE 1—continued

Ketone	Reagent	Derivative	Separation	Detection	Reference
Aliphatic aldehydes	DNP	DNPH	Flash exchange chromatography Mix DNPH with Celite, α-keto-glutaric acid, formaldehyde-DNPH Exchange at 250 °C Formaldehyde flushes the other carbonyls into the GC: Dinonyl phthalate on Celite		80
Aliphatic and aromatic carbonyls	DNP	DNPH	GC SF-96 on Chrom. W	FID	79
Carbonyls (cigarette smoke, flavouring agents)	DNP	DNPH	GC SE-30 on Chrom. G	FID	51
Aliphatic carbonyls (aroma concentrates of fruits)	DNP	DNPH	GC SE-30 on Chrom. G	FID	29
Aliphatic aldehydes Benzaldehyde	DNP	DNPH	GC SE-30 or F-60 on Chrom. W	FID ECD	46
C_1–C_{12} *n*-Aldehydes	Phenylhydrazine	PH	GC SE-30 on Chrom. W	FID	25
Aliphatic carbonyls (flavour compounds in foods)	Phenylhydrazine	PH	GC SE-30 on Chrom. W	FID	49
Di-*n*-hexyl ketone Di-*n*-heptyl ketone Androstan-17-one	Phenylhydrazine Pentafluorophenyl-hydrazine *N*-Aminopiperidine *N*-Aminohomo-piperidine	SB	GC F-60-Z or CNSi on Gas Chrom. P	FID MS	88

TABLE 1—continued

Ketone	Reagent	Derivative	Separation	Detection	Reference
Estrone (plasma)	Dimethyl sulfate Pentafluoro-phenylhydrazine	3-Methyl ether 17-Pentafluoro-phenylhydrazone	TLC on silica gel G Benzene GC XE-60, SE-30 or EPON 1001 on Gas Chrom. Q	Spectrofluori-metry ECD	1
OH- and OMe-Benzaldehydes	Tetra-*O*-acetyl-D-glucosyl-3-thio-semicarbazide 4-(Tetra-*O*-acetyl-β-D-glucopyranosyl)-3-thiosemicarbazide	Semicarbazones	TLC on silica gel $CHCl_3$/MeOH	Diazosulfanilic acid u.v. Spectrometry	70
Aliphatic and aromatic carbonyls	Semicarbazide	Semicarbazones	Pre-column reaction Injection block with diatomaceous earth/semicarbazide at 115 °C	FID	17
Aliphatic aldehydes (alkaline degradation of herqueinone)	Hydroxylamine-HCl	Oximes	GC Celite/di-2-ethyl-hexylphthalate		13
Benzaldehyde Salicylaldehyde	Hydroxylamine	Oximes	GC Silicone rubber grease on Celite		56
α-Keto-acids (urine, serum)	Hydroxylamine +BSA/TMCS	Oximes Oxime-TMS	GC OV-17 on Chrom. W	FID MS	81
Steroids	*N,N*-Dimethyl-hydrazine	*N,N*-Dimethyl-hydrazones	GC SE-30, QF-1 or CNSi on Gas Chrom. P	FID	37, 89

TABLE 1—continued

Ketone	Reagent	Derivative	Separation	Detection	Reference
Steroids	Methoxylamine-HCl +TMCS	MO MO-TMS	GC NGS or SE-30 on Gas Chrom. P	FID MS	24
Steroids (urine)	Methoxylamine-HCl +HMDS	MO MO-TMS	GC SE-30 on Gas Chrom. P	FID MS	28
Steroids (urine)	Methoxylamine-HCl +BSA/TMCS	MO MO-TMS	GC OV-1, OV-17 or SE-30 on Gas Chrom. P	FID	39
Steroids	Methoxylamine-HCl +BSA/TMCS	MO MO-TMS	GC OV-1 or OV-17 on Gas Chrom P TLC on silica gel G/Rhodamine 6G Benzene/EtOAc $CHCl_3$/MeOH	FID	43
Aldosterone	Methoxylamine-HCl +BSTFA +BSA/TMCS +Heptafluoro-butyric anhydride	MO MO-TMS MO-HFB	GC SE-30, OV-1, OV-17 or OV-22 on Gas Chrom. P	FID ECD MS	41
Steroids Keto-acids (urine)	Methoxylamine-HCl +BSA-TSIM +Heptafluoro-butyric anhydride	MO MO-TMS MO-HFB	GC SE-30, OV-1, OV-101, OV-17 on Gas Chrom. P	FID ECD	35

TABLE 1—continued

Ketone	Reagent	Derivative	Separation	Detection	Reference
Human urinary steroids	Methoxylamine-HCl	MO	GC SE-30 on Gas Chrom. P	FID MS	85
Adrenocortical steroid hormones	+TSIM +BSA	MO-TMS			
Steroids Endogenous cortico-steroids (rat adrenals, blood)	Methoxylamine-HCl +HMDS +BSA-d_{18}	MO MO-TMS MO-TMS-d_9	GC SE-30 on Gas Chrom. P	FID MS SIM	59
Estrogens (non-pregnant woman's blood)					
Metabolites of corti-costerone (rat liver)					
Steroids (urine)	Methoxylamine-HCl +BSA/TMCS	MO MO-TMS	GC Open-tube column SE-30 or OV-1	FID	76
Steroids (human urine)	*O*-Benzylhydroxyl-amine-HCl +TSIM	BO BO-TMS	GC Dexsil or SE-30 on Gas Chrom. P	FID MS Thermal conductivity	40
Steroids (urine of infants)	*O*-Benzylhydroxyl-amine-HCl +BSTFA	BO BO-TMS	GC SE-30 on Gas Chrom. P	FID MS	19

TABLE 2

Analysis of amines as Schiff bases

Amine	Reagent	Derivative	Separation	Detection	Reference
Biological amines	Acetone	SB	GC		9
Aminoalcohols	Butanone		SE-30 on Gas Chrom. P		
Aminophenols	+HMDS	SB-TMS			
Long chain amines	Acetone	SB	GC	FID	88
Diamines	1,1,1-Trifluoro-acetone		CNSi, SE-52 or F-60-Z on Gas Chrom. S		
Benzylamines	Cyclopentanone				
Amphetamine	Cyclohexanone				
	Cycloheptanone				
Biological amines	Acetone	SB	GC	FID	10
Catecholamines			F-60-Z or NGS on Gas Chrom. P		
Tryptamines					
Tryptamine	Acetone	SB	GC	FID	36
5-Hydroxytryptamine	+HMDS	SB-TMS	F-60-Z or NGS on Gas Chrom. P		
5-Methoxytryptamine					
Biological amines	Acetone	SB	GC	FID	12
	Cyclobutanone		F-60-Z or F-60 on Gas Chrom. P	MS	
	Cyclopentanone				
	Cyclohexanone				

TABLE 2—continued

Amine	Reagent	Derivative	Separation	Detection	Reference
Biological amines—*continued*					
	Cycloheptanone Diisopropyl-ketone +HMDS	SB-TMS			
Catecholamines	Pentafluoro-benzaldehyde +BSA	PFB PFB-TMS	GC SE-30 or OV-17	FID ECD MS	63
Amphetamine (urine)	Acetone	SB	GC SE-30 on Suprasorb	FID	78
Biological amines (adrenal extract)	Pentafluoro-benzaldehyde +BSA	PFB PFB-TMS	GC OV-17 on Gas Chrom. P	ECD SIM	59
Biological amines (rat adrenal extract)	Pentafluoro-benzaldehyde +BSA	PFB PFB-TMS	GC SE-30 or OV-17 on Gas Chrom. P	FID ECD SIM	52
Biological amines	Pentafluoro-benzaldehyde +BSA	PFB	GC SE-30, OV-1 or OV-17 on Gas Chrom. P	ECD SIM	53

TABLE 2—continued

Amine	Reagent	Derivative	Separation	Detection	Reference
Hydrazine (tobacco, tobacco smoke, tech. maleic hydrazide, pyrolysis products)	Pentafluoro-benzaldehyde	PFB Decafluoro-benzaldehyde azine	GC OV-17+QF-1 on Gas Chrom. Q TLC on silica gel Benzene TLC on Al_2O_3 Hexane/benzene	ECD	55
Aminophenols Benzidine *o*-Toluidine *β*-Naphthylamine *o*-Dianisidine	Benzaldehyde	SB	Reaction GC Reaction column: Me_2SiCl_2 on Chrom. W Analysis column: silicone QF-1 Inject amines, then titrate with benzaldehyde in heptane	FID	87
o-Dianisidine 2-Naphthylamine Aminophenols	Cyclohexanone Benzaldehyde 2-Heptanone	SB	GC Determine activation energy of SB-formation		86
Serotonin	*o*-Phthalaldehyde	SB	TLC on silica gel G $CHCl_3$/MeOH/aq. NH_3 Isopropanol/aq. NH_3	Spectro-fluori-metry	65

TABLE 3

Analysis of amines as isothiocyanates

Amine	Reagent	Derivative	Separation	Detection	Reference
Serotonin Methylserotonin Bufotenin 5-Methoxytryptamine	CS_2	NCS for prim. amines Dithiocarbamic acid for sec. amines	TLC on silica gel G Isopropanol/aq. NH_3	Dimethylamino-benzaldehyde in 1 N HCl	31
Amphetamine	CS_2	NCS	GC SE-30 or XE-60 on Chrom. W	FID ECD MS	3
Amphetamine	CS_2	NCS	GC SE-30 on Chrom. W	FID SIM	4
Amphetamine Mescaline STP Dopamine Norephedrine Noradrenaline	CS_2 +TSIM	NCS NCS-TMS	GC SE-30 or Versamid 900 on Chromosorb	SIM MS	5

TABLE 3—continued

Amine	Reagent	Derivative	Separation	Detection	Reference
Phenylethylamine Amphetamine Mescaline STP Dopamine Norephedrine Noradrenaline	CS_2 +TSIM	NCS NCS-TMS	GC SE-30 on Chromosorb	SIM	6
3-Methoxy-4-hydroxy- and 4-methoxy-3-hydroxy-phenylethylamines	CS_2 +Regisil-TMCS +trifluoro-acetic anhydride +BSA-d_{18}/TMCS-d_9	NCS NCS-TMS NCS-TFA NCS-TMS-d_9	GC OV-225 or OV-101 on Gas Chrom. Q	MS	67
Phenylethylamines Phenylisopropylamines Parnates Tryptamines	CS_2 +BSTFA/TMCS	NCS NCS-TMS	GC SE-30 or OV-225	MS	66
Phenylethylamines Phenylisopropylamines Tryptamines (urine)	CS_2 +BSTFA/TMCS +BSA-d_{18}/TMC-d_9	NCS NCS-TMS NCS-TMS-d_9	GC OV-225, OV-17 or SE-30 on Gas Chrom. Q	FID MS	68
3,4-Dimethoxyphen-ethylamine 3,4-Dimethoxyphenyl-isopropylamine 3-Methoxytyramine (urine)	CS_2	NCS	TLC on silica gel G $CHCl_3$, benzene/HOAc	$MeOH/H_2SO_4$ and u.v. irradiation with Spectrofluorimetry	67

REFERENCES

1. J. Attal, S. M. Hendeles and K. B. Eik-Nes, *Anal. Biochem.* **20**, 394 (1967).
2. M. L. Bastos and D. B. Hoffman, *J. Chromatogr. Sci.* **12**, 269 (1974).
3. H. Brandenberger and E. Hellbach, *Helv. Chim. Acta* **50**, 958 (1967).
4. H. Brandenberger, *Pharm. Acta Helv.* **45**, 394 (1970).
5. H. Brandenberger and D. Schnyder, *Z. Anal. Chem.* **261**, 297 (1972).
6. H. Brandenberger, *Proceedings of the International Symposium on Gas Chromatography Mass Spectrometry*, Elba, Italy (Ed. A. Frigerio), Tamburini Editore, Milano, 1972, p. 37.
7. H. Brandenberger, in *Clinical Biochemistry II* (Eds. Curtius, Roth), De Gruyter, New York, 1974, p. 1465.
8. A. Brantner, J. Vamos and A. Vegh, *Gyogyszereszet* **17**, 457 (1973); *C.A.* **80**, 90931h (1974).
9. E. Brochmann-Hanssen and A. Baerheim, *J. Pharm Sci.* **51**, 938 (1962).
10. C. J. W. Brooks and E. C. Horning, *Anal. Chem.* **36**, 1540 (1964).
11. D. A. Buyske, I. H. Owen, P. Wilder and M. E. Hobbs, *Anal. Chem.* **28**, 910 (1956).
12. P. Capella and E. C. Horning, *Anal. Chem.* **38**, 316 (1966).
13. J. Cason and E. R. Harris, *J. Org. Chem.* **24**, 676 (1959).
14. D. Cavallini, N. Frontali and G. Toschi, *Nature* **163**, 568; **164**, 792 (1949).
15. E. A. Corbin, *Anal. Chem.* **34**, 1244 (1962).
16. E. A. Corbin, D. P. Schwartz and M. Keeney, *J. Chromatogr.* **3**, 322 (1960).
17. D. A. Cronin, *J. Chromatogr.* **64**, 25 (1972).
18. M. B. Devani, C. J. Shishoo and B. K. Dadia, *J. Chromatogr.* **105**, 186 (1975).
19. P. G. Devaux, M. G. Horning, R. M. Hill and E. C. Horning, *Anal. Biochem.* **41**, 70 (1971).
20. J. H. Dhont and C. de Rody, *Analyst* 74 (1961).
21. G. M. Dyson and H. J. George, *J. Chem. Soc.* 1702 (1924).
22. R. Ellis, A. M. Gaddis and G. T. Currie, *Anal. Chem.* **30**, 475 (1958).
23. L. L. Engel, A. M. Neville, J. C. Orr and P. R. Ragatt, *Steroids* **16**, 377 (1970).
24. H. M. Fales and T. Luukkainen, *Anal. Chem.* **37**, 955 (1965).
25. E. Fedeli and M. Cirimeli, *J. Chromatogr.* **15**, 435 (1964).
26. L. F. Fieser and M. Fieser, *Organic Chemistry*, 1st edition, D. C. Heath, Boston, 1944, p. 615.
27. W. G. Galetto, R. E. Kepner, and A. D. Webb, *Anal. Chem.* **38**, 34 (1966).
28. W. L. Gardiner and E. C. Horning, *Biochem. Biophys. Acta* **115**, 524 (1966).
29. L. Gasco, R. Barrera and F. de la Cruz, *J. Chromatogr. Sci.* **7**, 228 (1969).
30. B. E. Gordon, F. Wopat, H. D. Burnham and L. C. Jones, *Anal. Chem.* **23**, 1954 (1951).
31. H. Gross and F. Franzen, *Biochem. Z.* **340**, 403 (1964).
32. C. G. Hammar, B. Holmstedt and R. Ryhage, *Anal. Biochem.* **25**, 532 (1968).
33. A. W. Hofmann, *Ber. dtsch. chem. Ges.* **1**, 25, 169 (1868).
34. Hopkin and Williams Limited, Chadwell Heath, Essex, England, Organic Reagents for Organic Analysis, 1950.
35. E. C. Horning and M. G. Horning, *J. Chromatogr. Sci.* **9**, 129 (1971).
36. E. C. Horning, M. G. Horning, W. J. A. Vanden Heuvel, K. L. Knox, B. Holmstedt and C. J. W. Brooks, *Anal. Chem.* **36**, 1546 (1964).
37. E. C. Horning and W. J. A. Vanden Heuvel, *Ann. Rev. Biochem.* **32**, 709 (1963).
38. E. C. Horning, W. J. A. Vanden Heuvel and B. G. Creech, *Methods of Biochemical Analysis XI* (Ed. D. Glick), Interscience, New York, 1963, p. 69.
39. E. C. Horning, M. G. Horning, N. Ikekawa, E. M. Chambaz, P. I. Jaakonmaki and C. J. W. Brooks, *J. Gas Chromatogr.* **5**, 283 (1967).
40. E. C. Horning, P. G. Devaux, A. C. Moffat, C. D. Pfaffenberger, N. Sakuchi and M. G. Horning, *Clin. Chim. Acta* **34**, 135 (1971).
41. E. C. Horning and B. F. Maume, *J. Chromatogr. Sci.* **7**, 411 (1969).
42. M. G. Horning, E. A. Boucher, A. M. Moss and E. C. Horning, *Anal. Lett.* **1**, 713 (1968).
43. M. G. Horning, A. M. Moss and E. C. Horning, *Biochem. Biophys. Acta* **148**, 597 (1967); *Anal. Biochem.* **22**, 284 (1968).

44. M. Hranisavljevic-Jakovljevic, I. Pejkovic-Tadic and A. Stojiljkovic, *J. Chromatogr.* **12**, 70 (1963).
45. D. Janne, R. Vihko and K. Sjovall, *Clin. Chim. Acta* **21**, 405 (1969).
46. H. Kallio, R. R. Linko and J. Kartaranta, *J. Chromatogr.* **65**, 355 (1972).
47. P. Karrer in *Lehrbuch der Organischen Chemie*, Vol. 13 (Ed. G. Thieme), Stuttgart, 1959, pp. 148 and 491.
48. E. Komanova, V. Knoppova and V. Koman, *J. Chromatogr.* **73**, 231 (1972).
49. J. Korolczuk, M. Daniewski and Z. Mielniczuk, *J. Chromatogr.* **88**, 177 (1974).
50. S. H. Koslow, F. Cattabeni and E. Costa, *Science* **176**, 177 (1972).
51. R. E. Leonard and J. E. Kiefer, *J. Gas Chromatogr.* **4**, 142 (1966).
52. J. C. Lhuguenot and B. F. Maume, *J. Chromatogr. Sci.* **12**, 411 (1974).
53. J. C. Lhuguenot and B. F. Maume, *Mass Spectrometry in Biochemistry and Medicine* (Eds. A. Frigerio and N. Castagnoli), Raven Press, New York, 1974, p. 111.
54. R. L. Lin and N. Narasimhachari, *Anal. Biochem.* **57**, 46 (1974).
55. Y. Liu, I. Schmeltz and D. Hoffmann, *Anal. Chem.* **46**, 885 (1974).
56. L. J. Lohr and R. W. Warren, *J. Chromatogr.* **8**, 127 (1962).
57. W. S. Lynn, L. A. Steele and E. Staple, *Anal. Chem.* **28**, 132 (1956).
58. I. Martin, *Chem. Ind.* 1439 (1958).
59. B. F. Maume, P. Bournot, J. C. Lhuguenot, C. Baron, F. Barbier, G. Maume, M. Prost and P. Padieu, *Anal. Chem.* **45**, 1073 (1973).
60. S. M. McElvain, *The Characterization of Organic Compounds*, Macmillan, New York, 1947.
61. D. F. Meigh, *Nature*, (*London*) **170**, 159 (1952).
62. L. D. Metcalfe, *J. Chromatogr. Sci.* **13**, 516 (1975).
63. A. C. Moffat and E. C. Horning, *Biochem. Biophys. Acta* **222**, 248 (1970).
64. J. D. Mold and M. T. McRae, *Tobacco Sci.* **1**, 40 (1957); *C.A.* **51**, 8381d (1957).
65. N. Narasimhachari and J. Plaut, *J. Chromatogr.* **57**, 433 (1971).
66. N. Narasimhachari and P. Vouros, *Anal. Biochem.* **45**, 154 (1972).
67. N. Narasimhachari and P. Vouros, *J. Chromatogr.* **70**, 135; N. Narasimhachari, J. Plaut and K. Leiner, *J. Chromatogr.* **64**, 341 (1972).
68. N. Narasimhachari and P. Vouros, *Biomed. Mass Spectrom.* **1**, 367 (1974).
69. S. Nesic, Z. Nikic and I. Pejkovic-Tadic and M. Hranisavljevic-Jakovljevic, *J. Chromatogr.* **76**, 185 (1973).
70. Z. Nowakowska and W. Wieniawski, *Acta Pol. Pharm.* **30**, 565, 571 (1973); *C.A.* **80**, 146466e, 146467f (1974).
71. J. W. Ralls, *Anal. Chem.* **32**, 332 (1960).
72. R. Reimendal and J. B. Sjovall, *Anal. Chem.* **45**, 1083 (1973).
73. R. G. Rice, J. Keller and J. G. Kirchner, *Anal. Chem.* **23**, 194 (1950).
74. J. Rosmus and Z. Deyl, *J. Chromatogr.* **6**, 187 (1961).
75. M. Roth and A. Hampai, *J. Chromatogr.* **83**, 353 (1973).
76. P. Sandra, M. Verzele and E. Vanluchene, *Chromatographia* **8**, 499 (1975).
77. D. P. Schwartz, O. W. Parks and M. Keeney, *Anal. Chem.* **34**, 669 (1962).
78. K. D. R. Setchell, J. D. H. Cooper, *Clin. Chim. Acta* **35**, 67 (1971).
79. R. J. Soukup, R. J. Scarpellino and E. Danielezik, *Anal. Chem.* **36**, 2255 (1964).
80. R. I. Stephens and A. P. Teszler, *Anal. Chem.* **32**, 1047 (1960).
81. H. J. Sternowsky, J. Roboz, F. Hutterer and G. Gaull, *Clin. Chim. Acta* **47**, 371 (1973).
82. H. M. Stevens and P. D. Evans, *Acta Pharmacol. Toxicol.* **32**, 525 (1973).
83. E. Sundt and M. Winter, *Anal. Chem.* **30**, 1620 (1958).
84. C. C. Sweeley, W. H. Elliot, I. Fries and R. Ryhage, *Anal. Chem.* **38**, 1549 (1966).
85. J.-P. Thenot and E. C. Horning, *Anal. Lett.* **5**, 21 (1972).
86. R. Toyoda, T. Nakagawa and T. Uno, *Bunseki Kagaku* **22**, 914 (1973); *C.A.* **80**, 59136d (1974).
87. T. Uno, T. Nakagawa and R. Toyoda, *Bunseki Kagaku* **21**, 993 (1972); *C.A.* **79**, 37728e (1973).

88. W. J. A. Vanden Heuvel, W. L. Gardiner and E. C. Horning, *Anal. Chem.* **36**, 1550 (1964).
89. W. J. A. Vanden Heuvel and E. C. Horning, *Biochem. Biophys. Acta* **74**, 560 (1963).
90. W. J. A. Vanden Heuvel, W. L. Gardiner and E. C. Horning, *J. Chromatogr.* **18**, 391 (1965).
91. T. Walle, *Acta Pharm. Suec.* **5**, 353 (1968).
92. A. D. Webb and R. E. Kepner, *Food Res.* **22**, 384 (1957).
93. L. A. Witting and B. S. Schweigert, *J. Am. Oil Chem. Soc.* **35**, 413 (1958).

CHAPTER 7

Cyclization

André Darbre

Department of Biochemistry, University of London King's College, Strand, London WC2R 2LS, United Kingdom

1 INTRODUCTION

The preparation of suitable derivatives for the purposes of chromatographic separation usually makes use of single protecting group donors, such as acylation (Chapter 3), silylation (Chapter 4) or esterification (Chapter 2) agents. When compounds are polyfunctional, it is possible to use reagents which specifically react with proximal groups; this may avoid multi-step operations, such as the esterification of a carboxyl group followed by acylation of a hydroxyl group elsewhere in the compound under investigation.

Cyclization may occur as a result of an intramolecular rearrangement between groups on the same molecule. This may be unexpected and a cautionary tale may be related here: when vitamin D_2 was gas-chromatographed it gave not one, but two peaks. These were found to correspond to isopyro- and pyrocalciferol formed by a thermal cyclization reaction above 210 °C. Similarly, vitamin D_3 gave isopyro- and pyrovitamin D_3.[1] Thus, when dealing with compounds possessing more than one reactive group, it is particularly important to ascertain by appropriate chemical or physicochemical methods the exact nature of the derivative represented by the TLC spot or the GLC peak on the chromatogram. GC–MS, which has developed enormously since its introduction about 1964, is one of the most powerful tools for this purpose, but GLC followed by trapping the derivative and its subsequent investigation by infrared spectroscopy, nuclear magnetic resonance, optical rotatory dispersion, etc., have also been used.

Intermolecular reactions may occur without addition or loss, for example, when 3 mol of formaldehyde combine to give paraldehyde. When amino acids condense to give diketopiperazines water is eliminated as shown in the following reaction.

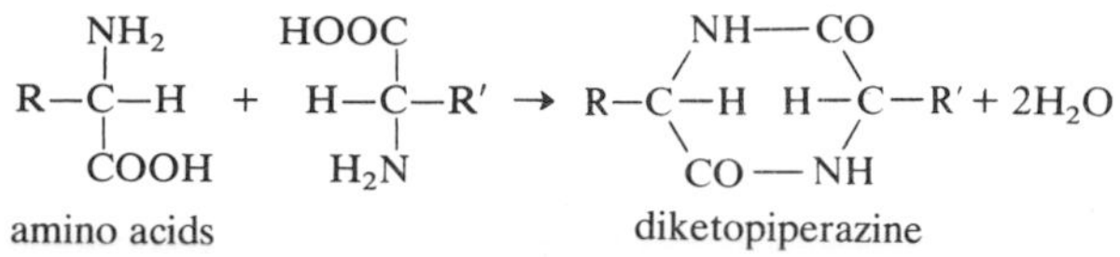

$$\begin{array}{ccc}
\underset{\underset{\text{COOH}}{|}}{\overset{\overset{\text{NH}_2}{|}}{\text{R}-\text{C}-\text{H}}} + \underset{\underset{\text{H}_2\text{N}}{|}}{\overset{\overset{\text{HOOC}}{|}}{\text{H}-\text{C}-\text{R}'}} \rightarrow & \text{R}-\overset{\text{NH}-\text{CO}}{\underset{\text{CO}-\text{NH}}{\text{C}-\text{H} \quad \text{H}-\text{C}-\text{R}'}} + 2\text{H}_2\text{O} \\
\text{amino acids} & \text{diketopiperazine}
\end{array}$$

A different form of cyclization involves the use of a bridging group. By using different types of bridging groups, it is possible to form cyclic compounds with rings of different sizes, and also to insert specific atoms or molecular groupings, which confer specific properties to aid in the subsequent analysis. A small molecule can thus be increased in molecular weight, to give a compound having less extreme volatility and having a greater molar response with a detector. A chromophoric group can be introduced for spectrometric analysis, radioactive or stable isotopes can be introduced for specific labelling and detection, and fragmentation-directing behaviour may be conferred for MS analysis. In addition, certain thermolabile groupings, such as the dihydroxyacetone side-chain of corticosteroids, may be stabilized for GLC separation by the formation of suitable cyclic derivatives.

In this chapter we are concerned with cyclization as a means of chemical derivatization which enables the cyclic derivative to be separated more efficiently and easily from other compounds in a mixture, and also to be determined quantitatively. Thus, as examples, neither cyclic ozonide reactive intermediates (see review[2]) nor cyclic osotriazoles formed from sugar osazones,[3] will be considered. Ring compounds resulting from chelation of metal ions are considered in Chapter 12.

In order to make a cyclic derivative, the parent compound must possess two functional groups of suitable reactivity with a spatial separation suitable for the formation of a ring. The cyclic derivative must be produced in high and reproducible yield, preferably by a simple method using essentially mild conditions, and one which is not unduly time consuming. Only one derivative should be produced from the original parent compound and it should be suitably stable for the conditions used during analytical separation. Thus, the possibility of ring–chain tautomerism should be borne in mind (see reviews[4,5]). Examples of this are shown in Fig. 1. Molecular rearrangements such as acyl

chain   ring

$$\text{CH}_2{=}\text{CHCH}_3 \;\rightleftharpoons\; \text{H}_2\text{C}\underset{\text{CH}_2}{\overset{\displaystyle|}{-\!\!-}}\text{CH}_2$$

$$\underset{\text{HCCOC}_6\text{H}_5}{\overset{\text{C}_6\text{H}_5-\overset{\displaystyle\|}{\text{C}}\text{COOH}}{}} \;\rightleftharpoons\; \text{C}_6\text{H}_5-\underset{\underset{\underset{\text{OH}}{|}}{\text{HC}-\text{C}-\text{C}_6\text{H}_5}}{\overset{\text{C}-\text{CO}}{\overset{\displaystyle\|}{}}}\overset{}{\underset{}{\text{O}}}$$

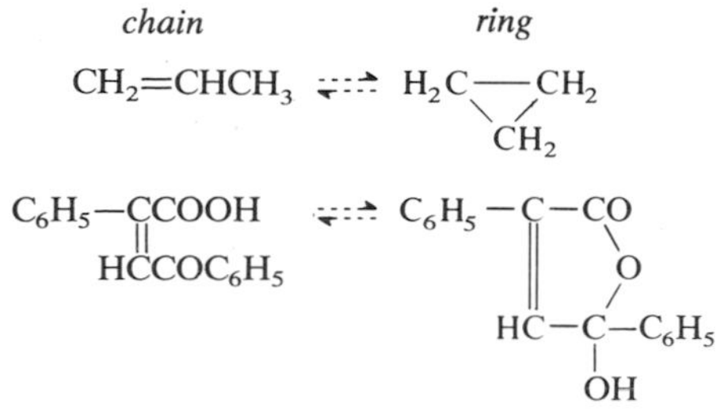

Fig. 1   Examples of tautomerism.

migration may occur when carrying out bridging reactions, particularly under the influence of strongly acidic or basic conditions. In separation processes where the parent compound is to be recovered the cyclization should be reversible.

2 ACETALS AND KETALS

2.1 General Introduction

Cyclic acetals and ketals are normally prepared from diol, triol and polyhydroxy compounds by reaction with an aldehyde or a ketone, usually to form a 1,3-dioxolane ring (Fig. 2). Other ring sizes may often be formed. Sulfydryl groups may also react. Thus, ketals are formed when cyclohexane-1, 2-dione is reacted with ethylene glycol, 2-mercaptoethanol and ethane-1,2-dithiol.[6] Note that condensation may lead to the introduction of an asymmetric centre into the cyclic derivative.

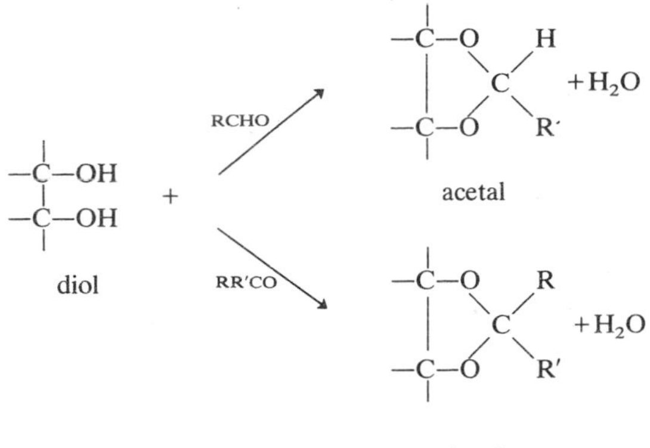

Fig. 2 Examples of acetal and ketal formation.

Acetal formation in aldoses and aldosides has been reviewed.[7] The reaction occurs by way of the mechanism shown in Fig. 3. The hemi-acetal formed is usually unstable and ring closure occurs to form the cyclic acetal. Optimum conditions occur with hydroxyl groups in *cis* positions with an O—O bond distance of 0.251 nm.[8] A wide variety of aldehydes and ketones have been used (Table 1). Isopropylidene derivatives (often referred to as acetonides) were shown to be important for mass spectrometry because of the intense fragment at M − 15, due to a loss of one of the *gem* methyl groups.[9,10]

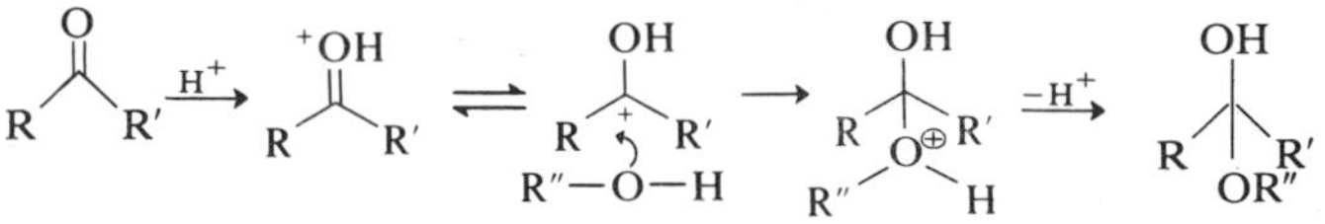

Fig. 3 Acid catalyst and hemi-acetal formation.

TABLE 1

Aldehydes and ketones used for acetal and ketal formation

Aldehydes	Ketones
Formaldehyde	Acetone
Acetaldehyde	2-Butanone
n-Propionaldehyde	Cyclopentanone
n-Butyraldehyde	Cyclohexanone
Glyoxal	Cyclohexan-1,2-dione
Benzaldehyde	Hexafluoroacetone
2-Furaldehyde	Trifluoromethylacetone
p-Anisaldehyde	*sym*-Chlorodifluoromethylacetone
p-Tolualdehyde	
3,7-Dimethyl-2,6-octadienal	
Cinnamaldehyde	

Many forms of chromatography have been used for the determination of acetals. Chemical modification of carbohydrate derivatives occurring on the GLC column has been reported.[11] Rearrangement of acetal and ketal groups occurred with benzylidene glycerols and with di-*O*-isopropylidene-D-mannitols. Benzylidene glycerol prepared by acid-catalysed condensation of glycerol with benzaldehyde gave rise to four GLC peaks, corresponding to the *cis* and *trans* isomers of 1,2- and 1,3-benzylidene glycerol. Only one derivative was stable when chromatographed: this was *cis*-1,3-*O*-benzylidene glycerol. The other three derivatives, when rechromatographed singly, each gave four GLC peaks thus showing that equilibrium conditions were not established.[11]

Acetals are formed in good yield. They are stable to many reagents such as: (i) under the basic conditions often used for esterification or etherification of hydroxyl groups; (ii) under a wide variety of oxidizing conditions with periodate, lead tetra-acetate, alkaline permanganate or silver oxide; (iii) under reducing conditions with hydrogen and platinum, palladium or Raney nickel catalysis, or with sodium borohydride or lithium aluminium hydride; (iv) under the influence of Lewis acids. They are sensitive to hydrolysis, but methylene acetals of aldoses are more stable than other acetals; further derivatization of other reactive groups on the molecule is thus possible.

The carbonyl reagent used is often a liquid, and it may be desirable to use diluents such as benzene, *p*-dioxane, ether or *N,N*-dimethylformamide. If the aldehyde is volatile a polymeric form may be used, such as the trimer paraldehyde instead of acetaldehyde; paraformaldehyde, polyoxymethylene, or formalin solution may be used instead of formaldehyde. Usually, hydrochloric acid (0.02–0.1 M) or sulfuric acid (0.01–0.5 M)[7] are used as catalysts, but ethanesulfonic acid,[12] zinc chloride[7] and cation exchange resins[13] have also been used. A dehydrating agent, such as anhydrous sodium sulfate, may be used and this also acts as a catalyst.[7] Water formed during the reaction may be removed by azeotropic distillation. The choice of catalyst can affect the product obtained, as shown when *O*-isopropylidene derivatives of D-xylose diethyl dithioacetal were prepared with acetone and anhydrous zinc chloride, copper sulfate or phosphorus pentoxide.[14]

2.2 Diols and triols

The reaction of glycerol with acetaldehyde was first reported in 1865.[15] Later studies[16] showed that condensation occurs to yield both 1,2- and 1,3-ethylidene glycerol derivatives (Fig. 4). Acetaldehyde introduces two asymmetric centres into the molecule and separation by GLC of the *cis* and *trans* pairs of 2-methyl-4-hydroxymethyl-1,3-dioxolane(**1**) and 2-methyl-5-hydroxy-1,3-dioxane(**2**) has been reported.[16]

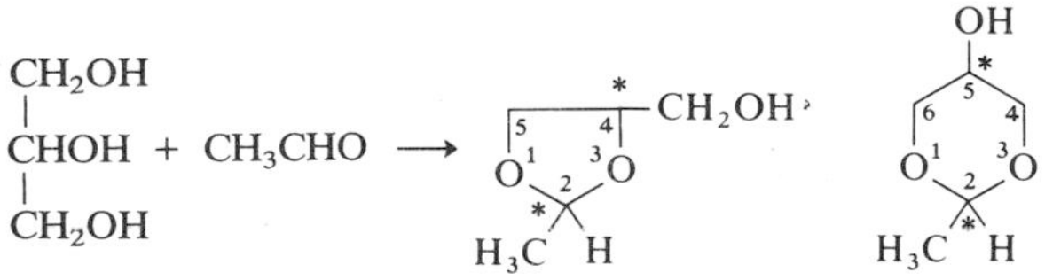

1 **2**

Fig. 4 Condensation of glycerol and acetaldehyde.

The isomerization of the acetal isomers is extremely sensitive to acid. The reaction of glycerol with acetone gave only the 5-membered cyclic ketal, 4-hydroxymethyl-2,2-dimethyl-1,3-dioxolane[17] (see p. 283). Benzylidene glycerols were also studied.[18] The acid-catalysed reaction of glycerol with *n*-hexadecanal led to the formation of both *cis* and *trans* isomers of 2-pentadecyl-4-hydroxymethyl-1,3-dioxolane and 2-pentadecyl-5-hydroxy-1,3-dioxane. The *cis* isomers were more thermodynamically stable[19] (see p. 289).

When uropygiols and diols were reacted with acetone, *threo* isomers gave the *trans*-substituted ring, whilst *erythro* isomers gave the *cis*-substituted ring.[20]

2.3 Carbohydrates

Acetal and ketal formation has been widely used in carbohydrate chemistry since 1895,[21,22] as intermediates in synthesis by selective blocking of hydroxyl groups and for structural studies (see reviews on tetritols, pentitols and hexitols,[23] cyclitols,[24] unsaturated sugars,[25] carbohydrates,[26] ketoses[27] and dithioacetals of sugars[28]). The formation of bi- or tricyclic ring systems with sugars is largely dependent on the availability of *cis*-1,2- and *cis*-1,3-diol groups. Thus, D-galactopyranose forms mainly the 1,2:3,4-di-*O*-isopropylidene derivative (**3**). *O*-isopropylidene and deuterio-isopropylidene pentoses and hexoses were studied by mass spectrometry.[9,10,29] Cyclohexanone dimethyl acetal was used for selective protection of sugars[30] as shown in Fig. 5.

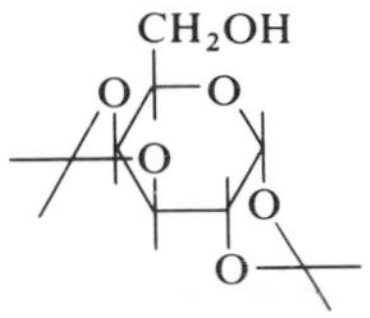

3

Methylene,[31] ethylidene,[32] di-*O*-methane sulfonyl,[32] butylidene,[33] benzylidene,[31,32,34–38] isopropylidene,[9,10,13,14,29,31,32,37,39,40] cyclopentylidene[8] and cyclohexylidene[8,41] derivatives of polyols have been reported (see pp. 282–289 for some examples).

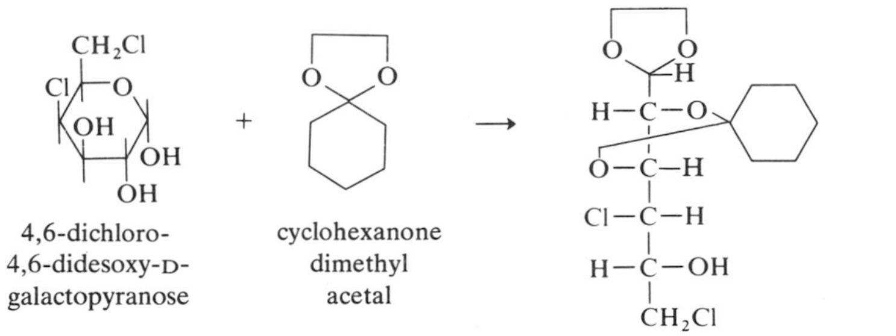

Fig. 5.

2.4 Steroids

Isopropylidene derivatives (commonly called acetonides) of 3,16α-oestradiol,[53] *cis* (but not *trans*) epimeric oestriols[54] and of 2β-3α-mercaptocholestanols which contain an *S*, *O*-ring[55] have been successfully separated. Some examples are given in the practical section (see p. 286).

Under normal GLC conditions 17-hydroxycorticosteroids undergo pyrolysis to C-19 ketosteroids. To avoid this problem a thermally stable C-20-21 side chain can be formed by cyclization (see also siliconides and boronates). The bis-methylenedioxy (formal) derivative (**5**) of 17-hydroxyketosteroids (**4**) was formed by reaction with formaldehyde using acid catalysis as shown in Fig. 6.

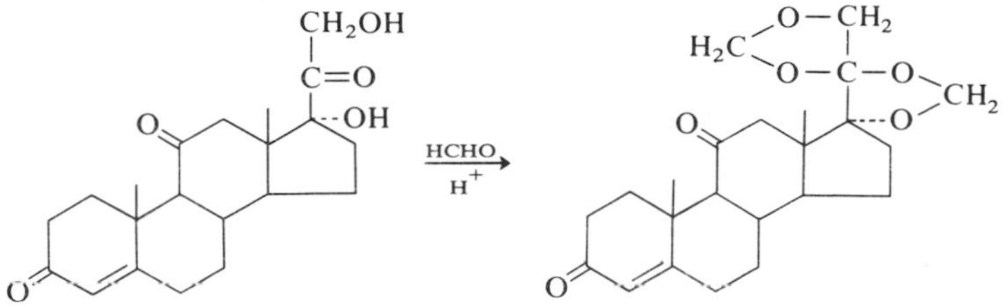

4 **5**

Fig. 6 Cyclic derivatives of corticosteroids.

The specificity of the reaction for the dihydroxyacetone function enables cortisone to be separated by gas chromatography from a mixture of steroids.[56] This is illustrated by an example in the practical section of the chapter (p. 283).

2.5 Lipids

Monoglycerides,[42] glyceryl ethers[43] and hydroxy fatty acids[44,45] have all been converted into isopropylidene derivatives for analysis by GLC. Positional and geometrical characterization of double bonds (see review[46]) has been achieved with unsaturated fatty esters[47] and alkenes[48] by preliminary oxidation with osmium tetroxide to form the diol with the subsequent formation of ispropylidene derivatives. An alternative method is to prepare a halohydrin,[49] and from this to form the hexafluoroacetone ketal.[50] It has been shown that ethylene chlorhydrin reacts with *sym*-chlorodifluoroacetone to form 2,2-bis (chlorodifluoromethyl)-1,3-dioxolane (**6**)[51] (see p. 289).

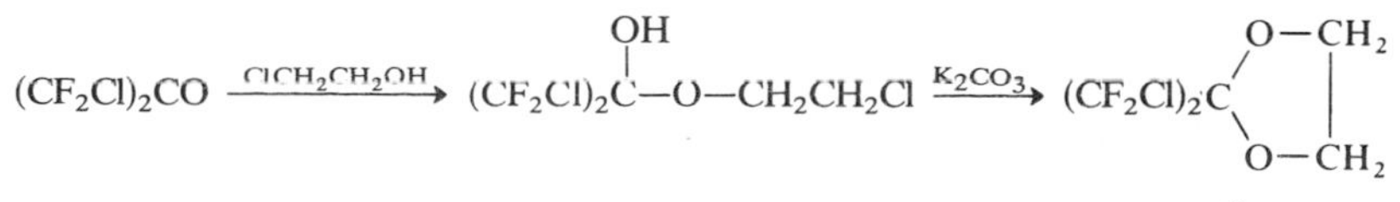

6

Synthesis of the ketal directly from the diol is not possible and the alkene must first be converted into the bromohydrin, (two products are possible when $R^1 \neq R^2$) before reacting with hexafluoroacetone[50] (see Practical Methods pp. 289–291). Addition of HOBr to the alkene is at least 97% *trans*-specific and the *cis*-alkene leads to formation of the *threo*-bromohydrin. This process is illustrated in Fig. 7. With hexafluoroacetone, *trans* addition via the hemiketal and subsequent ring closure leads to the *cis*-ketal.[50] The fragmentation patterns of polyunsaturated fatty acid derivatives have been studied by GC-MS.[52]

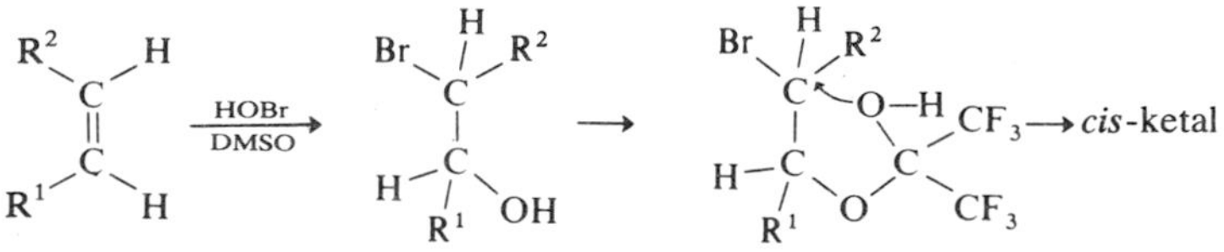

Fig. 7 Hexafluoroacetone ketal formation.

2.6 Miscellaneous

Dimethylketals of α-keto acids have been studied by GC–MS.[57] Acenaphthenequinone reacts with ethylene glycol to form acetal derivatives which have been studied by mass spectrometry.[58] Nucleosides and nucleotides have been converted to *O*-isopropylidene derivatives.[59,60] The GLC resolution of ($\pm$)-camphor was achieved by forming diastereomeric ketals from 2,3-butanediol.[61]

3 SILICONIDES

Siliconides are analogous in their structure to acetonides. Dichlorodimethylsilane in pyridine was used to prepare the cyclic dimethylsilyl derivative of 3β-acetoxy-16α-17α-dihydroxypregn-5-en-20-one,[62] and of 17β,18-hydroxy-oestradiol[63] at the *cis*-diol position. Cyclic silyl derivatives or siliconides of steroids have been prepared by reacting the dihydroxy acetone side chain of a steroid (**7**) such as cortisone, cortisol and betamethasone, with a new difunctional reagent, dimethyldiacetoxysilane (DMDAS) in the presence of triethylamine.[64] This is illustrated in Fig. 8. This reagent reacted with 20-oxo-17α-21-dihydroxy groups (**7**) and also with unhindered –OH groups to give a product with the structure R—O—$Si(CH_3)_2COCH_3$. It will not react with hindered groups such as 11β-OH (see Practical Methods section p. 291).

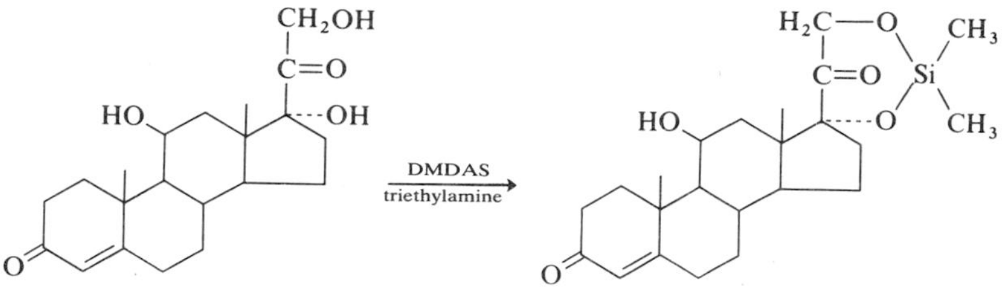

7

Fig. 8 Siliconide formation of corticosteroids.

The reagent was used in dilute solution to reduce the formation of intermolecular silyl bridges by cross-linking between free hydroxyl groups.

Dichlorodimethylsilane was used in pyridine solution to prepare dimethyl siliconides of anthranilic (**8**), salicylic (**9**) and thiosalicylic (**10**) acids[65] (see p. 292).

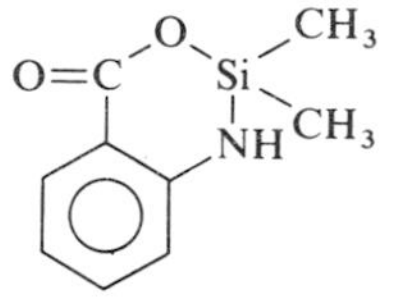

8

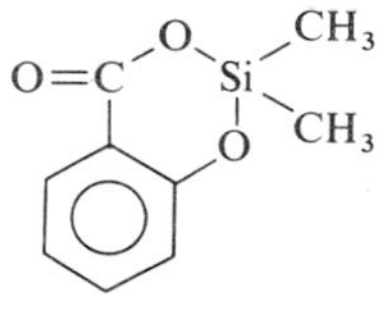

9

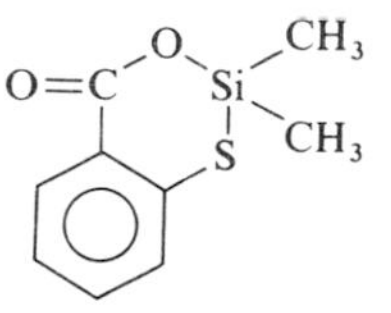

10

When tetramethoxysilane was reacted with C_2 to C_6 diols in the presence of *p*-toluenesulfonic acid, the products were different, depending on the molar ratios of diol to silane reagent[66] (Fig. 9). Triethyloxysilane was also studied. The ring size varied with the diol. The derivative from 1,6-dihydroxyhexane was distillable,[66] so it should be possible to analyse these compounds by GLC.

Molar ratio 1:1

$$Si(OCH_3)_4 + \quad \begin{matrix} CH_2OH \\ | \\ CH_2OH \end{matrix} \quad \rightarrow (CH_3O)_2 - Si \begin{matrix} O-CH_2 \\ | \\ O-CH_2 \end{matrix} + 2CH_3OH$$

Molar ratio 1:2

$$Si(OCH_3)_4 + 2 \begin{matrix} CH_2OH \\ | \\ CH_2OH \end{matrix} \quad \rightarrow Si \left[\begin{matrix} O-CH_2 \\ | \\ O-CH_2 \end{matrix} \right]_2 + 4CH_3OH$$

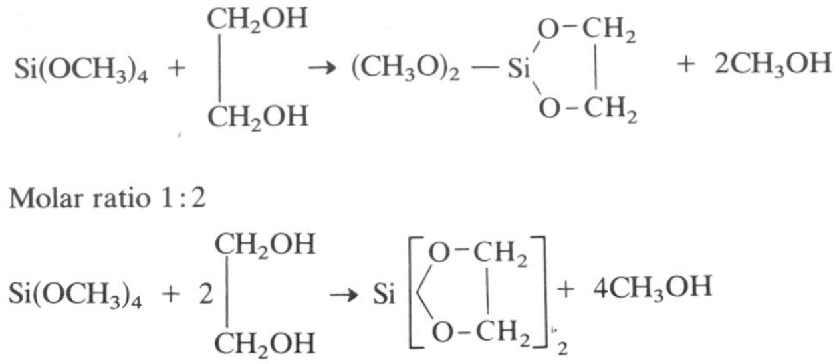

Fig. 9   Reactions of tetramethoxysilane with diols.

4   BORONATES

The interaction of boric acid with hydroxylated compounds in aqueous solution has been reviewed,[67,68] and the presence of benzeneboronic acid was utilized to alter the R_F values of sugars for paper chromatographic separations.[69] The preparation and usefulness of volatile alkaneboronates was first shown with cycloalkanediols. Cyclohexanediols were reacted with n-butylboronic acid or n-butylboroxine to yield cyclic derivatives with 5, 6 and 7 atoms (11, 12, 13) from the 1,2-, 1,3- and 1,4-diols respectively.[70] Only the cis-1,2-, trans-1,2-, cis-1,3- and cis-1,4-derivatives could be distilled; the trans-1,3- and trans-1,4-derivatives were not volatile. The use of GLC and the preparation of the more volatile methyl boronate derivatives were suggested. The parent diol was recovered by distillation after transesterification with ethylene glycol[70] (see p. 296).

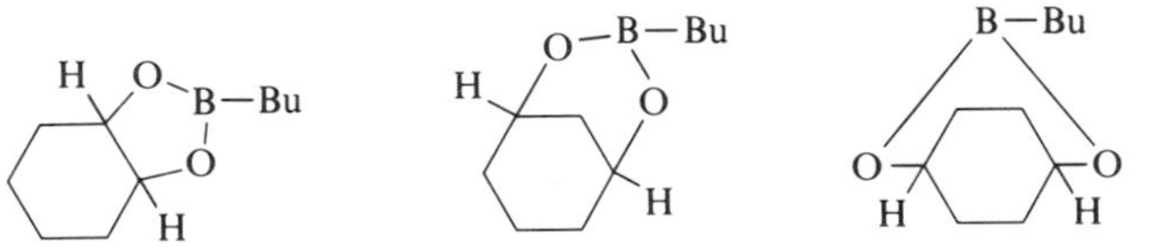

11   12   13

Alkaneboronate derivatives are stable in air but are rapidly hydrolysed with water,[71] moist organic solvents such as 1-butanol[69] or weak acid.[71] Phenylboronates of methyl glucose- and mannosepyranosides are stable in dry dioxane solution, but the parent glycoside has been rapidly regenerated on adding water or alcohol to the solution.[72] Recovery of the parent compound may be achieved by solvolysis (transesterification) of the boronate ring in some carbohydrate,[72-74] nucleoside[75] or steroid[76,77] derivatives at room temperature with anhydrous propane-1,3-diol. Indane-cis-1,2-diol forms a 'stable' ester and is itself effective in liberating many other α-diols.[78] Cyclic boronates form complexes with amines, and the stability of these complexes has been studied.[79]

4.1 Comparative studies

A comparison has been made by GC–MS of *n*-butyl and phenyl boronates of 1,2- and 1,3-aliphatic and aromatic diols.[78] Methyl boronates of corticosteroids were also compared with other derivatives by GC–MS.[80]

In an extensive GC–MS study it was concluded that all boronates of corticosteroids gave satisfactory derivatives.[76,77] They were formed in less than 5 min at room temperature and the reactions were similar in a variety of solvents. The retention times of the methyl and *tert*-butyl boronates were similar, while longer retention times were recorded for *n*-butyl which were greater than cyclohexyl which were greater than phenylboronates. The 17α,20-diols, 20,21-diols and 17α,20,21-triols gave single GLC peaks and could be submitted to TLC. Where additional hydroxyl groups remained these were acetylated or trimethylsilylated. Boronates of 17α,21-dihydroxy-20-oxo-steroids and of 20,21-ketals were too unstable for further derivatization in this way.

The analysis of methyl-, butyl- and phenyl boronates of per-*O*-TMS carbohydrates[81] and sphingosines[82] was carried out by GC–MS. Using trialkyl or triaryl boranes the derivatives of acetylacetone (**14**) and benzoin were prepared and analysed by GLC and MS[83] (see p. 298). This is shown in Fig. 10. Acetylacetone did not form butylboronate esters with *n*-butylboronic acid, although hop acids were successfully derivatized.[84]

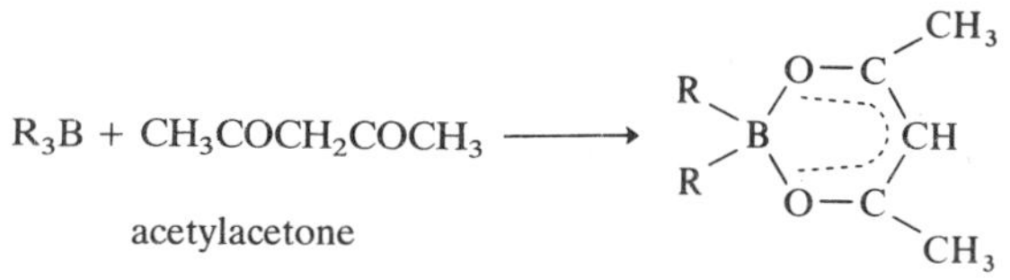

14

R = propyl, butyl, hexyl, phenyl, benzyl.

Fig. 10 Reaction of acetylacetone with trisubstituted boranes.

4.2 Methyl boronates

The methyl boronate derivatives show only a small mass increment, and they are more volatile and have shorter retention times than other alkyl boronates. An interest in the major metabolite of norepinephrine led to the determination of 3-methoxy-4-hydroxy-phenylethyleneglycol as its methyl- or *n*-butyl boronate[85,86] (see p. 294). These derivatives were more stable than the acyl derivatives for GC–MS determination. Sphingosines[82,87] and ceramides[88] have also been analysed as their methylboronates (see p. 294). The boronate derivative of 18-hydroxy-11-deoxycorticosterone resulted in a single compound for GC–MS analysis, whereas three products were obtained after trimethylsilylation. The methyl boronate derivatives had lower Kováts retention indices than the corresponding TMS derivatives.[87]

4.3 Butyl boronates

A wide range of compounds has been derivatized as butylboronates, such as 1:2-,[70,78,89,90] 1:3-,[70,78,89,90] 1:4-,[70] and 1:2-ene-diols[89] and α-OH, β-OH acids and β and γ-OH amines.[89,90]

The GLC separation of D-glucitol and D-galactitol butane boronates was completed in 6 min compared with 30–80 min under the same conditions for the acetate derivatives.[71] However, in order to obtain single peaks from carbohydrates by GLC other workers have found it necessary to trimethylsilylate free hydroxyl groups.[91,92]

Prostaglandins of the E and F series were successfully resolved by GC–MS analysis as their trimethylsilyl and *n*-butylboronic acid esters[93–96] (see p. 295). Prostaglandin $F_{1\alpha}$ was derivatized as shown in structure (**15**).[93]

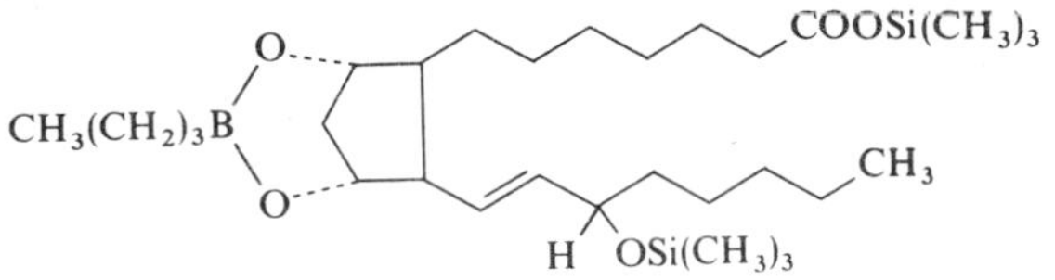

15

4.4 Phenyl boronates

Phenylboronates of steroids,[77,78,97] pentoses and 6-deoxyhexoses,[98] hexosides,[72] xylofuranosides,[99] sugar 1,2-, 1,3- and 1,4-diols,[100] other diols and triols,[101] nucleosides,[75,102] nucleotides,[102] 1,2- and 2,3-diamino, dihydroxy and aminohydroxy naphthalenes,[103] ethylene diamine, aminophenol and anthranilic acid,[104] chloral[105] and α-monoglycerides have been reported.[106] The phenylboronates of benzoin were prepared with triphenyl borane,[83] phenylboronic acid,[101,105] and phenylborodichloride.[101] Ten solid phenyl- and *p*-bromophenylboronates of diols and polyols have also been prepared.[107]

A new class of heterocyclic compounds, dihydrobenzoboradiazoles, were prepared by the reaction of boronic acids and their esters with *p*-phenylenediamine.[108] Phenylboronate derivatives were used for the novel synthesis of 6-*O*-methyl-D-glucose[109] and as a method of restricting isopropyl acetal formation in monosaccharides.[110] Monoalkenes were oxidized to diols and converted to phenylboronates for GC–MS analysis.[111] The production of characteristic hydrocarbon ions in a mass spectrometer from phenylboronates has been reported.[112] The hydrolysis of 1,2-, 1,3-, and 1,4-diol phenylboronates having respectively 5-, 6- and 7-membered rings was studied. The order of stability was 6- > 5- ≫ 7-membered rings. Involatile products were prepared from esters containing 8- or 9-membered rings.[113]

5 OTHER HETEROCYCLIC DERIVATIVES

5.1 Epoxides, episulfides

The simplest example of an α-epoxide or oxirane ring is ethylene oxide, e.g. **16**, R = H. When epoxides are regarded as derivatives of α-glycols, they are usually referred to as anhydro-derivatives as in carbohydrates. However, anhydro rings in sugars are not always formed between *cis*-hydroxyl groups (see reviews on oxirane derivatives of aldoses[14] and anhydro-derivatives of sugars[115] and alditols[116] and the mechanisms of epoxide reactions[117]). Epoxides may be formed by treatment of ethylenic compounds or *cis*-diols with peracids,[118] such as peracetic and performic acids.[119] Perbenzoic acid was used to prepare inositol anhydro derivatives.[120] In steroids an epoxide has been formed with *p*-nitroperbenzoic acid and the oxygen atom replaced by a sulfur to form the episulfide.[121] Alkenes may be epoxidized stereospecifically with *m*-chloroperbenzoic acid and the *cis*- and *trans*-oxiranes separated by GLC, the *cis* isomer usually having a longer retention time than the *trans* isomer.[122] Similar *cis–trans* determinations have been made with monounsaturated[123–125] and diene[123] fatty esters (see review[126]).

Epoxy derivatives may be analysed directly by TLC,[125,127] GLC,[122–124,127] or mass spectrometry.[125,127] Alternatively, the ring may be opened to form ketone[128] or α,β-hydroxy-*N*-dimethylamino[129] derivatives, or the ring may be split with periodic acid to give aldehyde and aldehyde fragments for analysis.[130] Selective oxirane ring opening in nucleosides was shown with HCl in dimethylformamide.[131]

Epoxides of fatty acids[132,133] and epoxyglycerides[134] in the presence of boron trifluoride react with ketones to give 1,3-dioxolane (ketal) derivatives: these are heat-stable for GLC. Derivatives were prepared with acetone, methyl ethyl ketone, methyl isobutyl ketone, 2-heptanone, cyclopentanone and cyclohexanone. Cyclopentanone gave derivatives which were best separated from the long chain common triglycerides[134] (see p. 299).

Terminal epoxides, 1,2-epoxylalkanes (**16**) may be reacted with boron trifluoride in methanol and the products (**17, 18**) examined by GLC.[135]

$$\underset{\mathbf{16}}{\mathrm{R{-}CH{-}CH_2}\ (\text{bridged by O})} + \mathrm{CH_3OH} \xrightarrow{\mathrm{BF_3}} \underset{\mathbf{17}}{\mathrm{RCH(OH)CH_2OCH_3}} + \underset{\mathbf{18}}{\mathrm{RCH(OCH_3)CH_2OH}}$$

5.2 Cyclic carbonates and thiocarbonates

In 1883 the preparation of a cyclic carbonate ester (**19**) from ethylene glycol and phosgene was reported.[136]

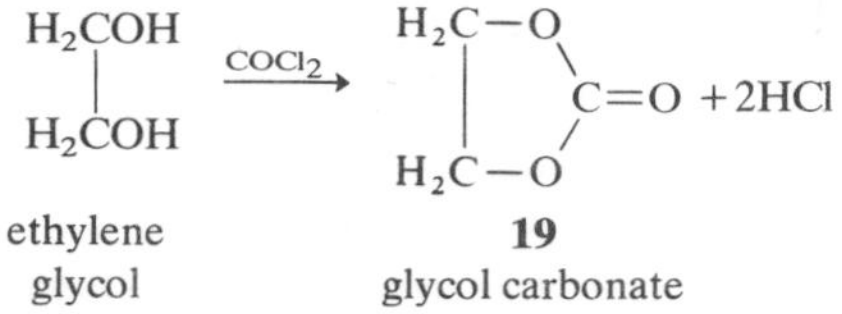

19

glycol carbonate

The carbonate and thiocarbonate derivatives of carbohydrates have been reviewed.[137] The most important characteristic of the carbonate group is its sensitivity to alkalis coupled with its relative stability to acids, a property largely complementary to that of the isopropylidene acetals. Cyclic carbonates may be formed with free sugars or derivatized sugars from chloroformic esters or, by treatment with phosgene either without solvent, or in the presence of inert solvents such as acetone, or organic bases such as pyridine or quinoline. In this way 1,2-isopropylidene-D-glucofuranose (**20**) was converted to 1,2-*O*-isopropylidene-D-glucofuranose-5,6-carbonate (**21**).[137]

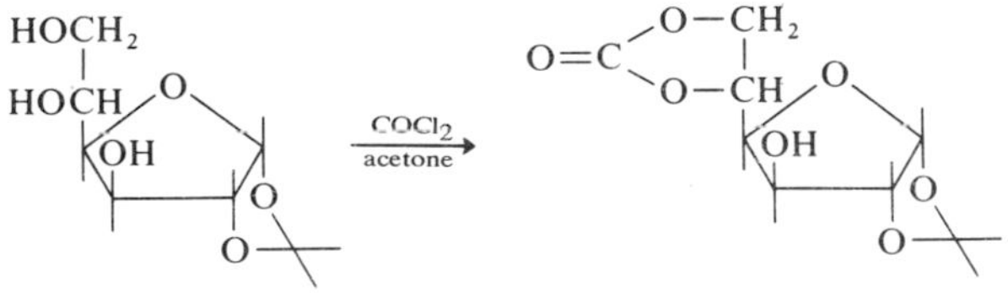

20 **21**

The furanose ring was preferred for some fused ring systems, so that from D-mannopyranose, D-mannofuranose 2,3-carbonate was formed with *cis*-hydroxyl groups.[138] Cyclization between *trans* groups is not usual. Cyclic carbonates of acyclic and cyclic 1,2-diols were examined by mass spectrometry.[139]

Treatment of the steroid hydroxy-xanthate derivatives with sodium borohydride in ethanol led to the formation of either the episulfide or the cyclic dithiocarbonate derivative[121] (see pp. 299–300). Reaction of cholestan-3β-ol-2α-ethyl xanthate with sodium borohydride or sodium ethoxide led to formation of cholestan-2α-3α-episulfide, whereas similar treatment of cholestan-3α-ol-2α-ethyl xanthate resulted in formation of the cyclic dithiocarbonate (Fig. 11). Only cholestan-2β,3α-trithiocarbonate was prepared successfully.[121]

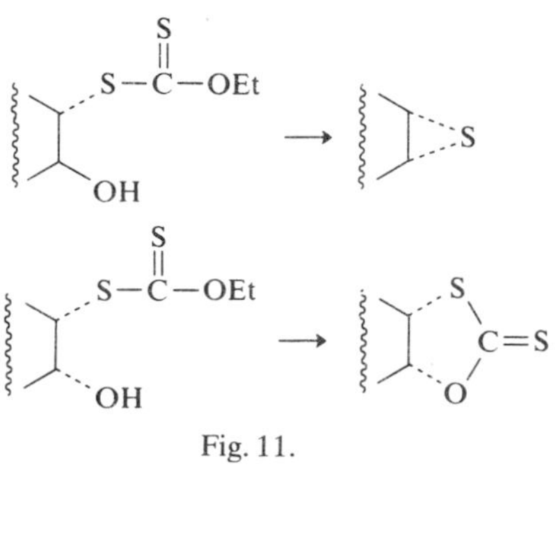

Fig. 11.

5.3 Cyclopropanes

Fatty acid methyl esters containing up to four double bonds (arachidonic acid) were converted almost quantitatively to their cyclopropane derivatives and were analysed on both polar and non-polar GLC columns by GC–MS.[140] (See p. 300.)

5.4 Quinoxalin-2-ols

Because of enolization and decomposition, keto acids often give multiple derivatives when esterified for GLC,[57] or when converted to 2,4-dinotrophenylhydrazones for paper chromatography.[141,142] The known reaction of *o*-dicarbonyl compounds with aromatic diamines was used to characterize α-keto acids as their quinoxalinol derivatives.[143] Single GLC peaks were obtained for the TMS derivatives of eleven α-keto acid quinoxalinol derivatives[144] (see p. 302). The proposed mechanism was as shown in Fig. 12.

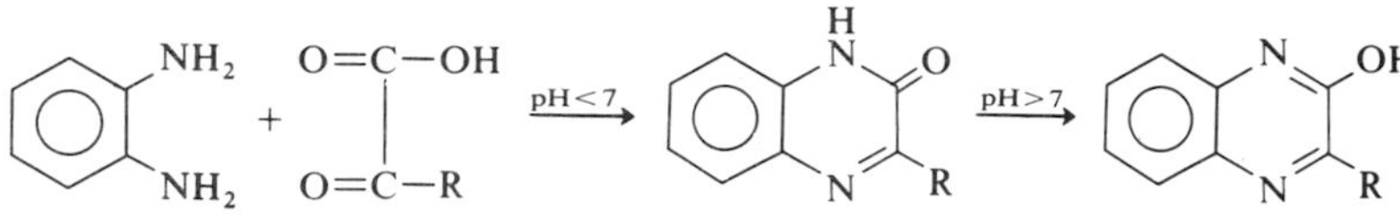

Fig. 12 Mechanism of reaction of α-ketoacids and *o*-phenylenediamine.

Oxaloacetic acid did not give the expected carboxymethylquinoxalinol, but gave methylquinoxalinol. The perdeuterio-TMS derivatives showed slightly shorter retention times than the unlabelled derivatives. The method was considered better than any of nine others which have been used, because the reaction was specific, the derivatives stable and crystalline, and because the chromatographic separation was easier since no free carboxyl group remained.

Pyruvic acid was condensed with *o*-phenylenediamine, and 4-chloro-, 4,5-dichloro-, and 3,4,5,6-tetrafluoro-*o*-phenylenediamine and it was shown by GC, MS and IR that the derivatives were *O*-silylated.[145] It had been previously shown by IR and NMR that pyruvic and α-ketoglutaric derivatives were *N*-silylated.[146] α-Keto acids were condensed with *o*-phenylenediamine[147,148] and 1,2-diamino-4-nitrobenzene[148,149] for their separation by paper chromatography and their determination in blood and urine[149] (see p. 301).

5.5 2,5-Diketopiperazines (piperazine-2,5-diones)

Amino acids[150] and amino acid esters (**22**) condense together in the presence of base to form 2,5-diketopiperazines (**23**)[151–2] (see review[153]), and these may be analysed by TLC.[154–6] In general, compounds with amide groups are not very volatile, and are therefore not very suitable for direct GLC analysis.

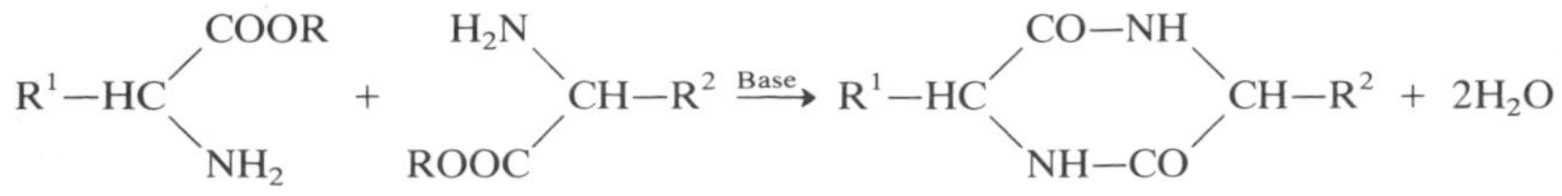

22 **23**

However, retention times for 13 amino acid DL-(*trans*)- and LL-(*cis*)-diketopiperazines have been reported for columns with 5% QF-1 or 3% EGS at 200–240 °C[156] and on 1% SE-30.[157] Forty diketopiperazine-TFA-derivatives were gas chromatographed on a column with 5% SE-30 at temperatures from 90 °C (*cis*-L-ala-L-ala) to 180° C (*cis*-L-phe-D-phe).[158] TMS-derivatives were chromatographed at lower temperatures than either methyl- or non-derivatized diketopiperazines.[159] Baseline separation between the *cis*- and *trans*-diketopiperazines was not always obtained.[158,159] Diketopiperazines may be prepared by the cyclization of dipeptide methyl esters in methanolic-ammonia for periods of 1–5 days[160] but racemization occurs.[154] Synthesis without racemization has been effected by heating the unblocked dipeptide or its hydrobromide salt in phenol just below the boiling point of phenol,[155] or by cyclizing the *t*-butyloxycarbonyl (*t*-BOC) dipeptide methyl ester[154] (see p. 303). Hydrogenolysis of eleven benzyloxycarbonyldipeptide methyl esters has been used to prepare diketopiperazines without racemization.[161]

5.6 2-Trifluoromethyl-oxazolin-5-ones from amino acids

The preparation of the azlactones or 2-trifluoromethyl-oxazolin-(5)-ones of *N*-TFA-alanine (**24**) and *N*-TFA-valine (**25**) was first described using simple heating of the amino acid with trifluoroacetic anhydride and distillation.[162] The glycine derivative was unstable and was not prepared but other methods of preparation were later described. The *N*-TFA-derivative of leucine may be reacted in pyridine with phosphorous oxychloride to give the cyclized product[163] (see Practical Methods p. 304). 2-H, 2-Methyl and 2-phenyl-oxazolin-5-ones were synthesized by the use of dicyclohexylcarbodiimide to dehydrate formyl-acetyl- and benzoyl-leucine respectively.[164] The position of the double bond has been confirmed at the 3-4 position.[165]

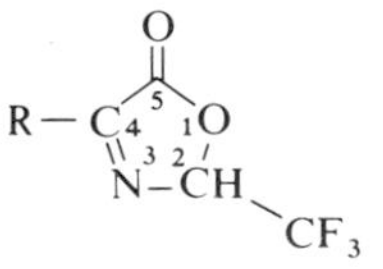

(**24**) R = methyl

(**25**) R = isopropyl

Ten amino acids as their 2-TFA-oxazolin-(5)-one derivatives have been separated by GLC.[166] They gave single peaks and they were more volatile than the corresponding *N*-TFA amino acid *n*-butyl esters[166,167] Because a new asymmetric carbon is formed at position 2, alloisoleucine having an asymmetric carbon in the side chain gave a mixture of two partly separated diastereomers.[166] The determination of phenylalanine in serum by this single-step reaction has been described[168] and this is covered in the practical section of this chapter. Racemization during peptide synthesis with dicyclohexylcarbodiimide has been attributed to the formation of oxazolone intermediates.[169] A series of papers has been published on the thermolysis of oxazolin-5-ones.[170]

The preparation of 2,2-bis-(trifluoromethyl)-4-substituted 1,3-oxazolidin-5-ones from α-amino acids and 2,2-bis(trifluoromethyl)-4-substituted 1,3-dioxolanes from hydroxy amino acids has been reported.[171] They were prepared from amino acids and trifluoroacetic anhydride. This work was extended to 26 α-amino, and α-imino acids and tri-functional α-amino acids using hexafluoroacetone in dimethylsulfoxide solution.[172] The more convenient use of *sym*-dichlorotetrafluoroacetone led to the preparation and GLC separation of 2,2-bis-(chlorodifluoromethyl)-4-subst.-1,3-oxazolidin-5-one derivatives of tyrosine and some of its iodinated analogues[173–4] (see p. 305).

5.7 Lactones

Lactones of branched chain γ- and δ-hydroxy aliphatic acids have been prepared and analysed by GC–MS. Some were synthesized by reacting a molar excess of methyl magnesium iodide with the appropriate keto acid[175] (see p. 306).

5.8 Thiazolidines

Several thiols and disulfides have been analysed qualitatively and quantitatively by treatment with pivaldehyde (2,2-dimethylpropanal) followed by GC–MS examination of the thiazolidine (**26**) or in some cases the Schiff base (**27**) derivatives.[176] Thiazolidines and the neopentylidene derivatives are formed rapidly in neutral or alkaline medium; strong base causes breakdown of the products, and all derivatives are hydrolysed by water.

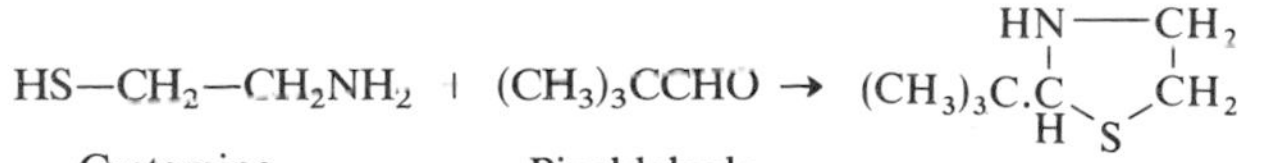

Cystamine Pivaldehyde

26

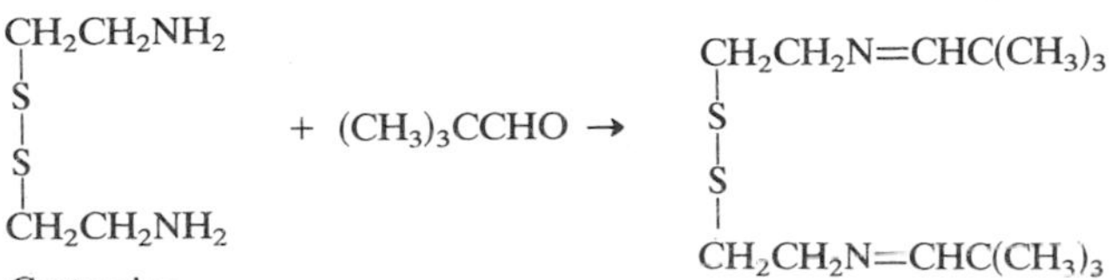

Cystamine

27

5.9 Cyclic sulfites

Cyclic sulfites of acyclic and cyclic 1,2-diols (**28**) have been examined by GLC and MS.[139] It was shown that there are two isomeric *meso*-pentane-2,4-diol cyclic sulfites.[177,178]

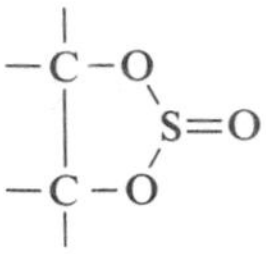

28

5.10 Organoarsenic compounds

The mass spectra of the following ring compounds (**29**–**31**) have been examined by mass spectrometry.[179]

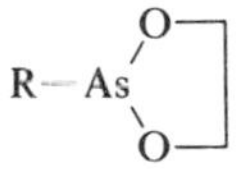

(**29**) 1,3,2-dioxarsolane

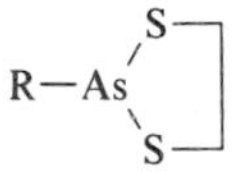

(**30**) 1,3,2-dithiarsolane

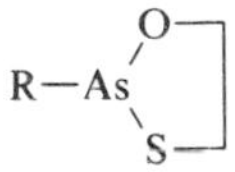

(**31**) 1,3,2-oxathiarsolane

5.11 Miscellaneous

Glyoxylic acid (**32**) is different to many of the α-keto acids studied in that it is not derivatized as the quinoxalinol. Instead it has been condensed with *N,N'*-diphenylethylenediamine (**33**) to form 1,3-diphenyl-imidazolidine-2-carboxylic acid (**34**) which may be trimethylsilylated for analysis by GLC[180] (see Practical Methods, Section 10.22).

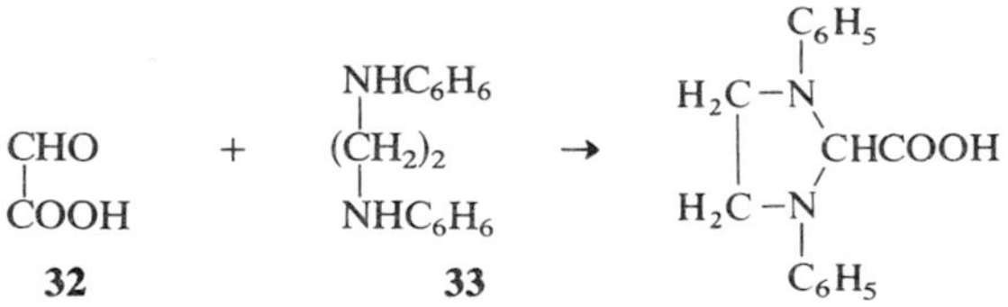

32 **33** **34**

Malonaldehyde has been condensed with guanidine to give 2-aminopyrimidine,[181] with urea to give 2-hydroxypyrimidine[182] and with arginine to give δ-*N*-(2-pyrimidinyl) ornithine[183] (see pp. 306–307).

A few simple α-amino acids have been converted to 2-morpholone derivatives (**35**) or 'Leuchs anhydrides' (**36**) by the reaction of propylene oxide on the amino acid.[184] These derivatives have not been used widely for GLC, because they are not applicable to all amino acids and polymerize on heating or in the presence of moisture or compounds with polar groups.

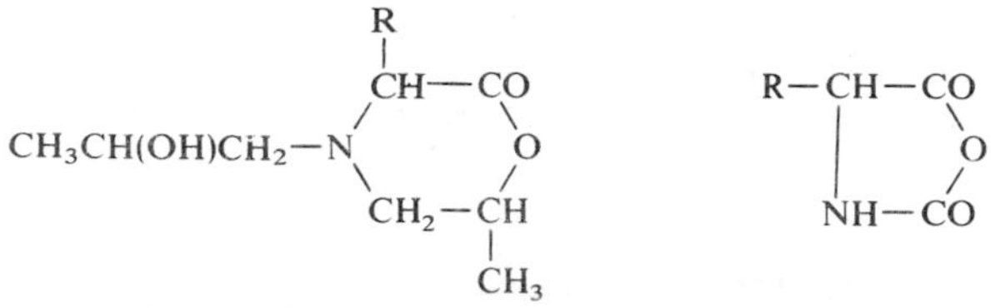

35 36

6 CYCLIC COMPOUNDS IN PROTEIN SEQUENCING

Hydantoins and more particularly thiohydantoin amino acid derivatives have become prominent over the past 25 years in connection with protein sequencing by stepwise degradation of the polypeptide chain. A macroreticular resin containing thiohydantoinyl groups has recently been reported.[185]

6.1 Phenylthiohydantoins

Edman[186,187] showed that phenylthiocarbamates of peptides could be cleaved by a mild chemical reaction releasing the *N*-terminal amino acid as its thiazolone derivative which was then rearranged to its more stable phenylthiohydantoin derivative. The reaction sequence is given in Fig. 13.

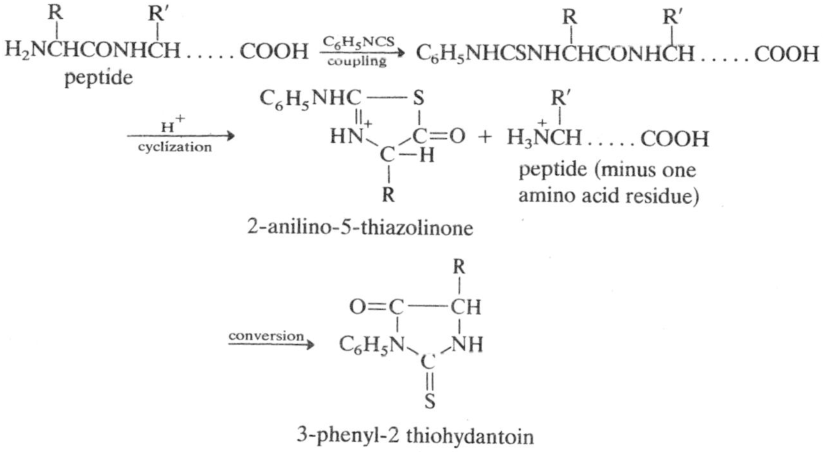

Fig. 13 Sequential degradation of a protein from the *N*-terminus by the Edman method.

The peptide is first reacted with phenylisothiocyanate at pH 8–9 at 40 °C, and in the presence of anhydrous trifluoroacetic acid cyclization then occurs with liberation of the *N*-terminal amino acid as the 2-anilino-thiazolinone. This ring compound rearranges in the presence of aqueous acid to give the amino acid phenylthiohydantoin.

The method has been automated[188] and developed for rapid manual sequencing of peptides.[189] Alternative procedures make use of solid-state attachment of the peptide to a resin for stepwise degradation,[190–191] or attachment of the reagent in the form of a thiocyanate resin.[192–193] Radioactive phenylisothiocyanate has been used.[194] Modified reagents have also been used, such as 4-sulfophenyl isothiocyanate,[195] *p*-chloro-,[196–197] *p*-bromo-[197–199], *p*-methoxy-,[196] and *p*-fluorophenylisothiocyanates.[196]

The thiazolinone derivatives are not very stable and their direct determination has not been widely used. By treatment with 1 M hydrochloric acid,[200] 5% acetic acid,[189] or 20% trifluoroacetic acid,[201–202] rearrangement of the ring leads to formation of the relatively stable phenylthiohydantoin amino acid derivatives which have been determined in a variety of ways. They are nearly all soluble in 1,2-dichloroethane-methanol (7:7, v/v).[202] However, their stability is greater in ethyl acetate than in pyridine, and metal columns cannot be used for GLC.[203] To obviate the problems of cyclization, *N*-phenyl thiourea derivatives of amino acids and short peptides have been analysed directly by MS.[204]

6.2 Methylthiohydantoins

The preparation of methylthiohydantoin amino acid derivatives has been described.[205] Methylisothiocyanate was introduced into the Edman degradative procedure in order to gain the advantage of releasing amino acid methylthiohydantoin derivatives which were more volatile than the phenyl derivatives for analysis by GLC. *N*-Methylthiourea derivatives of amino acids and short peptides have been analysed by mass spectrometry.[204,206]

6.3 Thiohydantoins

A method of sequencing from the carboxyl-terminus of a peptide by selective cleavage of the *C*-terminal amino acid as its thiohydantoin derivative was proposed in 1926.[207] The method was used to a limited extent[208–210] but detailed studies with peptides[211] and proteins[212] showed the possibilities for further development. The reaction steps are given in Fig. 14.

The peptide was reacted with ammonium thiocyanate in a mixture of acetic anhydride and acetic acid at 50 °C for a few hours. Cleavage was carried out with 12 M hydrochloric acid at 30 °C for 1.5 h and the thiohydantoin amino acid derived from the *C*-terminal of the peptide was identified. The method has also been applied in the solid-phase using polystyrene resins[213–214] and controlled-pore glass beads.[213–215] The amino acid thiohydantoins were identified by TLC,[212,216–218] GLC[219] and MS[219,220] or by hydrolysing and determining the parent amino acid.[212,219]

6.4 Iminohydantoins

A new method for protein sequencing from the carboxyl terminus which was claimed to be simple, rapid and efficient involved stepwise degradation of the

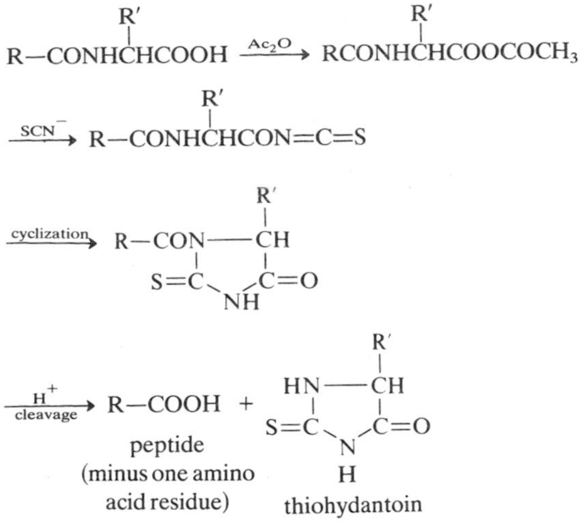

Fig. 14 Thiohydantoin method for the sequential degradation of protein from the carboxyl terminus.

peptide with the release of *C*-terminal amino acid as its iminohydantoin derivative.[221] The peptide carboxyl group was coupled in the presence of ethyl-dimethylaminopropyl-or dicyclohexyl carbodiimide to a thiuronium salt (*S*-alkylthiourea) to form a peptide acyl thiourea. Aqueous base at pH 10–11.5 resulted in formation of the peptide acyl cyanamide, which cyclized with the release of the *C*-terminal amino acid as its iminohydantoin derivative (Fig. 15).

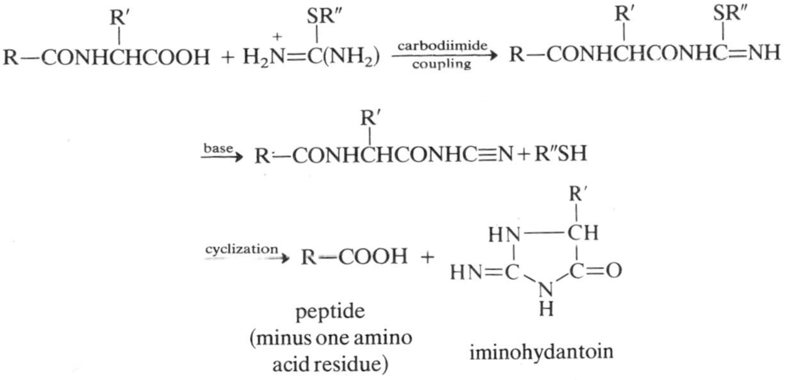

Fig. 15 Iminohydantoin method for the sequential degradation of protein from the carboxyl terminus.

The iminohydantoins of some amino acids were analysed by TLC and GLC. Coupling was better with ethyl-dimethylaminopropyl- rather than with dicyclohexyl-carbodiimide. *N*-Butyl thiouronium iodide was the most reactive salt of those tried and coupling was about 99% efficient in anhydrous pyridine at 50 °C in 10–15 min. Cleavage was effected at pH 10–11.5, irrespective of the bases studied, and aqueous pyridine was found to be convenient. Problems with proline, aspartic acid and glutamic acid still exist and many suggestions have been proposed for further investigations.[221]

6.5 Hydantoins

A method for the determination of the *N*-terminal amino acid in a protein involved carbamylation of α- and ε-NH_2 groups with excess potassium cyanate.[222] After heating with 6 M hydrochloric acid at 100 °C for 1 h the hydantoin of the *N*-terminal amino acid was isolated by ion-exchange chromatography (Fig. 16). The hydantoin was hydrolysed with 0.2 M sodium hydroxide or 6 M hydrochloric acid and the parent amino acid determined by conventional ion-exchange chromatography.[223]

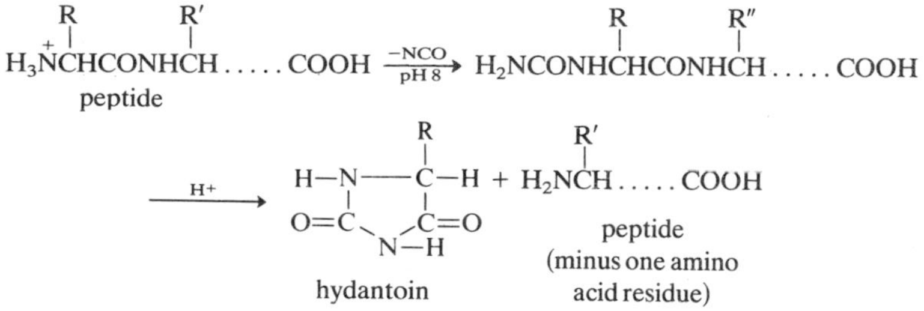

Fig. 16 Hydantoin method for the determination of the *N*-terminal amino acid of a protein.

It should be noted that diphenylhydantoin is an anticonvulsant drug and its determination, as well as that of its major metabolic derivative, 5-(*p*-hydroxyphenyl)-5-phenylhydantoin, was achieved by GLC after flash methylation.[224]

SECTION B—PRACTICAL METHODS

7 ACETALS AND KETALS

7.1 Methylenedioxy derivative of an anhydro sugar[31]

The sugar (0.36 g) was dissolved in 37% aqueous formaldehyde (2 ml) and concentrated hydrochloric acid (2 ml) and kept in a desiccator over concentrated sulfuric acid for 3 days. The solid residue was dissolved in ethyl acetate.

Comments: the authors prepared 1,6-anhydro-2,3:4,5-di-*O*-methylene-1(6)-thio-D-glucitol from 1,6-anhydro-1(6)-thio-D-glucitol.

7.2 Bismethylenedioxy derivative of a steroid[56]

Cortisone (1 mg) was dissolved in 0.5 ml of chloroform and 0.13 ml of 12 M hydrochloric acid, followed by 0.13 ml of 37% formalin solution. After stirring at 5 °C for 48 h, during which time the steroid dissolved, the solution was neutralized with a sodium hydroxide solution and dried over anhydrous sodium sulfate. Samples of the chloroform layer (1 μl) were analysed by GLC on 1% SE-30 at 235 °C.

Comments: Steroids with the dihydroxyacetone side chain reacted. Cortisone extracted from urine with ethyl acetate could be similarly derivatized. Hydrocortisone and some other related cortisones gave minor peaks in addition to one major peak.

7.3 Ethylidene derivatives[225]

7.3.1 USING ACETYLENE

Mercuric sulfate (1 g) was triturated in a mortar with 1 ml of concentrated sulfuric acid and the paste added to glycerol (23 g, 0.25 mol) in a two-necked flask with a stirrer and placed in an oil bath at 70 °C. After displacing the air with acetylene gas, bubbling with the gas under slight pressure was continued until acetylene (0.25 mol) was absorbed (over a period of about 2 h). The reaction product was dissolved in ether, filtered and neutralized with solid potassium carbonate.

7.3.2 USING PARALDEHYDE

Glycerol (27.6 g) was mixed with 1 : 1 aqueous sulfuric acid (0.5 ml) in a three-necked flask fitted with a reflux condenser and a stirrer with a seal, and heated to 100 °C in an oil bath. Paraldehyde (14.4 g) was added over a period of 30 min and heated for a further 3 h. The mixture was extracted with ether and the ether extract was dried over potassium carbonate.

Fractional distillation did not completely separate the isomers. Preparative GLC with 20% Carbowax-20 M or analytical GLC with 20% PDEAS were used.[16]

7.4 Isopropylidene derivatives

7.4.1 ISOPROPYLIDENES OF GLYCEROL[17]

Dry glycerol (10 g), 60 ml of a 0.3 M solution of hydrogen chloride in dry acetone and 4.0 g of powdered anhydrous sodium sulfate were stirred at room temperature for 12 h and neutralized with lead carbonate. The product was fractionally distilled from silver oxide (b.p. 82.5 °C at 11 mm Hg). GLC may be used to isolate 5-hydroxymethyl-2,2-dimethyl-1,3-dioxolane.

7.4.2 ISOPROPYLIDENES OF GLYCERYL ETHERS[43]

1-*O*-[Octadecyl]-glycerol (399 mg) was suspended in 10 ml of acetone, and 12 M perchloric acid (0.05 ml) was added with stirring. After 20 min water was carefully added until a faint turbidity appeared. The mixture was extracted with four volumes of ether, the ether extract was washed with water until acid-free, and evaporated to dryness *in vacuo*. The residue was dissolved in hexane and, if desired, the 2,3-*O*-isopropylidene glyceryl ether was purified by passing it down a 50 g silicic acid column and eluting with hexane-ether (9 : 1, by volume).

GLC analysis was carried out on 15% EGS at about 175 °C. For TLC analysis silica gel G plates were used and developed with petroleum ether–diethyl ether–acetic acid (90 : 10 : 1, by volume) and visualized by spraying with concentrated sulfuric acid and heating.

Comments: The method was used for glyceryl ethers and glyceryl ether phospholipids.

7.4.3 ISOPROPYLIDENES OF HYDROXY ACIDS[132]

For acids with vicinal hydroxyl groups, the hydroxy acid (0.2–0.3 g) was added to a 0.3 M solution of hydrogen chloride in 2,2-dimethoxypropane (2.2 g) and methanol (0.3 g) and this was kept in the dark at room temperature for 24 h. The solution was passed through a column of 3 g of basic, washed, activated alumina and eluted with hexane. The eluent was taken to dryness and the isopropylidene derivative analysed.

7.4.4 ISOPROPYLIDENES OF UNSATURATED FATTY ESTERS[47,52]

Preliminary oxidation. Osmium tetroxide (2 mg) and the unsaturated ester (1 mg) were added to a pyridine–dioxane mixture (1 : 8, by volume, 0.2 to 0.3 ml)[48] and allowed to stand at room temperature for 2 h. Sodium sulfite suspension (mix 4.5 ml of freshly prepared 1.27 M aqueous sodium sulfite and 1.5 ml of ethanol) was added and allowed to stand for a further 1.5 h. The mixture was centrifuged and the supernatant evaporated to dryness. The residue was dissolved in ether and evaporated.

Procedure. The resulting diol was condensed with acetone (1.0 ml or less) in the presence of anhydrous copper sulfate (50 mg) by heating in a reaction vial at 50 °C for 2 h, and analysed by GLC or GC–MS using a column of 3% OV-17.

7.4.5 ISOPROPYLIDENES OF UNSATURATED HYDROCARBONS[48]

Preliminary oxidation. A 1 : 8 by volume pyridine–dioxane mixture (0.2 ml) was added to 1 mg of hydrocarbon followed by 0.1 ml of dioxane containing 1 mg of osmium tetroxide, and allowed to stand for 2 h. A sodium sulfite suspension (made from 1.5 ml of freshly prepared 1.27 M aqueous sodium sulfite and 5 ml of ethanol) was then added. After 1 h, the mixture was centrifuged, and the supernatant taken to dryness. The residue was dissolved in ether and re-evaporated.

Procedure. The resulting diol was condensed with acetone (0.1–0.5 ml) in the presence of anhydrous copper sulfate (50 mg) by heating at 50 °C for 1 h.

Comments: The compounds Os $^{18}O_4$ or acetone-d_6 may be used for labelled derivatives.[47] Acetaldehyde was used for the preparation of 2-methyl-4, 5-dialkyl-1,3 dioxolanes.[48]

Preparation of oxygen-labelled osmium tetroxide (*Os* $^{18}O_4$)[47]. Labelled water (100 μl) containing 11.2% excess ^{18}O and 100 μl of dioxane were mixed and allowed to stand for 10 min, then a solution of osmium tetroxide (2 mg) in 100 μl of dioxane was added. Incorporation was reported to be 10–11% per oxygen.

CAUTION! *Osmium tetroxide is toxic.*

7.4.6 ISOPROPYLIDENE OF ANHYDRO SUGAR[31]

The sugar (0.9 g) was suspended in 50 ml of acetone containing concentrated sulfuric acid (0.5 ml) and stirred for 10 min until dissolved. After 20 h, the mixture was neutralized with solid sodium carbonate and filtered. The filtrate was evaporated to dryness and the residue recrystallized from methanol-water.

Comments: By this method, 1,6-anhydro-2,3:4,5-di-*O*-isopropylidene-1(6)-thio-D-glucitol was prepared from 1,6-anhydro-1(6)-thio-D-glucitol.

7.4.7 ISOPROPYLIDENE OF ALDOSES[10]

An aliquot of aldose solution was taken to dryness in a small reaction vial. The residue was shaken with 1.5 ml of acetone which was 0.1 M in concentrated sulfuric acid for 2 h, and neutralized with solid sodium bicarbonate. Samples of the solution were analysed by GLC on 3% XE-60 or 3% OV-225.

7.4.8 ISOPROPYLIDENES OF PENTOSES AND HEXOSES

Without boric acid.[110] L-Arabinose (1.5 g, 10 mmol) was stirred at room temperature for periods up to 24 h with dry acetone (100 ml) containing concentrated sulfuric acid (0.1 ml). At intervals samples were taken and neutralized with sodium carbonate for examination by TLC. The main product was 1,2:3,4-di-*O*-isopropylidene-α-L-arabinopyranose.

With boric acid. L-Arabinose (10 mmol) and boric acid (0.62 g) were stirred for 24 h with 100 ml of acid–acetone as above, and neutralized with sodium carbonate, filtered and concentrated to a syrup. Purification on a column of silica gel yielded 0.6 g of the di-acetal, 0.25 g of the 3,4- and 0.15 g of the 1,2-monoacetal.

Analysis was by TLC on Kieselgel HF_{254} (Merck) with a 3:1 mixture by volume of ethyl acetate–light petroleum (b.p. 60–80 °C) and a 4 : 1 mixture of chloroform–methanol. Visualization was achieved under u.v. illumination, or by spraying with 5% ethanolic sulfuric acid and heating to char.

7.4.9 ISOPROPYLIDENES OF OESTRIOLS[54]

The steroid (100 μg) was dissolved in 1 ml of dry acetone plus 0.5 ml of acetone previously saturated with dry hydrochloric acid gas at 0 °C. After 15 min at room temperature, the hydrochloric acid was neutralized by adding 0.6 M sodium bicarbonate solution. The solution was filtered, and the acetone evaporated *in vacuo*. The steroid–acetonide was extracted into ether.

Analysis was by GLC on a column with mixed 5% OV-210 and 2.5% OV-1. TLC was on Silica Gel G to which alcoholic ammonium bisulfate had been added, and was developed with a solution of ethanol in benzene (between 3:100 and 20:100 by volume).

Comments: *Trans* epimers did not react.

7.4.10 *O,S*-ISOPROPYLIDENE DERIVATIVE OF 3β-MERCAPTO-5α CHOLESTAN-2β-OL[55]

The mercaptosterol (383 mg) in acetone (30 ml) was refluxed with *p*-toluene sulfonic acid (60 mg) for 4 h and chromatographed on a column of Florisil (8 g). The acetonide was eluted with petroleum ether and petroleum ether–benzene 9:1 v/v), and recrystallized. It may be analysed by GLC.

7.4.11 ISOPROPYLIDENE-BENZENE BORONATES OF PENTOSES AND HEXOSES[110]

With benzeneboronic acid. The monosaccharide (10 mmol) was stirred with dry acetone (100 ml) containing benzeneboronic acid (10 mmol). Concentrated sulfuric acid (2 ml) was added and stirred for 24 h. The mixture was neutralized with anhydrous sodium carbonate, filtered and concentrated. Methanol (3 × 50 ml) was distilled from the residue and the syrup was fractionated on a column of silica gel.

Comments: This one-step reaction was used to make derivatives of glucose, mannose, galactose, arabinose and xylose.

D-glucose → 1,2-*O*-Isopropylidene-α-D-glucofuranose-3,5-benzeneboronate
D-mannose → 2,3-*O*-Isopropylidene -D-mannofuranose-5,6-benzeneboronate
D-galactose → 1,2-*O*-Isopropylidene-α-D-galactopyranose-3,4-benzeneboronate
L-arabinose → 1,2-*O*-Isopropylidene-β-L-arabinopyranose-3,4-benzeneboronate
D-xylose → 1,2-*O*-Isopropylidene-α-D-xylofuranose-3,5-benzeneboronate

These derivatives could be analysed by GLC after trimethylsilylation of unblocked hydroxyl groups.

7.5 Butylidene alditols[33]

The alditol (0.5 g) in concentrated hydrobromic acid (47%, 0.2 ml) was shaken with butyraldehyde (0.4 ml) for 2.5 h at room temperature, neutralized with aqueous sodium hydroxide and evaporated to a syrup under reduced pressure. It was extracted with light petroleum and dried first with calcium chloride and then with metallic sodium. It could be fractionated on neutral alumina (30 g), by elution with diethyl ether.

Analysis was by TLC on silica gel glass or pre-coated plastic plates, with butanone saturated with water, or benzene–methanol (9 : 1, by volume). GLC was carried out on a column of 7.5% Apiezon K at 182 °C. Hydroxy compounds were trimethylsilylated prior to GLC analysis.

Comments: the starting material 1-deoxy-D-glucitol gave a mixture of diacetals which varied with the time of reaction. After two days the main products were isomers of 2,4 : 5,6-di-*O*-butylidene-1-deoxy-D-glucitol, differing only in the configuration of the acetal carbon of the 5-membered ring.

7.6 Benzylidene derivatives

7.6.1 BENZYLIDENE OF GLYCEROL[18]

Air was bubbled continuously through a mixture of benzaldehyde (20 g), glycerol (22 g) and concentrated sulfuric acid (1 drop) at approx. 95 °C. Benzene (28 ml) was added and, the water (2.5 ml, about 70%) was distilled off azeotropically. The mixture was cooled, seeded with *cis*-1,3-*O*-benzylideneglycerol, and stored at 0 °C for 2 days. The crystalline product was dissolved in benzene, washed with aqueous ammonia and recovered. The *cis*- and *trans*-isomers could be separated by chromatography on alumina, or preferably by GLC.

Comment: two other methods were given for the preparation of 5–OH-2-phenyl 1,3-dioxanes.

7.6.2 BENZYLIDENE OF SUGARS[38]

α-Methylmannoside (20 g) was added to freshly distilled benzaldehyde (100 ml) and heated to 150–155 °C under carbon dioxide at about 0.45 atmospheres pressure. The sugar dissolved in 2–3 h, and the benzaldehyde was removed under reduced pressure. The residual syrup was poured hot into ethanol (200 ml) and a crystalline precipitate of 2,3 : 4,6-dibenzylidene-α-methylmannoside was obtained.

Comments: benzaldehyde introduced two new asymmetric centres into the sugar: two isomers were reported, with m.p. 97–98 °C and 181–182 °C.

Either TLC or GLC could be used for analysis.

7.6.3 BENZYLIDENE OF ALDITOLS[31]

The sugar (0.9 g) was dissolved in concentrated hydrochloric acid (2.5 ml) and benzaldehyde (2.5 ml) was added and stirred vigorously for 15 h. The mixture was diluted with chloroform, and the organic layer washed with 0.6 M aqueous sodium hydrogen carbonate, dried over anhydrous sodium sulfate, filtered and evaporated; the residue was crystallized from methanol–chloroform. Analysis was by TLC on Kieselgel G with ethyl acetate or ethyl acetate–carbon tetrachloride mixtures (1 : 1 or 1 : 3 or 1 : 5, by volume). For detection plates were sprayed with 0.1 M potassium permanganate: M sulfuric acid (1 : 1, by volume) and heated at 105 °C.[31]

7.6.4 BENZYLIDENE OF SUCROSE[36]

A solution of sucrose (2.5 g) in dry pyridine (50 ml) was treated with benzylidene bromide (2.8 ml) at 85 °C for 1 h, then a further 1 ml of benzylidene bromide was added and heated at 95 °C for 0.5 h. The 4, 6-*O*-benzylidenesucrose could be acetylated with acetic anhydride (5 ml) at 0 °C and allowed to stand at room temperature for 5 h. The solution was poured into ice-water and extracted with dichloromethane. The organic layer was washed with water and dried over anhydrous sodium sulfate. TLC and silica gel column chromatography were used to purify 1,2,3,3′,4′,6′,-hexa-*O*-acetyl-4,6-*O*-benzylidene sucrose.

7.7 Cyclopentylidene and cyclohexylidene derivatives

7.7.1 GENERAL METHODS

(a) Ethylene glycol (3.1 g), cyclopentanone (4.2 g) and about 10 μl of concentrated sulfuric acid were stirred under reflux at 100 °C for 10 h, cooled and extracted with ether. The ether was evaporated to give cyclopentylidene glycol (b.p. 153 °C).[8]

(b) A mixture of glycerol (2.0 g), cyclopentanone (4.0 g), gaseous hydrogen chloride (40 mg) and anhydrous sodium sulfate (0.8 g) was stirred at room temperature for 24 h, neutralized and the precipitated salts filtered off. A few mg of silver oxide were added and cyclopentylidene glycerol was obtained by vacuum distillation at 127–129 °C/10 mmHg.[8]

(c) D-Xylose (3 g), cyclohexanone (32 ml) and 0.4 ml of concentrated sulfuric acid were stirred at room temperature for 24 h, neutralized with sodium bicarbonate, and filtered. The cyclohexanone was evaporated under reduced pressure. The residue was crystallized from cold methanol. The product, 1,2;3,5-dicyclohexylidene-D-xylofuranose was recrystallized from ethanol, m.p. 105 °C.[41]

Comments: Various polyhydric alcohols and monosaccharides were converted to mono-, di- and tri-cyclopentylidene and cyclohexylidene derivatives.[8]

7.8 Cyclohexanone–ethylene acetal derivative of monosaccharides[30]

The monosaccharide was suspended (1 g) in 200 ml of toluene, cyclohexanone–ethylene acetal (7.0 ml) and 1 drop of concentrated sulfuric acid was added and the mixture refluxed for 6 h. After cooling 1 g of sodium bicarbonate was added and after stirring for 1 h, filtered and taken to dryness at less than 40 °C.

Analysis was performed on silica gel TLC plates Alu DC-Rolle F_{254} (Merck) with a toluene–ethanol (9: 1, by volume) as the eluent.

Comments: 4,6-Dichloro-4,6-didesoxy-D-galactopyranose was selectively blocked to yield the open chain derivative 4,6-dichloro-2,3-*O*-cyclohexylidene-4,6-didesoxy-D-galactose ethylene acetal (see p. 267).

7.9 Long chain cyclic acetals of glycerol[19]

A mixture of hexadecanal (0.96 g, 4 mmol), glycerol (0.92 g, 10 mmol), *p*-toluene sulfonic acid (0.5 g) and dry benzene (250 ml) was refluxed for 2 h in a three-necked flask fitted with a water entrainer, stirrer and inlet and outlet tubes for nitrogen gas.

The water formed was continuously removed by azeotropic distillation, and after 2 h the benzene was distilled off, the mixture was cooled to 0 °C and neutralized with 0.07 M aqueous potassium carbonate. It was extracted with three 150 ml portions of diethyl ether, the extracts washed with two 50 ml portions of water, then dried over anhydrous sodium sulfate and concentrated *in vacuo.* Yield 1.08 g (86%, mixture of 4 isomeric glycerol acetals).

After acetylation of the free hydroxyl group, isomers were analysed by GLC on 18% HiEff-2BP-EGS.

Preparative TLC was performed on Silica Gel H (Merck) 0.3 or 2.0 mm thick with hexane-diethyl ether (2 : 3, by volume). Opalescent bands of lipid could be seen by eye or sprayed with 2′,7′-dichlorofluorescein (0.2%) in ethanol and viewed under u.v. light.

7.10 Fluoro- and chlorofluoroketals from halohydrins

7.10.1 ETHYLENE CHLORHYDRIN AND ETHYLENE BROMOHYDRIN

(a) 2,2-Bis-(chlorodifluoromethyl)-1,-3-dioxolane was made from ethylene chlorhydrin[51] by stirring a mixture of *sym*-dichlorotetrafluoroacctonc (0.25 mol) and ethylene chlorhydrin (0.25 mol) for 15 min and then adding pentane (an inert solvent, 50 ml). Powdered potassium carbonate (0.25 mol) was carefully added over a period of 1 h. The exothermic reaction could be controlled by external cooling. When addition was complete, stirring was continued for 2 h, then the mixture was poured into 150 ml of distilled water, the organic layer removed and the aqueous solution extracted with two 80 ml portions of pentane. The combined pentane extracts were washed with water, dried over anhydrous magnesium carbonate, and concentrated under reduced pressure.

(b) Hexafluoroacetone ketals have been prepared from bromohydrins[50] as a method for characterization of double bonds. The derivatives gave characteristic fragmentation patterns on mass spectrometry. The alkene was converted to the bromohydrin, and this was reacted with hexafluoroacetone in the presence of tri-*n*-butylamine.

7.10.2 PREPARATION OF *CIS*-3-OCTENE BROMOHYDRIN[50]

Cis-3-octene (0.0313 mol) was weighed into a 250 ml round-bottomed flask. 50 ml of dry dimethylsulfoxide (DMSO) and 5.0 ml of 'wet' DMSO (0.0127 mol H_2O ml^{-1}) were stirred in vigorously, while *N*-bromosuccinimide (0.0626 mol) was added in small portions over 0.5 h. The orange reaction mixture was then stirred for a further 2 h and the mixture poured into a 500 ml separating funnel containing 100 ml of pentane and 200 ml of water. The pentane layer was removed and the aqueous layer extracted with 100 ml of pentane. The combined pentane extracts were extracted four times with 100 ml portions of water, the pentane extract dried over anhydrous magnesium sulfate, filtered and evaporated to dryness under reduced pressure. The residual sweet-smelling liquid was a mixture of *threo*-3-bromo-4-octanol and *threo*-4-bromo-3-octanol. It did not need further purification. Yield 6.7 g.

7.10.3 *CIS*-2,2-BIS(TRIFLUOROMETHYL)-4-BUTYL-5-ETHYL-1,3-DIOXOLANE[50]

The bromohydrin mixture (6.7 g) was transferred to a heavy-walled reaction tube with 10 ml of pentane, and tri-*n*-butylamine (7.45 ml) was added. The contents were frozen with liquid nitrogen and the tube was attached to a vacuum pump to remove air. Excess hexafluoroacetone* (0.0394 mol) was condensed into the reaction tube and the tube sealed *in vacuo*, and incubated in a safety container at 75 °C for 2 days. On cooling, crystals formed. The tube was frozen with liquid nitrogen, opened, and allowed to warm up to room temperature in a well-ventilated hood to allow excess hexafluoroacetone to escape. The mixture was washed into a separating funnel with 250 ml of water and 100 ml of pentane, separated, and the aqueous layer extracted twice more with 100 ml portions of pentane. The combined pentane layers were washed with two 100 ml portions of 0.5 M sulfuric acid and then with two 100 ml portions of water and dried over anhydrous magnesium sulfate. The excess pentane was removed *in vacuo* to give 9.82 g of light yellow liquid.

Analysis was by GLC on 15% Carbowax 20-M, by mass spectrometry, or by other physicochemical methods.

Comments: Scaling down the quantities is perfectly feasible. The use of a gas syringe (under a hood) would simplify the addition of small volumes of hexafluoroacetone gas to the reaction mixture. Tri-*n*-butylamine used here gave better results than potassium carbonate.[51] The addition of hexafluoroacetone to the bromohydrin was at least 97% *trans* specific, and with

---

* Instead of bubbling hexafluoroacetone gas into the pentane mixture at atmospheric pressure, a vacuum line of known volume with attached mercury manometer was filled with the gas and frozen in liquid nitrogen. The amount of gas used was calculated from the initial and final manometer readings.

cyclization of the hemiketal, an inversion of the configuration led to nearly quantitative *cis*-ketal formation (see p. 268).

7.10.4 BROMOHYDRIN FORMATION FROM ALKENE[49]

The alkene (1 mol) in dimethyl–sulfoxide under nitrogen was treated with water (2 mol) and cooled to about −10 °C. *N*-Bromosuccinimide (2 mol) was added with stirring. After a few minutes the solution became yellow and the exothermic reaction raised the temperature to 70 °C. This was unimportant unless volatility losses were likely, when external cooling was advisable. After stirring for a further 15 min, the reaction was quenched with water or dilute sodium bicarbonate solution, and the bromohydrin was extracted into ether.

8 SILICONIDES

8.1 Siliconides of steroids using dimethyldiacetoxysilane[64]

Dichlorodimethylsilane (1 mol) was refluxed with 2 mol of acetic anhydride for 4 h, and the acetyl chloride was slowly distilled off through a Dufton column. The temperature was increased and unreacted acetic anhyride was distilled off. The temperature was further raised and dimethyldiacetoxysilane distilled at 164–168 °C. Further redistillation gave a liquid boiling at 164–166 °C.

The reagent was first used for steroids with the 20-oxo-17α-21 dihydroxy side chain (see Section 3). The steroid in a reaction vial or a glass tube was treated with 20 μl of a dry hexane solution containing dimethyldiacetoxysilane (0.4 μl) and triethylamine (0.4 μl). The closed vial or sealed tube was heated at 40 °C for 2 h. Samples were analysed by GLC on a column of 1% OV-1 at 245 °C. The yields varied from 75–95% depending on the steroid. The derivatives were stable for 48 h in an unopened vial or sealed tube. A suitable internal standard was the preformed crystalline siliconide of 3β-acetoxy-16α,17α-dihydroxypregn-5-en-20-one.[62]

8.2 Siliconides of steroids using chlorosilanes[226]

(a) The steroid (100 μg) in pyridine (10 μl) was mixed with dichloro-dimethyl silane (10 μl) and benzene (20 μl) in a small reaction vial, and left at 40 °C for 2 h. Samples were analysed by GLC on 1% OV-1 at 240 °C. Thermally unstable derivatives are formed with cortisone.

(b) The steroid (10 μg) was dissolved in 1,1,3,3-tetramethyldisilazane (10 μl) and dimethylmethoxychlorosilane (10 μl of a 20% solution in benzene) and incubated in a sealed tube at 40 °C for 12 h. Samples were injected directly onto a GLC column, or evaporated *in vacuo* and dissolved in carbon disulfide for GLC on 1% OV-1 at 240 °C.

Dimethylmethoxychlorosilane may be prepared as follows. Methanol (23.2 g, 1.0 mol-equiv.) was cautiously stirred into dichlorodimethylsilane

(100 g) drop by drop over a period of 30 min. Hydrogen chloride was evolved. Stirring was continued for a further 30 min. On distillation through a Vigreux column, the fraction b.p. 77 °C (750 mm) was collected.

CAUTION! This must be done in an efficient fume hood.

Comments: Steroids, such as cortisol, with the dihydroxyacetone side chain, formed siliconides using this reagent, and the 11β-silyl ether formed at the same time. Siliconides are sensitive to moisture.

8.3. Siliconides of steroids with *cis*-diol groupings[62,63]

The dry steroid, 3β-acetoxy-16α,17α-dihydroxypregn-5-en-20-one, was allowed to react in a reaction vial with dichlorodimethylsilane (1 ml), pyridine (1 ml), and benzene (8 ml) for 1 h at room temperature.[62]

3-Methyl-18-hydroxy oestradiol-17β with dichlorodimethylsilane (50 μl) and pyridine (50 μl) was allowed to react for 1 h at room temperature. The 3-methyl oestradiol-17β, 18-dimethylsiliconide was analysed by GLC on 3% OV-1 or OV-17.[63]

8.4 Benzo-2,2-dimethyl-2-sila-1,3-dioxanone-(4) from salicylic acid[65]

A three-necked flask, fitted with condenser, stirrer and dropping funnel, was filled with dry benzene (200 ml) containing salicylic acid (20.7 g, 0.15 mol) and triethylamine (30.3 g, 0.3 mol). Dichlorodimethylsilane (19.2 g, 0.15 mol) was slowly added with vigorous stirring, and then refluxed for 3 h. Triethylamine hydrochloride was filtered off, the benzene was distilled off, and then the product distilled at 119 °C (4 mm Hg) m.p. 62 °C: the yield was 86%.

8.5 Benzo-2,2-dimethyl-2-sila-1-thia-3-oxanone-(4)- from thiosalicylic acid[65]

This was prepared as above, from thiosalicylic acid: b.p. 158 °C (10 mm Hg) m.p. 100 °C; yield 70%.

8.6 Benzo-2,2-dimethyl-2-sila-1-aza-3-oxanone-(4) from anthranilic acid[65]

Refluxing was carried out as before, but using anthranilic acid, and the benzene was distilled off. The residue in a small volume of benzene was precipitated with pentane or crystallized from carbon tetrachloride: m.p. 120 °C; yield 89%.

9 BORONATES

9.1 Boronic acid ester of monosaccharides[110]

The monosaccharide (10 mmol) was stirred with dry acetone (100 ml) containing boric acid (10 mmol) and concentrated sulfuric acid (2 ml) for 24 h at room temperature, neutralized with solid anhydrous sodium carbonate, filtered and concentrated for analysis.

9.2 Boronate derivatives of steroids[77]

The steroid (10 μmol) and the appropriate boronic acid (10 μmol) in ethyl acetate (1 ml) were allowed to stand in a stoppered tube for 5 min at room temperature. Analysis was by GLC or GC–MS on 1% OV-17 at 250 °C. TLC was on 'Chroma R sheet 500' (Mallinckrodt) with chloroform as the mobile phase.

For silylation, where necessary, the solution of boronate was evaporated to dryness, the residue dissolved in 0.1 ml of dry pyridine in a stoppered tube, hexamethyldisilazane (0.1 ml) and a trace of trimethylchlorosilane were added, and the mixture left at room temperature for 5 min. The solvent was evaporated off, and the silylated derivative extracted into cyclohexane for GLC.

Comments: Boronic acids kept best dry, under nitrogen gas. An excess of boronic acid could be used where other hydroxyl groups were absent. A slight excess was used to give a single GLC peak with 17α-,21α-hydroxy-20-ketones. Yields from 20,21-ketols were low, but could be improved by using up to a 3-molar excess of boronic acid. Unreacted (excess) boronic acids were eluted as their trimeric anhydrides (boroxines) and presented no problems.

9.3 Alkylboronates

Polyols or polyol methyl esters were reacted with 1 mg of methylboronic acid or *n*-butyl boroxine dissolved in 0.1 ml of pyridine at 100 °C for 15 min, and were analysed by GLC on 3% OV-17.

9.4 Alkyl- and phenylboronates of diols and hydroxyamines[90]

Boronic acids (methyl-, *n*-butyl-, cyclohexyl- and phenyl-) (0.95 mol) were added to a solution of the test compound (1 mg) in anhydrous dimethylformamide (100 μl) at room temperature. For hydroxyamines, boronic acids were used without additional solvent. Analysis was by GLC on 1% OV-17.

9.5 *n*-Butyl- and phenylboronates[78]

Equimolar amounts of the boronic acid and the diol were mixed in acetone solution. The cyclic ester formed in a few minutes at room temperature. Analysis was by GLC on 1% OV-17.

9.6 Methyl- and *n*-butylboronates of 3-methoxy-4-hydroxy-phenyl ethyleneglycol[86]

Solutions of 3-methoxy-4-hydroxyphenyl ethyleneglycol (5–40 μg) were taken to dryness in a reaction vial. 2,2-Dimethoxypropane as a water scavenger (0.3 ml) containing 300 μg of methyl- or *n*-butylboronic acid and 15 μg of phenanthrene as internal standard, was added, and left at room temperature for 15 min. One μl of the reaction mixture, together with 0.2 μl of *N,O*-bis-(trimethylsilyl trifluoroacetamide) was injected for GLC on 3% OV-101 with temperature programming from 130 to 230 °C at 15° min^{-1}.

Comments: Tailing GLC peaks were obtained when the free phenolic group was not silylated. Silylation *in situ* was used because prior treatment of the boronate with BSTFA caused solvolysis, with consequently decreased GLC peak size.

9.7 Methylboronates of sphingosines and ceramides[87,88]

Derivatives of sphingosines and ceramides were made in pyridine solution by the addition of 1.1 molar proportions of methylboronic acid. For ceramides derived from 2-hydroxyacids, 2.2 molar amounts of reagent were used. Reaction for 10 min at room temperature was usually sufficient. Analysis was by GLC or GC–MS with a capillary column. [Ceramides are *N*-acyl sphingosines, R—CH(OH)CH(NHCO—R)CH_2OH where R,R′ = alkyl, alkenyl or hydroxyalkyl.]

9.8 Butylboronates

9.8.1 BUTYLBORONATES OF CARBOHYDRATES

(a) The sugar or sugar alcohol (1 mg) and *n*-butylboronic acid (5 mg) were dissolved in 1 ml of pyridine.[71] It was sometimes necessary to boil the mixture to dissolve the sugar. Analysis was by GLC on 3% OV-17.

Comments: Butaneboronic acid could be contaminated with isobutaneboronic acid, which would result in multiple peaks. On 3% OV-17 at 92 °C isobutaneboronic acid gave $t_R = 7$ min and butaneboronic acid $t_R =$ 10 min.[37]

(b) General Method:[92] The sugar or alditol (1 mg) in dry pyridine (0.5 ml) was put in a reaction vial with an internal standard such as methyl palmitate. Butaneboronic acid (3 mg) was added and heated at 100 °C for 10 min. After cooling to room temperature, analysis, after trimethylsilylation if necessary, was by GLC on 1% ECNSS–M (temperature programme 115–190 °C) or 3% OV-225 (100–240 °C).

(c) Quantitative analysis of sugar mixtures:[92] sugars (0.1–10 μg of each sugar) were reacted in pyridine (0.5 ml) with a three-fold weight excess of butylboronic acid at 100 °C for 10 min, trimethylsilylated (see below) and diluted to 1 ml with pyridine in a volumetric flask. One μl aliquots were taken for GLC. Methylarachidate was a suitable internal standard. Silylation was done with either, (a) trimethylchlorosilane (0.1 ml) followed by hexamethyldisilazane (0.2 ml), or (b) *N,O*-bis (trimethylsilyl) acetamide (BSA) 0.1 ml of a 12.5% (v/v) solution in pyridine, or (c) as in (b) followed by trimethylchlorosilane (0.1 ml), or (d) 2,2-dimethoxy propane (5 μl) before trimethylsilylation.

(d) A solution of carbohydrate (about 200 μg) was taken to dryness in a reaction vial.[81] Dichloromethane (50–100 μl) was added and evaporated to remove water azeotropically. The appropriate boronic acid (about 20 mol excess) in pyridine was added and heated at 110 °C for 30 min. To the hot tube 50 μl of silylating reagent (prepared by mixing trimethylchlorosilane and BSTFA, 1 : 1, by volume) was added and left at 110 °C for 10 min. Analysis was by GLC on 0.05% OV-1 or OV-17 on glass beads, with oven temperature at 80 °C increasing by 8° min^{-1}. Eicosane was a suitable internal standard.

Comments: Methane-, butane- and benzene-boronic acids were used. Formation of the cyclic boronate was quantitative. No side-products were detected. The nature of the boronic acid derivative had no detectable influence on the reaction.

9.8.2 BUTYLBORONATES OF PROSTAGLANDINS[93]

Preliminary methoxime derivation. A mixture of 30 μg of each of prostaglandins E_1, E_2, F_1 and $F_{2\alpha}$ was put into a microtube of 100 μl volume with a rubber septum. Methoxime (20 μl) was added and left at room temperature overnight. This converted the E prostaglandins to the 9-methoxime, without altering the F prostaglandins. The mixture was evaporated to dryness with a stream of nitrogen gas.

Procedure: *n*-Butylboronic acid (30 μl) in dimethoxypropane (2.5 mg ml^{-1}) was added to the dry prostaglandins in a reaction vial and heated for 2 min at 60 °C. The prostaglandins F were converted to the 9,11-*n*-butylboronate derivatives. They were taken to dryness and trimethylsilylated by adding 20 μl of Tri-Sil-Z (Pierce) and heating for 5 min at 60 °C. Analysis was by GLC on 3% SE-30 at 220 °C, as quickly as possible to avoid solvolysis.

Comments: Dimethoxypropane was used both as a solvent and as a water scavenger. The GLC peaks for the prostaglandins E showed two derivatives arising from *syn–anti* isomerism about the methoxime group.[227,228] The F_α prostaglandins gave single peaks. The authors suggested that *n*-pentyl- or *n*-hexylboronates (for better separation) or halogenated alkylboronates (for electron-capture detection) might prove useful.[93] The reader is referred to Chapter 4 of this Handbook for alternatives to the proprietary silylation reagent used.

9.8.3 BUTYLBORONATES OF *CIS*, *TRANS*-1,3-CYCLOHEXANEDIOLS[70]

Separation of isomers. The mixture of *cis*, *trans*-1,3 cyclohexanediols (10 g, 86 mmol) was reacted with tri-*n*-butylboroxine (28.7 mmol) in a distillation flask. About 100 ml of benzene was added and the water of esterification removed by azeotropic distillation. The *cis* boronate esters were removed by distillation at 100–103 °C at 13 mm Hg.

The *cis* isomer was recovered by adding 0.1 ml of ethylene glycol and leaving to stand at room temperature for 30 min. After distilling off ethylene glycol boronate at 48–52 °C at 13 mm, *cis* 1,3-cyclohexanediol was recrystallized from ethyl acetate, m.p. 85–86 °C. The non-volatile *trans* 1,3-cyclohexanediol boronate ester was similarly transesterified and recovered.

9.8.4 BUTYLBORONATE OF 1,2-; 5,6-DIANHYDROGALACTITOL (DAG)[229]

Extraction from plasma: 10 ml of an isopropanol–chloroform mixture (9 : 1, by volume) was added to 1 ml of plasma containing 1–30 μg of DAG (an anti-neoplastic agent) and shaken for 10 min. Anhydrous potassium carbonate (1 g) was added and the mixture shaken for a further 10 min. After separation in the centrifuge the organic phase was removed and taken to dryness.

Esterification: The residue was dissolved in 100 μl of an acetone solution containing 250 μg of butaneboronic acid and 5 μg of *tetra*-(*O*-trimethylsilyl)-*meso*-erythritol(internal standard) and allowed to stand at room temperature for 5 min. Samples were analysed by GLC or GC–MS using a column of 3% SE-30 at 110 °C.

Comments: Because epoxide rings easily undergo hydration and aminolysis at high pH, the extraction from plasma cannot be carried out with ammonium carbonate as often used for drug extraction from biological fluids.

9.9 Phenylboronates

9.9.1 PHENYLBORONATES OF MONOSACCHARIDES[110]

The monosaccharide (10 mmol) was stirred with dry acetone (100 ml) containing benzeneboronic acid (10 mmol) and concentrated sulfuric acid (2 ml) for 24 h at room temperature. The mixture was neutralized with solid anhydrous sodium carbonate, filtered and concentrated for analysis (see Ref. 241).

9.9.2 PHENYLBORONATES OF ALCOHOLS, PHENOLS AND AMINES[100]

The diol, amino-alcohol or diamine (43 mmol) was refluxed for 4 h with phenylboronic acid (43 mmol) in 20 ml of dry acetone, under a calcium chloride drying tube. The ester product was purified by distillation under reduced pressure, by crystallization or by extraction into petroleum ether (b.p. 110–125 °C) and recrystallization.

Comments: GLC may also be used after further derivatization of other reaction groups, if these are present.

9.9.3 PHENYLBORONATES OF MONOALKENES[111]

The monoalkene (2 μl) was dissolved in 2 ml of dioxane and a solution of 60 mg of osmium tetroxide in 20 μl of dioxane-pyridine solution (1:8, by volume) added drop by drop. The reaction was allowed to stand for 15 min. A suspension of 14 g of disodium sulfite in 90 ml of water and 300 ml of methanol was added and stirring continued for 1 h at room temperature. The precipitate was filtered and washed with methanol. The filtrate and washings were evaporated to dryness on a rotary evaporator. The residue was dissolved in 30 ml of ether, dried with magnesium sulfate and filtered. The diol was reacted with 60 mg of phenylboronic acid in ether solution. After 15 min aliquots were analysed by GC–MS, using a capillary column coated with OV-61.

9.9.4 PHENYLBORONATES OF DIOLS AND TRIOLS[113] (USING PHENYLBORONIC ANHYDRIDE)

The diol or triol (25 mmol) was heated with phenylboronic anhydride (25 mmol) in boiling toluene until the azeotropic removal of water was complete. The remaining toluene was removed at 60 °C under reduced pressure. The esters were purified by distillation or by crystallization.

Comments: Phenylboronates of 1,2- 1,3- and 1,4-diols and some triols were studied.

9.9.5 PHENYLBORONATES OF CARBOHYDRATES (USING PHENYLBORONIC ANHYDRIDE)

Either, an equimolar mixture of a diol and phenylboronic anhydride were refluxed in toluene solution, with azeotropic removal of water and toluene; *or*, an equimolar mixture of a diol and phenylboronic anhydride was shaken at room temperature for a few minutes, the water layer was removed and the organic layer dried over magnesium sulfate. The product was purified by distillation (see Refs. 242, 243).

9.9.6 DIHYDROXY, DIAMINO AND HYDROXYAMINO NAPHTHALENES[103]

The 1,2- or 2,3-disubstituted naphthalene was refluxed with phenylboronic anhydride in benzene or toluene with azeotropic removal of water.

The product was purified by crystallization followed by vacuum sublimation.

9.10 Tri-*p*-bromobenzeneboronate of mannitol[107]

A solution of mannitol (1.2 g, 7.0 mmol in 10 ml of water) was added to a warm methanolic solution of *p*-bromobenzeneboronic acid (3.7 g, 18.0 mmol in 10 ml of methanol). After 10 min, the white precipitate was filtered off, washed with hot water and then cold methanol and dried at 105 °C. It was recrystallized from methanol-benzene mixture (1:1, by volume).

9.11 Dialkylboryl-acetylacetone (4-oxo-penten-(2)-yl-(2)-oxy dialkylborane)[83]

Triethylborane (0.25 mol) was added slowly to acetylacetone (0.66 mol) at room temperature. The reaction was speeded up by warming to 40–50 °C. The ether, unreacted acetylacetone and finally the diethylborane derivative, b.p. 91.5; 14 mm Hg were distilled off.

Comments: Dipropyl-, bibenzyl-, dibutyl-, dihexyl- and diphenyl-boryl acetylacetones were prepared. Benzoin and benzil were reacted similarly.

9.12 Diphenylboryl-acetylacetone[83]

Triphenylborane (13.9 g) was added slowly to acetylacetone (15 g) at 0 °C in an ice bath. After removal from the ice the temperature rose to about 60 °C. The diphenylboryl-acetylacetonate crystallized on cooling, m.p. 115–116°.

10 OTHER HETEROCYCLIC DERIVATIVES

10.1 Epoxides[122]

A solution of the alkene (0.1 g) was reacted with an equimolar quantity of *m*-chloroperbenzoic acid in 25 ml of dichloromethane at 6 °C overnight. The reaction mixture was extracted five times with 25 ml portions of 1 M sodium hydroxide, five times with distilled water and dried over anhydrous sodium sulfate. After filtration and evaporation to dryness under reduced pressure at 30 °C, the residue was redissolved in hexane for GLC on a column of 5% Apiezon L, Carbowax-20 M or DEGS. The *trans* isomers were found to have shorter retention times than the *cis* isomers.

10.2 Epoxides of monounsaturated and diene fatty acids

(a) The monounsaturated acid ester (3 μl) was mixed with 150 μl of peracetic acid and allowed to stand at room temperature for 2–3 h.[123,124] Double the quantity of peracetic acid was used for dienes and they were reacted for 4.5–5 h[123] or 6–8 h.[124] Samples were analysed directly or neutralized with sodium bicarbonate and extracted into petroleum ether. *Cis* and *trans* isomers were separated on a GLC column of 10% EGSS-X.

Comments: *Cis* monounsaturated compounds react in 3 h, but the *trans* isomers require 6 h.[124] Approximately 5% peracetic acid may be prepared as follows. Acetic anhydride (5 μl) was mixed with 1 ml of 30% hydrogen peroxide, stirred for 3 h at room temperature and allowed to stand for 16 h. The peracetic acid is stable at −25 °C for a minimum of 3 weeks.

(b) The monounsaturated acid ester (200 mg) was reacted with *m*-chloroperbenzoic acid (160 mg) in chloroform (20 ml) at room temperature for

3–4 h, and extracted into ether.[127] For analysis by TLC silica gel G was eluted with a mixture of light petroleum (b.p. 40–60 °C) and ether (4 : 1, by volume). The chromatogram was developed by spraying with an ethanolic solution of phosphomolybdic acid (10% by volume) and heating at 120 °C for 15 min. For GLC analysis a column of 10% DEGS or 3% Apiezon L was used.

Comments: All the 31 methyl epoxyoctadecanoate isomers were prepared and characterized.

10.3 Epoxyglyceride conversion to ketals[134]

The triglyceride sample (5–50 mg) was dissolved in 2.0 ml of iso-octane in a 10 ml volumetric flask or in a stoppered tube. A suitable ketone was made acidic with about 1 part per 100 of BF_3 etherate and 0.2 ml of this added to the iso-octane solution. The flask was shaken and allowed to stand at room temperature for 2 h. The reaction was stopped by adding 5 ml of 2 M aqueous NaCl and shaking. An aliquot (1–5 μl) of the organic layer was injected onto a GLC column of 3% OV-1 with a temperature programme from 260° to 374 °C.

Comments: Acetone, methyl ethyl ketone, methyl isobutyl ketone, 2-heptanone, cyclopentanone and cyclohexanone were studied. Cyclopentanone gave derivatives which were best separated from the long chain common triglycerides.[134]

10.4 Epoxide to isopropylidene conversion[132]

The following is a method for fatty acid epoxides. A solution of the fatty acid epoxide methyl ester (1 part) in acetone (1 part) was added to a well-stirred solution of boron-trifluoride etherate in acetone (1 : 200, by volume). An exothermic reaction resulted. After cooling, the mixture was diluted with water and the isopropylidene fatty acid methyl ester extracted into ether for analysis.

10.5 Episulfide of a steroid[121]

Cholestan-3β-ol-2α-ethyl xanthate (100 mg, 197 μmol) was dissolved in 25 ml of ethanol and excess sodium borohydride added (10 mg, 263 μmol). After 2 days at room temperature the mixture was diluted with water and the cholestan-2α,3α-episulfide extracted with *n*-hexane.

10.6 Carbonates of diols[139]

The diol (1 mol) was stirred with dry pyridine (1.5–2.0 mol) in chloroform (ethanol-free) at 0 °C and phosgene bubbled through the solution slowly for 1 h. The mixture was stirred for 1 h further at 0 °C and then 2–3 h at room temperature. The solution was transferred to a separating funnel and the organic phase washed with water until neutral and dried over anhydrous calcium sulfate. A sample was analysed by GLC on a column of 10% SE-30 or 5% Apiezon L. Cyclic sulfites of acyclic and cyclic 1,2-diols were examined by MS[139] and GLC.[139,178]

10.7 Carbonate derivatives of sugars[138]

6-*O*-Trityl-D-mannose (10 g) was added to pyridine (70 ml) and a solution of phosgene (2.4 g) in toluene (20 ml) added to the pyridine solution at 0 °C over a period of 30 min with vigorous stirring. The ice-bath was removed and after standing for 30 min the suspension was filtered. The filtrate was evaporated to a syrup, taken up in chloroform, passed through a silicic acid column and eluted with chloroform. Fractions containing 6-*O*-trityl-D-mannose-2, 3-carbonate were collected for analysis (see also Ref. 244).

10.8 Dithiocarbonate derivative of a steroid[121]

Cholestan-3α-ol-2α-ethyl-xanthate (28 mg, 55 μmol) was dissolved in 4 ml of ethanol and excess sodium borohydride added (15 mg, 390 μmol). After 4 days at room temperature, the reaction was diluted with water and the cholestan-2α(S), 3α(0)-dithiocarbonate extracted into ether for analysis.

10.9 Cyclopropane fatty acid methyl esters[140]

Preparation of zinc–copper couple (this should always be freshly prepared): Nearly boiling glacial acetic acid (10 ml) was stirred vigorously in a 25 ml two-necked flask and zinc dust added (2.0 g). After 1 min, a hot solution of glacial acetic acid (10 ml) containing cupric acetate monohydrate (0.4 g) was added, and the reaction stirred until the blue colour disappeared (about 1 min). The hot supernatant was decanted and in the same flask the zinc–copper couple washed five times with 20 ml portions of glacial acetic acid and then five times with 20 ml portions of dry ether.

Procedure: Dry ether (10 ml) was added to the flask containing the couple which had been fitted with a condenser and a dropping funnel. Diiodomethane (4 ml) was added and then the unsaturated fatty acid methyl ester (0.2 g) in 5 ml of ether added at a rate sufficient to keep the solution refluxing by the heat of reaction. When addition was complete the mixture was refluxed overnight in a nitrogen atmosphere. The ether solution was then decanted from the couple and washed three times with water before drying over anhydrous sodium sulfate. First the ether and then the diiodomethane were removed under reduced pressure (100 °C at 0.5 mm Hg). The cyclopropane fatty acid methyl ester may be purified by elution from a column of Florisil (20 cm × 1 cm) with 100 ml of petroleum ether–ether mixture (70:30, by volume). Analysis by

TLC on silica gel impregnated with silver nitrate was performed using petroleum ether–ether mixture (9:1, by volume) or, by GLC on 20% EGS and 2% phosphoric acid (polar) or 20% Apiezon L (non-polar) columns.

10.10 Nitroquinoxalinol derivatives of α-keto acids and condensation with 1,2-diamino-4-nitrobenzene

(a) Pyruvic acid (1.25 mmol, 110 mg) was dissolved in 10 ml of water and 1,2-diamino-4-nitrobenzene (DANB, 1.25 mol, 190 mg) dissolved in 10 ml of 2 M hydrochloric acid.[148] After 2 h at room temperature the quinoxalinol was filtered off, dried and recrystallized from hot ethanol–water, m.p. 245 °C. Paper chromatography was performed with the eluent ascending on large sheets of Whatman No. 4 paper using methanol, ethanol, or alcohol–water (9 : 1, or 7 : 3, by volume). The yellow spots may be seen in daylight but are accentuated by spraying with ethanolic potassium hydroxide (0.3 M). They appear as brown spots under u.v. light.

Comments: Derivatives of dimethylpyruvic, benzoylformic and α-ketoglutaric acids were also prepared. In addition derivatives were made by condensation with *o*-phenylenediamine but their chromatography and detection were less satisfactory.

(b) The detection of α-keto acids in blood and urine has been reported using quinoxalinol derivatives.[149] The method is as follows.

A sample of 6–10 ml of freshly drawn blood was added to 40 ml of freshly prepared metaphosphoric acid solution (5% w/v). After standing for 10 min, the mixture was centrifuged and the supernatant liquid removed into a separate tube. To this, 3 ml of 1,2-diamino-4-nitrobenzene solution (0.2% in 0.66 M hydrochloric acid) was added and after mixing allowed to stand at room temperature for 12–16 h. (A 10 ml sample of urine was treated with 1 ml of this reagent solution in the same way.) The mixture was extracted four times with 10 ml portions of ethyl acetate, and from the combined ethyl acetate extracts, the quinoxalinols were removed by extracting four times with 8 ml portions of sodium carbonate solution (0.5 M). The aqueous phase was adjusted to pH 4 with 10 M hydrochloric acid and extracted four times with 8 ml portions of ethyl acetate. The combined ethyl acetate extracts were evaporated to dryness at less than 40 °C, and the residue taken up in acetone for descending paper chromatography on Whatman No. 4 sheets, using as eluent ethanol-*n*-pentanol-(0.880) ammonia (5:8:6, by volume). The quinoxalinols were located as purple-brown spots under u.v. light. The pale yellow spots fade in daylight, but the colour can be enhanced by spraying with 1.25 M sodium hydroxide. Elution with 30% aqueous ethanol followed by measurement of the absorbance at 280 nm may be used for quantitative measurements.

Comments: Ethyl acetate must be washed before use with sodium carbonate solution (5% w/v) and redistilled twice, otherwise derivatives ‘streak’ during paper chromatography. The method is not generally applicable to aromatic α-keto acids, because these acids are unstable in solution and they do not form quinoxalinols as readily as the aliphatic acids (see next method).

10.11 Quinoxalinols from aromatic and aliphatic α-keto acids using condensation with *o*-phenylenediamine

(a) A sample of 10 mmol of α-keto acid (e.g. phenylpyruvic acid) was dissolved in 20 ml of 96% ethanol in a stoppered tube.[147] 0.8 M acetic acid (20 ml) containing 1.2 g of *o*-phenylenediamine was added and the mixture heated as 100 °C for 1 h. The quinoxalinol precipitated, and was washed with 1 : 1 aqueous ethanol and crystallized from a 3 : 7 mixture of aqueous ethanol. The product was analysed by paper chromatography using as eluent 20 : 1 by volume butanone: 0.1 M sodium hydroxide. The spots were detected under u.v. light at about 350 nm. Sensitivity limit was about 20 nmol. GLC may be used after trimethylsilylation (see next method).

Comments: This method was developed for aromatic acids. Lactones must first be converted to the acid form by boiling with ethanolic solutions of sodium hydroxide (15 mmol) for 5 min under nitrogen and then neutralizing with hydrochloric acid. Aqueous solutions of α-keto acids may be protected against oxidation with sodium hydrogen sulfide by adding 1 volume of 1 M aqueous sodium sulfide plus 1 M hydrochloric acid to 4 volumes of the α-keto acid solution. The method has also been applied to urine samples.

(b) The quinoxalinol derivatives of α-keto acids were prepared as described in the preceding method, and then silylated.[144] This was done by dissolving 1–2 mg of the sample in 100 μl of pyridine and reacting with 100 μl of BSTFA at 70 °C for 30 min. A sample of 1–2 μl was injected onto a GLC column of 3% Dexsil-300-GC, OV-1 or OV-17.

Comments: Oxaloacetic acid gave methylquinoxalinol and not the expected carboxymethylquinoxalinol.[144] TMS-quinoxalinols were prepared from 10 aliphatic α-keto acids and also phenylpyruvic acid[144] and separated by GLC. For GC–MS confirmation of structure 100 μl of $^{2}H_9$–BSA and 50 μl of $^{2}H_9$-TMCS may be used.

(c) A solution of the α-keto acid (10 mmol in 25 ml of water) was added slowly with stirring to a solution of *o*-phenylenediamine (37 mmol, 4 g) in 50 ml of 1.6 M acetic acid.[143] The solution darkened and the quinoxalinol crystallized out. After filtration, washing with water and drying in air, the product was recrystallized from benzene or aqueous acetone for analysis.

Comment: 3-Methyl-, -ethyl-, -*n*-propyl-, -isopropyl-, -*n*-butyl-, -isobutyl- and -*t*-butyl- derivatives were prepared.

(d) Sublimed *o*-phenylenediamine (60 mmol) was dissolved in 100 ml of 1.6 M acetic acid and mixed with a solution of 30 mmol of α-keto acid in 30 ml of water. The quinoxalinol precipitated out and was filtered off after 15 min, washed with water, dried and recrystallized from methanol. Suitable internal standards for biological samples were found to be 6-methyl-2-naphthol and *p*-nitrophenyl phenyl ether. After silylation the sample may be analysed by GLC with columns of 13% SE-30, 1% OV-17[146] or 3% OV-17[145] with a temperature programme from 100 to 200 °C. Silylation is best done by reacting 75–120 μl of pyridine solution of the quinoxalinol with 200 μl of BSA. Reaction is virtually immediate.[146] An alternative method involves reaction of the quinoxalinol with a 1 : 1 mixture by volume of acetonitrile: BSTFA at 20 °C for 30 min.[145]

10.12 2,5-Diketopiperazines (Piperazine-2,5-diones)

(a) The dipeptide (L-leucyl-L-tyrosine, 0.24 g) was heated with 10 g of phenol with stirring under nitrogen at 140–150 °C for 1 h.[155] The phenol was removed by sublimation under vacuum at room temperature. The residue was crystallized from ethanol–water, m.p. 292–293 °C.

A similar reaction with the dipeptide salt glycyl-L-histidine hydrobromide (0.6 g) involved heating as above for 5 h. The phenol was removed mainly by sublimation and residual phenol by ether extraction. Crystallization from acetone–water gave a sample with m.p. 244–245 °C (with decomposition).

(b) The *t*-BOC amino acid (2 mmol) was dissolved in methylene chloride (10 ml) at 0 °C and an amino acid methyl ester hydrochloride (2 mmol) and triethylamine (2 mmol) were added. After adding *N*-ethyl-*N'*-(3-dimethylaminopropyl) carbodiimide hydrochloride (2 mmol) the mixture was stored at −5 °C overnight.[154] The reaction mixture was washed with water, 0.33 M citric acid, sodium bicarbonate and water and the solution evaporated to dryness. The *t*-BOC dipeptide ester was recrystallized.

In a very similar procedure the *t*-BOC-dipeptide methyl ester (200 mg) was dissolved in formic acid (98%, 20 ml) and the solution kept at room temperature for 2 h. Excess formic acid was removed *in vacuo* and the crude dipeptide ester formate dissolved in *sec*-butanol (10-40 ml) and toluene (5–10 ml). The solution was boiled for 2–3 h and the solvent level maintained by adding further *sec*-butanol. The solution was concentrated to 5–10 ml and then cooled to 0 °C; the product was removed by filtration and recrystallized.[154]

To analyse these diketopiperazines by TLC silica gel plates with the eluents 2-butanol, chloroform-methanol-acetic acid (35 : 2 : 1, by volume) or 2-propyl ether-chloroform-acetic acid (6 : 3 : 1, by volume) were used.[155] The plates were dried thoroughly and sprayed with freshly diluted commercial chlorax bleach in water (10–15%). After spraying they were dried at room temperature under a hood for exactly 30 min, sprayed with ethanol and left for 10 min, then sprayed with a 1 : 1 mixture by volume of 0.06 M aqueous potassium iodide and a freshly prepared solution of 0.1 M *o*-toluidine in 1.6 M aqueous acetic acid. The deep blue spots given by N-protected compounds fade rapidly.[230] For analysis by GLC, columns of 5% QF-1 or 3% EGS at temperatures 200–240 °C have been used.[156]

TMS derivatives of these diketopiperazines (1–5 mg) may be prepared using a 1 : 3 mixture by volume of BSA: dry dimethylformamide and heating for 10 min at 80 °C. For GLC analysis a column of 3% SE-30 at 90 °C to 180 °C has been used.[159] TFA derivatives were prepared from the diketopiperazine (1–2 mg) reacted with 0.2 ml of trifluoroacetic anhydride at 90–100 °C for 5–30 min, depending upon its solubility. Columns of 1% EGSP-Z or 3% NGS at 100–200 °C have been used for GLC analysis.[158]

(c) Glycine (700 g) and 3500 ml of ethylene glycol were heated at 174–176 °C for 50 min with stirring.[150] After cooling overnight at 0 °C, the crystals of diketopiperazine were removed by filtration and washed with methanol. The product was recrystallized from methanol.

Comment: Although this reaction was performed on an enormous scale it could easily be adapted to a microprocedure.

10.13 2-Trifluoromethyl-oxazolidin-5-ones

(a) Alanine (0.98 g) was heated with trifluoroacetic anhydride (3.4 ml) for 2 h at 140 °C and the product distilled *in vacuo*. The azlactone fraction was redistilled with the addition of a few drops of trioctylamine b.p. 143 °C.[162]

(b) Trifluoroacetic anhydride (32 ml) was added carefully to 0.1 ml of an amino acid, with gentle warming. The mixture was refluxed at 80 °C for 20 min, then the bath temperature was raised to 120–130 °C for a further 20 min. The trifluoroacetic acid and trifluoroacetic anhydride were removed under reduced pressure. Low boiling oxazolones were freed of remaining trifluoroacetic acid by dissolving in ether and washing the ethereal solution with aqueous sodium bicarbonate solution. The ether layer was dried over anhydrous magnesium sulfate, filtered and the ether removed by distillation.[163]

(c) A TFA-derivative of an amino acid (0.1 mol) was dissolved in 30 g of pyridine and 40 ml of methylene chloride. The solution was cooled to −20 °C, and 9.2 g of freshly distilled phosphorous oxychloride added drop by drop with stirring. After 30 min at 0 °C, followed by 1 h at room temperature, methylene chloride was added and the solution washed with ice-water and 0.1 M ice-cold phosphoric acid. The methylene chloride layer was dried over anhydrous magnesium sulfate. The methylene chloride and then the oxazolidinone were distilled under reduced pressure.[163]

(d) An amino acid (1 mg) was heated with 0.1 ml of trifluoroacetic anhydride in a sealed tube or a Rotaflo stopcock[231] (see Chapter 1) at 150 °C for 10 min. The mixture was cooled to room temperature, evaporated under reduced pressure or in a stream of nitrogen gas, taking care to avoid volatility losses and 100 μl of ethyl acetate added. The product was analysed by GLC on a column of 5% OV-17 with temperature programming.[166]

Comment: This one-step reaction was applied to ten simple α-amino acids.

10.14 2-Trifluoromethyl ozazolidin-5-one of phenylalanine in serum[168]

A sample of 100 μl of serum in a centrifuge tube was mixed with 10 μl of a 5–8 mM aqueous solution of phenylglycine as the internal standard. Ethanol (500 μl) was added, the solution mixed thoroughly and centrifuged. A 400 μl portion of the deproteinized supernatant was transferred to a reaction vial and taken to dryness at 60–80 °C in a stream of nitrogen gas. Trifluoroacetic acid (250 μl) and trifluoroacetic anhydride (250 μl) were added and the closed vial was heated at 130–140 °C for 10 min. After cooling and evaporation to dryness in a stream of nitrogen, the residue was dissolved in 40 μl of ethyl acetate and 1 μl samples injected onto a GLC column of 3% OV-17 at 100 °C using a temperature programme up to 132 °C.

10.15 2,2-Bis-(trifluoromethyl)-4-substituted-1,3-oxazolidin-5-ones, the condensation of fluoroketones with amino acids[172]

Dry dimethylsulfoxide (15–20 ml) was added to the amino acid (20–30 mmol) in a three-necked flask with gas inlet, stirrer and condenser fitted with drying tube. A slow stream of hexafluoroacetone was passed through the liquid from a cylinder. The α-amino acid dissolved with gentle warming in 20–60 min and the reaction was stirred for 3–4 h, but 4–6 h was required for amino acids with additional reactive groups. The reaction mixture was then diluted with water and extracted with methylene chloride. The methylene chloride extract was washed twice with water, dried over sodium sulfate and evaporated *in vacuo*. The products were analysed by GLC.

Comments: An efficient fume-cupboard is essential. The hexafluoroacetone may be passed into water for disposal, or alternatively trapped with solid carbon dioxide at −70 °C.

10.16 2,2-Bis-(chlorodifluoromethyl)-4-methyl-1,3-oxazolidin-5 ones from amino acids[51]

A suspension of alanine (0.15 mol) was stirred in *sym*-dichlorotetrafluoroacetone and dimethylformamide (50 ml) added. The temperature rose to about 45 °C with the exothermic reaction. The reaction mixture was heated at 60 °C for 3 h, cooled and methylene chloride (85 ml) added. The reaction mixture was poured into 850 ml of water and extracted with methylene chloride. The combined methylene chloride extracts were washed with water, dried over anhydrous magnesium sulfate and concentrated under reduced pressure.

10.17 2,2-Bis-(chlorodifluoromethyl)-4-substituted-1,3-oxazolidin-5-ones from amino acids[173,174]

The amino acid or amino acid hydrochloride (2–5 μmol) in a reaction vial was completely dried by azeotropic removal of water with methylene chloride at 50 °C with nitrogen gas. A 120 μl portion of solvent consisting of a 1:100 dilution of pyridine in acetonitrile was added and in addition 30 μl of *sym*-dichlorotetrafluoroacetone. After the amino acid had dissolved reaction was complete in 5 min at 20 °C. However, a further period at 50 °C for 5 min was necessary in some cases. Further derivatization is possible, for example hydroxyl groups (e.g. of tyrosine) may be blocked by adding to the reaction mixture 20 μl of heptafluorobutyric anhydride and warming at 50 °C for 15 min. GLC analysis was done on a column of 5% SE-30 or 3% OV-17, using flame ionization or electron-capture detection.

Comments: Only tyrosine and its iodinated analogues were studied. After the first step for oxazolidone formation the reaction mixture may be taken to dryness, before derivatizing the hydroxyl groups, but this is not necessary.

10.18 Thiazolidines and Schiff Bases[176]

The thiol was dissolved in methanol (0.1 ml) and pivaldehyde (0.1 ml) added. Ion-exchange beads (AG1-X8 in bicarbonate form) were added until the solution was at pH 7–8. Molecular Sieve (3A) was added to remove water and after 10 min an aliquot was injected onto a GLC column of 5% SE-30 using a temperature programme from 80 to 250 °C. Biological samples were prepared by extraction with aqueous ethanol or trichloracetic acid and the extract was evaporated to dryness. The residues were esterified with 1.0 ml of a 1.25 M solution of hydrogen chloride in methanol for 30 min and then taken to dryness.

Comments: The reaction is dependent upon the pH. In pyridine the reaction is slow, and with triethylamine rapid. The basic ion-exchange resin is recommended. Cysteamine, cystamine, homocysteine lactone and S-containing amino acids were derivatized to yield either the thiazolidone or Schiff base.

10.19 Isocaprolactone[232]

Levulinic acid (1 mol) was dissolved in ether and methyl magnesium iodide (2 mol) added slowly. The addition complex was cooled and decomposed with sulfuric acid. The isocaprolactone was extracted into ether and washed with sodium bisulfite to remove free iodine. The product solution was evaporated to dryness before analysis. Lactones may be analysed by GC–MS on a column of 5% FFAP.[175]

10.20 Malonaldehyde conversion to 2-hydroxypyrimidine[182]

Urea was added to an aqueous solution of malonaldehyde (0.5 mM) to a concentration of 0.2 M. An equal volume of 12 M hydrochloric acid was added and allowed to stand at room temperature for 1 h. After evaporation under reduced pressure at about 60 °C, the product was converted to the free base by adding aqueous ammonium hydroxide to dissolve the residue and then taking to dryness. A 5 ml column of Dowex 50-H^+ was used to purify the product, eluting with 20 ml of 0.1 M acetic acid and then 20 ml of a mixture of 0.3 M acetic acid +0.1 M pyridine. The collected fraction was taken to dryness. The yield of 2-hydroxypyrimidine was about 90%.

GLC analysis: trimethylsilylation and injection on a column of 10% F60.

TLC analysis: Silica gel G, with distilled water eluent, sprayed with sodium fluoresceinate and detected under u.v. light.

Comments: The method was used to determine microgram quantities of malonaldehyde in tissue homogenates.[182] Malonaldehyde may be detected by its colour reaction with 2-thiobarbituric acid.[245]

10.21 Malonaldehyde reaction with arginine[183]

A solution of L-arginine (1 mmol) and 1,1,3,3-tetraethoxypropane (1.1 mmol) in 8 ml of 12 M hydrochloric acid was allowed to stand at 25 °C for 3 h, and then evaporated to dryness on a rotary evaporator. The δ-*N*-(2-pyrimidinyl)-L-ornithine which was formed, could be analysed at this stage (see below) or further purified. The residue was stored *in vacuo* over sodium hydroxide pellets overnight. It was then dissolved in 2 ml of water and loaded onto a column of Dowex 50-X8 (25 cm × 1 cm) equilibrated with 0.3 M acetic acid + 0.1 M pyridine solution. Elution with a linear gradient was produced by 120 ml of 0.3 M acetic acid + 0.1 M pyridine and 120 ml of 0.3 M acetic acid + 0.2 M pyridine. The effluent was monitored by its absorbance at 315 nm. The collected fractions were taken to dryness and the residue dissolved in 1 ml of 1 M hydrochloric acid and 9 ml of ethanol. δ-*N*-(2-Pyrimidinyl)-L-ornithine crystallized out, after the addition of 0.09 ml pyridine. TLC analysis was performed on Eastman sheet silica gel (type K 301 R) with an eluent of *n*-propanol-28% ammonium hydroxide (67:33, by volume) and the spots detected with ninhydrin. (GLC is possible with further derivatization.)

Comments: 1,1,3,3-Tetraethoxypropane is the ethyl acetal of malonaldehyde. α-*N*-Benzoyl-L-arginine was similarly converted to α-*N*-benzoyl-δ-*N*-(2-pyrimidinyl)-ornithine; its ethyl ester derivative is suitable for GLC.

10.22 1,3-Diphenylimidazolidine-2-carboxylic acid[180]

This reaction was proposed for glyoxylic acid (see Section 5.11).

The dry glyoxylic acid (2 mg, 20 μmol) was reacted with *N,N'*-diphenylethylenediamine (10 mg, 44 μmol) in pyridine (0.4 ml) by mixing and allowing to stand for 2 h at room temperature. BSA (0.5 ml) was added and after 2 h, an aliquot was taken for analysis on a column of 10% OV-101. The retention temperature was 264 °C, using a linear programme.

10.23 Cyclic sulfites of diols[139]

The diol (1 mol) and dry pyridine (1.5–2.0 mol) were stirred in chloroform (ethanol-free) at 0 °C and thionyl chloride (1 mol) was added drop by drop over 30 min. The mixture was stirred at room temperature for 1.5–2 h and then poured into a separating funnel. The organic phase was washed with water until neutral, then dried over anhydrous calcium sulfate. The product was analysed by GLC on a column of 10% SE-30 or 5% Apiezon L.

11 CYCLIC COMPOUNDS IN PROTEIN SEQUENCING

11.1 3-Phenyl-2-thiohydantoin-5-substituted (PTH) amino acids[233]

A sample of 10 mmol of the amino acid was dissolved in 50 ml of a 1 : 1 mixture by volume of pyridine and water. The solution was adjusted to about pH 9 with 1 M sodium hydroxide, and then heated to 40 °C and, with vigorous stirring, phenylisothiocyanate (2.4 ml) was added, maintaining the mixture at pH 9 by adding 1 M sodium hydroxide. The reaction was completed when alkali consumption ceased and this took about 30 min at 40 °C. The reaction was extracted several times with benzene to remove pyridine and excess phenylisothiocyanate. Molar hydrochloric acid was added equivalent to all the 1 M sodium hydroxide used during the reaction and this precipitated the phenylthiocarbamyl (PTC) amino acid. PTC-arginine was precipitated at about pH 7.0 and PTC-histidine at pH 3.5. The PTC-amino acid was suspended or dissolved in 30 ml of 1 M hydrochloric acid and refluxed for 2 h. The reaction mixture was taken to dryness *in vacuo* exhaustively to remove hydrogen chloride. The yield of PTH-amino acid was about 80–90%. The product was recrystallized from ethanol (histidine, isoleucine, leucine), water (arginine, hydroxyproline), ethanol–water (aspartic acid, asparagine, glutamic acid, lysine) or glacial acetic acid–water mixture (alanine, glycine, methionine, phenylalanine, proline, tryptophan, tyrosine). The preparations were dried *in vacuo* (1 mm Hg) over phosphorus pentoxide and KOH pellets for 48 h.

Some of the simpler derivatives may be gas chromatographed directly; others require trimethylsilylation or trifluoroacetylation.

Comments: The initial reaction is best carried out in a pH-stat and the quantities may be scaled down. The reader is referred to original papers for PTH derivatives of cysteic acid, *S*-carboxymethylcysteine and glutamine,[234] serine and threonine[235] and sarcosine and pipecolic acid.[236] The preparation and cleavage of PTC peptides has been described.[237]

11.2 3-Methyl-2-thiohydantoin-5-substituted (MTH) amino acids[238]

About 1 mg of amino acid in a 10 ml round bottomed flask was treated with 1 ml of water, 2 ml of pyridine and 5 μl of triethylamine. Nitrogen gas was passed through the solution for 1 min and then 5–10 ml of methyl isothiocyanate was added. The stoppered flask was shaken on a rotary mixer for 30 s. The flask was attached to a rotary evaporator and evacuated to 20 mm Hg and flushed with nitrogen. The flask was rotated at 40 °C for 30 min with continual nitrogen flushing and then the flask contents were evaporated to dryness. Ethyl acetate was added and taken to dryness to remove volatile substances. The product was cyclized to the thiazolinone by adding 0.5 ml of anhydrous trifluoroacetic acid and incubating at 40 °C for 5 min. The acid was removed on the rotary evaporator. The thiazolinone was converted to the MTH amino acid by adding 0.5 ml of 1 M hydrochloric acid and heating at 80 °C for 10 min. The MTH amino acid was extracted into ethyl acetate for analysis by GLC on a column of 1.5% OV-1 or 1.0% OV-17, at 160–260 °C (see also Ref. 246).

Comments: Peptides (0.05–0.5 μmol) are reacted in the same way. All solutions should have nitrogen bubbled through them before use and the reactions carried out under nitrogen. Details for the preparation of MTH amino acids[205] and of pipecolic acid and sarcosine[236] have been published. MTH derivatives are now commercially available. A method has been described for the GC–MS analysis of the TMS derivatives of the thiazolinones of asparagine and glutamine.[239]

11.3 Thiohydantoins

(a) A sample of 10 mmol of an amino acid was heated with 0.9 g of ammonium thiocyanate (recrystallized from ethanol), 10 ml of acetic anhydride (redistilled) and 1.3 ml of acetic acid at 100 °C for 30 min.[212] Excess water was added to hydrolyse any acetic anhydride remaining and then the mixture evaporated on a rotary evaporator. The residue was redissolved in 12 M hydrochloric acid and allowed to stand for 30 min at room temperature, the solution was taken to dryness and recrystallized from water.

GLC silylation procedure:[219] 500 μl of a 1:1 by volume pyridine:BSA mixture was added to about 1 mg of thiohydantoin and reacted at 50 °C for 10 min. This product was analysed on a column of 1% Dexsil-300-GC or 5% OV-17.

TLC analysis:[218] Polygram polyamide-6 u.v. 254 precoated plastic sheets were eluted with acetic acid–water (7:13, by volume) in the first dimension and chloroform-95% ethanol–acetic acid (20:10:3, by volume) in the second dimension.

Comments: The reader is referred to Ref. 212 for arginine, glutamic acid, *S*-aminoethyl cysteine, histidine and tryptophan derivatives. For details of protein sequencing see the original publications.

(b) When 10 mmol of amino acid had been dissolved in a mixture of 15 ml of acetic anhydride and 5 ml of acetic acid, 1.5 g of sodium thiocyanate was added in small portions with vigorous mixing.[216] The solution was heated at 80 °C for 30 min, taken to dryness under reduced pressure at less than 40 °C, and the residue dissolved in a minimum of concentrated hydrochloric acid and allowed to stand for 30 min at room temperature. Water was added and the mixture evaporated to dryness. The residue was crystallized from water.

11.4 Iminohydantoins[221]

11.4.1 PREPARATION OF *n*-BUTYL THIOURONIUM IODIDE

An excess of *n*-butyl iodide was reacted with thiourea by stirring in methanol at 50 °C for 20 min. The crystalline salt was filtered and washed with ethyl acetate. The purity was checked by TLC on cellulose plates with an eluent of *n*-butanol : pyridine : water : acetic acid (6 : 4 : 4 : 1, by volume). Thiourea and its derivatives were detected by spraying with 1% picric acid in ethanol, drying and spraying with 1 M KOH in ethanol.[240]

11.4.2 PROCEDURE FOR PREPARING THE IMINOHYDANTOINS

The *N*-acetylated amino acid or short peptide (0.25 M) is dissolved in anhydrous pyridine and reacted with dicyclohexylcarbodiimide (0.25 M) at 50 °C for 10–15 min. Alternatively, aqueous pyridine (up to 50% by volume) was used with ethyl-dimethylaminopropyl carbodiimide as the coupling agent. The product was fairly stable at neutral and mildly acidic pH. The product could be cleaved with aqueous base at pH 10–11.5, using a 1 : 1 mixture by volume of pyridine and aqueous 1 M trimethylamine at 50 °C for 10 min.

11.4.3 IDENTIFICATION OF THE IMINOHYDANTOIN

Analysis by TLC was performed on silica gel plates with eluent of ethyl acetate : methanol : 0.8 M acetic acid (4 : 2 : 1, by volume). The spray reagent was prepared by mixing 2.5 M sodium hydroxide, 10% sodium nitroprusside, 10% potassium ferricyanide and water (1 : 1 : 1 : 3, by volume). This kept for several weeks at 0 °C.[240] A purple-rose colouration was visible with about 40 nmol.[221]

For GLC analysis the sample was dried and trimethylsilylated with BSA : acetonitrile (1 : 1, by volume) under nitrogen at 125 °C for 30 min. Aliquots were injected onto a column of 1.5% OV-101.

Comments: Methylisothiouronium hemisulfate or *p*-chlorobenzylisothiouronium chloride may be used, but these are less reactive.[221]

11.5 Hydantoins[222]

The reagent solution was prepared from 1 ml of water, 0.3 ml of *N*-ethylmorpholine and 100 mg of potassium cyanate which were mixed and adjusted to pH 8 with about 0.1 ml of glacial acetic acid.

A sample of 0.5 μmol of an amino acid was warmed with 0.25 ml of the reagent solution in a reaction vial at 50 °C for 4 h, and then taken to dryness *in vacuo*. After adding 0.25 ml of 6 M hydrochloric acid the mixture was heated in a closed vial at 100 °C for about 30 min. The hydantoins which precipitated upon cooling were recrystallized from water (aspartic acid, glutamic acid, isoleucine, leucine, methionine, phenylalanine, tryptophan, tyrosine, valine).

If no precipitation occured, the hydrogen chloride was removed under reduced pressure and the residues triturated with warm absolute ethanol. The ethanol extract was taken to dryness and the hydantoin residue recrystallized from water (alanine, glycine, homocitrulline, proline, serine, threonine) or from ethanol (histidine). The product was analysed by TLC or GLC, after trimethylsilylation if necessary. Hydrolysis was possible with 6 M hydrochloric acid or 0.2 M sodium hydroxide to regenerate the parent amino acid.

Comments: See the original paper for the method used to determine the N-terminal amino acid in proteins.[222]

REFERENCES

1. H. Ziffer, W. J. A. Vanden Heuvel, E. O. A. Haahti and E. C. Horning, *J. Am. Chem. Soc.* **82**, 6411 (1960).
2. J. K. Sutherland, *Chem. Ind. (London)* 1607 (1961).
3. H. El Khadem, *Adv. Carbohydr. Chem.* **20**, 139 (1965).
4. P. R. Jones, *Chem. Rev.* **63**, 461 (1963).
5. R. Valters, *Russ. Chem. Rev.* **43**, 665 (1974).
6. R. H. Jaeger and H. Smith, *J. Chem. Soc.* 160 (1955).
7. A. N. De Belder, *Adv. Carbohydr. Chem.* **20**, 219 (1965).
8. V. M. Mićović and S. Stojiljković, *Tetrahedron* **4**, 186 (1958).
9. K. Biemann, H. K. Schnoes and J. A. McCloskcy, *Chem. Ind. (London)* 448 (1963).
10. S. Morgenlie, *Carbohydr. Res.* **41**, 285 (1975).
11. C. T. Bishop, F. P. Cooper and R. K. Murray, *Can. J. Chem.* **41**, 2245 (1963).
12. B. R. Baker and R. E. Schaub, *J. Am. Chem. Soc.* **77**, 5900 (1955).
13. K. Erne, *Acta Chem. Scand.* **9**, 893 (1955).
14. T. Van Es, *Carbohydr. Res.* **32**, 370 (1974).
15. T. Harnitzky and N. Menschutkine, *Liebig's Ann.* **136**, 126 (1865).
16. G. Aksnes, P. Albriktsen and P. Juvvik, *Acta Chem. Scand.* **19**, 920 (1965).
17. E. Fischer and E. Pfähler, *Ber. Dtsch. Chem. Ges.* **53**, 1606 (1920).
18. N. Baggett, J. S. Brimacombe, A. B. Foster, M. Stacey and D. H. Whiffen, *J. Chem. Soc.* 2574 (1960).
19. W. J. Baumann, *J. Org. Chem.* **36**, 2743 (1971).
20. E. O. A. Haahti and H. M. Fales, *J. Lipid Res.* **8**, 131 (1967).
21. E. Fischer, *Ber. Dtsch. Chem. Ges.* **28**, 1167 (1895).
22. A. Speier, *Ber. Dtsch. Chem. Ges.* **28**, 2531 (1895).
23. S. A. Barker and E. J. Bourne, *Adv. Carbohydr. Chem.* **7**, 138 (1952).
24. S. J. Angyal and L. Anderson, *Adv. Carbohydr. Chem.* **14**, 135 (1959).
25. R. J. Ferrier, *Adv. Carbohydr. Chem.* **20**, 67 (1965).
26. J. W. Berry, in *Advances in Chromatography*, Vol. 2, J. C. Giddings and R. A. Keller (Eds.), Marcel Dekker, New York, p. 271 (1966).
27. R. F. Brady Jr, *Adv. Carbohydr. Chem.* **26**, 197 (1971).
28. J. D. Wander and D. Horton, *Adv. Carbohydr. Chem.* **32**, 15 (1976).
29. D. C. DeJongh and K. Biemann, *J. Am. Chem. Soc.* **86**, 67 (1964).
30. H. Paulsen, H. Salzburg and H. Redlich, *Chem. Ber.* **109**, 3598 (1976).
31. J. Kuszmann, P. Sohár and G. Horváth, *Carbohydr. Res.* **50**, 45 (1976).
32. H. G. Jones, J. K. N. Jones and M. B. Perry, *Can. J. Chem.* **40**, 1559 (1962).

33. T. G. Bonner, D. Lewis and L. Yüceer, *Carbohydr. Res.* **49**, 119 (1976).
34. P. J. Garegg, L. Maron and C.-G. Swahn, *Acta Chem. Scand.* **26**, 518 (1972).
35. P. J. Garegg and C.-G. Swahn, *Acta Chem. Scand.* **26**, 3895 (1972).
36. R. Khan, *Carbohydr. Res.* **32**, 375 (1974).
37. E. J. Hedgley, D. Mérész, W. G. Overend and R. Rennie, *Chem. Ind.* (*London*) 938 (1960).
38. G. J. Robertson, *J. Chem. Soc.* 330 (1934).
39. N. A. Hughes and C. J. Wood, *Carbohydr. Res.* **49**, 225 (1976).
40. H. Arzoumanian, E. M. Acton and L. Goodman, *J. Am. Chem. Soc.* **86**, 74 (1964).
41. K. Heyns and J. Lenz, *Chem. Ber.* **94**, 348 (1961).
42. A. G. McInnes, N. H. Tattrie and M. Kates, *Am. Oil Chem. Soc*, **37**, 7 (1960).
43. D. J. Hanahan, J. Ekholm and C. M. Jackson, *Biochemistry* **2**, 630 (1963).
44. S. G. Batrakov, A. N. Ushakov and V. L. Sadovskaya, *Bioorganisheskaya Khim.* **2**, 1095 (1976).
45. O. D. Hensens, P. W. Khong and K. G. Lewis, *Aust. J. Chem.* **29**, 1549 (1976).
46. H. B. S. Conacher, *J. Chromatogr. Sci.* **14**, 405 (1976).
47. J. A. McCloskey and M. J. McClelland, *J. Am. Chem. Soc.* **87**, 5090 (1965).
48. R. E. Wolff, G. Wolff and J. A. McCloskey, *Tetrahedron* **22**, 3093 (1966).
49. D. R. Dalton, V. P. Dutta and D. C. Jones, *J. Am. Chem. Soc.* **90**, 5498 (1968).
50. B. M. Johnson and J. W. Taylor, *Anal. Chem.* **44**, 1438 (1972).
51. H. E. Simmons and D. W. Wiley, *J. Am. Chem. Soc.* **82**, 2288 (1960).
52. V. Dommes, F. Wirtz-Peitz and W.-H. Kunau, *J. Chromatogr. Sci.* **14**, 360 (1976).
53. M. N. Huffman, and M. H. Lott, *J. Biol. Chem.* **215**, 627 (1955).
54. J. C. Touchstone, M. Breckwoldt and T. Murawec, *J. Chromatogr.* **59**, 121 (1971).
55. T. Komeno, K. Tori and K. Takeda, *Tetrahedron* **21**, 1635 (1965).
56. M. A. Kirschner and H. M. Fales, *Anal. Chem.* **34**, 1548 (1962).
57. P. G. Simmonds, B. C. Pettitt and A. Zlatkis, *Anal. Chem.* **39**, 163 (1967).
58. A. I. Cohen, I. T. Harper and S. D. Levine, *J. Chem. Soc. Chem. Commun.* 1610 (1970).
59. H. T. Miles and H. M. Fales, *Anal. Chem.* **34**, 860 (1962).
60. E. J. Freyne, J. A. Lepoivre, F. C. Alderweireldt, M. J. O. Anteunis and A. De Bruyn, *J. Carbohydr. Nucleosides, Nucleotides* **3**, 113 (1976).
61. J. Casanova Jr and E. J. Corey, *Chem. Ind.* (*London*) 1664 (1961).
62. R. W. Kelly, *Tetrahedron Lett.* 967 (1969).
63. J. K. Findlay, L. Siekmann and H. Breuer, *Biochem. J.* **137**, 263 (1974).
64. R. W. Kelly, *J. Chromatogr.* **43**, 229 (1969).
65. M. Wieber and M. Schmidt, *Chem. Ber.* **96**, 1561 (1963).
66. R. C. Mehrotra and R. P. Narain, *Indian J. Chem.* **5**, 444 (1967).
67. J. Böeseken, *Adv. Carbohydr. Chem.* **4**, 189 (1949).
68. A. B. Foster, *Adv. Carbohydr. Chem.* **12**, 81 (1957).
69. E. J. Bourne, E. M. Lees and H. Weigel, *J. Chromatogr.* **11**, 253 (1963).
70. H. C. Brown and G. Zweifel, *J. Org. Chem.* **27**, 4708 (1962).
71. F. Eisenberg Jr, *Carbohydr. Res.* **19**, 135 (1971).
72. R. J. Ferrier, *J. Chem. Soc.* 2325 (1961).
73. R. J. Ferrier, D. Prasad, A. Rudowski and I. Sangster, *J. Chem. Soc.* 3330 (1964).
74. R. J. Ferrier and D. Prasad, *J. Chem. Soc.* 7429 (1965).
75. A. M. Yurkevich, I. I. Kolodkina, L. S. Varshavskaya, V. I. Borodulina-Shvetz, I. P. Rudakova and N. A. Preobrazhenski, *Tetrahedron* **25**, 477 (1969).
76. C. J. W. Brooks and D. J. Harvey, *Biochem. J.* **114**, 15P (1969).
77. C. J. W. Brooks and D. J. Harvey, *J. Chromatogr.* **54**, 193 (1971).
78. C. J. W. Brooks and J. Watson, *J. Chem. Soc. Chem. Commun.* 952 (1967).
79. A. Finch and J. C. Lockhart, *J. Chem. Soc.* 3723 (1962).
80. T. A. Baillie, C. J. W. Brooks and B. S. Middleditch, *Anal. Chem.* **44**, 30 (1972).
81. V. N. Reinhold, F. Wirtz-Peitz and K. Biemann, *Carbohydr. Res.* **37**, 203 (1974).

82. S. J. Gaskell and C. J. W. Brooks, *J. Chromatogr.* **122**, 415 (1976).
83. R. Köster and G. W. Rotermund, *Liebig's Ann. Chem.* **689**, 40 (1965).
84. S. J. Shaw, *Tetrahedron Lett.* 3033 (1968).
85. P. Biondi and M. Cagnasso, *Anal. Lett.* **9**, 507 (1976).
86. M. Cagnasso and P. A. Biondi, *Anal. Biochem.* **71**, 597 (1976).
87. S. J. Gaskell, C. G. Edmonds and C. J. W. Brooks, *Anal. Lett.* **9**, 325 (1976).
88. S. J. Gaskell, C. G. Edmonds and C. J. W. Brooks, *J. Chromatogr.* **126**, 591 (1976).
89. G. M. Anthony, C. J. W. Brooks, I. Maclean and I. Sangster, *J. Chromatogr. Sci.* **7**, 623 (1969).
90. C. J. W. Brooks and I. Maclean, *J. Chromatogr. Sci.* **9**, 18 (1971).
91. P. J. Wood and I. R. Siddiqui, *Carbohydr. Res.* **19**, 283 (1971).
92. P. J. Wood, I. R. Siddiqui and J. Weisz, *Carbohydr. Res.* **42**, 1 (1975).
93. C. Pace-Asciak and L. S. Wolfe, *J. Chromatogr.* **56**, 129 (1971).
94. R. W. Kelly, *Anal. Chem.* **45**, 2079 (1973).
95. R. W. Kelly and P. L. Taylor, *Anal. Chem.* **48**, 465 (1976).
96. C. Pace-Asciak, *J. Am. Chem. Soc.* **98**, 2348 (1976).
97. J. W. Brooks and J. Watson, in *Gas Chromatography 1968*, C. L. A. Harbourne and R. Stock (Eds.), Proc. 7th Int. Symp. Copenhagen, Institute of Petroleum, London (1969), p. 129.
98. M. L. Wolfrom and J. Solms, *J. Org. Chem.* **21**, 815 (1956).
99. R. J. Ferrier, D. Prasad and A. Rudowski, *J. Chem. Soc.* 858 (1965).
100. J. M. Sugihara and C. M. Bowman, *J. Am. Chem. Soc.* **80**, 2443 (1958).
101. M. Pailer and H. Huemer, *Monatsch. Chem.* **95**, 373 (1964).
102. J. J. Dolhun and J. L. Wiebers, *J. Am. Chem. Soc.* **91**, 7755 (1969).
103. R. Hemming and D. G. Johnston, *J. Chem. Soc.* 466 (1964).
104. M. Pailer and W. Fenzl, *Monatsch. Chem.* **92**, 1294 (1961).
105. R. L. Letsinger and S. B. Hamilton, *J. Org. Chem.* **25**, 592 (1960).
106. G. Kresze and F. Schäuffelhut, *Z. Anal. Chem.* **229**, 401 (1967).
107. H. G. Kuivila, A. H. Keough and E. J. Soboczenski, *J. Org. Chem.* **19**, 780 (1954).
108. R. L. Letsinger and S. B. Hamilton, *J. Am. Chem. Soc.* **80**, 5411 (1958).
109. E. J. Bourne, I. R. McKinley and H. Weigel, *Carbohydr. Res.* **25**, 516 (1972).
110. B. E. Stacey and B. Tierney, *Carbohydr. Res.* **49**, 129 (1976).
111. W. Blum and W. J. Richter, *Helv. Chim. Acta* **57**, 1744 (1974).
112. I. R. McKinley and H. Weigel, *J. Chem. Soc. Chem. Commun.* 1022 (1970).
113. R. A. Bowie and O. C. Musgrave, *J. Chem. Soc.* 3945 (1963).
114. N. R. Williams, *Adv. Carbohydr. Chem.* **25**, 109 (1970).
115. J. Defaye, *Adv. Carbohydr. Chem.* **25**, 181 (1970).
116. S. Soltzberg, *Adv. Carbohydr. Chem.* **25**, 229 (1970).
117. R. E. Parker and N. S. Isaacs, *Chem. Rev.* **59**, 737 (1959).
118. D. Swern, *Org. React.* **7**, 378 (1953).
119. D. Swern and G. B. Dickel, *J. Am. Chem. Soc.* **76**, 1957 (1954).
120. M. Nakajima, I. Tomida, N. Kurihara and S. Takei, *Chem. Ber.* **92**, 173 (1959).
121. D. A. Lightner and C. Djerassi, *Tetrahedron* **21**, 583 (1965).
122. L. M. McDonough and D. A. George, *J. Chromatogr. Sci.* **8**, 158 (1970).
123. E. A. Emken, *Lipids* **6**, 686 (1971).
124. E. A. Emken, *Lipids* **7**, 459 (1972).
125. R. T. Alpin and L. Coles, *J. Chem. Soc. Chem. Commun.* 858 (1967).
126. E. H. Pryde and J. C. Cowan, in *Topics in Lipid Chemistry*, Vol. 2, F. D. Gunstone (Ed.), Logos Press, London (1971) p. 1.
127. F. D. Gunstone and F. R. Jacobsberg, *Chem. Phys. Lipids* **9**, 26 (1972).
128. G. W. Kenner and E. Stenhagen, *Acta Chem. Scand.* **18**, 1551 (1964).
129. H. Audier, S. Bory, M. Fétizon, P. Longevialle and R. Toubiana, *Bull. Soc. Chim. Fr.* 3034 (1964).
130. E. A. Emken and H. J. Dutton, *Lipids* **9**, 272 (1974).

131. R. Mengel and H. Wiedner, *Chem. Ber.* **109**, 1395 (1976).
132. J. L. O'Donnell, L. E. Gast, J. C. Cowan, W. J. De Jarlais and G. E. McManis, *J. Am. Oil Chemists' Soc.* **44**, 652 (1974).
133. S. M. Osman and G. A. Qazi, *Fette Seifen Anstrichmittel* **78**, 326 (1976).
134. J. A. Fioriti, M. J. Kanuk and R. J. Sims, *J. Chromatogr. Sci.* **7**, 448 (1969).
135. C. R. Glowacki, P. J. Menardi and W. E. Link, *J. Am. Oil Chemists' Soc.* **47**, **(7)** 225 (1970).
136. J. Nemirowsky, *J. Prakt. Chem.* **28**, 439 (1883).
137. L. Hough, J. E. Priddle and R. S. Theobald, *Adv. Carbohydr. Chem.* **15**, 91 (1960).
138. A. S. Perlin, *Can. J. Chem.* **42**, 1365 (1964).
139. P. Brown and C. Djerassi, *Tetrahedron* **24**, 2949 (1968).
140. W. W. Christie and R. T. Holman, *Lipids* **1**, 176 (1966).
141. F. A. Isherwood and R. L. Jones, *Nature* (*London*) **175**, 419 (1955).
142. K. Schwarz, *Arch. Biochem. Biophys.* **92**, 168 (1961).
143. D. C. Morrison, *J. Am. Chem. Soc.* **76**, 4483 (1954).
144. U. Langenbeck, H. U. Möhring and K.-P. Dieckmann, *J. Chromatogr.* **115**, 65 (1975).
145. A. Frigerio, P. Martelli, K. M. Baker, and P. A. Biondi, *J. Chromatogr.* **81**, 139 (1973).
146. N. E. Hoffman and T. A. Killinger, *Anal. Chem.* **41**, 162 (1969).
147. K. H. Nielsen, *J. Chromatogr.* **10**, 463 (1963).
148. D. J. D. Hockenhull and G. D. Floodgate, *Biochem. J.* **52**, 38 (1952).
149. K. W. Taylor and M. J. H. Smith, *Analyst* **80**, 607 (1955).
150. H. F. Schott, J. B. Larkin, L. B. Rockland and M. S. Dunn, *J. Org. Chem.* **12**, 490 (1947).
151. F. Weygand, A. Prox. E. C. Jorgensen, R. Axén and P. Kirchner, *Z. Naturforsch. Teil B* **18**, 93 (1963).
152. K. Blau, in *Biomedical Applications of Gas Chromatography*, Vol. 2, H. A. Szymanski (Ed.), Plenum, New York (1967), p. 1.
153. J. P. Greenstein and M. Winitz, *Chemistry of the Amino Acids*, Vol. 2, Wiley, New York (1961), p. 793
154. D. E. Nitecki, B. Halpern and J. W. Westley, *J. Org. Chem.* **33**, 864 (1968).
155. K. D. Kopple and H. G. Ghazarian, *J. Org. Chem.* **33**, 862 (1968).
156. J. W. Westley, V. A. Close, D. N. Nitecki and B. Halpern, *Anal. Chem.* **40**, 1888 (1968).
157. S. Tamura, A. Suzuki, Y. Aoki and N. Otake, *Agr. Biol. Chem. Tokyo* **28**, 650 (1964).
158. G. P. Slater, *J. Chromatogr.* **64**, 166 (1972).
159. A. B. Mauger, *J. Chromatogr.* **37**, 315 (1968).
160. E. Fisher, *Dt. Chem. Ges.* **39**, 2893 (1906).
161. G. P. Slater, *Chem. Ind.* (*London*) 1092 (1969).
162. F. Weygand, and U. Glöckler, *Chem. Ber.* **89**, 653 (1956).
163. F. Weygand, W. Steglish and H. Tanner, *Liebig's Ann. Chem.* **658**, 128 (1962).
164. I. Z. Siemion and K. Nowak, *Rocz. Chem.* **34**, 1479 (1960).
165. F. Weygand, W. Steglich, D. Mayer and W. von Philipsborn, *Chem. Ber.* **97**, 2023 (1964).
166. O. Grahl-Nielsen and E. Solheim, *J. Chem. Soc. Chem. Commun.* 1092 (1972).
167. O. Grahl-Nielsen and E. Solheim, *J. Chromatogr.* **69**, 366 (1972).
168. O. Grahl-Nielsen and B. Møvik, *Biochem. Med.* **12**, 143 (1975).
169. F. Weygand, A. Prox, L. Schmidhammer and W. König, *Angew. Chem. Int. Ed. Eng.* **2**, 183 (1963).
170. S. Götze and W. Steglich, *Chem. Ber.* **109**, 2335 (1976).
171. F. Weygand, *Z. Anal. Chem.* **205**, 406 (1964).
172. F. Weygand, K. Burger and K. Engelhardt, *Chem. Ber.* **99**, 1461 (1966).
173. P. Hušek, *J. Chromatogr.* **91**, 475 (1974).
174. P. Hušek, *J. Chromatogr.* **91**, 483 (1974).
175. E. Honkanen, T. Moisio and P. Karvonen, *Acta Chem. Scand.* **23**, 531 (1969).
176. E. Jellum, V. A. Bacon, W. Patton, W. Pereira Jr and B. Halpern, *Anal. Biochem.* **31**, 339 (1969).
177. J. G. Pritchard and R. L. Vollmer, *J. Org. Chem.* **28**, 1545 (1963).

178. P. C. Lauterbur, J. G. Pritchard and R. L. Vollmer, *J. Chem. Soc.* 5307 (1963).
179. P. Frøyen and J. Møller, *Acta Chem. Scand. Part B* **29**, 61 (1975).
180. R. A. Chalmers and R. W. E. Watts, *Analyst* **97**, 951 (1972).
181. D. J. Brown, *The Pyrimidines*, Interscience, New York, (1962), p. 32.
182. M. Hamberg, W. G. Niehaus Jr and B. Samuelsson, *Anal. Biochem.* **22**, 145 (1968).
183. T. P. King, *Biochemistry* **5**, 3454 (1966).
184. S. Coussement and M. Renard III, *Chromatography and Methods of Immediate Separation*, Association of Greek Chemists, Athens (1966), p. 151.
185. A. Sughi and N. Ogawe, *Chem. Pharm. Bull.* **24**, 1349 (1976).
186. P. Edman, *Acta Chem. Scand.* **4**, 283 (1950).
187. P. Edman, *Acta Chem. Scand.* **10**, 761 (1956).
188. P. Edman and G. Begg, *Eur. J. Biochem.* **1**, 80 (1967).
189. R. B. Meagher, *Anal. Biochem.* **67**, 404 (1975).
190. R. A. Laursen, *J. Am. Chem. Soc.* **88**, 5344 (1966).
191. A. Dijkstra, H. A. Billiet, A. H. Van Doninck, H. Van Velthuyzen, L. Maat and H. C. Beyerman, *Recl. Trav. Chim. Pays-Bas* **86**, 65 (1967).
192. G. Manecke and G. Günzel, *Naturwiss.* **55**, 1 (1968).
193. L. M. Dowling and G. R. Stark, *Biochemistry* **8**, 4728 (1969).
194. J. Silver and L. Hood, *Anal. Biochem.* **67**, 392 (1975).
195. C. Largman and G. Moss, *Anal. Biochem.* **69**, 70 (1975).
196. G. J. M. Van der Kerk, C. W. Pluygers and G. de Vries, *Org. Synth.* **45**, 19 (1965).
197. A. Murai and Y. Takeuchi, *Bull. Chem. Soc. Jpn* **48**, 2911 (1975).
198. F. Weygand and R. Obermeier, *Eur. J. Biochem.* **20**, 72 (1971).
199. H. Tschesche and E. Wachter, *Eur. J. Biochem.* **16**, 187 (1970).
200. C. L. Zimmerman, E. Appella and J. J. Pisano, *Anal. Biochem.* **75**, 77 (1976).
201. B. Wittmann-Liebold, Hoppe-Seyler's *Z. Physiol. Chem.* **354**, 1415 (1973).
202. B. Wittmann-Liebold, H. Graffunder and H. Kohls, *Anal. Biochem.* **75**, 621 (1976).
203. J. J. Pisano, T. J. Bronzert and H. B. Brewer, *Anal. Biochem.* **45**, 43 (1972).
204. T. Fairwell, S. Ellis and R. E. Lovins, *Anal. Biochem.* **53**, 115 (1973).
205. V. M. Stepanov and V. F. Krivstov, *J. Gen. Chem. USSR* **35**, 53, 556, 988 (1965).
206. S. Ellis, T. Fairwell and R. E. Lovins, *Biochem. Biophys. Res. Commun.* **49**, 1407 (1972).
207. P. Schlack and W. Kumpf, Hoppe-Seyler's *Z. Physiol. Chem.* **154**, 125 (1926).
208. S. G. Waley and J. Watson, *J. Chem. Soc.* 2394 (1951).
209. V. H. Baptist and H. B. Bull, *J. Am. Chem. Soc.* **75**, 1727 (1953).
210. S. W. Fox, T. L. Hurst, J. F. Griffith and O. Underwood, *J. Am. Chem. Soc.* **77**, 3119 (1955).
211. G. R. Stark, *Biochemistry* **7**, 1796 (1968).
212. L. D. Cromwell and G. R. Stark, *Biochemistry* **8**, 4735 (1969).
213. A. Darbre and M. Rangarajan, in *Solid-Phase Methods in Protein Sequence Analysis*, R. A. Laurson (Ed.), Proc. 1st Int. Conf., Pierce Chemical Co., Rockford, Ill. (1975), p. 131.
214. M. Rangarajan and A. Darbre, *Biochem. J.* **157**, 307 (1976).
215. M. Williams and B. Kassell, *FEBS Lett.* **54**, 353 (1975).
216. S. Yamashita, *Biochem. Biophys. Acta* **229**, 301 (1971).
217. S. Yamashita and N. Ishikawa, in *Chemistry and Biology of Peptides*, J. Meinhofer (Ed.), 3rd Proc. American Peptide Symp., Ann Arbor Science Publishers, Ann Arbor, Ill. (1972), p. 701.
218. M. Rangarajan and A. Darbre, *Biochem. J.* **147**, 435 (1975).
219. M. Rangarajan, R. E. Ardrey and A. Darbre, *J. Chromatogr.* **87**, 499 (1973).
220. T. Suzuki, S. Matsui and K. Tuzimura, *Agr. Biol. Chem. Tokyo* **36**, 1061 (1972).
221. G. E. Tarr in *Solid-Phase Methods in Protein Sequence Analysis*, R. A. Laursen (Ed.), Proc. 1st Int. Conf., Pierce Chemical Co., Rockford, Ill. (1975), p. 139.
222. G. R. Stark and D. G. Smyth, *J. Biol. Chem.* **238**, 214 (1963).
223. C. H. W. Hirs, S. Moore and W. H. Stein, *J. Biol. Chem.* **219**, 623 (1956).
224. A. Estas and A. Dumont, *J. Chromatogr.* **82**, 307 (1973).

225. H. S. Hill, A. C. Hill and H. Hibbert, *J. Am. Chem. Soc.* **50**, 2242 (1928).
226. R. W. Kelley, *Steroids* **13**, 507 (1969).
227. K. Gréen, *Chem. Phys. Lipids* **3**, 254 (1969).
228. F. Vane and M. G. Horning, *Anal. Lett.* **2**, 357 (1969).
229. T. Kimira, L. A. Sternson and T. Higuchi, *Clin. Chem.* **22**, 1639 (1976).
230. D. E. Nitecki and J. W. Goodman, *Biochemistry* **5**, 665 (1966).
231. A. Darbre, *Lab. Pract.* **20**, 726 (1971).
232. P. K. Porter, *J. Am. Chem. Soc.* **45**, 1086 (1923).
233. P. Edman, *Acta Chem. Scand.* **4**, 277 (1950).
234. P. Edman and K. Lauber, *Acta Chem. Scand.* **10**, 466 (1956).
235. V. M. Ingram, *J. Chem. Soc.* 3717 (1953).
236. H. C. Beyerman, L. Maat, A. Sinnema and A. Van Veen, *Recl. Trav. Chim. Pays-Bas* **87**, 11 (1968).
237. H. Fraenkel-Conrat, J. I. Harris and A. L. Levy, in *Methods of Biochemical Analysis*, Vol. 2, D. Glick (Ed.), Interscience, New York (1955), p. 388.
238. D. E. Vance and D. S. Feingold, *Anal. Biochem.* **36**, 30 (1970).
239. M. A. Young and D. M. Desiderio, *Anal. Lett.* **8**, 1 (1975).
240. D. Waldi, in *Thin-layer Chromatography*, E. Stahl (Ed.), Springer-Verlag, Berlin (1962), p. 496.
241. D. S. Robinson, J. Eagles and R. Self, *Carbohydr. Res*, **26**, 204 (1973).
242. I. R. McKinley and H. Weigel, *Carbohydr. Res.* **31**, 17 (1973).
243. E. J. Bourne, I. R. McKinley and H. Weigel, *Carbohydr. Res.* **35**, 141 (1974).
244. H. Komura, T. Yoshino and Y. Ishido, *Carbohydr. Res.* **40**, 391 (1975).
245. F. Berheim, M. L. C. Berheim and K. M. Wilbur, *J. Biol. Chem.* **174**, 257 (1948).
246. W. M. Lamkin, N. S. Jones, T. Pan and D. N. Ward, *Anal. Biochem.* **58**, 549 (1974).

CHAPTER 8

Microreactions

May N. Inscoe

Organic Chemicals Synthesis Laboratory, Agricultural Environmental Quality Institute, U.S. Department of Agriculture, Research Center East, Beltsville, Maryland 20705, U.S.A.

and

Graham S. King and Karl Blau

Bernhard Baron Memorial Research Laboratories, Queen Charlotte's Maternity Hospital, Goldhawk Road, London W6 0XG.

1 INTRODUCTION

The term 'microreaction' could, strictly speaking, be applied to any chemical reaction or derivatization carried out on a small enough scale, and indeed this handbook is full of examples. In this chapter the term 'microreaction' will be restricted to such microscale reactions as are sufficiently vigorous to cause some fairly extensive rearrangement or cleavage of the reacting substance or substances, as opposed to the usual types of derivatization, which simply modify protic functions such as hydroxyl or amino. Such a vigorous reaction may have several advantages; for example it may give structural information via the characterization of the carbon skeleton of the product. Microreactions may be designed to remove specific groupings and so lead to products which are more volatile, or have better chromatographic properties, or are adapted to certain modes of detection.

These reactions are scaled down to the micro or submicro level not only for greater ease and convenience, but also to enable reagents to be used in vast excess in order to achieve complete reaction. Apart from this, many reactions may be violent or difficult to control on a large scale, but can be performed safely on a microscale because the amount of heat generated can be readily absorbed or dissipated.

There are many analytical problems in which amounts of sample material are severely limited. For example, it may be necessary to analyse for a few nanograms of drugs,[1,2] biologically important acids,[3] or prostaglandins[4–6] in biological fluids, gibberelins in plants,[7] or maleic hydrazide (1,2-dihydro-3,6-pyrazinedione) in cigarette smoke;[8] or the extract from 78 000 insects may have yielded no more than a few micrograms of active pheromone with which to make a structural identification.[9] Such qualitative and quantitative analyses of biologically active materials, drug levels and pesticide residues and structural determinations of natural products are but some of the areas where microchemical reactions for chromatography are essential. Confirmation of identity in residue analyses and classification by functional group are other important applications. Microchemistry in conjunction with chromatography is being applied with increasing frequency in many areas to achieve economy of time and materials. Almost any chemical reaction could be adapted into a microprocedure for structural analysis. It is not within the scope of this chapter to cover the whole field of organic microchemistry but the aim is to point out a few useful techniques and reactions which will give an idea of the range of analytical methods which may be applied.

Many of the methods currently in use were devised to solve specific problems, particularly those encountered in working with biological materials. Accordingly, descriptions of micro methods may be found widely scattered throughout the scientific literature in publications of natural product chemistry, clinical chemistry, and biochemistry. Other papers may be found in literature dealing with specific subjects such as amino acids and proteins, carbohydrates, lipids and steroids, or the more general ones concerned with analytical chemistry and chromatography. While it is difficult to define what should be considered a microreaction, in this chapter it is taken as a reaction requiring no more than 10 mg of sample; many of the procedures require only a few nanograms or even picograms.

Even though many common chemical reactions can be used for micro work, it must be emphasized that a microprocedure is not merely a scaled-down version of a macro method. When only a few milligrams or micrograms of sample are available, it is crucial to avoid loss of material and prevent contamination; the entire procedure must be devised with this in mind. Since losses or contamination arise particularly during sample transfers, these must be kept to a minimum. In general the precautions discussed in Chapter 1 are most important. With any microprocedure it is important to perfect the reaction technique and the required micromanipulations with known substances before using valuable experimental samples. Even the most straightforward reaction may occasionally lead to unanticipated problems. It is usually important to run blanks and controls. Ideally, reactions used for microderivatization will be completed rapidly, and will form stable products, with no side-reactions and no subsequent interference from excess reagents or other reaction products. Since such ideal reactions are rare, it is often necessary to devise methods to achieve reproducible results even when these conditions are not met.

With some reactions it is possible to take advantage of the separations achieved in the chromatographic process and apply the reaction mixture directly onto the column or TLC plate. Elimination of solvent by using the reagent itself as the solvent is sometimes possible, as in some silylation and acyl imidazole reactions. In a gas chromatographic procedure for the determination of maleic hydrazide residues in tobacco, bis(trimethylsilyl)acetamide (BSA) serves as both the extracting and also the derivatization agent;[10] better results were obtained by direct treatment of tobacco powder with the BSA followed by injection of the reaction mixture than were obtained with the usual extraction methods. A procedure called 'derivative formation in a solid matrix' which has been used for confirmation of pesticide residue identities, is simple and is reported to give high yields.[11–13] The sample in a small amount of solvent is introduced onto a microcolumn packed with alumina inpregnated with an appropriate reagent (sulfuric acid, acetic anhydride, potassium hydroxide, etc.). After reaction, with heating if necessary, the products are eluted with a suitable solvent and analysed by GLC.

Bitner and co-workers[14] have described an integrated system for pre-injection microreactions and have demonstrated its value for transesterification and ozonolysis reactions. An electronically controlled reactor unit is connected to a pyrolysis 'needle-gun' assembly: the needle may be inserted into a GLC injection port. Sample and reactants are injected into the system, heated and then vaporized into the GLC in a stream of helium; the reactor may also be filled with ozone and the ozonolysis products examined. Zinc oxide granules were used to trap carboxylic acids formed in the pyrolysis.

We will now consider two forms of chemical modification—classical microchemical reactions prior to analysis by gas chromatography and reaction chromatography.

2 REACTION CHROMATOGRAPHY

2.1 Reaction gas chromatography (RGC)

The term 'reaction gas chromatography' (RGC) is used to describe the occurrence of a reaction during the chromatographic process where the eluted compounds differ from those in the original sample before injection. The reaction may occur pre-column, on-column, or post-column, although post-column reactions are not as important from the point of view of derivatizations and are not within the scope of this book. A reaction that occurs before sample injection is not an RGC reaction, but one taking place in a pre-column reactor that is part of the chromatographic system is an RGC process, even when the products are held up in a trap to allow the reaction to go to completion before introduction on to the column. There are inevitably some techniques which fall into the 'grey' area between these two categories. In this Handbook we can discuss only briefly the subject of RGC: in many ways it might be more appropriately considered in a publication on GLC techniques. A number of reviews have appeared on RGC[21–23] and the reader is also recommended to refer to the book by Ma and Ladas.[15]

As with derivatization reactions, the purpose of RGC may be to make volatile products from less volatile or non-volatile materials (e.g. methyl esters from salts of carboxylic acids), to make volatile materials non-volatile (subtractive GLC), to change the properties of the material to improve chromatographic separations (e.g. esters from acids) or to derive structural information (peak-shifting techniques). The reagents may be contained in a pre-column reactor, or they may be mixed with the sample before injection, injected immediately after the sample, or be in the carrier gas. Reactions involving hydrogen and a catalyst are the most common in RGC.[16,17] Hydrogenation of the double bonds in a compound yields the saturated analogue, which is often easier to identify or quantify. On-line catalytic hydrogenation may be used for unsaturated functions in a wide variety of compounds. The normal technique is to pack a small (less than 1 cm) length of column with a good palladium or platinum catalyst and use hydrogen as a carrier gas; the injected compounds are rapidly hydrogenated during their passage over the catalyst. Other catalysts have been used for specific reductions (Table 1); these include zinc dust and metal hydrides. A platinum black catalyst has also been used for pyrolytic dehydrogenation in a pre-column reactor. A detailed discussion of the GC techniques involved is given by Beroza and Sarmiento[16] or Ma and Ladas.[15] At higher reaction temperatures hydrogenolysis may occur.[18,19] This means that during passage through the on-line reactor substituents are removed and the product is generally the saturated carbon skeleton of the original compound. However, when an oxygen function is present at an end carbon atom (e.g. carboxylic acid, aldehyde, primary alcohol) the end carbon may also be removed. Paradoxically, at high temperatures some aromatization of cyclic compounds by dehydrogenation may occur. Carbon skeleton chromatography is particularly valuable when used in conjunction with mass spectrometry,[20] to establish product structures.

Siggia and co-workers[34,37,39] have developed a pre-column fusion technique, suitable for a variety of reactions. The apparatus is a small pyrolysis unit with a stainless steel loop between the heated zone and column. The sample is introduced into the pyrolyser as a mixture with the fusion reagent, the mixture is heated and the product vapours condense in the loop which is cooled in liquid nitrogen. The next stage is to remove the sample container and flash-heat the loop to drive the products into the GC column. Some pre-column reactions can be performed in simple straight tubes containing a suitable reagent: for example a stainless steel tube packed with a 1:9 mixture of powdered potassium hydroxide and 60–80 mesh glass beads has been used for pre-column hydrolytic cleavage of carbamates to alcohols, as shown in the following equation:[62]

$$R^1NHCOOR^2 + KOH \rightarrow R^2OH + R^1NHCOOK$$

TABLE 1
Examples of reactions used in conjunction with gas chromatography

Reagent(s)	Reaction	References
A. Examples of Reaction Gas Chromatography		
H_2 gas and catalyst (platinum/palladium type)	Hydrogenation of double bonds	16, 17, 24–30
	Hydrogenation (carbon skeleton)	18, 19, 28, 31
Zn dust and H_2 gas	Hydrogenolysis to parent hydrocarbons	32
Lithium borohydride	Reduction of carbonyl compounds	24–26
Carbohydrazide	Reduction of azo, nitro, and sulfonate compounds by pre-column fusion	34
Alkali (NaOH, KOH, or soda-lime)	Saponification of esters	25, 26, 38
	Pre-column fusion reactor	37
	Formation of amines or ammonia from amides, nitriles, or urea compounds by pre-column fusion.	39
	Conversion of arylsulfonic acids to phenols.	40
Basic compounds (NH_3 or organic amines)	Release of free bases from nitrogenous drugs injected as salts	41
Phosphoric acid	Conversion to amides to nitriles	58, 59
B. Examples of Subtraction Techniques		
Lithium aluminium hydride	Subtraction of oxygen-containing compounds (alcohols, aldehydes, ketones, esters, epoxides)	24, 33
Zinc oxide	Subtraction of carboxylic acids, partial subtraction of alcohols and phenols	35, 36
o-Dianisidine	Subtraction of aldehydes	35
Benzidine	Subtraction of aldehydes, ketones, epoxides	35
Hydroxylamine	Subtraction of carbonyl compounds	42
Phosphoric acid	Subtraction of epoxides	35
	Subtraction of bases	57
Boric acid	Subtraction of alcohols (1° and 2°), 3° alcohols not affected	24, 33, 35
C. Co-injection or 'In-Syringe Reaction' Methods		
Tetramethylammonium hydroxide, tetrabutyl-ammonium hydroxide, trimethylphenylammonium hydroxide	On-column alkylation of acids amides, amino acids, barbiturates, etc., by injection of solution of quaternary salt (Chapter 5)	43–53

TABLE 1—continued

Reagent(s)	Reaction	References
MeOH + NaOH	On-column 'transesterification' of carbamates (Chapter 2)	54
Methyl iodide and silver oxide	Methylation of acids and alcohols (Chapter 5)	55
Diazomethane	Methylation of acids (Chapter 5)	26, 56
α-Ketoglutaric acid	Regeneration of carbonyl compounds from dinitrophenylhydrazones	60, 61
Trimethylsilylation	Formation of TMS derivatives of acids, alcohols, phosphates, etc. (Chapter 4)	63, 64

One of the commonest RGC techniques is component subtraction. This simply involves the introduction of a subtraction reagent into the top of the GLC column or the insertion of a subtraction loop (Fig. 1). As an example a small plug of zinc oxide will subtract acids from the injected sample and boric acid will remove primary and secondary alcohols; other examples are given in Table 1.

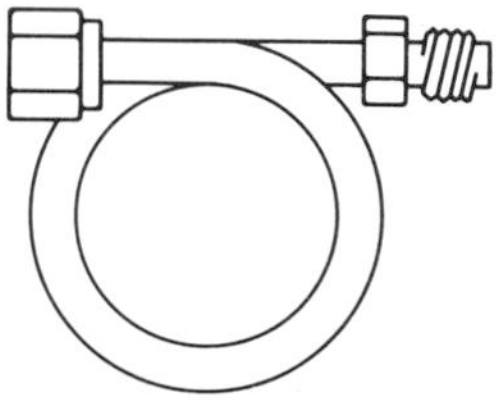

Fig. 1 A 15 cm subtraction loop (Bierl, Beroza and Ashton, Ref. 35. Reproduced by kind permission of *Mikrochimica Acta*—Springer Verlag).

2.1.1 SUBTRACTION OF EPOXIDES BY TREATMENT WITH PHOSPHORIC ACID (GLC)[35]

Phosphoric acid reacts with epoxides to form glycols, which are more polar and are therefore less volatile than the starting compound. Subtraction of epoxides is accomplished by means of a short 15 cm length of column tubing (Fig. 1) containing 100–200 mg of column packing treated with 5% by weight of 85% phosphoric acid. The phosphoric acid is deposited on the support by evaporation from an aqueous solution. The rest of the phosphoric acid loop is filled with the packing used in the analytical column, and a similar 'blank' loop for reference is filled with the same packing. The packing is held in place in the normal manner using plugs of glass wool. Although most reaction loops can be mounted either before or after the analytical column, it is probably preferable to place the phosphoric acid loop after the column to avoid contamination of the column with the acid or the reaction products. The compound to be examined is dissolved in a suitable solvent, such as hexane. The sample size injected is about 60 μg, but amounts ranging from 1 μg to 1 mg have been used. The resulting chromatogram is compared with the reference chromatogram obtained when the 'blank' loop is used. The phosphoric acid loop subtracts more than 95% of the epoxide components from the carrier gas stream; subtraction of other compounds is generally less than 30% except with a few reactive ethers such as methyl benzyl ether or 1,2-dimethoxyethane. Higher concentrations of acid also subtract methyl ketones and methyl ethers.

2.2 Reaction thin layer chromatography (RTLC)

Reaction thin layer chromatography (RTLC) is useful chiefly because of its simplicity and rapidity. Many of the reactions used for the detection of developed spots have also been successfully used as RTLC reactions before development. Occasionally the reaction is performed after the first development in a two stage elution process. Generally the sample is applied to the plate and then over-spotted or sprayed with the reagent. The reagent can also be applied to the plate as a band through which the material must pass during development, this has been called 'elatography'.[65] The reagent can also be included in the elution solvent itself (e.g. bromine[66]). It is sometimes possible to incorporate the reagent into the thin layer and an example of this is the use of zinc with nitro compounds;[67] a distinction should be made between such a procedure, in which a derivative is formed, and procedures which involve impregnation of the layer with a complexing reagent to improve the separation characteristics of the layer (e.g. boric acid or silver nitrate). Examples of reactions that have been used for RTLC are given in Table 2; the subject was reviewed by Beroza.[23] A particularly interesting technique has been described by Tumlinson and co-workers[91] and this is presented in Section 2.2.1.

2.2.1 RTLC OF KETONES ELUTED FROM A GAS CHROMATOGRAPH[91]

It is possible to form carbonyl derivatives such as 2,4-dinitrophenylhydrazones on a TLC plate as the compounds emerge from the exit port of a gas chromatograph. In this procedure, the plate is positioned so that the starting line of the plate is under the column exit and the plate is close but not touching. A drop (about 10 μl) of 2,4-dinitrophenylhydrazine (2,4-DNP) reagent is spotted onto the starting line of the plate just before elution of a GC peak, and the emerging material is allowed to impinge on the moist spot. Additional reagent is added to the spot as needed to keep it moist while elution continues. Several components of a mixture are treated on a single plate by repositioning the plate after each peak has emerged and repeating the process. The plate is then developed without further treatment. With this procedure

TABLE 2

Examples of reactions in thin layer chromatography

Reagent(s)	Reaction	References
H_2 and catalyst	Hydrogenation	68, 69
Lithium aluminium hydride	Reduction of carbonyl compounds and esters	70
Sodium borohydride	Reduction of carbonyl compounds	68, 71–73
Zinc and HCl	Reduction of nitro compounds	67, 74
Stannous chloride	Reduction of nitro compounds	75
Titanium trichloride	Reduction of nitrosamines or 1,3-dihydro-7-nitro-2-1,4-benzodiazepin-2-one and metabolites	76, 77
Halogens	Halogenation of organic compounds	66, 68, 69, 78–83
HCl	Hydrolysis of esters, plasmalogens, glycosides etc.	68, 84–88
Nitric acid	Nitration	75
Nitrous acid or nitric oxide	Diazotization	75, 89
Phosphoric acid	Subtraction of epoxides	90
Acetic anhydride or acetyl chloride	Acetylation (Chapter 3)	68, 71
Chromic oxide	Oxidation of alcohols	68, 70, 71
2,4-Dinitrophenyl-hydrazine	Carbonyl derivatization (Chapter 6)	68, 71, 91–95
Fluorodinitrobenzene	Amino acid protection (Chapter 10)	96, 97
Dansyl chloride	Amine derivatization (Chapter 9)	98
Potassium hydroxide	Saponification	68, 70, 71
Sodium methoxide	Transesterification (Chapter 2)	99, 100

50 μg or less of a compound can be derivatized. The 2,4-DNP reagent is prepared by dissolving 50 g 2,4-dinitrophenylhydrazine in 600 ml of 85% phosphoric acid, diluting with 395 ml 95% ethanol and filtering.

Another interesting technique using RTLC has been described by Bierl and co-workers.[90] This is subtraction of epoxides with phosphoric acid and is described below.

2.2.2 SUBTRACTION OF EPOXIDES WITH PHOSPHORIC ACID (TLC)[90]

Precoated silica gel plates are washed with diethyl ether and 5 μl of 1 M phosphoric acid is placed at each spot to be used for epoxide subtraction, with an adjacent untreated area left beside each spot to allow comparison. After a few minutes, another 5 μl aliquot of the phosphoric acid is added to each original spot and the plate allowed to dry for 20 min. The sample compound (20–100 μg) is spotted onto the phosphoric acid, with a duplicate sample on the adjacent blank space. After a reaction time of one hour the plate is developed with the appropriate eluent. The reaction products (chiefly diols) remain at or near the origin and the original compound is 'subtracted' and does not appear at its associated R_f position. By limiting the reaction time to five minutes it is often possible to determine the epoxide configuration. Unhindered (*cis*) epoxides react rapidly and are subtracted almost completely, whereas hindered (*trans*) epoxides react more slowly and are subtracted only partially with this shorter reaction time.

2.3 Reaction liquid chromatography (RLC)

Derivatization reactions have not been used as an integral part of liquid chromatographic procedures to any great extent, unless one includes ion-exchange columns or columns with optically active substrates for resolution of racemic mixtures. There are several examples of the use of a column or microcolumn as a reactor (e.g. for methylation of acids with diazomethane[101] or reduction of carbonyl compounds with sodium borohydride[102]) without combining the reaction with a liquid chromatographic separation. When reactant-containing columns (e.g. periodic acid for cleavage of glycols or epoxides,[103] dinitrophenylhydrazine[103] or semicarbazide[104] for carbonyl derivatization) are used in combination with a chromatographic column, the procedure may be regarded as being reaction chromatography. With the development of high-performance liquid chromatographic techniques, new applications in this area may be expected, especially in conjunction with fluorescence detectors.

3 MICROCHEMICAL REACTIONS PRIOR TO CHROMATOGRAPHY

3.1 Reactions involving carbon to carbon multiple bonds

Location of the position of one or more double bonds in a molecule is a frequent analytical problem, especially in the field of lipid and fatty acid analysis and petrochemical gas chromatography. If we consider the case of the two monounsaturated C-18 fatty acids in Fig. 2, there is a fairly obvious problem of identification. The two isomers will be difficult to separate unless a very high resolution capillary column is used, and if they are separated chromatographically there is still the problem of identification. Unless authentic standards are available this is almost impossible, even using powerful techniques such as GC–MS, because their mass spectra will be very similar. To solve this type of problem it is necessary to take advantage of the relative lability of the double bond to oxidation, and to examine the products of such chemical reactions by suitable techniques.

Oxidative cleavage of a double bond may be achieved in a number of ways, most of which involve 1,2-diol intermediates. An exception is ozonolysis which is probably one of the most widely used analytical methods for such cleavages.

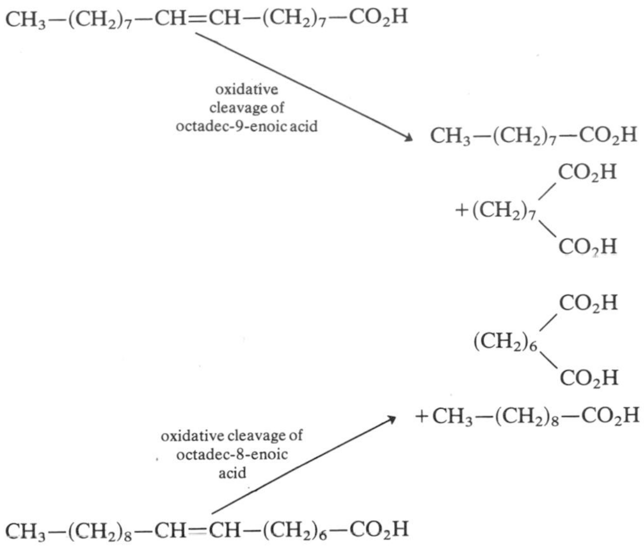

Fig. 2 Oxidative cleavage of unsaturated C-18 acids.

The intermediate ozonide which is formed by the addition of one or more molecules of ozone to the double bond has not been well defined and several structures have been proposed.[105,106] Ozonides are stable only at very low temperatures and decompose readily into carbonyl compounds and hydroperoxides; to convert an alkene function into two components (as illustrated in Fig. 3) an ozone generator is used to produce ozone which is passed into the compound in a suitable solvent at a low temperature. A reducing agent is then introduced to convert the ozonolysis products into carbonyl compounds. Beroza and Bierl[107] have described a micro-ozonolysis unit which is illustrated in Fig. 4. Commercial versions of this apparatus are now available.

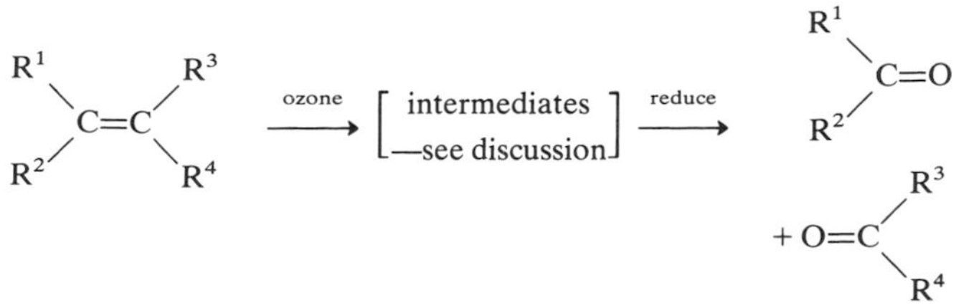

Fig. 3 Ozonolysis of an alkene.

3.1.1 MICRO-OZONOLYSIS FOR LOCATION OF UNSATURATION USING ANALYSIS BY GAS–LIQUID CHROMATOGRAPHY.[108]

In this procedure, frequently used in pheromone identification studies, triphenylphosphine is the reducing agent, and the resulting carbonyl compounds are determined by gas chromatography. Ozone is generated by passing

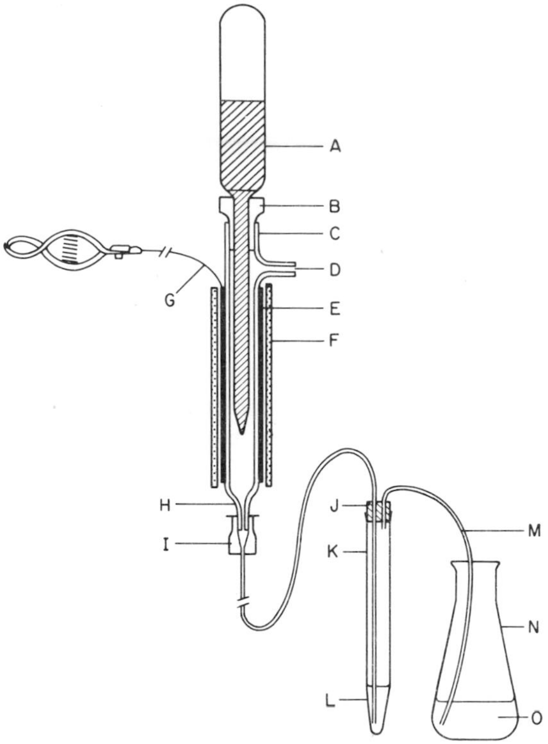

Fig. 4 A cross-sectional view of the micro ozone generator (Beroza and Bierl, Ref. 107. Reproduced by kind permission of *Mikrochimica Acta*—Springer Verlag). A—Metal tip of a Tesla® coil, machined to 3 mm outside diameter, this is the inner electrode of the ozonizer; B—Teflon® adapter; C—Glass tube with a side-arm inlet (D) for oxygen and nitrogen inlet; E—Outer electrode of aluminium foil; F—Rubber tubing insulation; G—Copper wire (20 gauge) for connection to ground; H, I—Luer fitting for Teflon® tubing which is connected to . . . K—A reaction vessel containing . . . L—The sample dissolved in a solvent; N, O—Solution to absorb excess ozone.

oxygen at 10 ml min^{-1} through the micro-ozonizer shown in Fig. 4. The reaction vessel is a tapered tube, made by sealing the small end of a medicine dropper. After passing through the reaction solution, the gases are passed into a few millilitres of indicating solution (5% potassium iodide in 5% aqueous sulfuric acid, with a little added starch) to signal completion of ozonization by the blue colour produced with excess ozone. The sample (1–25 μg) in 100 μl of solvent is placed in the reaction-tube and cooled to −70 °C in a xylene/solid carbon dioxide bath. The ozone-oxygen mixture from the ozonizer is bubbled through the solution until the indicator turns blue (10–15 s). The reaction mixture is then purged with nitrogen for 15 s to remove reagent gases. Powdered triphenylphosphine (about 1 mg) is dropped into the solution, the tube is stoppered, agitated to dissolve the powder and allowed to warm up to room temperature. A 2 μl aliquot of the reaction solution is then injected into the gas chromatograph for the analysis or extracted in a suitable solvent system, derivatized with suitable carbonyl group reagents (Chapter 6) and then analysed.

Carbon disulfide is a suitable solvent for most ozonolyses, but peaks for carbonyl fragments smaller than valeraldehyde are usually obscured by the solvent peak when this is used, and the products are chromatographed directly. For the smaller fragments (C_2 to C_5), redistilled pentyl acetate is more satisfactory.

Ozonolysis is probably the most widely used oxidative cleavage method, closely followed by permanganate and peroxide oxidations.[109–111,119] Under vigorous conditions the intermediate diol is split into two acids using permanganate, but milder conditions allow isolation of the 1,2-diol which has the *cis* configuration. It is probably wiser to use specific reagents for these reactions rather than to rely upon mild or vigorous conditions to achieve a particular transformation. Periodate oxidation of an alkene to a diol followed by permanganate cleavage gives carboxylic acid products which are determined as their corresponding methyl esters,[112,113] although almost any other suitable method of esterification could also be used (Chapter 2). Osmium tetroxide has been used extensively for *cis*-hydroxylation of double bonds and the diol product may be alkylated to form the dimethoxy compound,[114] silylated to give the TMS–ether[115–117] or bridged to form an isopropylidene derivative.[118] Hydrogen peroxide has also been used as a reagent for converting the alkene into an epoxide which is then either hydrolysed to a diol, oxidized to a mixture of ketones[110] or converted into a dimethylamino alcohol.[111] Diols are themselves easily cleaved as already discussed. Many of these techniques are rather specialized and the reader is referred to the original papers for the precise practical details. Having made this point, periodic acid oxidations are widely applicable and the practical details are of some interest.

3.1.2 CLEAVAGE OF EPOXIDES WITH PERIODIC ACID WITH ANALYSIS BY GAS–LIQUID CHROMATOGRAPHY[90]

Periodic acid cleaves epoxides to aldehyde or ketone fragments; these can be derivatized for identification but may also be determined by direct gas chromatography of the reaction solution. It is therefore necessary to be sure that the solvent peak will not mask the oxidation products. With methylene dichloride, products equal to or larger than heptanal may be observed, while carbon tetrachloride is used for products with lesser retention times. Using 1,8-dibromooctane as a solvent, products from C_2 to C_{12} emerge before the solvent peak.

Periodic acid as purchased (H_5IO_6) does not give complete reaction; HIO_4 obtained by vacuum-drying the ground material to constant weight over phosphorus pentoxide gives satisfactory results. The reaction is performed in a small screw-cap bottle (35 mm by 12 mm) and the sample (1–100 μg) in 100 μl of solvent is placed in the bottle and the vacuum-dried reagent (340 times the sample weight plus 2 mg extra) is added and the bottle sealed. After shaking for five minutes at room temperature, the mixture is centrifuged for 1 min and an aliquot of the clear solution (1–10 μl) is injected into the gas chromatograph.

One of the main reasons for not oxidizing a double bond through to its cleavage products is that it may be necessary to find out whether the geometry of substitution is *cis* or *trans*. If a mass spectrometer is available, an *O,O*-isopropylidene derivative could solve this problem. An investigation of the value of this approach was tried using oleic and elaidic acids.[118] The two acids were stereospecifically hydroxylated using osmium tetroxide and the isopropylidene-methyl ester derivatives were examined by GC–MS. The spectra of these geometrical isomers were measurably different, and comparison with authentic standards allowed assignment of the double bond geometry.* Mass spectrometry, to some extent, removes the need to cleave a double bond even for simply determining its position in a chain. The oxygen atoms of protected diols direct the fragmentation in the mass spectrometer source, and the observed ions establish the original double bond position.[115,116] An example is shown in Fig. 5. Methylation of the 1,2-diol has also been used as an alternative to silylation; this has been done using methyl iodide and sodium

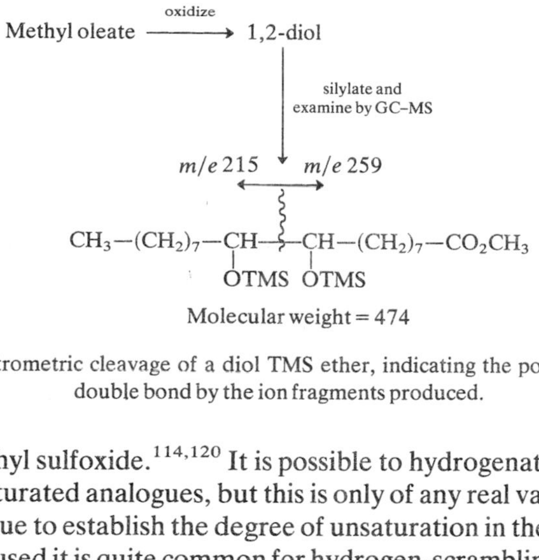

Fig. 5 Mass spectrometric cleavage of a diol TMS ether, indicating the position of the original double bond by the ion fragments produced.

hydride–dimethyl sulfoxide.[114,120] It is possible to hydrogenate multiple bonds to their fully saturated analogues, but this is only of any real value when used as a 'shift' technique to establish the degree of unsaturation in the molecule. Even if deuterium is used it is quite common for hydrogen-scrambling to occur on the surface of the catalyst, and so even isotopic hydrogenation is not impressive as a means of identifying the position of an unsaturation. A method for microhydrogenation is given in Section 3.3.2.

* It seems reasonable to assume that butyl boronate derivatives of diols (Chapter 7) could be just as useful as the isopropylidene compounds.

3.2 Reactions involving the carbonyl group

Chemical modification of carbonyl groups prior to chromatography has been undertaken for a wide variety of analytical reasons. It would only rarely seem to be worthwhile to attempt cleavage adjacent to or reduction of a carbonyl function simply to prove its presence in a molecule. Other methods such as the formation of an oxime derivative with, for example, methoxyamine or ethoxyamine would probably achieve this objective using some sort of shift technique (see Section 3.8, p. 341). In specific cases, reduction has proved necessary to improve an analytical procedure; this may be seen in the GC analysis of aldoses, some small peptides and hydroxy fatty acids.

Aldoses are troublesome to analyse by simple derivative formation followed by GC because their anomeric forms cause multiple peaks. The carbonyl function may be blocked by oxime formation or reduced using borohydride. After reduction the resultant alcohol is acylated or silylated together with the rest of the alditol hydroxy groups. The use of borohydride reduction is more convenient in some ways because it is quite simple and removes the need for double derivatization. Alditol acetates have found wide application as the derivatives of choice for these compounds. The procedure of Sawadeker[121] has been widely adopted; this involves reduction of the aldoses in dilute sodium hydroxide with sodium borohydride at room temperature. Excess borohydride is destroyed by the addition of acetic acid. An effective way of removing excess boric acid (a by-product of the reduction) is by the addition of methanol to the dry residue and evaporation of the methyl borate produced in a stream of nitrogen. For complete removal, this procedure may have to be repeated several times. The alditol products are acetylated with acetic anhydride, often with the addition of pyridine; alditol acetates are readily soluble in methylene dichloride. Crowell and Burnett[122] have reported poor results with the methanol procedure for removing boric acid and have recommended prolonged acetylation under reflux overnight without removal of boric acid; it seems that the presence of the boric acid slows down the rate of acetylation. Deuteriated sodium borohydride and D_2O are now readily available and there is no reason why isotope shift techniques using GC–MS could not be applied to this reduction, to determine the number of carbonyl groups in a molecule.

3.2.1 REDUCTION OF ALDOSES WITH BOROHYDRIDE FOLLOWED BY ACETYLATION[122]—METHOD NUMBER 1.

The carbohydrate rhamnose (200 mg) is added to an aqueous aldose mixture as an internal standard; the aldose mixture in the original publication resulted from the hydrolysis of 0.35 g of wood pulp by the method of Saeman *et al.*[123] The aqueous mixture is concentrated to about 10 ml at 60 °C in a rotary evaporator. The sugars are reduced with 0.4 g of sodium borohydride, dissolved in 10 ml water and after one hour at room temperature the excess reagent is destroyed by the addition of acetic acid until gas evolution ceases.

The solution is evaporated to dryness and dehydrated by the addition of 200 ml of methanol and re-evaporating to dryness. The residue is acetylated by refluxing with 10 ml of a 1:1 mixture of acetic anhydride in pyridine for about 12 h. The mixture is evaporated to a syrupy residue, water (2 ml) is added and again the mixture is evaporated to dryness to remove excess acylation reagents. The residue is dissolved in 10 ml of methylene chloride and 1 μl of this solution is injected into the gas chromatograph.

Comment: This is hardly a microreaction nor is it particularly rapid, but it is included because it illustrates the principles of alditol acetate formation and shows one method for dealing with the boric acid reduction product. It could undoubtedly be scaled down and simplified. Other acylation procedures for alditols are described in Chapter 3.

3.2.2 REDUCTION OF ALDOSES WITH BOROHYDRIDE FOLLOWED BY ACETYLATION[124]—METHOD NUMBER 2

Approximately 2 ml of an aqueous neutralized polysaccharide hydrolysate solution is treated with 1 ml of molar ammonium hydroxide containing 2 mg of sodium borohydride. The mixture is stirred at room temperature for one hour. Excess borohydride is destroyed with acetic acid and the resultant solution evaporated almost to dryness at 40–45 °C in a rotary evaporator. A suitably small quantity of an internal standard such as inositol is added at this stage and the residue evaporated fully to dryness. The boric acid reduction by-product is removed as its volatile methyl ester by boiling the residue with 2 ml of a 5:1 methanol: benzene solution containing 25 μl of glacial acetic acid for 5 min; this solution is then evaporated in a stream of nitrogen at 70–80 °C. This procedure is repeated four times omitting the benzene. The final residue is pumped at high vacuum at room temperature and then acetylated with 1.5 ml of acetic anhydride at 100 °C for one hour (under nitrogen, with agitation if possible). Excess anhydride is removed *in vacuo* at 40 °C and the final traces removed in high vacuum. The residue is extracted with 3–4 ml of methylene chloride by stirring for 30 min, the supernatant is evaporated in a stream of nitrogen, pumped at high vacuum for several hours, diluted with 200 μl of methylene chloride and used for the analysis using a 3% OV-225 column (2 m long).

Comment: The acetic anhydride may be replaced with a suitable perfluoracyl anhydride for a more volatile derivative or for ECD gas chromatography. (Chapter 3).

3.2.3. 'IN-SYRINGE' REDUCTION OF ALDEHYDES AND KETONES WITH BOROHYDRIDE[125]

The reducing mixture (0.5 μl) of saturated ethanolic sodium borohydride is spread by plunger action over the wall of a 10 μl syringe. One microlitre of an ether solution of the sample (approximately 1:100) is introduced into the syringe and left to react for 2 min. The resultant solution is injected directly into the GC.

Comment: This is only useful for volatile alcohols (C_6–C_{12}) but the simplicity of the technique suggests other possibilities for further reaction. The method was also used for esterification with 51% BF_3 in methanol for five minutes and for oxidation and hydrolysis, which will be discussed later. The concept of the syringe as reaction vessel is obviously capable of wide application.

The reduction of small peptides with lithium aluminium hydride prior to silylation and examination by GC–MS is another example of chemical modification used to refine the precision of an analysis.[127] A typical method would involve enzymic cleavage of a peptide to smaller di-, tri- and tetra-peptides, followed by esterification of this mixture with methanolic-HCl and acetylation with acetic anhydride to increase the compounds' solubilities in the reduction solvent which is 1,2-dimethoxyethane. After reduction with $LiAlH_4$, the product is silylated with TMS-diethylamine to protect only the hydroxy groups and the derivatives are examined by GC–MS. For small peptides this approach appears to give a good quality, rapid GLC analysis; the mass spectra of these amino-alcohols are also informative and readily interpreted. A similar approach has been applied to hydroxylated stearic acids by Esselman and Clagett.[126] In this case it is not clear whether the authors found any advantages in analysing the resultant silylated polyhydroxyoctadecanes as opposed to the derivatized parent hydroxy acids.

3.2.4 REDUCTION OF HYDROXY FATTY ACIDS WITH LITHIUM ALUMINIUM HYDRIDE[126]

A saturated ethereal solution of lithium aluminium hydride is added dropwise to a solution of the TMS derivatives (*O*-TMS-ether-esters) of hydroxy acids in diethyl ether until no more hydrogen is evolved (about 1 min at 25 °C). Dilute hydrochloride acid is added and the mixture extracted with ether. The ether solution is evaporated to dryness and the new hydroxyl group silylated by adding more BSA; this final solution is diluted with ether before analysis by GLC.

Of the many reagents available for the reduction of carbonyl groups, some would seem to be particularly suitable for microreductions prior to chromatography. Diborane, although a highly reactive material, should be easy to generate and use on a microscale, but in many cases borohydride could be just as effective. Sodium amalgam and platinum metal catalysts are also easily used on a small scale.

Analogous to the alditol acetate method for sugar analysis, the carbonyl group of aldoses has also been modified by conversion into a nitrile function.[128,129] This is achieved by formation of the oxime with hydroxylamine and subsequent dehydration and acetylation with acetic anhydride.

3.2.5 CONVERSION OF NEUTRAL ALDOSES TO ALDONITRILE ACETATES[128]

A dried mixture of monosaccharides containing 1 mg of each sugar is dissolved in pyridine (0.6 ml), treated with 12 mg of dry hydroxylamine hydrochloride and heated in a reaction vial at 90 °C for 30 min. After cooling, acetic anhydride (1.8 ml) is added and the heating continued for another 30 min. The cooled solution is evaporated to dryness under vacuum at below 40 °C. The residue is dissolved in 0.1 ml of dry chloroform and an aliquot injected into the gas chromatograph.

3.2.6 IODOFORM REACTION OF METHYL KETONES

One specific microreaction which should find a useful application in GC analysis has been mentioned by Crippen;[130] this is the well-known iodoform reaction. This specific test for the CH_3CO-group is simple to carry out and both the iodoform itself[131] and the carboxylic acid product could be detected by chromatographic methods (Fig. 6).

$$R-\underset{\underset{O}{\|}}{C}-CH_3 + 3NaOI \rightarrow \underset{\substack{\text{yellow}\\ \text{precipitate}}}{CHI_3} + RCOONa + 2NaOH$$

Fig. 6 The iodoform reaction.

Aliphatic aldehydes have been oxidized to the corresponding acids using silver oxide and the resulting acids esterified with diazomethane (Chapter 2), followed by gas chromatographic analysis of the methyl esters.[132]

Frequently carbonyl compounds are isolated from complex mixtures using condensation derivatives such as dinitrophenylhydrazones, semicarbazides or complexes with Girard's reagent-T. Methods have been developed for regenerating the parent carbonyl compound prior to GC analysis.[133–6] One of the simplest and most interesting of these[133] involves preparing a mixture of the derivative with 2-oxoglutaric acid and sodium bicarbonate in a capillary tube which is sealed at one end, inserting the open end of the tube into the GC injection port through a septum, and heating the mixture rapidly. The liberated carbonyl compound then passes into the GC column.

3.3 Reactions at benzylic carbon atoms

Vigorous oxidative degradation, apart from splitting multiple bonds and forming quinone structures may also cleave aromatic compounds at benzylic carbon atoms. The acid or ketone oxidation products can be analysed chromatographically. An example of this is the analysis of terodiline as benzophenone[137] using ECD gas liquid chromatography (Fig. 7). In this case the sensitivity of the ECD to benzophenone is an important consideration, especially when combined with the chromatographic simplicity of its detection.

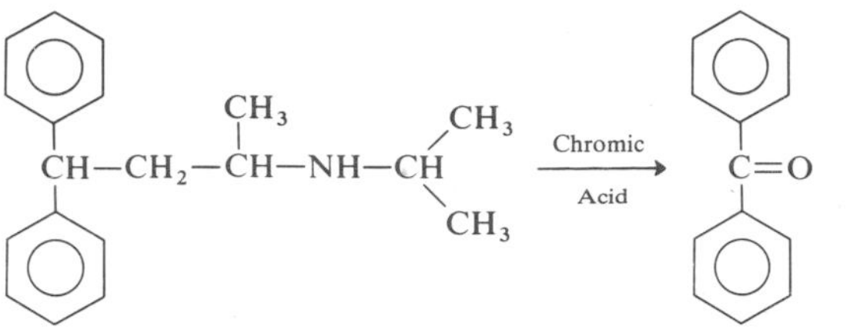

Fig. 7 Oxidation of terodiline to benzophenone.

3.3.1 OXIDATION OF TERODILINE WITH CHROMIC ACID[137]

The oxidation reagent is prepared by diluting 3 g of concentrated sulfuric acid to 50 ml with glacial acetic acid and then adding 167 mg of chromium trioxide. The mixture is agitated to dissolve the oxide and only stored for a maximum of 24 h.

The details of extraction of terodiline from serum is fairly intricate and involves ion-pair techniques. The reader is referred to the original paper for details of this procedure; the oxidation technique is described here because it is more generally applicable. The sample residue in a glass-stoppered tube is treated with 1 ml of the oxidation mixture (the sample contains about 100 ng of terodiline). After agitating briefly the mixture is allowed to stand at room temperature for about 30 min. A 5 ml portion of 5 M sodium hydroxide and 4 ml of water are added, the resultant mixture is cooled to about 10–15 °C and 0.1 ml of heptane added. The mixture is shaken for 15 min and the aqueous phase removed; a trace of anhydrous sodium sulfate is used to dry the heptane solution, which is used for GLC analysis.

Oxidation of this general type has been applied to the compounds bromopheniramine, chloropheniramine and amitriptyline (Fig. 8).

In the cases of bromo- and chloropheniramine alkaline permanganate, oxidation was used[138] and for amitriptyline, ceric sulfate in 5.4 M sulfuric acid was satisfactory.[139] In both cases an ECD gas chromatography method was chosen.

Benzylic carbon atoms are also sensitive to hydrogenolysis over a palladium-charcoal catalyst and this has been used as the first stage in an ECD–GC assay of nicotine.[140] Nicotine is not easily or cleanly derivatized by electron-capturing reagents but by using hydrogenation a product is obtained which can be acylated with a perfluoracyl reagent and measured by ECD (Fig. 9). It seems that there is a potentially useful application for the new pentafluorobenzyl-chloroformate reagent (see Section 3.5) for the analysis of this tertiary amine.

The two examples of terodiline and nicotine illustrate one of the dangers of using degradative methods to obtain an easily analysable product for quantitative work as opposed to structural elucidation. In Fig. 7 it may be seen that any compound with a diphenyl alkane structure will give benzophenone, and in Fig.

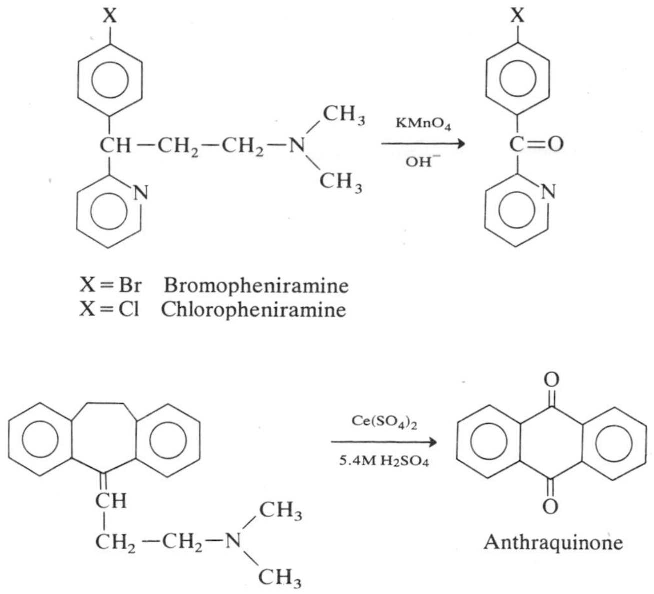

Amitriptyline

Fig. 8 Oxidation to aromatic ketones.

9 compounds with any other pattern of unsaturation or easily hydrogenolysed groups might interfere.

3.3.2 GAS CHROMATOGRAPHIC ANALYSIS OF NICOTINE USING ECD[140]

A solution of 2 mg of nicotine hydrochloride in 3 ml of water is added to 5 mg of a 10% palladium-charcoal catalyst in a 20 ml glass vessel and the mixture hydrogenated at a pressure of 1.38 bar for 4 h at room temperature. The

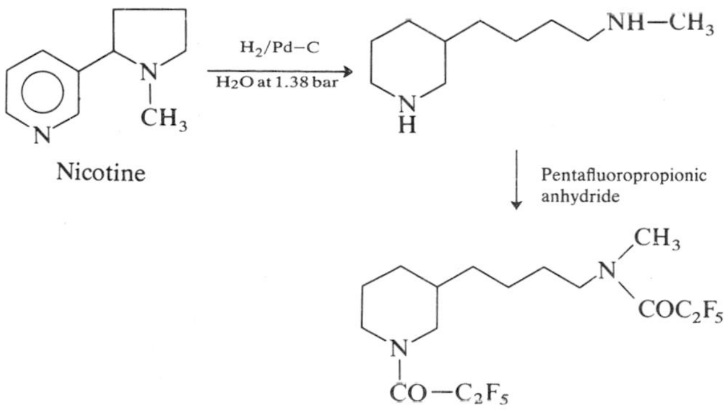

Fig. 9 Hydrogenolysis of nicotine.

catalyst is removed by filtration through a sintered glass filter, washed with water (2 ml) and the filtrate collected in a 15 ml centrifuge tube. The filtrate is evaporated to dryness under reduced pressure and desiccated at 70 °C under vacuum for 1 h. The residue may then be treated with a perfluoroacylation reagent as described in Chapter 3. The original authors used 0.4 ml of pentafluoropropionic anhydride at 70 °C for 2 h, and then evaporated the excess reagent before adding heptane and performing several alkaline and acid washes.

3.4 Organochlorine residue analysis

Because of the importance of accurate identification of pesticide residues in crops and the similarity of many of the commercial agents used, identification systems using GC–ECD retention data before and after chemical modification have been developed. For example *p,p′*-DDT has been treated with basic reagent to remove the elements of HCl; endrin has been rearranged with BF_3; heptachlor is acetylated in the allylic position using silver acetate in acetic acid; and nitration has been applied to the polychlorinated biphenyls. An excellent review on this specialized field of GC analysis has appeared by Cochrane and Chau.[141]

3.5 Reactions involving tertiary amines

Tertiary amines often pose a difficult problem for analysis by GLC because of their polarity; analysis by condensed-phase chromatography may not necessarily be so difficult. Improved chromatographic separation is often achieved by the addition of basic materials to the eluent or stationary phase, for example by the addition of ammonia to a TLC eluent or by treating the GLC stationary phase with an alkali. There is, however, a problem with high-sensitivity gas chromatography. Alkali-flame detectors which are specific for nitrogen should give a good limit of detection but the techniques are still new and to some extent dependent upon the design and reliability of the GC system. When considering the use of an electron-capture detector there is the problem of derivatization: obviously a tertiary amine has no replaceable hydrogen atom to allow insertion of a halogen-containing function. Recently a novel procedure has been described by Hartvig and co-workers[142–144] for derivatizing tertiary amines with pentafluorobenzyl chloroformate to form electron-capturing carbamates.

The very high sensitivity of the ECD to pentafluorobenzyl groups is discussed in Chapter 3, and so this approach opens up the possibility of detecting tertiary amines at nanogram and picogram levels. One of the stimuli for the development of pentafluorobenzyl carbamate derivatives was the number of important antihypertensive and psychoactive drugs with tertiary amino functions.

The earliest approach[142] used ethyl chloroformate to react with the amine to form a carbamate as in the following equation:

$$R{-}N\begin{matrix}CH_3\\ CH_3\end{matrix} + ClCOOC_2H_5 \rightarrow R{-}N\begin{matrix}CH_3\\ COOC_2H_5\end{matrix} + CH_3Cl$$

The product of this reaction may then be hydrolysed to a secondary amine and derivatized as follows:

$$R{-}N\begin{matrix}CH_3\\ COOC_2H_5\end{matrix} + H_2O + HBr \rightarrow R{-}\overset{+}{N}H\begin{matrix}CH_3\\ H\end{matrix}\ Br^- + CO_2 + C_2H_5OH$$

A detailed investigation of the scope of this demethylation was undertaken[144] using a number of diphenylpropylamine-type compounds, and the results were fairly variable. In some cases carbamates resulted, in others deamination occurred, and in general there were several side-reactions, but the nature of the alkyl groups on the nitrogen had little effect upon carbamate formation. Subsequent investigations established that pentafluorobenzyl chloroformate was the most useful reagent because the hydrolysis step is not necessary, as may be seen from the following equation:

$$R{-}N\begin{matrix}CH_3\\ CH_3\end{matrix} + ClCOOCH_2{-}C_6F_5 \rightarrow R{-}N\begin{matrix}CH_3\\ COOCH_2C_6F_5\end{matrix} + CH_3Cl$$

It was found that base catalysis had a pronounced effect upon the reaction using *N,N*,1-trimethyl-3,3-diphenylpropylamine as a test compound (Fig. 10).

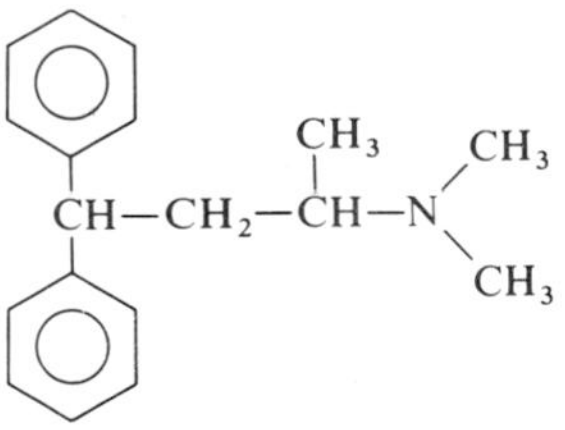

Fig. 10 *N,N*-1-trimethyl-3,3-diphenylpropylamine.

Anhydrous sodium carbonate was found to be the best catalyst[146] and it is possible to detect as little as 3 pg of this tertiary amine after only 30 min of reaction time. Although their ECD and GC properties are very useful the one difficult problem in using these derivatives is removal of the excess reagent: this has been overcome by treating the reaction mixture with alcoholic alkali before examination by GLC.

3.5.1 PREPARATION OF PENTAFLUOROBENZYL CARBAMATE FROM TERTIARY AMINE[145]

Pentafluorobenzyl chloroformate (20 μl) is added to a solution of the amine (0.5–1.0 mg ml^{-1}) in heptane (200 μl) contained in a tapered 3 ml reaction vial; 10 mg of anhydrous sodium carbonate is added and the capped tube heated at 100 °C for 30 min (or longer if necessary). The cooled heptane solution is treated with 500 μl of 0.5 M potassium hydroxide in a 3:1 methanol:water solvent; the mixture is shaken for several minutes and diluted with 1 ml of water; the organic phase is removed and dried over a trace of anhydrous sodium carbonate and an aliquot used for GLC analysis.

Tertiary amines have also been analysed as their Hofmann degradation products[147,148] but this type of procedure is apparently not always quantitative and has not been widely adopted. It also seems likely that side-reactions could be a problem.

3.5.2 HOFMANN DEGRADATION OF AMINES (USING EXHAUSTIVE METHYLATION)[147]

The free amine (less than 10 μg) dissolved in chloroform is placed in a small tapered vial, the chloroform is evaporated and replaced with 10 μl of benzene and 4 μl of methyl iodide. The mixture is allowed to stand for 5 min at room temperature, a small particle of moist silver oxide is added and the tube warmed at about 70 °C until the solvent has evaporated. The residue is dried *in vacuo* and dissolved in 10 μl of carbon disulfide, and a small aliquot is examined by GLC. Breakdown of the Hofmann products presumably occurs in the GLC flash-heater.

3.6 Quaternary ammonium compounds

Quaternary ammonium salts usually decompose on heating and so their decomposition products can be detected by gas chromatography. In many cases condensed-phase chromatography is an ideal solution for quaternary salts and the reader is referred to Chapter 14 on ion-pair extraction. The thermal decomposition products of quaternary ammonium halides may be tertiary amines (from long-chain compounds[149]), dimethylamine esters from acetylcholine esters[150] or Hofmann-type elimination products in the presence of a base (see above). The majority of GC methods for the analysis of quaternary amines have relied on pyrolysis. One exception is the dequaternization process using sodium benzenethiolate in anhydrous butanone (methyl ethyl ketone);[151] the resulting tertiary amines are analysed by GC using a phenyldiethanolamine succinate stationary phase especially suited for their separation. The technique is involved and in general it seems advisable to consider simpler analytical approaches such as ion-pair chromatography.

3.7 Sulfur-containing compounds

Many smaller sulfur-containing molecules may be chromatographed directly by GC using suitable chromatographic conditions. These include compounds such as hydrogen sulfide, dimethylsulfide, dimethylsulfoxide and alkanethiols. An excellent review and tabulation of these has been presented by Ronkainen and colleagues.[152] One of the most intractable problems in the analysis of thiols, sulfides, and sulfoxides is their ease of oxidation and this is associated with the poor volatility of most sulfoxides and sulfones.

Thiols, apart from their sensitivity to oxidation, may for the most part be treated with protecting groups in the same way as alcohols. Pentafluorobenzyl bromide (Chapter 5) has been used as a reagent for treating thiols to volatilize them for electron-capture detection.[153] Desulfurisation of thiols, sulfides and disulfides using Raney nickel is a possibility but does not appear to have found much application apart from the analysis of thiophene and thionaphthene.[154] Because of the sensitivity of the sulfide function to oxidation, Bowman and Beroza[155] have advocated oxidation to the sulfone before analysis to ensure a single detectable compound. However, it should be remembered that sulfones are often not very volatile. This oxidation may be performed using *m*-chloroperbenzoic acid (which is now commercially available) in chloroform.

3.7.1 OXIDATION OF SULFOXIDES AND SULFIDES TO SULFONES[155]

Add 10–100 mg of *m*-chloroperbenzoic acid to 5 ml of chloroform containing the sulfur compound. The precise quantity of oxidant can only be determined experimentally and should be sufficient for complete reaction but not such an excess that it is difficult to remove. The oxidation is allowed to proceed at room temperature for 40 min and then the solution is introduced onto the top of a small alumina column. The column is eluted with 30 ml of chloroform and the eluate concentrated as necessary for the analysis.

Sulfonic acids may be esterified in a number of ways and very often the methods used for carboxylic acids (Chapter 2) are perfectly satisfactory. However, sulfonic esters and sulfonamides (even after alkylation) are not very volatile and there are several alternative methods for dealing with these acid functions. One of these is microfusion of the sodium or potassium salts with the corresponding hydroxide, which is best performed in a microreactor at the front-end of the GC column.[40] This is really reaction-GC and has been discussed already. Such a fusion process is sufficiently drastic to be virtually a form of pyrolysis, and in some circumstances it could result in a complex product mixture. An interesting method for increasing the volatility of the sulfonic acid group by conversion to the corresponding sulfonyl halide has been described. Both chlorides[156] and fluorides[157] have been used.

3.7.2 CONVERSION OF SULFONIC ACIDS INTO SULFONYL CHLORIDES[156,157]

Add 10 μl of phosphorus pentachloride to approximately 1 mg of the sodium sulfonate in a reaction vial, cap the vial and heat to 100 °C for 1 h.

Shake the vial occasionally to ensure mixing. Cool the tube and add 30 μl of benzene, recap and shake the vial, allow to separate into two layers, remove the supernatant liquid into another small vial and add 5 μl of water. Cap and shake the vial vigorously to hydrolyse excess phosphorus chlorides, remove the benzene layer into another small vial containing a trace of anhydrous sodium sulfate and use this solution of sulfonyl chloride for the analysis.

By heating the wet benzene extract with 2 mg of potassium fluoride dihydrate at about 100 °C in a reaction vial for 1.5 h, halogen exchange occurs to give sulfonyl fluoride in the benzene phase.

Thiocarbamates, analogous to normal carbamates, are readily hydrolysed with 0.5 M sulfuric acid to the corresponding amine,[158] as shown in the following equation:

$$2 \quad \overset{R}{\underset{R'}{\diagdown}} N-C \overset{\diagup S}{\underset{\diagdown SH}{}} + H_2SO_4 \rightarrow 2 \quad \overset{R}{\underset{R'}{\diagdown}} \overset{+}{NH_2} + 2CS_2$$
$$SO_4^{2-}$$

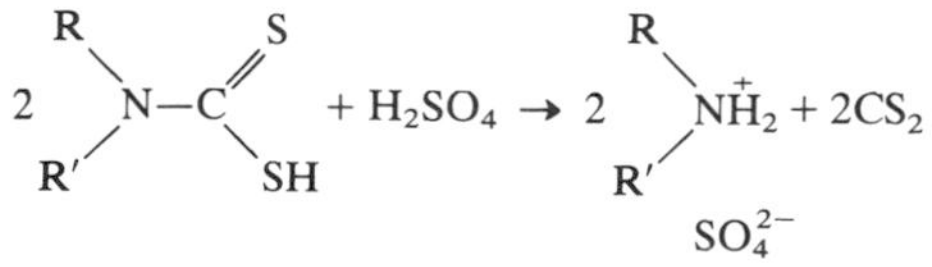

The hydrolysis may be carried out on a small scale in a reaction vial at about 100 °C followed by a simple acid/base extraction. Alternatively it is possible to esterify these compounds to their S-alkyl esters using diazoalkanes.[159]

Thioesters such as acylthioesters of Co-enzyme-A are reduced by sodium borohydride to the corresponding alcohol;[160] borohydride does not reduce normal esters and so this is a fairly specific reaction:

$$R-\overset{\overset{O}{\|}}{C}-S-\boxed{CoA} \xrightarrow{NaBH_4} R-CH_2-OH + \boxed{CoA}-SH$$

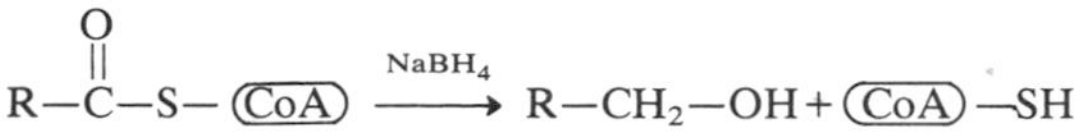

3.8 Miscellaneous methods

3.8.1 HYDROGENOLYSIS

An interesting further example of hydrogenolysis has been provided by McCloskey and Law.[161] Cyclopropane fatty acids are almost indistinguishable from unsaturated fatty acids by GC–MS, and so the compound is hydrogenated as the methyl ester over a platinum catalyst to give a mixture of methyl-substituted fatty acid esters (Fig. 11). This mixture can be analysed by GC–MS.

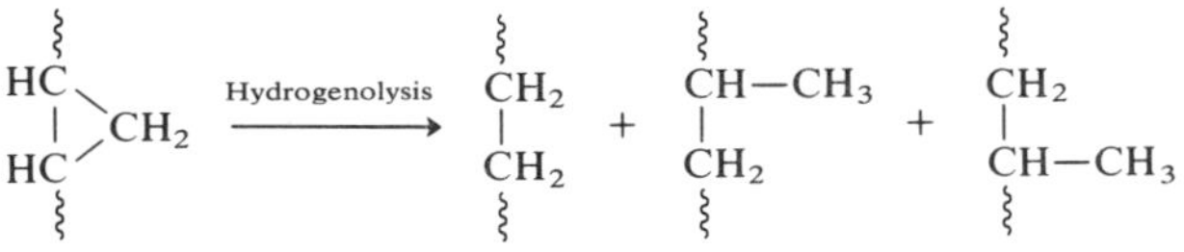

Fig. 11 Hydrogenolysis of cyclopropanes.

3.8.2 'SHIFT' TECHNIQUES

One of the most valuable structural elucidation techniques in microchemical modification for GC and GC–MS is the 'shift technique'. This may be applied in GC alone by observing the change in retention index on e.g. substituting a propionyl group for an acetyl group and so establishing the number of protectable functions in the molecule. In GC–MS the technique is even more powerful[162] and mass shifts may be observed on moving from $(CH_3)_3$-Si-groups to $(CD_3)_3Si$-, from CF_3CO- to C_2F_5CO- or from $=N{-}OCH_3$ to $=N{-}OC_2H_5$ etc. A brief discussion of the shift rule in GC–MS is given in Waller's book on biochemical mass spectrometry.[163]

3.8.3 SAPONIFICATION (AQUEOUS ALKALINE HYDROLYSIS)

Saponification is a fairly general reaction which has many applications as shown in the tables of Chapter 1. An illustrative example is given here. Many lepidopteran sex pheromones are acetates of long chain unsaturated alcohols. In studies to establish the structure of an unknown material, the presence of such a compound is usually demonstrated by the loss of activity and formation of the alcohol accomplished by alkaline hydrolysis and the regeneration of the original material upon acetylation.

3.8.4 MILD SAPONIFICATION OF A PHEROMONE ACETATE, FOLLOWED BY REACETYLATION[164]

The purified pheromone (40 ng) in 20 μl of ether is treated with 2 μl of a saturated solution of potassium carbonate in methanol. After 10 min at room temperature an aliquot of the mixture is taken for analysis by GLC. The solvent is then removed from the reaction mixture with a stream of nitrogen and 5 μl of a mixture of acetic anhydride and pyridine (40:60 by volume) is added to the residue. The mixture is allowed to stand for 15 min at room temperature and then analysed again by GLC.

A more vigorous saponification would involve simply heating a residue with molar aqueous alkali at about 60 °C in a reaction vial for a suitable period and then isolating either acidic, basic or neutral products by micro acid/base extraction with a suitable solvent system. An extremely simple procedure has been described for hydrolysis of esters in a GLC injection syringe; the product alcohol may be analysed directly.

3.8.5 'IN-SYRINGE' HYDROLYSIS OF ESTERS[125]

The hydrolysis reagent consists of saturated ethanolic sodium hydroxide; 0.5 μl of this is drawn into a dry 10 μl microsyringe and spread onto the wall by plunger action. One microlitre of a solution of the ester in ether (approx. 1:100) is introduced into the syringe and left there for 15 min, after which time the entire contents are injected into the gas chromatograph. The alcohols investigated were smaller than decanol.

4 CONCLUSION

Although the procedures described are very varied, the scope of such microreactions and the ingenuity shown in devising some of these microprocedures indicates a wealth of approaches that may be applied to problems of derivatization and structural identification by chromatography. It can be seen that the availability of only small amounts of material for analysis need not be an obstacle to its characterization. Even where the availability of material is not a limiting factor, some of the microprocedures described in this chapter may be able to give valuable information more quickly and more economically than more conventional methods.

REFERENCES

1. N.-E. Larsen and J. Naestoft, *J. Chromatogr.* **109**, 259 (1975).
2. W. Dünges, G. Naundorf and N. Seiler, *J. Chromatogr. Sci.* **12**, 655 (1974).
3. E. Grushka, H. D. Hurst and E. Kitka, *J. Chromatogr.* **112**, 673 (1975).
4. E. Oswald, D. Parks, T. Eling and B. J. Corbett, *J. Chromatogr.* **93**, 47 (1974).
5. S. Nicosia and G. Galli, *Anal. Biochem.* **61**, 192 (1974).
6. A. J. F. Wickramasinghe and R. S. Shaw, *Biochem. J.* **141**, 179 (1974).
7. G. Schneider, S. Jaenicke and G. Sembdner, *J. Chromatogr.* **109**, 409 (1975).
8. Y.-Y. Liu and D. Hoffmann, *Anal. Chem.* **45**, 2270 (1973).
9. B. A. Bierl, M. Beroza and C. W. Collier, *J. Econ. Entomol.* **65**, 659 (1972).
10. A. F. Haberer, W. S. Schlotzhauer and O. T. Chortyk, *J. Agric. Food Chem.* **22**, 328 (1974).
11. A. S. Y. Chau and K. Terry, *J. Ass. Off. Anal. Chem.* **55**, 1228 (1972).
12. A. S. Y. Chau and K. Terry, *J. Ass. Off. Anal. Chem.* **57**, 394 (1974).
13. A. S. Y. Chau, *J. Ass. Off. Anal. Chem.* **57**, 585 (1974).
14. E. D. Bitner, V. L. Davison and H. J. Dutton, *J. Am. Oil Chem. Soc.* **46**, 113A (1969).
15. T. S. Ma and A. S. Ladas, *Organic Functional Group Analysis by Gas Chromatography*, Academic Press, London, 1976.
16. M. Beroza and R. Sarmiento, *Anal. Chem.* **38**, 1042 (1966).
17. T. L. Mounts and H. J. Dutton, *Anal. Chem.* **37**, 641 (1965).
18. M. Beroza and R. Sarmiento, *Anal. Chem.* **35**, 1353 (1963).
19. M. Beroza and R. Sarmiento, *Anal. Chem.* **36**, 1744 (1964).
20. R. M. Silverstein, R. G. Brownlee, T. E. Bellas, D. L. Wood and L. E. Browne, *Science* **159**, 889 (1968).
21. M. Beroza and R. A. Coad, *J. Gas Chromatogr.* **4**, 199 (1966).
22. M. Beroza and M. N. Inscoe, in L. S. Ettre and W. H. McFadden (Eds) *Ancillary Techniques of Gas Chromatography*, Wiley–Interscience, New York, 1969, pp. 89–144.
23. M. Beroza in E. Sz. Kovats (Ed.) *Chimia (Supplement)—Column Chromatography*, Chimia, Stuttgart, 1970, pp. 92–103.
24. F. E. Regnier and J. C. Huang, *J. Chromatogr. Sci.* **8**, 267 (1970).
25. F. Drawert, W. Heimann, R. Emberger and R. Tressl, *Chromatographia* **2**, 57 (1969).
26. R. Tressl, F. Drawert and U. Prenzel, *Chromatographia* **6**, 7 (1973).
27. D. A. Cronin and J. Gilbert, *J. Chromatogr.* **87**, 387 (1973).

28. R. E. Kepner and H. Maarse, *J. Chromatogr.* **66**, 229 (1972).
29. G. Stanley and B. H. Kennett, *J. Chromatogr.* **75**, 304 (1973).
30. D. Henneberg, U. Heinrichs and G. Schomburg, *J. Chromatogr.* **112**, 343 (1975).
31. G. T. Pearce, W. E. Gore, R. M. Silverstein, J. W. Peacock, R. A. Cuthbert, G. N. Lanier and J. B. Simeone, *J. Chem. Ecol.* **1**, 115 (1975).
32. Z. Stransky, J. Gruz and E. Ruzicka, *J. Chromatogr.* **59**, 158 (1971).
33. N. Propenko, A. Rabinovich, N. Dubrova and M. Dementyeva, *J. Chromatogr.* **69**, 47 (1972).
34. P. C. Rahn and S. Siggia, *Anal. Chem.* **45**, 2336 (1973).
35. B. A. Bierl, M. Beroza and W. T. Ashton, *Mikrochim. Acta* 637 (1969).
36. V. L. Davison and H. J. Dutton, *Anal. Chem.* **38**, 1302 (1966).
37. S. P. Frankoski and S. Siggia, *Anal. Chem.* **44**, 507 (1972).
38. G. Lundquist and C. E. Meloan, *Anal. Chem.* **43**, 1122 (1971).
39. S. P. Frankoski and S. Siggia, *Anal. Chem.* **44**, 2078 (1972).
40. S. Siggia, L. R. Whitlock and J. C. Tao, *Anal. Chem.* **41**, 1387 (1969).
41. N. D. Greenwood and H. E. Nursten, *J. Chromatogr.* **92**, 323 (1974).
42. Y. G. Osokin, V. S. Feldblum and S. I. Kryukov, *Neftekhimia* **6**, 333 (1966).
43. B. Halpern, W. E. Pereira, M. D. Solomon and E. Steed, *Anal. Biochem.* **39**, 156 (1971).
44. W. L. Clapp, R. J. Valis, S. R. Kramer and J. W. Mercer, *Anal. Chem.* **46**, 613 (1974).
45. E. B. Solow, J. M. Metaxas and T. R. Summers, *J. Chromatogr. Sci.* **12**, 256 (1974).
46. W. J. A. VandenHeuvel and V. F. Gruber, *J. Chromatogr.* **112**, 513 (1975).
47. J. W. Schwarze and M. N. Gilmour, *Anal. Chem.* **41**, 1686 (1969).
48. B. S. Middleditch and D. M. Desiderio, *Anal. Lett.* **5**, 605 (1972).
49. R. H. Hammer, B. J. Wilder, R. R. Streiff and A. Mayersdorf, *J. Pharm. Sci.* **60**, 327 (1971).
50. U. Langenbeck and J. E. Seegmiller, *J. Chromatogr.* **80**, 81 (1973).
51. K. M. Williams and B. Halpern, *Anal. Lett.* **6**, 839 (1973).
52. J. J. Rosenfeld, B. Bowins, J. Roberts, J. Perkins and A. S. Macpherson, *Anal. Chem.* **46**, 2232 (1974).
53. I. Gan, J. Korth and B. Halpern, *J. Chromatogr.* **92**, 435 (1974).
54. H. A. Moye, *J. Agric. Food Chem.* **19**, 452 (1971).
55. C. B. Johnson and E. Wong, *J. Chromatogr.* **109**, 403 (1975).
56. D. B. de Oliveira znd W. E. Harris, *Anal. Lett.* **6**, 1107 (1973).
57. J. Fryčka and J. Pospišil, *J. Chromatogr.* **67**, 366 (1972).
58. K. Yasuda and K. Nakashima, *Bunseki Kagaku* **17**, 1536 (1968).
59. D. Gaede and C. E. Meloan, *Anal. Letters* **6**, 71 (1973).
60. J. W. Ralls, *Anal. Chem.* **36**, 946 (1964).
61. H. Halvarson, *J. Chromatogr.* **57**, 406 (1971).
62. A. S. Ladas and T. S. Ma, *Mikrochim. Acta* 853 (1973).
63. G. G. Esposito, *Anal. Chem.* **40**, 1902 (1968).
64. P. M. Wise and R. H. Hanson, *Anal. Chem.* **44**, 2393 (1972).
65. A. Becker, *Z. Anal. Chem.* **174**, 161 (1960).
66. J. W. Copius-Peereboom and H. W. Beekes, *J. Chromatogr.* **17**, 99 (1965).
67. S. K. Yasuda, *J. Chromatogr.* **13**, 78 (1964).
68. M. H. A. Elgamal and M. B. E. Fayez, *Z. Anal. Chem.* **226**, 408 (1967).
69. H. P. Kaufmann and T. H. Khoe, *Fette, Seifen, Anstrichm.* **64**, 81 (1962).
70. J. M. Miller and J. G. Kirchner, *Anal. Chem.* **25**, 1107 (1953).
71. C. Mathis, *Ann. Pharm. Franc.* **23**, 331 (1965).
72. L. L. Smith and J. C. Price, *J. Chromatogr.* **26**, 509 (1967).
73. J. Polesuk and T. S. Ma, *J. Chromatogr.* **57**, 315 (1971).
74. S. K. Yasuda, *J. Chromatogr.* **71**, 481 (1972).
75. J. Polesuk and T. S. Ma, *Mikrochim. Acta* 352 (1969).
76. H. Klus and H. Kuhn, *J. Chromatogr.* **109**, 425 (1975).
77. H. Schütz, *J. Chromatogr.* **94**, 159 (1974).
78. J. W. Copius-Peereboom and H. W. Beekes, *J. Chromatogr.* **9**, 316 (1962).

79. J. Polesuk and T. S. Ma, *Mikrochim. Acta* 662 (1971).
80. A. M. Gardner, *J. Ass. Off. Anal. Chem.* **54**, 517 (1971).
81. C. Schütz and H. Schütz, *Arzneimittelforschung* **23**, 428 (1973).
82. M. Wilk and U. Brill, *Arch. Pharm.* **301**, 282 (1968).
83. M. Wilk, U. Hoppe, W. Taupp and J. Rochlitz, *J. Chromatogr.* **27**, 311 (1967).
84. M. Baggiolini and B. Dewald, *J. Chromatogr.* **30**, 256 (1967).
85. H.-C. Curtis and M. Müller, *J. Chromatogr.* **32**, 222 (1968).
86. H. H. O. Schmid and H. K. Mangold, *Biochim. Biophys. Acta* **125**, 182 (1966).
87. H. H. O. Schmid, P. C. Bandi and K. K. Sun, *Biochim. Biophys. Acta* **231**, 270 (1971).
88. V. E. Vas'kovskii and V. M. Dembitskii, *J. Chromatogr.* **115**, 645 (1975).
89. A. Fono, A.-M. Sapse and T. S. Ma, *Mikrochim. Acta* 1098 (1965).
90. B. A. Bierl, M. Beroza and M. H. Aldridge, *Anal. Chem.* **43**, 636 (1971).
91. J. H. Tumlinson, J. P. Minyard, P. A. Hedin and A. C. Thompson, *J. Chromatogr.* **29**, 80 (1967).
92. P. J. Holloway and S. B. Challen, *J. Chromatogr.* **25**, 336 (1966).
93. P. Froment and A. Robert, *Chromatographia* **4**, 173 (1971).
94. I. Wilczynska, *Chem. Anal. (Warsaw)* **17**, 21 (1972).
95. E. D. Stedman, *Analyst* **94**, 594 (1969).
96. G. Pataki, *J. Chromatogr.* **16**, 541 (1964).
97. G. Pataki, J. Borko, H. C. Curtius and F. Tancredi, *Chromatographia*, **1**, 406 (1968).
98. J. F. Lawrence and G. W. Laver, *J. Ass. Off. Anal. Chem.* **57**, 1022 (1974).
99. J. R. Swarthout and R. J. Gross, *J. Am. Oil Chem. Soc.* **41**, 378 (1964).
100. K. Oette and M. Doss, *J. Chromatogr.* **32**, 439 (1968).
101. D. P. Schwartz and R. S. Bright, *Anal. Biochem.* **61**, 271 (1974).
102. D. P. Schwartz and R. Reynolds, *Microchem. J.* **20**, 50 (1975).
103. D. P. Schwartz, J. L. Weihrauch and L. H. Burgwald, *Anal. Chem.* **41**, 984 (1969).
104. M. D. Booth and B. Fleet, *Analyst* **94**, 844 (1969).
105. R. Criegee, *Rec. Chem. Progr.* **18**, 111 (1967).
106. P. R. Story, J. A. Alford, W. C. Ray and J. R. Burgess, *J. Am. Chem. Soc.* **93**, 3044 (1971).
107. M. Beroza and B. A. Bierl, *Mikrochim. Acta* 720 (1969).
108. M. Beroza and B. A. Bierl, *Anal. Chem.* **39**, 1131 (1967).
109. G. Bergström, B. Kullenberg, S. Ställberg-Stenhagen and E. Stenhagen, *Arkiv. Kemi* **28**, 453 (1968).
110. G. W. Kenner and E. Stenhagen, *Acta Chem. Scand.* **18**, 551 (1964).
111. H. Audier, S. Bory, M. Fetizon, P. Longevialle and R. Toubiana, *Bull. Soc. Chim. Fr.* 3034 (1964).
112. E. von Rudolf, *Can. J. Chem.* **34**, 1413 (1956).
113. D. T. Downing and R. S. Greene, *Lipids* **3**, 96 (1968).
114. W. G. Niehaus and R. Ryhage, *Anal. Chem.* **40**, 1840 (1968).
115. G. Eglington, D. H. Hunneman and A. McCormick, *Org. Mass Spectrom.* **1**, 593 (1968).
116. J. A. McCloskey, R. N. Stillwell and A. M. Lawson, *Anal. Chem.* **40**, 233 (1968).
117. C. J. Argoudelis and E. G. Perkins, *Lipids* **3**, 379 (1968).
118. J. A. McCloskey and M. J. McClelland, *J. Am. Chem. Soc.* **87**, 5090 (1965).
119. R. C. Crippen, *The Identification of Compounds with the Aid of Gas Chromatography*, McGraw-Hill, New York, 1973, pp. 248–251.
120. J. C. M. Schogt and P. Haverkamp-Bagemann, *J. Lipid Res.* **6**, 466 (1965).
121. J. S. Sawardeker, J. H. Sloneker and A. Jeanes, *Anal. Chem.* **37**, 1602 (1965).
122. E. P. Crowell and B. B. Burnett, *Anal. Chem.* **39**, 121 (1967).
123. J. F. Saeman, W. E. Moore, R. L. Mitchell and M. A. Millet, *Tappi* **37**, 336 (1964).
124. L. J. Griggs, A. Post, E. R. White, J. A. Finkelstein, W. E. Moeckel, K. G. Holden, J. E. Zarembo and J. A. Weisbach, *Anal. Biochem.* **43**, 369 (1971).
125. K. M. Fredricks and R. Taylor, *Anal. Chem.* **38**, 1961 (1966).
126. W. J. Esselman and C. O. Claggett, *J. Lipid. Res.* **10**, 234 (1969).

127. K. Biemann in G. R. Waller (Ed.), *Biochemical Applications of Mass Spectrometry*, Wiley–Interscience, New York, 1972, pp. 423–425.
128. R. Varma, R. S. Varma and A. H. Wardi, *J. Chromatogr.* **77**, 222 (1973).
129. R. Varma, R. S. Varma, W. S. Allen and A. H. Wardi, *J. Chromatogr.* **86**, 205 (1973).
130. R. C. Crippen, *The Identification of Compounds with the Aid of Gas Chromatography*, McGraw-Hill, New York, 1973, p. 266.
131. R. D. Stephens and C. H. V. Middelem, *J. Agric. Food Chem.* **14**, 149 (1966).
132. J. C. M. Schogt, P. H. Bagemann and J. H. Recourt, *J. Lipid Res.* **2**, 142 (1967).
133. J. W. Ralls, *Anal. Chem.* **36**, 946 (1964).
134. L. A. Jones and R. J. Monroe, *Anal. Chem.* **37**, 935 (1965).
135. D. F. Gadbois, P. G. Schauerer and F. J. King, *Anal. Chem.* **40**, 1362 (1968).
136. W. L. Stanley, R. M. Ikeda, S. H. Vannier and L. A. Rolle, *J. Food Sci.* **26**, 43 (1961).
137. J. Vessman and S. Stromberg, *Acta Pharm. Suec.* **6**, 505 (1969).
138. R. B. Bruce, J. E. Pitts and F. M. Pinchbeck, *Anal. Chem.* **40**, 1246 (1969).
139. P. Hartvig, S. Strandberg and B. Näslund, *J. Chromatogr.* **118**, 65 (1976).
140. L. Neelakantan and H. B. Kostenbauder, *Anal. Chem.* **46**, 452 (1974).
141. W. P. Cochrane and A. S. Y. Chau, *Advances in Chemistry*, Vol. XX, 1971, pp. 11–26.
142. P. Hartvig and J. Vessman, *Anal. Lett.* **7**, 223 (1974).
143. J. Vessman, P. Hartvig and M. Molander, *Anal. Lett.* **6**, 699 (1973).
144. P. Hartvig and J. Vessman, *Acta Pharm. Suec.* **11**, 115 (1974).
145. P. Hartvig, W. Handl, J. Vessman and C. M. Svahn, *Anal. Chem.* **48**, 390 (1976).
146. P. Hartvig and J. Vessman, *J. Chromatogr. Sci.* **12**, 722 (1974).
147. H. B. Hucker and J. K. Miller, *J. Chromatogr.* **32**, 408 (1968).
148. H. V. Street, *J. Chromatogr.* **73**, 73 (1972).
149. D. E. Schmidt, P. I. A. Szilagyi and J. P. Green, *J. Chromatogr.* **7**, 248 (1969).
150. P. I. A. Szilagyi, D. E. Schmidt and J. P. Green, *Anal. Chem.* **40**, 2009 (1968).
151(a) I. Hanin and D. J. Jenden, *Biochem. Pharmacol.* **18**, 837 (1969).
151(b) D. J. Jenden, I. Hanin and S. I. Lamb, *Anal. Chem.* **40**, 125 (1968).
152. P. Ronkainen, J. Denslow and O. Leppänen, *J. Chromatogr. Sci.* **11**, 384 (1973).
153. F. K. Kawahara, *Anal. Chem.* **40**, 1009 (1968).
154. J. Undeova, M. Hrivnac, M. Dodova, R. Staszewsi and J. Janak, *J. Chromatogr.* **40**, 359 (1969).
155. M. C. Bowman and M. Beroza, *J. Ass. Off. Anal. Chem.* **52**, 1231 (1969).
156. J. B. Himes and I. J. J. Dowbak, *J. Gas Chromatogr.* **3**, 194 (1965).
157. J. S. Parsons, *J. Gas Chromatogr.* **5**, 254 (1967).
158. C. Bighi, *J. Chromatogr.* **14**, 348 (1964).
159. F. I. Onuska and W. R. Boos, *Anal. Chem.* **45**, 967 (1973).
160. F. J. Barron and L. A. Mooney, *Anal. Chem.* **40**, 1742 (1968).
161. J. A. McCloskey and J. H. Law, *Lipids* **2**, 225 (1967).
162. K. Biemann and G. Spiteller, *J. Am. Chem. Soc.* **84**, 4578 (1962).
163. K. Biemann in G. R. Waller (Ed.), *Biochemical Applications of Mass Spectrometry*, Wiley–Interscience, New York, 1972, p. 266.
164. B. F. Nesbitt, P. S. Beevor, D. R. Hall, R. Lester and V. A. Dyck, *Insect Biochem.* **6**, 105 (1976).

CHAPTER 9

Fluorescent Derivatives

Nikolaus Seiler and Lothar Demisch

Max-Planck-Institut für Hirnforschung
und
Zentrum der Psychiatrie, Klinikum der Universität
Frankfurt am Main, West Germany.

1 INTRODUCTION

It was realized early on that the fluorescence exhibited by certain substances was a valuable property for their analysis because it was measurable at lower concentrations than optical absorbance, because it was linear with concentration over a wide concentration range, and because its relatively high specificity permitted determination even in the presence of other substances. The development of reliable and sensitive spectrofluorimeters enabled this analytical potential to be realized, and numerous analytical methods for the determination of compounds with native fluorescence have been developed.[1–3] Many methods were also developed where a compound of interest was converted to a fluorescent product by a specially devised chemical reaction to enable it to be determined by the sensitive technique of fluorimetry, and some of these reactions will be discussed. Our main concern, however, is with the application of reagents which are designed to make fluorescent derivatives of compounds of interest, in order to permit their determination or their detection by fluorimetric methods, especially allied to chromatographic (or electrophoretic) separation processes.

1.1 The choice of fluorigenic reagents

The most important reagents with fluorigenic groups used for labelling compounds are summarized in Fig. 1.

Sensitivity is normally one of the most important aspects of fluorescence methods. High fluorescence efficiency of the derivatives is therefore necessary. With all methods in which a reagent reacts with a certain functional group, their specificity is normally limited by the efficiency of the separation methods. From

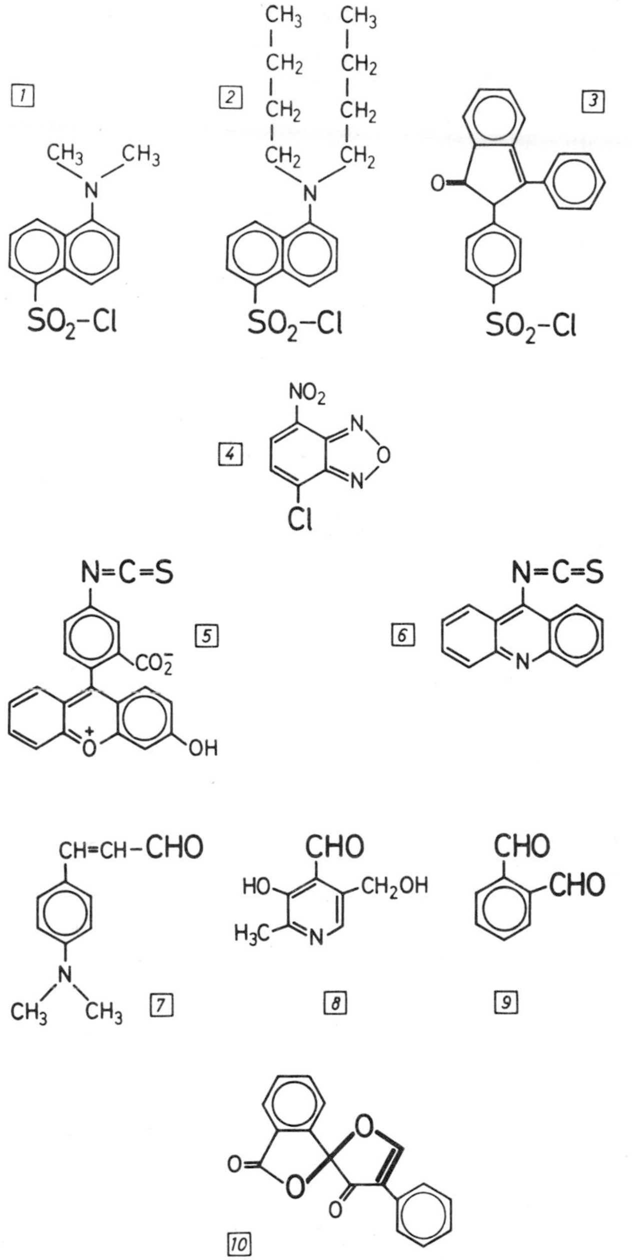

Fig. 1 Reagents for fluorescence labelling of amino groups. 1. 5-dimethylaminonaphthalene-1-sulfonyl chloride (Dns—Cl); 2. 5-di-*n*-butylaminonaphthalene-1-sulfonyl chloride (Bns—Cl); 3. 2-*p*-chlorosulfophenyl-3-phenyl indone (Dis—Cl); 4. 4-chloro-7-nitrobenzo[*c*]-1,2,5-oxadiazole (Nbd—Cl); 5. fluorescein isothiocyanate; 6. 9-isothiocyanatoacridine; 7. *p*-dimethylamino-cinnamaldehyde; 8. pyridoxal; 9. *o*-phthaldialdehyde; 10. fluorescamine.

a practical viewpoint, any reagent for fluorescent derivative formation should fulfil some or all of the following prerequisites:

1. Rapid quantitative reaction under mild conditions, preferably in water or at least in aqueous media.
2. Formation of relatively non-polar derivatives, in order to permit their isolation and concentration by extraction into organic solvents.
3. Specificity for a certain functional group.
4. Excess of the reagent should be easily separable from its reaction products. The derivatives should have favourable chromatographic properties.
5. High fluorescence efficiency; emission at a wavelength long enough to avoid the generally bluish background fluorescence of solvents and adsorbents, but within the range of high sensitivity of the available detectors.
6. High molar extinction coefficient at a suitable wavelength, for instance at one of the intense mercury lines; no overlapping of excitation and emission bands.
7. If the reagent is to be used for end-group determination of peptides and proteins, or if conjugates of certain compounds are to be studied, the derivatives must be stable to hydrolytic cleavage.

The history of fluorescence labelling of small molecules for the purpose of their identification and quantitative determination began with the application of 5-dimethylaminonaphthalene-1-sulfonyl chloride (Dns—Cl) as a fluorescent end-group reagent in protein chemistry by Gray and Hartley[4] in 1963. This was the first successful attempt to replace the non-fluorescent 2-naphthalene-sulfonyl chloride of Fischer and Bergell[5,6] and Sanger's 2,4-dinitrofluorobenzene[7,8] by a fluorescent label, and increased the sensitivity of end-group determinations by more than an order of magnitude.[9,10] Thin-layer chromatography as a separation method, quantitative methods, and application of this reagent to amine analysis were all introduced by Seiler and co-workers.[11–13]

Although the measurement of fluorescence depolarization and decay are still powerful tools in the study of the physical chemistry of proteins,[14] the application of fluorescence labelling of proteins has lost some of its importance, since improved instrumentation allows the measurement of proteins' native fluorescence. Most of the fluorescent reagents developed for this purpose reacted with the amino groups of proteins, and since these same reagents were mainly applied for microanalytical purposes, few fluorescent labels are applied for functional groups other than primary or secondary amino groups, and they are rarely used.

Measurement of low concentrations of compounds by means of fluorimetry is normally the main purpose of fluorescence labelling of compounds without other physical properties that allow sensitive detection. Where there is fluorescence quenching by known or unknown components of the sample solutions, or a lacking of specificity of the fluorescence measurements, the application of alternative methods for the determination of fluorescent derivatives may be necessary. In such cases fluorescence may be used to monitor the chromatographic separations, while the alternative quantitation methods are applied. Radioactivity measurement and mass spectrometry are frequently used. If a reagent is selected for a certain analytical purpose, besides its reactivity and the fluorescence and chromatographic characteristics of its derivatives, its accessibility to radioactive labelling and its suitability for qualitative and quantitative mass spectrometry should therefore also be considered.

2 5-DIMETHYLAMINONAPHTHALENE-1-SULFONYL CHLORIDE (Dns—Cl)

2.1. History

G. Weber introduced 5-dimethylaminonaphthalene-1-sulfonylchloride (Dns—Cl) (Fig. 1-1) in 1952 as a reagent for the preparation of fluorescent conjugates of proteins, and used it for the study of protein structure by fluorescence polarization measurements. It is particularly useful for proteins with a molecular weight of 10 000–200 000.[15–17] Shore and Pardee[18] showed the value of Dns-conjugates for the study of energy transfer from tryptophan-containing proteins. The protein fluorescence almost overlaps the dye absorption. Subsequently Dns-Cl was used for the preparation of fluorescent antibodies in numerous studies,[19] and for the study of active centres of enzymes. The stoichiometric inhibition of α-chymotrypsin by reaction with Dns—Cl[20] suggested a method for the identification of the constituent amino acid in question, namely degradation of the fluorescent enzyme, and identification of the fluorescent fragment. From this, Gray and Hartley developed an end-group method with Dns—Cl as reagent,[4,9,21] analogous to the method of Sanger,[7,8] and used high-voltage paper electrophoresis to separate the dansylated amino acids. Subsequently Seiler and Wiechmann developed two-dimensional TLC systems for this purpose,[11] and Boulton and Bush[22] paper chromatographic systems. With the introduction of the polyamide sheets by Woods and Wang,[23] the method became a routine one, and found broad application. The dansylation procedure was first applied to the identification of biogenic amines in 1965.[12] Methods for the quantitative estimation of the Dns-derivatives were first published in 1966,[13] and with minor variations, are still in use.[10,24–29]

Preparation of Dns—Cl[10,17]

5-Dimethylaminonaphthalene-1-sulfonic acid (2 g, purified by crystallization from water containing a few drops of sulfuric acid, using charcoal) are ground in a mortar with 3.5 g of PCl_5 for 3–5 min. After HCl evolution has ceased, cracked ice and water are added and the mixture is vigorously stirred to wash out phosphoric acid. If the reaction product does not crystallize, it is extracted with ethyl acetate. The ethyl acetate solution is washed with sodium bicarbonate, and then with water until neutral. After drying with Na_2SO_4, and evaporation of the solvent, Dns—Cl usually crystallizes. Traces of phosphoric or other acids increase the rate of decomposition of Dns—Cl. Dns—Cl is also available commercially.

2.2 Purification of Dns—Cl (including commercial Dns—Cl)

About 200mg of Dns—Cl in 2–3 ml of toluene are applied to a column (2 × 20 cm) of silica gel H (E. Merck, Darmstadt, Germany), or its equivalent, prepared by filling a suitable glass column with a slurry of silica gel in toluene. On elution with toluene Dns—Cl is eluted ahead its derivatives. The solvent is evaporated *in vacuo* giving orange crystals, m.p. 69 °C; easily soluble in alcohols, ketones, aromatic hydrocarbons and chloroform; less readily soluble in aliphatic hydrocarbons and very slightly soluble in water. The fluorescence efficiency of Dns—Cl is low.

2.3 Reactions of Dns—Cl

As an aromatic sulfonyl chloride, Dns—Cl reacts with primary and secondary amino groups even at a slightly alkaline pH; at higher pH with phenols, imidazoles, and slowly even with alcohols. Derivatives with barbiturates and purines are known.[9,10,24,30] Thiol compounds form the corresponding disulfides.[10,24]

The reaction with amines is formulated in Fig. 2. Polyfunctional molecules react completely under the usual conditions: *O,N*-bis-Dns-tyramine, *N,N*-bis-Dns-histamine, tri-Dns-spermidine, tetra-Dns-spermine and tri-Dns-norepinephrine are the products, when excess of the reagent is used. Amino acids react first with the amino group. The Dns-amino acids, tend, however, to react with excess Dns-Cl to form mixed anhydrides, especially at high pH. Under the usual reaction conditions anhydrides of α-amino acids partially

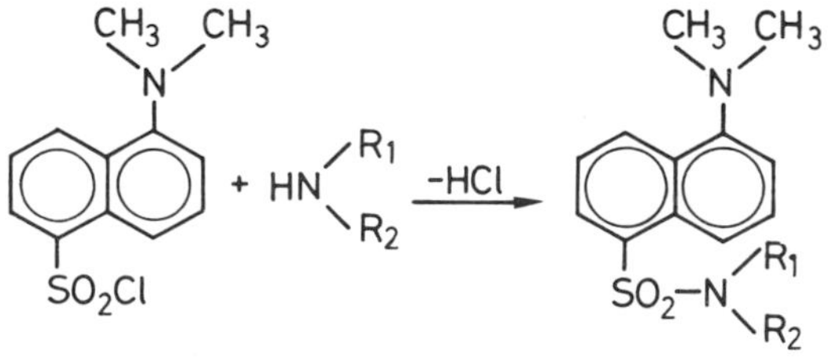

Fig. 2 Reaction of Dns—Cl with an amine.

beak down to CO, Dns—NH_2, the aldehyde with one carbon less than the parent amino acid and dimethylaminonaphthalene-sulfonic acid[31] (Fig. 3). γ-Amino acids under the same conditions form the γ-lactams[25,32–34] (Fig.

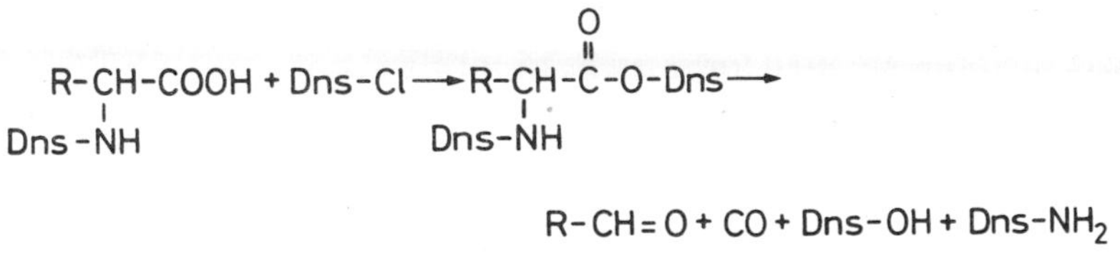

Fig. 3 Decomposition of α-amino acids with excessive Dns—Cl.

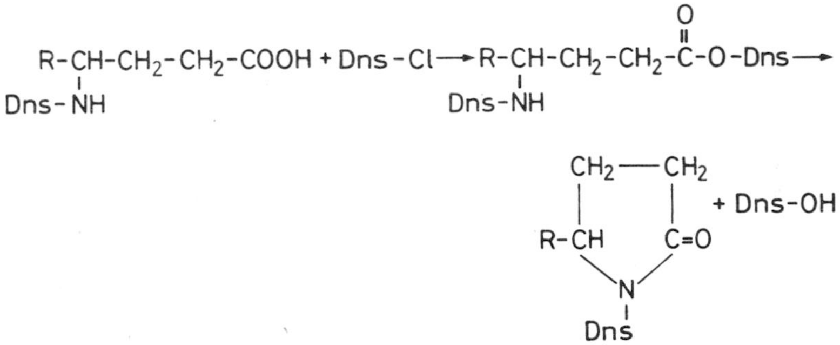

Fig. 4 Formation of Dns-2-oxopyrrolidine from Dns-γ-aminobutyrate with excess of reagent.

4). Steric hindrance as with the branched chain aliphatic amino acids, favours decomposition. Amino acids with secondary amino groups (proline, hydroxyproline), and β-amino acids are relatively stable. These breakdown reactions with excess Dns—Cl are the main obstacle in the determination of free amino acids in tissues and body fluids. The lactam formation can be exploited for the assay of γ-amino-butyric acid.[32,35,36]

With primary amines side-reactions are unusual, and it is normally possible to obtain stoichiometric amounts of the reaction products. At high pH and elevated temperatures tertiary amines may be attacked to a significant extent, forming the Dns-derivative of secondary amines by elimination of an alkyl or aryl residue:[37] Dns-dimethylamine may be formed by this direct electrophilic attack or by a mechanism analogous to that of the Bucherer reaction.[38]

The rate of dansylation increases with increasing pH, but is paralleled by an increased rate of hydrolysis of the Dns-Cl. Optimal conditions are those under which the reactive groups most effectively compete for the limited amount of reagent available. Labelling of most amino acids, amines and phenols is optimal at pH 9.5–10 (at room temperature). At pH values below 8 the unreactive protonated form of the amino groups predominates.

Dns—Cl is only very slightly soluble in water and so dansylation is mostly done in acetone–water mixtures. Other organic solvents, e.g. dioxane, have been used. For the dansylation of barbiturates, ethyl acetate has been used.[30]

2.4 Procedures for the use of Dns—Cl

2.4.1 GENERAL PROCEDURE FOR THE DERIVATIVE FORMATION WITH AMINES[10]

Tissue extracts in 0.2 M perchloric acid (100–500 μl) or similar amounts of sample solutions are used for assay in 15 ml glass-stoppered centrifuge tubes. Samples with internal standards are prepared by addition of known amounts of the compounds to be determined, corresponding approximately to the amounts of these substances in the samples. External standards are carried through the procedure in similar amounts. In addition two or three blanks (containing the appropriate amount of perchloric acid) are prepared. Three times the sample volume of a solution of Dns—Cl in acetone or dioxane (10 mg ml^{-1}) are added to each sample. The mixture is saturated with $Na_2CO_3.10H_2O$. The stoppered centrifuge tubes are stored overnight at room temperature. Shaking of the tubes is advisable, especially during the initial phase of the reaction. By sonicating the tubes in an ultrasonic cleaning-bath, the reaction can be completed within 2–3 h at room temperature. Since Dns-derivatives are light-sensitive, unnecessary exposure to bright light should be avoided.

In order to remove excess reagent (which can be recognized by its yellow colour) 5 mg of proline dissolved in 20 μl of water is added to each sample, if amines or γ-aminobutyric acid are to be determined. After sonification for about 3 min or storage for 30 min, the Dns-amine derivatives are extracted with 3–5 volumes of toluene, (i.e. 5–10 ml) and separated by centrifugation.

For the estimation of amino acids the excess of Dns—Cl must be minimized, because of the partial breakdown of the Dns-amino acids and the need to separate dimethylaminonaphthalene sulfonic acid from the amino acid derivatives. If this acid is formed in large amounts, it may cause serious difficulties. Usually the excess of reagent is kept as small as possible. Excess reagent can be removed from the reaction mixture by extraction with *n*-heptane or toluene. Side-products of the dansylation together with Dns-amine derivatives are also removed, at least partially, if *n*-heptane is used as solvent; and nearly completely, if toluene is used. Toluene extraction may, however, remove traces of the Dns-amino acids as well. The extraction of Dns—NH_2 is not complete under these conditions. After neutralization (for example with NaH_2PO_4) most of the Dns-amino acids can be extracted from the water phase with ethyl acetate. Exceptions are the derivatives of aspartate, arginine, taurine, mono-Dns-lysine and orithine. Dns-glutamate also remains in the aqueous layer, but *N*-Dns-pyrrolidonecarboxylic acid, the reaction product of glutamate[34] with excess reagent, is extractable. Other non-extractable compounds of importance are *O*-Dns-choline and *O*-Dns-bufotenine.

2.4.2 RECOMMENDED PROCEDURE FOR PEPTIDES[9]

A solution of the peptide (0.5–5 nmol) is transferred to a small test tube. After drying *in vacuo*, the peptide is redissolved in 15 μl of 0.2 M $NaHCO_3$ and dried a second time. The peptide is then dissolved in 15 μl of water. The pH at this point should be 8.5–9.0, and if correct, 15 μl of Dns—Cl (5 mg ml^{-1} of acetone) are added. The reaction is allowed to proceed for 1 h at 37 °C or 2 h at room temperature.

For the identification of the labelled amino acid, the solutions are dried *in vacuo*, and 100 μl of 6.1 M HCl is added to each tube. The tubes are sealed and hydrolysed for 6–18 h at 105 °C. HCl is removed from the opened tubes *in vacuo* over NaOH pellets. Most of the material in the peptide hydrolysates is a mixture of sodium chloride and Dns-OH, with small amounts of Dns-amino acids and Dns-NH_2. Two extractions with water-saturated ethyl acetate (100 μl) removes the neutral Dns-amino acids almost completely, Dns-aspartate and glutamate approximately 80%, moderate amounts of mono-Dns-lysine and *O*-Dns-tyrosine, and only traces of Dns-histidine, arginine, and cysteic acid.

2.4.3 REACTION OF BARBITURATES IN ETHYL ACETATE[30]

Fifty nmol of a barbiturate dissolved in 100 μl of ethyl acetate are refluxed for 2 h with 300 nmol Dns—Cl and 1–5 mg solid K_2CO_3. Care has to be taken that the reaction mixture is not overheated locally, because thermal decomposition of Dns-derivatives is severe under these circumstances. Aliquots of the ethyl acetate solution can be immediately applied to thin-layer plates or columns for chromatographic separation. Other compounds than barbiturates can be reacted with Dns—Cl in a similar way.

2.5 Properties of Dns-derivatives

2.5.1 GENERAL PROPERTIES

Derivatives of the usual amino acids (now commercially available) and a considerable number of amine, aminophenol[24] and phenoaldehyde derivatives[39] have been prepared on a preparative scale. The Dns-derivatives are pale yellow to yellow crystalline solids, mostly readily soluble in organic solvents, and only slightly soluble in water. The NH-group of the derivatives of ammonia and of primary amines can be deprotonated;[13] the solubility of these compounds in water therefore increases with increasing pH.

2.5.2 STABILITY TO HYDROLYSIS

Dns-amides are stable to acid hydrolysis, and fairly stable even to hydrolysis with bases. However, transfer of dansyl groups to other bases occurs in mild alkali; this reaction must be born in mind when hydrolysis is carried out in alkaline media. In Table 1 some data concerning the stability of Dns-derivatives to hydrolysis are summarized. Phenolic esters are much less stable than the sulfonamides, therefore it is possible quantitatively to split off the *O*-Dns from *O*,*N*-Dns-aminophenols with methanolic KOH. This is of considerable analytical interest because the *N*-Dns-derivatives of tyramine,

TABLE 1

Stability of Dns-derivatives to hydrolysis. (A) Hydrolysis with 6.1 M HCl; 105 °C; 18 h (according to Gray[9]) (B) Hydrolysis with 5 M methanolic KOH; 30 min, 50 °C (according to Seiler and Deckardt[40])

Derivative	Percent remaining (A)	Percent remaining (B)	Derivative	Percent remaining (B)
Dns-glycine	82		Dns-dimethylamine	100
Dns-α-alanine	93		Dns-piperidine	100
Dbs-β-alanine		49	Dns-ethanolamine	100
Dns-serine	65		bis-Dns-cystamine	60
Dns-threonine		22	bis-Dns-putrescine	100
Dns-valine	100		tri-Dns-spermidine	100
Dns-leucine	100		tetra-Dns-spermine	100
Dns-isoleucine	100	78	Dns-aniline	96
Dns-proline	23		Dns-benzylamine	100
Dns-hydroxyproline		6	Dns-β-phenylethylamine	100
Dns-aspartate		8	Dns-tryptamine	100
Dns-tryptophan	0	82	bis-Dns-histamine	0
bis-Dns-lysine		80	Dns-phenol	0
O,N-bis-Dns-tyrosine		0	*O,N*-bis-Dns-tyramine	100[a]
Dns-ammonia		100	*O,N*-bis-Dns-synephrine	100[a]
Dns-methylamine		100	*O,N*-bis-Dns-5-hydroxytryptamine	0

[a] Recovered as *N*-Dns-derivative.

synephrine etc. have much higher fluorescence efficiencies than the corresponding *O,N*-bis-Dns-derivatives. Imidazoles and alcoholic esters are unstable, both in acidic and alkaline media. Dns-derivatives of imidazoles, can, however, be selectively cleaved with formic acid.[41]

2.5.3 SPECTRAL CHARACTERISTICS

The spectral properties of Dns-derivatives are dependent on structural features, but are profoundly changed by solvents, adsorption to active surfaces, and pH. As a rule, fluorescence emission maxima of compounds with primary amino groups are at shorter wavelengths than those with secondary amino groups (Table 2). Imidazoles and the dansylated lactams fluoresce at relatively long wavelength, comparable with phenol-esters. For instance, Dns-γ-aminobutyric acid has its activation maximum (in methanol) at 354 nm, and its fluorescence maximum at 534 nm, whereas the corresponding lactam (Dns-2-oxo-pyrrolidine) is maximally activated at 367 nm and emits at 567 nm.[32] Consequently Dns-amine and amino acid derivatives fluoresce greenish to yellow whereas phenols, imidazoles and 2-oxo-pyrrolidines show an orange fluorescence.

TABLE 2

Fluorescence efficiency of some Dns- and Bns-derivatives.[a] (Emission maxima in parenthesis)

Derivative	Solvent: Ethyl acetate	Solvent: Water
ε-Dns-lysine		0.026
Dns-proline		0.053
Dns-glycine		0.065
Dns-glutamic acid		0.066
Dns-tryptophan	0.54 (510 nm)	0.068 (578 nm)
bis-Dns-cystine		0.087
Dns-methionine		0.088
Dns-valine		0.091
Dns-methylamine	0.61 (502 nm)	
Dns-dimethylamine	0.53 (510 nm)	
Dns-piperidine	0.59 (508 nm)	
Dns-pyrrolidine	0.59 (508 nm)	
Bns-methylamine	0.72 (500 nm)	
Bns-dimethylamine	0.64 (505 nm)	
Bns-piperidine	0.68 (505 nm)	
Bns-pyrrolidine	0.68 – (503 nm)	

[a] The values of the Dns-amino acids are from Chen[42] and those of the amine derivatives from Seiler *et al.*[43]

The quantum yields of amine- and amino acid-derivatives are of the same order of magnitude (Table 2). Phenol- and imidazole-derivatives, however, have much lower fluorescence efficiencies. In Table 3, the spectral characteristics of an amine derivative, of an aminophenol- and a phenol-derivative are compared. The interesting point is that the aminophenol-derivative (*O,N*-bis-Dns-*p*-tyramine) shows much lower fluorescence efficiency than the analogous amine derivative, although it bears two apparently independent fluorophores. But we observe only one activation and fluorescence maximum, and the usual two absorption maxima, which are between the corresponding bands of the

TABLE 3

Comparison of the spectral characteristics of Dns-β-phenylethylamine. *O,N*-bis-Dns-*p*-tyramine and Dns-phenol in methanol as solvent[10,13]

	Absorption maximum/ nm	$\varepsilon_{(\lambda)}$	Fluorescence excitation/ nm	Fluorescence emission/ nm	Relative fluorescence efficiency
Dns-β-phenyl-ethylamine	336 252	4 530 14 100	360	521	100
O,N-bis-Dns-*p*-tyramine	344 254	7 880 26 800	362	525	58
Dns-phenol	350 255	4 050 14 100	375	540	10

amine- and the phenol-derivative (Table 3). Similar observations can be made e.g. with the Dns-histamine derivatives. Since the two fluorophores are not connected by a system of conjugated double bonds, and the spectral characteristics are not changed by dilution, one might asume an intramolecular interrelation of the π-electron systems of the fluorophores.

If in acidic solutions the dimethylamino group of the Dns-derivatives is protonated, fluorescence efficiency decreases to a low level.[13] Protonation may be at least partially the reason for the decreased fluorescence efficiency of Dns-derivatives adsorbed to active silica gel, since spraying with bases increases fluorescence efficiency very considerably, besides stabilizing the Dns-derivatives against (most probably light-induced) oxidative degradation.[10,13,24] In strongly alkaline media the amido group of the Dns-derivatives of ammonia and of primary amines are ionized, and a hypsochromic shift of the absorption and fluorescence bands is observed.[13]

Fluorescence efficiency of Dns-derivatives is strongly solvent dependent: increased fluorescence efficiency and a hypsochromic shift of the emission maximum are observed with decreasing dielectric constant. In Table 4 are summarized data reported by Chen[42] for Dns-tryptophan. It is interesting to note in this connection, that addition of about 8 mmol cyclohepta-amylose to the aqueous media greatly enhances fluorescence efficiency.[44]

Oxygen significantly quenches fluorescence of Dns-derivatives. The selection of the appropriate solvent for extraction and subsequent measurement by fluorimetry is therefore of considerable practical importance. Flushing of the solvent with nitrogen, or distillation in nitrogen atmosphere (under reduced pressure) for removal of oxygen is advisable.

TABLE 4

Fluorescence efficiency and emission maximum of Dns-D,L-tryptophan in solvents of different dielectric constants (according to Chen[42])

Solvent	Solvent dielectric constant/ debye	Emission maximum/ nm	Fluorescence efficiency
Water	78.5	578	0.068
Glycerol	42.5	553	0.18
Ethylene glycol	37.0	543	0.36
Propylene glycol	32.0	538	0.37
Dimethylformamide	36.5	517	0.59
Ethanol	25.8	529	0.50
n-Butanol	19.2	519	0.50
Acetone	21.5	513	0.35
Ethyl acetate	6.1	510	0.54
Chloroform	5.1	508	0.41
Dioxane	3.0	500	0.70

2.6 Applications

2.6.1 SEPARATION PROCEDURES

Most current chromatographic procedures have been applied to the separation of Dns-derivatives, except GC, where the volatility of the Dns-derivatives is too low (N. Seiler, unpublished work). High-voltage paper electrophoresis was the first method introduced for the identification of dansylated amino acids and for the separation of peptides.[4,9] and subsequently electrophoretic separations on thin layers were also reported.[45,46] These methods are, however, unsuitable for the separation of amine derivatives. TLC on silica gel has been widely used for the separation of amino acid, peptide and amine derivatives. A large variety of solvent systems of varying polarity is now available, which allow the separation of even complex mixtures of amino acids,[10,28,47,48] and amines.[10,24,49] Fingerprinting or identification of Dns-peptides can also be carried out on silica gel thin-layers.[41,50,51] For the separation of Dns-peptides, electrophoresis in polyacrylamide gel[52] and on cellogel strips[53] was recently recommended. Alumina,[54] Kieselgur[55] and cellulose thin-layers[56,57] have also been tried.

TLC on activated layers may be advantageously applied to a broad spectrum of separation methods, from pure adsorption chromatography to pure partition chromatography. There are few separation problems that cannot be solved by selection of the appropriate adsorbent and solvent systems. Furthermore, the background fluorescence of these layers is generally low, and the amount of extractable impurities, which interfere with mass spectrometric methods (see p. 359) can be kept to a minimum, using suitable binders.

Paper chromatography of Dns-derivatives[22,57–59] is no longer significant, but TLC on polyamide sheets is widely used. These methods are similar in that only partition chromatography can normally be used. In these systems derivatives of amines and amino acids move together, so that they cannot be separated as easily from each other as on active layers. The limited choice of solvents suitable for separation of Dns-derivatives on polyamide sheets is attested by the fact that the original solvents of Woods and Wang[23] have been only slightly modified for the separation of the protein amino acids[9] and for the amino acids and amine derivatives of a variety of tissues.[27,60–62] The polyamide sheets are good for end-group determinations of peptides and proteins because of the high detection sensitivity (small spots, high fluorescence efficiency), stability of the Dns-derivatives and rapid chromatographic development. Disadvantages are the restriction to partition chromatography, the relatively high background-fluorescence, the extraction of compounds that interfere with mass spectrometry, and especially the limited sample capacity of the layers which severely limits the applicability of polyamide sheets to the analysis of tissues and biological fluids.

For improvement of separations and especially to allow automated procedures, column chromatographic methods have more recently been proposed for the separation of Dns-derivatives.[63–71] Major progress can be expected from automated column chromatography of the Dns-derivatives of amines and amino acids in the picomole range.

2.6.2 QUANTITATIVE ESTIMATION

For the quantitative assay of Dns-derivatives four different methods are at our disposal: optical absorption, fluorimetry, quantitative mass spectrometry and the application of radioactive Dns–Cl.

Colorimetry. The intense absorption bands of Dns-derivatives in the range of 250–255 nm and 335–350 nm[10] (see Table 3) are suitable for the quantitative measurement of Dns-derivatives by absorptimetry, but have apparently been used only by Engelhardt *et al.*[67] for the monitoring of Dns-amino acids in the effluent of an HPLC column.

Fluorimetry. Fluorimetry of chromatographically separated Dns-derivatives is the most obvious method for quantitative measurement of Dns-derivatives. Extraction of the derivatives from the adsorbent, and *in situ* fluorescence scanning of the chromatograms are equally suitable, as is the monitoring of fluorescence in column-effluents. Principles and applications of the procedures for the estimation of Dns-derivatives on thin-layer chromatograms were first reported by Seiler and Wiechmann,[10,13,24] and for paper chromatograms by Boulton.[72] Some modifications for the assay of Dns-amino acids were subsequently published.[61,73–78] Extraction procedures and *in situ* scanning are equally reliable and sensitive, and both methods can be applied to unidimensional and bidimensional TLC or electrophoresis. Reproducibility is normally within ±5%.

Suitable solvents should be used for the extraction: amino acid derivatives are best extracted from silica gel layers with methanol or with a mixture of methanol: 25% NH_3 (95:5)[13] or with chloroform:methanol:acetic acid (7:2:2).[77] Peptides are extractable with acetone:water (1:1),[78] while amine derivatives are extractable with less polar solvents, e.g. ethyl acetate, benzene:acetic acid (99:1), benzene:triethylamine (95:5)[13,24] or with dioxane, if as well as fluorescence, radioactivity is to be measured by liquid scintillation counting.[79] Chloroform is suitable for the extraction of Dns-amino acids from polyamide sheets.[61] Excitation and emission wavelengths are, if possible, adjusted to the individual Dns-derivatives; however, for almost every compound, activation can generally be achieved using the 365 nm mercury line.

The theoretical basis of *in situ* fluorescence scanning and its practical application has been described.[80,263,264] Since Dns-derivatives are labile on active layers, spraying with a solution of triethanolamine in propanol-2 (1:4) is advantageous,[13] since it increases the fluorescence and stability of the Dns-derivatives. For the evaluation of two-dimensional thin-layer chromatograms an apparatus was devised which allows two-dimensional fluorimetry.[73,74] Fluorescence intensity is linear with the amount of Dns-derivative up to at least 10 nmol ml^{-1} and 1 nmol/spot, respectively.

Quantitative mass spectrometry. Mass spectrometry is almost indispensable for the identification of small amounts of unknown compounds. Chromatographic criteria alone are insufficient for unambiguous identification. Other physical methods normally require more than nanomole quantities of material.

Since electron impact mass spectra of Dns-derivatives usually include the molecular ions, these derivatives are quite useful for the mass spectrometric identification of unknown compounds, although only a few typical fragments characteristic of the labelled molecule are formed (Fig. 5). The main fragment ion (m/e 170 or 171) is that formed by cleavage of the S—C bond of the sulfonyl group.[33,81–85]

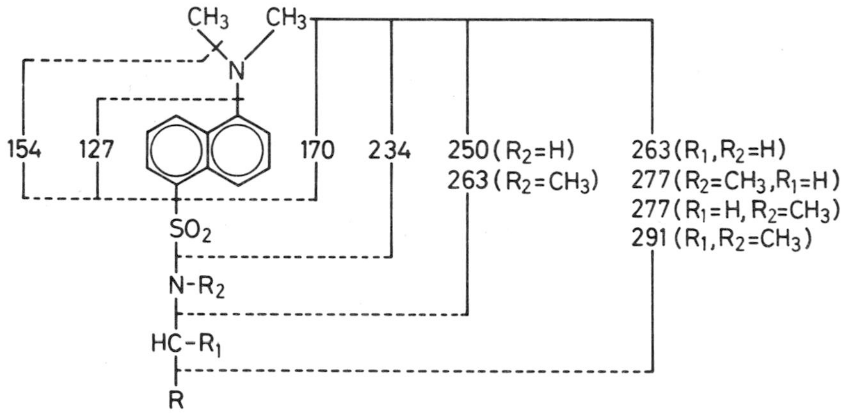

Fig. 5 Fragmentation of Dns-derivatives induced by electron impact.

The quantitative evaluation of the molecular ion or of a typical fragment-ion is much more specific than fluorimetry. Although underivatized compounds can be determined with a method called integrated ion-current monitoring,[86] it is often advantageous to use Dns-derivatives.[87] The background at the mass range of the molecular ions of Dns-derivatives is normally very low, so that erroneous peak identification is much less likely than with low molecular weight compounds.

The principle of the procedure is that the sample together with a standard is evaporated from the direct probe of the mass spectrometer. The ion currents of the molecular ions (or fragment ions) of sample and standard are alternately recorded against time during the evaporation process, and subsequently integrated. The ratio of the integrated ion currents of sample and standard is a measure of the amount of sample, essentially independent of changes in instrumental sensitivity. For detailed descriptions of suitable procedures see Refs. 87 and 88. Picomole amounts of Dns-derivatives can be measured with these methods.

Radioactive Dns–Cl. [*N*-Methyl-^{14}C]Dns—Cl (specific radioactivity 10–30 Ci mol^{-1}) and [G-^{3}H]Dns—Cl (specific radioactivity 3–10 Ci mmol^{-1}) are available from several commercial sources. The labelled reagents can be applied just like the non-labelled Dns—Cl. Their application allows the substitution of fluorimetry by the more convenient and automated liquid scintillation counting. The fluorescence of the derivatives is used only for visualization of the spots on the chromatograms. For quantitative evaluation the adsorbent is scraped off, or the layer is cut out with a razor in the case of polyamide sheets, and placed directly into the counting vials. 0.5–1 ml of methanol is added, and after about 15 min, 10 ml of the liquid scintillator. Since only small amounts are usually measured, counting efficiency is high, and it is not necessary to make corrections for different amounts. In comparison with fluorimetry, the sensitivity is about one order of magnitude greater if the reagent with the highest available specific radioactivity is used. Detection sensitivity can be further increased by preparing autoradiographs of the chromatograms, which can be evaluated by microscope photometry.[89]

For quantitative derivatization, Dns—Cl has to be employed in large excess, so that even the use of microtechniques for the preparation of the derivatives[60] is expensive. Users of labelled Dns—Cl are therefore frequently tempted to use too little reagent.[90] More important, the *specificity* of the method is not increased by the use of one of the labelled reagents, as compared with fluorimetry. Nevertheless both ^{14}C and ^{3}H-Dns—Cl have frequently been used for the identification or estimation of amino acids and of amines.[26,27,60,89–93] A considerable improvement of the application of labelled Dns—Cl lies in the use of double isotope techniques.[94–97] The procedure is that a known amount of the compound to be determined is added to the tissue sample in the form of its ^{14}C-labelled specimen (with high specific radioactivity) as internal standard. The compound is now isolated by the usual procedure and then derivatized with [G-^{3}H]Dns—Cl. The Dns-derivative is purified extensively by the usual chromatographic procedures and the radioactivity of the two isotopes is measured. The amount of non-radioactive compound present in the sample can be calculated from the $^{3}H/^{14}C$-ratio. Incomplete derivate formation or losses during the isolation procedure do not influence the result, as long as the isolated spot is homogeneous. The coefficient of variation for the isotope ratio in blanks containing only ^{14}C-4-aminobutyrate was 6.2%.[95]

2.6.3 APPLICATIONS

Dns—Cl is the most widely used fluorescent reagent. Beside its use in the physical chemistry of protein structure, its applications range from the determination of small amounts of amines, amino acids and phenolic compounds in tissues and biological fluids to the quantitative assay of drugs and drug metabolites; from the evaluation of peptide sequences to the characterization of active centres of enzymes. Since it is impossible to review all methods and modes of application in detail within this chapter, the most typical aspects of the use of Dns—Cl are summarized in Table 5.

TABLE 5

Applications of Dns-Cl

Type of application	Compound	Animal and tissue	Reference
End-group analysis	Peptides, proteins		9
N-terminal amino acid			29
			98
			99
			100
Aminoacyl residues	Peptides, proteins		101
Formyl- and acetyl-groups	Proteins		102
C-terminal amino acid	Peptides, proteins		103
			104
Sequence analysis	Peptides, proteins		21
Edman degradation			105
			106
			107
	chymotrypsinogen,		108
	immunoglubulin G		109
			110
Fingerprinting of peptides	Protein digests and peptide mixtures		50
			51
			76
			111
			112
Indentification of peptides	Different peptides		41
Determination of molecular weight	Histones		52
			53
Recovery of biologically active peptides			113
Labelling with cycloheptaamylose	Mucopoly-saccharides		114
Dns-complex	Proteins		115
			116
Determination/identification of amino acids		Superfusates of rat cerebral cortex	117
		Rat optic nerve	92
		Rat dorsal ganglia	118
		Invertebrate neurons	119
		Axonal transport	120
	Histidine		121
	4-Aminobutyrate	Mouse tissues	35
		Trout brain	122
		Mouse brain	79
		Mouse liver	123
		Torpedo electric organ	124
		Enzymatic method	125

TABLE 5—continued

Type of application	Compound	Animal and tissue	Reference
Characterization of *t*-RNA			91
Determination/ identification of amines			24
General methods			25
Specific methods	β-Phenylethylamine	Rat brain	126
	p-Tyramine	Rat brain	126
			127
			128
			129
	Tryptamine	Octopus brain	129
		Rat brain	126
	3,4-Dimethoxy-β-phenylethylamine	Human urine	130
			131
			132
	Serotonin	Mouse brain	133
		Toad brain	133
	Bufotenin	Toad brain	133
			134
	Putrescine, spermidine, spermine	Rat brain	135
		Trout brain	136
	Acetylputrescine	Mouse tissue	137
	Cadaverine	Snail brain	138
		Mouse brain	139
	Piperidine	Snail brain	140
		Mouse brain	141
	Cyclohexylamine		142
	Hexosamines		143
Determination/ identification of phenolic compounds	Estrogens		144
			145
	Hydroxybiphenyls		146
	Pyridoxalphosphate		147
Determination/ identification of drugs	Narcotics		148
	Barbiturates		30
			69
	Chlorpromazine metabolites		149
			150
			151
			54
	Cannabinoids		152
			153

TABLE 5—continued

Type of application	Compound	Animal and tissue	Reference
Determination of enzyme activities	Carboxylic ester hydrolase		154
	Cholinesterase		155
	Acetyl-CoA:1,		156
	4-diaminobutane		157
	N-acetyltransferase		

3 5-DI-*n*-BUTYLAMINONAPHTHALENE-1-SULFONYL CHLORIDE (Bns—Cl)

5-Di-*n*-butylaminonaphthalene-1-sulfonyl chloride (Bns—Cl) (Fig. 1-2) is an analogue of Dns—Cl. It is somewhat more stable in storage. Its reaction with amines is slightly slower. The fluorescence activation maxima of its derivatives are shifted a little towards longer wavelengths, the emission maxima towards shorter wavelengths (Fig. 6) and its fluorescence efficiency is about 10% higher compared with Dns-derivatives (Table 2).[43] Its uses are the same as those of Dns—Cl. It can be used, for instance, for the analysis of end-groups of proteins and peptides. A suitable solvent system has been reported for the complete separation on polyamide sheets of the Bns-derivatives for all the amino acids normally found in proteins.[158]

Bns—Cl was synthesized to overcome difficulties in the estimation of piperidine in tissues.[159] It was hoped to improve quantitative mass spectrometry (see p. 395) by increasing the molecular weight of the reagent, and by

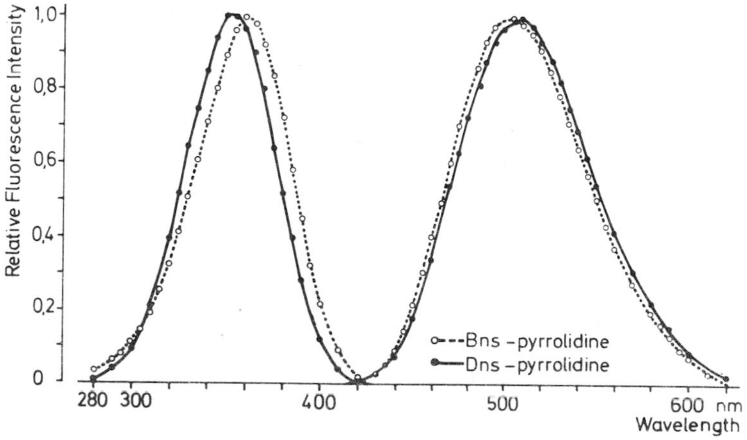

Fig. 6 Activation and fluorescence spectra of Dns- and Bns-pyrrolidine in ethyl acetate solution.

decreasing its polarity. These expectations were fulfilled: Bns-derivatives are much more lipid soluble than Dns-derivatives. In consequence they can be separated on thin-layer chromatograms with less polar systems.[158,159] More important is that they are extracted from reaction mixtures and adsorbents with less polar solvents than the corresponding Dns-derivatives. This means that all amino acid Bns-derivatives can be extracted from reaction mixtures with ethyl acetate, and that the samples extracted for quantitative mass spectrometry contain less impurities. It means too, that Bns-derivatives are eluted from reversed-phase columns at higher methanol concentrations, so that increased fluorescence gives about three-fold better detection sensitivity.

The greatest advantage of the Bns-derivatives is their characteristic fragmentation pattern on electron impact (Fig. 7).[43] Bns-derivatives are preferentially cleaved in the butyl side chain, forming a base peak at $(M-43)^+$, which retains the complete information of the derivatized molecule. This is in sharp contrast with the fragment formation of Dns-derivatives (see Fig. 5). Besides the $(M-43)^+$ ion, the molecular ion $(M^{+\cdot})$ is observed with about the same abundance as in case of Dns-derivatives, so that derivate formation with Bns—Cl allows not only the determination of smaller amounts, but facilitates the identification of molecular ions in mass spectra of mixtures of compounds by means of the two characteristic ions $M^{+\cdot}$ and $(M-43)^+$, whose intensities are observed in a fixed ratio. In contrast with Dns—Cl, Bns—Cl can also be used for the determination of the biologically important amines methylamine and dimethylamine.[43] Apart from the fact that Bns-derivatives are difficult to crystallize, they have no disadvantage in comparison with Dns-derivatives, and can therefore be considered an advance on Dns-derivatives.

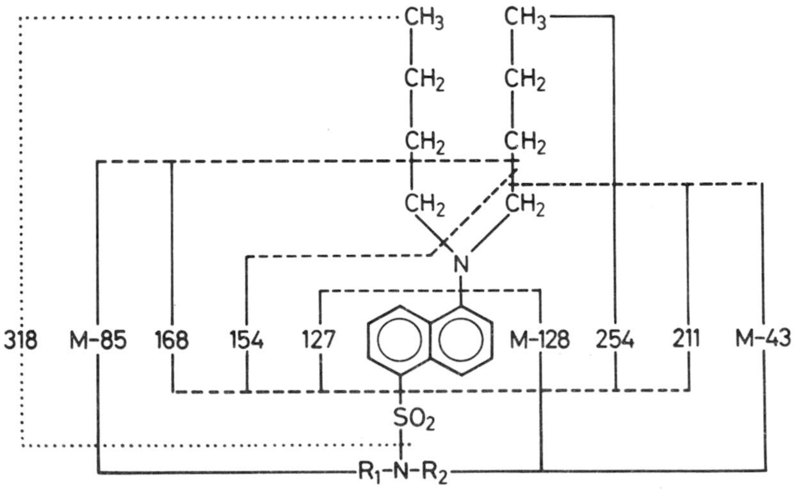

Fig. 7 Fragmentation of Bns-derivatives induced by electron impact.

4 6-METHYLANILINONAPTHALENE-2-SULFONYL CHLORIDE (Mns—Cl)

6-Methylanilinonaphthalene-2-sulfonyl chloride (Fig. 8) (Mns—Cl) was introduced by Cory *et al.*[160] for *N*-terminal protein labelling. Its characteristics as a reagent are quite similar to Dns—Cl and Bns—Cl, and so are the

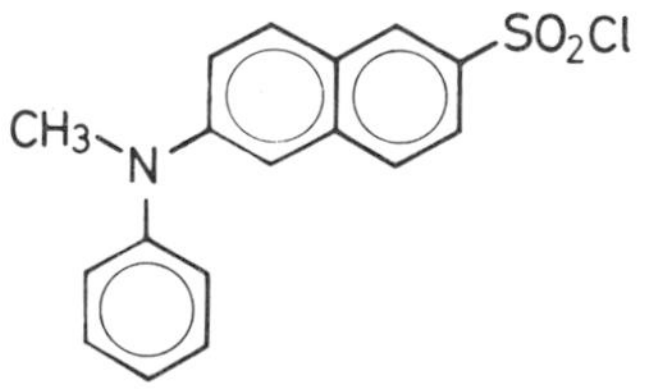

Fig. 8 6-Methylanilinonaphthalene-2-sulfonyl chloride (Mns—Cl).

procedures for derivate formation and application. Mns-derivatives have considerably higher molar extinction coefficients than the corresponding Dns-derivatives. For Mns—NH_2 $\varepsilon_{255} = 4.2 \times 10^4$ and $\varepsilon_{321} = 2.3 \times 10^4$, as measured in *n*-propanol,[29] the corresponding values for Dns–NH_2 are: $\varepsilon_{252} = 1.3 \times 10^4$ and $\varepsilon_{333} = 0.43 \times 10^4$, in methanol.[10] Therefore the Mns-derivatives fluoresce more intensely. The maximum of fluorescence emission is at shorter wavelength than that of the corresponding Dns-derivatives, around 450 nm, with an excitation maximum at about 321 nm, whereas Dns-derivatives fluoresce maximally at wavelengths above 500 nm.

5 2-*p*-CHLOROSULFOPHENYL-3-PHENYL INDONE (Dis—Cl)

2-*p*-Chlorosulfophenyl-3-phenyl indone (Dis—Cl) (Fig. 1-3) is in some respect similar to Dns—Cl. It reacts with primary and secondary amines, amino acids, imidazoles, phenols and even with alcohols to the corresponding amides and esters, respectively.[161] The sulfonamides are stable to hydrolytic cleavage with 6 M HCl. Dis—Cl is, therefore, a suitable reagent for the determination of the amino-end groups of peptides and proteins. The procedure[162] is quite similar to the dansylation method of Gray and Hartley.[9]

Dis—Cl forms orange to red derivatives which can be detected on chromatograms with about the same sensitivity as the dinitrophenyl derivatives. However, Dis-derivatives can be rearranged to strongly green fluorescing 1-phenyl-3-*p*-sulfophenyl-isobenzofuranic derivatives, by reaction with alcoholic KOH or with sodium ethylate solution (Fig. 9).[163] This increases detection sensitivity, permitting detection of amino acids in the range 0.1–1 pmol.

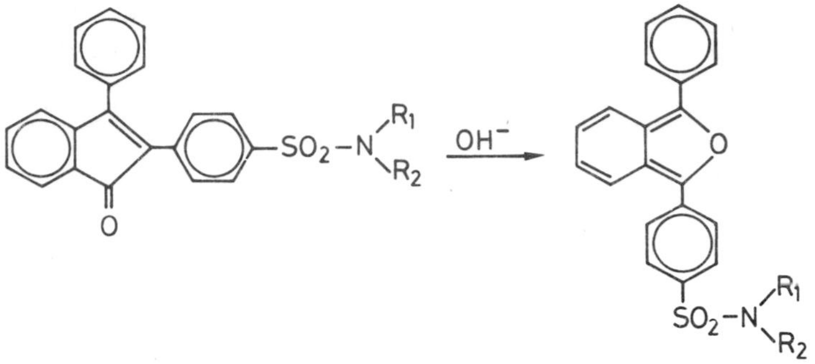

Fig. 9 Formation of 1-phenyl-3-*p*-sulfophenylisobenzofuranic derivative from sulfophenyl-3-phenyl-indone.

5.1 Synthesis

Direct sulfonation of 2,3-diphenylindone leads to an almost quantitative conversion into orange 2-*p*-chlorosulfonphenyl-3-phenyl indone.[161]

5.2 Reaction

Dis-derivatives are prepared by mixing equal volumes of the amino acid solution in 0.1 M $NaHCO_3$ with the solution of Dis—Cl (1 mg ml^{-1} of acetone). If sulfonyl chloride is precipitated, acetone is added until a clear solution is obtained.[162] After leaving the solution for 3 h in a closed tube at room temperature, it is evaporated to dryness. The residue is dissolved in 1 ml of acetone: methanol (1 : 1). Aliquots of this solution are applied to TLC plates.

5.3 Applications

For the use of end-group determinations Ivanov and Yukhnovska[163] worked out TLC systems for the separation of the Dis-derivatives of the usual amino acids. After development the chromatograms are dried at 105 °C and then sprayed with a solution of sodium ethylate (5 g per 100 ml of 96% ethanol). The fluorescent spots are immediately visualized under a u.v. lamp (365 nm).

Direct (*in situ*) fluorescence measurement of the fluorescent spots is not feasible. Durko *et al.*[164] recommend the extraction of the chromatographically separated Dis-derivatives with acetone, evaporation of the extract and addition of 5 ml sodium ethoxide solution 25 min before fluorescence measurement. For the pyridoxamine derivative, excitation was 410 nm, and emission was measured at 480 nm.

6 4-CHLORO-7-NITROBENZO[*c*]-1,2,5-OXADIAZOLE (Nbd—Cl)

4-Chloro-7-nitrobenzo[*c*]-1,2,5-oxadiazole (4-chloro-7-nitrobenzofurazan; Nbd—Cl) was noticed as a possible fluorescent label during the examination of benzofurazans as potential antileukemic agents and blockers of thiol groups.[165] Nbd—Cl is prepared by nitrating 4-chlorobenzofurazan[166] obtained from 2,6-dichloraniline via dichloronitrosobenzene.[167] It is also commercially available. Nbd—Cl is a stable, non-fluorescent pale-yellow solid; m.p. 97 °C; readily soluble in organic solvents, somewhat more soluble in water than Dns—Cl.

6.1 Reactions (Fig. 10)

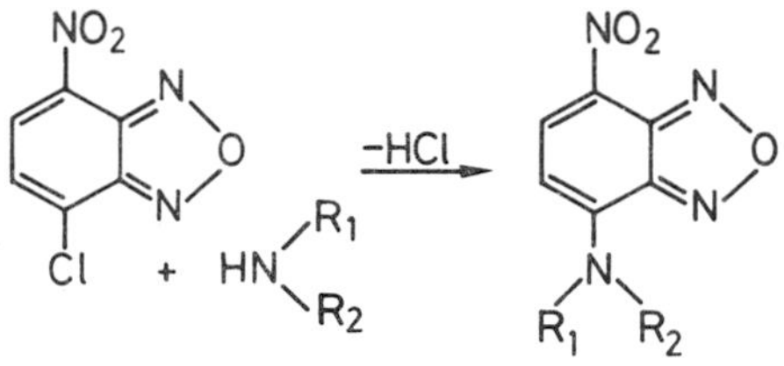

Fig. 10 Derivative formation with Nbd—Cl.

As an aryl halide with an activated halogen, Nbd—Cl is closely related to dinitrofluorobenzene. In aqueous solutions or in organic solvents it reacts with primary and secondary amino groups, and less readily with phenolic OH and thiol groups under alkaline conditions. Thiol-containing compounds (*N*-acetyl cysteine, glutathione, proteins) dissolved in 50 mM sodium citrate buffer, pH 7, containing 1 mM EDTA, react rapidly with Nbd—Cl.[168]

For derivate formation with amines the following procedure was recommended:[169] 25–500 μl of the amine solution (1–20 μg amine) is mixed with 4 volumes of a 0.5 mg/ml solution of Nbd—Cl in methanol. 50–100 μl of a 0.1 M $NaHCO_3$ solution is added. The reaction is completed by heating to 55 °C for 60 min. Yields of Nbd-piperidine and Nbd-dimethylamine were 95%. For TLC aliquots of the reaction mixture are spotted onto silica gel G plates or polyamide sheets.

For the reaction with amino acids and peptides a mixture of 2.0 ml ethanol, 0.1 ml Nbd—Cl solution in ethanol (1.4 mg ml^{-1}), 0.1 ml ethanol saturated with sodium acetate, and 0.1 ml amino acid solution (400–600 μg ml^{-1}), is heated to 75 °C for 20 min.[170]

Lawrence and Frei[171] recommend a two-phase system of water: isobutylmethylketone, but formation of by-products under these conditions seems to be worse than in methanol.[169] Ethyl acetate and chloroform have also been suggested as reaction media: for the estimation of methamphetamine 100 ml of the 24-h urine were adjusted to pH 9, and then extracted with 50 ml of ethyl acetate. The ethyl acetate phase was concentrated to 10 ml. Nbd—Cl (5 mg) was added, together with 1 drop of 2 M NaOH, and left at room temperature for 4 h.[172] Amphetamine and morphine were derivatized in the same way, using chloroform instead of ethyl acetate.[173,174]

Nbd-derivatives can be extracted after dilution of the reaction mixtures with water, using ethyl acetate or dichloromethane. Removal of excess reagent can be achieved by chromatography on silica gel:[169] the reaction mixture is applied to a silica gel column (1.25 cm × 60 cm; particle size 10–40 μm). With cyclohexane : ethylacetate (1 : 1) the non-fluorescent Nbd—Cl is the first compound to be eluted.

6.2 Properties of Nbd-derivatives

A thorough study of the chemical and physical properties of the Nbd-derivatives has not yet been published. They are stable against hydrolysis with 6 M HCl for 4 h at 110 °C,[170] so that end-groups of peptides and proteins can be determined with Nbd—Cl. Nbd-derivatives are stable in solutions and on chromatograms, if they are protected from irradiation.

In contrast with most other fluorescent labels, the absorption maximum of the Nbd-derivatives is in the visible region. Unfortunately activation and emission bands overlap. In Table 6 some spectral characteristics of several types of Nbd-derivatives have been summarized. Fluorescence activation and emission maxima of a large number of amino acid-derivatives have been published by Fager *et al.*[170]

Quantum yields of Nbd-derivatives do not seem to have been determined. According to Klimisch and Stadler[169] about 50 pmol Nbd-dimethylamine ml^{-1} (ethyl acetate) were measurable by fluorimetry. Ghosh and Whitehouse[165] found the detection sensitivity of Nbd- and Dns-glycine to be comparable.

6.3 The use of Nbd—Cl

Although the reagent is non-fluorescent, it nevertheless interferes with fluorescence measurements, so that the excess has to be removed before fluorimetry. Some authors recommend the measurement of total fluorescence, after removal of excess reagent, without TLC separation of the reaction products. Morphine and amphetamine in blood and urine were determined in this way.[173,174] However, since by-products of the reaction are inevitable, such procedures may give erroneous values if small amounts are to be determined. Usually chromatographic procedures have to be applied before quantitative measurement. For the separation of Nbd-amine derivatives only TLC has been used, either on silica gel G plates or on polyamide sheets.[169,172,175] *In situ* fluorimetric estimation on polyamide sheets showed linearity of peak areas with the amount of material in the range of 70–700 pmoles.[169] Solvents for the separation of Nbd-amino acids were reported by Fager *et al.*,[170] using silica gel

TABLE 6

Spectral characteristics of NBD derivatives

	Ethyl acetate		50 mM sodium citrate (with 1 mM EDTA), pH 7.0		Methanol		
Compound	Absorption maximum nm	Emission maximum nm	Absorption maximum nm	Emission maximum nm	Absorption maximum nm	Colour (on silica gel)	Reference
N-Acetyl-*S*-NBD-cysteine			425	545			168
N,S-bis-NBD-Cysteine methylester	464	512					165
S-NBD-Glutathion			420	540			168
Phosphorylase b			430	525			168
N-NBD-Cyclohexylamine			475	545			168
S-NBD-Thioglycollic acid	464	512					165
O-NBD-*p*-Nitrophenylether	464	524					165
O-NBD-*N*-Acetyltyrosine	464	512					165
N-NBD-Glycine	464	512					165
N-NBD-Methylamine	470(464)	524(512)					165, 169
N-NBD-Dimethylamine	464(464)	522(512)					165, 169
N-NBD-Diethylamine	475	527					169
N-NBD-Pyrrolidine	476	523					169
N-NBD-Piperidine	473	531					169
N-NBD-Benzylamine	464	504				red orange	165
N-NBD-Aniline	464	524					165
N-NBD-Ephedrine					476	orange	175
N-NBD-Methamphetamine					478	orange	175
N-NBD-Norephedrine					464	yellow	175
N-NBD-Methylphenidate					475	orange	175
N-NBD-Amphetamine					464	yellow	175

G plates. Chloroform : methanol (7 : 3) gave the best separations. Mass spectrometry proved to be suitable for the identification of Nbd-derivatives.[176]

7 ISOTHIOCYANATES

Isocyanates and isothiocyanates react with primary amines to give urea and thiourea derivatives respectively. 1,2-Benzanthrylisocyanate was the first fluorescent probe for protein study.[177] Subsequently a number of other cyanates were used for the same purpose (fluoresceinisocyanate, 3-phenyl-7-isocyanatocoumarin and 1-isocyanatoanthracene, among others[178]). However, isocyanates react quite readily with water and alcohols to give urethanes. For this reason they were replaced by the less reactive isothiocyanates. Only three isothiocyanates have so far been used for chromatography, and isocyanates seem not to have been used at all.

8 9-ISOTHIOCYANATOACRIDINE

The reaction of 9-isothiocyanatoacridine (Fig. 1-6) with primary and secondary aliphatic amines leads to several fluorescent products. The fluorescence of one of these products could be related to the amount of amine present.[179] On chromatographic evidence the fluorophore is a cyclization product (2-alkylamino-1,3-thiazino-(4-5,6-kl) acridine) (Fig. 11), which is formed by photooxidation from the initially formed thiourea derivative.[180]

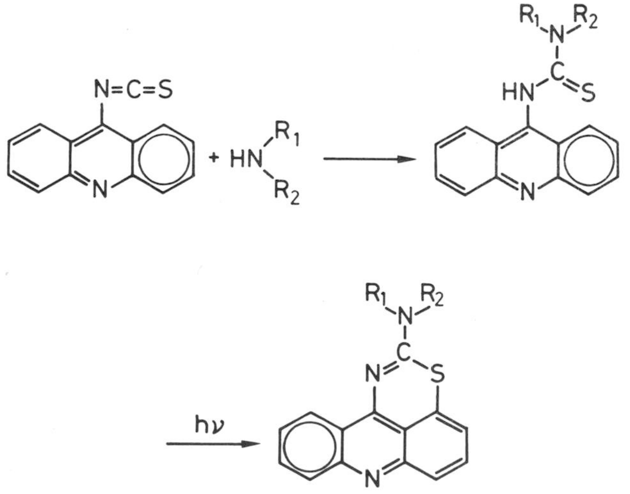

Fig. 11 Derivative formation with 9-isothiocyanatoacridine.

8.1 Preparation

9-Isocyanatoacridine is prepared from 9-chloroacridine by replacement of the chloro group by the isothiocyanate group of silver thiocyanate.[181]

8.2 Procedure[180]

All amines with $pK_a > 9.3$ were successfully determined with the following procedure, while no compound with a $pK_a < 5.3$ could be measured. In practice 5 ml of a toluene solution of the amine (free base) is mixed with 2 ml of a solution of 9-isothiocyanatoacridine (80 mg/100 ml of toluene), and stored overnight at room temperature. This is followed by a 'limited morning exposure to indirect sunlight'. Experiments to maximize the yield of cyclized product by exposure of the reaction mixture to u.v. light in a photochemical reactor were unsuccessful. Under these conditions the required excess of reagent was readily converted into acridone, resulting in excessive background fluorescence. While water interferes with the method, limited amounts of water can be tolerated to permit analysis of aqueous solutions of amines.[179] Quantification is by fluorimetry of the toluene reaction mixture (excitation at 295–310 nm; emission at 500–525 nm). When the reaction mixture was mixed with acidic ethanol (0.5 ml of the toluene solution with 4.5 ml of ethanol and 0.1 ml of conc. HCl), there was a pronounced increase in the fluorescence of the amine reaction solution compared to the blank, at the cost of a decrease in the stability of the derivative.

A few TLC separations have been carried out with the amphetamine and *n*-butylamine derivatives. Linearity of *in situ* fluorescence measurement was found over a range of 3–25 μg (recovery 98.9%; standard deviation 9.9%).

9 FLUORESCEIN ISOTHIOCYANATE

Fluorescein isothiocyanate (Fig. 1-5) was proposed by Maeda *et al.* as a fluorescent end-group reagent[182] analogous to phenylisothiocyanate in the Edman procedure.[183] A procedure for the determination of free amino acids, by formation of fluorescent thiohydantoins with fluorescein isothiocyanate was reported by Kawauchi *et al.*[184] This reagent does not seem to have been used for estimation of amines.

9.1 Preparation

Nitrofluorescein is prepared by the method of Coons and Kaplan[185] from resorcinol and 4-nitrophthalic acid. The two isomers formed are separated by fractional crystallization. The nitro group is reduced according to McKinney *et al.*[186] Finally fluorescein isothiocyanate is obtained from aminofluorescein by treatment with thiophosgen.[187] The hydrochloride is an orange powder.

9.2 The use of fluorescein isothiocyanate[184]

The reaction of fluorescein isothiocyanate with an amino acid is formulated in Fig. 12. In the first step the fluorescein isothiocarbamyl amino acid is formed by addition of the amino acid to the isothiocyanate group. This reaction product could be used for the assay of the amino acids. However, the thiocarbamyl amino acid is usually converted into the corresponding thiohydantoin by acidification.

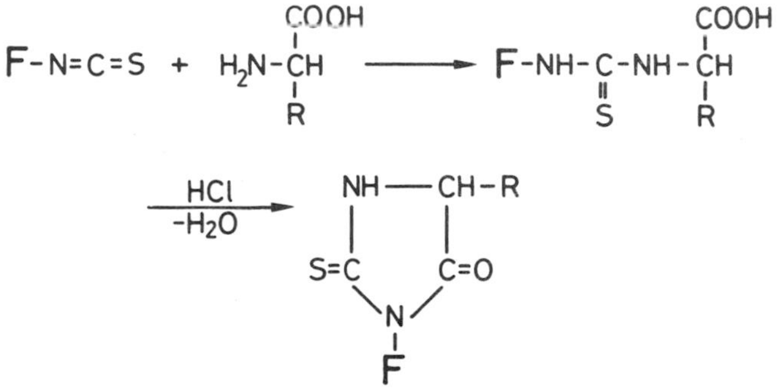

Fig. 12 Formation of the fluorescent thiothydantoin from fluoresceinisothiocyanate.

Fluorescein isothiocyanate (0.3 mmol) is dissolved in 5 ml of acetone with a trace of pyridine. This solution (0.05 ml) is added to an amino acid solution (0.5 ml, about 10 μmol) in 0.2 M carbonate–bicarbonate buffer, pH 9, and allowed to react at 25 °C for 4 h. The reaction is stopped by acidifying with glacial acetic acid to pH 4.5, and the precipitate formed is dissolved in about 0.5 ml of acetone. To this solution 0.2 ml of 6 M HCl is added to convert the fluorescein thiocarbamyl derivative to yield fluorescein thiohydantoin. The solutions of the fluorescent thiohydantoins are immediately used for TLC. On silica gel G unreacted fluoresceinisothiocyanate runs behind the solvent front and aminofluorescein-HCl, a decomposition product, stays at the origin in the solvent systems devised by Kawauchi *et al.*[184]

Sulfhydryl groups seem to react with fluorescein isothiocyanate; phenolic groups (of tyrosine) react to a small extent under the reaction conditions. Fluoresceinthiohydantoins in amounts exceeding 1 nmol are visible as yellow spots on TLC plates, and show up as intense greenish-yellow fluorescent spots under u.v. light.

4-Dimethylamino-1-naphthyl isothiocyanate, another fluorescent Edman reagent, is used analogously.[265] It is commercially available.

10 FLUORESCAMINE

Fluorescamine (4-phenylspiro[furan-2(3H),1′-phthalan]-3,3′dione) (Fig. 1-10) is a non-fluorescent compound; m.p. 154 °C. It is readily soluble in acetone, dioxane, tetrahydrofuran, dimethylsulfoxide, etc. Its solutions in acetone or acetonitrile are stable, for at least 12 weeks. It is only slightly soluble in water, but it is hydrolysed in aqueous media to nonfluorescent products. A 0.5 M solution in acetone is commercially available.

10.1 History

In 1962 the formation of a fluorescent product from phenylalanine and ninhydrin in the presence of a peptide was used by McCaman and Robins for the specific estimation of phenylalanine in blood.[188] Nearly ten years later Udenfriend and co-workers showed that phenylacetaldehyde, which is formed by oxidative decarboxylation of phenylalanine by ninhydrin, reacts with the excess of ninhydrin and with a primary amino group to give the fluorescent product.[189] The elucidation of this mechanism, together with the structure of the fluorescent reaction product[190] led to the synthesis of fluorescamine.[191] The fluorophore formed by fluorescamine is identical with that obtained by reacting phenylacetaldehyde, ninhydrin and a primary amine.

10.2 Reaction and reaction conditions

Compounds with nucleophilic functional groups (primary and secondary amines, alcohols, water, etc.) are capable of reacting with fluorescamine, but only primary amines form fluorescent products. Fluorescamine is therefore a specific reagent for compounds with primary amino groups. It has this specificity in common with reagents that form Schiff bases, like pyridoxal and *p*-dimethylaminocinnamaldehyde.

The reaction of fluorophore formation is shown in Fig. 13. In a reversible, rapid reaction the primary amine is added to the double bond of the reagent.

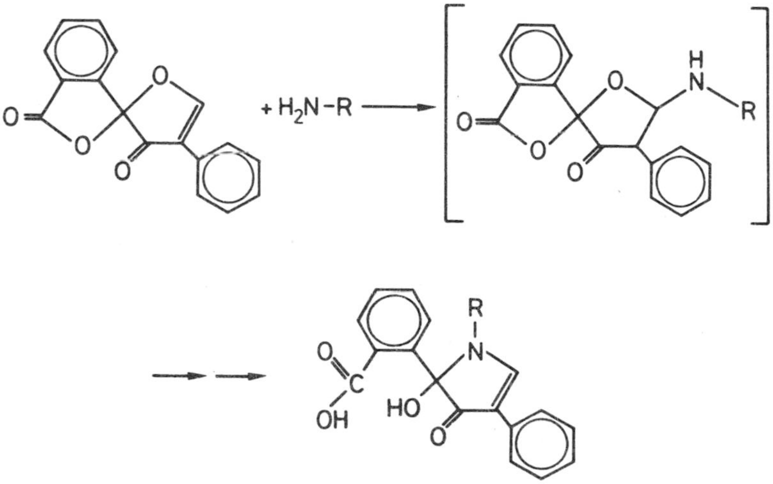

Fig. 13 Derivative formation with fluorescamine.

The addition product is then transformed in a multistep rearrangement to the fluorophore.[192] For the reaction of aliphatic primary amines and peptides a pH of 8–8.5 is adequate, whereas proteins and amino acids are normally reacted at pH 9. At higher pH hydrolytic cleavage of the reagent competes with the formation of the fluorophore, so that yields of the fluorophore decrease. Buffer and salt concentrations, and organic solvents in the reaction mixture influence fluorophore formation as well, but their influence is less pronounced than that of pH.

The estimation of amines, amino acids, peptides and proteins is done by essentially the same procedure.[193,194] Three volumes of a solution of a primary amine, peptide or protein in 0.05 M phosphate buffer pH 8–9.5 (or a 0.05 M sodium borate buffer pH 8–9.5) are mixed under vigorous stirring with 1 volume of fluorescamine solution (28 mg/100 ml of acetone, or 56 mg/100 ml of dioxane). (Standardized conditions of stirring are essential, for reproducible fluorophore formation.) Fluorescence measurement is carried out 5–30 min after mixing. Yields of the fluorophore are 80–90%. There is a linear relationship between the concentration of the amine and the observed fluorescence up to 25–30 nmol ml^{-1}.

10.3 Properties of the fluorophore

The reaction products of some amines (ethylamine, *n*-butylamine, benzylamine, 3,4-dimethoxy-β-phenylethylamine, aniline) and of some peptides (pentaglycine, leucylalanine, among others) with fluorescamine were synthesized as crystalline solids.[193] Solutions of these compounds are stable in neutral and mildly alkaline medium, in the dark, even for days. In acidic solutions fluorescence rapidly deteriorates by a sequence of reactions. On the other hand, fluorescence efficiency is highest at acidic pH, due to the highly fluorescent protonated form of the fluorophore (Fig. 14). In freshly prepared fluorophore solutions the fluorescence intensity decreases in two distinct steps with increasing pH, reflecting the presence of two dissociable acidic functions in the molecule, with pK values of 3.8 and 11.6 (Fig. 14).

Fluorescence measurements are normally carried out in the range of pH 4.5–10.5, i.e. in the range of maximal stability of the fluorophore (Fig. 14), corresponding to the molecule with the dissociated carboxyl group. In this range absorption maximum is at 390 nm, and fluorescence emission at 475 nm. Quantum yields of amine and peptide derivatives in ethanol are normally in the range of 0.2–0.34, although some remarkable exceptions exist.[193]

10.4 Applications of fluorescamine

Fluorescamine has found its main application in the estimation of amines, amino acids or peptides, after separation by ion-exchange column chromatography,[195,196] especially in automated amino acid analysis, where the reagent

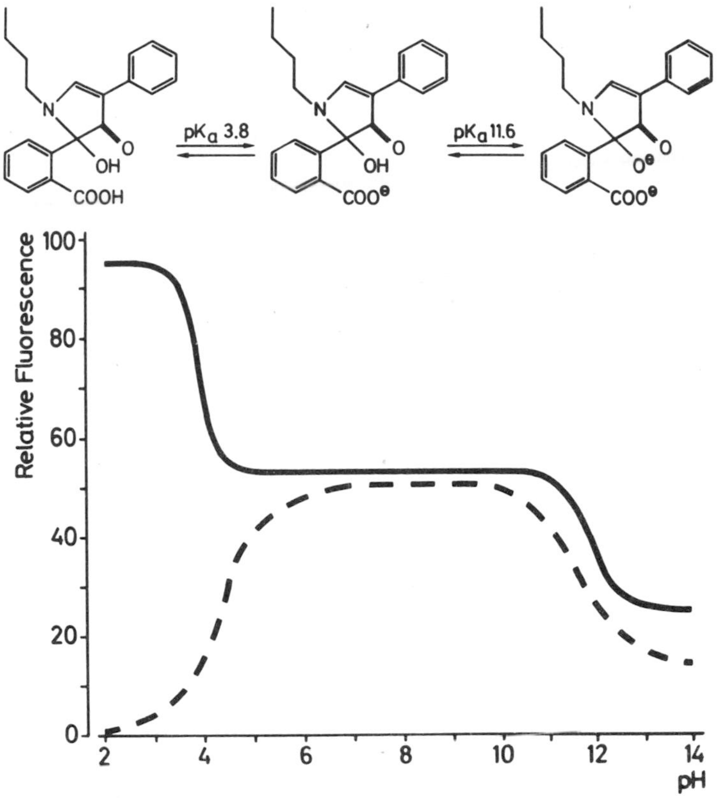

Fig. 14 Fluorescamine fluorophore: pH-dependence of fluorescence efficiency(———) and stability of fluorescence (– – – – – –). (According to S. De Bernardo *et al.*[193])

(15 mg Fluorescamine/100 ml of acetone) is continuously added to the column effluent buffered with sodium borate (0.16 M; pH 9.6).[195] With this method primary amino acids could be estimated in the picmole range. Fluorescence was linear up to 2.5 nmol, the precision being the same as that obtained in the nanomole range with ninhydrin. Proline and hydroxyproline do not yield fluorescent products, but can be oxidatively decarboxylated with *N*-chlorosuccinimide to the corresponding imino acids, which are hydrolysed to primary amines, and reacted with fluorescamine.[197]

Fluorescamine can be used for the estimation of proteins[194,198] and their visualization in gel-electrophoretograms.[199] Fortunately fluorescamine does not give a fluorophore with ammonia.

A more recent application of fluorescamine is its use as spray reagent for the estimation of amines separated by TLC.[200] The TLC plate was dried and then sprayed first with a solution of 10% triethylamine in methylene chloride, air dried for several seconds, and then sprayed with a solution of 0.5 mg ml^{-1} fluorescamine in acetone. Finally the dried plate was resprayed with the triethylamine solution. Although this method allows the visualization of about 500 pmol of an amino acid or peptide, it is not quantitative. It should be kept in mind that fluorescamine quenches fluorescence, so that the measured fluorescence intensity is not only dependent on the amount of amine, acid or peptide, but also on the excess of reagent.

So far there have only been preliminary reports on the chromatography of fluorophores made with fluorescamine *before* their separation. Imai *et al.*[201] ran the fluorophores from some amines (dopamine, norepinephrine and their respective 3-*O*-methylation products, putrescine, spermidine and spermine), and from some peptides on silica gel plates. They showed that about 250–500 pmol of these derivatives were detectable. Quantitative evaluation of the fluorescent spots was, however, not satisfactory. Imai,[202] therefore, worked out an HPLC system for the catecholamine derivatives, using Hitachi gel 3011 with methanol −0.1 M Tris buffer pH 8.0 (7:3) for separation. The fluorophores were measurable at the 100 pmol level. Similar results were reported by Samejima[203] for the fluorophores of putrescine, spermidine and spermine. He used a reversed phase column (chemically bonded octadecylsilane) and a gradient of methanol in 0.1 M sodium borate buffer pH 8. In all these cases the fluorophores were formed by the usual procedure, namely by mixing the reagent with the sample solution in buffer. Aliquots of the reaction mixture were immediately applied to the TLC plates, or to the HPLC columns.

11 PYRIDOXAL AND PYRIDOXAL-5-PHOSPHATE

Pyridoxal and pyridoxal-5-phosphate are the only *naturally occurring* compounds, which have been used as fluorescent labels. These compounds form Schiff bases with primary amino groups. The reduction of the C=N-bond with sodium borohydride ($NaBH_4$) leads to the formation of fluorescent pyridoxyl derivatives.[204,205] Both pyridoxal and pyridoxal-5-phosphate are commercially available.

11.1 Reaction

The reaction of pyridoxal with a primary amine and the reduction of the Schiff base to the pyridoxyl derivative is shown in Fig. 15. Since only primary amino groups form Schiff bases, the reaction should be specific for compounds with primary amino groups. However proline and hydroxyproline reacted as well, and as the authors note,[205] even secondary amines can be bought to reaction, especially in non-aqueous solvents. A ketoenimine is assumed to be an intermediate of the reaction sequence in this case. Histidine reacts first with the primary amino group to a Schiff base. This compound is, however rearranged to a non-reducible imidazopyridine derivative. The following procedure has been suggested for the reaction of the reagents with amino acids and

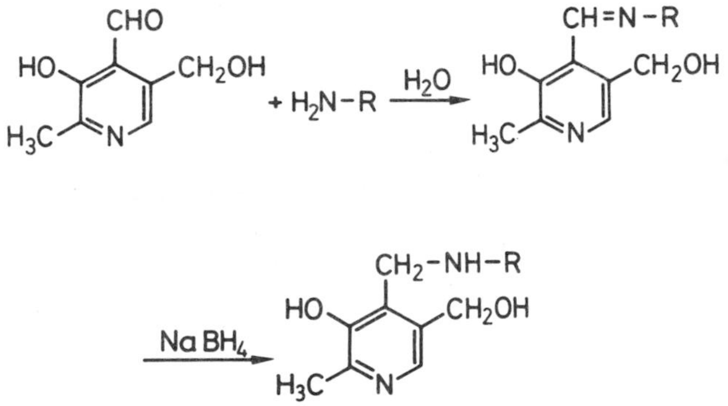

Fig. 15 Derivatization reactions with pyridoxal.

primary amines: amino acids and amines are dissolved in 0.5 M phosphate buffer pH 9.3. This solution is mixed with the same volume of a freshly prepared 0.4 M pyridoxal solution in the same buffer. (Pyridoxal: amino acids = 5 : 1). After 30 min at 8 °C, 100 mg of $NaBH_4$ dissolved in 1 ml of 0.1 M NaOH is added. The excess of $NaBH_4$ is decomposed by acidification to pH 1–2 with hydrochloric acid.

The yield of α-amino acid derivatives is >90%. Lysine and ornithine form dipyridoxyl derivatives. Recovery of the proline and hydroxyproline derivatives is about 25%.

For derivate formation with amines, pyridoxal-5-phosphate is preferred to pyridoxal, in order to allow ion-exchange chromatographic separation. The reaction mixtures are immediately applied to ion-exchange columns. Other separation methods have not been used.

11.2 Properties of pyridoxal derivatives

Pyridoxyl derivatives are stable at 0 °C for weeks, if protected from light. Fluorescence decays within minutes, if the solutions are irradiated especially with light at the wavelength of the absorption maximum.

Pyridoxylamino acids exhibit spectral characteristics similar to that of pyridoxamine: absorption maximum at 255 nm and 328 nm; fluorescence emission at 400 nm. Fluorescence efficiency is pH-dependent, with maximum fluorescence at pH 5.28 (except for the histidine derivative, which shows maximal fluorescence at pH 12). Molar extinction coefficients and fluorescence efficiencies are, with few exceptions, the same for all pyridoxyl amino acids. Between 10 and 100 pmol of an amino acid can be determined in the effluent of an amino acid analyser.[205]

The pyridoxal method has been used only for ion-exchange chromatography. The pyridoxyl residue increases the retention time, but with few exceptions the amino acid derivatives are eluted from the column in the same order as the free amino acids. For fluorescence measurement one part of the effluent is mixed with 49 parts of sodium citrate buffer, pH 5.28. One of the advantages of the pyridoxal method is that one could substitute sodium borotritide ($NaBT_4$), which is available with specific activities of about 116 Ci $mmol^{-1}$. Thus radioactively labelled pyridoxyl derivatives could be obtained, allowing the detection of about 0.1 pmol of an amine or amino acid. However, since the handling of $NaBT_4$ requires special precautions, and since the price of $NaBT_4$ with high specific radioactivity is high, this method has not yet been applied to actual analytical problems.

12 *o*-PHTHALDIALDEHYDE (OPT)

o-Phthaldialdehyde (Fig. 1-9) is a widely used commercially available reagent. It has been used for the assay of histamine[206,207] and spermidine.[208] A number of biologically important compounds,[209–211] among others glutathione, arginine, agmatine and 5-hydroxy- and 5-methoxy-indole derivatives, form fluorescent derivatives with *o*-phthaldialdehyde. The condensation with *o*-phthaldialdehyde has been used, therefore, for the demonstration of some of these compounds in tissues,[212–215] and, after their separation, on chromatograms.[216–219] Recently Roth[220] showed that nearly all amino acids can form fluorescent condensation products with *o*-phthaldialdehyde in the presence of strong reducing agents. The chemical reactions involved in the formation of the fluorophores are unknown.

12.1 Reaction conditions and applications of OPT

For the estimation of histamine the sample is dissolved in 2.3 ml of dilute NaOH (pH 12.4–12.7) and then mixed with 0.1 ml of OPT solution (2 mg ml^{-1} in methanol) at 0 °C under continuous bubbling with N_2 for 40 min. The condensation reaction is stopped by acidification to pH 2.5–3.5. Fluorescence is stable for 2–5 h. Fluorescence is read after 10–20 min (excitation at 350 nm, emission at 440 nm). This procedure allows the estimation of about 2 ng of histamine. The reaction conditions minimize the interference of spermidine: 10 ng histamine equal 8 μg of spermidine. The level of fluorescence increases in a linear fashion up to 0.5 $\mu g\ ml^{-1}$ of histamine. Addition of $SrCl_2$ (final concentration 0.1 M) or $CdCl_2$ (final concentration 1 mM) with *o*-phthaldialdehyde prevents spermidine fluorescence, but not that of histamine. Certain anions (iodide, bromide, acetate, citrate, etc.) markedly reduce fluorescence, whereas chloride and nitrate are relatively inert. Addition of formaldehyde (final concentration 1 M) a few minutes after *o*-phthaldialdehyde, or boiling for 2 h, will abolish the fluorescence of histamine, but not that of spermidine.[207] An automated procedure (using the Auto-Analyzer), based on the condensation of *o*-phthaldialdehyde has been devised for the estimation of histamine.[221]

For spermidine assay (for instance in column effluents) 2.3 ml of the sample solution in 1 mM NaOH (pH 11.0–11.6) are mixed with 0.1 ml of a 2 mg ml^{-1} *o*-phthaldialdehyde solution and heated for 2 min at 100 °C. The reaction is terminated by acidification to pH 3.5, followed by cooling in ice. Fluorescence is read at 400 nm (activation at 350 nm). The minimum detectable amount of spermidine is about 10 ng ml^{-1}. Linearity extends to 2.5 μg ml^{-1} [222] of spermidine. Similar reaction conditions can be applied for the assay of various indole derivatives.[223]

For amino acids the following procedure has been suggested: 100 μl of the sample solution are mixed with 3 ml reagent. Fluorescence is activated at a wavelength of 340 nm; emission is measured at 455 nm within 5–25 min of mixing.[220] The reagent is prepared as follows: 1.5 ml of OPT solution (10 mg ml^{-1} of ethanol) are mixed with 90 ml of 0.05 M sodium borate buffer pH 9.5. To this solution 1.5 ml of a solution of 2-mercaptoethanol (5 μl ml^{-1} of ethanol) are added. The reagent must be freshly prepared each day. This reaction forms the basis of a method for automated amino acid analysis.[224] The effluent of the ion-exchange column (25 ml h^{-1}) is mixed with the reagent (0.8 g OPT, 10 ml ethanol, 200 μl 2-mercaptoethanol per litre 0.1 M sodium borate buffer pH 10) which is pumped at 30 ml h^{-1}. Reaction is completed in a Teflon® delay coil. With a conventional amino acid analyser connected to an Amino Microfluorometer, amino acids could be determined with good precision at the 0.5 nmol level.

If OPT is used as a spray reagent, the TLC plates are first sprayed with acetic acid in acetone (1–2 drops per 10 ml), or with NaOH in methanol (1–2 drops of 10 M NaOH per 10 ml), to optimize the pH for the condensation. After a few minutes in the oven, the plates are sprayed with a 0.2 mg ml^{-1} solution of OPT in acetone and viewed under u.v. light.[216]

13 CONDENSATION REACTIONS WITH FORMALDEHYDE, GLYOXYLIC ACID, *p*-DIMETHYLAMINOCINNAMIC ALDEHYDE AND BENZOPYRONE

Certain substituted arylalkylamines and indolic compounds (catecholamines, tyramines, mescaline, 3,4-dimethoxy-β-phenylethylamine, tryptamine, bufotenin, tryptophan, tyrosine, indoleacetic acid) can be condensed with formaldehyde, glyoxylic acid and other aliphatic aldehydes, to fluorescent reaction products. The condensation with these aldehydes is almost exclusively carried out after the chromatographic separation of the free amines or amino acids. These reagents, like OPT, are therefore not fluorescent labels in a strict sense, but a short description of their use is given below, since some of these procedures have proved to be useful for the sensitive determination of naturally occurring compounds of great importance.

13.1 Formaldehyde and glyoxylic acid

Acidic solutions of formaldehyde in ethanol were introduced by Prochazka as a spray reagent for indole derivatives.[225] Exposure of freeze-dried tissue sections to formaldehyde gas was introduced by Falck and Hillarp as a histochemical method for the fluorescence microscopic demonstration of catecholamines and of serotonin.[226,227] This method has found wide application. It was especially useful for the localization of adrenergic and serotinergic neurones.[228] Catecholamines and serotonin and some related amines can also be visualized on chromatograms by exposure to formaldehyde vapour.[216,229–231] Substituted arylakylamines under these conditions usually form dihydroisoquinolines, and indoleamines dihydroharmans (Fig. 16). Isoquinolines and harmans might, however, be formed from certain amines either by dehydration (in case of amines with a hydroxyl group in the side chain such as norepinephrine), or by dehydrogenation, respectively. The fluorescence characteristics of amine derivatives formed in this way strongly depend on the structural features of the amines, and on environmental factors.[232]

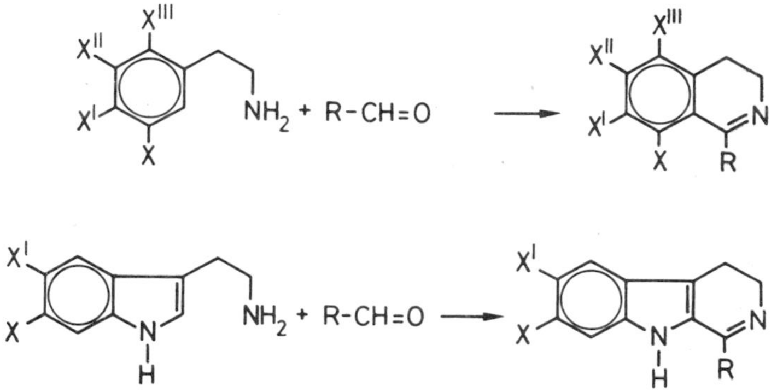

Fig. 16 Condensation reaction of an aliphatic aldehyde (formaldehyde: R = H) with a β-phenylethylamine and with an indoleamine (X′, X″ etc. = substituents).

For the sensitive determination of a variety of arylalkylamines, including mescaline, and of some aromatic amino acids, the following procedure has been suggested:[233] 1 ml of the sample solution in methanol is mixed with 1 ml ammonia solution (30 ml conc. NH_3 + 70 ml methanol) and 1 ml formaldehyde solution (10 ml 35% aqueous formaldehyde + 90 ml methanol) and heated in a water bath of 95 °C for 25 min. After addition of 0.1 ml of 6 M HCl the samples are evaporated by continued heating for 70 min. The dry residue is dissolved in 3 ml of methanol. Fluorescence can be measured immediately, but is stable for at least 24 h. Fluorescence of mescaline was linear over a range of 0.01–10 $\mu g\ ml^{-1}$ (excitation at 365 nm; emission at 505 nm).

For visualization and quantitative determination of compounds separated by TLC, the silica gel plate is first sprayed with an 1 : 1 mixture of 35% aqueous formaldehyde and 25% ammonia, followed by heating at 100 °C and 60 min. The plate is then sprayed with 6 M HCl, or with a modified Prochazka reagent (35% aqueous formaldehyde + 6 M HCl + ethanol (1:2:1)), and again heated for 60 min at 100 °C. In comparison with exposure to formaldehyde vapour, this procedure not only extended the number of amines and amino acids that can be measured, but at the same time increased detection sensitivity, By *in situ* fluorescence scanning 5 ng of tryptophan was measurable (excitation at 390 nm, emission at 504 nm) with a standard deviation of ±3%.[233]

During the last years there have been attempts to substitute other aldehydes. Glyoxylic acid seems to be the most promising both for the histochemical localization of certain biogenic amines,[234,235] and as spray reagent.[236] Spraying with a 0.6 M solution of glyoxylic acid in ethanol, followed by heating at 100 °C for 45 min allows a comparably sensitive detection of arylakylamines, indoleamines and of aromatic amino acids, when used as described above.

13.2 *p*-Dimethylaminocinnamaldehyde

p-Dimethylaminocinnamaldehyde (Fig. 1-7) was used as spray reagent for demonstration of various amines.[237–242] It condenses with primary amines, forming Schiff bases with a suitable system of conjugated double bonds (Fig. 17). β-Phenylethylamine was determined in blood and urine by heating the chloroform solution with this reagent to 85 °C for 15 min, followed by fluorimetric measurement (excitation at 470 nm; emission at 520 nm).[243] The

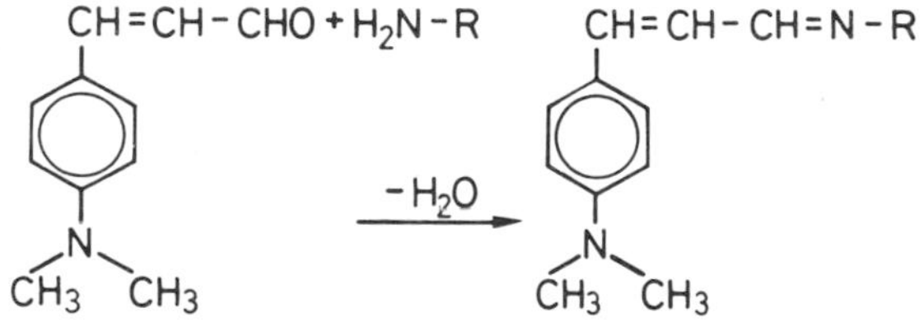

Fig. 17 Condensation of *p*-dimethylaminocinnamaldehyde with an aliphatic primary amine.

structure of the reaction product with indolic compounds is unknown. Baumann studied the applicability of *p*-dimethylaminocinnamaldehyde to the quantitative measurement of tryptophan derivatives separated by TLC.[244] Spraying of the chromatograms with a solution of 2 g reagent in a mixture containing 100 ml of 6 M HCl and 100 ml of ethanol, followed by drying for 2 min at 60 °C was adequate for visualization of the spots. However, instead of fluorimetry, densitometry was used for the quantitative evaluation of the chromatograms.

13.3 ω-Formyl-*o*-hydroxyacetophenone and benzopyrone

ω-Formyl-*o*-hydroxyacetophenone and benzo-γ-pyrone form with primary and secondary aliphatic and aromatic amines β-aminovinyl-*o*-

hydroxyphenylketones. Kostka[266] has utilized this reaction for the derivative formation of amines, and their sensitive detection on TLC. The chromatographic characteristics of a number of enamines were studied using ethyl acetate : benzene (1 : 5), chloroform : xylene (4 : 1) and acetone : xylene (1 : 9) and silica gel G layers. 0.1–1.0 g of the derivatives can be detected in the u.v.

14 FLUORESCENT REAGENTS FOR CARBONYL COMPOUNDS

Aldehydes and ketones readily react with hydrazine and hydrazines, respectively, to the corresponding hydrazones, so it seems reasonable to devise fluorescent hydrazines for the fluorescence labelling of carbonyl compounds, by analogy with the well-known coloured 2,4-dinitrophenylhydrazine. However, this has not been widely attempted, and mono-Dns-hydrazine (Fig. 18) seems to be the only such reagent used for actual analytical problems, i.e. for the estimation of ketosteroids.[245,246]

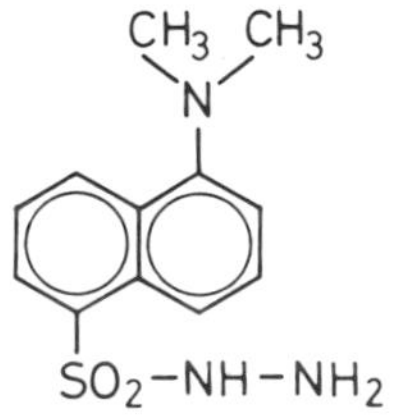

Fig. 18 Mono-Dns-hydrazine.

14.1 Preparation and properties of mono-Dns-hydrazine[24]

Mono-Dns-hydrazine is prepared by drop by drop addition of a concentrated solution of Dns—Cl to a solution of a large excess of hydrazine monohydrate in acetone. The rate of addition of Dns—Cl is matched to the progress of the reaction, which can be followed by the disappearance of the yellow colour of Dns—Cl. The reaction mixture is diluted with water and the product is extracted with ethyl acetate. Pale yellow crystals, m.p. 126 °C, are obtained on evaporation to dryness, soluble in practically all organic solvents but only slightly soluble in water.

14.2 Derivative formation and use of mono-Dns-hydrazine[247]

The dry residue of the steroid extract, in a glass-stoppered centrifuge tube is treated with ethanolic HCl (concn. 0.65 ml l^{-1}) followed by 0.2 ml of the mono-Dns-hydrazine solution (2 mg ml^{-1} of ethanol). The tube is heated for 10 min in a boiling water-bath and then cooled to room temperature. A solution of 5 mg sodium pyruvate per ml ethanol (0.2 ml) is mixed with the contents of the tube to react with the excess of mono-Dns-hydrazine. After 15 min at room temperature 3 ml of 0.5 M aqueous NaOH is added and the steroid derivatives are extracted with 6 ml of diethyl ether. If the amounts are sufficient, and if a suitable method is available for the separation of the reaction products from the excess of the reagent, aliquots of the reaction mixture can be applied to chromatographic separation immediately after completion of the reaction. Quantitation is by the same methods used for other Dns-derivatives. In ethanol excitation is at 340 ml and emission is at 525 nm. Fluorescence increases linearly over the range of 0.1–1 $\mu g\ ml^{-1}$.

14.3 Pyridoxamine

The results with pyridoxal as a reagent for primary amino groups[204,205] suggest that the analogous reaction of pyridoxamine with carbonyl compounds should be promising. The Schiff base formed from pyridoxamine and an aldehyde or ketone could be reduced with $NaBH_4$ to the pyridoxyl derivative, as with the reaction product of pyridoxal and amino acids (see p. 576), followed by fluorimetry after its separation from other fluorescent compounds. No attempts to use this approach appear to have been reported.

15 SULFHYDRYL REAGENTS

A large number of fluorescent sulfhydryl-directed reagents have been developed: 4-Chloro-7-nitrobenzo-[*c*]-1,2,5-oxadiazole (Nbd—Cl),[168,248] bis-fluorescein cystine,[249,250] fluorescein mercuric acetate,[251,252] mercurochrome,[253] iodoacetamide naphthylamine sulfonate,[254] *N*-(3-pyrene)-maleimide,[255] *N*-(1-anilinonaphthyl-4)-maleimide,[256] *N*-[*p*-(2-benzimidazolyl)phenyl]-maleimide,[257] bis-(5-dimethylaminonaphthalene-1-sulfonyl cystine,[249,258] 5-dimethylamino-naphthalene-1-sulfonyl-aziridine[259] and *S*-mercuric-*N*-(5-dimethylaminonaphthalene-1-sulfonyl)-cysteine.[260] None of these reagents has been used for the labelling of low molecular weight sulfhydryl compounds in connection with chromatographic separations. However, at least some of these reagents should in principle, be applicable.

The reaction of Nbd—Cl has already been mentioned (see p. 367). Maleimides react by addition of the SH-group to the double bond, forming a covalent linkage, the fluorescent disulfide reagents form mixed disulfides with sulfhydryl compounds, and *S*-mercuric-*N*-(5-dimethylaminonaphthalene-1-sulfonyl)-cysteine reacts according to Fig. 19, forming a mercury-bridged mercaptide. This reaction was used for the titration of glutathione and cysteine.[260]

$$\text{Dns-Cys-SHg}^{+} + \text{HS-R} \longrightarrow \text{Dns-Cys-S-Hg-S-R} + \text{H}^{+}$$

16 CONCLUSION

The term 'fluorescent label' has not always been used in the strict sense in this chapter: reagents have been discussed that lead to the formation of derivatives useful only after the prior separation of non-derivatized compounds. Examples of this type of reagent are the various aldehydes (with the exception of pyridoxal). On the other hand, many reagents, which are used quite analogously, have not even been mentioned, as for instance ethylenediamine for the assay of catecholamines,[178] α,γ-anhydroaconitic acid for determination of tertiary amines,[261] or 8-anilinonaphthalene-1-sulfonate[262] and related reagents for the detection of lipids and lipoproteins. Although selection has to a certain extent been arbitrary, the emphasis has been put on those reactions which are useful for quantitative determination. Mass spectrometry, GC–MS and SIM are increasingly taking over the role of specific detection methods, whereas sensitive quantitative chemical methods are still needed.

None of the available reagents fulfills the ideal requirements of a fluorescent label. The major defects are side-reactions and lack of specificity towards any given functional group. Nevertheless, the use of fluorescent labels has considerably extended our ability to determine small amounts of certain compounds, especially in protein chemistry and in the biochemistry of amines. With these new methods, interest is expanding from a small number of biologically active amines (catecholamines, serotonin, histamine) for which sensitive methods were available, to the whole spectrum of biogenic amines in general. With the available fluorescent reagents we are restricted to the study of compounds with primary and secondary amino groups. The analytical methods for tertiary amines are still quite inadequate, even if we include the possibilities offered by GLC. In the present state of development only a small number of the available reagents is actually employed, and only a few functional groups are in practice used for derivative formation. However, with the experience gathered during the last decade, it may be hoped that the planned synthesis of fluorescent labels will extend and improve our analytical capabilities, especially as far as sensitivity, reliability and automation are concerned.

REFERENCES

1. S. Udenfriend, *Fluorescence Assay in Biology and Medicine*, Academic Press, New York, Vol. I, 1962; Vol. II, 1969.
2. R. A. Passwater, *Guide to Fluorescence Literature*, Plenum Press, New York, 1967.
3. G. G. Guilbault, Ed., *Fluorescence, Theory, Instrumentation and Practice*, Marcel Dekker, New York, 1967.
4. W. R. Gray and B. S. Hartley, *Biochem. J.* **89**, 59 (1963).
5. E. Fischer and P. Bergell, *Ber. Dtsch. Chem. Ges.* **35**, 3779 (1902).
6. E. Fischer and P. Bergell, *Ber. Dtsch. Chem. Ges.* **36**, 2592 (1903).
7. F. Sanger, *Biochem. J.* **39**, 507 (1945).
8. F. Sanger and H. Tuppy, *Biochem. J.* **49**, 463 (1951).

9. W. R. Gray, *Methods Enzymol.* **25**, 121 (1972).
10. N. Seiler, *Methods Biochem. Anal.* **18**, 259 (1970).
11. N. Seiler and J. Wiechmann, *Experientia* **20**, 559 (1964).
12. N. Seiler and M. Wiechmann, *Experientia* **21**, 203 (1965).
13. N. Seiler and M. Wiechmann, *Z. Anal. Chem.* **220**, 109 (1966).
14. L. Brand and J. R. Gohlke, *Ann. Rev. Biochem.* **41**, 843 (1972).
15. G. Weber, *Biochem. J.* **51**, 155 (1952).
16. G. Weber, *Adv. Protein Chem.* **8**, 415 (1953).
17. D. J. R. Laurence, *Methods Enzymol.* **6**, 174 (1957).
18. V. G. Shore and A. B. Pardee, *Arch. Biochem. Biophys.* **62**, 355 (1956).
19. A. H. Coons, *Immunofluorescence, Public Health Reports, U.S.* **75**, 937 (1960).
20. B. S. Hartley and V. Massey, *Biochim. Biophys. Acta* **21**, 58 (1956).
21. W. R. Gray, *Methods Enzymol.* **25**, 333 (1972).
22. A. A. Boulton and I. E. Bush, *Biochem. J.* **92**, 11 (1964).
23. K. R. Woods and K.-T. Wang, *Biochim. Biophys. Acta* **133**, 369 (1966).
24. N. Seiler and M. Wiechmann, in *Progress in Thin-Layer Chromatography and Related Methods* (A. Niederwieser and G. Pataki, Eds), Vol. 1, p. 94, Ann Arbor-Humphrey Science Publishers, Ann Arbor, London, 1970.
25. N. Seiler, *J. Chromatogr.* **63**, 97 (1971).
26. V. Neuhoff, in *Micromethods in Molecular Biology* (V. Neuhoff Ed.) p. 87, Chapman and Hall, London, 1973.
27. N. N. Osborne, *Prog. Neurobiol.* **1**, 301 (1973).
28. A. Niederwieser, *Methods Enzymol.* **25**, 60 (1972).
29. J. Rosmus and Z. Deyl, *Chromatogr. Rev.* **13**, 163 (1971).
30. W. Dünges and H. W. Peter in *Methods of Analysis of Anti-Epileptic Drugs*, Excerpta Medica Intern. Congress Series No. 286, p. 126, Amsterdam, 1972.
31. D. J. Neadle and R. J. Pollit, *Biochem. J.* **97**, 607 (1965).
32. N. Seiler and M. Wiechmann, Hoppe-Seyler's *Z. Physiol. Chem.* **349**, 588 (1968).
33. N. Seiler, H. H. Schneider and K.-D. Sonnenberg, *Z. Anal. Chem.* **252**, 127 (1970).
34. N. Seiler, M. Wiechmann, H. A. Fischer and G. Werner, *Brain Res.* **28**, 317 (1971).
35. N. Seiler and M. Wiechmann, Hopper-Seyler's *Z. Physiol. Chem.* **350**, 1493 (1969).
36. N. Seiler in *Research Methods in Neurochemistry* (N. Marks and R. Rodnight, Eds) Vol. 3, p. 409, Plenum Press, New York, 1975.
37. M. Wiechmann, unpublished observations.
38. A. Rieche and H. Seeboth, *Angew. Chem.* **70**, 52, 312 (1958).
39. H. Ockenfels, H. Thomas and E. Schmitz, *Z. Naturforsch. Teil B* **25**, 922 (1970).
40. N. Seiler and K. Deckardt, *J. Chromatogr.* **107**, 227 (1975).
41. Z. Tamura, T. Nakajima, T. Nakayama, J. J. Pisano and S. Udenfriend, *Anal. Biochem.* **52**, 595 (1973).
42. R. F. Chen, *Arch. Biochem. Biophys.* **120**, 609 (1967).
43. N. Seiler, T. Schmidt-Glenewinkel and H. H. Schneider, *J. Chromatogr.* **84**, 95 (1973).
44. T. Kinoshita, F. Iinuma and A. Tsuji, *Biochem. Biophys. Res. Commun.* **51**, 666 (1973).
45. M. S. Arnott and D. N. Ward, *Anal. Biochem.* **21**, 50 (1967).
46. B. P. Sloan, *J. Chromatogr.* **42**, 426 (1969).
47. R. S. Fager and C. B. Kutina, *J. Chromatogr.* **76**, 268 (1973).
48. H. T. Nagashawa, P. S. Fraser and J. A. Elberling, *J. Chromatogr.* **44**, 300 (1969).
49. N. Seiler and M. Wiechmann, *J. Chromatogr.* **28**, 351 (1967).
50. J. Langner, Hoppe-Seyler's *Z. Physiol. Chem.* **347**, 275 (1966).
51. R. S. Atherton and A. R. Thompson, *Biochem. J.* **111**, 797 (1969).
52. T. Kato and M. Makoto, *Anal. Biochem.* **66**, 515 (1975).
53. M. O. Creighton and J. R. Trevithick, *Anal. Biochem.* **50**, 255 (1972).
54. I. S. Forrest, S. D. Rose, L. G. Brookes, B. Halpern, V. A. Bacon and I. A. Silberg, *Agressologie* **11**, 127 (1970).

55. K. Igarashi, I. Izumi, K. Hara and S. Hirose, *Chem. Pharm. Bull.* **22**, 451 (1974).
56. R. L. Munier and A. M. Drapier, *Chromatographia* **5**, 306 (1972).
57. R. L. Munier, C. Thommegay and A. M. Drapier, *Chromatographia* **1**, 95 (1968).
58. A. A. Boulton, *Abstr.* 2nd Intern. Neurochemical Meeting, Oxford, 1965.
59. G. L. Moore and R. S. Antonoff, *Anal. Biochem.* **39**, 260 (1971).
60. G. Briel, V. Neuhoff and M. Maier, Hoppe-Seyler's *Z. Physiol. Chem.* **353**, 540 (1972).
61. J. Airhart, S. Sibiga, H. Sanders and E. A. Khairallah, *Anal. Biochem.* **53**, 132 (1973).
62. P. Teichgräber, I. Krautschick, R. Arnold and D. Biesold, *Pharmazie* **29**, 186 (1974).
63. G. Nota, G. Marino, V. Buonocore, and A. Ballio, *J. Chromatogr.* **46**, 103 (1970).
64. R. W. Frei and J. F. Lawrence, *J. Chromatogr.* **83**, 321 (1971).
65. N. Seiler and H. H. Schneider, *J. Chromatogr.* **59**, 367 (1971).
66. Z. Deyl and J. Rosmus, *J. Chromatogr.* **69**, 129 (1972).
67. H. Engelhardt, J. Asshauer, U. Neue and N. Weigand, *Anal. Chem.* **46**, 336 (1974).
68. T. Seki and H. Wada, *J. Chromatogr.* **102**, 251 (1974).
69. W. Dünges, G. Naundorf and N. Seiler, *J. Chromatrogr. Sci.* **12**, 655 (1974).
70. S. R. Abbott, A. Abu-Shumays, K. O. Loeffler and I. S. Forrest, *Res. Commun. Chem. Pathol. Pharmacol.* **10**, 9 (1975).
71. M. M. Abdel-Monem and K. Ohno, *J. Chromatogr.* **107**, 416 (1975).
72. A. A. Boulton, *Methods Biochem. Anal.* **16**, 327 (1968).
73. V. V. Nesterov, B. G. Belinsky and L. G. Senyutenkova, *Biochimiia* **34**, 666 (1969).
74. V. A. Spivak, V. M. Orlov, V. V. Scherbukhin and Ja. M. Varshavsky, *Anal. Biochem.* **35**, 227 (1970).
75. V. A. Spivak, V. V. Scherbukhin, V. M. Orlov and Ja. M. Varshavsky, *Anal. Biochem.* **39**, 271 (1971).
76. V. A. Spivak, M. I. Levjant, S. P. Katrukha and Ja. M. Varshavsky, *Anal. Biochem.* **44**, 503 (1971).
77. C. Gros and B. Labouesse, *Eur. J. Biochem.* **7**, 463 (1969).
78. J. P. Zanetta, G. Vincendon, P. Mandel and G. Gombos *J. Chromatogr.* **51**, 441 (1970).
79. N. Seiler and B. Knödgen, Hoppe-Seyler's *Z. Physiol. Chem.* **352**, 97 (1971).
80. R. Klaus, *J. Chromatogr.* **16**, 311 (1964).
81. J. Reisch, H. Alfes, N. Jantos and H. Möllmann, *Acta Pharm. Suec.* **5**, 393 (1968).
82. C. R. Creveling, K. Kondo and J. W. Daly, *Clin. Chem.* **14**, 302 (1968).
83. G. Marino and V. Buonocore, *Biochem. J.* **110**, 603 (1968).
84. N. Seiler, H. H. Schneider and K.-D. Sonnenberg, *Anal. Biochem.* **44**, 451 (1971).
85. H. Egge, H. Ockenfels, H. Thomas and F. Zilliken, *Z. Naturforsch. Teil B* **26**, 229 (1971).
86. A. E. Jenkins and J. R. Majer, in *Mass Spectrometry*, Butterworth, London, 1968, p. 253.
87. N. Seiler and B. Knödgen, *Org. Mass Spectrom.* **7**, 97 (1973).
88. D. A. Durden, B. A. Davis and A. A. Boulton, *Biomed. Mass Spectrom.* **1**, 83 (1974).
89. M. Weise and G. M. Eisenbach, *Experientia* **28**, 245 (1972).
90. V. Neuhoff and M. Weise, *Arzneim Forsch.* **20**, 368 (1970).
91. V. Neuhoff, F. von der Haar, E. Schlimme and M. Weise, Hoppe-Seyer's *Z. Physiol. Chem.* **350**, 121 (1969).
92. L. Casola and G. DiMatteo, *Anal. Biochem.* **49**, 416 (1972).
93. J. P. Brown and R. N. Perham, *Eur. J. Biochem.* **39**, 69 (1973).
94. A. J. Kennedy and M. J. Voades, *J. Neurochem.* **23**, 1093 (1974).
95. S. R. Snodgrass and L. L. Iversen, *Nature (London) New Biol.* **241**, 154 (1973).
96. P. M. Beart and S. R. Snodgrass, *J. Neurochem.* **24**, 821 (1975).
97. M. H. Joseph and J. Halliday, *Anal. Biochem.* **64**, 389 (1975).
98. G. Schmer, Hoppe-Seyler's *Z. Physiol. Chem.* **348**, 199 (1967).
99. K. Worowski, *Wiad. Chem.* **27**, 789 (1973).
100. L. Casola, G. Di Matteo, G. Di Prisco and F. Cervone, *Anal. Biochem.* **57**, 38 (1974).
101. G. Rapoport, M. F. Glatron and M. M. Lecadet, *C. R. Acad. Sci. Ser. D* **265**, 639 (1967).
102. G. Schmer and H. Kreil, *Anal. Biochem.* **29**, 186 (1969).

103. P. Nedkov and N. Genov, *Biochim. Biophys. Acta* **127**, 544 (1966).
104. B. Mesrob and V. Holeysovsky, *Collect. Czech. Chem. Commun.* **32**, 1976 (1967).
105. G. Pataki, *Helv. Chim. Acta* **50**, 1069 (1967).
106. C. J. Bruton and B. S. Hartley, *J. Mol. Biol.* **52**, 165 (1970).
107. M. E. Percy and B. M. Buchwald, *Anal. Biochem.* **45**, 60 (1972).
108. B. S. Hartley, *Nature (London)*, **201**, 1284 (1964).
109. P. J. Piggot and E. M. Press, *Biochem. J.* **104** 616 (1967).
110. P. Wahl, *Biochim. Biophys. Acta* **175**, 55 (1969).
111. W. R. Gray and B. S. Hartley, *Biochem. J.* **89**, 379, (1963).
112. G. Schmer and H. Kreil, *J. Chromatogr.* **28**, 458 (1967).
113. Z. Tamura, T. Tanimura and H. Yoshida, *Chem. Pharm. Bull.* **15**, 252 (1967).
114. S. G. Whitney and E. A. Grula, *Biochim. Biophys. Acta* **158**, 124 (1968).
115. T. Kinoshita, F. Iinuma and A. Tsuki, *Anal. Biochem.* **61**, 632 (1974).
116. T. Kinoshita, F. Iinuma and A. Tsuji, *Anal. Biochem.* **66**, 104 (1975).
117. K. Crowshaw, S. J. Jessup and P. W. Ramwell, *Biochem. J.* **103**, 79 (1967).
118. N. N. Osborne, P. H. Wu and V. Neuhoff, *Brain Res.* **74** 175 (1974).
119. N. N. Osborne and V. Neuhoff, *Naturwissenschaften* **60**, 78 (1973).
120. P. J. Roberts, P. Keen and J. F. Mitchell, *J. Neurochem.* **21**, 199 (1973).
121. V. Neuhoff, G. Briel and A. Maelicke, *Arzneim. Forsch.* **21**, 104 (1971).
122. N. Seiler, M.-J. Al-Therib and K. Kataoka, *J. Neurochem.* **20**, 699 (1973).
123. N. Seiler and N. Eichentopf, *Biochem. J.* **152**, 201 (1975).
124. N. N. Osborne, H. Zimmermann, M. J. Dowdall and N. Seiler, *Brain Res.* **88**, 115 (1975).
125. R. H. Strang, *J. Neurochem.* **25**, 27 (1975).
126. A. A. Boulton, S. R. Philips and D. A. Durden, *J. Chromatogr.* **82**, 137 (1973).
127. A. A. Boulton and J. R. Majer, in *Research Methods in Neurochemistry* (N. Marks and R. Rodnight, Eds), Vol. 1, p.341, Plenum, New York, 1972.
128. S. Axelsson, A. Björklund and N. Seiler, *Life Sci.* **13**, 1411 (1973).
129. A. V. Jurio and S. R. Philips, *Brain Res.* **83**, 180 (1975).
130. C. R. Creveling and J. W. Daly, *Nature* (*London*) **216**, 190 (1967).
131. G. Braun, Dissertation, University of Bonn, 1974.
132. W. H. Vogel and D. Ahlberg, *Anal. Biochem.* **18**, 563 (1967).
133. N. Seiler and K. Bruder, *J. Chromatogr.* **106**, 159 (1975).
134. S. Axelsson, A. Björklund and N. Seiler, *Life Sci.* **10**, Part I, 745 (1971).
135. N. Seiler and T. Schmidt-Glenewinkel, *J. Neurochem.* **24**, 791 (1975).
136. N. Seiler and U. Lamberty, *J. Neurochem.* **20**, 709 (1973).
137. N. Seiler, M.-J. Al-Therib and B. Knödgen, Hoppe-Seyler's *Z. Physiol. Chem.* **354**, 589 (1973).
138. H. Dolezalova, M. Stepita-Klauco and N. Seiler, *Brain Res.* **67**, 349 (1974).
139. H. Dolezalova, M. Stepita-Klauco and R. Fairweather, *Brain Res.* **77**, 166 (1974).
140. H. Dolezalova, E. Giacobini, N. Seiler and H. H. Schneider, *Brain Res.* **55**, 242 (1973).
141. H. Dolezalova and M. Stepita-Klauco, *Brain Res.* **74**, 182 (1974).
142. T. Takahashi, M. Watanabe, M. Asahina and T. Yamaha, *J. Pharm. Soc. Jpn* **92**, 393 (1972).
143. A. A. Galoyan, B. K. Mesrob and V. Holeysovsky, *J. Chromatogr.* **24**, 440 (1966).
144. L. P. Penzes and G. W. Oertel, *J. Chromatogr.* **51**, 325 (1970)
145. G. W. Oertel and L. P. Penzes, *Z. Analyt. Chem.* **252**, 306 (1970).
146. M. Frei-Häusler and R. W. Frei, *J. Chromatogr.* **79**, 207 (1973).
147. I. Durko and A. A. Boulton, Abstr. 1st Intern. Meeting of the Intern. Soc. Neurochem., Strasbourg, 1967.
148. I. K. Ho, H. H. Loh and E. L. Way, *Proc. West. Pharmacol. Soc.* **14**, 183 (1971).
149. P. N. Kaul, M. W. Conway and M. L. Clark, *Nature* (*London*) **226**, 372 (1970).
150. P. N. Kaul, M. W. Conway, M. K. Ticku and M. L. Clark, *J. Pharm. Sci.* **59**, 1745 (1970).
151. P. N. Kaul, M. W. Conway, M. K. Ticku and M. L. Clark, *J. Pharm. Sci.* **61**, 581 (1972).

152. I. S. Forrest, D. E. Green, S. S. Rose, G. C. Skinner and D. M. Torres, *Res. Commun. Chem. Pathol. Pharmacol.* **2**, 787 (1971).
153. W. W. Just, G. Werner and M. Wiechmann, *Naturwiss.* **59**, 222 (1971).
154. N. Seiler, L. Kamenikova and G. Werner, *Collect. Czech. Chem. Commun.* **34**, 719 (1969).
155. N. Seiler, L. Kamenikova and G. Werner, Hoppe-Seyler's *Z. Physiol. Chem.* **348**, 768 (1967).
156. N. Seiler and M.-J. Al-Therib, *Biochim. Biophys. Acta* **354**, 206 (1974).
157. N. Seiler, U. Lamberty and M.-J. Al-Therib, *J. Neurochem.* **24**, 797 (1975).
158. N. Seiler and B. Knödgen, *J. Chromatogr.* **97**, 286 (1974).
159. N. Seiler and H. H. Schneider, *Biomed. Mass Spectrom.* **1**, 381 (1974).
160. R. P. Cory, R. R. Becker, R. Rosenbluth and T. Isenberg, *J. Am. Chem. Soc.* **90**, 1643 (1968).
161. Ch. P. Ivanov, *Monatsh. Chem.* **97**, 1499 (1966).
162. Ch. P. Ivanov and Y. Vladovska-Yukhnovska, *J. Chromatogr.* **71**, 111 (1972).
163. Ch. P. Ivanov and Y. Vladovska-Yukhnovska, *Biochim. Biophys. Acta* **194**, 345 (1969).
164. I. Durko, Y. Vladovska-Yukhnovska and Ch. P. Ivanov, *Clin. Chim. Acta* **49**, 407 (1973).
165. P. B. Ghosh and M. W. Whitehouse, *Biochem. J.* **108**, 155 (1968).
166. A. J. Boulton, P. B. Ghosh and A. R. Katritzky, *J. Chem. Soc. B*, 1004 (1966).
167. A. J. Boulton, P. B. Ghosh and A. R. Katritzky, *Tetrahedron Lett.* **25**, 2887 (1966).
168. D. J. Birkett, N. C. Price, G. K. Radda and A. G. Salmon, *FEBS Lett.* **6**, 346 (1970).
169. H.-J. Klimisch and L. Stadler, *J. Chromatogr.* **90**, 141 (1974).
170. R. S. Fager, C. B. Kutina and E. W. Abrahamson, *Anal. Biochem.* **53**, 290 (1973).
171. J. F. Lawrence and R. W. Frei, *Anal. Chem.* **44**, 2046 (1972).
172. D. Clasing, H. Alfes, H. Möllmann and J. Reisch, *Z. Klin. Chem. Klin. Biochem.* **7**, 648 (1969).
173. J. Montforte, R. J. Bath and I. Sunshine, *Clin. Chem.* **18**, 1329 (1972).
174. F. Van Hoof and A. Heyndrick, *Anal. Chem.* **46**, 286 (1974).
175. J. Reisch, H. J. Kommert, H. Alfes and H. Möllmann, *Z. Analyt. Chem.* **247**, 56 (1969).
176. J. Reisch, H. Alfes, H. J. Kommert, N. Jantos, H. Möllmann and E. Clasing, *Pharmazie* **25**, 331 (1970).
177. H. J. Creech and R. N. Jones, *J. Am. Chem. Soc.* **62**, 1970 (1940).
178. S. Udenfriend, *Fluorescence Assay in Biology and Medicine*, Academic Press, New York, p. 215.
179. J. E. Sinsheimer, D. D. Hong, J. T. Stewart, M. L. Fink and J. H. Burckhalter, *J. Pharm. Sci.* **60**, 141 (1971).
180. A. DeLeenheer, J. E. Sinsheimer and J. H. Burckhalter, *J. Pharm. Sci.* **62**, 1370 (1973).
181. P. Kristan, *Chem. Zvesti.* **15**, 641 (1961).
182. H. Maeda, N. Ishida, H. Kawauchi and K. Tuzimura, *J. Biochem.* (*Tokyo*) **65**, 777 (1969).
183. P. Edman, *Acta Chem. Scand.* **4**, 283 (1950).
184. H. Kawauchi, K. Tuzimura, H. Maeda and N. Ishida, *J. Biochem.* (*Tokyo*) **66**, 783 (1969).
185. A. H. Coons and M. H. Kaplan, *J. Exp. Med.* **91**, 1 (1950).
186. R. M. McKinney, J. T. Spillane and G. W. Pierce, *J. Org. Chem.* **27**, 3986 (1962).
187. J. L. Riggs, R. J. Seiwald and J. H. Burckhalter, *Am. J. Pathol.* **34**, 1081 (1958).
188. M. W. McCaman and E. Robins, *J. Lab. Clin. Med.* **59**, 885 (1962).
189. K. Samejima, W. Dairman and S. Udenfriend, *Anal. Biochem.* **42**, 222 (1971).
190. M. Weigele, J. F. Blount, J. P. Tengi, R. C. Czajkowski and W. Leimgruber, *J. Am. Chem. Soc.* **94**, 4052 (1972).
191. M. Weigele, S. L. DeBernardo, J. P. Tengi and W. Leimgruber, *J. Am. Chem. Soc.* **94**, 5927 (1972).
192. S. Stein, P. Böhlen and S. Udenfriend, *Arch. Biochem. Biophys.* **163**, 400 (1974).
193. S. De Bernardo, M. Weigele, V. Toome, K. Manhart, W. Leimgruber, P. Böhlen, S. Stein and S. Udenfriend, *Arch. Biochem. Biophys.* **163**, 390 (1974).

194. P. Böhlen, S. Stein, W. Dairman and S. Udenfriend, *Arch. Biochem. Biophys.* **155**, 213 (1973).
195. S. Stein, P. Böhlen, J. Stone, W. Dairman and S. Udenfriend, *Arch. Biochem. Biophys.* **155**, 202 (1973).
196. A. M. Felix and G. Terkelsen, *Arch. Biochem. Biophys.* **157**, 177 (1973).
197. M. Weigele, S. De Bernardo and W. Leimgruber, *Biochem. Biophys. Res. Commun.* **50**, 352 (1973).
198. C. Schwabe, *Anal. Biochem.* **53**, 484 (1973).
199. W. L. Ragland, J. L. Pace and D. L. Kemper, *J. Int. Res. Commun.* **1**, 7 (1973).
200. A. M. Felix and M. H. Jimenez, *J. Chromatogr.* **89**, 361 (1974).
201. K. Imai, P. Böhlen, S. Stein and S. Udenfriend, *Arch. Biochem. Biophys.* **161**, 161 (1974).
202. K. Imai, *J. Chromatogr.* **105**, 135 (1975).
203. K. Samejima, *J. Chromatogr.* **96**, 250 (1974).
204. N. Lustenberger, H. W. Lange and K. Hempel, *Angew. Chem.* **84**, 255 (1972).
205. H. W. Lange, N. Lustenberger and K. Hempel, *Z. Anal. Chem.* **261**, 337 (1972).
206. P. A. Shore, A. Burckhalter and V. H. Cohen, *J. Pharmacol. Exp. Ther.* **127**, 182 (1959).
207. R. Hakanson, A. L. Rönnberg and K. Sjölund, *Anal. Biochem.* **47**, 356 (1972).
208. L. T. Kremzner, *Anal. Biochem.* **15**, 270 (1966).
209. E. A. Carlini and J. P. Green, *Brit. J. Pharmacol.* **20**, 264 (1963).
210. A. H. Anton and D. F. Sayre, *J. Pharmacol. Exp. Ther.* **166**, 285 (1969).
211. R. P. Maickel and F. P. Miller, *Anal. Chem.* **38**, 1937 (1966).
212. L. Juhlin and W. B. Shelley, *J. Histochem. Cytochem.* **14**, 525 (1966).
213. B. Ehinger and R. Thunberg, *Exp. Cell. Res.* **47**, 116 (1967).
214. R. Hakanson and C. Owman, *Life Sci.* **6**, 759 (1967).
215. R. Hakanson, L. Juhlin, C. Owman and B. Sporrong, *J. Histochem. Cytochem.* **18**, 93 (1970).
216. D. Aures, R. Fleming and R. Hakanson, *J. Chromatogr.* **33**, 480 (1968).
217. D. Turner and S. L. Wightman, *J. Chromatogr.* **32**, 315 (1968).
218. W. B. Shelley and L. Juhlin, *J. Chromatogr.* **22**, 130 (1966).
219. L. Edvinsson, R. Hakanson, A. L. Rönnberg and F. Sundler, *J. Chromatogr.* **67**, 81 (1972).
220. M. Roth, *Anal. Chem.* **43**, 880 (1971).
221. D. P. Evans, J. A. Lewis and D. S. Thomson, *Life Sci.* **12** (Part II), 327 (1973).
222. R. Hakanson and A. L. Rönnberg, *Anal. Biochem.* **54**, 353 (1973).
223. F. P. Miller and R. P. Maickel, *Life Sci.* **9** (Part I), 747 (1970).
224. M. Roth and A. Hampai, *J. Chromatogr.* **83**, 353 (1973).
225. Z. Prochazka, *Chem. Listy* **47**, 1643 (1953).
226. A. Carlsson, B. Falck and N. A. Hillarp, *J. Histochem. Cytochem.* **10**, 348 (1962).
227. B. Falck, *Acta Physiol. Scand. Suppl.* **197**, 1 (1962).
228. U. Ungerstedt, *Acta Physiol. Scand. Suppl.* **367**, 1 (1971).
229. D. Aures, A. Björklund and R. Hakanson, *Z. Anal. Chem.* **243**, 564 (1968).
230. E. J. Cowles, G. M. Christensen and A. C. Hilding, *J. Chromatogr.* **35**, 389 (1968).
231. A. Björklund, B. Falck and R. Hakanson, *J. Chromatogr.* **40**, 186 (1969).
232. A. Björklund, B. Falck and R. Hakanson, *Acta Physiol. Scand. Suppl.* **318**, 1 (1968).
233. N. Seiler and M. Wiechmann, Hoppe-Seyler's *Z. Physiol. Chem.* **337**, 229 (1964).
234. S. Axelsson, A. Björklund and O. Lindvall, *J. Histochem. Cytochem.* **20**, 435 (1972).
235. S. Axelsson, A. Björklund, B. Falck, O. Lindvall and L.-A. Svensson, *Acta Physiol. Scand.* **87**, 57 (1973).
236. S. Axelsson, A. Björklund and O. Lindvall, *J. Chromatogr.* **105**, 211 (1975).
237. J. Harley-Mason and A. A. P. G. Archer, *Biochem. J.* **69**, 60 (1958).
238. R. A. Heacock and M. E. Mahon, *J. Chromatogr.* **17**, 338 (1965).
239. H. Tanimukai, *J. Chromatogr.* **30**, 155 (1967).
240. B. Heller, N. Narasimhachari, J. Spaide, L. Haskovec and H. E. Himwich, *Experientia* **26**, 503 (1970).
241. P. Baumann, P. Scherer, W. Krämer and N. Matussek, *J. Chromatogr.* **59**, 463 (1971).

242. P. Baumann and N. Narasimhachari, *J. Chromatogr.* **86**, 269 (1973).
243. H. Spatz and N. Spatz, *Biochem. Med.* **6**, 1 (1972).
244. P. Baumann, *J. Chromatogr.* **109**, 313 (1975).
245. R. Chayen, R. Dvir, S. Gould and A. Harell, *Israel J. Chem.* **8**, 157 (1970).
246. V. Graef, *Z. Klin. Chem. Klin. Biochem.* **8**, 320 (1970).
247. R. Chayen, R. Dvir, S. Gould and A. Harell, *Anal. Biochem.* **42**, 283 (1971).
248. G. Allen and G. Lowe, *Biochem. J.* **133**, 679 (1973).
249. C. Wu and L. Stryer, *Proc. Natl. Acad. Sci. U.S.* **69**, 1104 (1972).
250. J. R. Bunting and R. E. Cathou, *J. Mol. Biol.* **77**, 223 (1973).
251. E. Karush, N. R. Klinman and R. Marks, *Anal. Biochem.* **9**, 100 (1964).
252. J. R. Heitz and B. Anderson, *Arch. Biochem. Biophys.* **127**, 637 (1968).
253. J. K. Weltman, A. R. Frackelton Jr, R. P. Szaro and R. M. Dowben, *Biophys. Soc. Abstr.* **12**, 277a (1972).
254. E. Hudson and G. Weber, *Biochemistry* **12**, 4154 (1973).
255. J. L. Weltman, R. P. Szaro, A. R. Frackelton Jr, R. M. Dowben, J. R. Bunting and R. E. Cathou, *J. Biol. Chem.* **248**, 3173 (1973).
256. Y. Kanaoka, M. Machida, M. Machida and T. Sekine, *Biochim. Biophys. Acta* **217**, 563 (1973).
257. Y. Kanaoka, M. Machida, K. Ando and T. Sekine, *Biochim. Biophys. Acta* **207**, 269 (1970).
258. H. C. Cheung, R. Cooke and L. Smith, *Arch. Biochem. Biophys.* **142**, 333 (1971).
259. W. H. Scouten, R. Lubcher and W. Boughman, *Biochim. Biophys. Acta* **336**, 421 (1974).
260. P. C. Leavis and S. S. Lehrer, *Biochemistry* **13**, 3042 (1974).
261. A. B. Groth and M. E. Dahlen, *Acta Chem. Scand.* **21**, 291 (1967).
262. C. Gitler, *Anal. Biochem.* **50**, 324 (1972).
263. N. Seiler, G. Werner and M. Wiechmann, *Naturwiss.* **20**, 643 (1963).
264. N. Seiler and H. Möller, *Chromatographia* **2**, 470 (1969).
265. H. Ichikawa, T. Tanimura, T. Nakajima and Z. Tamura, *Chem. Pharm. Bull.* **18**, 1493 (1970).
266. K. Kostka, *J. Chromatogr.* **49**, 249 (1970).

CHAPTER 10

Dinitrophenyl and Other Nitrophenyl Derivatives

David J. Edwards

University of Pittsburgh School of Medicine, Department of Psychiatry, Western Psychiatric Institute and Clinic, 3811 O'Hara Street, Pittsburgh, Penn. 15261, U.S.A.

Nitrophenyl derivatives are often used for the separation and analysis of amines, phenols, sulfhydryls and carbonyl compounds. The 2,4-dinitrophenyl (DNP) derivatives and 2,4,6-trinitrophenyl (TNP) derivatives in particular have found widespread application in methods of chemical analysis as well as of protein modification. While this chapter will not attempt to review all the uses of these derivatives, it is hoped that, instead, it will provide the reader with sufficient information for the possible application of nitrophenyl derivatives to particular problems in his own laboratory. Therefore, attention will be focused upon the specificities of the derivatizing reagents, properties of the corresponding derivatives and practical methods for their preparation, separation and detection.

1 PROPERTIES OF NITROPHENYL DERIVATIVES

The DNP derivatives are relatively stable and may, therefore, be separated by thin-layer chromatography, paper electrophoresis or gas chromatography. The bright yellow colour of the DNP-amines facilitates their chromatographic separation and identification. The high molar extinction coefficient of amines ($\varepsilon \sim 1.6 \times 10^4$ $[\text{mol/l}]^{-1}$ cm^{-1} at 360 nm[1]) is often used to quantify these compounds spectrophotometrically. The spectrum for DNP-phenylethylamine is illustrated in Fig. 1, displaying the strong absorption band near 360 nm and a somewhat weaker band at 260 nm which is characteristic of DNP-amines. The DNP moiety is also highly electron-capturing, enabling nanogram quantities of DNP-derivatives to be analysed by a gas chromatograph equipped with an electron-capture detector.[2] Its relatively high molecular weight (mol. wt. = 167) and polar nature of the DNP moiety is used to advantage in preparing DNP-derivatives for the separation and analysis of volatile amines.

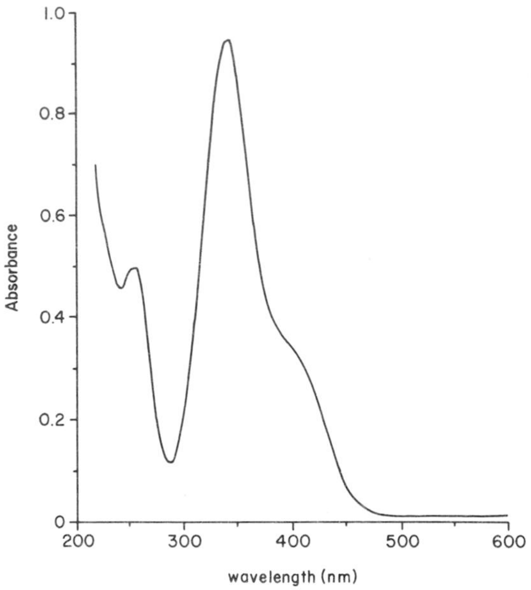

Fig. 1 Absorption spectrum of DNP-phenylethylamine in ethanol. ε (at 346 nm) = 1.69×10^4 $[\text{mol/l}]^{-1}$ cm^{-1}.

The TNP-derivatives have similar properties to those of the DNP-derivatives, but, being more polar, are not so amenable to separation by gas chromatography. Moreover, the TNP-amines are generally more labile and photosensitive than the corresponding DNP-amines.[3] Therefore, TNP-derivatives have been limited to use in the colorimetric analysis of amino acids[4] and peptides,[5] and in protein modification.[6–8]

While DNP- and TNP-derivatives account for the vast majority of the applications of nitrophenyl derivatives, other derivatives have been used in particular applications where somewhat different properties of the products were desirable. For example, 2,6-dinitro-4-trifluoromethyl (DNT) and 2-nitro-4-trifluoromethyl (MNT) derivatives of primary amines have shorter retention times on gas chromatography than the corresponding DNP-derivatives, and, additionally, these derivatives have a somewhat greater sensitivity than DNP-derivatives towards electron-capture detection.[9] Nitrobenzyl and dinitrobenzyl derivatives are useful in separations carried out by high-pressure liquid chromatography (HPLC), since they may be detected by the ultraviolet absorption of these chromophores at 254 nm ($\varepsilon \sim 5 \times 10^3$ $[\text{mol/l}]^{-1}$ cm^{-1}).[10] It would appear that many of these derivatives will find wider application in the future as reagents having a variety of specificities become commercially available.

Physical constants, including the melting points and molar extinction coefficients, have been tabulated in the literature for DNP-amino acids,[1]

DNP-aliphatic amines[11,66] and for TNP-amino acids.[3] Mass spectra have been published for DNP-amino acids[12] and for DNP-derivatives of aliphatic amines.[13]

2 REAGENTS

2.1 1-Fluoro-2,4-dinitrobenzene (FDNB)

By far the most commonly used reagent for preparing DNP-derivatives is FDNB. This reagent was originally used by Sanger[14] in 1945 to identify the free amino groups of insulin. The amino group reacts with FDNB by the following reaction, which is usually carried out in an alkaline solution:

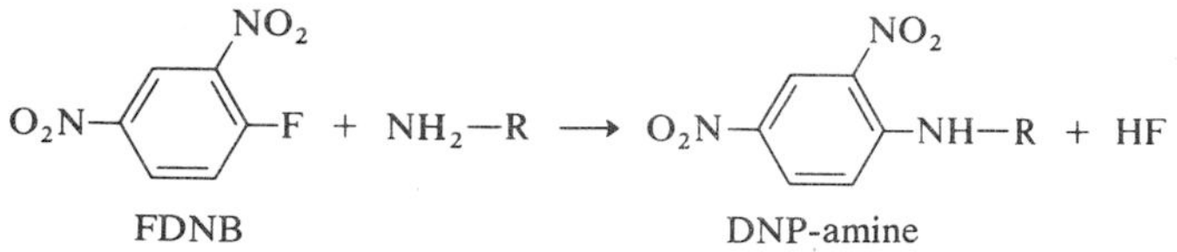

'Sanger's reagent', as FDNB is often called, has been widely employed for the identification of the *N*-terminal amino acid residues of proteins. The procedure involves reaction of the protein with FDNB, the hydrolysis of the protein to individual amino acid residues, and identification of the residue(s) which are dinitrophenylated. Alternatively, the *N*-terminal amino acid residue of small peptides may be determined by a substraction method in which the amino acid compositions of the peptide, before and after dinitrophenylation, are compared. For more details in the use of FDNB for *N*-terminal analysis, the reader should consult reviews devoted to this subject.[15,16]

Nowdays, the use of FDNB for *N*-terminal analysis has been largely replaced by dansylation (Chapter 9) and Edman degradation procedures. However, FDNB remains to be useful as a group specific reagent to study binding sites and tertiary structure of proteins.[6,8]

Precautions should be taken for amino acid residues which have other functional groups which become dinitrophenylated by FDNB. For example, FDNB dinitrophenylates the phenolic group of tyrosine, the imidazole group of histidine, the sulfhydryl group of cysteine, and, of course, the ε-amino group of lysine. Since *O*-DNP-tyrosine is colourless,[16] it does not interfere with detecting DNP-amino acids by their yellow colour. If it is necessary to obtain derivatives containing DNP on only the amino group, the DNP group can be selectively removed from the phenolic, imidazole, and sulfhydryl groups by thiolysis with 2-mercaptoethanol.[17]

FDNB has also been the most commonly used reagent for the preparation of DNP-derivatives of low molecular weight compounds, such as amino acids and biogenic amines. However, amines which contain no ionizable functional groups besides the amino group, when dinitrophenylated, are not easily separable from the excess reagent.* For this reason, an ionizable dinitrophenylating reagent, such as 2,4-dinitrobenzenesulfonic acid, is more convenient for derivatizing these amines (see below). On the other hand, FDNB is the reagent of choice for preparing the DNP-derivatives of amino acids, since an excess of reagent may be readily removed from the product by solvent extraction under the same alkaline conditions at which the reaction was carried out. The aqueous solution may be subsequently acidified to render the DNP-amino acids soluble in organic solvents. By using the appropriate organic solvent, the derivatives may be extracted to eliminate salts which may interfere with their chromatography. As already mentioned, the side chain functional groups of several amino acids also react with FDNB, such that the di-DNP-derivatives of these amino acids may be formed. The specificity of FDNB is summarized and compared with that of other nitrophenylating reagents in Table 1.

TABLE 1

Specificity of nitrophenyl derivatizing agents

Derivatizing agent	Functional groups reacted	Reference
FDNB	Primary and secondary amines, alcohols (aliphatic and phenolic), sulfhydryls, imidazole	6, 7
DNBS	Primary amines, secondary amines	19, 29
TNBS	Primary amino groups, sulfhydryls (does not react with imino groups of histidine or proline or hydroxy groups of tyrosine, threonine, water, and alcohol)	3, 6, 7, 48
DNPH	Aldehydes, ketones	23
	quinones	49

2.2 2,4-Dinitrobenzenesulfonic acid (DNBS)

DNBS is another reagent useful for preparing DNP-derivatives. It was developed by Eisen and co-workers[19] in 1953. Like FDNB, it was originally used for reacting with proteins, but has now found many applications for preparing derivatives of small molecules. DNBS offers the advantage of being

* Excess FDNB may be hydrolysed by strong alkali. An alternative technique which may be used to avoid problems caused by an excess reagent is to add glycine to react with the excess reagent. The resulting DNP-glycine can then be extracted by acqueous solutions and thus be separated from DNP-amines which remain in the organic layer.[18]

water soluble. Therefore, certain alkylamines and arylamines whose DNP-derivatives are soluble in organic solvents may be conveniently dinitrophenylated with DNBS. After the derivatives are formed, they may then be easily separated from an excess of reagent by solvent extraction (usually at the same pH at which the reaction was carried out). Another important advantage of DNBS is its greater specificity for amino groups. Whereas FDNB reacts with sulfhydryl, imidazole and hydroxyl groups in addition to amino groups, DNBS is far more specific for amino groups (Table 1). This reagent is useful for preparing the *N*-DNP-derivatives of amino alcohols, for example, since the hydroxyl group does not react with DNBS. DNBS does react with secondary amines, though at a slower rate than with primary amines.[20] It should be kept in mind that DNBS generally reacts more slowly than does FDNB, so that longer reaction times or more strongly alkaline solutions may be required in order to complete the reaction.

2.3 2,4,6-Trinitrobenzenesulfonic acid (TNBS)

TNBS is the reagent most commonly used for preparing TNP-derivatives, although 1-chloro-2,4,6-trinitrobenzene has also been used.[3] Trinitrophenylation with TNBS occurs more readily than does dinitrophenylation using the analogous reagent, DNBS. The reaction of TNBS with amines is generally carried out at room temperature. Moreover, the reaction rate at neutral pH is sufficiently rapid to permit modification of proteins under non-denaturating conditions. The specificity of TNBS is similar to that of DNBS and is, therefore, more selective for amino groups than is FDNB. TNBS does react to some extent with sulfhydryl groups, but the resulting *S*-TNP-derivatives are unstable in alkaline solution.[6] TNBS reacts very weakly or not at at all with the hydroxyl groups of tyrosine, threonine and serine, the imidazole group of histidine, the imino group of proline and the guanidino group of arginine.[3] It has recently been reported that large quantities of reducing sugars and other carbonyl compounds can interfere with spectrophotometric methods using TNBS.[21]

2.4 2,4-Dinitrophenylhydrazine (DNPH)

DNPH is often used as a reagent for preparing derivatives of aldehydes and ketones and the DNP-hydrazones which are formed may be separated by chromatography on TLC.[22] Since most DNP-hydrazones are solids, their melting points may be used as a convenient procedure for the identification of unknown aldehydes and ketones.[23] Derivatization is generally carried out in aqueous ethanolic sulfuric acid solution.[23] The reaction is usually complete within 10 min at room temperature and the derivatives crystallize out of solution.

2.5 Other nitrophenylating reagents

DNT and MNT derivatives may be prepared by reaction with α,α,α-trifluoro-3,5-dinitro-4-chlorotoluene and α,α,α,4-tetrafluoro-3-nitrotoluene, respectively.[9] Although these reagents have been used in only a few cases, these derivatives appear to offer some advantages in gas chromatography, especially with electron-capture detection. However, like FDNB, these reagents lack specificity for primary amines. In particular, their reaction with alcohols has been noted.[9] It would seem that sulfonic acid analogues of these reagents might be useful for specifically introducing DNT and MNT moieties onto primary amino groups, in the same way that DNBS is a useful analogue of FDNB for introducing the DNP moiety.

3 PREPARATION OF NITROPHENYL DERIVATIVES

Table 2 summarizes the conditions which have been used for the preparation of DNP or TNP-derivatives of various compounds and should be consulted in conjunction with this section. Some of these procedures were designed for the preparation of milligram or even gram quantities of a derivative of a pure compound. When these procedures are used for chemical analysis, the reaction conditions should be appropriately modified in order to prevent too large an excess of reagent or interference by other compounds in the sample. For example, FDNB and DNBS react with water to produce dinitrophenol as a side product. If formed in large amounts, dinitrophenol may interfere with the separation of DNP-amino acids.[24] The amount of dinitrophenol formed may be minimized by reducing the reagent to sample ratio or the reaction time or temperature as much as possible without reducing the yield of the desired derivatives.

3.1 DNP-derivatives formed with FDNB

Detailed procedures for the use of FDNB in the preparation of particular DNP-amino acids have been given elsewhere.[1,16,24,25] The reaction is usually carried out in either aqueous or alcoholic solution at alkaline pH (usually pH 9). Following completion of the reaction, the reaction mixture is extracted with peroxide-free ether in order to remove dinitrophenol. (Traces of peroxide will oxidize DNP-methionine to the sulfone.) The aqueous solution is then acidified and the ether-soluble DNP-amino acids are extracted with peroxide-free ether. The DNP-amino acids remaining in the aqueous phase (including DNP-histidine, DNP-arginine, DNP-cysteic acid, ε-DNP-lysine and *O*-DNP-tyrosine) may be extracted with butanol or *sec*-butanol: ethylacetate (1 : 1).[24]

A longer reaction time may be required in order to dinitrophenylate both functional groups of histidine completely.[24] Amino acids contained in peptides

TABLE 2

Methods for the preparation of nitrophenyl-derivatives

Reagent	Compounds derivatized	Methods	References
FDNB	Amino acids	1–2 mg of the amino acids are dissolved in 1 ml 2% $NaHCO_3$ and mixed with 2 ml ethanol containing 0.05 ml FDNB. Shake (in the dark) at room temperature for 2 h. (5 h for aspartic and glutamic acids.) Dilute with 5 ml H_2O and extract twice with peroxide-free ether. Acidify the aqueous layer with 2 drops of 6 M HCl and extract the DNP-amino acids with 2 ml portions of ether.	16
FDNB		Amino acids (extracted from 0.5 ml serum) in ethanol/H_2O are added to 2 ml phosphate (pH 8.8), 1 ml 5% FDNB in ethanol and 3 ml ethanol; shake at 40 °C for 2 h. Extract 3 times with 30 ml ether. Add 15 ml 10% HCl to aqueous layer (+buffer wash of ether layer) and extract DNP-amino acids 3 times with 20 ml EtOAc. Dry over anhydrous Na_2SO_4	26
FDNB		Amino acids (extracted from approximately 50 mg of tissue) dissolved in 500 μl borate buffer, pH 10, and 50 μl of 10% FDNB in absolute ethanol is added. The reaction is carried out at 60 °C for 30 min. Add 300 μl of 5 M HCl. The excess FDNB and dinitrophenol are removed by extraction 3 times with 1 ml heptane : bromobenzene (80 : 20) and the DNP-amino acids are then extracted 3 times with 400 μl 1 M HCl, combined and dried *in vacuo.*	27
FDNB	Aliphatic amines	1 g of the amine is dissolved in 50 ml of 2.5% borate solution. FDNB (2 ml in 25 ml of *p*-dioxane) is added and the mixture is heated on a steam bath for 1 h. Add 50 ml of 2 M NaOH and heat for 1 h to hydrolyse the excess reagent.	11
FDNB	Phenols	10 μl of each phenol in 4 ml acetone is added to 0.1 ml of a saturated sodium methoxide solution in methanol and 1 ml of FDNB (1% w/v in acetone). The solution is refluxed for 30 min. 25 ml of 2.5% NaOH is added and the DNP-ethers are extracted into 25 ml of chloroform.	28

TABLE 2—continued

Reagent	Compounds derivatized	Methods	References
DNBS	Aliphatic amines	1 ml of amine solution and 1 ml of 0.25 M DNBS in saturated sodium tetraborate are heated in a boiling water bath for 40 min. DNP amines are extracted into 4 ml benzene.	20
DNBS	Phenylethyl-amines	50 μl of solution of amines (100–200 ng of each) and 50 μl 0.25 M DNBS (purified as in Ref. 20) are heated in a boiling water bath for 15 min. After the tubes are cooled, DNP-amines are extracted with 400 and 200 μl portions of benzene.	30
DNBS	Amino alcohols	0.5 mmol of amino alcohol is mixed with 100% excess DNBS, 2.5 ml of 0.5 M Na_2CO_3 and 2.5 ml H_2O and heated in a sealed tube at 100 °C for 1 h. Extract derivatives with chloroform.	29
TNBS	Amino acids	75 mg amino acid, 200 mg TNBS, and 200 mg $NaHCO_3$ are dissolved in 10 ml H_2O and allowed to stand at room temperature (in the dark) for 2 h.	3
DNPH	Ketones and aldehydes	To 0.5 g of the carbonyl compound in 20 ml of 95% ethanol, add DNPH reagent (freshly prepared according to text) and allow to stand at room temperature 5–10 min.	23
DNPH	Ketones and aldehydes	The carbonyl compounds (from a urine extract) are dissolved in 0.1 ml of 0.2% (w/v) DNPH in ethyl acetate and evaporated to dryness under N_2 at 35–40 °C. Add 1.0 ml of 0.03% (w/v) trichloroacetic acid in abs. benzene. After 30 min at 35–40 °C, evaporate the reaction mixture to dryness under N_2 at 35–40 °C.	31

or proteins can, of course, be dinitrophenylated only on their side chains. If these peptides or proteins are hydrolysed after dinitrophenylation, mono-DNP-derivatives will be released, including ε-DNP-lysine, *im*-DNP-histidine and *O*-DNP-tyrosine. Methods for the preparation of authentic samples of these derivatives are given by Mills and Beale.[16] The α-amino group is usually protected by an acetyl group, which may be removed after the DNP-derivative is formed.

Procedures were developed by Ikekawa *et al.*[26] and by Shank and Aprison[27] for the dinitrophenylation of amino acids in serum and tissues, respectively. In the method of Ikekawa *et al.*[26] ethanol/H_2O is added to the serum in order to deproteinize it prior to derivatization. Dinitrophenylation is then carried out in phosphate buffer (pH 8.8). By the method of Shank and Aprison,[27] tissues are homogenized in 10 volumes of 5% trichloroacetic acid. The homogenates are centrifuged for 20 min at 23 000 *g* and the supernatants (500 μl) are extracted 4 times with 1 ml of ether to remove the trichloroacetic acid and are dried. This extract is resuspended in borate buffer and dinitrophenylated.

Day *et al.*[11] used FDNB to dinitrophenylate both primary and secondary aliphatic amines containing 1–4 carbon atoms. These include methylamine, ethylamine and diethylamine. Excess reagent was hydrolysed in strong base (pH 14) by heating for 1 h in a steam bath. However, the use of DNBS as the dinitrophenylating reagent obviates the need for destroying the excess reagent (see below) and is, therefore, the reagent of choice for dinitrophenylating amines other than amino acids.

Cohen *et al.*[28] have also used FDNB to dinitrophenylate phenols. This method was used to make DNP-derivatives of 36 different phenols.

3.2 DNP-derivatives formed with DNBS

Since DNBS is less reactive than FDNB, dinitrophenylation with this reagent must usually be carried out in a boiling-water bath. However, the greater specificity of DNBS and its water solubility has made this reagent highly useful for the preparation of DNP-derivatives in certain circumstances. For example, Crawhall and Elliott[29] used DNBS to prepare the *N*-DNP-derivatives of several amino alcohols. The reaction was carried out in a carbonate solution at 100 °C. Smith and Jepson[20] subsequently used DNBS in borate solution to dinitrophenylate various aliphatic amines and amino alcohols. Edwards and Blau[30] modified this method to derivatize nanogram quantities of phenylethylamines, including phenylethylamine, phenylethanolamine, tyramine, octopamine, and normetanephrine. It was found that the amount of dinitrophenol formed could be minimized without affecting the amount of derivative formed by reducing the reaction time to 15 min.

The DNBS reagent may be prepared by using the technical grade of either the free acid or the sodium salt,[20] although we have found the salt form to be more satisfactory.[30] A 0.25 M solution of DNBS is prepared by dissolving 67.5 g of the sodium salt (mol. wt. = 270.15) per litre of saturated sodium tetraborate. This solution is then washed in a separatory funnel four times with chloroform in order to remove any impurity present in the practical grade of DNBS. This washing step may not be necessary if one uses a reagent grade of DNBS (98% pure), which is now commercially available (Eastman). This reagent can be conveniently prepared in a large batch and stored frozen in aliquots.[30]

3.3 TNP-derivatives

TNBS is a highly reactive trinitrophenylating reagent. The reaction with amino groups is normally carried out at pH 7.5 to 8.5 at room temperature.[3] Amino groups are usually quantitatively derivatized in 2 h. Kinetic studies indicate that only the unprotonated amino group is the reactive species.[3]

TNBS is often used as a spray reagent for locating amino-containing compounds on paper. In this case, TNBS may be prepared as a 0.01% solution in acetone: ethanol (1:1 v/v), containing 1% v/v pyridine.[20]

3.4 2,4-Dinitrophenylhydrazones

A preparative scale procedure for DNP-hydrazones of aldehydes and ketones is given by Shriner *et al.*[23] To a solution containing 0.5 g of a carbonyl compound in 20 ml of 95% ethanol is added the DNPH reagent, which has been freshly prepared as described below. The solution is allowed to sit at room temperature until the derivative has precipitated out (usually 5–10 min).

The DNPH reagent is prepared by adding 2 ml conc. H_2SO_4 to 0.4 g DNPH in a 25 ml flask. Then 3 ml water is added dropwise, and finally, 10 ml of 95% ethanol is added.[23]

Treiber and Oertel[31] have presented a procedure for forming the DNP-hydrazones of ketosteroids in urine extracts. The reaction is carried out at 35–40 °C for 30 min in 1 ml of 0.03% (w/v) trichloroacetic acid in benzene. This procedure was found to be suitable for derivatizing microgram quantities of ketosteroids.

4 SEPARATION OF NITROPHENYL DERIVATIVES

Derivative formation may be carried out in order to facilitate the separation of a mixture of compounds as well as to aid in their detection. The properties of the DNP group are often useful for both the separation and for the quantitation of compounds. For example, a mixture of amino acids, such as might be obtained from either protein hydrolysates or from physiological fluids, may be derivatized as a mixture before the DNP-amino acids are separated. In contrast, TNP-amino acids are less stable than the corresponding DNP-derivatives; and, therefore, trinitrophenylation is more commonly carried out solely for the purpose of detection, usually in the place of ninhydrin, of amino acids and peptides which have been previously separated by ion-exchange chromatography or by paper chromatography.

The differences in chemical properties of compounds before and after derivative formation may also be taken advantage of for isolating a group of similar compounds from complex mixtures, such as may be encountered in biological samples. For example, Edwards and Blau[30] isolated phenylethylamines from biological tissues by a two-step extraction scheme. First, the free amines were extracted into butanol and subsequently back into dilute acid. Then the amines were dinitrophenylated, enabling the derivatives of interest to be extracted into benzene. This two-step purification scheme produced a high degree of clean-up of tissue samples, and served to eliminate large quantities of impurities which may contaminate the electron-capture detector used in these analyses.

DNP-derivatives may be separated by a variety of techniques, including paper chromatography, thin-layer chromatography, high-voltage paper electrophoresis, silica gel column chromatography and gas chromatography. The application of these techniques to the separation of DNP-derivatives is given in Section 6. The particular separation technique which should be selected depends upon a number of factors: (1) nature of the compounds of interest and of impurities present; (2) quantities of these compounds; (3) whether a quantitiatve or qualitative analysis is needed; (4) availability of equipment; and (5) whether recovery of the compound after its analysis is desired.

Paper chromatography has commonly been used for the separation of DNP-amino acid mixtures. Normally, chromatography of the ether-soluble DNP-amino acids and the water soluble DNP-amino acids are carried out separately. The ether-soluble derivatives may be separated by two-dimensional chromatography, using organic solvents (e.g. toluene) in the first direction and phosphate buffer in the second direction. On the other hand, the water soluble DNP-amino acids may be separated one-dimensionally in butanol-acetic acid-water.[16] Solvent systems which are useful for the separation of DNP-amino acids have been reviewed by Biserte *et al.*[32] and by Mills and Beale.[16] Paper chromatography has the advantage of being rapid and simple and requires little equipment. However, this method is relatively insensitive and nonquantitative. Usually, 10 μg of each derivative must be used in order to be visible. Another disadvantage of this technique is its inability to separate all the DNP-amino acids; the DNP-leucines and the DNP-dicarboxylic amino acids are particularly difficult to separate.[16]

Thin-layer chromatography (TLC) allows DNP-derivatives to be separated in a manner very similar to that attained by paper chromatography, but offers the advantages of being more rapid, having better resolution, and permitting smaller quantities of compounds to be detected. As little as 0.2 μg of DNP-derivatives can be seen on TLC plates.[24] For the separation of DNP-amino acids by TLC, the reader is referrred to several reviews for details of solvent systems, R_f values, etc.[16,24,25] Chromatography may be carried out on either cellulose[16,33] or, silica gel plates.[24,25] As for paper chromatography, ether-soluble and water-soluble DNP-amino acids are generally separated by TLC using different solvent systems. An example of two-dimensional TLC of the ether-soluble DNP-amino acids on silica gel G is shown in Fig. 2. Wang and co-workers[34,35] succeeded in separating 31 DNP-amino acids on polyamide layer chromatography. Smith and Jepson[20] separated DNP-derivatives of aliphatic amines such as methylamine, ethylamine and propylamine by TLC on

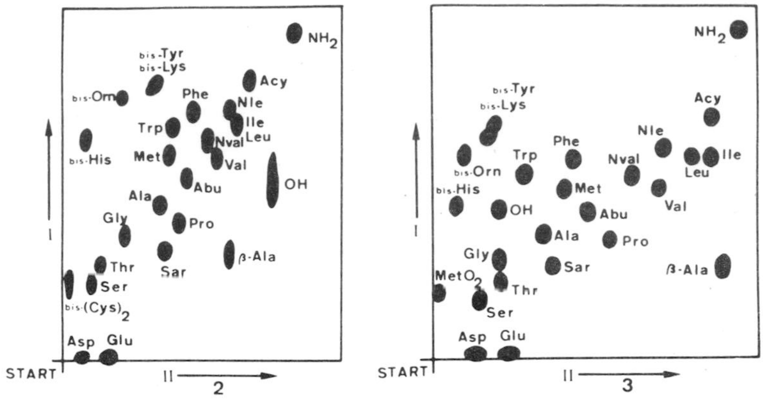

Fig. 2 Two-dimensional TLC of DNP-amino acids on silica gel G. Solvent systems are as follows: 1. toluene-pyridine-2-chloroethanol–0.8 M ammonia (10 : 3 : 6 : 6); 2. chloroform-benzyl alcohol-acetic acid (70 : 30 : 3); 3. benzene-pyridine-acetic acid (40 : 10 : 1).[25] Abbreviations include: *bis*-, bis-DNP-derivatives; NH_2, 2,4-dinitroaniline, and *OH*, 2,4-dinitrophenol. (From *Chromatography*, Erich Heftmann, © 1975 by Litton Educational Publishing, Inc. Reprinted by permission of Van Nostrand Reinhold Co. and Dr A. Niederwieser.[69])

silicic acid plates (adsorption chromatography) and on celite/glycerol plates (partition chromatography). TLC has also been successfully used for the chromatography of DNP-derivatives of several neutral dipeptides.[67,69]

TLC has been a widely used technique for the separation of DNP-hydrazones. Adsorbents which have been used include silica gel G,[36,37] alumina,[37,38] and zinc carbonate.[39] Since the adsorptive properties of DNP-hydrazones undergo a change in the presence of base and certain inorganic compounds, such as zinc carbonate, the composition of the adsorbent and the solvent can have a marked effect on the chromatographic behaviour of these derivatives. An extensive review of adsorbents and solvent systems which have been used for the TLC of DNP-hydrazones is given by Stahl and Jork.[22]

High-voltage paper electrophoresis has also been used to separate DNP-amino acids.[40] Although several of the amino acid derivatives are not separated electrophoretically, this technique appears useful if combined with chromatography in the second direction. The DNP-derivatives of aspartic acid and glutamic acid, which are difficult to resolve chromatographically, are readily separated by high-voltage paper electrophoresis using a pH 6.5 buffer containing pyridine : acetic acid : water (25 : 1 : 225).

A quantitative, automated procedure for the analysis of DNP-amino acids has been developed by which the DNP-amino acids are pumped through a hydrated silica gel column. These derivatives are then quantitated by measuring their absorption in a colorimeter equipped with a flow cell.[41] This method is reproducible and quantitative, but requires as much as 20 h for a single analysis. In addition, the eluate may be recovered by fractional collection, if desired.

Gas chromatography offers the advantages of being both rapid and providing a very high resolution: although the instrument is relatively expensive, gas chromatographs have come to be standard equipment in many laboratories. The method depends upon the compounds separated being volatile. Although DNP-derivatives of aliphatic amines and phenylethylamines are sufficiently volatile and have good gas chromatographic properties, DNP-derivatives of amino acids and hydroxylated amines must be further derivatized before they may be gas chromatographed.[30,42] Usually, the DNP-amino acids are separated as their methyl esters.[42,43] On the other hand, hydroxylated amines may be volatilized by trimethylsilylation.[30] Crosby and Bowers[9] found that DNT and MNT-derivatives of primary amines are more volatile than the corresponding DNP-derivatives and, therefore, are better suited for gas chromatography. Gas chromatography has also been used successfully for the separation of DNP-hydrazone derivatives of carbonyl compounds.[44] However, gas chromatography has been used far less frequently than TLC for the separation of these derivatives, perhaps owing to the relatively low volatility of these compounds and, consequently, the somewhat high column temperatures (250 °C) which must be used. They may also show thermal decomposition.

High-pressure liquid chromatography (HPLC) has a resolving power similar to gas chromatography and has the advantage of permitting the rapid separation of non-volatile compounds at room temperature. Several nitrophenylating reagents are commercially available for reacting with a variety of functional groups, including carboxylic acids and amines.[10] For example, the *p*-nitrobenzyl esters of palmitic and stearic acids may be easily separated and detected by HPLC.[10]

Ito *et al.*[68] have utilized an elution centrifuge to separate a mixture of DNP-amino acids by countercurrent chromatography. This procedure results in a relatively high efficiency (approximately 10 000 theoretical plates) but requires about 54 h to complete.

5 DETECTION

Detection of nitrophenyl derivatives usually takes advantage of the particular properties of the nitrophenyl moiety. The high molar extinction coefficient of these compounds facilitates the detection of small quantities of these derivatives which have been separated by paper chromatography, TLC, high-voltage paper electrophoresis, or column chromatography. Only a few micrograms or less of DNP-amino acids may be required for them to be visually detected by their yellow colour on paper chromatograms or TLC plates. Nitrophenyl derivatives can be detected with greater sensitivity by using a fluorescent indicator in the TLC plates. As little as 0.06 μg of a DNP-amino acid can be detected under u.v. light by observing its quenching of the fluorescent indicator. In addition, this technique would enable the colourless

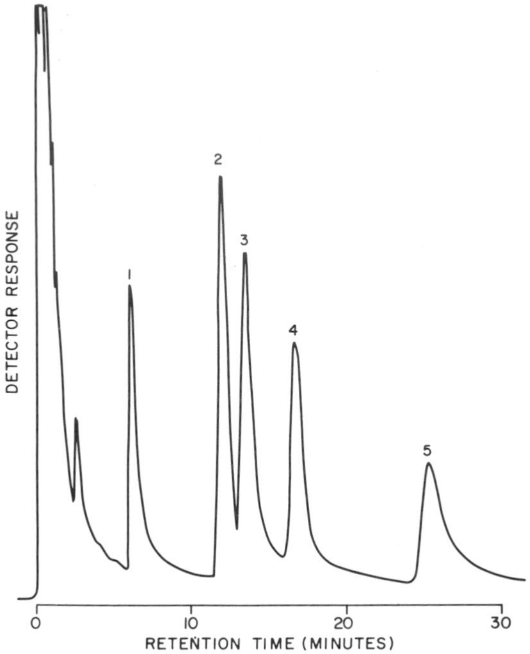

Fig. 3 Gas–liquid chromatograph of DNP-amines with electron-capture detection. Peaks represent the *N*-DNP, *O*-trimethylsilyl-derivatives of the following: 1. phenylethylamine (10 ng); 2. octopamine (10 ng); 3. *p*-tyramine (10 ng); 4. normetanephrine (20 ng); and 5. 3,4-dimethoxyphenylethylamine (20 ng). The column was 1% OV-17, $6' \times \frac{1}{8}''$, operated at 230 °C. Reproduced from Ref. 30 with permission from Academic Press, Inc.

O-DNP-derivative of tyrosine to be visualized.[25] Since DNP-hydrazones undergo a colour change at alkaline pH, the chromatograms may be sprayed with a base such as pyridine or ethanolamine to intensify the spots.

The high molar extinction coefficient of nitrophenyl groups makes nitrophenyl derivatives attractive for the measurement of a variety of compounds by HPLC.[10] DNP-amines, for example, can be detected by their u.v. absorption at 360 nm.[45] As this technique becomes more widely used, one would expect to see a greater application of nitrophenyl derivative formation.

The sensitivity of the DNP moiety to electron-capture detection has been taken advantage of for the quantitation of nanogram or even picogram quantities of DNP-derivatives which have been separated by gas chromatography.[2] Edwards and Blau[30] have used electron-capture detection to measure as little as 0.1 ng of the DNP-derivatives of several biologically important phenylethylamines (Fig. 3). In addition to its sensitivity, the electron-capture detector is also highly specific for certain halogen- and nitro-containing compounds. While most components of biological tissues have a low response towards the electron-capture detector, the specific introduction of an electron-capturing group enhances the response only of those compounds that now carry this group. Under the best conditions, the specificity which can be achieved can be as great as the specificity of the reagent used (see Chapter 2).

The use of radioactively labelled FDNB has also been used as a sensitive method for the determination of DNP-derivatives. Both ^{3}H- and ^{14}C-labelled FDNB of high specific activities are now commercially available. Beale and Whitehead[46] were able to determine the *N*-terminal groups of as little as 1 μg of a protein by dinitrophenylating with ^{3}H-FDNB. The ^{3}H-counts measured in the DNP-amino acids separated by paper chromatography gave a measure of the amount of each amino acid which had been dinitrophenylated in the protein sample. ^{14}C-DNP-amino acids were added to the sample to correct for losses during the procedure. Kesner *et al.*[41] developed a system by which the eluate from a silica gel column containing radioactively labelled DNP-amino acids could be continuously collected on filter paper, dried and monitored for radioactivity (Fig. 4). Radioactive FDNB may also be used to quantitate small

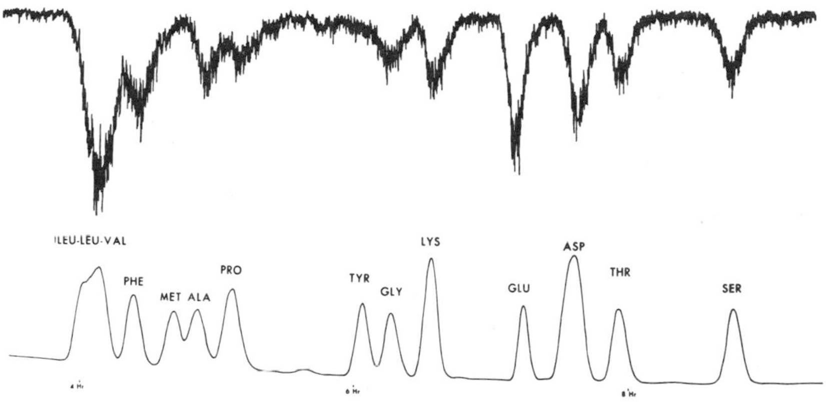

Fig. 4 Separation with simultaneous radioactive counting of a mixture of labelled amino acids derived from ^{14}C-algal protein hydrolysate. The lower photometric trace is due to a synthetic mixture of unlabelled DNP-amino acids added to the sample. Reproduced from Ref. 41 with permission from Academic Press, Inc.

amounts of derivatives separated by TLC or paper chromatography. The unlabelled derivative can be added to the sample as a carrier so that the labelled derivative may be readily located. However, caution must be exercised to keep the total amount of DNP-derivative in the eluted spot below the amount which would cause quenching during scintillation counting.[47]

Mass spectrometry used in combination with paper chromatography or other separation techniques could potentially be used for the simultaneous analysis of various metabolites and their deuterated analogues.[12] Such a procedure could be useful for studying the metabolism of a wide variety of compounds, even if present in trace amounts, and for determining the concentrations of these compounds in small tissue samples.

TABLE 3

Applications of nitrophenyl derivatives for chromatography

Compounds derivatized	Reagent	Derivative formed	Chromatographic techniques	Applications	References
Amino acids	FDNB	DNP-amino acids	Paper chromatography	*N*-terminal analysis of proteins; amino acid content of proteins and peptides	15, 16, 32
Amino acids	FDNB	DNP-amino acids	Thin-layer chromatography (polyamide)	Separation of standards	34, 35, 50
Amino acids	FDNB	DNP-amino acids and DNP-amino acid methyl esters	Thin-layer chromatography (silica gel G)	*N*-terminal analysis of proteins and peptides	15, 24
Amino acids	FDNB	DNP-amino acids	Thin-layer chromatography (Chrom AR sheets [Mallinckrodt])	Determination of concentration and specific radioactivity of amino acids in nervous tissue extracts	27
Amino acids	FDNB	DNP-amino acids	High voltage paper electrophoresis	Separation of standards	32, 40
Amino acids	FDNB	DNP-amino acids	Cellulose TLC/ electrophoresis	Separation of standards	33
Amino acids	FDNB	DNP-amino acids	Automated silica gel column	Separation of DNP-amino acids and DNP-peptides derived from a Pronase digest of ribonuclease, and from labelled ^{14}C-algal protein hydrolysate.	41
Amino acids	FDNB	DNP-amino acid methyl esters	Gas chromatography (1% SE-30)	Separation of amino acids present in hydrolysate of gramacidin A	43
Amino acids	FDNB	DNP-amino acid methyl esters	Gas chromatography (SE-30, QF-1, and PhSi)	Separation of standards	42
Amino acids	FDNB	DNP-amino acid methyl esters	Gas chromatography (1.5% SE-30)	Analysis of free amino acids in nervous tissue	51–53

TABLE 3—continued

Compounds derivatized	Reagent	Derivative formed	Chromatographic techniques	Applications	References
Amino acids	FDNB	*N*-DNP, *O*-TMS-amino acid methyl esters	Gas chromatography (1.5% SE-30, 1.5% XE-60, 1% XE-61, 1% SE-52, and 1.5% QF-1)	Analysis of free amino acids in serum	12
Aliphatic amines	FDNB	DNP-amines	Paper chromatography (reversed phase)	Analysis of piperidine and pyrrolidine formed *in vitro* in brain tissue	54, 55
Aliphatic amines (primary and secondary)	FDNB	DNP-amines	TLC (silica gel G) and gas chromatography (2% DEGS)	Separation of standards (C_1–C_4 amines)	13
Aliphatic amines (primary and secondary)	^{3}H-FDNB	DNP-amines	TLC (silica gel G)	Analysis of volatile amines in mouse brain	47
Aliphatic amines (primary and secondary); aromatic amines	FDNB	DNP-amines	Gas chromatography with electron-capture detection) (1% NPGSe, 1% PDEAS, 1% Versamide 900, 2% SE-30, 0.1% NPGSe+0.8% SE-30, and 1% Apiezon L)	Determination of diethylamine and amphetamine in urine	56
Aliphatic amines (secondary)	FDNB	DNP-amines	High-performance liquid chromatography	Separation of DNP-dimethylamine, DNP-diethylamine and DNP-pyrrolidine standards	45
Aliphatic amines, amino alcohols	DNBS	DNP-amines	Celite and alumina columns; TLC (silicic acid and celite-glycerol plates)	Separation of urinary amines	20
Aliphatic amines	DNBS	DNP-amines	Gas chromatography (10% SE-30;[60] 1% XE-60 and 2% DEGS[59])	Separation of aliphatic amines in whole blood and urine	57–59

TABLE 3—continued

Compounds derivatized	Reagent	Derivative formed	Chromatographic techniques	Applications	References
Phenylalkylamines	DNBS	*N*-DNP, *O*-TMS-amines	Gas chromatography (with electron-capture detection) (1% OV-17; 1% OV-25; 1% OV-210)	Determination of nanogram amounts of phenylethylamines in brain and liver samples	30, 60, 61
Aromatic amines	FDNB	DNP-amines	Gas chromatography (with electron-capture detection) (XE-60)	Analysis of herbicidal carbamate and urea derivatives hydrolysed to aromatic amines	62
Peptides	^{3}H-FDNB	DNP-peptides	TLC	Double isotope assay for angiotensin I	63
Phenols	FDNB	*O*-DNP-ethers	Gas chromatography (with electron-capture detection) (1% XE-60 + 0.1% Epikote 1001)	Analysis of insecticidal carbamates hydrolysed to free phenols	28, 64
Ketones	DNPH	DNP-hydrazones	TLC	Determination of 11-deoxy-17-ketosteroids in human urine	31
Ketones and aldehydes	DNPH	DNP-hydrazones	TLC	Separation of aldehyde and ketone standards	22, 36–39
Ketones and aldehydes	DNPH	DNP-hydrazones	Gas chromatography (10% SF-96; 3% Dexsil 300 GC)	Separation of standards	44, 70, 71
Ketones and aldehydes	DNPH	DNP-hydrazones	High-pressure liquid chromatography	Separation of standards	72, 73
Keto acids	DNPH	DNP-hydrazones	Sephadex LH-20	Separation of standards	65
Quinones	DNPH	DNP-hydrazones	TLC	Separation of standards	49

6 APPLICATIONS

In Table 3 is compiled a variety of applications of nitrophenyl derivatives which have so far appeared in the literature. Although the majority of these applications involve the chromatography of amino acids, it is clear that these derivatives are also useful in the separation and analysis of many other compounds.

REFERENCES

1. K. R. Rao and H. A. Sober, *J. Am. Chem. Soc.* **76**, 1328 (1954).
2. R. A. Landowne and S. R. Lipsky, *Nature (London)* **199**, 141 (1963).
3. T. Okuyama and K. Satake, *J. Biochem.* **47**, 454 (1960).
4. H. H. Brown, *Clin. Chem.* **14**, 967 (1968).
5. R. Delaney, *Anal. Biochem.* **46**, 413 (1972).
6. G. E. Means and R. E. Fenney, *Chemical Modification of Proteins*, Holden-Day, San Francisco, 1971, pp. 118–123.
7. G. R. Stark, *Adv. Prot. Chem.* **24**, 265 (1970).
8. L. A. Cohen, *Ann. Rev. Biochem.* **37**, 695 (1968).
9. D. G. Crosby and J. B. Bowers, *J. Agric. Food Chem.* **16**, 839 (1968).
10. Regis Lab Notes, No. 18, Regis Chemical Co. (1975).
11. E. W. Day Jr, T. Golab and J. R. Koons, *Anal. Chem.* **38**, 1053 (1966).
12. M. H. Studier and L. P. Moore, *Biochem. Biophys. Res. Commun.* **40**, 894 (1970).
13. A. Zeman and I. P. G. Wirotama, *Z. Anal. Chem.* **247**, 155 (1969).
14. F. Sanger, *Biochem. J.* **39**, 507 (1945).
15. J. L. Bailey, *Techniques in Protein Chemistry*, Elsevier, Amsterdam, 1967, pp. 163–221.
16. G. L. Mills and D. Beale, *Chromatographic and Electrophoretic Techniques*, Interscience, New York (Ed. I. Smith), 1969, pp. 170–188.
17. S. Shaltiel, *Biochem. Biophys. Res. Commun.* **29**, 178 (1967).
18. M. J. Kolbezen, J. W. Eckert and B. F. Bretschneider, *Anal. Chem.* **34**, 583 (1962).
19. H. N. Eisen, S. Belman and M. E. Carsten, *J. Am. Chem. Soc.* **75**, 4583 (1953).
20. A. D. Smith and J. B. Jepson, *Anal. Biochem.* **18**, 36 (1967).
21. W. C. Burger, *Anal. Biochem.* **57**, 306 (1974).
22. E. Stahl and H. Jork, *Thin-Layer Chromatography: A Laboratory Handbook*, Springer-Verlag, New York (Ed. E. Stahl), 1969, pp. 206–258.
23. R. L. Shriner, R. C. Fuson and D. Y. Curtin, *The Systematic Identification of Organic Compounds: A Laboratory Manual*, Wiley, New York, 1964, pp. 253–254.
24. G. Pataki, *Techniques of Thin-Layer Chromatography in Amino Acid and Peptide Chemistry*, Humphrey Science Publ. Inc., Ann Arbor, 1969, pp. 126–143.
25. M. Brenner, A. Niederwieser and G. Pataki, *Thin-Layer Chromatography: A Laboratory Handbook*, Springer-Verlag, New York (Ed. E. Stahl), 1969, pp. 756–772.
26. N. Ikekawa, O. Hoshino, R. Watanuki, H. Orimo, T. Fujita and M. Yoshikawa, *Anal. Biochem.* **17**, 16 (1966).
27. R. P. Shank and M. H. Aprison, *Anal. Biochem.* **35**, 136 (1970).
28. I. C. Cohen, J. Norcup, J. H. A. Ruzicka and B. B. Wheals, *J. Chromatogr.* **44**, 251 (1969).
29. J. C. Crawhall and D. F. Elliott, *Biochem. J.* **61**, 264 (1955).
30. D. J. Edwards and K. Blau, *Anal. Biochem.* **45**, 387, (1972).
31. L. Treiber and G. W. Oertel, *Clin. Chim. Acta*, **17**, 81 (1967).
32. G. Biserte, J. W. Holleman, J. Holleman-Dehove and P. Sautière, *J. Chromatogr.* **2**, 225 (1959); **3**, 85 (1960).

33. R. L. Munier and G. Sarrazin, *J. Chromatogr.* **22**, 347 (1966).
34. K.-T. Wang, J. M. K. Huang and I. S. Y. Wang, *J. Chromatogr.* **22**, 362 (1966).
35. K.-T. Wang and I. S. Y. Wang, *J. Chromatogr.* **24**, 460 (1966).
36. G. A. Byrne, *J. Chromatogr.* **20**, 528 (1965).
37. E. Denti and M. P. Luboz, *J. Chromatogr.* **18**, 325 (1965).
38. A. Jart and A. J. Bigler, *J. Chromatogr.* **23**, 261 (1966).
39. C. F. Beyer and T. E. Kargl, *J. Chromatogr.* **65**, 435 (1972).
40. S. Fittkau, *J. Chromatogr.* **18**, 331 (1965).
41. L. Kesner, E. Muntwyler, G. E. Griffin and P. Quaranta, *Methods in Enzymology*, Vol. XI, Enzyme Structure, Academic Press, New York, 1967, pp. 94–108.
42. J. J. Pisano, W. J. A. VandenHeuvel and E. C. Horning, *Biochem. Biophys. Res. Comm.* **7**, 82 (1962).
43. S.-I. Ishii and B. Witkop, *J. Am. Chem. Soc.* **85**, 1832 (1963).
44. R. J. Soukup, R. J. Scarpellino and E. Danielczik, *Anal. Chem.* **36**, 2255 (1964).
45. G. B. Cox, *J. Chromatogr.* **83**, 471 (1973).
46. D. Beale and J. K. Whitehead, *Tritium in the Physical and Biological Sciences I*, International Atomic Energy Agency, Vienna, 1962, pp. 179–190.
47. R. Nixon, *Anal. Biochem.* **48**, 460 (1972).
48. R. B. Freedman and G. K. Radda, *Biochem. J.* **108**, 383 (1968).
49. P. Juvvik and B. Sundby, *J. Chromatogr.* **76**, 487 (1973).
50. K.-T. Wang, Y.-T. Lin and I. S. Y. Wang, *Adv. Chromatogr.* **11**, 73 (1974).
51. M. H. Aprison, W. J. McBride and A. R. Freeman, *J. Neurochem.* **21**, 87 (1973).
52. W. J. McBride, A. R. Freeman, L. T. Graham, Jr. and M. H. Aprison, *Brain Res.* **59**, 440 (1973).
53. J. E. Smith, J. D. Lane, P. A. Shea, W. J. McBride and M. H. Aprison, *Anal. Biochem.* **64**, 149 (1975).
54. Y. Kasé, M. Kataoka and T. Miyata, *Life Sci.* **6**, 2427 (1967).
55. Y. Yamanishi, Y. Kasé, T. Miyata and M. Kataoka, *Life Sci.* **9**, 409 (1970).
56. T. Walle, *Acta Pharm. Suec.* **5**, 367 (1968).
57. S. Baba, I. Hashimoto and Y. Ishitoya, *J. Chromatogr.* **88**, 373 (1974).
58. S. Baba, I. Hashimoto, Y. Ishitoya and Y. Fukuoka, *Clin. Chim. Acta* **62**, 309 (1975).
59. Y. Ishitoya, S. Baba and I. Hashimoto, *Clin. Chim. Acta* **46**, 55 (1973).
60. D. J. Edwards and K. Blau, *J. Neurochem.* **19**, 1829 (1972).
61. D. J. Edwards and K. Blau, *Biochem. J.* **132**, 95 (1973).
62. I. C. Cohen and B. B. Wheals, *J. Chromatogr.* **43**, 233 (1969).
63. R. I. Gregerman and M. Kowatch, *J. Clin. Endocrinol.* **32**, 110 (1971).
64. I. C. Cohen, J. Norcup, J. H. A. Ruzicka and B. B. Wheals, *J. Chromatogr.* **49**, 215 (1970).
65. R. V. Winchester, *J. Chromatogr.* **78**, 429 (1973).
66. A. Oikawa, S. Noguchi, T. Matsushima and M. Inouye, *J. Biochem.* **50**, 157 (1961).
67. C. Martel and D. J. Phelps, *J. Chromatogr.* **115**, 633 (1975).
68. Y. Ito, R. L. Bowman and F. W. Noble, *Anal. Biochem.* **49**, 1 (1972).
69. A. Niederwieser, *Chromatography: A Laboratory Handbook of Chromatographic and Electrophoretic Methods* (3rd edition), Van Nostrand Reinhold, New York, 1975, pp. 393–465.
70. Y. Hoshika and Y. Takata, *J. Chromatogr.* **120**, 379 (1976).
71. C. Bachmann, R. Baumgartner, H. Wick and J. P. Colombo, *Clin. Chim. Acta* **66**, 287 (1976).
72. L. J. Papa and L. P. Turner, *J. Chromatogr. Sci.* **10**, 747 (1972).
73. M. A. Carey and H. E. Persinger, *J. Chromatogr. Sci.* **10**, 537 (1972).

CHAPTER 11

Derivatives of Inorganic Anions for Gas Chromatography

William C. Butts,

Group Health Hospital, 200 15th Ave. E., Seattle, Washington, 98112, U.S.A.

1 INTRODUCTION

The application of gas chromatography to the analysis of inorganic anions is a new and as yet relatively unexplored field. This is due in large part to the non-volatility of the simple anions and the failure to find a universal derivative comparable to the volatile β-diketone chelates of inorganic cations. In spite of these difficulties there have been some innovative and successful developments in the derivatization and chromatography of inorganic anions. Rodriguez-Vazquez[1] and Anvaer and Drugov[2] have reviewed many of these methods; Table 1 summarizes essentially the entire scope of the work done to date with simple anions. As these existing techniques are developed and new derivatives are discovered, the analysis of anions by gas chromatography should become an increasingly frequent and useful method.

An important advantage of this type of anion analysis is the excellent sensitivity and specificity of available detectors[3,4] to many of the elements making up the common anions. The selectivity of electron-capture for halogens and nitro-compounds, flame-photometry for sulfur, phosphorus and silicon,[5] microcoulometric detector for halides, phosphorus and sulfur, and alkali flame ionization for nitrogen, phosphorus, and sulfur offer potentially superior techniques for many anions.

The two largest classes of anions which have been successfully analysed by gas chromatography are the halides and the oxyanions. In the latter category phosphate and the substituted phosphates have been studied most extensively. Because of the concentration of effort on these anions, most of the quantitative procedures in Appendix II involve them.

2 HALIDE DERIVATIVES

Although halogen-containing compounds have been routinely analysed by gas chromatography for many years, it was not until 1968 that Fresen, Cox and

TABLE 1

Anions determined by gas chromatography

Anion	Derivative[a]	Sample matrix	Stationary phase[b]	Detector[c]	References
AsO_3^{3-}	$(TMS)_3AsO_3$	H_2O	SE-30, OV-17	FID, MS	21
	$(TMS)_3AsO_3$	solid salt	SE-30	FID	45
AsO_4^{3-}	$(TMS_3AsO_4$	H_2O	SE-30, OV-17	FID, MS	20, 21
	$(TMS)_3AsO_4$	solid salt	SE-30	FID	45
$(CH_3)_2AsO_2^-$	$(CH_3)_2AsI$	H_2O	DC-200	EC	45
	$(CH_3)_2AsO_2(TMS)$	solid salt	SE-30	FID	45
	$(CH_3)_2AsCH_2CH_2CN$	solid salt	OV-17	FID	45
BO_3^{3-}	$(TMS)_3BO_3$	H_2O	SE-30, OV-17	FID, MS	21
Br^-	$BrCH_2CH_2OH$	H_2O	EGS	FID	13
	CH_3Br	H_2O	Poropak Q	TC	11
	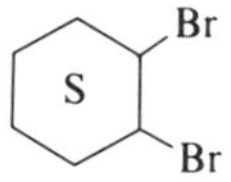	blood	OV-17	FID	14
	$Br(CH_2)_6CH_3$	H_2O	Apiezon L	EC, FID, MS	12
CN^-	CNBr	H_2O	Poropak Q	EC	47
	CNCl	blood, urine	Halcomid M-18	EC	46
CO_3^{2-}	$(TMS)_2CO_3$	H_2O	SE-30, OV-17	FID, MS	20, 21
	CO_2	solid salt	Poropak Q	TC	42
	CO_2	H_2O	Silica gel	TC	43
$C_2O_4^{2-}$	$(TMS)_2C_2O_4$	H_2O	SE-30, OV-17	FID, MS	20, 21
Cl^-	$ClCH_2CH_2OH$	H_2O	EGS	FID	13
	CH_3Cl	H_2O	Poropak Q	TC	11
	$Cl(CH_2)_6CH_3$	H_2O	Apiezon L	EC, FID, MS	12
F^-	$(C_2H_5)_3SiF$	various	DC-200	FID	8
	$(C_2H_5)_3SiF$	H_2O	DC-550	FID	9
	$(CH_3)_3SiF$	teeth	DC 200/50	FID	7
	$(CH_3)_3SiF$	serum, urine	DC 200/50	FID	6, 48
	CH_3F	H_2O	Poropak Q	TC	11
I^-	ICH_2CH_2OH	H_2O	EGS	FID	13
	CH_3I	H_2O	Poropak Q	TC	11
	ICH_2COCH_3	milk	PDS	EC	15
	$I(CH_2)_6CH_3$	H_2O	Apiezon L	EC, FID, MS	12
IO_3^-	I_2	H_2O	DNP	TC	16
NO_2^-	$C_6H_5NO_2$	H_2O	Apiezon M	EC	44
NO_3^-	$C_6H_5NO_2$	H_2O	Apiezon M	EC	44
PO_3^{3-}	$(TMS)_3PO_3$	H_2O	SE-30, OV-17	FID, MS	20, 21
PO_4^{3-}	$(TMS)_3PO_4$	H_2O	DC-430	TC	19(a)
	$(TMS)_3PO_4$	Nucleic Acid	DC-430	TC	19(c)
	$(TMS)_3PO_4$	Nucleic Acid	NPGS	TC	19(b)
	$(TMS)_3PO_4$	Nucleic Acid	OV-101	FID	22
	$(TMS)_3PO_4$	H_2O	SE-30, OV-17	FID, MS	20, 21
	$(TMS)_3PO_4$	H_2O	SE-30	TC	23
	$(TMS)_3PO_4$	H_2O	OV-225	FID, FPD	24
S^{-2}	H_2S	solid salt	Poropak Q	TC	42
	H_2S	H_2O	Silica gel	TC	43
SCN^-	CNBr	H_2O	Poropak Q	EC	47
SO_3^{2-}	SO_2	solid salt	Poropak Q	TC	42
	SO_2	H_2O	Silica gel	TC	43
SO_4^{2-}	$(TMS)_2SO_4$	H_2O	SE-30, OV-17	FID, FPD, MS	20, 21
SiO_4^{4-}	$(TMS)_4SiO_4$	solid salt	SE-30	FID	17, 18, 49
$Si_2O_7^{6-}$	$(TMS)_6Si_2O_7$	solid salt	SE-30	FID	17, 18, 49

TABLE 1—continued

Anion	Derivative[a]	Sample matrix	Stationary phase[b]	Detector[c]	References
$Si_3O_9^{6-}$	$(TMS)_6Si_3O_9$	solid salt	SE-30	FID	18
$Si_3O_{10}^{8-}$	$(TMS)_8Si_3O_{10}$	solid salt	SE-30	FID	18, 49
$Si_4O_{12}^{8-}$	$(TMS)_8Si_4O_{12}$	solid salt	SE-30	FID	18
VO_4^{3-}	$(TMS)_3VO_4$	H_2O	SE-30, OV-17	FID, MS	20, 21

[a] TMS = $Si(CH_3)_3$.
[b] EGS = ethylene glycol succinate; PDS = 1,5-pentanediol succinate; DNP = dinonylphthlate; NPGS = neopentylglycol succinate.
[c] FID = flame ionization; MS = mass spectrometer; EC = electron-capture; TC = thermal conductivity; FPD = flame photometric.

Witter[6] were among the first to form a haloderivative for the purpose of analysing for a halide. They determined fluoride in a variety of biological materials by extracting and reacting the fluoride with trimethylchlorosilane in benzene (Appendix I, 7) to form volatile trimethylfluorosilane. Munksgaard and Bruun[7] successfully applied this method to the quantitative analysis of fluoride in tooth enamel. Ranfft[8] and Bock and Strecker[9] subsequently developed methods for the determination of fluoride in a variety of matrices using the analogous triethylchlorosilane to form triethylfluorosilane (Appendix I, 8). The unique volatility of trialkylsilyl compounds has made the use of such derivatives extremely popular in the application of gas chromatography to many classes of compounds[10] including anions, as will be further emphasized in the following section on oxyanions.

The most general method reported for the gas chromatographic analysis of halides involves their reaction with tetraalkylammonium ions, and the subsequent thermal elimination of the volatile alkyl halides (Appendix I, 3 and 4). Fluoride acts anomalously to the other halides, forming methylfluoride from its tetramethylammonium salt, but decomposing to hydrogen fluoride when a longer alkyl chain is involved. MacGee and Allen[11] formed tetramethyl, ethyl and propyl ammonium halide salts on a cation exchange resin. Matthews *et al.*[12] employed direct extraction into toluene-undecanol solutions of tetraheptylammonium carbonate. In both studies thermal elimination of the alkyl halides occurred upon injection into the heated inlet of the chromatograph. Figure 1 illustrates the results achieved by Matthews.

Three other studies deserve mention, each involving the reaction of halides with a simple organic molecule to form a volatile product. Rüssel[13] bubbled ethylene oxide through solutions of HCl, HBr and HI to form the corresponding 2-haloethanol derivatives (Appendix I, 2). The three halides can be determined simultaneously in a single sample. Archer[14] has reported a quantitative method for the determination of bromide in blood in which bromide is

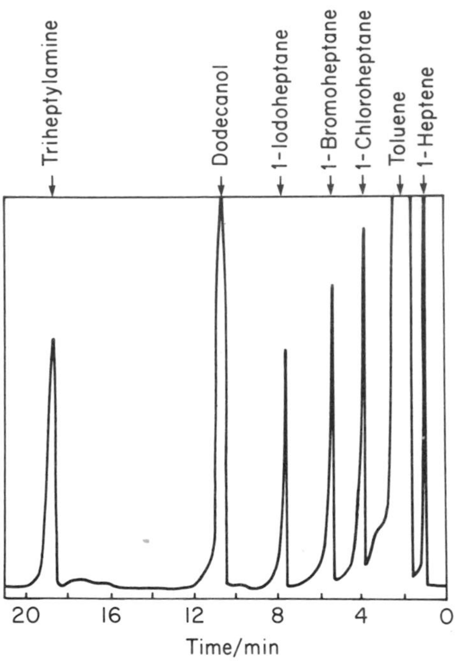

Fig. 1 Gas chromatographic separation of the *n*-heptyl derivatives of chloride, bromide and fluoride. Triheptylamine and heptene are reaction by-products; the solvent is 10% dodecanol in toluene. (Matthews *et al.*, Ref. 24).

oxidized to bromine and extracted into a solution containing cyclohexene to form 1,2-dibromocyclohexane (Appendix I, 5). Grys[15] used the oxidation of iodide by iodate to iodine followed by reaction with acetone to form monoiodoacetone (Appendix I, 6) to determine iodide in milk. Pennington[16] used the same initial reaction (Appendix I, 1) to determine iodate; the liberated iodine was measured directly by gas chromatography.

3 ANION ESTERS

The closest approach to a general technique for the gas chromatographic analysis of anions has been that involving the formation of the trimethylsilyl esters of several oxyanions. Lentz[17] was the first to report such an analysis using the reaction between trimethylchlorosilane and several silicates to form volatile trimethylsilyl silicates (Appendix I, 10). Wu *et al.*[18] employed the same technique successfully to separate five silicates, including two with cyclic structures, and confirmed the stoichiometry of the individual derivatives by mass spectrometry.

Apparently working independently of Lentz, Hashizume and Sasaki[19] formed the trimethylsilyl derivatives of orthophosphate using a similar reaction. The use of such derivatives to analyse phosphates and substituted phosphates has received considerable attention and will be further discussed in the following sections. Butts[20] and Butts and Rainey[21] further extended the use of trimethylsilyl esters to the simultaneous gas chromatographic separation and mass spectral analysis of carbonate, oxalate, borate, phosphite, phosphate, arsenite, arsenate, sulfate and vanadate. These derivatives were formed by the direct reaction of the ammonium salt of the anion with bis-(trimethylsilyl)-trifluoroacetamide containing 1% trimethylchlorosilane (Appendix I, 9). Complete reaction coincided with dissolution of the sample and required the ammonium salt of the anion; with the exception of phosphate, the other anions gave extremely low yields when present as the sodium or potassium salt. Dimethylformamide was used as a solvent to increase the solubility of most of the salts. A typical chromatogram using SE-30 as the stationary phase is shown in Fig. 2.

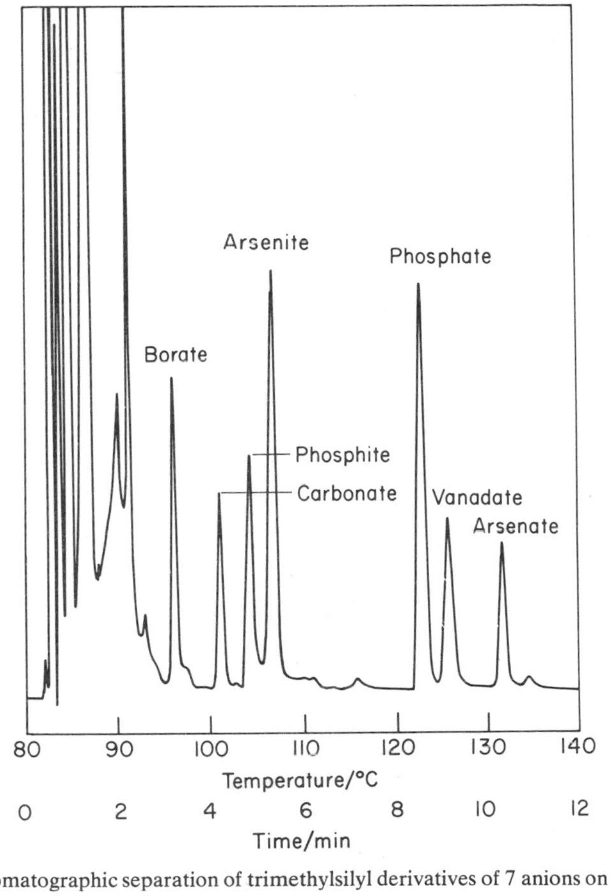

Fig. 2 Gas chromatographic separation of trimethylsilyl derivatives of 7 anions on SE-30. (Butts and Rainey, Ref. 21).

Biborate, bicarbonate, bisulfate and dihydrogen-phosphate gave the same products as borate, carbonate, sulfate and monohydrogen-phosphate respectively. Sulfite was oxidized to sulfate during the analysis procedure; nitrate and chromate produced a discoloured reaction mixture and no chromatographic peaks; and bismuthate, stannate and tungstate were apparently insoluble in the reagent.

4 PHOSPHATE

Phosphate is unique among the oxyanions that have been subjected to analysis by gas chromatography because of its frequent association with organic compounds and because of the stability and ease of formation of its esters.

Three refinements have been made on the earlier work discussed above, leading to methods for the quantitative chromatographic analysis of phosphate as its trimethylsilyl ester. Pacáková and Nekvasil[22] used ion-exchange, neutralization, and drying by evaporation or lyophilization to prepare nucleic acid hydrolysates for derivatization. Reaction gas chromatography was employed by Wiese and Hanson[23] in which ammonium chloride was added to an aqueous phosphate sample and the solution injected into a heated pre-column where the water was vaporized leaving ammonium phosphate in the pre-column. Bis-(trimethylsilyl)-trifluoroacetamide was subsequently injected into the pre-column and after a two-minute delay the reaction products were swept into the analytical column.

A third method of isolating and silylating phosphate was reported by Matthews *et al.*[24] The phosphate was removed from the aqueous phase and concentrated fifty-fold by a single solvent extraction with a quaternary alkyl ammonium salt (Adogen-464-HCO_3^-) in toluene–octanol. Derivatization was readily accomplished in the organic phase by reaction with bis-(trimethylsilyl)-trifluoroacetamide. Although Matthews obtained excellent quantitative results with a flame photometric detector and Pacáková with a flame ionization detector, both authors cautioned about peak inversion with large sample volumes.

5 SUBSTITUTED PHOSPHATES

Numerous methods for the gas chromatographic analysis of organophosphates have been developed in parallel with the methods for inorganic phosphate. Many of the silylation techniques will simultaneously derivatize phosphate, organophosphates, and numerous non-phosphate organic compounds. Three categories of organophosphate compounds (pesticides, alkyl phosphates, and carbohydrate phosphates) have been analysed by gas chromatography. The phosphate-containing pesticides have been the subject of several reviews[25] and are discussed in books such as that by Gudzinowicz.[26] The reader is referred to these sources for further discussion. In general the pesticides are fully substituted phosphates and require no further derivatization.[27] Shafik *et al.*[28,29] and Moye[30] discuss derivatization of these pesticides by transesterification.

Methyl esters and trimethylsilyl esters have been used for the separation of a series of alkyl phosphates. Following the original work of Hardy in 1964,[31] Saey (1973),[32] Horton (1972),[33] Daemen and Dankelman (1973),[34] and Brignocchi (1973)[35] have prepared and chromatographed the methyl esters of mono- and dialkyl (C_4–C_{16}) phosphates using diazomethane. Although diazomethane is potentially hazardous and requires special apparatus for its use, Daemen obtained results for the gas chromatographic analysis of the alkylphosphoric acid content of commercial surfactants that agreed exceptionally well with the results obtained by potentiometric titration. Boyden and Clift[36] have prepared trimethylsilyl esters of mono- and dibutyl phosphate using a mixture of trimethylchlorosilane and hexamethyldisilazane. Harvey and Horning[37b] more recently have prepared the silyl derivatives of a series of alkylphosphonates using bis-(trimethylsilyl)-trifluoroacetamide and trimethylchlorosilane in acetonitrile with a ten-minute reaction time at room temperature. Although this technique has not been reported for alkyl phosphates, its use should be considered because of the simplicity of technique and the proven ability of the reagent to derivatize organophosphate compounds.

Table 2 contains retention data as methylene unit values for the compounds discussed above as well as for the third class of organophosphates, the carbohydrate phosphates and nucleotides. The use of retention data in the form of methylene unit values or Kovats retention indices is encouraged because it affords interlaboratory comparability. Laboratories using the same stationary phases should be able easily to match the values in Table 2 to within 0.5 methylene units.

Because of the widespread importance of carbohydrate phosphates and nucleotides in biological systems, numerous efforts have been made to analyse them. Wells *et al.*[38] first formed volatile monosaccharide phosphate derivatives by treating the phosphoric acid group with diazomethane and then silylating the sugar with trimethylchlorosilane and hexamethyldisilazane. Hashizume and Sasaki[19a] and Hancock[39] used the same reaction to volatilize several nucleotides. Butts[20] later used bis-(trimethylsilyl)-trifluoroacetamide containing 1% trimethylchlorosilane in an equal volume of pyridine to derivatize 13 different mononucleotides with a reaction time of 16 h at 75 °C. Lawson *et al.*[40] in the gas chromatography–mass spectrometry studies of silylated nucleotides and Butts (unpublished data) have observed that the apparent yield is highly variable and dependent on numerous factors including previous use of the column. At present these methods for nucleotides must be considered strictly qualitative.

TABLE 2
Retention data as methylene unit (MU) values for major peaks of derivatives of organophosphates

Parent organophosphorus compound[a]	Derivative[b]	MU Value SE-30	MU Value OV-17
1. mono-2-ethylhexylphosphoric acid	Me		16.8
2. mono-*n*-decylphosphoric acid	Me		19.9
3. mono-*n*-dodecylphosphoric acid	Me		22.0
4. mono-*n*-tetradecylphosphoric acid	Me		23.9
5. bis-2-ethylhexylphosphoric acid	Me		22.0
6. di-*n*-decylphosphoric acid	Me		27.6
7. di-*n*-dodecylphosphoric acid	Me		31.8
8. di-*n*-tetradecylphosphoric acid	Me		35.7
9. methylphosphonic acid	TMS	11.35	12.40
10. ethylphosphonic acid	TMS	12.05	13.10
11. phenylphosphonic acid	TMS	15.30	17.15
12. 1-phenylmethylphosphonic acid	TMS	16.50	18.20
13. methylene diphosphonic acid	TMS	18.00	18.95
14. phosphonoacetic acid	TMS	15.40	16.50
15. phosphonopropionic acid	TMS	16.55	17.65
16. uridine-2′-monophosphate (3′ same MU)	TMS	28.28	
17. uridine-5′-monophosphate	TMS	28.96	
18. adenosine-2′-monophosphate	TMS	30.06	
19. adenosine-3′-monophosphate	TMS	30.28	
20. inosine-5′-monophosphate	TMS	30.40	
21. deoxyadenosine-5′-monophosphate	TMS	30.92	
22. adenosine-5′-monophosphate	TMS	31.14	
23. xanthosine-5′-monophosphate	TMS	31.34	
24. guanosine-2′-monophosphate (3′ same MU)	TMS	31.59	
25. guanosine-5′-monophosphate	TMS	31.79	
26. deoxyguanosine-5′-monophosphate	TMS	32.24	
27. orthophosphoric acid	TMS	12.70	13.45
28. pyrophosphoric acid	TMS	16.60	18.00
29. glycoaldehyde phosphate	MO–TMS	14.80	16.25
30. α-glycerophosphate	TMS	18.05	18.65
31. β-glycerophosphate	TMS	17.65	18.25
32. 2-phosphoglyceric acid	TMS	18.15	19.20
33. 3-phosphoglyceric acid	TMS	18.45	19.45
34. glycerol-1,2-diphosphate	TMS	22.25	23.35
35. glycerol-1,3-diphosphate	TMS	22.55	23.75
36. 2,3-diphosphoglyceric acid	TMS	22.40	23.85
37. 2-phosphoenolypyruvic acid	TMS	15.95	17.35
38. glyceraldehyde phosphate	MO–TMS	17.55	18.70
39. dihydroxyacetone phosphate	MO–TMS	17.75	19.00
40. erythrose-4-phosphate	TMS	18.95	19.70
41. 2-deoxyribose-5-phosphate	TMS	20.00	21.00
	MO-TMS	20.75	21.55
42. xylose-1-phosphate	TMS	22.00	22.50
43. xylose-5-phosphate	TMS	21.70	22.40
	MO–TMS	21.80	22.30
44. ribose-1-phosphate	TMS	20.60	22.00

TABLE 2—continued

Parent organophosphorus compound[a]	Derivative[b]	MU Value SE-30	MU Value OV-17
45. ribose-5-phosphate	TMS	21.65	22.20
	MO–TMS	21.85	21.95
46. ribulose-5-phosphate	TMS	21.75	22.35
	MO–TMS	21.70	21.95
			22.25
47. ribulose-1,5-diphosphate	TMS		26.90
48. 2-deoxyglucose-6-phosphate	TMS	22.75	23.35
			23.70
	MO-TMS	22.85	23.35
49. 2-deoxy-6-phosphogluconic acid	TMS	23.50	23.65
50. glucose-1-phosphate	TMS	23.45	23.85
		23.80	24.45
51. glucose-6-phosphate	TMS	24.45	24.80
		24.85	24.95
	MO–TMS	24.05	24.00
52. 1-phosphoglucuronic acid	TMS	23.75	25.00
53. glucosamine-6-phosphate	TMS	24.15	25.00
54. 6-phosphogluconic acid	TMS	25.10	24.75
55. mannose-1-phosphate	TMS	22.60	23.25
		23.35	
56. mannose-6-phosphate	TMS	23.60	23.70
			24.35
	MO–TMS	24.00	23.75
57. galactose-1-phosphate	TMS	23.30	22.78
58. galactose-6-phosphate	TMS	24.00	24.00
		24.35	24.35
59. 1-phosphogalacturonic acid	TMS	24.55	26.10
60. fructose-1-phosphate	TMS	23.05	22.85
		23.35	22.30
		23.60	23.50
61. fructose-6-phosphate	TMS	23.35	23.25
	MO–TMS	23.95	23.80
62. fructose-1,6-diphosphate	TMS	27.30	27.65
		27.60	28.05
63. sedoheptulose-7-phosphate	TMS	25.30	24.75
		25.45	24.95
	MO–TMS	26.45	26.05
64. mannitol-1-phosphate	TMS	24.45	23.75
65. sorbitol-6-phosphate	TMS	24.55	23.60
66. myoinositol-2-phosphate	TMS	25.55	25.05
67. inositol-3-phosphate	TMS	23.65	22.70
68. glycerol-1-(myoinositol)-1-phosphate	TMS	29.10	28.00

[a] Compounds 1–8 taken from Daemen and Dankelman, Ref. 34; MU values calculated by multiplying author's Kovats indices by 0.01. Compounds 9–15 taken from Harvey and Horning, Ref. 37(b); Compounds 16–26 taken from Butts, Ref. 20; Compounds 27–68 taken from Harvey and Horning, Ref. 37(a).

[b] Me = methyl ester; TMS = trimethylsilyl; MO–TMS = methoxime–trimethylsilyl.

Several efforts have been made to chromatograph the silyl derivatives of carbohydrate phosphates. The most successful and extensive has been that of Harvey and Horning.[37a] Simple trimethylsilyl derivatives were formed by reaction with either bis-(trimethylsilyl)-trifluoroacetamide or trimethylsilylimidazole. The silyl derivatives of the keto and aldehyde phosphates were less stable than the other carbohydrate phosphates and it was necessary to convert the carbonyl group into a methoxime derivative by reaction with methoxylamine hydrochloride. Methoxime derivatives can also be used to reduce the number of peaks formed by the α and β anomers of some sugar phosphates, but on more polar phases, such as OV-17, multiple peaks may be seen for the *syn* and *anti* isomers of the methoxime. As with the nucleotides, this is at present a qualitative method.

6 MASS SPECTROMETRY

The mass spectra of the trimethylsilyl derivatives of several oxyanions (Figs 3 and 4) were obtained by Butts and Rainey[21] using combined gas chromatography–mass spectrometry. These spectra possess several common characteristics, particularly the presence of a large M−15 peak (loss of a methyl group) and peaks at m/e 73[$Si(CH_3)_3^+$], 75, 147 and 207. These peaks are typical of trimethylsilyl derivatives in general and are silicon-containing fragments with m/e 147, resulting from a rearrangement of two TMS groups. All of the spectra in Figs 3 and 4 have a metastable peak at m/e 36.3 indicating the transition m/e 147 → m/e 73. Zinbo and Sherman[41] have reported a very similar spectrum for the trimethylsilyl derivative of phosphate, and in addition used deuterium-labelled derivatives to confirm the fragmentation pattern. Wu *et al.*[18] observed a similar fragmentation process with the silyl derivatives of a series of silicate anions. These spectra contained a significant M−15 peak and a base peak of m/e 73, along with a large peak at m/e 147. Ions of low abundance but structural significance occurred from successive loss of 88 m/e units [$(CH_3)_2CH_2SiO$] and/or 162 m/e units [$(CH_3)_3Si—O—Si(CH_3)_3$].

Several studies including those of Lawson *et al.*[40] and Harvey and Horning[37] have been made of the mass spectra of organic phosphates including alkylphosphonates, carbohydrate phosphates and nucleotides. These spectra show features of the phosphate derivatives discussed above, as well as fragments characteristic of the organic group. A most interesting feature is the appearance of a peak at m/e 387 identified as tetrakis-(trimethylsilyl)-phosphate which is formed by an ion–molecule reaction with organophosphates but is not present in the spectrum of phosphate itself.[41]

7 SPECIFIC ANION DERIVATIVES

In addition to the two categories (halides and trimethylsilyl esters) already discussed there are a number of specific derivatives developed for individual anions.

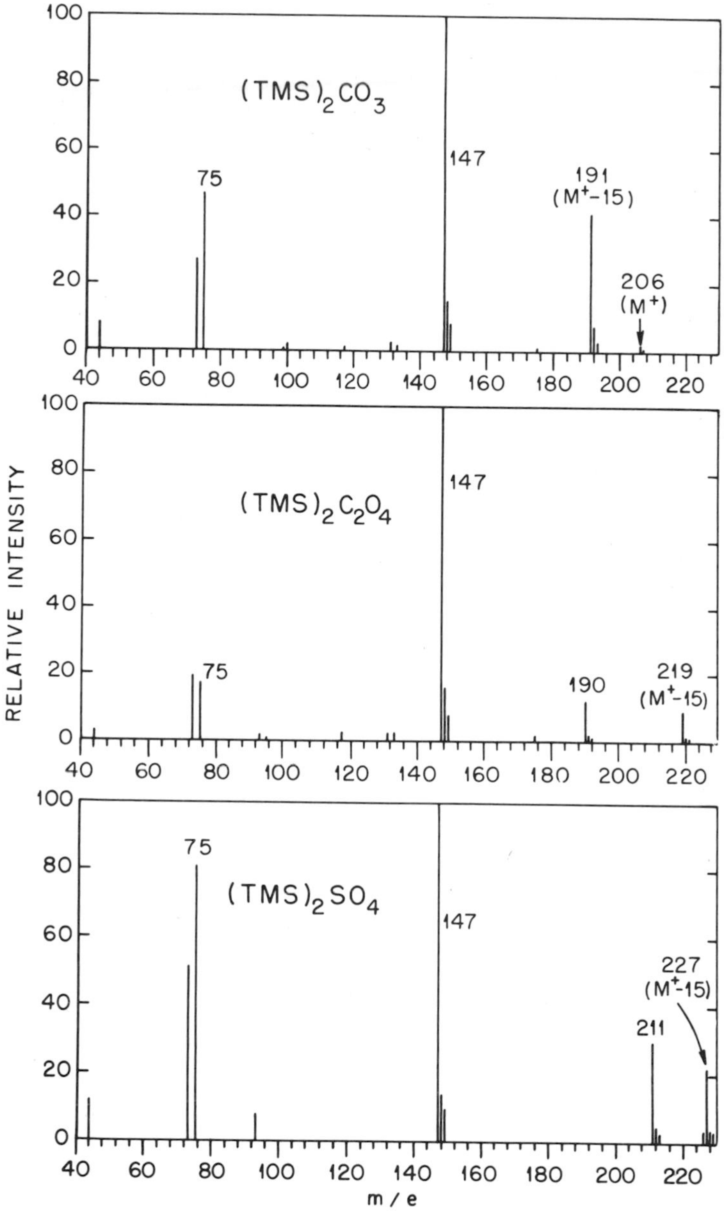

Fig. 3 Mass spectra of trimethylsilyl derivatives of divalent anions. (Butts and Rainey, Ref. 21).

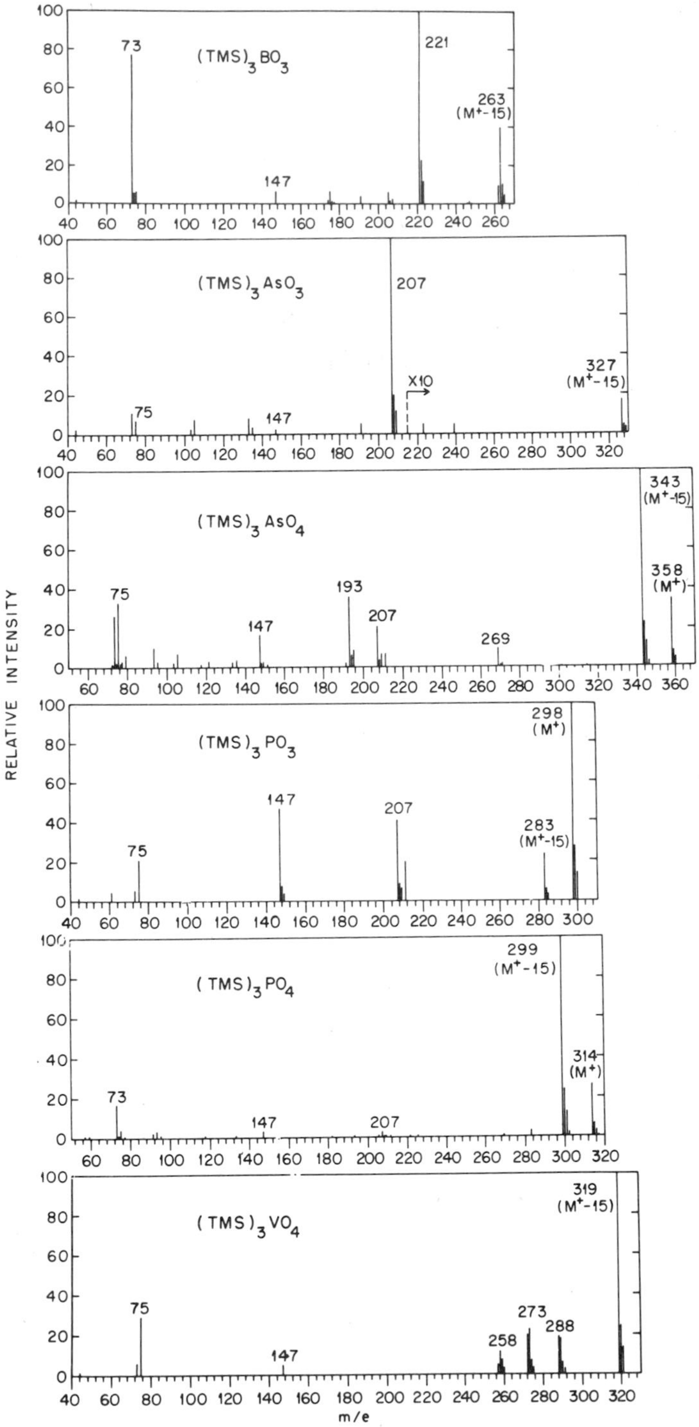

Fig. 4 Mass spectra of trimethylsilyl derivatives of trivalent anions. (Butts and Rainey, Ref. 21).

Sulfite, sulfide and carbonate will react with strong acid to liberate volatile SO_2, H_2S and CO_2 respectively; these gases may be separated by gas–solid chromatography. Birk *et al.*[42] and McDonald[43] have designed reaction chambers to permit these reactions (Appendix I, 11 and 14) to take place on-line to a gas chromatograph. Both solid and liquid samples can be analysed; sensitivity is limited largely by the type of thermal conductivity detector used.

Nitrate has been determined gas chromatographically[44] after conversion to nitrobenzene (Appendix I, 12). Nitrite can also be determined in this manner by first oxidizing it to nitrate with potassium permanganate. Excellent sensitivity can be obtained using an electron-capture detector.

Soderquist, Crosby and Bowers[45] investigated several methods (Appendix I, 9 and 13) for the derivatization of cacodylic acid [$(CH_3)_2AsO_2H$] and concluded that the formation of iododimethylarsine was more reliable than silylation which was not reproducible or cyanoethylation which was too slow. Iododimethylarsine is readily extractable into hexane, volatile, and because of its strong electron-capturing property can be detected in nanogram amounts with an electron-capture detector.

Cyanide can be rendered volatile by converting it to cyanogen halide. Valentour, Aggarwal and Sunshine[46] used Chloramine T to form cyanogen chloride from cyanide separated from blood by microdiffusion. Nota and Palombari[47] formed cyanogen bromide from cyanide or thiocyanate by reaction with bromine water. Thiocyanate can be determined in the presence of cyanide by removing the cyanide by reaction with formaldehyde. Excellent sensitivity can be obtained for both cyanogen chloride and cyanogen bromide using electron-capture detection.

8 APPENDIX I: REACTIONS

The following are reported reactions employed in derivatization of inorganic anions. (Underlined compound is the species chromatographed.)

8.1 Halogens

1. IO_3^- $\qquad IO_3^- + 5I^- + 6H^+ \rightarrow \underline{3I_2} + 3H_2O$

2. Cl^-, Br^-, I^- $\qquad X^- + \underbrace{CH_2-CH_2}_{O} + H^+ \rightarrow \underline{XCH_2CH_2OH}$

3. F^-, Cl^-, Br^-, I^- $\qquad X^- + (CH_3)_4N^+ \rightarrow (CH_3)_4NX$
   $(CH_3)_4NX \xrightarrow{\Delta} \underline{CH_3X} + (CH_3)_3N$

4. Cl^-, Br^-, I^- $\qquad X^- + R_4N^+ \rightarrow R_4NX$
   $R_4NX \xrightarrow{\Delta} \underline{RX} + R_3N$
   where $R = CH_2CH_3$, $CH_2CH_2CH_3$, or $(CH_2)_6CH_3$

5. Br^- $6Br^- + 2\,MnO_4^- + 8H^+ \rightarrow 3Br_2 + 2MnO_2 + 4H_2O$

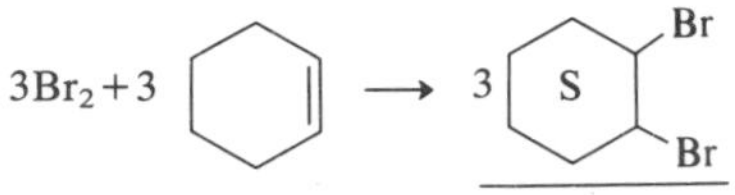

6. I^- $5I^- + IO_3^- + 6H^+ \rightarrow 3I_2 + 3H_2O$

$3I_2 + 3CH_3COCH_3 \rightarrow \underline{3ICH_2COCH_3} + 3H^+ + 3I^-$

7. F^- $(CH_3)_3SiCl + H_2O \rightarrow (CH_3)_3SiOH + H^+ + Cl^-$

$F^- + (CH_3)_3SiOH + H^+ \rightarrow \underline{(CH_3)_3SiF} + H_2O$

8. F^- $F^- + (CH_3CH_2)_3SiCl \rightarrow \underline{(CH_3CH_2)_3SiF} + Cl^-$

8.2 Oxyanions

9. CO_3^{2-}, $C_2O_4^{2-}$, BO_3^{3-}, PO_3^{3-}, AsO_3^{3-}, PO_4^{3-}, AsO_4^{3-}, SO_4^{3-}, VO_4^{3-}, $(CH_3)_2AsO_2^-$

$Y^{-n} + H^+ + n\,BSTFA \xrightarrow{TMCS} \underline{(TMS)_n\,Y} + n\,MSTFA$

where

BSTFA = bis-(trimethylsilyl)-trifluoroacetamide

TMCS = trimethylchlorosilane

MSTFA = mono-(trimethylsilyl)-trifluoroacetamide

10. SiO_4^{4-}, $Si_2O_7^{6-}$, $Si_3O_9^{6-}$, $Si_3O_{10}^{8-}$, $Si_4O_{12}^{8-}$

$M_pY + n\,HCl \rightarrow H_nY + pM^{2+} + n\,Cl^-$

$H_nY + n(CH_3)_3SiCl \rightarrow \underline{[(CH_3)_3Si]_nY} + n\,H^+ + n\,Cl^-$

where M = Mg, Fe, Zn, Ca, Na

11. SO_3^{2-}, CO_3^{2-} $YO_3^{2-} + 2H^+ \rightarrow H_2YO_3$

$H_2YO_3 \rightarrow \underline{YO_2} + H_2O$

12. NO_2^-, NO_3^- $3\,NO_2^- + 2MnO_4^- + 2H^+ \rightarrow 3\,NO_3^- + 2MnO_2 + H_2O$

$3NO_3^- + 3C_6H_6 + H^+ \rightarrow \underline{3C_6H_5NO_2} + H_2O$

13. $(CH_3)_2AsO_2H$

(a) $(CH_3)_2AsO_2H + 3HI \rightarrow \underline{(CH_3)_2AsI} + 2H_2O + I_2$

(b) $(CH_3)_2AsO_2H + 4H_2 \rightarrow (CH_3)_2AsH + 2H_2O$

$(CH_3)_2AsH + CH_2{=}CHCN \rightarrow \underline{(CH_3)_2AsCH_2CH_2CN}$

8.3 Sulfide, cyanide and thiocyanate

14. S^{2-} $S^{2-} + 2H^+ \rightarrow \underline{H_2S}$

15. CN^-, SCN^- (a) $CN^- + Br_2 \rightarrow \underline{BrCN} + Br^-$

(b) $SCN^- + 4Br_2 + 4H_2O \rightarrow \underline{BrCN} + SO_4^{2-} + 7Br^- + 8H^+$

16. CN^- $CN^- + CH_3C_6H_4SO_2NClNa + 2H^+ \rightarrow$

$\underline{ClCN} + CH_3C_6H_4SO_2NH_2 + Na^+$

9 APPENDIX II: PROCEDURES

The following procedures are intended to contain sufficient detail so that they can be used without reference to another source. The reader should bear in mind that such methods are flexible and he is encouraged to vary sample and reagent volumes to suit his needs. Similarly the stationary phases and operating temperatures may be intelligently varied, particularly if one uses the McReynolds or Rohrschneider constants in choosing a column.

As with any sensitive gas chromatographic method, care must be taken to avoid contamination or other artifacts such as those produced by decomposition during derivatization or in the chromatographic system. In particular, an all-glass chromatographic system, and especially glass columns, should be used with the following procedures. Commercially available glass vials with Teflon®-lined septum caps are suitable reaction vessels for many of these procedures.

9.1 Fluoride in teeth[7]

The sample is dissolved in 0.5 M perchloric acid and 5 μl of this solution added to a polyethylene tube containing 50 μl of benzene in which is dissolved 50 μg trimethylchlorosilane and 65 ng 2-methylbutane as internal standard. The sample is agitated vigorously at 4 °C for 20 min and the benzene (upper) layer separated and injected (6–8 μl aliquot) into the gas chromatograph to determine the trimethylfluorosilane formed. A freshly prepared standard of fluoride in 0.5 M perchloric acid is treated similarly. Less than 1 ng fluoride per sample has been determined using a flame ionization detector. Column: 2 m $\times$ 0.125 in (0.32 cm) i.d., 20% DC-200/50 on 80/100 AW Chromosorb P; temperature: 80 °C.

9.2 Fluoride in organic or inorganic matrices[8]

Organic material: Ash 20 g of air-dried sample in a nickel crucible for 3 h, or preferably overnight, at 550 °C. After cooling add 5 g of NaOH and heat for 3 min over a Bunsen burner. Lower the flame and continue heating until the entire sample becomes liquid. Cool, dissolve in H_2O, and add 10 ml of conc. HCl.

Inorganic material: Dissolve 2.5 g of sample in 10 ml conc. HCl and add 10 ml H_2O. Materials which are not completely soluble are fused with NaOH as above.

Derivatization: The test solutions and appropriate standard solution (NaF dissolved in 10 ml HCl and 10 ml H_2O) are transferred to a flask and 5 ml of reagent solution (0.75 g triethylchlorosilane in 100 ml tetrachloroethylene) added. The mixture is vigorously shaken for 45 min and the lower phase separated and analysed for triethylfluorosilane by gas chromatography. Concentrations as low as 15 $\mu g\ g^{-1}$ have been determined with a flame ionization detector. Column: 1.8 m × 0.2 cm, 20% DC-200 on Gaschrom Q; temperature: 60 °C.

9.3 Bromide in blood[14]

Mix 2 ml of blood or 2 ml of standard KBr solution (0.25, 0.50, 0.75 and 1 mg bromide ml^{-1} of 0.5% w/v aqueous NaCl) with 2 ml of H_2O and add 10 ml of 20% trichloroacetic acid solution. Mix thoroughly and filter the blood samples through a Whatman No. 541 filter-paper. Transfer 7 ml of each filtrate or acidified standard solution to stoppered test-tubes and add 2 ml of 2.5 M H_2SO_4, 50–60 mg of $KMnO_4$, and 2 ml of cyclohexene reagent (70 mg of 1,6-dibromohexane and 2 ml of cyclohexene diluted to 100 ml with cyclohexane). Stopper and shake the tube vigorously for 30 s, add 100–120 mg Na_2SO_3 and mix gently until decolourized. Allow the layers to separate and inject 2 μl of the upper phase into the chromatograph. The peak areas of the 1,6-dibromohexane added as internal standard and the 1,2-dibromohexane formed in the reaction are graphically used to determine the blood bromide concentrations. Blood specimens containing 0.1–1.0 mg bromide ml^{-1} have been analysed using a flame ionization detector. Column: 2 m × 2 mm i.d., 2.5% OV-17 on 80/100 Gaschrom Q; temperature: 70–170 °C at 10 °C min^{-1}.

9.4 Chloride, bromide and iodide[13]

The hydrohalic acids are formed by passing a solution of the halides through a column containing a strongly acidic ion-exchanger. Ethylene oxide is passed (2 bubbles s^{-1}) for 15 min through 10 ml of the acid cooled to 4 °C. Alternatively, 1 ml of liquid ethylene oxide may be added to the acid. After 1 h, the sample is diluted to 15 ml and 1 μl aliquot injected into the chromatograph. A 0.1 M solution in HCl, HI and HBr has been analysed by this method. Column: 2 m, 12% EGS on Gaschrom P; temperature: 100 °C.

9.5 Chloride, bromide and iodide[12]

A 0.1 M solution of tetraheptylammonium chloride in 10% (v/v) undecanol-toluene is repeatedly contacted with aqueous solutions of Na_2CO_3 until chloride cannot be detected in the aqueous phase with $AgNO_3$. 1 ml of the resulting solution of tetraheptylammonium carbonate is used to extract 30 ml of aqueous solution containing the halides. The tetraheptylammonium halides are converted to 1-haloheptanes on injection into a gas chromatograph with an inlet temperature of 225 °C. Linear responses have been obtained with a flame ionization detector for aqueous halide concentrations in the approximate range 1–50 $\mu g\ ml^{-1}$. Column: 10 ft × 0.125 in o.d. (3 m × 0.32 cm), 5% Apiezon L on 80/100 Chromsorb W (HP); temperature: 50–145 °C at 5 °C min^{-1} then 145–250 °C at 20 °C min^{-1}.

9.6 Iodide in milk[15]

Milk samples (12 ml) are mixed with 12 ml of water and 3 ml of 5 M H_2SO_4. This mixture is filtered after 15 min when protein precipitation is complete. Portions (4.5 ml) of the filtrate are mixed with 0.5 ml of 5 M acetone and 0.05 ml of 100 ppm KIO_3 solution. After allowing 30 min for the iodine to be converted to iodoacetone, the mixture is extracted with 5 ml of hexane and an aliquot of the hexane injected into the gas chromatograph. Standard iodide solutions are treated similarly to obtain response factors. Using a nickel-63 electron capture detector a detection limit of 1 p.p.b. iodide and a relative standard deviation of 2.5% have been obtained. Column: 1.5 m × 6 mm o.d., 5% PDS on 80/100 Varaport 30; temperature: 125 °C.

9.7 Iodate[16]

Add 2 ml of KI solution (15.5 mg ml^{-1}) to 5 ml of the solution containing iodate. Two drops of 3 M HCl are added and the solution shaken for 30 s. The liberated iodine is extracted by shaking with 1 ml of CCl_4 for 1 min. A 2 μl aliquot of the CCl_4 solution is injected into the chromatograph. Aqueous solutions containing 1 mg iodate ml^{-1} have been analysed with a thermal conductivity detector. Column: 6 ft × 0.25 in o.d. (1.8 m × 0.64 cm), 20% DNP on 40/60 AW-Chromosorb W; temperature: 165 °C.

9.8 Phosphate[24]

Two hundred millilitres of an aqueous phosphate solution are contacted for 10 min with 4 ml of 0.1 N Adogen-464 (a quaternary alkylammonium salt) as the bicarbonate salt in 7% (v/v) octanol in toluene. After separation of the phases by centrifugation, equal volumes of the organic layer and bis-(trimethylsilyl)-trifluoroacetamide containing 1% trimethylchlorosilane are mixed in capped vials and allowed to react for 15 min at room temperature. A 1–5 μl aliquot is injected into the chromatograph; samples and standard solutions of sodium phosphate are treated identically. Less than 20 ng of phosphate have been readily detected with flame ionization and flame photometric detectors. Column: 6 ft × 0.25 in o.d. (1.8 m × 0.64 cm), 5% OV-225 on 80/100 Chromosorb W (HP); temperature: 95 °C.

9.9 Phosphate and bases in nucleic acids[19]

About 10 mg of nucleic acid previously dried for 2 h at 60 °C over phosphoric anhydride *in vacuo* is accurately weighed and hydrolysed with 0.1 ml of 70% $HClO_4$ at 100 °C for 1 h. The hydrolysate is mixed with 0.3 ml of 6 M KOH and shaken for 3 min in an ice-bath. When heat evolution ceases, 0.25 ml of 6 M HCl and then 0.25 ml of 28% aqueous ammonia are added. The mixture is evaporated to dryness and dried for 1 h at 100 °C under reduced pressure. Add 0.3 ml hexamethyldisilazane, 0.15 ml trimethylchlorosilane and 0.1 ml pyridine to the residue and heat at 150 °C for 1 h. After cooling, 10 ml pyridine containing phenanthrene as an internal standard are added and an aliquot of the mixture is injected into the chromatograph. Standard solutions of the bases and ammonium phosphate need only be dried and derivatized. Column: 75 cm × 3 mm i.d., 5% DC-430 on 80/100 AW-silanized Celite 545; temperature: 80–220 °C at 8 °C min^{-1}.

9.10 Phosphate in nucleic acids and nucleotides[22]

The dried residue in procedure 9.9 is heated at 150 °C for 10 min with a twenty-fold molar excess of bis-(trimethylsilyl)-trifluoroacetamide or bis-(trimethylsilyl)-acetamide containing naphthalene as an internal standard. The solution is analysed directly by gas chromatography. Column: 180 cm × 4 mm i.d., 5% OV-101 on 60/80 Chromosorb W (HP); temperature: 155 °C.

9.11 Mono-, di-, and tributyl phosphate[36]

The phosphate sample (100 mg) is weighed into a 10 ml volumetric flask to which is added 3 ml trimethylchlorosilane, 3 ml hexamethyldisilazane and 50 mg dodecane as internal standard. The flask is made up to volume with diethyl ether and one hour is allowed for the reaction to go to completion. Response factors are determined by treating pure standards in a similar manner. Column: 6 ft × 0.375 in, (1.8 m × 0.95 cm) 3% SE-30; temperature: 125 °C (10 min) to 185 °C at 4 °C min^{-1}, and 120 °C (10 min) to 160 °C at 2 °C min^{-1}.

9.12 Alkylphosphonates[37b]

The phosphonate (1 mg) is dissolved in 0.1 ml acetonitrile, 0.1 ml bis-(trimethylsilyl)-trifluoroacetamide and 0.05 ml trimethylchlorosilane. Derivatization is complete after 15 min at room temperature and an aliquot is then injected directly into the gas chromatograph. Column: 12 ft (3.6 m) × 4 mm, 1% SE-30 or OV-17 on 100/120 AW Gaschrom P; temperature: 90–170 °C.

9.13 Carbohydrate phosphates[37a]

TMS derivatives. A 1 mg sample of the carbohydrate phosphate, as its cyclohexylammonium salt or as the free phosphate isolated by DEAE-Sephadex chromatography, is added to 0.1 ml of trimethylsilylimidazale (TSIM) in 0.1 ml of acetonitrile and allowed to react for 5 min at room temperature.

Methoxime-TMS derivatives. A 1 mg sample of the carbohydrate phosphate and 1 mg of methoxylamine hydrochloride are added to 0.2 ml of pyridine and kept at room temperature for 1 h. One ml of TSIM is added to the mixture and heated at 60 °C for 5 min.

Aliquots of the reaction mixtures can be injected directly into the gas chromatograph. Column: 12 ft × 4 mm i.d., 1% SE-30 or OV-17 on 100/200 silanized AW Gaschrom P; temperature: 140–280 °C at 2 °C min^{-1}.

9.14 Nucleotides[20b]

Add 1–2 mg of nucleotides to 0.2 ml pyridine and 0.3 ml of bis-(trimethylsilyl)-trifluoroacetamide containing 1% trimethylchlorosilane. Derivatization is completed by heating at 75 °C for 3 h. Any inorganic phosphate present will also be derivatized by this procedure. Aliquots of the solution are injected directly into the chromatograph. Column: 6 ft × 0.25 in o.d. (1.8 m × 0.64 mm), 5% SE-30 on 80/100 Chromosorb W (HP); temperature: 100–300 °C at 10 °C min^{-1}.

9.15 Borate, carbonate, phosphate, phosphite, sulfate, arsenate, arsenite and vanadate[21]

A sample (5–10 mg) of the ammonium salts of the anions is added to 200 μl each of dimethylformamide and bis-(trimethylsilyl)-trifluoroacetamide and allowed to react overnight at room temperature. If the anion is not in the form of ammonium salt, the available salt is placed on a 13 cm × 1 cm^2 column of Dowex 50W-X8 cation-exchange resin in the NH_4^+ form and eluted with 15 ml of distilled water. The collected eluent is taken to dryness and derivatized as above. An aliquot of the reaction mixture is injected directly into the chromatograph. Column: 12 ft × 0.25 in o.d. (3.6 m × 0.64 mm), 3% OV-17 on 80/100 Chromosorb W (HP) or 6 ft × 0.25 in o.d. (1.8 m × 0.64 mm), 5% SE-30 on 80/100 Chromosorb W (HP); temperature: 70–150 °C at 5 °C min^{-1}.

9.16 Silicate anions in silicate rock[18]

The crystalline solid is ground to pass a 100 mesh sieve. Twenty grams of this powder is added to a mixture of 125 g ice, 150 ml concentrated HCl, 300 ml 2-propanol and 200 ml hexamethyldisilazane which has previously been stirred for 1 h. The entire reaction mixture is then stirred an additional 16 h at room temperature. After filtration the silazane (upper) layer is separated, washed with H_2O, stirred with 2.5 g Amberlyst 15 for 1 h, filtered and stripped to a pot temperature of 115 °C. The liquid residue in the distillation flask is analysed by gas chromatography. Column: 12 ft × 0.125 in i.d. (3.6 m × 0.32 mm), 3% SE-30 on 80/100 Chromosorb W; temperature: 70–170 °C at 20 °C min^{-1}, then 170–240 °C at 8 °C min^{-1}, then 240–300 °C at 2 °C min^{-1}.

9.17 Nitrate and nitrite[44]

Nitrate: Five ml of aqueous sample is mixed vigorously with 5 ml of benzene. Fifteen ml of $H_2SO_4 . H_2O$ (3:1) is added and mixed; after relieving the pressure the flask is heated at 75 °C for 5 min. The flask is again shaken and cooled to room temperature. The benzene solution is diluted to give a nitrobenzene concentration of about 7 ng μl^{-1} and 2-nitrotoluene is added as internal standard to a concentration of about 14 ng μl^{-1}. The benzene solution is then analysed by gas chromatography. Standards are treated the same as samples.

Nitrate plus nitrite: One ml of 0.1 M $KMnO_4$ is added to 1–4 ml of sample and the total volume made up to 5 ml with H_2O. The rest of the procedure is as above.

Nitrite and nitrate have been determined at levels of 0.1 to 50 ppm in aqueous solution using an electron-capture detector. Column: 4 ft × 0.25 in o.d. (1.2 m × 0.64 mm), 6.9% Apiezon M on 60/80 Diataport S; temperature: 120 °C.

9.18 Cyanide in blood, urine and other aqueous solutions[46]

Cyanide is separated from the sample using conventional microdiffusion cells with 0.1 M NaOH in the centre well. One ml of this NaOH solution is mixed with 2.0 ml hexane, 20 ml 1.0 M NaH_2PO_4, and 1 ml of a solution containing 250 mg Chloramine T/100 ml H_2O. The mixture is shaken, placed in an ice-bath for 5 min, shaken again, and replaced in the ice-bath. After the hexane layer separates, an aliquot is injected to measure the ClCN formed in the reaction. As little as 25 ng CN^- ml^{-1} sample can be determined using an electron-capture detector. Column: 6 ft × 0.25 in (1.8 m × 0.64 mm), 7% Hallcomid M-18 on 90/110 Anakrom ABS; temperature: 55°C.

9.19 Cyanide and thiocyanate in water[45]

The water sample is cleared of suspended material by centrifugation. 3 ml of the sample and 0.5 ml of 20% (w/w) H_3PO_4 are added to a 5 ml flask. Bromine water is added until the yellow colour persists. The flask is shaken occasionally for 5 min, then the excess bromine is removed by addition of 0.2 ml of a 5% aqueous solution of phenol. The solution is made up to 5 ml with water and an aliquot injected directly into the chromatograph to measure the BrCN formed. A standard curve is prepared by analysing solutions of KSCN as above. This method gives the total of SCN^- and CN^- present. To determine only SCN^-, neutralize the water sample, add a volume of 40% formaldehyde equal to 2% of the sample volume and treat as above. Water samples containing 0.02–0.50

ppm CN^- or SCN^- have been analysed with an electron-capture detector. Column: 1 m×3 mm i.d., 80/100 Poropak Q; temperature: 90 °C or 130 °C.

REFERENCES

1. J. A. Rodriguez-Vazquez, *Anal. Chim. Acta* **73**, 1 (1974).
2. B. I. Anvaer and Y. S. Drugov, *Zh. Anal. Khim.* **26**, 1180 (1971).
3. D. F. S. Natusch and T. M. Thorpe, *Anal. Chem.* **45**, 1184A (1973).
4. F. Eisenberg, in *Analytical Chemistry of Phosphorus Compounds*, Wiley–Interscience, New York, 1972, p. 69.
5. R. W. Morrow, J. A. Dean, W. D. Shults and M. R. Guerin, *J. Chromatogr. Sci.* **7**, 572 (1969).
6. J. A. Fresen, F. H. Cox and M. J. Witter, *Pharm. Weekblad* **103**, 909 (1968).
7. E. C. Munksgaard and C. Bruun, *Arch. Oral Biol.* **18**, 735 (1973).
8. K. Ranfft, *Z. Anal. Chem.* **269**, 18 (1974).
9. R. Bock and S. Strecker, *Z. Anal. Chem.* **266**, 110 (1973).
10. A. E. Pierce, in *Silylation of Organic Compounds*, Pierce Chemical Company, Rockford, Illinois, U.S.A., 1968.
11. J. MacGee and K. G. Allen, *Anal Chem.* **42**, 1672 (1970).
12. D. R. Matthews, W. D. Shults and J. A. Dean, *Anal. Lett.* **6**, 513 (1973).
13. H. A. Rüssel, *Angew. Chem. Int. Ed. Engl.* **9**, 374 (1970).
14. A. W. Archer, *Analyst (London)* **97**, 428 (1972).
15. S. Grys, *J. Chromatogr.* **100**, 43 (1974).
16. S. N. Pennington, *J. Chromatogr.* **36**, 400 (1968).
17. C. W. Lentz, *Inorg. Chem.* **3**, 574 (1964).
18. F. F. H. Wu, J. Gotz, W. D. Jamieson and C. R. Masson, *J. Chromatogr.* **48**, 515 (1970).
19. (a) T. Hashizume and Y. Sasaki, *Anal. Biochem.* **15**, 199 (1966); (b) **23**, 316 (1967); (c) **24**, 232 (1968).
20. (a) W. C. Butts, *Anal. Lett.* **3**, 29 (1970); (b) *J. Chromatogr. Sci.* **8**, 474 (1970).
21. W. C. Butts and W. T. Rainey, *Anal. Chem.* **43**, 538 (1971).
22. V. Pacáková and J. Nakvasil, *J. Chromatogr.* **91**, 459 (1974).
23. P. M. Wiese and R. H. Hanson, *Anal. Chem.* **44**, 2393 (1972).
24. D. R. Matthews, W. D. Shults, M. R. Guerin and J. A. Dean, *Anal. Chem.* **43**, 1582 (1971).
25. W. Thornburg, *Anal. Chem.* **43**, 145R (1971); **45**, 151R (1973); **47**, 157R (1975).
26. B. J. Gudzinowicz, in *Gas Chromatographic Analysis of Drugs and Pesticides*, Marcel Dekker, New York, 1967, p. 407.
27. J. N. Sieber and J. C. Markle, *Environ. Contam. Toxicol.* **7**, 72 (1972).
28. M. T. Shafik and H. F. Enos, *J. Agric. Food Chem.* **17**, 1186 (1969).
29. M. T. Shafik, D. E. Bradway, H. F. Enos and A. R. Vobs, *J. Agric. Food Chem.* **21**, 625 (1973).
30. H. A. Moye, *J. Agric. Food Chem.* **21**, 621 (1973).
31. C. J. Hardy, *J. Chromatogr.* **13**, 372 (1964).
32. J. C. Saey, *J. Chromatogr.* **87**, 57 (1973).
33. A. D. Horton, *J. Chromatogr. Sci.* **10**, 125 (1972).
34. J. M. H. Daemen and W. Dankelman, *J. Chromatogr.* **78**, 281 (1973).
35. A. Brignocchi, *Anal. Lett.* **6**, 523 (1973).
36. J. W. Boyden and M. Clift, *Z. Anal. Chem.* **256**, 351 (1971).
37. (a) D. J. Harvey and M. G. Horning, *J. Chromatogr.* **76**, 51 (1973); (b) **79**, 65 (1973).
38. W. Wells, T. Katazi, C. Bentley and C. Sweeley, *Biochim. Biophys. Acta* **82**, 408 (1964).
39. (a) R. L. Hancock, *J. Gas Chromatogr.* **4**, 363 (1966); (b) **6**, 431 (1968).
40. A. M. Lawson, R. N. Stillwell, M. M. Tacker, K. Tsuboyama and J. A. McCloskey, *J. Am. Chem. Soc.* **93**, 1014 (1971).
41. M. Zinbo and W. R. Sherman, *Tetrahedron Lett.* **33**, 2811 (1969).

42. J. R. Birk, C. M. Larsen and R. G. Wilbourn, *Anal. Chem.* **42**, 273 (1970).
43. K. L. McDonald, *Anal. Chem.* **44**, 1298 (1972).
44. D. J. Glover and J. C. Hoffsommer, *J. Chromatogr.* **94**, 334 (1974).
45. C. J. Soderquist, D. G. Crosby and J. B. Bowers, *Anal. Chem.* **46**, 155 (1974).
46. J. C. Valentour, V. Aggarwal and I. Sunshine, *Anal. Chem.* **46**, 924 (1974).
47. G. Nota and R. Palombari, *J. Chromatogr.* **84**, 37 (1973).
48. O. Backer-Dirks, J. M. P. A. Jongeling-Eijndhoven, T. D. Flissebaalje and J. Gedalia, *Caries Res.* **8**, 181 (1974).
49. I. B. Smith and C. R. Mason, *Can. J. Chem.* **49**, 683 (1971).

CHAPTER 12

The Gas–Liquid Chromatography of Metal Ions via Chelation and Non-Chelation Techniques

Paul Mushak

Department of Pathology, School of Medicine, University of North Carolina at Chapel Hill, Chapel Hill, N.C., 27514, U.S.A.

1 INTRODUCTION

This section is chiefly concerned with the use of GLC techniques for the quantitative evaluation of metal ions, to include both cationic species and those metals presented for analysis as the oxy- or other anionic molecular forms. The major divisions of dicussion in this section centre about chelation and non-chelation GLC techniques, with major emphasis being appropriately given to the former. A number of criteria must be met in the satisfactory utilization of GLC for metal ion analysis. Firstly, the derivative(s) of the ion must be thermally stable, chromatographically volatile and sufficiently inert to permit minimal interaction with column walls, column packing and other components of the chromatographic ensemble. Secondly, the selected routes to formation of suitable metal derivatives should give the desired product for analysis in high yield, should be reproducible, and should be consistent with the sample treatment needed for the satisfactory isolation of the metal ion from its matrix.

2 CHELATION TECHNIQUES IN THE GAS–LIQUID CHROMATOGRAPHIC ANALYSIS OF METALS

2.1 Chelating ligands

2.1.1 β-DIKETONES AND THEIR DERIVATIVES

To date, the class of metal chelating agents which have been most studied with respect to GLC analysis of their metal derivatives has been the β-diketone group; this also includes their halo- and alkyl-substituted derivatives as well as the corresponding thio- and amine forms. Structurally, these agents exist to varying degrees as keto–enols, with a configuration that permits hydrogen bonding:

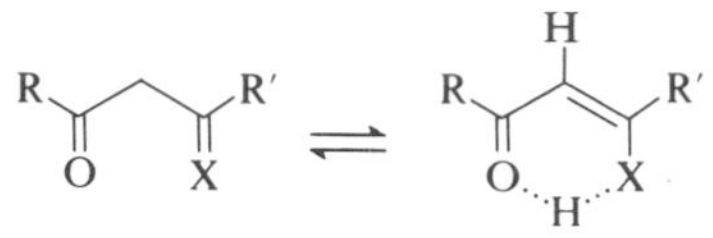

R,R′ = haloalkyl, alkyl, aryl, alkaryl . . .
X = O,S,HN—R″ . . .

The extent of enolization among the various β-diketones is variable, from 80% for pentane-2,4-dione[1] to 92–100% for the fluorinated derivatives.[2]

A newer class of β-diketone derivatives, developed for the analysis of metals by GLC, are the β-ketoamines, deriving from both mono- and bifunctional amines, and the variably substituted ketonic substrates.

2.1.2 PREPARATION OF β-DIKETONES AND THEIR DERIVATIVES

The classic method for synthesis of β-diketones is via the Claisen condensation, in which a ketone possessing at least one α-hydrogen atom is caused to react with an appropriately substituted ester, using a variety of strong bases to promote the condensation,[3] usually in benzene or diethyl ether solution.

$$R-\overset{O}{\overset{\|}{C}}-CHR_1R_2 + C_nH_mF_{(2n+1)-m}-\overset{O}{\overset{\|}{C}}-O-R_3 \xrightarrow{B^-} R-\overset{O}{\overset{\|}{C}}-CR_1R_2-\overset{O}{\overset{\|}{C}}-C_nH_mF_{(2n+1)-m}$$

R = alkyl, aryl
R_1,R_2 = H, alkyl, aryl
R_3 = ethyl
$n = 1, 2, \ldots$
$m = 0, 1, 2 \ldots$
$B^- = OR^-, NH_2^-, H^-$

Isolation and purification of the diketones involve a variety of conventional procedures such as extraction and distillation, as well as isolation via a metal derivative such as the copper (II) complex and subsequent decomposition by acid to liberate the free ligand in comparatively pure form. The Claisen reaction gives highest yields of diketones where the ketonic reactant possesses a methyl group. Reduced yields occur with increasing substitution because of cleavage side-reactions.[3] In Table 1 are shown various β-diketones of interest in metal analysis which are prepared by Claisen condensation, together with pertinent physical data.

An alternative, less commonly employed synthetic route entails acid-catalysed acylation of a ketone, via the use of boron trifluoride, with an

TABLE 1

Properties of selected fluorinated β-diketones

Compound	Density/g ml⁻	B.p. or m.p./°C (Pressure/mm Hg)	Ref.
Trifluoroacetylacetone, H(tfa)	1.271	107 (760)	4
Hexafluoroacetylacetone, H(hfa)	1.480	70.0–70.2 (760)	5
Heptafluorodimethyloctanedione, H(fod)	1.292	33 (2.7)	6
		56.5–64 (12–14)	7
		43–49 (2–4)	8
Octafluorohexanedione, H(ofhd)	1.538	85 (760)	7
Decafluoroheptanedione, H(dfhd)	1.592	99–105	9
1,1,1-Trifluoro-3-methyl-2,4-pentanedione	1.220	123 (755)	4
1,1,1-Trifluoro-2,4-hexanedione	1.220	124 (755)	4
1,1,1-Trifluoro-6-methyl-2,4-heptanedione	1.130	78 (64)	4
1,1,1-Trifluoro-4-phenyl-2,4-butanedione, H(bta)		39–40.5 m.p.	4
2-Thenoyltrifluoroacetone		42.5–43.2 m.p.	4
1,1,1,2,2,3,3-heptafluoro-4,6-heptanedione	1.365	39–42 (15–16)	10, 11
1,1,1-Trifluoro-5,5-dimethyl-2,4-hexandione		138–141 (760)	11
1,1,1,2,2-Pentafluoro-3,5-hexanedione	1.373	111–112 (631)	12

appropriate acid anhydride:[3]

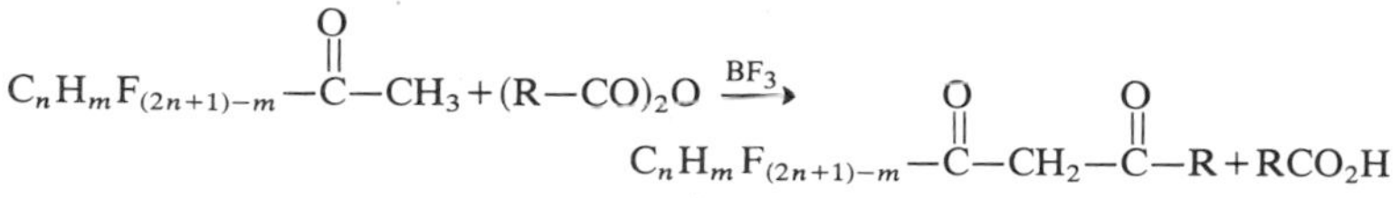

R = alkyl, aryl
$n = 1, 2, \ldots$
$m = 0, 1, \ldots$

In this manner, fluoroacetone reacted with acetic anhydride at low temperatures, with BF_3-saturated acetic acid as solvent, yields 1-fluoropentane-2,4-dione.

A procedure for the insertion of fluorine at the γ position of a β-diketone involves treatment of the sodium salt with perchloryl fluoride:[13]

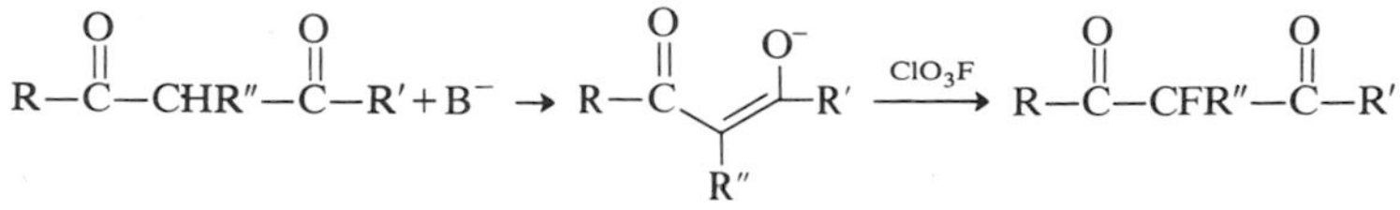

$R = R' =$ alkyl, aryl
$R'' =$ alkyl, H
$B^- = CH_3O^-, C_2H_5O^-$

Thio β-diketones are usually obtained directly from the oxo-precursor via the action of anhydrous hydrogen sulfide and hydrogen chloride at low temperatures:

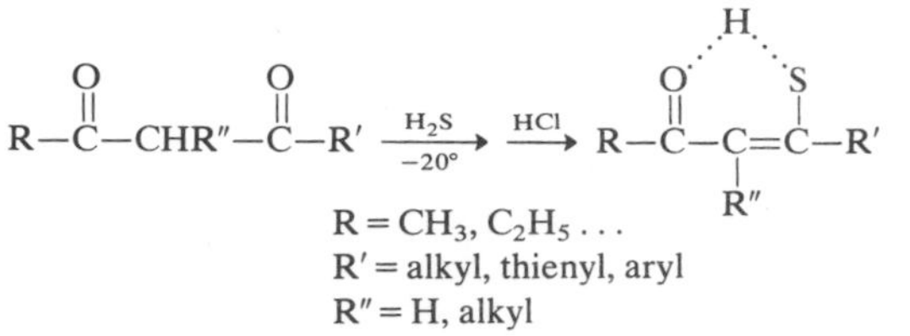

$$R-\overset{\overset{\displaystyle O}{\|}}{C}-CHR''-\overset{\overset{\displaystyle O}{\|}}{C}=R' \xrightarrow[-20°]{H_2S} \xrightarrow{HCl} R-\overset{\overset{\displaystyle O\cdots H\cdots S}{|}}{C}-\underset{\underset{\displaystyle R''}{|}}{C}=\overset{}{C}-R'$$

R = CH$_3$, C$_2$H$_5$ ...
R′ = alkyl, thienyl, aryl
R″ = H, alkyl

A second synthetic route is the Claisen condensation between an *O*-methyl thionic ester and a methyl ketone using sodium amide as the condensing agent:[14]

$$R-\overset{\overset{\displaystyle O}{\|}}{C}-CH_3 + R'-\overset{\overset{\displaystyle S}{\|}}{C}-O-CH_3 \xrightarrow{NH_2^-} R-\overset{\overset{\displaystyle O\cdots H\cdots S}{|}}{C}-CH=\overset{}{C}-R'$$

R = alkyl, haloalkyl, aryl ...
R′ = alkyl, haloalkyl, aryl

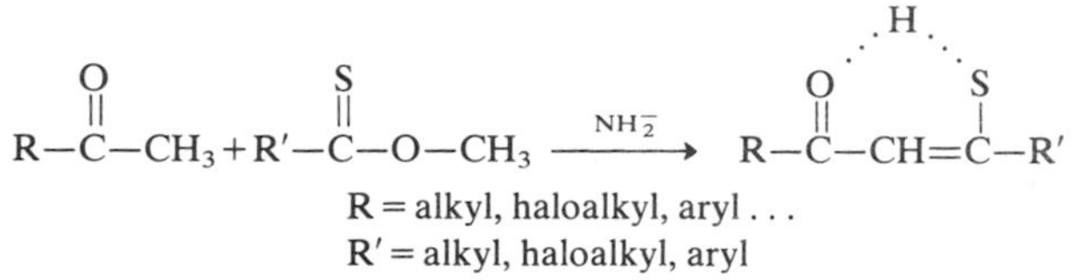

Various thio-β-diketones described in the literature are presented in Table 2.

TABLE 2

Sulfur derivatives of selected β-diketones

Compound	B.p. or m.p./°C (Pressure/mm Hg)	Ref.
1,1,1-Trifluoro-4-mercaptopent-3-en-2-one	50 (20)	15
1,1,1-Trifluoro-4-mercapto-4-phenylbut-3-en-2-one	95 (5)	16
1,1,1-Trifluoro-4-(2-thienyl)-4-mercaptobut-3-en-2-one	62–64 m.p.	15
1,1,1-Trifluoro-4-mercapto-4-(4-methylphenyl)-but-3-en-2-one	Dec.	17
1,1,1-Trifluoro-4-mercapto-4-(4-methoxyphenyl)-but-3-en-2-one	Dec.	17
1,1,1-Trifluoro-4-mercapto-4-(4-bromophenyl)-but-3-en-2-one	110 (2)	17
1,1,1,5,5,5-Hexafluoro-4-mercaptopent-3-en-2-one	75–78 (12)	18
1,1,1-Trifluoro-4-(2-furyl)-4-mercaptobut-3-en-2-one	102 (1)	19
1,1,1,2,2-Pentafluoro-5-mercapt-6,6-dimethylhept-4-en-3-one		19
1,1,1-Trifluoro-4-mercapto-5,5-dimethylhex-3-en-2-one		19

The major route to the preparation of various mono- and diamine derivatives of β-diketones involves the condensation of an appropriate amine or diamine, in the correct stoichiometric ratio, with the appropriate diketone:

$$R-NH_2 + nR''-\overset{\overset{\displaystyle O}{\|}}{C}-CH_2-\overset{\overset{\displaystyle O}{\|}}{C}-R'''$$

or

$$R'\overset{\displaystyle NH_2}{\underset{\displaystyle NH_2}{\diagdown}}$$

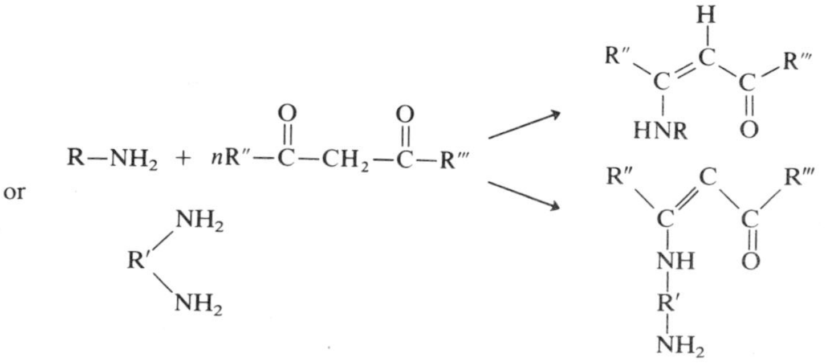

Various amine derivatives which have been developed recently for metal analysis by GLC are listed in Table 3.

TABLE 3

Selected amine derivatives of β-dicarbonyl compounds

Compound	B.p. or m.p./°C	Ref.
4-Imino-2-pentanone	43	20
4-Methylamino-2-pentanone	40–41	21
4-Amino-1,1,1-trifluoropent-3-en-2-one	87–88	22
N,N'-Ethylenebis(acetylacetoneimine)	104	23
N,N'-Ethylenebis(trifluoroacetylacetoneimine)	156	24
N,N'-Propylenebis(trifluoroacetylacetoneimine)	52	24
N,N'-Butylenebis(trifluoroacetylacetoneimine)	—	25

2.1.3 PROPERTIES OF β-DIKETONES AND THEIR DERIVATIVES

Substitution of various groups for α- and/or β-hydrogens in β-diketones markedly influences the properties of the ligand, especially with regard to ligand acidity and reactivity of the carbonyl function(s) to nucleophilic reagents. Fluorinated β-diketones are considerably more acidic than the parent compounds, the acidity increasing with increasing fluorination, because of the strong electron-withdrawing effect of fluorine (Table 4).

TABLE 4

Effect of fluorine on acidity of pentane-2,4-dione

	pK_a	Ref.
Pentane-2,4-dione	8.9	26
1,1,1-Trifluoropentane-2,4-dione	6.7	27
1,1,1,5,5,5-Hexafluoropentane-2,4-dione	4.6	5

With regard to lability to nucleophilic agents such as water and the lower aliphatic alcohols, hexafluoroacetylacetone, as a somewhat extreme example, yields the corresponding tetraol[28] and dihemiketal[29] with water and methanol respectively. In the case of thiodiketones, the lability of these ligands to oxidation and geminal-thiol formation may be a practical drawback. Monothioacetylacetone, for example, is stable in non-polar solvents for up to six hours, but decomposition occurs over a 24-h period in benzene, hexane, carbon tetrachloride, ether, chloroform, methanol and water.[30] These agents are best prepared for storage as the lead chelate, from which they may be liberated by treatment with dry hydrogen sulfide just before use.[30]

Where α- and β-substituents in a thio-β-diketone are not equivalent, the nature of the substituents apparently determines whether these agents are predominantly visualized as the *O*- or *S*-enolic form. Thus, recent spectral studies indicate that, while the *O*-enol is the dominant form in thiotrifluoroacetylacetone, the thioenolic configuration is evident in thioacetylacetone and thioacetylpivalylmethane:[31]

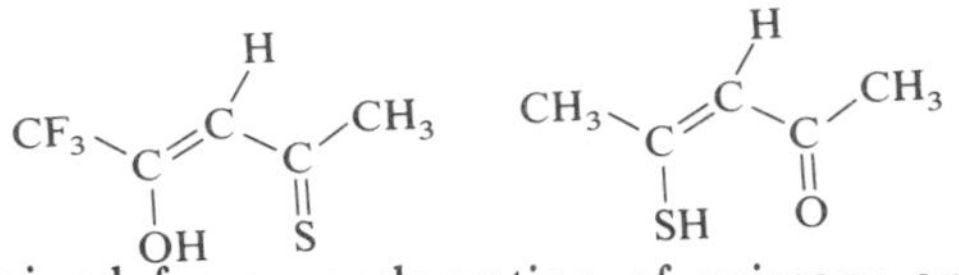

β-ketoamines, obtained from condensation of primary amines with β-diketones or β-keto-aldehydes can exist in three tautomeric forms:

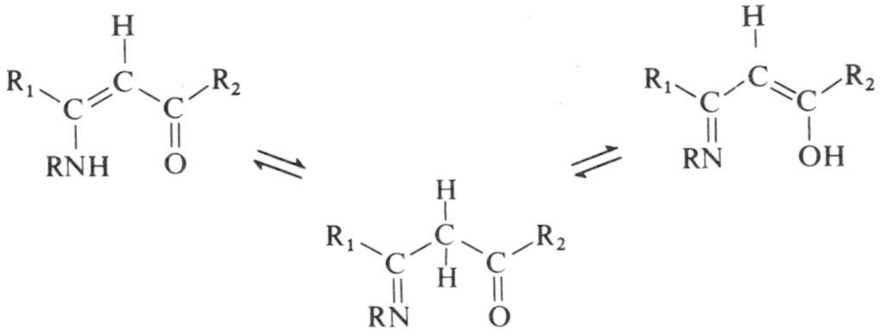

I II III

Recent n.m.r studies, carried out in non-polar solvents, indicate that the ketoamine structure I predominates (95 mol %).[32] This has been rationalized on the basis of greater resonance stabilization.

In asymmetrical β-ketoamines such as:

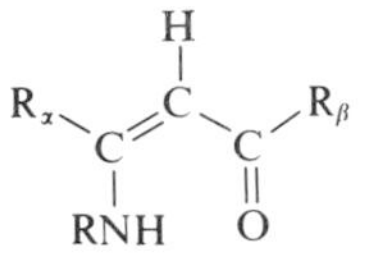

$R_\alpha \neq R_\beta$

the nature of the substituents determine the structure. Thus, benzoylacetone and trifluoroacetylacetone react to place trifluoromethyl and phenyl groups bound to the free carbonyl (R_β = phenyl, CF_3) whereas with formylacetone, the formyl moiety enters into reaction with the amine reagent ($R_\alpha = H$).[33]

2.1.4 MISCELLANEOUS CHELATING AGENTS

Dithiocarbamates, prepared via the action of perchlorobicycloalkyl primary and secondary amines on carbon disulfide, have been reported in connection with preliminary GLC studies of various groups of metals.[34] These reagents have found little analytical application. *N,N*-dialkyldithiocarbamates have been studied in some detail in our own laboratory. Diethyldithiocarbamates of various arsenicals show promise as derivatives in arsenic analysis with various media (report in press). Esters of phosphorodithioic acid form strong complexes with selected metals and a preliminary report of GLC studies of some of these has appeared.[35] The GLC properties of complexes between metals and amine derivatives of salicylaldehyde have been investigated.[36]

2.2. Metal chelates of interest in metal analysis

This section deals with metal derivatives which are currently in use in metal analysis by GLC, as well as those which offer promise for this application.

2.2.1 SYNTHESIS OF METAL β-DIKETONATES

Several general methods are employed to prepare metal derivatives of the various β-diketones and their derivatives.[37] The most widely employed synthetic route is to treat a salt of the metal with free ligand. The salt may be the nitrate, acetate, etc., and the reaction is carried out in aqueous methanolic or ethanolic solution, buffered at a slightly alkaline pH, with the ligand being added directly. Since the resulting chelate usually comes out of solution, it is simply isolated for further conventional purification via recrystallization, column chromatography or sublimation. In the case of ternary complex synthesis, the metal, the diketone and a neutral adduct agent, the latter ligating agent is either added together with the diketone, or contained in an extracting solvent for preliminary isolation.[38]

A second preparative route is the direct action of the diketone with a metal chloride (anhydrous suspension) in a non-polar solvent such as carbon tetrachloride. The mixture is then refluxed, with the evolution of hydrogen chloride.[39]

Various metal β-diketonates of interest in metal analysis by GLC are listed in Table 5, with some of their physical properties.

2.2.2 PROPERTIES OF METAL β-DIKETONATES.

In metal β-diketonate complexes, both oxygen atoms of the dioxo-ligands, or mixed atoms of the thio- and amine derivatives, coordinate to the metal, yielding a 6-membered metallocycle. The general 'chelate effect' serves to promote formation of the metallocycle in preference to polymerization, driven by a favourable entropy change, since chelation consumes one molecule of ligand but usually releases two or more solvating molecules such as water from the coordination sphere of the metal, which thus increases the degree of disorder in the reaction system.[54]

A specific factor operating with β-diketonates of many metals is the formation of complexes where six π electrons are available for delocalization in a metallocyclic structure, bearing a theoretical analogy to the resonance-stabilized six π-electron carbocycle benzene. While the question of aromaticity in certain metal β-diketonates is still a controversial one,[55] a number of these complexes do undergo reactions characteristic of aromatic nuclei, such as electrophilic substitution,[56] and demonstrate spectrochemical evidence for the existence of delocalized electronic systems within the ring. The reader is directed to reviews by Collman[56] and Fackler[55] for more detailed discussion.

Many metal ions, on chelation with β-diketones and their derivatives, yield complexes which have so far been of limited application in analysis of metal ions by GLC. Many divalent ions typically react to furnish coordinatively

TABLE 5

Metal β-diketonates of interest in GLC of metals

Chelate	M.p./°C	Sublimation/ °C (Pressure/ mm Hg)	Colour	Ref.
Al(III) of				
trifluoroacetylacetone	121–122	48	White	40
hexafluoroacetylacetone	73–74		White	41
benzoyltrifluoroacetone	173–174	88	White	42
2-thenoyltrifluoroacetone	203–205	125	White	42
Be(II) of				
trifluoroacetylacetone	112	38	White	43
hexafluoroacetylacetone	70–71		White	44
benzoyltrifluoracetone	143–144	78	White	42
2-thenoyltrifluoroacetone	169–170	73, dec.	White	42
Co(II) of				
trifluoroacetylacetone (5/2 H_2O)			Red-brown	45
hexafluoroacetylacetone (2 H_2O)	172–174			41
benzoyltrifluoroacetone	158	55	Yellow	42
2-thenoyltrifluoroacetone	200–215, dec.	69	Yellow	42
Co(III) of				
trifluoroacetylacetone	129–129.5, *cis*		Green	40
	158–158.5, *trans*		Green	40
hexafluoroacetylacetone	94–95		Green	46
Cr(III) of				
trifluoroacetylacetone	112–114, *cis*		Violet	40
	154.5–155, *trans*		Violet	
hexafluoroacetylacetone	84–85	25	Green	41
Cu(II) of				
trifluoroacetylacetone	189		Blue-violet	4
hexafluoroacetylacetone	95–98		Purple	47
hexafluoroacetylacetone.H_2O	126–128		Green	41
hexafluoroacetylacetone.$2H_2O$	134–136		Green	47
benzoyltrifluoroacetone	243–244		Green	4
2-thenoyltrifluoroacetone	242–243	112, dec.	Green	42
decafluoro-2,4-heptanedione	73–83	50 (0.1)	Purple	9
Ce(III) of				
trifluoroacetylacetone	130–131		Yellow	43
Ce(IV) of				
2-thenoyltrifluoroacetone	181		Brown	48
Dy(III) of				
2-thenoyltrifluoroacetone	193		Yellow	48
Dy(III) of				
dimethylheptafluorooctane-4,6-dione	180–188			6
Eu(III) of				
trifluoroacetylacetone	132–134		White	43
2-thenoyltrifluoroacetone	180		Pink	48
dimethylheptafluorooctane-4,6-dione	205–212, dec.			6
Eu(IV) of				
2-thenoyltrifluoroacetone	164–165		White	49
2-thenoyltrifluoroacetone.H_2O	59–67			6
Er(III) of				
2-thenoyltrifluoroacetone	125		Pink	48
dimethylheptafluorooctane-4,6-dione	158–164			6

TABLE 5—continued

Chelate	M.p./°C	Sublimation/ °C (Pressure/ mm Hg)	Colour	Ref.
Fe(III) of				
trifluoroacetylacetone	114, *trans*		Red	40
hexafluoroacetylacetone	49	35	Red	41
benzoyltrifluoroacetone	128–129	89, dec.	Red	42
2-thenoyltrifluoroacetone	159–160	135 (79)	Red	42
Fe(II) of				
trifluoroacetylacetone		115–120(10^{-3})	Red	50
hexafluoroacetylacetone		45–50 (10^{-3})	Red	50
Ga(III) of				
trifluoroacetylacetone	128.5–129.5, *trans*		White	40
Gd(III) of				
trifluoroacetylacetone	133–135		White	43
2-thenoyltrifluoroacetone	122		Grey	48
dimethylheptafluorooctane-4,6-dione	203–213, dec.			6
Hf(IV) of				
trifluoroacetylacetone	125–128	115 (0.05)	White	51
2-thenoyltrifluoroacetone	220–223		White	51
decafluoroheptanedione	70 (b.p. at 0.02 mm Hg)			9
Ho(III) of				
2-thenoyltrifluoroacetone	135		Brown	48
dimethylheptafluorooctane-4,6-dione	172–178			6
In(III) of				
trifluoroacetylacetone	118–120		Ivory	40
La(III) of				
trifluoroacetylacetone	169			43
hexafluoroacetylacetone	120–125		White	43
2-thenoyltrifluoroacetone	135		Brown	48
dimethylheptafluorooctane-4,6-dione	215–230, dec.			6
Nd(III) of				
trifluoroacetylacetone	133–134		Pink	43
hexafluoroacetylacetone	117		Pink	43
2-thenoyltrifluoroacetone	181–182		Pink	48
dimethylheptafluorooctane-4,6-dione	210–215, dec.		Pink	6
Ni(II) of				
trifluoroacetylacetone		150(0.5)	Green	42
hexafluoroacetylacetone.$2H_2O$	207–208		Green	41
benzoyltrifluoroacetone	223–224	98	Green	42
2-thenoyltrifluoroacetone	291–295	70	Yellow	42
Rh(III) of				
trifluoroacetylacetone	148.5–149, *cis*		Yellow	40
hexafluoroacetylacetone	114–115		Yellow	41
Sc(III) of				
trifluoroacetylacetone	106–107	90–95	White	43
dimethylheptafluorooctane-4,6-dione	25			6
Monothio Chelate Analogues				
Cd(II) of thiotrifluoroacetylacetone	197		Yellow	16
Co(III) of thiotrifluoroacetylacetone	131		Dark brown	16
Cu(II) of thiotrifluoroacetylacetone	202		Red-brown	16
Hg(II) of thiotrifluoroacetylacetone	146, dec.		Yellow	16
Ni(II) of thiotrifluoroacetylacetone	153		Brown	16

TABLE 5—continued

Chelate	M.p./°C	Sublimation/ °C (Pressure/ mm Hg)	Colour	Ref.
Pd(II) of thiotrifluoroacetylacetone	154		Orange	16
Pt(II) of thiotrifluoroacetylacetone	144		Red	16
Zn(II) of thiotrifluoroacetylacetone	101		Yellow	16
Ni(II) of thiohexafluoroacetylacetone	25			52
Cd(II) of thiothenoyltrifluoroacetone	213, dec.		Yellow-brown	16
Co(II) of thiothenoyltrifluoroacetone	229		Blue-brown	26
Co(III) of thiothenoyltrifluoroacetone	235		Brown	16
Cu(II) of thiothenoyltrifluoroacetone	230–231		Olive-brown	26
Fe(III) of thiothenoyltrifluoroacetone	165		Black	16
Hg(II) of thiothenoyltrifluoroacetone	178		Yellow	16
Pd(II) of thiothenoyltrifluoroacetone	236		Brown-red	26
Pt(II) of thiothenoyltrifluoroacetone	218		Brown	16
Ni(II) of thiothenoyltrifluoroacetone	235–237		Brown-red	26
Zn(II) of thiothenoyltrifluoroacetone	177–178		Yellow-green	26
Ketoamine Chelate Analogues				
Cu(II) of				
N-Methylsalicylaldimine	158			36
4-Imino-2-pentanone	185–186			36
4-Methylimino-2-pentanone		120		36
Ni(II) of				
4-Imino-2-pentanone	246			36
Bisacetylacetone-ethylenediimine of				
Cu(II)		175 (0.1)		53
Ni(II)		170 (0.1)		53
Pd(II)		200 (0.1)		53
Bisacetylacetone-propylenediimine of				
Cu(II)		170 (0.1)		53
Ni(II)		165 (0.1)		53
Pd(II)		190 (0.1)		53
Bistrifluoroacetylacetone-ethylenediimine of				
Cu(II)		170 (0.1)		53
Ni(II)		165 (0.1)		53
Pd(II)		190 (0.1)		53
Bistrifluoroacetylacetone-propylenediimine of				
Cu(II)		165 (0.1)		53
Ni(II)		165 (0.1)		53
Pd(II)		185 (0.1)		53
Pt(II)		190 (0.1)		53

unsaturated species which react further to form hydrates, polymeric material or other compounds which complicate isolation of the complex from an analytical medium, and interfere during chromatography.[57] A detailed discussion of the chemistry of such complexes has been presented.[58] More recent work indicates that the use of adduct agents can improve the extraction and chromatographic properties, since such agents compete with the processes of hydration (solvation) and polymerization to yield the more easily handled derivatives.[59] An alternative approach has been to employ β-diketones so substituted that steric hindrance prevents the approach of agents which result in solvation or polymerization.

2.3 Quantitative GLC analysis of metal ions by chelation techniques

Successful applications of GLC to the analysis of various metal ions are discussed in some detail. This treatment is not intended to be an exhaustive review of all GLC studies of metal ions, but gives a selection of actual analytical applications involving a variety of media.

2.3.1 GENERAL CHROMATOGRAPHIC CONSIDERATIONS

Packings: The factors of stability and polarity of various metal β-diketonates have narrowed the choice of stationary phase(s) in packings for metal analysis to the silicone gums and oils and the hydrocarbon high-vacuum greases: Apiezon L, SE-30, SE-52, QF-1, DC-oil series, etc.[60] The support must be as inert as possible, and this requires properly deactivated diatomaceous earths, the usual treatment being silanization with e.g. dichlorodimethylsilane and hexamethyldisilazane. The problem of on-column loss of analyte may be particularly troublesome if one is carrying out determinations in the microgram range or less. Most columns, for metal analysis work are of borosilicate glass or PTFE, glass columns being more widely used because they are cheaper and can be used at higher operating temperatures.

Detectors: While all of the conventional GLC detectors have been used in GLC of various metal derivatives, those reports dealing with analytical application in the trace of ultra-trace range have usually employed FID or EC detectors. The very high sensitivity possible with the EC detector when used for the analysis of halogenated materials has been exploited in those metal analysis studies where fluorinated chelate derivatives are employed.[37] Despite the value of FID and EC detectors in metal analysis by GLC, these detection methods are relatively non-specific in that they also give peaks with any organic impurities present. More specific but less widely employed detectors are mass spectrometers,[61] flame-photometric detectors[62] and microwave emission units.[63,64]

2.4 Quantitative GLC analysis of various metal ions

2.4.1 ALUMINIUM

Morie and Sweet[65] analysed mixtures of aluminium with gallium and indium by solvent extraction-gas chromatography techniques using trifluoroacetyl acetone as chelating agent. Aqueous solutions were acetate-buffered at pH 4–7 and extracted into benzene using 4-h extraction time. Using a glass column packed with 0.5% DC-550 on silanized glass beans and an FID, milligram levels of aluminium yielded recoveries of 71–100% with a relative precision of

2%. In a related report,[66] aluminium as well as copper and iron were extracted from acetate-buffered aqueous media using a solution of trifluoroacetylacetone (0.1 M) in chloroform. At a pH of 4.5–5.5, 99.7% of aluminium can be removed in the 0.5 mM range. Aluminium and other metals in alloys can be determined by the procedure of Scribner *et al.*[67] NBS alloys (nickel/copper) were suitably treated and diluted samples were analysed for aluminium content using trifluoroacetylacetone as chelating agent. Moshier and Schwarberg[68] measured aluminium, copper and iron in NBS alloys. Employing a glass column packed with 12% Tissuemat E on Gas Pack F and operated at 105°C, aluminium and iron could be detected simultaneously, while analysis of copper required a modified procedure. Relative deviations of 1.83 and 1.08 were found for the two NBS alloy samples analysed. Nanogram levels of aluminium have been detected in rat liver.[69] Wet-ashing of the liver samples via sulfuric/nitric acid mixture was followed by sample neutralization, pH adjustment to 4.5 with buffer and extraction with a benzene solution (0.5%) of trifluoroacetylacetone. An instrument equipped with an EC (^{63}Ni) detector and a column (glass) of 3% OV-17 on Chromosorb G was used.

An interesting application of GLC in aluminium determination is that of Genty *et al.*[70] who combined solvent extraction and GLC for detecting this metal in uranium. Aluminium (and chromium) in uranyl nitrate solutions were extracted with a benzene solution (0.1 M) of trifluoroacetylacetone after acetate buffering. An EC detector and a glass column filled with glass beads coated with 0.2% DC-710 permitted a detection limit of 0.1 ppm aluminium and linear calibration data in the range 8×10^{-5}–10^{-6} micrograms. Extraction efficiency was 100% at pH 5.0 and with a one-hour extraction time. No effect of a variety of added ionic elements was seen. Reference uranium solutions containing 0.2 ppm aluminium gave GLC values of 0.18 ± 0.03 ppm.

A troublesome medium for metal analysis, sea water, has been analysed for aluminium by GLC. Lee and Burrell[71] extracted sea-water samples (30 ml) with a toluene solution (0.1 M) of trifluoroacetylacetone and achieved detection of aluminium in the picogram range when EC detection and a column containing a mixture of DC-710 and Carbowax 20 M as stationary phase were employed. In a related but more extensive investigation of GLC analyses of metal ions in various waters, Minear and Palesh[72] successfully applied chelation-extraction and GLC to the analysis of aluminium and a variety of other metals—beryllium, chromium and copper. Using trifluoroacetylacetone as chelant, and an instrument equipped with an electron-capture (^{3}H) detector, sensitivity observed was in the range 10^{-6}–10^{-7} g ml^{-1}. The packings 5% QF-1 on Variport 30 and 15% Carbowax 20 M on Chromosorb W permitted analysis of aluminium, copper and chromium, but not of iron.

2.4.2 BERYLLIUM

A number of reports deal with beryllium analysis by GLC in a variety of media and, in fact, GLC techniques applied to beryllium have been more extensive than with any other element, with the possible exception of chromium. A rapid micro-analytical GLC procedure for beryllium in biological fluids has been published by Sievers and his co-workers,[73] who investigated urine, whole blood, liver homogenates and plant extracts. Direct treatment with trifluoroacetylacetone in sealed glass ampoules minimized element loss or contamination via conventional ashing procedures. With Teflon® columns packed with 5% SE-52 on 60/80 mesh Gas Chrom Z and an electron-capture detector, a sensitivity of 10^{-10} g Be/50 microlitre sample was reported. Average recovery from all media was 93.5% and no interferences by common anions and cations was observed. A later valuable report from the same laboratory,[74] on samples from laboratory animals fed radioisotopic beryllium, indicated that beryllium incorporated into tissues *in vivo* can indeed be removed via the direct action of the chelating agent. Few investigators who study applicability of new methods for metal analysis take the added experimental pains to establish recovery and precision data relative to *in vivo* incorporation of a particular element into biological matter, but the widespread practice of *in vitro* 'spiking' to evaluate quantitative factors is a crude simulation of the tightness of metal binding in the biological matrix and may not be valid where the samples are only partially mineralized.

In a rather extensive survey of beryllium measurement in various media—blood, urine, organs, plants, food and sewage—Kaiser and his co-workers[75] employed GLC in conjunction with both direct reaction of reagent and prior ashing techniques. A ^{63}Ni-equipped electron-capture detector and silanized glass columns packed with 5% SE-52 on Gas Chrom Z were used. Recoveries of beryllium in all media were monitored using ^{7}Be as radiotracer to optimize the isolation conditions. Ashing consisted of the usual wet-ashing techniques (nitric or nitric/hydrofluoric acids) as well as low-temperature decomposition using an oxygen RF-generated high energy plasma. The minimum amount of beryllium detectable was 10^{-11} g, using samples of less than one gram. Relative standard deviations were considerably better for the direct reaction method than with the conventional ashing methods.

In environmental investigations, Ross and Sievers,[76] studying water and particulate matter gathered on air sampling filters, employed the EC detector as well as mass-spectrometric techniques for beryllium detection. Strips of the air filter were mineralized using a low-temperature asher, the residues digested in strong acid and the resulting solutions adjusted to pH 5.5–6.0; further sample work-up involved sample extraction with a dilute solution (0.164 M) of trifluoroacetylacetone in benzene. Using a glass column packed with 2.8% W-98 silicone on Diatoport S at 110 °C, relative standard deviations of 3.0% were obtainable. Interfering metal ions were complexed with EDTA. A double-focusing mass spectrometer was employed as a specific detector, using a fragment of bis(trifluoroacetylacetonato) beryllium(II) which is both intense and relatively free of interference. The overall sensitivity of both the MS and ECD's is such that levels are routinely analysed which are two orders of magnitude below the AEC-recommended beryllium concentration of 0.01 μg Be/m^3 air.

Foremen *et al.*,[77] analysing human and rat urine samples for beryllium via GLC, also employed the direct and prior ashing procedures. Detailed study showed either method to be satisfactory while a survey of column materials and packings indicated that Teflon® columns packed with SE-52 gave the best results. With EC detection, the minimal amount of element detectable in urine was 1.0 ng ml^{-1} using EDTA as masking agent and a 0.05 M benzene solution of trifluoroacetylacetone.

Eisentraut *et al.*[78] successfully evaluated beryllium in lunar, terrestrial and meteoric samples using GLC. Lunar samples were those procured from the Apollo 11 and 12 missions. Chips of material were powdered in a diamond mortar and fused with sodium carbonate, followed by dilute HCl treatment and transfer to polyethylene bottles. After pH adjustment of the solutions to about 4.0 and a final adjustment to pH 5.0 with acetate buffer, both EDTA and chelant (trifluoroacetylacetone) in benzene were added, and the samples heated to 95 °C for short periods. For GLC analysis a tritium-foil EC detector was employed in an instrument equipped with a Teflon® column containing 10% SE-30 on 70/80 mesh Gas Chrom Z. Using peak height measurement, the reported sensitivity was 4×10^{-14} g beryllium.

2.4.3 CHROMIUM

Ross and Sievers[79] measured the chromium content of high- and low-carbon steels (NBS 106B and 107A) using several methods of sample preparation. Samples were solubilized by heating in mantles or in a radio-frequency field. Small amounts (2–4 mg) of alloy were reacted directly with trifluoroacetylacetone in the presence of nitric acid. Where residue still remained, this was dissolved in 70% nitric acid and evaporated to dryness followed by a repetition of the whole process. The resulting red solutions were extracted with benzene followed by removal of excess chelant via alkaline reextraction (dilute sodium hydroxide). Precision data included a relative error of 1.4–1.7% when using a Teflon® column packed with 15% SE-52 on Anakrom ABS and an EC detector.

Chromium in biological media has been determined in a number of laboratories: this element is difficult to measure owing to low levels of occurrence in biological samples and its role in carbohydrate metabolism is of increasing interest. Hansen *et al.*[80] utilized the direct reaction route of a number of fluorinated diketones with whole blood or plasma: hexane solutions of the chelants were warmed with the samples in sealed tubes. Tracer recovery studies involving ^{51}Cr administered to laboratory animals furnished recoveries of the element in whole blood and plasma of 82 and 98% respectively when using trifluoroacetylacetone. Other agents, hexafluoroacetylacetone and heptafluorodimethyloctanedione, gave recoveries of up to 50%. For these studies, glass columns packed with 5% Dow-Corning LSX on Gas Chrom P were used with an EC detector. Determination of chromium down to a level of 5 ng ml^{-1} was claimed. Chromium in serum which has first been wet-ashed for liberation of the element was determined by Savory *et al.*[81] Serum was ashed in a mixture of sulfuric, nitric and perchloric acids and worked up in the usual manner to provide a buffered medium of pH 6.0. Chelation–extraction was in a dilute solution of trifluoroacetylacetone in benzene, on a reciprocal shaker block heated at 70 °C. Comparative studies using atomic absorption spectrometry furnished a coefficient of correlation (R) of 0.96 for a group of serum samples, while calibration data demonstrated linearity up to 1.0 ppm (100 $\mu g\ dl^{-1}$) chromium. A detection limit of 0.001 $\mu g\ dl^{-1}$ Cr in the analyte medium (benzene) was achieved. GLC was done on glass columns packed with 5% QF-1 on 60/80 WAW Chromosorb in an instrument equipped with a ^{63}Ni EC detector. A related study is that of Booth and Darby,[82] who measured soft tissue and serum levels of chromium. Wet-ashing was carried out in Vycor flasks followed by pH adjustment of the digests to 6.0 using acetate buffer. Chelation-extraction was carried out using a benzene solution of trifluoroacetylation in a reaction time of one hour at 70 °C. Lindane served as an internal standard. Recovery data for chromium in serum and liver homogenate were 94 and 88–104% respectively, with a relative standard deviation for liver of about 17.0%.

Wolf *et al.*[61] employed direct reaction and ashing techniques for chromium assessment in plasma and serum using a double-focusing mass spectrometer and an EC detector. As with beryllium, mass spectral quantitation was achieved using a relatively clean fragment of chromium(III)trifluoroacetylacetonate. Comparable data were obtained with the two means of detection.

Chromium in lunar dust and rocks obtained in the Apollo 11 and 12 experiments has been measured by GLC using as preliminary reference matter the USGS rock standards BCR-I, DTS-I and PCC-I. Small samples (less than one milligram) were fused with carbonate, taken up with hydrochloric acid and reacted directly with trifluoroacetylacetone. An EC detector was employed.

2.4.4 COPPER

In addition to analyses of copper in admixture with other ions as noted above, GLC techniques have been applied to the assessment of copper in tap water and in nickel–copper alloys.[84] In the latter cases, copper was chromatographed as the bis(acetyl-pivalyl-methane)-ethylenediimine complex. For copper-content of the alloys, samples were solubilized in the usual manner, the pH adjusted to 6–7 and chelation-extraction carried out via a hexane solution of the β-ketoamine ligand. Using a chromatographic unit equipped with an FID, as little as 10^{-9} g of copper could be observed, with linearity over the range 10^{-8}–10^{-9} g copper. By means of the butylenediimine derivative of trifluoroacetylacetone, Uden *et al.*[85] studied copper levels in liver, kidney and lung of animals fed the element. Best chromatographic data were obtained using 2% Dexsil 300 GC on 100/120 mesh Chromosorb 750. It was noted that, in comparison to atomic absorption spectrometric values, GLC determination gave lower levels of copper.

2.4.5 NICKEL

Barrett and co-workers[52] studied the use of monothiotrifluoroacetylacetone in the analysis of nickel in media alloys, tea and fats. In the case of alloys, dissolution in 6 M HCl and HNO_3 was followed by pH adjustment with buffer, ammonia addition and the extraction of the resulting solution with a dilute hexane solution of the chelant. Nickel in tea samples involved use of both wet- and dry-ashing sample preparations. In the former case, final residues were taken up in dilute HNO_3 while dry-ash residues were taken up in deionized water. Fats were initially dried on a hot plate and then dry-ashed at 540 °C. Further work-up was essentially that for alloy samples. For all nickel determination, a Teflon® column packed with Silicone E-350 on 60/80 mesh Universal B was employed in a unit equipped with a tritium EC detector. Temperature programming from 140–170 °C gave the best chromatographic separation. While copper seriously interferes, it can be removed readily by preliminary sample treatment using hydrogen sulfide.

2.4.6 OTHER METALS

Heptafluorodimethyloctanedione has been employed for the assessment of cobalt in vitamin B_{12}, liver, blood and urine.[86] Benzene solutions of the chelating agent were heated with the various media in sealed tubes at 75 °C, after first rendering the materials alkaline and adding hydrogen peroxide. After removal of the excess of chelant from the tubes by alkali washing, aliquots were introduced into a gas–liquid chromatograph equipped with an EC detector. A detection limit of 4.4×10^{-11} g of cobalt was reported.

NBS samples of Mesabi iron ore were employed to demonstrate the utility of iron analysis by GLC, using heptafluorodimethyloctanedione.[87] In these experiments, small samples (less than one mg) were reacted in sealed capillary tubes with neat reagent and the tubes then crushed in the injection port of a modified gas chromatograph. Good correspondence of GLC-determined iron values with the assay data were observed. As a detector, thermal conductivity was used at the end of a Teflon® column packed with 10% SE-30 on Gas Chrom Z.

Uranium and thorium in aqueous solution have been determined via their ternary complexes with hexafluoroacetylacetone and the neutral extraction synergist, dibutyl sulfoxide.[88] Extractions were carried out using benzene solutions of chelant and synergist, the concentrations of the agents being 2–5 times that of the combined metal levels. Stainless-steel columns packed with 17.8% QF-1 on 100/120 mesh Chromosorb W gave the best separations. The detection limits were 0.4 mg Th ml^{-1} and 0.6 mg U ml^{-1} using thermal conductivity detection.

The yttrium and cerium group lanthanides were studied by Burgett and Fritz,[89] using decafluoroheptane-3,5-dione as chelant and dibutyl sulfoxide as neutral synergist. Solutions of the lanthanides (aqueous, pH 5.5) were extracted with cyclohexane solutions of combined chelant and synergist, these being present in amounts three times that of the total metal concentrations. Recovery studies showed near-quantitative and quantitative extraction of the yttrium and cerium groups, respectively, under these conditions.

3 NON-CHELATION TECHNIQUES IN METAL ION GAS–LIQUID CHROMATOGRAPHY

3.1 Metal halides

GLC determinations of metals via the halides are quite troublesome because of the chemical lability of these derivatives, both in chromatographic systems and in the various analytical solutions. The halide derivatives most widely employed for various groups of metals are the fluorides and chlorides; the fluorides are more volatile than the corresponding chloride derivatives but tend to be more reactive.

Conversion of metal ions to their respective halides involves various halogenating techniques. For chlorination, direct reaction of the metals with chlorine gas or hydrogen chloride has been employed as well as reaction of a metal oxide with a halohydrocarbon at elevated temperature. A less general route is the passage of a mixture of hydride over a heavy metal chloride such as gold(III) chloride at high temperature. GLC studies involving metal chlorides and fluorides require the use of highly inert column packings and an inert chromatographic assembly, particularly the injector port, column walls and detector interiors. An early successful analysis of a mixture of the elements germanium and silicon, as reported by Phillips and Timms,[90] involved the hydrolysis of alloys of these two species with magnesium. The resulting hydrides were converted to the chlorides via passage over gold(III) chloride and chromatographed on a column 13% Silicone 702 on Celite using a gas-density balance for detection. A fairly general approach for the GLC analysis of metals and their alloys, carbides, oxides, sulfides and salts is the procedure of Juvet and Fisher[91] in which these derivatives are reacted *in situ* with fluorine using a reactor-system interfaced with a gas chromatograph. Uranium, selenium, technicium, tungsten, molybdenum and rhenium were measured quantitatively via this scheme. Column packing consisted of 15% Kel-F oil No. 10 on 40/60 mesh Chromosorb T in a Teflon® column. Removal of moisture and reactive organic matter in packings is first carried out by column conditioning with fluorine or chlorine trifluoride.

Sie and co-workers[92] investigated the analysis of silicon and tin in various alloys. Small amounts (1–50 mg) of material were heated in a quartz tube at 600–900 °C, followed by passage of chlorine. A trapping column of Haloport-F (30/80 mesh) coated with 15% Kel-F collected the desired chlorides followed by release into an analytical column similarly constructed and held at 75 °C. Using a gas-density balance and Teflon®-coated filaments, a sensitivity of 50 ppm and recoveries of 97 ± 5% for silicon were achieved.

Germanium in powdered coal may be determined by GLC via generation of the chloride by the action at high temperature of hydrogen chloride on a mixture of powdered coal and Chromosorb (1 : 1) that has been wetted with sulfuric acid.[93] Analysing 1.5 g samples and using activated charcoal AR-3 as packing, a sensitivity of 3.3 g ton^{-1} with an error of ±6% was reported.

Titanium may also be measured via GLC. Sievers *et al.* determined this element in oxide mixtures via reaction with carbon tetrachloride at elevated temperature to yield the tetrachloride. Reactants were sealed in a small capillary, heated and the capillary was inserted directly into a modified gas chromatograph. The capillaries were crushed and the reaction products swept directly into the carrier gas stream. By means of a standard bauxite sample, the relative error of analysis was determined to be 1.1%, while the response was linear over the range 0.02–0.17 mg of titanium oxide. The column, stainless steel, was packed with 15% Histowax on 60/80 mesh Gas Pak F heated at 77 °C.

Uranium, as the hexafluoride, may be chromatographically determined by employment of suitably modified instrumentation. Hamlin *et al.*[95] employed nickel-plated surfaces in the chromatographic train and a katharometer constructed with nickel filaments. As packing, Kel-F 300 low-density moulding powder coated with Kel-F 40 oil was used.

In the procedure of Becker *et al.*,[96] tin in zirconium–tin alloys was converted to the chloride via the action of chlorine. A detection limit of 4 μg and quantitative recoveries were noted, the values in the case of standard samples being in good agreement with data from X-ray fluorescence and polarographic determinations.

A recent analytical report by Russian workers, concerning GLC determination of selenium and tellurium, involved the reaction of xenon difluoride with the oxides of these elements to furnish the fluorides.[97] Teflon® columns packed with 20% Kel-F 10 on Polychrom I and operated at 78 °C were used in conjunction with a katharometer.

3.2 Miscellaneous non-chelation methods

3.2.1 DIRECT ANALYSES

Inorganic mercury determination by GLC has been under active investigation in several laboratories. In all cases, the inorganic form is reacted with various organic and organometallic reagents to form an organomercurial which is sufficiently stable and volatile to be evaluated by GLC. Jones and Nickless[98] reported a procedure for inorganic mercury in various media entailing the action of sodium benzenesulfinate (Peters reaction) on the element in suitably treated samples. The phenylmercury formed was extracted into toluene as the chloride, and an aliquot of the toluene layer was injected into a GLC equipped with an EC detector and a column packed with 5% ethyleneglycol adipate polyester on Supasorb at 185 °C. The detection limit was 2×10^{-10} g of the chloride, while amounts of mercury in the range 0.05–50 ppm could be determined. In a related study, the above authors have explored the use of the organosilicon salt, sodium 2,2′-dimethyl-2-silapentane-5-sulfonate for conversion of inorganic mercury to the chromatographic methyl derivative.[99] This reagent is quite stable in strong acids and high temperature, permitting quantitative recovery data over the range 0.003–10 ppm of mercury. Satisfactory analyses for mercury were achieved with wet-ashed fish and sediment samples.

A variety of organometallic reagents were successfully employed in our laboratory[100] for the determination of inorganic mercury in various media. Best results were achieved with methyl(pentacyano)cobalt(III) and tetraphenylboron(III), generating respectively methyl and phenyl mercury. These studies entailed the use of a stationary phase chemically bonded to the support which is commercially available (Durapak Carbowax 400, low K'), and this permits mercurial analysis for protracted periods of time without the necessity of periodic column treatments. Detection limits of mercury, as the generated organomercurial, were in the range of 10–50 ppb from e.g. urine, blood and tissue homogenates.

The propensity of selenium to form volatile piazselenol derivatives have prompted a number of reports describing the GLC determination of this element. In the method of Young and Christian,[101] 2,3-diaminonaphthalene reacts with selenium at pH 2.0 to furnish the hexane-extractable piazselenol. As little as 5×10^{-10} g of selenium could be detected using an EC detector, while calibration data demonstrated linearity over the range 0.1–1.0 μg. The procedure was applied to human blood and urine as well as to river water and gave levels which agree well with neutron activation analysis. Nakashima and Toei,[102] similarly, reacted selenium, as selenious acid, with 4-chloro-*o*-phenylendiamine at pH 0–1 and extracted the derivative into toluene. Linearity extended up to 0.80 μg, while the limit of detection was 0.04 μg. As a means of assessing the selenium content of tellurium metal, Shimoishi first reacted the sample with *aqua regia* followed by 4-nitro-*o*-phenylenediamine and, finally, by extraction into toluene. Using small amounts of sample (several milligrams), as little as 10 ng of selenium could be determined. Common ions even in large excess did not appear to interfere. In a related report,[104] Shimoishi studied sea water, without prior concentrating of sample volumes, and cited a detection limit of 2 ng ml^{-1} organic extractant.

Ruthenium was determined as the thiosemicarbazide after a prior TLC separation, and extraction from the TLC plate with ethanol-pyridine.[105] A reproducible decomposition of the complex in the injector port of a gas chromatograph permitted quantitation of the element over the range 0.2–2.0 ng.

Schwedt and Rüssel[106] described a chromatographic method for inorganic arsenic, in which arsenic in biological materials was first extracted into chloroform as the dithiocarbamate and then caused to react with a phenyl-Grignard compound. The resulting triphenylarsenic(III) was then measured using an instrument equipped with a FID. Concentrations of arsenic as low as 2.0 $\mu g\ g^{-1}$ medium were determined.

3.2.2 INDIRECT ANALYSES

Aluminium as well as aluminium carbide were determined indirectly by treatment of the metal with hydrochloric acid and measuring the generated hydrogen or methane by GLC.[107] Similarly, metallic aluminium in thin films could be assessed via chromatographic measurement of the hydrogen generated on treatment with strong base. In this case 60–300 μg of aluminium could be measured using thermal conductivity detection.

Aqueous solutions of palladium were determined indirectly via treatment of the metal in perchloric acid with a mixture of propene/propane (1 : 1) and GLC analysis of the residual hydrocarbon, propene entering into π-complex formation with palladium(II).[108]

The hydrogen liberated from decomposition of uranium and thorium hydrides at elevated temperatures was quantitatively determined to furnish an indirect measurement of these elements.[109]

Very small amounts of magnesium metal were determined by measuring chromatographically the amount of hydrogen liberated on treatment with sulfuric acid.[110] Obviously, this approach is of value with any metallic species capable of liberating hydrogen from acid solution.

4 COMPARATIVE DISCUSSION OF METAL ION ANALYSES BY GAS–LIQUID CHROMATOGRAPHY

The foregoing discussion clearly demonstrates the increasing value of GLC in the quantitative determination of a variety of metallic ions. While early work dealt primarily with a limited number of ions—those which formed highly stable, volatile and easily characterized derivatives—progress in our understanding of coordination chemistry has permitted an expansion in the number of metal ions which may be measured in this way. GLC of metal ions at this stage of study involves, and will continue to involve, the chelate derivatives, particularly the β-diketonates. No doubt, some use will continue to be made, in special cases, of halide and other derivatives of metals where laboratories have the requisite instrumentation and expertise to cope with the special problems which arise. Any rigorous critique of a general method of analysis must take into account parallel developments in competitive procedures; in the case of metal ion analysis by GLC, one must also consider the considerable developments simultaneously occurring in atomic absorption spectrometry and neutron activation analysis. Furthermore, GLC of metal ions must be viewed in the light of new demands which may be placed upon any technique for ion analysis, a particular example of this being multi-element analysis in the shortest possible time using limited amounts of sample. In this regard, GLC analysis for certain ions can certainly compete with atomic absorption spectrometry and other micro techniques, even with the advent of microsampling accessories such as the carbon-rod and carbon-cup furnace methods. In the case of beryllium, the chromatographic approach appears to be superior in terms of sensitivity and sample requirements to any competitive procedure.

A unique use of GLC in metal ion analysis, using mass spectrometry as a specific detector, is the assessment of isotope distribution ratios for a given chromatographible species. This was found to be of particular value in extraterrestrial sample analysis, few other methods are intrinsically capable of furnishing this type of data.

As to the question of simultaneous multi-element analysis of a given sample, few procedures admittedly exist for permitting this via GLC. However, the present existence of various groups of chelants which may be successfully used with different classes of metal ions should prompt continuing study of GLC as a technique for multi-element analysis.

REFERENCES

1. A. Gero, *J. Org. Chem.* **19**, 469 (1954).
2. J. L. Burdett and M. T. Rogers, *J. Am. Chem. Soc.* **86**, 2105 (1964) and references cited therein.
3. C. R. Hauser, F. W. Swamer and J. T. Adams in *Organic Reactions* (R. Adams *et al.* Eds) Vol. VIII, Wiley, New York, 1954, p. 59.
4. J. C. Reid and M. Calvin, *J. Am. Chem. Soc.* **72**, 2948 (1950).
5. R. L. Belford, A. E. Martell and M. Calvin, *J. Inorg. Nucl. Chem.* **2**, 11 (1956).
6. C. S. Springer Jr, D. W. Meek and R. E. Sievers, *Inorg. Chem.* **6**, 1105 (1967).
7. W. G. Scribner, *Annual Report, Contract No. AF-33(615)-1093*, September, 1967.
8. B. P. Pullen, M.S. Dissertation, University of Tennessee, 1967.
9. R. W. Moshier and R. E. Sievers, *The Gas Chromatography of Metal Chelates*, Pergamon, New York, 1965.
10. G. K. Schweitzer, B. P. Pullen and Y. H. Fang, *Anal. Chim. Acta* **43**, 332 (1968).
11. H. I. Schlesinger, H. C. Brown, J. J. Katz, S. Archer and R. A. Lad, *J. Am. Chem. Soc.* **75**, 2446 (1953).
12. J. D. Park, H. A. Brown and J. D. Lacher, *J. Am. Chem. Soc.* **75**, 4753 (1953).
13. C. E. Inman, R. E. Oesterling and E. A. Tyezkowski, *J. Am. Chem. Soc.* **80**, 6533 (1958).
14. E. Uhlemann and P. H. Thomas, *J. Prakt. Chem.* **34**, 180 (1966).

15. R. K. Y. Ho, S. E. Livingstone and T. N. Lockyer, *Aust. J. Chem.* **19**, 1179 (1966).
16. R. K. Y. Ho, S. E. Livingstone and T. N. Lockyer, *Aust. J. Chem.* **21**, 103 (1968).
17. R. K. Y. Ho and S. E. Livingstone, *Aust. J. Chem.* **21**, 1781 (1968).
18. E. Bayer, H. P. Miller and R. E. Sievers, *Anal. Chem.* **43**, 2012 (1971).
19. C. S. Saba and T. R. Sweet, *Anal. Chim. Acta* **69**, 478 (1974).
20. A. Combes and C. Combes, *Bull. Soc. Chim. Fr.* **7**, 778 (1892).
21. H. F. Holtzclaw Jr, J. P. Collman and R. M. Alice, *J. Am. Chem. Soc.* **80**, 1100 (1958).
22. S. Portnoy, *J. Org. Chem.* **30**, 3377 (1965).
23. J. Beretka, B. O. West and M. T. O'Connor, *Aust. J. Chem.* **17**, 192 (1964).
24. A. E. Martell, R. L. Belford and M. Calvin, *J. Inorg. Nucl. Chem.* **5**, 170 (1958).
25. P. C. Uden and K. Blessel, *Inorg. Chem.* **12**, 352 (1973).
26. E. W. Berg and K. P. Reed, *Anal. Chim. Acta* **36**, 372 (1966).
27. M. Calvin and K. W. Wilson, *J. Am. Chem. Soc.* **67**, 2003 (1945).
28. B. G. Schultz and E. M. Larsen, *J. Am. Chem. Soc.* **71**, 3250 (1949).
29. K. Sato, Y. Kodama and K. Arakawa, *Nippon Kagaku Zasshi* **87**, 821 (1966).
30. R. Belcher, W. I. Stephen, I. J. Thomson and P. C. Uden, *J. Inorg. Nucl. Chem.* **33**, 1851 (1971).
31. R. Belcher, W. I. Stephen, I. J. Thomson and P. C. Uden, *J. Inorg. Nucl. Chem.* **34**, 1017 (1972).
32. G. O. Dudek and E. P. Dudek, *J. Am. Chem. Soc.* **86**, 4283 (1964).
33. G. O. Dudek and G. P. Volpp, *J. Am. Chem. Soc.* **85**, 2697 (1963).
34. I. Schupan, K. Ballschmiter and G. Tölg, *Z. Anal. Chem.* **255**, 116 (1971).
35. T. J. Cardwell and P. S. McDonough, *Inorg. Nucl. Chem. Lett.* **10**, 283 (1974).
36. M. Miyazaki, T. Imanori, T. Kumugi and Z. Tamura, *Chem. Pharm. Bull.* **14**, 117 (1966).
37. P. Mushak, M. T. Glenn and J. Savory in *Fluorine Chemical Review* (P. Tarrant, Ed.), Vol. VI, Dekker, New York, 1973, Ch. 2.
38. E. R. Melby, N. J. Rose, E. Abrumson and J. C. Caris, *J. Am. Chem. Soc.* **86**, 5117 (1964).
39. M. L. Morris, R. W. Moshier and R. E. Sievers, *Inorg. Syn.* **9**, 50 (1967).
40. R. C. Fay and T. S. Piper, *J. Am. Chem. Soc.* **85**, 500 (1963).
41. M. L. Morris, R. W. Moshier and R. E. Sievers, *Inorg. Chem.* **2**, 411 (1963).
42. E. W. Berg and J. T. Truemper, *J. Phys. Chem.* **64**, 487 (1960).
43. R. A. Staniforth, Doctoral Dissertation, Ohio State University, 1943.
44. R. D. Hill, Thesis, University of Manitoba (1962).
45. R. H. Holm and F. A. Cotton, *J. Inorg. Nucl. Chem.* **15**, 63 (1960).
46. H. Veening, W. E. Bachman and D. M. Wilkinson, *J. Gas Chromatogr.* **5**, 248 (1967).
47. J. A. Bertrand and R. J. Kaplan, *Inorg. Chem.* **5**, 489 (1966).
48. D. Purushothan, V. R. Rao and Bh. S. V. Rao, *Anal. Chim. Acta* **33**, 182 (1965).
49. R. G. Charles and E. P. Riedel, *J. Inorg. Nucl. Chem.* **29**, 715 (1967).
50. D. A. Buckingham, R. C. Gorges and J. T. Henry, *Aust. J. Chem.* **20**, 281 (1967).
51. E. M. Larsen, G. Terry and J. Leddy, *J. Am. Chem. Soc.* **75**, 5107 (1953).
52. R. S. Barrett, R. Belcher, W. I. Stephen and P. C. Uden, *Proc. Soc. Anal. Chem.* **10**, 167 (1973).
53. R. Belcher, K. Blessel, T. Cardwell, M. Pravica, W. I. Stephen and P. C. Uden, *J. Inorg. Nucl. Chem.* **35**, 1127 (1971).
54. F. A. Cotton and G. Wilkinson, *Advanced Inorganic Chemistry: A Comprehensive Text*, 2nd Edn, Wiley, New York, 1966.
55. J. P. Fackler Jr, *Prog. Inorg. Chem.* **7**, 361 (1966).
56. J. P. Collman, *Angew. Chem. Intern. Ed.* **4**, 132 (1965).
57. R. Belcher, R. J. Martin, W. I. Stephen, D. E. Henderson, A. Kamalizad and P. C. Uden, *Anal. Chem.* **45**, 1197 (1973).
58. D. P. Graddon, *Coord. Chem. Rev.* **4**, 1 (1969).
59. C. A. Burgett and J. S. Fritz, *J. Chromatogr.* **77**, 265 (1973).
60. J. A. Rodriguez-Vazquez, *Anal. Chim. Acta*, **73**, 1 (1974).

61. W. R. Wolf, M. L. Taylor, B. M. Hughes, T. O. Tiernan and R. E. Sievers, *Anal. Chem.* **44**, 616 (1972).
62. R. S. Juvet Jr and S. P. Cram, *Anal. Chem.* **42**, 14R (1970).
63. H. Kawaguchi, T. Sakomoto and A. Mizuike, *Talanta*, **20**, 321 (1973).
64. R. M. Dagnall, T. S. West and P. Whitehead, *Analyst* **98**, 647 (1973).
65. G. P. Morie and T. R. Sweet, *Anal. Chem.* **37**, 1552 (1965).
66. G. P. Morie and T. R. Sweet, *Anal. Chim. Acta*, **34**, 314 (1966).
67. W. G. Scribner, W. J. Treat, J. D. Weis and R. W. Moshier, *Anal. Chem.* **37**, 1136 (1965).
68. R. W. Moshier and J. E. Schwarberg, *Talanta* **13**, 445 (1966).
69. M. Miyazaki and H. Kaneko, *Chem. Pharm. Bull.* **18**, 1933 (1970).
70. C. Genty, H. Horin, P. Malherbe and R. Schott, *Anal. Chem.* **43**, 235 (1971).
71. M. L. Lee and D. C. Burrell, *Anal. Chim. Acta* **66**, 245 (1973).
72. R. A. Minear and C. M. Palesh, 165th. ACS National Meeting, Dallas, Texas, April, 1973.
73. M. S. Black and R. E. Sievers, *Anal. Chem.* **45**, 1773 (1973) and earlier work cited therein.
74. M. L. Taylor and E. L. Arnold, *Anal. Chem.* **43**, 1328 (1971).
75. G. Kaiser, E. Grallath, P. Tschöpel and G. Tölg, *Z. Anal. Chem.* **259**, 257 (1972).
76. W. D. Ross and R. E. Sievers, *Environ. Sci. Technol.* **6**, 155 (1972).
77. J. K. Foreman, T. A. Gough and E. A. Walker, *Analyst* **95**, 797 (1970).
78. K. J. Eisentraut, D. G. Johnson, M. F. Richardson and R. E. Sievers, 161st National ACS Meeting, Los Angeles, Calif., March–April, 1971.
79. W. D. Ross and R. E. Sievers, *Anal. Chem.* **41**, 1109 (1969).
80. L. C. Hansen, W. G. Scribner, T. W. Gilbert and R. E. Sievers, *Anal. Chem.* **43**, 349 (1971).
81. J. Savory, P. Mushak, F. W. Sunderman Jr, R. H. Estes and N. O. Roszel, *Anal. Chem.* **42**, 294 (1970).
82. G. H. Booth Jr and W. J. Darby, *Anal. Chem.* **43**, 831 (1971).
83. W. R. Wolf and R. E. Sievers, 161st National ACS Meeting, Los Angeles, Calif., March–April, 1971.
84. R. Belcher, A. Khalique and W. I. Stephen, unpublished data cited in Ref. 60.
85. P. C. Uden, D. E. Henderson and C. A. Burgett, *Anal. Lett.* **7**, 807 (1974).
86. W. D. Ross, W. G. Scribner and R. E. Sievers, in *Gas Chromatorgraphy* (N. Sock, Ed.), Institute of Petroleum, London, 1970.
87. R. E. Sievers, J. W. Connolly and W. D. Ross, *J. Gas Chromatogr.* **5**, 241 (1967).
88. R. F. Sieck, J. J. Richard, K. Iverson and C. V. Banks, *Anal. Chem.* **43**, 913 (1971).
89. C. A. Burgett and J. S. Fritz, *Talanta* **20**, 363 (1973).
90. C. S. G. Phillips and P. L. Timms, *Anal. Chem.* **35**, 505 (1963).
91. R. S. Juvet Jr and R. L. Fisher, *Anal. Chem.* **38**, 1860 (1966).
92. S. T. Sie, J. P. A. Bleumer and G. W. A. Rijnders, *Sep. Sci.* **3**, 165 (1968).
93. M. L. Sazonov, J. E. Almova, M. S. Selinkina and A. A. Zhuhovitskii, *Khim. Tverd. Topl.* **3**, 64 (1968).
94. R. E. Sievers, G. Wheeler Jr and W. D. Ross, *Anal. Chem.* **38**, 306 (1966).
95. A. G. Hamlin, G. Iverson and T. R. Phillips, *Anal. Chem.* **35**, 2037 (1963).
96. J. H. Becker, J. Chevallier and J. Spitz, *Z. Anal. Chem.* **247**, 301 (1967).
97. N. N. Aleinikov, D. N. Sokalov, L. K. Golubena, B. L. Korsunskii and F. I. Dukovitskii, *Izv. Akad. Nauk SSSR, Ser. Khim.* 2614 (1973).
98. P. Jones and G. Nickless, *J. Chromatogr.* **76**, 285 (1973).
99. P. Jones and G. Nickless, *Proc. Soc. Anal. Chem.* **10**, 270 (1973).
100. P. Zarnegar and P. Mushak, *Anal. Chim. Acta* **69**, 389 (1974).
101. J. W. Young and G. D. Christian, *Anal. Chim. Acta* **65**, 127 (1973).
102. S. Nakashima and K. Toei, *Talanta* **15**, 1475 (1968).
103. Y. Shimoishi, *Bull. Chem. Soc. Jpn* **44**, 3370 (1971).
104. Y. Shimoishi, *Anal. Chim. Acta* **64**, 465 (1973).
105. K. Ballschmiter, *J. Chromatogr. Sci.* **8**, 491 (1970).
106. G. Schwedt and H. A. Rüssel, *Chromatographia* **5**, 242 (1972).

107. A. B. Nersesyants, E. L. Zaklarov and Z. A. Bystrova, *Zavod. Lab.* **36**, 1043 (1970).
108. T. Koga and T. Hara, *Bull. Chem. Soc. Jpn* **39**, 1353 (1966).
109. E. A. Schaefer and J. O. Hibbitts, *Nucl. Sci. Abstr.* **20**, 2280 (1966).
110. V. D. Hogan and F. R. Taylor, *Anal. Chem.* **40**, 1387 (1968).

CHAPTER 13

Derivatives for Chromatographic Resolution of Optically Active Compounds

B. Halpern

Department of Chemistry, University of Wollongong, N.S.W., Australia

1 INTRODUCTON

Extensive research has been carried out into the application of chromatographic techniques to the resolution of asymmetric compounds, and several reviews have been published.[1,2] Two general methods are available. The first uses acid-base reactions, condensations or complex-forming reactions with chiral reagents to form diastereoisomeric mixtures which are separated by chromatography. This is a simple extension of the classical resolution procedure in which adsorption or partition differences are exploited to separate the diastereoisomers. The second technique uses stereoselective sorption of enantiomers on a chiral stationary phase and is a more difficult experimental process. Both methods have been used with all of the common modes of chromatographic separation, but, at the time of writing, publications on the uses of column adsorption chromatography, thin-layer chromatography, liquid and ion-exchange chromatography and countercurrent distribution have been quite limited. The application of gas–liquid chromatography to the separation of volatile optical isomers has, however, been extensive.[2–4] For this reason this chapter has been restricted to gas–liquid chromatographic separations, but for the sake of completeness a bibliography of non-gas–liquid chromatographic methods has been included.

The aim of the chapter is to provide, in an easily usable form, answers to practical questions such as: how do I resolve a racemate by gas chromatography? What reagent do I use? How do I make it? What chromatographic conditions do I use to separate the enantiomers or diastereoisomers?

1.1 Explanation of terms used in this chapter

The term 'resolution' is used to describe the separation of configurational isomers. 'Resolution factor' ($r_{II/I}$) stands for the ratio of retention volumes of configurational isomers. The R, S nomenclature is used throughout and refers to the absolute configuration of asymmetric compounds, and the *d*, *l* nomenclature is used to refer to the sign of rotation of chiral compounds.

1.2 Selection of a resolving agent for the analysis of an enantiomeric mixture

From the following procedures it may be seen that the formation of volatile diastereoisomers from an enantiomeric mixture can offer a general method for separating the mixture by gas chromatography. To be of value in this application, the diastereoisomer-forming reaction must be essentially quantitative with either conservation or complete inversion of the optical centre purity. The derivatives must be thermally and sterically stable during the chromatographic separation process. For preparative applications a high yield reconversion of the diastereoisomers to the optically pure enantiomers is an additional prerequisite. Many factors influence the degree of separation of diastereoisomers by GC but the nature of the resolving agent is very important and certainly far more significant than the stationary liquid phase.[4] The following factors should be considered; they affect the usefulness of the derivatizing reagent and the degree of resolution of the resultant diastereoisomers.

1. Chromatography on an optically inactive phase can not differentiate between enantiomers (*ld* can not be separated from *dl* and *dd* can not be separated from *ll*), hence the GC resolution of diastereoisomers will not lead to sterically pure enantiomers unless an optically pure resolving agent is used.

2. Conformational immobility of the groups attached to the asymmetric centre of the resolving agent enhances separation[68] and, in particular, resolution might be greatly increased if the asymmetric centre is part of a ring system.[69]

3. A large size differential in the groups attached to the asymmetric carbon atom of the reagent can give a more effective reagent.[68]

4. The distance between the asymmetric carbon atoms of the reagent and the compound to be resolved should be minimized and if possible kept to less than 3 atoms.[70,71]

5. The nature of the central polar group between the asymmetric carbon atom of the reagent and the compound to be resolved may have an influence on the resolution factor ($r_{II/I}$). It has been shown that amides (—CONH—) give better results than esters (—CO_2—) and that a central carbonyl group is not essential for separation.[71]

6. The presence of an additional polar or polarizable group close to the functional group is desirable as it facilitates hydrogen bonding or π-orbital overlap between the diastereoisomer formed and the stationary phase. Resolving agents containing aromatic rings or acyl groups are likely to fulfil these requirements for multiple interactions. A chiral compound which is devoid of such groups is unlikely to be a good resolving agent.

7. Finally the reagent should be easy to prepare or should be commercially available in a state of high optical purity. It should be stable under the conditions of use and of storage (usually at low temperature). For preparative applications, a low cost or any easy recovery of the compound for re-use is also important.

2 APPLICATIONS

Gas chromatographic separation of enantiomers or diastereoisomers lends itself to both analytical and preparative applications; so far no practical method for the preparative resolution of chiral compounds has yet been published. Among the most important analytical applications are the following.

2.1 The determination of optical purity

The gas chromatographic separation of enantiomers and diastereoisomers provides the basis for a convenient method for the determination of optical purity. The advantages of this analysis over conventional polarimetric measurement are that optical and chemical impurities are separable during chromatography and that a steric purity measurement can be done on a mixture at the microgram level. Of the two available methods, resolution of enantiomers on optically active stationary phases is to be preferred, as in this technique derivatization of the sample is done with a symmetrical reagent and alteration of the real enantiomer ratio during preparation of the sample is less likely. Some analytical work using this method has been published[84–86,133] In practice it is difficult to implement such an analysis. Differential interaction of the solutes with the stationary phase are very small and the use of a high efficiency capillary or open tubular column is essential. Unfortunately the stationary phases presently available have a limited thermal and optical stability[77] and the gas chromatography at prudent temperatures leads to very long analysis times. Developments in column technology will probably change this picture. Recent work on long hydrocarbon chain optically active phases[133] has secured operating temperatures as high as 190 °C. Most of the examples of GC resolution cited in the literature have made use of the separation of diastereoisomers.[1–3] This approach involves the conversion of enantiomers (A_+ and A_-) into a mixture of diastereoisomers (A_+B_+ and A_-B_+) by reaction with a chiral reagent B_+. The reagent must be optically pure or the amount of antipodic impurity in B_+ must be known. This can normally be determined by running a reagent blank with an optically pure standard. Kinetic resolution during derivatization must be avoided and stability of the diastereoisomers during chromatography must be established. In most cases these prerequisites can be checked by running a control experiment with a racemic mixture of A. Under conditions which permit baseline separation of the two diastereoisomeric peaks and with electronic integration of the GC ouput known enantiomeric compositions have been determined experimentally with an absolute error of less than 0.6%.[11] The technique is compatible with radio chromatography and it has been applied to optical purity measurements of ^{14}C labelled amino acids.[87] Other useful applications are determination of the correct value of the specific rotation of resolved chiral compounds[88–90,90a] and monitoring the progress of a classical resolution process.[16,91] A modification of this technique suitable for the determination of nanogram amounts has been described.[92] It involves the use of a mixture of deuteriated *l*-reagent and unlabelled *d*-reagent to form a diastereoisomeric mixture. A gas chromatograph is used to separate the diastereoisomers and a mass spectrometer is used to monitor the steric purity of amino acids by selected ion monitoring (mass fragmentography).

2.2 The determination of configuration of asymmetric compounds by gas chromatography

It was first pointed out by Gil-Av that the order of emergence of diastereoisomeric 2-*n*-alkyl-α-alkanoyloxypropionates from a gas chromatographic column could serve as a basis for the assignment of configuration for chiral 2-*n*-alkanols.[93] It has been shown since that similar useful empirical relationships between elution order and configuration can be formulated for the GC behaviour of other closely related compounds, such as α-amino acids.[14,94] The use of GC retention correlations for the assignment of configuration depends on the consistency of the order of elution of diastereoisomers within a homologous series. In general this has been shown to hold, with the possible exception of the first member of a homologous series, the case where a structural change has altered the relative size of substituents at the asymmetric carbon atom, or where a ring structure is formed. The advantages of the method over optical rotary dispersion, circular dichroism and nuclear magnetic resonance methods are increased sensitivity of the GC detector, the simplicity of interpretation of the chromatograms and its applicability to the analysis of mixtures. Since the assignment of configuration depends on the order of elution only, resolution factors can be minimal and the optical purity of the resolving agent is not so important. The procedure can be adapted to the analysis of natural products where usually only one enantiomer is available. In these cases the compound (*l*) is coupled with racemic (*d***l**) and optically pure reagent (*l**) prior to analysis. An assignment is made which is based on the elution order of the (*d***l*) and (*l***l*) diastereoisomers of the unknown. This is possible because the gas chromatograph does not distinguish (*d***l*) from *l***d*) and (*d***d*) from *l***l*). The technique has been used, to esablish the configuration of a new unsaturated α-amino acid isolated from a New Guinea mushroom;[95] to confirm the configuration of amino acids in the peptide antibiotics gramicidin and vernamycin;[96] to demonstrate the presence of R-β-amino isobutyric

TABLE 1

The order of gas chromatographic elution of diastereoisomers derived from homologous or closely related compounds

Diastereoisomer	First GC peak	Second GC peak	Known exceptions
N-TFA-dipeptide esters	S*S	S*R	TFA-alanyl-alanine methyl ester
			TFA-prolyl-amino acid methyl ester
N-TFA-prolylamino acid esters	S*R	S*S	Lysine and ornithine derivatives
N-TFA-prolyl-1-methylalkyl amides	S*R	S*S	Cyclic amines (SS<SR)
N-TFA-prolyl-1-amino-1-phenylethanes	S*R	S*S	
N-TFA-amino acid-2-alkyl esters	RS*	SS*	TFA-phenyl glycine-2-octyl ester
α-Chloralkanoyl-amino acid esters	S*R	S*S	
2,5-Diketopiperazines	S*R	S*S	Cyclo-alanyl-alanine
			Proline containing cyclo peptides
α-Acetoxyalkanoic acid-2-alkyl esters	RS*	SS*	
α-Hydroxyalkanoic acid-2-alkyl esters	RS*	SS*	
2-Phenylpropionic acid-2-alkyl esters	S*S	SS*	3-Hydroxy fatty acids
Menthyl carbonates	R*S	R*R	
Drimanoic acid alkyl esters	R*S	R*R	
Chrysanthemic acid alkyl esters	R*S	R*R	
3β-Acetoxy-Δ^5-etienates	S*S	S*R	
Alkyl-*N*-(1-phenylethyl) carbamates	S*S	RR*	Aryl/alkyl carbinols
α-Alkylphenylacetyl-2-methylamino-1-phenylpropane	SS*	RS*	

*S or R refers to the absolute configuration of the resolving agent.

acid in urine[97] and to identify S-alloisoleucine in the blood of patients suffering from maple syrup urine disease.[98] The method has been useful for stereochemical characterization of small amounts of isoprenoid acids[39,45] and α-alkylphenylacetic acids.[40,43,44] It has also been applied to chiral compounds containing more than one asymmetric centre such as aldonic acids,[40,52] steroidal ketones[38] and amines.[99] Such diastereoisomeric derivatives may display more than two chromatographic peaks, but regular relationships between retention volumes and the stereochemistry at the carbon atom closest to the functional group can be developed. In a combination of resolution procedures, the gas chromatography of diastereoisomeric α-phenylbutyric acid amides has been used in conjunction with Horeau's kinetic resolution method,[100,101] to determine the absolute configuration of various steroidal and terpene alcohols.[102]

A consistent correlation of the order of emergence with configuration has also been found for enantiomeric amines and amino acids, separated on optically active phases.[103,104]

2.3 Biochemical and geochemical applications

One of the most fundamental considerations of living organisms is the optical purity of the monomers which make up biopolymers (proteins, nucleic acids), and as a consequence metabolic processes usually exhibit some steric discrimination. Gas chromatography of diastereoisomeric derivatives is ideally suited to the study of such processes, as the steric purity of microgram quantities of a substrate can be determined without isolation or rigorous purification. Biochemical applications of the GC technique include stereoselective hydrolysis of diastereoisomeric 2-chlorpropionyl amino acid,[105] leucyl amide[106] and phenylalanine alkyl ester[107] substrates by proteolytic enzymes; steric analysis of 2-hydroxy acids derived from brain cerebrosides[54] and steric analysis of hydroperoxides formed by lipoxygenase-catalysed conversion of linoleic acid.[108] Other studies cover the mechanism of microsomal $\omega 2$-hydroxylation of fatty acids[56–58] and determination of the chirality of samples of 13-hydroxy-9,11-octadecadienoic acid derived from human arterial lipids.[23] The technique has also been used to demonstrate stereospecific action of microorganisms in soils.[109] The method has also been of value in pharmaceutical applications[110,111] since it is known that the antipode of some drugs have different pharmacological effects. Geochemical applications have been stimulated by the proposition that sterically-specified information in a macromolecule is strong evidence for the presence of life on a planet.[112–114] Since the direct determination of optical activity requires at least 10 μg of pure material, gas chromatography has become the analytical method of choice for life detection experiments. In geochemical studies, investigations have centred around the lunar samples, meteorites and ancient sediments. Analysis of lunar samples from the Apollo 11 and 12 missions showed no evidence of any biomolecules indigenous to the lunar surface. The analysis of meteorites has provided evidence for the presence of extraterrestrial carbon compounds by the presence of racemic protein and nonprotein amino acids in the Murchison[115] and Murray[116] meteorites. The conversion of S-amino acids of biological origin to an equilibrium mixture of R and S amino acids has been used as a reliable indicator of age for ancient sediments less than 400 000 years old.[117,118] Gas chromatography of the methyl and menthyl esters of individual isoprenoid fatty acids isolated from a geological source of the Eocene period showed that the diastereoisomer composition was compatible with a chlorophyll origin.[119]

2.4 Racemization during peptide synthesis

The separation of diastereoisomeric dipeptides by gas chromatography greatly facilitates the study of racemization during peptide synthesis.[60] The carbobenzoxy function (Cbz) which is commonly used for amino group protection in peptide synthesis is not suitable for direct GC analysis and in order to monitor the degree of racemization, it is necessary to replace the Cbz group by a trifluoroacetyl group. The analysis of peptides containing more than two amino acid residues requires a preliminary degradation with methanolic hydrogen chloride. The GC analysis of dipeptides as their *N*-trifluoroacetyl methyl ester derivatives has been used to evaluate 16 peptide coupling methods.[120] An alternative approach using GC resolution of 2-chlorpropionyl amino acid methyl esters has been used to study the effect of the penultimate amino acid on the degree of racemization of phenylalanine, when acylated phenylalanyl dipeptides are coupled with glycine.[121] In this case the recovered phenylalanine has to be converted to a volatile diastereoisomeric derivative by acylation and esterification.[121] Racemization during amide bond formation can also be studied by the GC of diastereoisomeric *N*-*t*-butyloxycarbonyl(BOC)-amino acid amides.[122] The advantage of these model compounds for this purpose is that the reaction mixture can be analysed directly, as the protecting group need not be changed prior to analysis. The GC of BOC-dipeptide methyl esters, prepared by the Merrifield method, has confirmed the absence of racemization during solid phase peptide synthesis.[123]

3 THE MECHANISM OF THE RESOLUTION PROCESS

Mechanistic studies with model diastereoisomers show that chromatographic behaviour is a function of the differential interactions between the stationary liquid phase and the conformations of the solvated diastereoisomers. With polar compounds such as alcohols, amides and esters, intramolecular and intermolecular interactions such as hydrogen bonding, dipoles and polarization are the predominant factors which determine orientation and conformation.[2,4,28a] For a pair of polar diastereoisomers such as *cis*- and *trans*-1,2-cyclopentandiol, the isomer with the higher degree of intramolecular hydrogen bonding (*cis*) has the lower retention volume on a polar phase.[124–126] The steric environment of a functional group also influences the degree to which it can interact with a stationary phase.[126] Karger has used the concept of differential accessibility to a key polar group to account for the chromatographic behaviour of diastereoisomers.[50,127]

In a series of model diastereoisomers such as the α-acetoxypropionates of secondary alcohols, the correct order of elution can be predicted on the basis of greater interaction of the ester function of the SS diastereoisomer with the stationary phase. An increase in the size differential of the groups attached to the asymmetric carbon centre will increase the resolution factor. The improved resolution obtainable with α-bromoalkanoic acid compared with α-chloroalkanoic acid[128] diastereoisomers and 3,3-dimethylbutyl ester compared with 2-butyl ester[129] diastereoisomers agree with such expectations. Karger *et al.* have also shown that the incorporation of one or both of the asymmetric carbons into a ring system, such as in *N*-TFA-*S*-prolyl-2-methylindoline, can result in very large resolution factors ($r_{II/I}$ 1.35).[69] Systematic structural variation studies with diastereoisomeric model esters show that the distance between the asymmetric carbon atoms is an important factor in separation and that the distance between the asymmetric carbons should not be more than three atoms.[70] Studies with *meso* and racemic polar and apolar diastereoisomers show that chromatographic behaviour can be explained on the basis of molecular shape.[70] Surprisingly, the introduction of a methylene group between the two asymmetric centres of these compounds leads to better resolution.[70] Finally Feibush has shown that the order of emergence, and the resolution factors of a series of model esters can be predicted from the relative 'size' or 'bulk' of substituents at the asymmetric carbons.[130] Similar studies with optically active phases suggest that separation of enantiomers is the result of transient cyclic hydrogen-bonded association complexes between solute and solvent.[103,131,132] The presence of a bulky substituent at the asymmetric centre and in close proximity to a functional group can give excellent resolution of antipodes.[76,82,83] The topic of resolution on optically-active stationary phases has been reviewed[135] recently and this includes a full discussion of the factors affecting separation.

4 RESOLUTION OF OPTICAL ISOMERS BY GAS CHROMATOGRAPHY OF DIASTEREOISOMERS

4.1 Amino acids

α-Amino acids can be converted into a mixture of diastereoisomers by derivatizing with an asymmetric agent at either of the two functional groups linked to the α-carbon atom. For successful gas chromatography the remaining functional group and the side chain of polyfunctional amino acids must also be derivatized.

4.1.1 REACTION OF THE α-AMINO GROUP WITH AN ASYMMETRIC REAGENT

N-TFA-prolylchloride (1),[5] α-chloroisovaleryl chloride (2),[6] menthylchloroformate (3)[7] and teresantalinyl chloride (4)[8] have been employed as acylating agents for this purpose but only (1) has been used for a sufficiently large number of amino acids to be of general value.

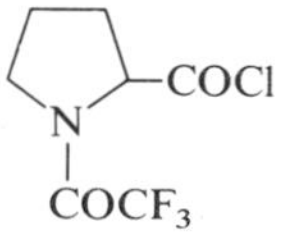

(1)

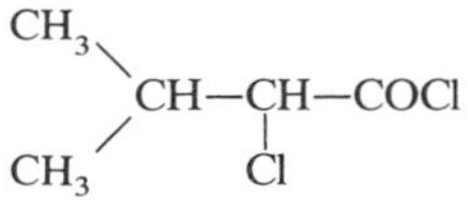

(2)

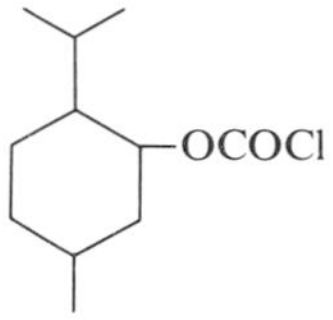

(3)

(4)

In all cases the carboxylic acid function of the amino acid has to be protected, usually by esterification. To resolve the common polyfunctional amino acids it is necessary to derivatize the hydroxyl groups of serine, threonine[9] and tyrosine[10] and the indole function of tryptophan[10] to form the *O*- or *N*-trimethylsilyl compounds. The thiol function of cysteine may be incorporated into a thiazolidine ring.[9] Racemization has not been observed and is not expected in this type of coupling reaction. The *N*-TFA-prolyl chloride reagent has been used to determine the enantiomeric purity of amino acids with an absolute error of less than ±0.7%.[11]

4.1.2 PREPARATION OF TFA-*l*-PROLYL CHLORIDE (1)

Trifluoroacetic anhydride (4 ml) dissolved in anhydrous ether (20 ml) is added to *l*-proline (1.15 g, dried in vacuum at 60 °C for 24 h) and cooled to −10 °C. After 10 min the flask is removed from the cooling-bath and the reaction mixture is kept at room temperature for 2 h. The ether and the unreacted anhydride are removed in a vacuum at room temperature. A solution of thionyl chloride (10 ml) in dry benzene (15 ml) is added to the residue at 0 °C and the reaction mixture is kept at room temperature for a further 2.0 h. The benzene and unreacted thionyl chloride are then removed under vacuum at below 40 °C, the residue is dissolved in dichloromethane (5 ml) and the solvent is then removed. This process is repeated twice more. The residue is transferred quantitatively to a standard flask with dry methylene chloride and made up to 100 ml. This solution contains 1 mmol of *N*-TFA-*l*-prolylchloride in 10 ml. The reagent can be stored over molecular sieve (Linde Type 3A) and is stable for several months at 0 °C.

4.1.3 PREPARATION OF *l*-α-CHLOROISOVALERYL CHLORIDE (2)[12]

d-Valine (1 g) is treated with a mixture of conc. hydrochloric acid (3 ml) and conc. nitric acid (1 ml) at 40–50 °C. Deamination occurs with liberation of nitrogen and the *d*-valine is converted to optically pure *l*-α-chloroisovaleric acid. The mixture is extracted with ether washed with water and dried over sodium sulfate, the solvent is then removed under vacuum. The residue is treated with thionyl chloride (2 ml) at 60 °C for 1 h. Excess thionyl chloride is removed under vacuum and the acid chloride (b.p. 113 °C) is redissolved in dry methylene chloride or benzene (10 ml) for immediate use without further purification.

4.1.4 PREPARATION OF *l*-MENTHYL CHLOROFORMATE (3)[7]

l-Menthol (15.63 g, 0.1 mol) and quinoline (12.9 g, 0.1 mol) are added to a solution of phosgene (20 g, 0.2 mol) in toluene (100 ml) at 0 °C, and the mixture is stirred overnight. The quinoline hydrochloride is filtered off, and excess phosgene is removed by bubbling dry nitrogen through the filtrate. The solution is transferred to a 100 ml standard flask containing calcium carbonate and stored at 0 °C. Under these conditions, the solution, which contains 1 mmol of the reagent ml^{-1}, can be kept for several months without any noticeable decomposition.

4.1.5 PREPARATION OF *l*-TERESANTALINYL CHLORIDE (4)[8]

l-Teresantalic acid (2 mg) in dry toluene (0.1 ml) is treated with redistilled thionyl chloride (0.2 ml) at 60 °C for 1 h. Excess thionyl chloride is removed in a stream of dry nitrogen and the acid chloride is dissolved in dry toluene for immediate use without further purification.

4.1.6 PREPARATION OF *N*-TFA-*l*-PROLYL(*d,l*)AMINO ACID METHYL AND 1-BUTYL ESTERS[5,11]

Methanol (1 ml) is cooled in a dry-ice–alcohol bath and thionyl chloride (0.1 ml) is added drop by drop. This solution is then added to the amino acid (0.1 mmol) and the resulting solution is heated under reflux for 30 min. Alternatively the amino acid (0.1 mmol) is refluxed for 3 h with 5 ml of anhydrous 1-butanol which is 3 M in anhydrous HCl. The reaction mixtures are concentrated under vacuum and the last traces of the reagent solution are removed by adding dry methanol or butanol (0.5 ml) respectively and evaporating under vacuum. The residue is dissolved in *N*-TFA-*l*-prolyl chloride (1) reagent (0.11 mmol, 1.1 ml) and the solution is neutralized with triethylamine (0.22 mmol, 0.03 ml) at below 10 °C. After washing with water (2 ml), hydrochloric acid (2 ml, 1 M), sodium bicarbonate solution (10%, 2 ml) and water (2 ml), the organic layer is dried over sodium sulfate and concentrated under vacuum. A sample of the solution is then analysed by gas chromatography (Table 2).

4.1.7 PREPARATION OF *N*-TFA-*l*-PROLYL(*d,l*)AMINO ACID METHYL AND 1-BUTYL ESTERS OF TRYPTOPHAN AND β-HYDROXY-AMINO ACIDS[9,10]

(a) The amino acid (0.1 mmol) is esterified with thionyl chloride (0.1 ml) and methanol (1 ml) or butanol–hydrogen chloride as described in the previous section and the excess reagent and solvent are removed. Hexamethyldisilazane (0.5 ml) is then added and the suspension is refluxed till a clear solution results. The excess reagent is removed under vacuum and the residue is condensed with a solution of (1) in methylene chloride (0.1 mmol in 1 ml). The reaction mixture is then processed as described in the previous section, and a sample is analysed by gas chromatography (Table 2).

(b) The *N*-TFA-*l*-prolyl-amino acid ester is prepared as described in Procedure 4.1.6. The methylene chloride is removed under vacuum and the residue is treated with pyridine (0.5 ml) and BSTFA with 1% TMCS (0.5 ml) at room temperature for 15 min.[10] A sample of the reaction mixture is analysed by gas chromatography (Table 2).

4.1.8 PREPARATION OF *l*-α-CHLOROISOVALERYL(*d,l*)AMINO ACID METHYL ESTERS[6]

The dry amino acid methyl ester (1 mg) prepared as described in Procedure 4.1.6 is treated with *l*-α-chloroisovaleryl chloride in methylene chloride (0.5 ml) prepared as in Procedure 4.1.3. The suspension is cooled to 0 °C and triethylamine is added to bring the pH to 8–9. After washing and drying over sodium sulfate a sample is analysed by gas chromatography (Table 3).

4.1.9 PREPARATION OF *l*-MENTHYLOXYCARBONYL(*d,l*)AMINO ACID METHYL ESTERS[7]

The amino acid (1 mmol) is esterified with the thionyl chloride–methanol reagent described in Procedure 1.6 and the solvent and excess reagent are removed under vacuum. An excess of *l*-menthyl chloroformate (3) in toluene (1.1 ml, 1.1 mmol) and pyridine (0.2 ml) are added and the mixture is left at room temperature for 30 min. After washing and drying a sample is analysed by gas chromatography (Table 3).

4.1.10 PREPARATION OF *l*-TERESANTALINYL(*d,l*)AMINO ACID METHYL AND BUTYL ESTERS[8]

The amino acid methyl ester (1 mg) or butyl ester prepared as in Procedure 4.1.6 is dissolved in tetrahydrofuran (0.8 ml) and pyridine (0.2 ml) and *l*-teresantalinyl chloride solution (0.2 ml, see p. 466) prepared as described in

TABLE 2

Gas chromatographic separation for diastereoisomeric TFA-prolyl-amino acid methyl and butyl esters[5,10]

Amino acid	Column conditions	Amino acid derivative	Derivative formed by method 4.1.3–4.1.9	$r_{II/I}$
Alanine	O	methyl ester	6	1.14
Valine	O	methyl ester	6	1.15
Leucine	O	methyl ester	6	1.09
Proline	O	methyl ester	6	1.11
Threonine	P	*O*-TMS	7(a)	1.23
Serine	P	*O*-TMS	7(a)	1.22
Hydroxyproline	P	*O*-TMS	7(a)	1.21
Aspartic acid	P	methyl ester	6	1.08
Glutamic acid	P	methyl ester	6	1.15
Methionine	P	methyl ester	6	1.14
Cysteine	P	thiazolidine-4-carboxylic acid	(9)	1.56
Phenylalanine	Q	methyl ester	6	1.04
Phenylalanine	R	*n*-butyl ester	6	1.06
Tyrosine	R	*O*-TMS-butyl ester	7(b)	1.04
Tryptophan	R	*N*-TMS-butyl ester	7(b)	not stated
Ornithine	S	Di-TFA-prolyl	6	1.10
Lysine	S	Di-TFA-prolyl	6	1.16
Allothreonine	P	*O*-TMS	7(a)	1.16
β-Hydroxyvaline	P	*O*-TMS	7(a)	1.21
Homoserine	P	*O*-TMS	7(a)	1.23
β-Hydroxyglutamic acid	P	*O*-TMS	7(a)	1.10
Penicillamine	P	5,5-dimethyl-thiazolidine-carboxylic acid	(9)	1.16
DOPA	R	*O*-TMS-butyl ester	7(b)	1.05

Column O: 5 ft × $\frac{1}{8}$ in packed with 5% SE-30 on Chromosorb W, run isothermally at 176 °C and with a nitrogen flow of 28 ml min^{-1}.
Column P: 5 ft × $\frac{1}{8}$ in packed with 0.5% EGA on Chromosorb W, run isothermally at 185 °C and with a nitrogen flow of 46 ml min^{-1}.
Column Q: 5 ft × $\frac{1}{8}$ in packed with 0.5% EGA on Aeropak 30 run isothermally at 220 °C and with a nitrogen flow of 30 ml min^{-1}.
Column R: 100 ft × 0.02 in Scot column coated with Dexsil 300 GC, temperature 225 °C, and with a helium flow of 1 ml min^{-1}.
Column S: 5 ft × $\frac{1}{4}$ in packed with 5% OV-1 on Supelcoport, run isothermally at 270 °C and with a nitrogen flow of 30 ml min^{-1}.
The order of emergence of *S*-prolyl derivatives for amino acids has been found to be r_{SR} before r_{SS} with the exception of lysine and ornithine,[2] where the subscripts refer respectively to the configuration of the proline and the asymmetric carbon to which the amine function is attached.

TABLE 3

Gas chromatographic separation for diastereoisomeric *N*-acyl-amino acid methyl esters

Amino acid	α-Chloroiscvaleryl derivatives prepared by method 4.1.8[6]			Menthyloxycarbonyl derivatives prepared by method 4.1.9[7]			Teresantalinyl ester derivatives prepared by method 4.1.10[8]			
	Column	temperature	$r_{II/I}$	Column	temperature	$r_{II/I}$	Column	temperature	$r_{II/I}$	$r_{II/I}$ (butyl ester)
Alanine	W	161	1.13	Y	170	1.14	Z	155	1.10	1.07
Valine	X	161	1.22	Y	170	1.06	Z	155	1.05	1.03
Leucine	X	161	1.15	Y	170	1.10	Z	155	1.05	1.03
Isoleucine	X	161	1.18				Z	155	1.03	1.04
Proline	X	185	1.17				Z	155	1.20	1.04
Aspartic acid							Z	155	1.05	1.01
Methionine							Z	155	1.09	1.04
Threonine	W	145	1.11							
Glutamic acid							Z	155	1.07	1.06
Phenylalanine	W	185	1.08	Y	200	1.10	Z	155	1.04	1.03
Norvaline							Z	155	1.08	1.05
Norleucine							Z	155	1.05	1.04
Ethionine							Z	155	1.03	1.01

Column W: 5 ft × $\frac{1}{8}$ in packed with 5% FFAP on chromosorb W nitrogen flow 30 ml min^{-1}.
Column X: 5 ft × $\frac{1}{8}$ in packed with 5% SE-30 on chromosorb W nitrogen flow 30 ml min^{-1}.
Column Y: 5 ft × $\frac{1}{8}$ in packed with 5% QF-1 on Aeropack 30 nitrogen flow 30 ml min^{-1}.
Column Z: 3 m × 3 mm packed with 0.5% PEGA on chromosorb W nitrogen flow 30 ml min^{-1}.

Procedure 4.1.5 is added at room temperature. After 30 min a sample of the solution is analysed by gas chromatography (Table 3).

4.1.11 REACTION OF CARBOXYL GROUP WITH AN ASYMMETRIC REAGENT

Amino acids can also be resolved as esters of 2-butanol,[13,14] 2-octanol,[13] 3,3-dimethyl-2-butanol[15] or menthyl alcohol.[16,17,22] In all cases the α-amino group has to be blocked by reaction with, for example, an acylation reagent such as trifluoroacetic anhydride. All protein amino acids with the exception of arginine, histidine and cystine can be resolved as their *N*-TFA-2-butyl esters on polar capillary columns. Some diastereoisomeric pairs such as those of serine, threonine, cysteine and hydroxyproline are only poorly resolved. A superior but more time-consuming procedure consists of forming an *O*- or *S*-acetyl-*N*-TFA-amino acid-2-butyl ester from these polyfunctional amino acids which successfully resolved on a Carbowax 20 M column.[15]

Diastereoisomeric *N*-TFA amino acid alkyl esters can be resolved on packed columns if a suitable asymmetric alcohol is used. The menthyl[16] and 3,3-dimethyl-2-butyl[18] esters are satisfactory but are not recommended for the steric analysis of amino acids as there is some indication that the rate of diastereoisomer formation can be significantly different for the two enantiomers.

Experimental Procedure. The following procedure for resolving 2-butanol is preferable to the brucine monophthalate method.[19]

4.1.12 (a) PREPARATION OF *d*-2-BUTANOL[20]

d-Valine (0.25 mol, 29 g), *d,l*-2-butanol (0.58 mol, 43 g), *p*-toluenesulfonic acid (60 g), benzene (35 ml) and toluene (15 ml) are refluxed in a Dean and Stark apparatus until a clear solution results (~30 h). The reaction mixture is cooled and filtered to remove traces of unchanged amino acid. After evaporation of the solvent and excess alcohol the residual oil is set aside for several hours. The residue is then diluted with anhydrous ether (100 ml), filtered, washed with ether and dried. The *p*-toluene sulfonate of *d*-valine-*d*-2-butyl ester can be obtained in 98% optical purity after six consecutive crystallizations from benzene. The ester (0.1 mol) is hydrolysed with aqueous sodium hydroxide (0.2 mol) at 25 °C. The active alcohol is isolated by ether extraction followed by fractional distillation.

(b) PREPARATION OF *d*-3,3-DIMETHYL-2-BUTANOL[20]

Optically pure *d*-3,3-dimethyl-2-butanol can be obtained by fractional crystallization of *d*-alanine-*d*-3,3-dimethyl-2-butyl ester *p*-toluene sulfonate salt using the same procedure as described for the resolution *d*-2-butanol.

4.1.13 PREPARATION OF *N*-TFA-AMINO ACID-2-BUTYL ESTERS[13,14]

An 8 M solution of hydrogen chloride in butanol is prepared from 3 ml of *d*-2-butanol and this is added to 5 mg or less of amino acid or amino acid mixture.

The mixture is refluxed for 1 h under anhydrous conditions. The excess reagent is removed under vacuum at 60 °C, and this is repeated twice more with 2 ml of methylene chloride. The residue is dissolved in anhydrous methylene chloride (0.5 ml) cooled to below 0 °C, and trifluoroacetic anhydride (0.1 ml) is added. After 2 h at room temperature the mixture is concentrated under vacuum at below 30 °C. Before gas chromatography the derivative is dissolved in methylene chloride or chloroform (Table 3). The method gives better than 95% yield of ester, but strict exclusion of moisture and removal of all excess 2-butanol is essential for reasonable yields of the *O*-TFA, and *S*-TFA derivatives of threonine, serine and cysteine.

4.1.14 PREPARATION OF *N*-TFA-AMINO ACID-2-OCTYL,[13] 3,3-DIMETHYL-2-BUTYL[15] AND MENTHYL ESTERS[16,22]

The preparation of these esters can be carried out as described in Procedure 4.1.13. There are alternative procedures:

(a) The amino acid (25 mg), menthol (60 mg) and *p*-toluene sulfonic acid (60 mg) are refluxed in benzene (3.5 ml) and toluene (1.5 ml) in a Dean and Stark apparatus until a clear solution results. The solvents are removed and the residue is dissolved in ethyl acetate (1 ml), triethylamine (0.15 ml) and methyl trifluoroacetate (0.3 ml). After 2 h at 25 °C the solution is washed, dried over sodium sulfate and a sample is analysed by gas chromatography.

(b) Acidic or basic amino acids (5 mg) are first esterified with dry HCl-methanol reagent (3 ml) at 80 °C for 1 h. After removal of the excess reagent under vacuum, menthol (2 g) is added and the sample is trans-esterified by passing hydrogen chloride gas through the reaction mixture at 105 °C for 3 h. The menthyl esters are then *N*-trifluoroacetylated and a sample is analysed by gas chromatography.

(c) Hydroxy amino acids (1 mg) are first derivatized as described in (a) or (b) and are then treated with hexamethyldisilazane-trimethylchlorosilane reagent in anhydrous pyridine (0.5 ml) for 5 min at room temperature. A sample is then analysed by gas chromatography. The reader is referred to the original papers cited in Refs. 13, 15–18, and 22 for the GC retention volumes and the recommended chromatographic conditions for this group of derivatives.

4.2 Amines

Enantiomeric amino compounds may be converted to diastereoisomeric amides by acylation with a wide variety of suitable chiral reagents.

4.2.1 REACTION OF THE AMINO GROUP WITH AN ASYMMETRIC REAGENT

N-TFA–prolyl chloride (1),[5] drimanoyl chloride (5),[23] *trans*-chrysanthemoyl chloride (6)[23] and α-phenylbutyric anhydride (7)[24] have been used as acylating agents for this purpose.

TABLE 4

Gas chromatographic separation for diastereoisomeric TFA-amino acid 2-butyl esters prepared by method 4.1.13

	Column conditions	Amino acid derivative	$r_{II/I}$
Alanine	S	mono-TFA	1.04
Valine	S	mono-TFA	1.07
α-Aminobutyric	S	mono-TFA	1.04
Isoleucine	S	mono-TFA	1.07
Norvaline	S	mono-TFA	1.05
Leucine	S	mono-TFA	1.06
Norleucine	S	mono-TFA	1.05
Serine	S	di-TFA	1.02
Proline	S	mono-TFA	1.04
α-Amino octanoic	S	mono-TFA	1.02
Cysteine	S	di-TFA	1.00
Hydroxyproline	S	di-TFA	1.02
Aspartic acid	S	di-2-butyl ester	1.01
Threonine	S	mono-TFA	1.02
Methionine	S	mono-TFA	1.02
Phenylalanine	S	mono-TFA	1.03
Glutamic acid	S	di-2-butyl ester	1.03
Ornithine	T	di-TFA	1.01
Lysine	T	di-TFA	1.02
Threonine	U	*O*-acetyl[15]	1.04
Serine	U	*O*-acetyl[15]	1.05
Hydroxyproline	U	*O*-acetyl[15]	1.03
Cysteine	U	*S*-acetyl[15]	0.95
Tyrosine	V	*O*-acetyl[15]	1.05
Tryptophan	V	di-TFA[15]	1.05

Column S: 150 ft × 0.02 in Carbowax 1540, He at 8 ml min^{-1} at 100 °C for 25 min, then programmed to 140 °C at 1° min^{-1}.
Column T: 150 ft × 0.02 in Ucon LB550-X, He at 10.5 ml min^{-1} at 63–140 °C at 2° min^{-1}.
Column U: 150 ft × 0.02 in Carbowax 20 M, He at 10 ml min^{-1} at 150 °C isothermally.
Column V: 150 ft × 0.02 in DEGS, He at 7.5 ml min^{-1} at 180 °C isothermally.

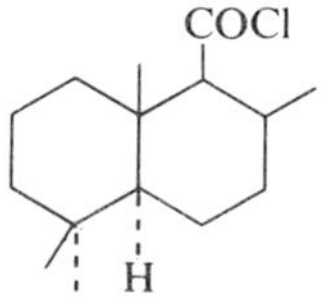

(5)

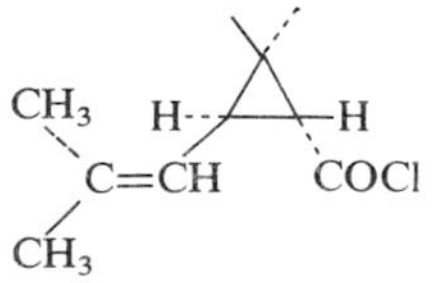

(6)

$$(CH_3-CH_2-\underset{\displaystyle C_6H_5}{\underset{|}{C}H}-CO)_2O$$

(7)

Experimental Procedure. The preparation of *N*-TFA–prolyl chloride (1) has been described in Procedure 4.1.2.

4.2.2 PREPARATION OF DRIMANOYL CHLORIDE (5) AND *TRANS*-CHRYSANTHEMOYL CHLORIDE (6)[23]

d-Drimanoic acid[23] (2 mg) or *d*-*trans*-chrysanthemic acid (2 mg) are dissolved in dry toluene (0.1 ml) and are treated with thionyl chloride (0.2 ml) at 60 °C for 1 h. Excess reagent is removed in a stream of nitrogen and the acid chloride is redissolved in dry toluene for immediate use.

4.2.3 PREPARATION OF α-PHENYLBUTYRIC ANHYDRIDE (7)[25,26]

l- or *d*-Phenylbutyric acid (2.8 g) is suspended in anhydrous ether (20 ml) and oxalyl chloride (0.63 ml) is added over 5 min at 0 °C. The reaction mixture is stirred at room temperature for several hours. The sodium chloride is filtered off, and the ether is evaporated in the cold. The anhydride ($\alpha_D = +139$ °C) is used without further purification.

4.2.4 PREPARATION OF *N*-TFA-*l*-PROLYL (*d,l*) AMIDES[5,27,28,28a]

In a typical experiment the racemic amine (0.1 mmol) and *N*-TFA-*l*-prolyl chloride reagent (1.2 ml) (1) are mixed in a screw-capped vial and adjusted to pH 9 with triethylamine. The mixture is kept at room temperature for 15 min and then washed successively with water, 20% citric acid solution, water, saturated sodium bicarbonate, and finally water. The organic layer is dried over anhydrous magnesium sulfate, filtered and evaporated to dryness. The residue is dissolved in ethyl acetate (1 ml) for analysis by GC (Tables 5, 6 and 7).

4.2.5 PREPARATION OF *d*-DRIMANOYL (*d,l*) AMIDES[23]

The amine (1 mg) is dissolved in dry toluene (20 μl) and is treated with a 3 molar excess of freshly prepared drimanoyl chloride (5) dissolved in dry toluene (0.3 ml). The reaction mixture is heated at 60 °C for 1–2 h. The crude product can be analysed directly by GC (Table 8).

4.2.6 PREPARATION OF *d*-*TRANS*-CHRYSANTHEMOYL (*d,l*) AMIDES[23]

d-*trans*-Chrysanthemoyl chloride (1 mmol), triethylamine (2 mmol) and the amine (0.2 mmol) are dissolved in chloroform (2 ml) and the solution is refluxed for 10 min. The reaction mixture can be analysed directly by GC (Table 7).

4.2.7 PREPARATION OF *d*-α-PHENYLBUTYRYL (*d,l*) AMIDES[24,134]

The amine (0.1 mmol) is treated with *d*-α-phenylbutyric acid anhydride (0.1 mmol) in toluene (0.5 ml) for 10 min at room temperature. The reaction

TABLE 5

Gas chromatographic separation of *N*-TFA-prolyl-amides prepared by method 4.2.4 from asymmetric aliphatic and heterocyclic amines[29]

Amine	Column	Separation temperature	Retention time of diastereoisomers (minutes) TFA-pro(*l*)amine	TFA-pro(*d*)amine	$r_{II/I}$
2-Aminobutane	M	140	11.7	12.4	1.06
2-Amino-3-methyl butane	M	140	12.4	14.0	1.13
2-Amino pentane	M	140	15.2	17.6	1.16
2-Amino-4-methyl pentane	M	140	12.8	15.8	1.23
2-Amino octane	M	155	16.1	19.25	1.20
2-Methyl pyrrolidine	N	210	3.6	3.2	1.11
2-Methyl piperidine	N	210	4.0	3.6	1.11
3-Methyl piperidine	O	230	61.6	57.1	1.08
2-Ethyl piperidine	N	210	4.7	4.0	1.17
2-Propyl piperidine	N	210	5.0	4.4	1.13
Salsolidine	N	250	16.7	19.1	1.14
1-Methyl-1,2,3,4-tetrahydroisoquinoline	N	210	10.7	12.0	1.12
3-Methyl-1,2,3,4-tetrahydroisoquinoline	N	210	14.0	12.0	1.17
2-Methyl indoline	N	210	9.8	7.7	1.27
2-Methyl-1,2,3,4-tetrahydroquinoline	N	210	7.3	5.9	1.25
trans-Decahydroquinoline	O	230	123.4	149.4	1.21

Column M: 15 ft × $\frac{1}{4}$ in packed with 0.75% DEGS/0.25% EGSS-X on Chromosorb W (acid washed); nitrogen flow 67 ml min^{-1}.
Column N: 5 ft × $\frac{1}{8}$ in packed with 5% DCLSX-3-0295 on DMCS-treated Chromosorb W; nitrogen flow 30 ml min^{-1}.
Column O: 10 ft × $\frac{1}{4}$ in packed with 20% W/WHI-EEF 4B on DMCS-treated Chromosorb P; nitrogen flow 39 ml min^{-1}.

TABLE 6

Gas chromatographic separation of *N*-TFA-prolyl-amides prepared by method 4.2.4 from asymmetric 1-phenylisopropylamine drugs[31]

Chemical structure

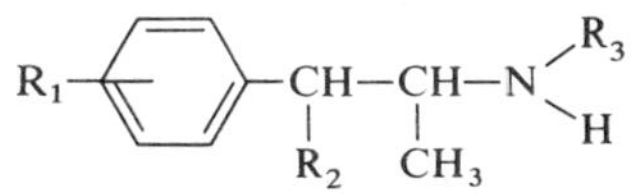

Drug	R_1	R_2	R_3	Retention time of diastereoisomers/min TFA-pro(*l*)amide	TFA-pro(*d*)amide	$r_{II/I}$
Amphetamine	H	H	H	29	32	1.10
N-Methylamphetamine	H	H	CH_3	52	55	1.06
N-Ethylamphetamine	H	H	CH_2CH_3	58	63	1.09
N-Propylamphetamine	H	H	$CH_2CH_2CH_3$	71	75	1.06
N-Butylamphetamine	H	H	$CH_2CH_2CH_2CH_3$	94	98	1.05
N-Cyanoethylamphetamine	H	H	$CH_2{-}CH_2{-}CN$	155	160	1.04
p-Chloroamphetamine	Cl	H	H	56	65	1.16
Norfenfluramine	*m*CF_3	H	H	26	27.5	1.06
Fenfluramine	*m*CF_3	H	CH_2CH_3	46	50	1.09
N-Allylnorfenfluramine	*m*CF_3	H	$CH_2{-}CH{=}CH_2$	58.5	62	1.06
N-Propynynorfenfluramine	*m*CF_3	H	$CH{-}C{\equiv}CH$	59.7	60.5	1.01
Norephedrine	H	OH	H	70	62	1.13
Norpseudoephedrine	H	OH	H	64	71	1.11
Ephedrine	H	OH	CH_3	105	98	1.07
Pseudoephedrine	H	OH	CH_3	101	105	1.04
N-Ethylnorephedrine	H	OH	CH_2CH_3	121	114	1.06
N-Ethylnorpseudoephedrine	H	OH	CH_2CH_3	113	118	1.04

Column: 2 m × $\frac{1}{8}$ in packed with 3% SE-30 on DMCS-treated Chromosorb G. Run isothermally at 170 °C using a nitrogen flow rate of 25 ml min^{-1}.

TABLE 7

Gas chromatographic separation of *N*-TFA-prolyl-amides[28] and chrysanthemoyl-amides[30] derived from 1-phenyl, 1-naphthyl and 1,2-diphenyl-ethylamines

Amine	TFA-*N*-prolylamides[a] prepared by method 4.2.4 $r_{II/I}$	Chrysanthemoylamides[b] prepared by method 4.2.6 $r_{II/I}$
1-Phenylethylamine	1.20	N.S.
1-Phenylbutylamine	1.18	1.04
2-Methyl-1-phenylpropylamine	1.27	1.05
3-Methyl-1-phenylbutylamine	1.16	1.05
N-Isobutyl-1-phenylethylamine	1.04[c]	—
1-(4′-Bromophenyl)ethylamine	1.29	—
1-(4′-Methoxyphenyl)ethylamine	1.27	—
1-(2′,3′-Xylyl)ethylamine	1.30	N.S.
1-(α-Naphthyl)ethylamine	1.50	N.S.
1-(β-Naphthyl)ethylamine	1.48	N.S.
1,2-Diphenylethylamine	1.10	1.08
1-Phenyl-2-(4′-tolyl)ethylamine	1.07	1.07
1-Phenyl-2-(2′-tolyl)ethylamine	1.08	1.08
2-(4′-Chlorophenyl)-1-phenylethylamine	1.06	1.09
2-Phenyl-1-(4′-tolyl)ethylamine	1.12	1.07
1-(4′-Isopropylphenyl)-2-phenylethylamine	1.17	1.08
2-Phenyl-1-(4′-tert-butylphenyl)-ethylamine	1.20	1.07
1,2,2-Triphenylethylamine	1.11[c]	1.07
2-Phenyl-1-(2′-thiophenyl)-ethylamine	1.07	1.07

[a] Column: 3 m × 3 mm packed with 3% SE-30 on DMCS-treated Chromosorb W; column temperatures 170, 200, 220 or 250 °C using a nitrogen flow of 50 ml min^{-1}.
[b] Column: 5 m × 3 mm packed with 10% QF-1 on DMCS-treated Chromosorb W; run isothermally at 230 °C and using a nitrogen flow of 50 ml min^{-1}.
[c] Column: 3 m × 3 mm packed with 2% QF-1 on DMCS-treated Chromosorb W; conditions as for column [a].

mixture is diluted with ethyl acetate before analysis by GC (Table 8). Methods and data are also available for α-phenylpropionamides and α-chlorophenylacetamides.[134]

4.3 Alcohols

Asymmetric alcohols can be converted into a mixture of diastereoisomers by reaction with suitable chiral reagents to form diastereoisomeric carbonates, esters and carbamates. Many successful separations have been reported using a variety of asymmetric reagents such as acyl chlorides, isocyantes and chloroformates.

TABLE 8

Gas chromatographic separation of drimanoyl-amides[23] and α-phenylbutyryl-amides[24,134]

	Drimanoylamides Procedure 4.2.5 Retention index values		α-Phenylbutyrylamide Procedure 4.2.7 Retention index values	
Amine	Acyl *l*-amine	Acyl *d*-amine	Acyl *l*-amine	Acyl *d*-amine
α-Phenylethylamine	2515	2535	2060	2030
α-Methylphenylethylamine	2630	2630		
N-Methyl-α-phenylethylamine			2120	2105
Amphetamine	2620	2600	2115	2125
Chloroamphetamine			2305	2320
3-Methoxy-4,5-methylenedioxy amphetamine			2615	2635
Methylamphetamine			2160	2195

Column: 5 m × 3 mm packed with 1% SE-30 or 1% OV-17 on Gas Chrom. Q.
The retention relationships in all the amide derivatives are that the diastereoisomers from the R-amines are eluted before the diastereoisomers derived from the S-amines.

4.3.1 REACTION OF HYDROXYL GROUP WITH AN ASYMMETRIC REAGENT

Menthyl chloroformate (3),[7] drimanoyl chloride (5),[23] *trans*-chrysanthemoyl chloride (6),[23] 3β-acetoxy-Δ^5-etienic acid chloride (8)[32] and 1-phenylethyl-isocyanate (9)[33] and 2-phenylpropionyl chloride (12)[36] have been used for this purpose.

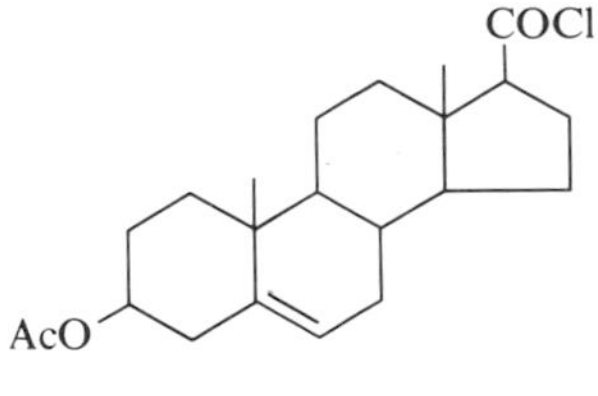

(8)

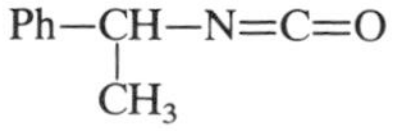

(9)

Experimental procedure. The preparation of menthylchloroformate (3), drimanoyl chloride (5), *trans*-chrysanthemoyl chloride (6) and 2-phenyl-proprionyl chloride (12) are described in Procedures 4.1.4 and 4.6.4, respectively.

4.3.2 PREPARATION OF 3β-ACETOXY-Δ^5-ETIENIC ACID CHLORIDE[34]

A solution of 3β-acetoxy-Δ^5-etienic acid (2.9 g) in thionyl chloride (10 ml) is kept at room temperature for 4 h. After removal of the excess reagent in vacuum the acid chloride is obtained as pale yellow crystals. It is used without further purification.

4.3.3 PREPARATION OF *d*-1-PHENYLETHYLISOCYANATE[33]

d-1-Phenylethylamine (25 g) is added to dry toluene (250 ml) which has been saturated with hydrogen chloride. After a further addition of toluene (175 ml) the suspension is refluxed for 4 h while phosgene is bubbled through the reaction mixture. The resulting solution is allowed to cool to room temperature and dry nitrogen is bubbled through the solution to remove excess phosgene. The solution is then fractionated under vacuum to yield the product. b.p. 74–76 °C, $[\alpha]_D^{26}+12.9°$ (c, 1.0905, chloroform).

4.3.4 PREPARATION OF DIASTEREOISOMERIC MENTHYL CARBONATES[7]

A toluene solution of *l*-menthylchloroformate (3) (0.11 ml, 0.11 mmol, prepared as in 4.1.4) and pyridine (10 μl) is added to the asymmetric alcohol (0.1 mmol) and the reaction is allowed to stand for 30 min at room temperature. After washing and drying over sodium sulfate a part of the solution is analysed by gas chromatography (Table 9).

4.3.5 PREPARATION OF DRIMANOIC ACID AND *TRANS*-CHRYSANTHEMANOIC ACID ALKYL ESTERS[23]

The alcohol (1 mg) in dry toluene (20 μl) is treated with a freshly prepared acid chloride solution (5) or (6) (3 mol excess, 0.3 ml) prepared as in 4.2.2 at 60 °C for 1–2 h. The reaction mixture is concentrated under vacuum and a part of the sample is analysed by GC. The reader is referred to the original papers[23,24] for GC retention volumes and the recommended chromatographic conditions for the separation of the diastereoisomers derived from 2-octanol, menthol, isomenthol, borneol, isoborncol, fenchol and pantolactone.

4.3.6 PREPARATION OF 3β-ACETOXY-Δ^5-ETIENTIC ACID ESTERS[31]

The alcohol (18 μmol), 3β-acetoxy-Δ^5-etienic acid chloride (50 μmol) (8) and pyridine (0.01 ml) are refluxed in benzene (1.0 ml) for 1 h. A sample of the reaction mixture is used directly for GC analysis (Table 9).

4.3.7 PREPARATION OF ALKYL-*N*-(1-PHENYLETHYL)-CARBAMATES[32]

The alcohol (1–1.5 mg) and *d*-1-phenylethylisocyanate (9) (1.5 mg) are heated in a sealed ampoule at 110 °C for 7 h. Methanol (0.5 ml) is added to convert the excess of the reagent to methyl *N*-(1-phenylethyl)-carbamate. A part of the solution is then analysed by GC (Table 10).

4.3.8 PREPARATION OF 2-PHENYLPROPIONIC ACID ALKYL ESTERS[36]

The alcohol (100 μg) is added to *d*-2-phenylpropionyl chloride solution (12) (80 μl). Dry pyridine (20 μl) is added and the mixture is left for 2 h at room temperature in a desiccator. The solution is evaporated to dryness and the

TABLE 9

Gas chromatographic resolution of phenylalkylcarbinols as menthyl carbonates[7,32] and 3β-acetoxy-Δ^5-etienates[32]

Alcohol	Menthyl carbonates (Procedure 4.3.4) retention times of diastereoisomers/min from			3β-Acetoxy-Δ^5-etienates (Procedure 4.3.6) retention times of diastereoisomers/min from		
	l-alcohol	*d*-alcohol	$r_{II/I}$	*l*-alcohol	*d*-alcohol	$r_{II/I}$
Phenylmethylcarbinol	11.0	12.0	1.09	16.0	17.5	1.09
Phenylethylcarbinol	9.5	11.0	1.16	19.0	21.5	1.13
Phenyl-*n*-propylcarbinol	12.5	14.0	1.12	22.0	24.0	1.09
Phenyl-*n*-butylcarbinol	19.0	20.5	1.08	27.5	29.0	1.05
Phenylcyclohexylcarbinol	38.5	41.5	1.08	93.5	98.5	1.05
Methyl-*n*-hexylcarbinol	9.0	10.0	1.11	9.0	9.5	1.06

Column: 6 ft $\times \frac{1}{8}$ in packed with 3% OV-17 on Gas Chrom Q with a nitrogen flow of 20 ml min^{-1}, and a separation temperature of 150° and 260 °C, respectively.

TABLE 10

Gas chromatographic separation of asymmetric alkyl-*N*-(1-phenylethyl)-carbamates[3 33,35] prepared by Procedure 4.3.3

Alkyl-*N*-(1-phenylethyl) carbamate derived from	Retention time of diastereoisomers min^{-1} from		
	d-alcohol	*l*-alcohol	$r_{II/I}$
2-Butanol	43.7	44.7	1.01
3-Methyl-2-butanol	27.9	28.6	1.03
2-Pentanol	32.2	33.1	1.03
3,3-Dimethyl-2-butanol	29.5	30.5	1.03
3-Methyl-2-pentanol	40.4	41.3	1.02
3-Methyl-2-hexanol	54.1	56.2	1.04
2-Octanol	77.9	82.6	1.06
2-Methyl-3-octanol	75.7	78.4	1.04
3-Penten-2-ol	40.3	41.0	1.02
1-Hexen-3-ol	45.7	46.4	1.02
1-Hepten-3-ol	62.2	64.2	1.03
1-Octen-3-ol	85.3	89.1	1.04
Menthol[b]	34.6	36.8	1.06
Phenylmethylcarbinol[b]	56.5	53.5	1.06
2-Dodecanol[c]	4.4	4.9	1.11
2-Tetradecanol[c]	8.3	9.2	1.11
2-Hexadecanol[c]	15.6	17.5	1.12
2-Eicosanol[c]	6.2	6.7	1.08
2-Tetracosanol[c]	16.9	18.5	1.09

[a] Column A: 150 ft × 0.02 in s.s. Column coated with 31 mg of Carbowax 20 M using a helium flow of 9.8 ml min^{-1} and a separation temperature of 170 °C.
[b] Column B: 50 ft × 0.02 in s.s. Column coated with 33 mg of OV-225 using a helium flow of 17.3 ml min^{-1} and a separation temperature of 170 °C.
[c] Column C: 1.8 m × 3 mm, packed with 1% QF-1 on Gas Chrom. Q using a helium flow of 30 ml min^{-1} and a separation temperature of 160 and 200 °C.

residue dissolved in chloroform; a sample is analysed by GC. The diastereoisomers derived from 3-hexanol, 3-heptanol, 3-octanol, 2-octanol and 2-eicosanol can be resolved ($r_{II/I}$ 1.05–1.13) on a glass column packed with 5% QF-1 on Gas-Chrom-Q run isothermally at 120°, 160° and 232 °C respectively.[36]

4.4 Ketones

Gas chromatographic separation of the diastereoisomeric acetal derived from camphor and 2,3-butanediol is of historical interest only;[37] attempts to use such acetalization for optical resolution of other ketones have been unsuccessful.[38] Cyclic ketones with an asymmetric carbon α to a carbonyl function react with 2,2,2-trifluoro-1-phenylethylhydrazine (10) to form diastereoisomeric hydrazones which can be resolved by GC.[38] Since hydrazone formation may be accompanied by some racemization, the method is not recommended for the steric analysis of ketones, if a hydrogen is present alpha to the asymmetric carbon.

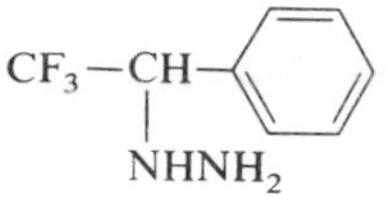

(10)

4.4.1 PREPARATION OF 2,2,2-TRIFLUORO-1-PHENYLETHYLHYDRAZINE[38]

(a) d,l-*2,2,2-Trifluoro-1-phenylethanol.* 2,2,2-Trifluoroacetophenone (50 g) in 250 ml of ether is added gradually to a stirred suspension of lithium aluminium hydride (23.5 g) in 500 ml of ether. The mixture is refluxed for 3 h, left stirring overnight at room temperature, and the alcohol is isolated in the normal way. The product is distilled; yield 42.3 g; b.p. 58 °C at 4 mm Hg.

(b) d,l-*2,2,2-Trifluoroethyl-1-phenylethyl* p-*toluene-sulfonate.* A solution of sodium hydroxide is added to a mixture of *p*-toluenesulfonyl chloride (50 g) and *d,l*-2,2,2-trifluoro-1-phenylethanol (42.3 g) in 250 ml of acetone. The mixture is stirred for 5 days at room temperature, the solvent is evaporated, and the residue partitioned between ether and water; the ether layer is separated, and the aqueous layer extracted with two 50 ml portions of ether. The combined ethereal layers are washed with dilute ammonia and water, and dried over anhydrous magnesium sulfate. Evaporation of the ether gives white crystals, m.p. 114–115 °C (from hot acetone); yield 62 g.

(c) d,l-*2,2,2-Trifluoro-1-phenylethylhydrazine.* The above *p*-toluenesulfonate (62 g) and hydrazine hydrate (140 ml, 99%) are refluxed for 12 h. The solution is cooled and extracted with four 50 ml portions of ether. The combined ether layers are washed with saturated sodium bicarbonate solution and water, and dried over anhydrous magnesium sulfate. Evaporation of the solvent gives a yellow oil, which is distilled in vacuum; yield 15 g b.p. 50 °C at 0.03 mm Hg.

(d) *Resolution of* d,l-*2,2,2-trifluoro-1-phenylethylhydrazine.* 5α-Androstan-17-one (3.3 g) and *d,l*-2,2,2-trifluoro-1-phenylethylhydrazine (3.42 g) are dissolved in 15 ml of a solution prepared by dissolving sodium acetate (2 g) and acetic acid (1 ml) in alcohol (100 ml) and the mixture is boiled under reflux for 6 h. The solution is kept at 5 °C overnight, and the crystals filtered and washed with ethanol. Recrystallization from hot ethanol yields 1.5 g of the hydrazone, m.p. 122–124 °C. $[\alpha]_D^{27}$ +40.9° (c. 0.420 chloroform).

(e) The optically pure steroid hydrazone (1.5 g) is dissolved in a mixture of water (15 ml), ethanol (15 ml), and sulfuric acid (6 ml). The mixture is refluxed for 2 h and set aside overnight. The solvent is partly evaporated, and the steroid ketone extracted with four 20 ml portions of ether. The aqueous layer is made alkaline with potassium carbonate and extracted with four 20 ml portions of ether. The combined ethereal solutions are dried over anhydrous magnesium sulfate, filtered, and evaporated to yield a yellow oil which is distilled in vacuum (b.p. 50 °C at 0.03 mm Hg). $[\alpha]_D^{26}$ +36.2° (c. 0.915 ethanol).

4.4.2 PREPARATION OF 2,2,2-TRIFLUORO-1-PHENYLHYDRAZONES[38]

In a typical experiment the ketone (0.1 mmol) and *d*-2,2,2-trifluoro-1-phenylethylhydrazine are dissolved in 0.5 ml of a mixture of ethanol, sodium acetate and acetic acid (100 : 2 : 1). The solution is kept at room temperature for 2 h in the case of unhindered ketones or is refluxed for 2 h in the case of sterically hindered ketone. A sample of the solution is then analysed by GC (Table 11).

4.5 Aliphatic and alicyclic acids

Aliphatic and alicyclic acids can be converted into a mixture of diastereoisomeric esters[39] or amides[40] which can be resolved by gas chromatography. The difficulty of separation increases with the distance of the asymmetric carbon from the carboxylic acid group. The method has proved of value for the stereochemical characterization of small amounts of isoprenoid acids.

4.5.1 REACTION OF CARBOXYL GROUP WITH AN ASYMMETRIC ALCOHOL

l-Menthyl esters have been found to give enhanced separation when used to form isoprenoid acid diastereoisomers,[39] whilst the *cis* and *trans* isomers of chrysanthemic acid could be resolved best as the 2-octyl esters.[41,42]

4.5.2 PREPARATION OF ALIPHATIC ACID-2-OCTYL ESTERS[41]

The acid (30 mg), pyridine (1 ml) thionyl chloride (1 ml) and *d* or *l*-2-octanol (1 ml) are heated on a boiling water bath for 20 min. After cooling to room temperature a small sample of the solution is analysed by GC. The reader is referred to the original publications[41,42] for the retention volumes and the recommended chromatographic conditions for the separation of chrysanthemic acid derivatives.

4.5.3 PREPARATION OF ALIPHATIC ACID-MENTHYL ESTERS[43,44]

The acid (20 mg) and thionyl chloride (1 ml) are refluxed for 20 min and the excess of thionyl chloride is evaporated under vacuum. *l*-Menthol (16 mg), pyridine (10 mg) and benzene (2 ml) are added to the residue and the reaction mixture is allowed to stand for 3 h. A small sample of the solution is analysed by GC (Table 12). The reader is referred to the original publications[39,45–47] for the GC retention volumes and the recommended chromatographic conditions for the separation of 3,7,11,15-tetramethylhexadecanoic (phytanic), 2,6,10,14-tetramethylpentadecanoic (pristanic), 5,9,13-trimethyltetradecanoic, 3,7,11-trimethyldodecanoic (farnesanic), 2,6,10-trimethylundecanoic-*l*-menthyl esters and other isoprenoid menthyl ester diastereoisomers.

TABLE 11

Gas chromatographic separation of 2,2,2-trifluoro-1-phenylethylhydrazones prepared by Procedure 4.4.2

Hydrazone derived from	Separation temperature/°C	Retention time of diastereoisomers/min		$r_{II/I}$
		1st peak	2nd peak	
2-Methylcyclopentanone	140	7.5	8.3	1.11
2-Methylcyclohexanone	140	12.0	13.1	1.09
2-Ethylcyclohexanone	140	17.3	18.2	1.03
3-Methylcyclopentanone	140	10.1	10.1	1.00
Menthone	150	12.0	13.0	1.08
Camphor	150	13.2	13.9	1.05
2-*n*-Butylcyclohexanone	160	11.7	12.2	1.04
1-Methyl-*cis*-bicyclo[4,2,0]octan-5-one	160	12.2	14.0	1.15
2-Heptylcyclopentanone	170	16.4	19.1	1.16
trans-Decal-1-one	170	13.3	14.7	1.11
Furopelargone	180	21.4	22.3	1.04
Methyl *cis*-2-acetyl-1-methylcyclobutylacetate	180	6.7	7.7	1.15
1,1,4,4,6-Pentamethyldecal-7-one	190	12.2	14.9	1.22
1-Acetoxy-5α,9β-dimethyl-*trans*-decal-6-one	230	7.8	8.7	1.12
5α-Androstan-17-one	250	11.9	12.4	1.04
5α-Androst-2-en-17-one	250	16.5	17.3	1.05
5α-Androstan-3-one	250	19.4	19.4	1.00
3β-Chloro-5α-androstan-17-one	260	31.8	33.4	1.05
17β-Acetoxy-5α-androstan-1-one	270	10.8	11.2	1.04
17β-Acetoxy-2α-methyl-5α-androstan-3-one	280	11.8	13.1	1.11
17β-Acetoxy-4α-methyl-5α-androstan-3-one	280	13.8	14.8	1.07

Column: 6 ft × $\frac{1}{8}$ in packed with 5% OV-17 on Chromosorb W using a nitrogen flow of 30 ml min^{-1}.

TABLE 12

Gas chromatographic separation of menthylisopropyl-4-substituted phenylacetates prepared by Procedure 4.5.3[43]

R—Ph—CH—COOH (with $CH(CH_3)_2$ on CH)

R=	Column	$r_{II/I}$
H	1	1.17
CH_3	2	1.15
OCH_3	2	1.16
F	1	1.16
Cl	1	1.16
Br	1	1.14
3,4-$O-CH_2-O$	2	1.15

Column 1: 1 m × 0.2 mm packed with 5% DGAP using separation temperature of 120 °C, and a nitrogen flow of 48 ml min^{-1}. Column 2: 1.7 m × 0.2 mm packed with 5% DEGS using a separation temperature of 170° C and a nitrogen flow of 50 ml min^{-1}.

4.5.4 REACTION OF THE CARBOXYL GROUP WITH AN ASYMMETRIC AMINE

2-Alkyl phenylacetic acids have been resolved as amides of desoxyephedrine[40] (11). The gas chromatographic resolution factor ($r_{II/I}$) increases with the branching of the alkyl side chain.

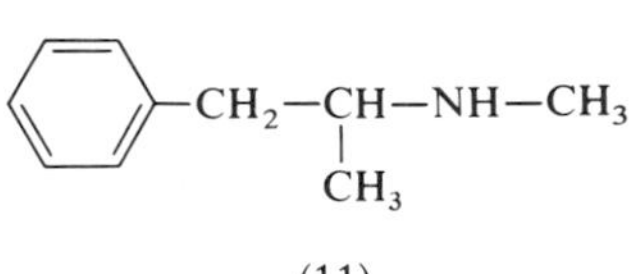

(11)

4.5.5 PREPARATION OF *d,l*-2-ALKYLPHENYLACETYL-*d*-2-METHYLAMINO-1-PHENYLPROPANES[40]

The alkylphenyl acetic acid (1 mmol) is treated with dicyclohexylcarbodiimide (1 mmol) in tetrahydrofuran (3 ml). After cooling *d*-2-methylamino-1-phenylpropane [(11), 1 mmol] is added and the suspension is stirred for 24 h. The mixture is diluted with ethyl acetate and the dicyclohexylurea formed is filtered off. A part of the solution is analysed by GC (Table 13).

TABLE 13

Gas chromatographic separation of 2-alkylphenylacetyl-2-methylamino-1-phenyl-propanes prepared by Procedure 4.5.5

Ph—CH(R)—COOH R =	$r_{II/I}$
Methyl	1.13
Ethyl	1.06
n-Propyl	1.10
Isopropyl	1.01
n-Butyl	1.03
Isobutyl	1.06
t-Butyl	1.82
Cyclopentyl	1.04

Column: 5 ft $\times \frac{1}{8}$ in packed with 1% HI-EFF-8AP on Gas Chrom Q using a separation temperature of 200 °C and a nitrogen flow of 35 ml min^{-1}.

4.6 Hydroxy fatty acids

Hydroxy fatty acids can be converted into a mixture of diastereoisomers by derivatizing either of the two functional groups with an asymmetric reagent. In general gas chromatographic resolution of the diastereoisomeric mixture is greatly improved if the remaining functional group is also derivatized.

4.6.1 REACTION OF THE CARBOXYL GROUP WITH AN ASYMMETRIC REAGENT

Hydroxy fatty acids can be resolved as esters of asymmetric alcohols.[48] The gas chromatographic resolution factor ($r_{II/I}$) increases with the branching of the alcohol used for esterification.[49,50] The hydroxyl function should be blocked by acetylation but some diastereoisomeric 2-hydroxy-propionates have been separated.[50,51] The method is best suited for the steric analysis of 2-hydroxy acids. Attempts to resolve diastereoisomeric esters of other positional hydroxy fatty acid isomers have been unsuccessful. Diastereoisomeric esters derived from 2-hydroxy acids carrying additional alcohol and/or acid functions, such as aldonic acids, can be separated. Since aldonic acids can be prepared by oxidation of aldoses, the method has been used to determine the stereochemistry of carbon 2 of aldoses.[49,52]

Experimental procedure. *d*-2-Butanol, *d*-2-pentanol,*d*-3-methyl-2-butanol, *d*-3,3-dimethyl-2-butanol and *d*-menthol can be prepared in over 98% optical purity by the procedure outlined in Section 4.1.12.[20]

4.6.2 PREPARATION OF ACYLATED HYDROXY FATTY ACID-2-ALKYL ESTERS[48]

The hydroxy acid (0.3 g), optically active alcohol (3 g), *p*-toluene sulfonic acid (0.3 g) and benzene (10 ml) are refluxed in a Dean and Stark apparatus until all of the water formed is removed. After removal of the solvent the residue is refluxed with an excess of the acyl chloride or the corresponding anhydride for 4 h. The excess of the reagent is distilled off, the residue is dissolved in benzene, and after washing with aqueous sodium carbonate (10%) and water a sample is injected into the GC. (Table 14).

TABLE 14

Separation of diastereoisomeric acetylated hydroxy acid alkyl esters[48,49,52] prepared by Procedure 4.6.2

Acetylated hydroxy acid	2-butyl esters $r_{II/I}$	3-methyl-2-butyl ester $r_{II/I}$	3,3-dimethyl-2-butyl ester $r_{II/I}$
Lactic acid	1.06	1.11	1.14
Glyceric acid	1.03	1.06	1.07
Malic acid	1.02	1.04	1.04
Threonic acid	1.03	1.06	1.09
Erythronic acid	1.02	1.04	1.06
Mandelic acid	1.02	1.05	1,07
Xylonic acid	1.03	1.05	1.09
Lyxonic acid	1.01	1.03	1.04
Ribonic acid	1.02	1.05	1.08
Arabonic acid	1.03	1.06	1.13

Column: 150 ft × 0.02 in packed with Carbowax 20 M, using a helium flow of 6.5–8.7 ml min^{-1} and separation temperatures of 98°, 150°, 160° and 175 °C respectively.

Polyhydroxy, lactone forming, acids such as aldonic acids must be acylated first before diastereoisomer formation is attempted.[49,52,53]

4.6.3 REACTION OF THE HYDROXYL GROUP OF HYDROXY FATTY ACIDS WITH AN ASYMMETRIC REAGENT

Menthylchloroformate (3)[7] *N*-1-phenylethylisocyanate (9)[3] and 2-phenylpropionyl chloride (12)[36] have been used as acylation reagents for this purpose.

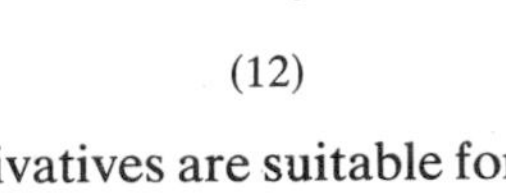

(12)

Menthyloxycarbonyl derivatives are suitable for the resolution of 2-hydroxy fatty acid methyl esters[7,54,55] and *N*-(1-phenylethyl) urethanes have been used for steric analysis of ω2-hydroxy acids*.[35,56–58] Attempts to resolve other positional hydroxy acid isomers with these reagents have been unsuccessful.[36] The 2-phenylpropionyl chloride reagent (12) has been developed for optical analysis of 3-hydroxy, ω4-hydroxy, ω3-hydroxy and ω2-hydroxy fatty acids.[36]

Experimental procedure. Preparation of the menthylchloroformate reagent (3) and *N*-1-phenylethylisocyanate (9) has been described in Sections 4.1.4 and 4.3.3 respectively.

4.6.4 PREPARATION OF (*d*)-2-PHENYL PROPIONYL CHLORIDE (12)[36]

d,l-2-Phenyl propionic acid (3 g, 20 mmol) and (*d*)-1-phenylethylamine (2.42 g, 20 mmol) were dissolved in acetone (80 ml) and the mixture heated to 70 °C for 5 min. After cooling to −20 °C, 2.8 g of the diastereoisomeric salt are obtained. After three recrystallizations the salt (0.3 g) is decomposed with 2 M hydrochloric acid and the (*d*)Δ2-phenyl propionic acid is extracted into ether. The extract is washed, dried over sodium carbonate and evaporated. The pale yellow syrup (α_D = +88 °C, in benzene, 0.2 g) is heated with thionyl chloride (0.24 ml) at 70 °C for 30 min. After removal of the excess reagent under vacuum the residue is dissolved in benzene and the solution is evaporated to remove the last traces of thionyl chloride. The product is dissolved in dry benzene (2.4 ml) to give a standard solution of (12) which contains 0.5 μmol ml^{-1}.

4.6.5 PREPARATION OF MENTHYLOXYCARBONYL DERIVATIVES OF 2-HYDROXY FATTY ACID METHYL ESTERS[55]

The 2-hydroxy acid (0.1 mmol) is treated with methanolic hydrogen chloride (1.25 M, 0.5 ml) and after 3 h at room temperature, the solvent is removed under vacuum. An excess of *l*-menthyl chloroformate (0.11 ml, 0.11 mmol) (3)[7] and pyridine (0.02 ml) are added and the reaction mixture is allowed to stand for 0.5 h. After washing and drying over sodium sulfate a sample is injected into the gas chromatograph (Table 15).

4.6.6 PREPARATION OF *N*-(1-PHENYLETHYL)URETHANES OF HYDROXY FATTY ACID METHYL ESTERS[54]

The methyl ester (3 μmol) is treated with *d*-1-phenylethylisocyanate (9) (30 μmol in 0.1 ml of toluene)[33] at 120 °C for 3 h in an inert atmosphere. The solvent is removed in vacuum and the residue dissolved in ethyl acetate (1 ml); a portion of this solution is analysed by GC. The diastereoisomeric urethanes have been separated on a 1.8 m column packed with 1% QF-1 on Chrom Q, at 210 °C. The separation factors for methyl-17-hydroxyoctadecanoate, methyl 19-hydroxyeicosanate and methyl-21-hydroxydocosanate are 1.08.[35]

* The carbon atom of the terminal methyl group of the fatty acid is designated 'ω1'.

TABLE 15

Gas chromatographic separation of diastereoisomeric menthyloxycarbonyl-2-hydroxycarboxylic acid methyl esters[7,35,59] prepared by method 4.6.5

Acid	Separation temperature[a] °C	$r_{II/I}$
Lactic	190	1.09
Lactic	170[b]	1.09
Malic	190	1.07
2-Hydroxyglutaric	190	1.02
2-Hydroxyvaleric	190	1.12
2-Hydroxyisovaleric	170[b]	1.14
2-Hydroxycaproic	190	1.07
2-Hydroxyisocaproic	170[b]	1.15
2-Hydroxycaprylic	210	1.10
2-Hydroxycapric	210	1.10
2-Hydroxylauric	240	1.08
2-Hydroxymyristic	240	1.09
2-Hydroxypalmitic	240	1.07
2-Hydroxystearic	240	1.08
3-Phenyllactic	200[b]	1.14
2-Hydroxyheptanoic	141[b]	1.16
2-Hydroxysebacic	200[b]	1.10

[a] Column: 10 ft × 4 mm packed with 1.5% OV-210 on Gas Chrom Q.
[b] Column: 5 ft × $\frac{1}{8}$ in packed with 5% QF-1 on Aeropack 30 using a nitrogen flow of 30 ml min^{-1}.

4.6.7 PREPARATION OF 2-PHENYLPROPIONATE DERIVATIVES OF HYDROXY FATTY ACID METHYL ESTERS[36]

The hydroxy acid methyl ester (100 μg) is treated with a benzene solution of *d*-2-phenylpropionyl chloride (12), (0.08 ml, 0.5 μmol ml^{-1}) and pyridine (0.02 ml). The solution is left for 2 h and the solvent is removed under vacuum. The residue is dissolved in chloroform (0.1 ml) and this solution is applied to a TLC plate (250 μm thick, Silica Gel G, activated at 120 °C for 1 h before use). The plate is developed with chloroform (stabilized with 1% ethanol) and the compounds are detected under u.v. light after spraying with 2′,7′-dichlorofluorescein. The TLC spots are eluted from the plate and the silica gel washed with ethyl acetate; a sample is injected into the gas chromatograph. The diastereoisomers derived from the 17, 16, 15, 14, 3 and 2-hydroxy- 18:0 fatty acid methyl esters and the 3-hydroxy-10:0, 13:0, 14:0, 16:0 and 18:0 fatty acid methyl esters have been separated ($r_{II/I}$ 1.03–1.08) on a glass column packed with 5% QF-1 on Gas Chrom Q run isothermally at 230° and 200 °C respectively.

4.7 Dipeptides

Volatile diastereoisomeric dipeptide derivatives can be resolved on capillary or packed columns.[60] For successful chromatographic resolution, it is necessary to derivatize all of the protic groups; this is usually done by trifluoroacetylation of the amino function and methylation of the carboxylic acid group. The silylation of hydroxyl functions in the dipeptide is necessary for efficient resolution of dipeptides containing hydroxy amino acid residues.[61] An alternative approach involves conversion of the dipeptides to 2,5-dioxopiperazines[62] which can be resolved by gas chromatography[63,64] with or without further reaction to form *N*-methyl[64] or trifluoroacetyl derivatives.[65]

4.7.1 PREPARATION OF TRIFLUOROACETYL-DIPEPTIDE METHYL ESTERS

The dipeptide (1 mg) is esterified with thionyl chloride–methanol (1 ml, 1:9) as described in 4.1.6 and the solvent and excess reagent are removed under vacuum. The residue is dissolved in methanol (0.5 ml) then methyltrifluoroacetate (0.2 ml) and triethylamine (0.1 ml) are added. After 3 h the solution is diluted with ethyl acetate (10 ml), washed and dried. After concentrating to 0.5 ml, a portion is analysed by gas chromatography.[60]

4.7.2 PREPARATION OF TRIFLUROACETYL-*O*-SILYL-DIPEPTIDE METHYL ESTERS[61]

The *N*-TFA-peptide methyl ester (1 mg) is refluxed with hexamethyldisilazane (0.5 ml); ammonia is evolved. After 10 min the excess reagent is removed under vacuum and the residue dissolved in dry ethylacetate (0.2 ml); a sample is analysed by gas chromatography.[60,61]

4.7.3 PREPARATION OF 2,5-DIOXOPIPERAZINES

The crude dipeptide ester, its acetate or formate salt (2 mg) is dissolved in *sec*-butyl alcohol (1–2 ml) and toluene (0.5–1 ml). The solution is boiled for 2–3 hours and the solvent level is maintained by the addition of fresh *sec*-butyl alcohol. After concentrating the solution to 0.5 ml a sample is analysed by GC (Table 16).

4.7.4 PREPARATION OF *N*-METHYLATED 2,5-DIOXOPIPERAZINES[64]

The dioxopiperazine (5 mg) is dissolved in dimethylformamide (0.3 ml), methyl iodide (0.025 ml) is added followed by sodium hydride (6 mg). The mixture is stirred at room temperature for 3 h and filtered. A sample is analysed by GC (Table 16).

4.7.5 PREPARATION OF TRIFLUOROACETYL DERIVATIVES OF 2,5-DIOXOPIPERAZINES[65]

The 2,5-dioxopiperazine (1 mg) is dissolved in trifluoroacetic anhydride (0.2 ml) in a sealed container and heated at 90–100° for 5–30 min till a

TABLE 16

Gas chromatographic separation of diastereoisomeric 2,5-dioxopiperazines

2,5-Dioxopiperazine	Underivatized[a] prepared by method 4.7.3[63] $r_{II/I}$	*N*-methylated[b] prepared by method 4.7.4[64] $r_{II/I}$	*N*-trifluoroacetylated[c] prepared by method 4.7.5[65] $r_{II/I}$
c-Ala-ala	1.00		1.16
c-Ala-val	1.00	1.30	1.07
c-Ala-ile			1.08
c-Ala-leu		1.20	1.05
c-Val-val	1.07		1.10
c-Val-ile			1.09
c-Val-leu		1.33	1.13
c-Leu-leu	1.04		1.17
c-Ile-ile			1.11
c-Ile-leu			1.10
c-Ala-phe	1.10		1.33
c-Val-phe			1.12
c-Ile-phe			1.24
c-Leu-phe	1.10		1.13
c-Ala-tyr			1.40
c-Ala-trp			1.28
c-Phe-phe	1.08		1.11
c-Ala-phegly	1.10		1.04
c-Ser-phe			1.46
c-Pro-ala	1.18		1.46
c-Pro-leu	1.17		
c-Pro-pro	1.10		

[a] Column: 5 ft × $\frac{1}{8}$ in packed with 5% QF-1 on Aeropak, run isothermally at 210° or 240 °C with a nitrogen flow of 30 ml min^{-1}.

[b] Column: 6 ft × 3.4 mm packed with 5% EGS on Gas Chrom Q run isothermally at 205 °C with a helium flow of 40 ml min^{-1}.

[c] Column: 5 ft × 3 mm packed with 5% SE-30 on DMCS treated acid washed Chromosorb W, run isothermally at 90°, 100°, 130°, 150° and 180 °C with a helium flow of 25 ml min^{-1}.

homogeneous solution results. A sample of the solution is then analysed by GC (Table 16).

4.8 Derivatives which increase electron-capture sensitivity for the gas chromatographic analysis of diastereoisomeric mixtures

A few papers describing the resolution of diastereoisomers using derivatives with high electron affinity have been published. The derivatives used were selected for their improved separation capabilities or greater volatility[66,67] rather than their electron capturing properties. Most of the amine[28a,67a] and amino acid analysis has received most attention and in all cases derivatization involved replacement of the *N*-trifluoroacetyl function with a *N*-perfluoroacetyl group. Pentafluoropropionic (13) and heptafluorobutyric (14) anhydrides have been used as acylating agents for this purpose.

4.8.1 PREPARATION OF *N*-PERFLUOROACYL-*l*-PROLYLCHLORIDE (15) OR (16)

l-Proline (1.15 g) is treated with pentafluoropropionic anhydride (13) or heptafluorobutyric anhydride [(14), 4 ml] dissolved in anhydrous ether (20 ml) at −10 °C. The reaction is then carried out as described in section 4.1.2.[5,11] The solution containing 1 mmol of the reagent (15) or (16) in 10 ml, can be stored for several months at 0 °C.

4.8.2 PREPARATION OF *N*-PERFLUOROACYL-*l*-PROLYL-AMINO ACID ALKYL ESTERS[66]

The amino acid (0.1 mmol) is esterified with thionyl chloride–methanol (1+9, 1 ml), HCl-isopropanol (3 N in anhydrous HCl, 1 ml), HCl-butanol (3 N, 1 ml) or HCl-menthol (1 ml) as described in Section 4.1. The dry residue is treated with the *N*-perfluoroacyl-*l*-prolyl chloride reagent (15) or (16), (0.11 mmol in 1.1 ml) as described in Section 4.1.6, and a sample of solution is analysed by GC (Table 17).

TABLE 17

Gas chromatographic separation of diastereoisomeric *N*-perfluoroacylprolyl-amino acid alkyl esters prepared by method 4.8.2[66]

	Pentafluoropropionyl derivatives				Heptafluorobutyryl derivatives			
Amino Acid	methyl $r_{II/I}$	butyl $r_{II/I}$	isopropyl $r_{II/I}$	menthyl esters $r_{II/I}$	methyl $r_{II/I}$	butyl $r_{II/I}$	isopropyl $r_{II/I}$	menthyl esters $r_{II/I}$
Alanine	1.14	1.18	1.14	1.08	1.15	1.18	1.15	1.09
Valine	1.16	1.18	1.14	1.10	1.17	1.20	1.16	1.10
Leucine	1.10	1.12	1.10	1.06	1.11	1.15	1.11	1.09
Proline	1.09	1.11	1.08	1.05	1.08	1.10	1.08	1.06

Column: 5.5 ft × $\frac{1}{4}$ in packed with 5% OV-1 on Supelcoport. Column temperature was 250 °C for menthyl esters and 185 °C for the other ester derivatives.

4.8.3 PREPARATION OF *N*-PERFLUOROACYL-(*d,l*)-AMINO ACID-2-ALKYL ESTERS[67]

A suspension of the amino acid (10 μmol) in *d*-2-alcohol (3 N in anhydrous HCl, 3 ml) is heated at 100 °C for 3 h. The solvent is then evaporated in a stream of dry nitrogen at 100 °C. The dry residue is treated with a solution of the anhydride (13) or (14) (0.2 ml) in methylene chloride (1.8 ml) at 100 °C for 15 min. The samples are concentrated at room temperature, care being taken to avoid loss of volatile amino acid derivatives and a sample of the residue is analysed by GC. The diastereoisomers derived from isovaline, *N*-methylalanine, *β*-amino-*n* or isobutyric acid and pipecolic acid can be resolved as *N*-pentafluoropropionyl and *N*-heptafluorobutyryl amino acid 2-hexyl, 2-heptyl and 2-octyl esters on a 250 ft × 0.02 in capillary column coated with Ucon 75H-90 000.

5 GC RESOLUTION OF ENANTIOMERS ON OPTICALLY ACTIVE STATIONARY PHASES

Chromatographic resolution which does not require prior formation of diastereoisomers is especially attractive in terms of speed and overall sensitivity of the assay. Gas chromatographic resolution of amino acid enantiomers on asymmetric stationary phases was first demonstrated by Gil-Av using a glass capillary column coated with *N*-TFA-isoleucine lauryl ester.[72] Subsequently it was shown that TFA-dipeptide cyclohexyl ester phases[73–75] have better thermal stability and higher separation factors; recently *N*-docosanoyl-*t*-butyramides[133] have improved this still further. These enantiomer resolutions on chiral phases are believed to occur by way of formation of transient diastereoisomeric hydrogen-bonded association complexes between the phase and the solute. ^{13}C and ^{1}H nuclear magnetic resonance studies of hydrogen bonding in chiral ureide–amide systems suggest that the only significant hydrogen-bond interaction is between the ester carbonyl of the phase and the amide proton of the solute.[76] At the time of writing, this technique has been used for the resolution of amine and amino acid enantiomers only. The main problem is the limited thermal and optical stability of some of the stationary phases[77] which leads to extremely long analysis times at the safe operating temperatures, and difficulty in quantitation of the results due to excessive stationary phase bleeding. Optically active stationary phases are becoming increasingly available commercially, and before embarking on their synthesis readers are recommended to consult catalogues of firms specializing in gas chromatographic materials to see whether suitable resolving stationary phases are listed.

5.1 Amino acids

N-TFA-dipeptide cyclohexyl esters,[73–75] carbonyl-bis (amino acid ester),[76] *N*-TFA-isoleucine lauryl ester[72] and *N*-lauroyl-valyl-*t*-butyramide[77,78] stationary phases give optimum GLC resolution for *N*-TFA amino acid esters. Most of the separation factors of enantiomers have been in the range of 1.06 to 1.20 and for complete resolution of the optical isomers a capillary column of high efficiency is used. To avoid partial decomposition of cysteine, serine and threonine derivatives during chromatography, the use of specially prepared high-efficiency glass capillaries is recommended.[77] It should be noted that untreated glass columns coated with any of the dipeptide stationary phases may deteriorate after one to two days use.[77] The recently developed phase *N*-docosanoyl-L-valine *t*-butyramide[133] is usable up to 190 °C and gives efficient separations of *N*-TFA amino acid isopropyl esters.

5.1.1 PREPARATION OF GLASS CAPILLARY COLUMNS[77]

A 1% solution of 3-aminopropyl triethoxysilane in methylene chloride is forced through the column (0.3 mm × 40 m) and the ends are then sealed. The column is heated to 120–130 °C and afterwards the excess reagent is removed by washing with methylene chloride. The column is subsequently coated by the static method[79] with a solution of 0.2% stationary phase in methylene chloride.

5.1.2 PREPARATION OF *N*-TFA-DIPEPTIDE CYCLOHEXYL ESTER STATIONARY PHASES.[80]

(a) *N-TFA-Amino acid cyclohexyl ester hydrogen chloride*: The selected C-terminal amino acid (5 g) in dry cyclohexanol (150 ml) is heated to 100 °C in an oil bath whilst dry hydrogen chloride is passed through the reaction mixture for 3 h or until a clear solution results. After removal of the excess alcohol in vacuum at below 50 °C the residue is triturated with petroleum ether.

(b) *Dipeptide cyclohexyl ester trifluoroacetate*: The amino acid cyclohexyl ester hydrogen chloride (10 mmol), 1-hydroxybenztriazole (10 mmol, 1.35 g) and *N*-ethyl morpholine (4 mol, 1.1 ml) are added to a solution of the *N*-*t*-butyloxycarbonyl derivative of the selected *N*-terminal amino acid (10 mmol) in dry tetrahydrofuran (30 ml) at 0 °C. Dicyclohexylcarbodiimide (2.06 g, 10 mmol), in tetrahydrofuran (15 ml) is added at 0° C and the reaction mixture is kept for 1 h at room temperature. After removal of the precipitated urea, the solvent is evaporated in vacuum and the residue is dissolved in ethyl acetate (250 ml). After washing with 2 N citric acid, $NaHCO_3$, and water the solvent is removed and the *t*-butyloxycarbonyl residue group is removed with trifluoroacetic acid (10 ml) in methylene chloride (10 ml). After removal of the solvent and addition of anhydrous ether the salt can be obtained in a solid form.

(c) *N-TFA-dipeptide cyclohexyl esters*: The product from (b) above is treated with trifluoroacetic anhydride (10 ml) in methylene chloride (40 ml) at 0 °C. After one hour at room temperature, the reagent and solvent are evaporated.

5.1.3 PREPARATION OF CARBONYL-BIS-LEUCINE ISOPROPYL ESTER STATIONARY PHASE[76]

d-Leucine (6 g) is refluxed with dry 2-propanol (200 ml) for 5 h whilst dry hydrogen chloride is passed through the reaction mixture. Excess 2-propanol is removed and the solid leucine isopropyl ester hydrochloride is neutralized with sodium bicarbonate solution. The solution is extracted with ether and the solvent removed, benzene, pyridine (2 mol) and one equivalent of phosgene (~12% w/w in benzene) are added and the solution is refluxed for 2 h. After filtration, and removal of the solvent in vacuum the crude product is purified by silica gel chromatography in chloroform.

5.1.4 PREPARATION OF AMINO ACID *t*-BUTYL ESTERS[80]

The amino acid mixture (300 mg) is dissolved in absolute dioxane (10 ml) and concentrated sulfuric acid (0.7 ml) in a Parr pressure bottle (500 ml capacity). Liquid isobutene (10 ml) is added and the bottle is assembled in a hydrogenation apparatus. After 5 h the pressure is released and the reaction mixture is poured into a ice-cold mixture of ether (50 ml) and sodium hydroxide (40 ml, 1 N). The aqueous layer is extracted 3 times with ether and the combined dried ether extracts evaporated to dryness. The residue is treated as described in Section 5.1.6.

5.1.5 PREPARATION OF AMINO ACID ISOPROPYL ESTERS[80]

The amino acid (5 mg) is dissolved in isopropanol-hydrogen chloride (3 N, 3 ml) in a tube sealed with a Teflon®-lined screw cap. The solution is heated to 100 °C for 50 min and then the isopropanol–HCl is removed using a rotary evaporator or a stream of nitrogen. The residue is treated as described in Section 5.1.6.

5.1.6 PREPARATION OF TFA AND PENTAFLUOROPROPIONYL (PFP) AMINO ACID *t*-BUTYL AND ISOPROPYL ESTERS[80]

The residues from Sections 5.1.4 or 5.1.5 are diluted with methylene chloride (2 ml) and trifluoroacetic or pentafluoropropionic anhydride (1 ml) is added. After 1 h at room temperature the excess solvent and reagent are removed. The residue is dissolved in methylene chloride (1 ml) and an aliquot is analysed by GC (Table 8). The TFA-amino acid isopropyl esters of alanine, aminobutyric acid, valine, norvaline, leucine, isoleucine and *t*-leucine can also be resolved on other dipeptide phases.[80]

5.2 Amines

Chiral ureide stationary phases in liquid crystal form have been used to resolve *N*-TFA, *N*-PFP and *N*-heptafluorobutyryl (HFB) derivatives of asymmetric amines.[82,83] The solid form of carbonyl-bis(L-valine isopropyl ester) gave enhanced resolution for amides compared to the liquid form of the stationary phase. The high degree of order present in smectic liquid crystals of carbonyl-bis-(leucine isopropyl ester) assists the separation process[82,83] on packed columns. The preparation of a chiral ureide stationary phase is described in Section 5.1.3.

5.2.1 PREPARATION OF *N*-TFA, PFP AND HFB DERIVATIVES OF AMINES

Methods of preparation are described in the chapter on acylation (Chapter 3); GC properties of enantiomers are listed in Table 19.

TABLE 18

Gas chromatographic separation of TFA and PFP-amino acid isopropyl esters on *N*-TFA-valyl-valine cyclohexyl ester stationary phase[81]

Amino acid	TFA derivative $r_{II/I}$	PFP derivative $r_{II/I}$
Alanine	1.12	1.11
Valine	1.09	1.09
Threonine	1.10	1.09
Isoleucine	1.18	1.18
Leucine	1.12	1.12
Serine	1.08	1.07
Proline	1.04	1.05
Cysteine	1.08	1.08
Aspartic acid[a]	1.06	1.00
Methionine[a]	1.12	1.11
Phenylalanine[a]	1.12	1.13
Glutamic acid[a]	1.09	1.09

Column: 400 ft × 0.02 in, run at 110 °C with helium excess pressure 10 psi (0.7 bar). Retention times varied between 41 and 409 min.

[a] Column: 10 ft × 0.02 in run at 110 °C with helium pressure 5 psi (0.35 bar).

TABLE 19

Gas chromatographic separation of perfluoroacetyl-amides on carbonyl-bis-(valine alkyl ester)[82]

Amine derivative	Methyl ester phase $r_{II/I}$		Ethyl ester phase $r_{II/I}$		Isopropyl ester phase $r_{II/I}$	
N-TFA-2-amino octane			1.04[a]	(1.00)[b]		
N-TFA-2-amino ethylbenzene	1.06[a]	(1.07)[b]	1.17	(1.09)	1.11[a]	(1.42)[c]
N-PFP- 2-amino ethylbenzene	1.07	(1.07)	1.26	(1.10)	1.11	(1.86)
N-HFB-2-amino ethylbenzene	1.06	(1.07)	1.34	(1.10)	1.13	(2.11)
N-TFA-2-amino-3-phenyl propane			1.00	(1.00)	1.00	
N-TFA-2-amino-4-phenyl propane			1.108	(1.00)	1.19	

Column: 6 ft × $\frac{1}{8}$ in i.d. packed with 5% carbonyl-bis-amino acid alkyl ester on DMCS treated Chromosorb G.

[a] Temperature 385 K (smectic phase).

[b] Temperature 398 K (isotropic liquid phase).

[c] Temperature 370 K (smectic phase).

6 BIBLIOGRAPHY OF NON-GAS CHROMATOGRAPHIC METHODS

Paper chromatography

C. E. Dalgliesh, *J. Chem. Soc.* 3940 (1952).
L. E. Ruhland, E. Work, R. F. Denman and D. S. Hoare, *J. Am. Chem. Soc.* **77**, 4844 (1955).
S. F. Contractor and J. Wragg, *Nature* **208**, 71 (1965).
D. R. Galpin and A. C. Huitric, *J. Pharm. Sci.* **57**, 447 (1968).
D. R. Galpin and A. C. Huitric, *J. Org. Chem.* **33**, 921 (1968).
L. E. Fellows and E. A. Bell, *Phytochemistry* **9**, 924 (1970).

Column chromatography

H. Krebs, J. A. Wagner and J. Diewald, *Chem. Ber.* **89**, 1875 (1958).
Th. Wieland and H. Bende, *Chem. Ber.* **98**, 504 (1965).
H. Krebs and W. Schumacher. *Chem. Ber.* **99**, 1341 (1966).
Chr. G. Kratchanov and M. Il. Popova, *J. Chromatogr.* **37**, 297 (1968).
Chr. G. Kratchanov, M. Il. Popova, Tz. Obretenov and N. Ivanov, *J. Chromatogr.* **43**, 66 (1969).
M. Il. Popova and Chr. G. Kratchanov. *J. Chromatogr.* **72**, 192 (1972).
G. Blaschke, *Angew. Chem.* **83**, 547 (1971).
M. Il. Popova, Chr. G. Kratchanov and M. Kuntcheva. *J. Chromatogr.* **87**, 581 (1973).
E. Blasius, K.-P. Janzen and G. Klautke, *Z. Anal. Chem.* **277**, 374 (1975).
H. Furukawa, E. Sakakibara, A. Kamei and K. Ito, *Chem. Pharm. Bull.* (*Tokyo*) **23**, 1625 (1975).

Thin-layer chromatography

D. E. Nitecki, B. Halpern and J. W. Westley, *J. Org. Chem.* **33**, 864 (1968).
J. W. Westley, V. A. Close, D. E. Nitecki and B. Halpern, *Anal. Chem.* **40**, 1888 (1968).
W. Freytag and K. H. Ney, *J. Chromatogr.* **41**, 473 (1969).
G. Simonneaux, A. Meyer and G. Jaouen, *Chem. Commun.* 69 (1975).
E. Taschner, A. Chimiak, J. F. Biernat, T. Sokolowska, C. Wasielewski and B. Rzeszotarska. Proc. 5th European Peptide Symposium (H. C. Beyerman, Ed) Noordwijk 1966, North Holland Publ. Co., Amsterdam, 1967, p. 195.

Counter-current distribution

K. Bauer, H. Falk and K. Schlögl, *Monatsh. Chem.* **99**, 2186 (1968).

REFERENCES

1. M. Raban and K. Mislow, *Topics in Stereochemistry* Vol. 2 (N. L. Allinger and E. L. Eliel, Eds) Wiley–Interscience, New York, 1967, pp. 199–230.
2. E. Gil-Av and D. Nurok. *Adv. Chromatogr.* **10**, 99 (1974).
3. J. W. Westley, *Chemistry and Biochemistry of Amino Acids, Peptides and Proteins* (B. Weinstein, Ed.) Dekker, New York, 1971, pp. 1–21.
4. B. L. Karger, *Anal. Chem.* **39[8]**, 24A (1967).
5. B. Halpern and J. W. Westley, *Biochem. Biophys. Res. Comm.* **19**, 361 (1965).
6. B. Halpern and J. W. Westley, *Chem. Commun.* 246 (1965).
7. J. W. Westley and B. Halpern, *J. Org. Chem.* **33**, 3978 (1968).

8. T. Nambara, J. Goto, K. Taguchi and T. Iwata, *J. Chromatogr.* **100**, 180 (1974).
9. B. Halpern and J. W. Westley, *Tetrahedron Lett.* 2283 (1966).
10. H. Iwase, *Chem. Pharm. Bull.* **23**, 217 (1975).
11. W. Bonner, *J. Chromatogr. Sci.* **10**, 159 (1972).
12. M. Renard, *Bull. Soc. Chim. Biol.* **28**, 497 (1946).
13. E. Gil-Av, R. Charles-Sigler, G. Fischer and D. Nurok, *J. Gas Chromatogr.* **4**, 51 (1966).
14. G. E. Pollock, V. I. Oyama and R. D. Johnson, *J. Gas Chromatogr.* **3**, 174 (1965).
15. G. E. Pollock and A. H. Kawauchi, *Anal. Chem.* **40**, 1356 (1968).
16. B. Halpern and J. W. Westley, *Chem. Commun.* 421 (1965).
17. S. V. Vitt, M. B. Saprovskaya, J. P. Gudkova and V. M. Belikov, *Tetrahedron Lett.* 2575 (1965).
18. G. S. Ayers, R. E. Monroe and J. H. Mossholder, *J. Chromatogr.* **63**, 259 (1971).
19. A. W. Ingersoll in *Organic Reactions* Vol. II. (R. Adams, Ed.) Wiley, New York, 1944, p. 376.
20. B. Halpern and J. W. Westley, *Aust, J. Chem.* **19**, 1533 (1966).
21. H. Iwase and A. Murai, *Chem. Pharm. Bull.* **22**, 8 (1974).
22. M. Hasegawa and I. Matsubara, *Anal. Biochem.* **63**, 308 (1975).
23. C. J. W. Brooks, M. T. Gilbert and J. D. Gilbert, *Anal. Chem.* **45**, 896 (1973).
24. M. T. Gilbert, J. D. Gilbert and C. J. W. Brooks, *Biomed. Mass Spectrom.* **1**, 274 (1974).
25. J. D. Gilbert and C. J. W. Brooks, *Anal. Lett.* **6**, 639 (1973).
26. R. Weidmann and A. Horeau, *Bull. Soc. Chim.* 117 (1967).
27. W. E. Pereira, Jr. and B. Halpern, *Aust. J. Chem.* **25**, 667 (1972).
28. A. Murano, *Agric. Biol Chem.* **37**, 981 (1973).
28a. R. W. Souter, *J. Chromatogr.* **108**, 265 (1975).
29. J. W. Westley and B. Halpern, *Gas Chromatography* 1968. (S. L. A. Harbourn, Ed.) Institute of Petroleum, London, 1969, p. 119.
30. A. Murano and S. Fujiwara, *Agric. Biol. Chem.* **37**, 1977 (1973).
31. A. H. Beckett and B. Testa, *J. Chromatogr.* **69**, 285 (1972).
32. M. W. Anders and M. J. Cooper, *Anal. Chem.* **43**, 1093 (1971).
33. W. Pereira, V. A. Bacon, W. Patton and B. Halpern, *Anal. Lett.* **3**, 23 (1970).
34. C. Djerassi and J. Staunton, *J. Am. Chem. Soc.* **83**, 736 (1961).
35. M. Hamberg, *Chem. Phys. Lipids* **6**, 152 (1971).
36. S. Hammarström and M. Hambert, *Anal. Biochem.* **52**, 169 (1973).
37. J. Casanova and E. J. Corey, *Chem. Ind.* (*London*) 1664 (1961).
38. W. E. Pereira, M. Salomon and B. Halpern, *Aust. J. Chem.* **24**, 1103 (1971).
39. I. MacLeen, G. Eglington, R. G. Ackman and S. N. Hooper, *Nature* (*London*) **218**, 1019 (1968).
40. B. Halpern and J. W. Westley, *Chem. Commun.* 237 (1967).
41. A. Murano, *Agric. Biol. Chem.* **36**, 2203 (1972).
42. C. J. W. Brooks, J. D. Gilbert and M. T. Gilbert, *Mass Spectrometry in Biochemistry and Medicine* (A. Frigerio and N. Castagnoli, Eds) Raven Press, New York 379 (1974).
43. M. Miyakado, N. Ohno, Y. Okuno, M. Hirana, K. Fujimoto and H. Yoshioka, *Agric. Biol. Chem.* **39**, 267 (1975).
44. A. Horeau and J. P. Guetté, *C.R. Acad. Sci.* (*Paris*) *Ser. C* **267**, 257 (1968).
45. R. E. Cox, J. R. Maxwell, G. Eglington and C. R. Pillinger, *Chem. Commun.* 1639 (1970).
46. R. G. Ackman, S. N. Hooper, M. Kates, A. K. Sen Gupta, G. Eglington and I. MacLeen, *J. Chromatogr.* **33**, 256 (1969).
47. R. G. Ackman, R. E. Cox, G. Eglington, S. A. Hooper and J. R. Maxwell, *J. Chromatogr. Sci.* **10**, 392 (1972).
48. E. Gil-Av, R. Charles-Sigler, G. Fischer and D. Nurok, *J. Gas Chromatogr.* **4**, 51 (1966).
49. G. E. Pollock and D. A. Jermany, *J. Gas Chromatogr.* **6**, 412 (1968).
50. B. L. Karger, R. L. Stern, H. C. Rose and W. Keane, in *Gas Chromatography 1966* (A. B. Littlewood Ed.) Institute of Petroleum, London, 1967, p. 240.

51. E. Gil-Av and D. Nurok, *Proc. Chem. Soc.* 146 (1962).
52. G. E. Pollock and D. A. Jermany, *J. Chromatogr. Sci.* **8**, 296 (1970).
53. R. L. Whistler and M. L. Walfrom, *Methods Carbohydr. Chem.* **2**, 25 (1968).
54. S. Hammarström, *FEBS. Lett.* **5**, 192 (1969).
55. R. G. Annett and P. K. Stumpf, *Anal. Biochem.* **47**, 638 (1972).
56. M. Hambert and I. Björkhem, *J. Biol. Chem.* **246**, 7411 (1971).
57. M. Hambert and I. Björkhem, *J. Biol. Chem.* **246**, 7417 (1971).
58. I. Björkhem and M. Hamberg, *Biochem. Biophys. Res. Commun.* **47**, 333 (1972).
59. M. Hamberg, *Anal. Biochem.* **43**, 515 (1971).
60. F. Weygand, A. Prox, L. Schmidhammer and W. König, *Angew. Chem. Int. Ed. Engl.* **2**, 183 (1963).
61. F. Weygand, A. Prox, E. C. Jorgensen, R. Axén and P. Kirchner, *Z. Naturforsch.* **186**, 93 (1963).
62. D. E. Nitecki, B. Halpern and J. W. Westley, *J. Org. Chem.* **33**, 864 (1968).
63. J. W. Westley, V. A. Close, D. Nitecki and B. Halpern, *Anal. Chem.* **40**, 1888 (1968).
64. A. B. Mauger, R. B. Desai, I. Rittner and W. J. Rzeszotarski, *J. Chem. Soc. Perkin Trans 1* 2146. (1972).
65. G. P. Slater, *J. Chromatogr.* **64**, 166 (1972).
66. H. Iwase and A. Murai, *Chem. Pharm. Bull.* **22**, 1455 (1974).
67. G. E. Pollock, *Anal. Chem.* **44**, 2368 (1972).
67a. S. Martin, M. Rowland and N. Castagnoli, Jr., *J. Pharm. Sci.* **62**, 821 (1973).
68. B. L. Karger, R. L. Stern, H. C. Rose and W. Keane, *Gas Chromatography 1966* (A. B. Littlewood, Ed.) Institute of Petroleum, London, p. 240 (1967).
69. B. L. Karger, R. L. Stern, W. Keane, B. Halpern and J. W. Westley, *Anal. Chem.* **39**, 228 (1967).
70. B. Feibush and L. Spialter, *J. Chem. Soc. Perkin Trans.* **2** 106 (1971).
71. B. L. Karger, S. Herliczek and R. L. Stern, *Chem. Commun.* 625 (1969).
72. E. Gil-Av, B. Feibush and R. Charles-Sigler in *Gas Chromatography 1966* (A. B. Littlewood, Ed.) Institute of Petroleum, London p. 227 (1967).
73. W. Parr, C. Yang, E. Bayer and E. Gil-Av, *J. Chromatogr. Sci.* **8**, 591 (1970).
74. S. Nakaparksin, P. Birrell, E. Gil-Av and S. Oro, *J. Chromatogr. Sci.* **8**, 177 (1970).
75. W. Parr and P. J. Howard, *Angew. Chem.* **84**, 586 (1972).
76. C. H. Lochmüller, J. M. Harris and R. W. Souter, *J. Chromatogr.* **71**, 405 (1972).
77. W. A. Koenig and G. J. Nicholson, *Anal. Chem.* **47**, 951 (1975).
78. B. Feibush, *Chem. Commun.* 544 (1971).
79. J. Bouche and M. Verzele, *J. Gas Chromatogr.* **6**, 501 (1968).
80. W. Parr and P. Y. Howard, *Anal. Chem.* **45**, 711 (1973).
81. W. Parr, J. Pleterski, C. Yang and E. Bayer, *J. Chromatogr. Sci.* **9**, 141 (1971).
82. C. H. Lochmüller and R. W. Souter, *J. Chromatogr.* **88**, 41 (1974).
83. C. H. Lochmüller and R. W. Souter, *J. Chromatogr.* **87**, 243 (1973).
84. E. Gil-Av, R. Z. Korman and W. Weinstein, *Biochim. Biophys. Acta*, **211**, 101 (1970).
85. S. Nakaparksin, E. Gil-Av and J. Oro, *Anal. Biochem.* **33**, 374 (1970).
86. E. Bayer, E. Gil-Av, W. A. König, S. Nakaparksin, J. Oro and W. Parr, *J. Am. Chem. Soc.* **92**, 1738 (1970).
87. A. Barooshian, M. J. Lautenschleger and W. G. Harris, *Anal. Biochem.* **44**, 543 (1971).
88. D. M. Feigl and H. S. Mosher, *Chem. Commun.* 615 (1965).
89. D. M. Feigl and H. S. Mosher, *J. Org. Chem.* **33**, 4242 (1968).
90. J. P. Guetté and A. Horeau, *Tetrahedron Lett.* 3049 (1965).
90a. A. Horeau, *Tetrahedron* **31**, 1307 (1975).
91. M. Belikov, T. F. Savaleva and E. N. Safanova, *Izv. Akad. Nauk. S.S.S.R. Otd. Khim. Nauk.* 1461 (1971).
92. B. Halpern, J. W. Westley, I. Wredenhagen and J. Lederberg, *Biochem. Biophys. Res. Commun.* **20**, 710 (1965).

93. E. Gil-Av, R. Charles-Sigler, G. Fischer and D. Nurok, *J. Gas Chromatogr.* **4**, 51 (1966).
94. J. W. Westley and B. Halpern, in *Gas Chromatography 1968* (A. B. Littlewood, Ed) Institute of Petroleum, London 1969, p. 119.
95. R. Rudzats, E. Gellert and B. Halpern, *Biochem. Biophys. Res. Commun.* **47**, 290 (1972).
96. R. Charles-Sigler and E. Gil-Av, *Tetrahedron Lett.* 4231 (1966).
97. E. Solem, E. Jellum and L. Eldjarn, *Clin. Chim. Acta* **50**, 393 (1974).
98. B. Halpern and G. E. Pollock, *Biochem. Med.* **4**, 352 (1970).
99. W. E. Pereira and B. Halpern, *Aust. J. Chem.* **25**, 667 (1972).
100. A. Horeau, *Tetrahedron Lett.* 506 (1961).
101. A. Horeau and H. B. Kagan, *Tetrahedron* **20**, 2431 (1964).
102. C. J. W. Brooks and J. D. Gilbert, *Chem. Commun.* 194 (1973).
103. B. Feibush, E. Gil-Av and T. Tamari, *J. Chem. Soc. Perkin Trans. 2*, 1197 (1972).
104. H. Rubinstein, B. Feibush and E. Gil-Av, *J. Chem. Soc. Perkins Trans. 2*, 2094 (1973).
105. B. Halpern, J. Ricks and J. W. Westley, *Anal. Biochem.* **14**, 156 (1966).
106. B. Halpern, J. Ricks and J. W. Westley, *Chem. Commun.* 679 (1966).
107. B. Halpern, J. Ricks and J. W. Westley, *Aust. J. Chem.* **20**, 389 (1967).
108. A. J. Markovetz, P. K. Stumpf and S. Hammarström, *Lipids I* 159 (1972).
109. B. Halpern, J. W. Westley, P. J. Anderson and J. Lederberg, *Anal. Biochem.* **17,** 179 (1966).
110. E. Gordis, *Biochem. Pharmacol.* **15**, 2124 (1966).
111. L. Gunn, *Biochem. Pharmacol.* **16**, 863 (1967).
112. T. L. V. Ulbricht, *Comparative Biochemistry* Vol. 4, Part B, (M. Florkin and H. S. Mason, Eds) Academic Press, New York, 1962, pp. 1–25.
113. J. Lederberg, *Nature* (*London*) **207**, 9 (1965).
114. B. Halpern, *Appl. Opt.* **8**, 1349 (1969).
115. K. Kvenvolden, J. Lawless, K. Pering, E. Peterson, J. Flores, C. Ponnamperuma, I. R. Kaplan and C. Moore, *Nature* (*London*) **228**, 923 (1970).
116. J. Lawless, K. Kvenvolden, E. Peterson, C. Ponnamperuma and C. Moore, *Science* **173**, 626 (1971).
117. J. Wehmiller and P. E. Hare, *Science* **173**, 907 (1971).
118. K. Kvenvolden, E. Peterson and F. S. Brown, *Science* **169**, 1079 (1970).
119. I. Maclean, G. Eglington, K. Douraghi-Zadeh, R. G. Ackman and S. N. Hooper, *Nature* (*London*) **218**, 1019 (1968).
120. A. Prox. F. Weygand, W. König and L. Schmidhammer, *Proc. 6th European Peptide Symposium* Athens, 1963 (L. Zervas, Ed.) Pergamon, New York, p. 139.
121. S. Lande and R. A. Landowne, *Tetrahedron* **22**, 3085 (1966).
122. B. Halpern, L. F. Chew and J. W. Westley, *Anal. Chem.* **39**, 399 (1967).
123. B. Halpern, L. F. Chew, V. Close and W. Patton, *Tetrahedron Lett.* 5163 (1968).
124. D. Nurok, G. I. Taylor and A. M. Stephen, *J. Chem. Soc. B* 291 (1968).
125. C. H. DePuy and P. R. Story, *Tetrahedron Lett.* 20 (1959).
126. Y. Gault and W. Felkin, *Bull. Soc. Chim. Fr.* 742 (1965).
127. H. C. Rose, R. L. Stern and B. L. Karger, *Anal. Chem.* **38**, 469, (1966).
128. R. L. Stern, B. L. Karger, W. Keane and H. D. Rose, *J. Chromatogr.* **39**, 17 (1969).
129. J. W. Westley, B. Halpern and B. L. Karger, *Anal. Chem.* **38**, 469 (1966).
130. B. Feibush, *Anal. Chem.* **43**, 1098 (1971).
131. E. Gil-Av, B. Feibush and R. Charles-Sigler, *Tetrahedron Lett.* 1009 (1966).
132. B. Feibush and E. Gil-Av, *J. Gas Chromatogr.* **5**, 257 (1967).
133. R. Charles, U. Beitler, B. Feibush and E. Gil-Av, *J. Chromatogr.* **112**, 121 (1975).
134. M. T. Gilbert, J. D. Gilbert and C. J. W. Brooks, *Biomed. Mass Spectrom.* **1**, 274 (1974).
135. C. H. Lochmüller and R. W. Souter, *J. Chromatogr.* **113**, 283 (1975).

CHAPTER 14

Ion-pair Extraction and Ion-pair Chromatography

G. Schill, R. Modin, K. O. Borg and B.-A. Persson

Department of Analytical Pharmaceutical Chemistry, Biomedical Centre, University of Uppsala, Box 574, S-751 23 Uppsala, Sweden

1 INTRODUCTION

Ion-pair extraction is a form of derivatization in which the compound of interest forms an ionic bond with a counter-ion of opposite charge selected to confer solubility in an organic solvent on the resulting ion-pair. The conditions are chosen to maximize the partition ratio in favour of the solvent phase, and to prevent dissociation of the ion-pair. It is described as an 'ion-pair' rather than as a salt because the counter-ion is usually present in non-stoichiometric excess. Although initially the technique was used to extract basic and acidic compounds from aqueous solution with counter-ions of opposite charge (ion-pair *extraction*), it is clear that immobilizing one or other of the two immiscible phases and passing the other over the resulting stationary phase leads to a classical partition chromatographic system. This chapter deals with both ion-pair extraction and ion-pair chromatography.

In recent years there has been a significant growth in the application of liquid–liquid chromatography using ion-pair techniques for the isolation and analysis of readily ionizable organic compounds. The technique has found particular use in the separation of compounds such as carboxylic acids and amines,[1,2] types of compounds which are ionized by the transfer of a hydrogen ion and are termed 'protolytes'. A protolyte is a compound which may be ionized by the transfer of a proton, the process may thus be termed 'protolysis'. Weak protolytes have high pK_a or pK_b, and strong protolytes have low pK_a and pK_b values. Ionic substances such as sulfonates and quaternary ammonium compounds have also been analysed by this method.[3,4] Due to the fairly favourable chromatographic properties of these ion-pair systems (which are explained below), some very efficient separation systems have been developed. Although most applications employ column chromatography, some interesting and useful thin-layer techniques have also been described.[5,6]

A major advantage of the ion-pair technique is the possibility of selecting counter-ions which have a high response to specific detectors such as those using photometry or fluorimetry, and this permits determination of submicrogram amounts of compounds which otherwise have a low detector response. This has been valuable in batch determination methods for ammonium compounds, by extraction with counter-anions of high molar absorbance like bromothymol blue,[7] dipicrylamine[8] and picrate,[9] or with highly fluorescent anions such as anthracene sulfonate[10] and dimethoxyanthracene sulfonate.[11] This way of increasing sensitivity has also been utilized in chromatographic separations and determinations. Acetyl choline and choline were isolated from biological material as picrate ion-pairs,[12,13] alkylammonium compounds were separated on a column containing naphthalene-2-sulfonate in the stationary phase,[14] and carboxylates were chromatographed as ion-pairs with dimethylprotriptyline.[15]

2 ION-PAIR EXTRACTION

The factors which govern the process of ion-pair extraction are the initial concentrations of the ion and of the counter-ion in their respective phases, the pH of the aqueous phase, the nature of the organic solvent, and the specific interactions of the various ions with one another and with the two phases. To see how these factors can be manipulated to optimize the extraction, we must analyse the physicochemical aspects of the system. In the following treatment a cationic species in aqueous solution is extracted with an anionic counter-ion; the same treatment of course applies to the converse situation.

2.1 Extraction constant

A cation, Q^+, for example a quaternary ammonium salt or a protonated amine, can be extracted from aqueous solution into an organic solvent as an ion-pair, QX, after addition of a counter-ion X^-. Without the addition of the counter-ion the partition of the salts of the cationic compounds is almost entirely in favour of the aqueous phase. The extraction can be illustrated by Eqn (1).

$$Q^+_{aq} + X^-_{aq} \leftrightarrows QX_{org} \qquad (1)$$

The equilibrium is characterized by the extraction constant, E_{QX}, which is defined by Eqn (2):

$$E_{QX} = \frac{[QX]_{org}}{[Q^+][X^-]} \quad (2)$$

The magnitude of E_{QX} depends on the properties of the ions, Q^+ and X^-, and on the nature of the organic solvent.

Table 1 illustrates the influence of the anionic nature on the extraction of the tetrabutylammonium cation into chloroform. The listed extraction constants extend over ten orders of magnitude from the hydrophilic inorganic anions to the very hydrophobic bromothymol blue and dipicrylamine. The possibility of selecting an anion within such a broad range gives the ion-pair technique a high degree of flexibility, and this permits extraction of compounds with widely differing properties.

TABLE 1

Extraction constants of tetrabutylammonium ion pairs (organic phase, $CHCl_3$)

Class	Anion component	log E_{QX}	Ref.
Inorganic	Cl^-	−0.11	(16)
	Br^-	1.29	(16)
	I^-	3.01	(16)
	ClO_4^-	3.48	(16)
Sulfonic acid	Toluene-4-sulfonic acid	2.33	(17)
	Naphthalene-2-sulfonic acid	3.45	(17)
	Anthracene-2-sulfonic acid	5.11	(10)
Phenol	Picric acid	5.91	(16)
	2,4-Dinitro-1-naphthol	6.45	(17)
Carboxylic acid	Acetic acid	−2.12	(18)
	Benzoic acid	0.39	(18)
	Salicylic acid	2.42	(18)
	Benzylpenicillin	2.03	(19)
Acid dye	Methyl orange	5.47	(10)
	Bromothymol blue	8.0	(10)
	Dipicrylamine	9.6	(17)

2.2 Side reactions

Equation (1) expresses the fundamental equilibrium of ion-pair extraction, but in many cases the ions or the extracted ion-pair or all of them can participate in other equilibria which have to be taken into consideration when the optimum extraction conditions have to be calculated. All processes that involve other equilibria of ions in the *aqueous* phase will tend to reduce the effective extraction constant, so reducing the extraction yield. Examples of such processes are protolysis (hydrogen ion transfers), association in the aqueous phase with different ions and with itself (polymerization), and micelle formation.[17,21] Correspondingly, some other equilibria in the *organic* phase tend to improve the extraction yield. Processes of this kind include dimerization and dissociation of the ion-pair. When the ion-pair components are weak protolytes, they can be subjected to protolysis in the aqueous phase and also be transferred to the organic phase in uncharged form. For example, considering the anion X^-, which is the anion of a weak acid and which is partially associated in the aqueous phase:

$$X^- + H^+ \underset{K'_{HX}}{\rightleftharpoons} HX \overset{k_{d(HX)}}{\rightleftharpoons} HX_{org} \tag{3}$$

K'_{HX} is the apparent acid dissociation constant, and $k_{d(HX)}$ is the partition coefficient between the two phases. Thus the conditional extraction constant will vary with hydrogen ion concentration (pH), and this will affect the partition ratio of Q^+. Dimerization appears at higher concentrations, and is in most cases of minor importance in analytical work.[22] Dissociation, on the other hand, is a typical low-concentration effect, and has to be considered for extractions in the sub-microgram scale.[23] Side reactions in the organic phase by the formation of adducts, can contribute to more efficient extraction systems, particularly with protolytes which act both as counter-ions and adduct-forming agents. The extraction of a cation, Q^+, by an acid, HX, with such properties is illustrated by Eqn (4):

$$Q^+ + n\,HX_{org} = QX(HX)_{n-1\,org} + H^+ \tag{4}$$

The extent of the extraction depends on the value of n, the concentration of HX and on the pH of the aqueous phase. Since n is usually greater than unity, the concentration dependence is much more pronounced than in a simple ion-pair extraction.[24] Adduct-forming properties have been observed among organic phosphoric and carboxylic acids.[18,25] Ion-pairing and adduct-forming agents have also been used in the extraction of anions, e.g. the extraction of penicillins with dodecylamine.[19]

2.3 Organic solvent effects

The efficiency of the extraction depends on hydration of the ion-pair components in the aqueous phase and on the hydrogen-bonding ability and polarity of the ion-pair in the organic phase. The components of an ion-pair bind each other by electrostatic forces, but hydrogen-bonding can also be important. An increase in hydrogen-bonding causes a lowering of the polarity of the ion-pair, and gives a higher extraction efficiency.[26]

The influence of the counter-ion and of the organic solvent on the extraction of ions with different hydrogen-bonding abilities is illustrated in Fig. 1, where extraction constants are plotted for tricyclic ammonium compounds with different anions and organic solvents. In this study high extraction constants

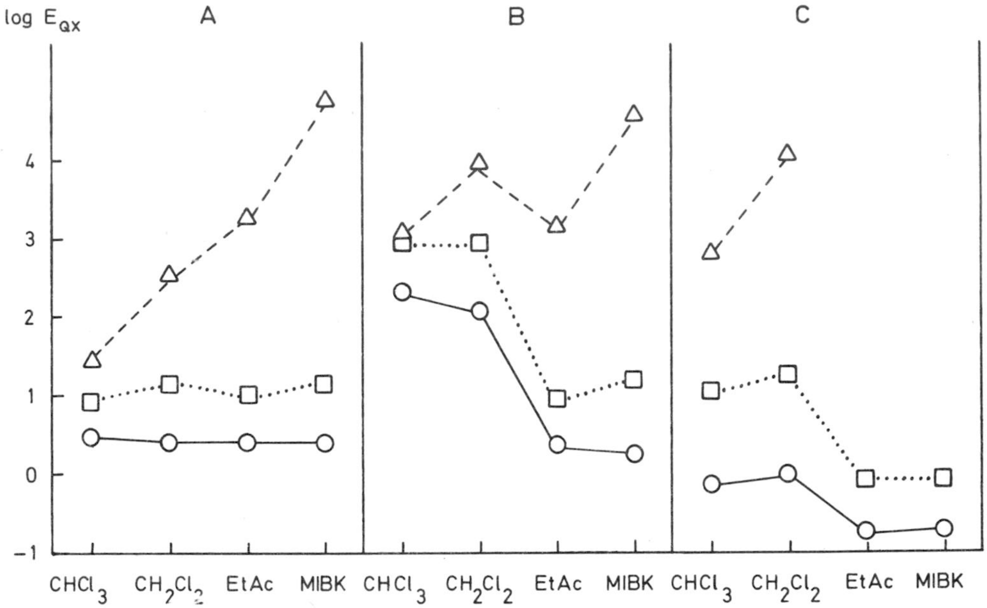

Fig. 1 Extraction of inorganic anions as ion-pairs into different organic solvents. A, nortriptyline ion-pairs; B, amitriptyline ion-pairs; C, *N*-methylamitriptyline ion-pairs; ○ = chloride; □ = bromide; △ = perchlorate.

were obtained when the hydrogen-bonding was strong between the ion-pair components and between the ion-pair and the solvent.

The influence of the polarity of the solvent and of hydrogen-bonding is further illustrated in Fig. 2, which shows extraction constants for picrate ion-pairs with three quaternary ammonium ions.[16,27] Low polarity solvents such as carbon tetrachloride and benzene give more than 10^3 times lower extraction constants than methylene chloride and 1-pentanol. Considering extraction of trimethylethylammonium and of its hydroxy derivative (choline), the weakly hydrogen-bonding solvents $CHCl_3$ and CH_2Cl_2 give greater selectivity than the hydrogen-accepting ethyl acetate and methylisobutyl ketone.

Lipophilic alcohols are good extractants for ion-pairs, because of their ability to act both as proton donors and proton acceptors. Their properties can be modified by dilution with an inert or less hydrogen-bonding solvent such as hexane or chloroform. The solvation effect of the alcohol in such mixtures can often be expressed as an adduct formation with fixed stoichiometry.[28] Extraction of an ion-pair by an organic phase that contains an alcohol (ROH) can be illustrated by Eqn (5):

$$HA^+_{aq} + X^-_{aq} + n\,ROH_{org} = HAX(ROH)_{n,\,org} \qquad (5)$$

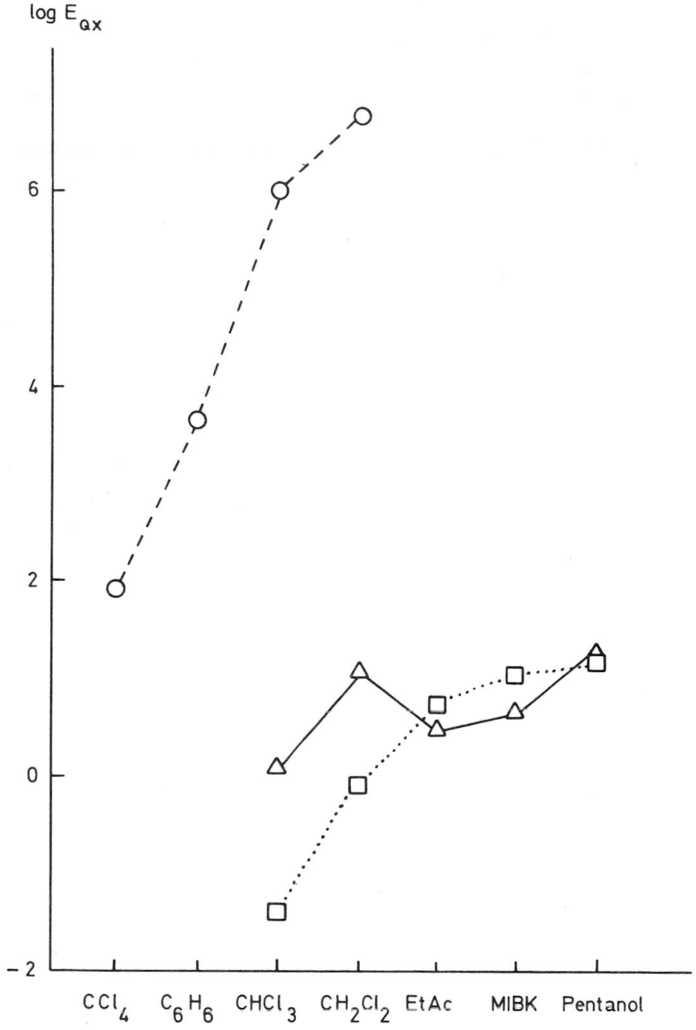

Fig. 2 Extraction of picrate ion-pairs into different organic solvents. ○ = tetrabutylammonium; △ = trimethylethylammonium; □ = choline.

Equation (6) shows the equilibrium:

$$E_{HAX,ROH} = \frac{[HAX(ROH)_n]_{org}}{[HA^+][X^-][ROH]^n_{org}} \quad (6)$$

When only ion-pairs solvated by ROH are present in the organic phase, Eqn (6) can be transformed into

$$\log E_{HAX} = \log E_{HAX,ROH} + n \log ROH_{org} \quad (7)$$

A relationship of this kind with a constant value of *n* has been found in numerous cases.[29,30] The effect of different kinds of lipophilic alcohols is demonstrated in Fig. 3. Alcohols with straight chains give higher extraction constants than branched chain alcohols. The slope, *n*, is however about 2 in all cases.

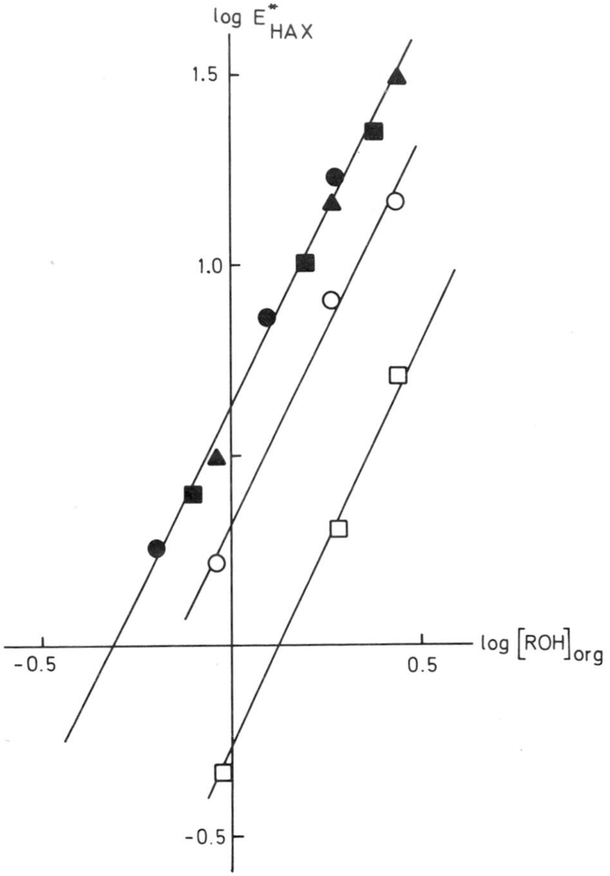

Fig. 3 Conditional extraction of chlorpromazine chloride with cyclohexane—lipophilic alcohol as the organic phase.[29] ▲ = 1-pentanol; ■ = 1-hexanol; ● = 1-octanol; ○ = 3-pentanol; □ = 2-methyl-1-butanol.

The stoichiometry of the alcohol adducts changes with the hydrogen-bonding ability of the ion-pair.[2] An example is given in Fig. 4, which shows the extraction of the secondary amine alprenolol, and of its hydroxy derivative, as perchlorates. The slope is 2 for the amine and 3 for the hydroxy-derivative, and the extraction selectivity decreases with increasing alcohol content of the organic phase.

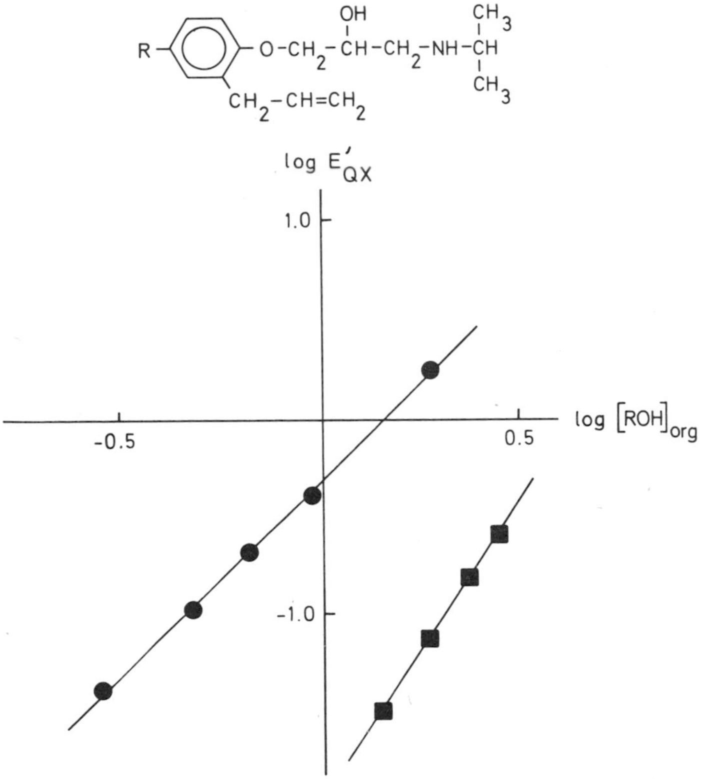

Fig. 4 Conditional extraction constant of alprenolol and 4-hydroxy-alprenolol with cyclohexane-1-pentanol-[2]. ● = alprenolol; ■ = 4-hydroxyalprenolol.

2.4 Ion-pair structure and extraction constant

If we look at the relationship between the structure of the ion-pair and the extraction constant, we see that the extraction constant increases with increasing number of alkyl or aryl groups, and there is, as a rule, within a homologous series a linear relationship between the number of methylene groups and the logarithm of the extraction constant (Fig. 5). With methylene chloride or chloroform as organic solvent, log *E* increases 0.5–0.6 by the addition of every CH_2. Similar relationships have been found with a wide variety of organic solvents.[31] In ion-pairs such as alkylammonium ions with hydrogen-accepting anions the extraction constant usually increases in the following order: quaternary < primary < secondary < tertiary using $CHCl_3$ and CH_2Cl_2 for extraction. The order may change with hydrogen-accepting solvents, as seen in Fig. 6.

Highly hydrophilic substituents, i.e. hydroxy, carboxylic and amino groups, decrease the extraction constant by 1–2 log units when the organic phase is

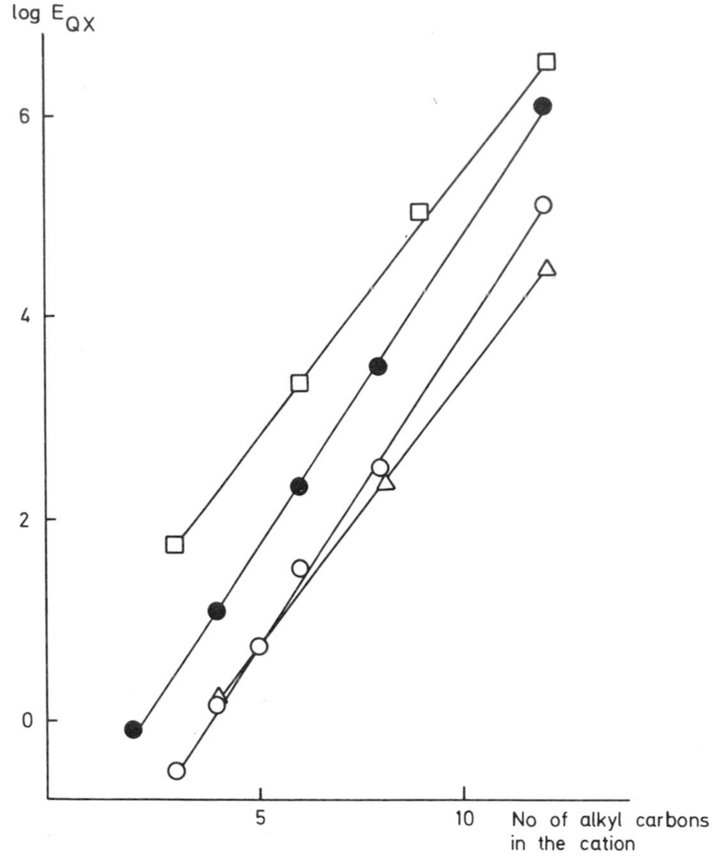

Fig. 5 Extraction constants of ion-pairs between alkylammonium ions and picrate.[16] ○ = primary amine; ● = secondary amine; □ = tertiary amine; △ = quaternary ammonium ion.

chloroform or a similar solvent,[7] but the decrease is considerably lower where the organic phase is strongly hydrogen-bonding e.g. pentanol.[32]

Masking of strongly hydrophilic groups by alkylation or acetylation has a very drastic effect on the extraction constant, as demonstrated in Table 2. Methylation of one hydroxyl in morphine increases its extraction constant by 2.5 log units, and acetylation of both hydroxyls gives an increase of 4 log units.

The variation of the extraction constant with anions has been demonstrated in Table 1 and Figs 1 and 6. Extraction constants increase according to the series $Cl^- < Br^- < I^- < ClO_4^-$ because of decreasing hydration in the aqueous phase, but the relative magnitudes also depend on hydrogen-bonding between the anion and cation.

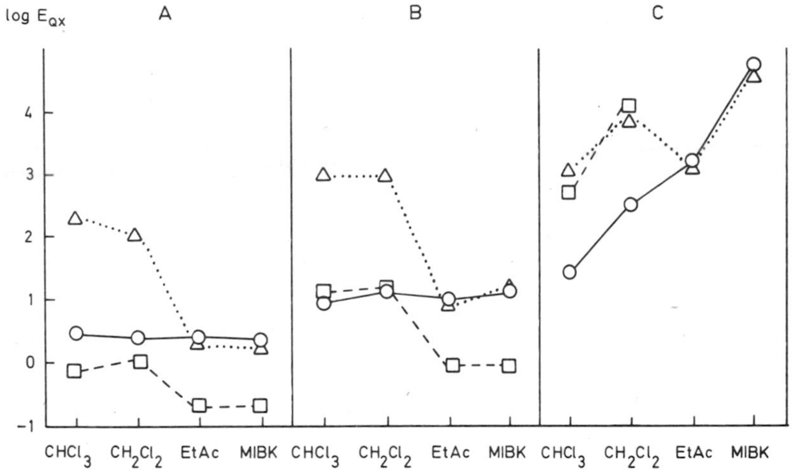

Fig. 6 Extraction of nortriptyline, amitriptyline and *N*-methyl-amitriptyline as ion-pairs into different organic solvents. A, chloride ion-pairs; B, bromide ion-pairs; C, perchlorate ion-pairs; ○ = nortriptyline; △ = amitriptyline; □ = *N*-methylamitriptyline.

TABLE 2

Extraction constants of morphine derivatives[7] (organic phase, CH_2Cl_2; anion, bromothymol blue, HB^-)

Cation (HA^+)	log E_{HAHB}	Functional groups
Morphine	4.4	ArOH; ROH
Codeine	6.8	$ArOCH_3$; ROH
Diacetylmorphine	8.5	$ArOCOCH_3$; $ROCOCH_3$
Dihydrocodeinone	7.6	$ArOCH_3$; R=O

The extraction constant of organic anions usually increases in the series carboxylate < sulfonate < sulfate. Hydroxylation *ortho* to an aromatic carboxyl, sulfonate or sulfate group will increase the constant by about two log units because of intramolecular hydrogen bonding (see benzoate–salicylate).[18] while *meta* hydroxylation decreases the constant by 2 log units.

3 ION-PAIR PARTITION CHROMATOGRAPHY

3.1 Chromatographic principles

Ion-pair partition chromatography can be done both in conventional chromatography and in the reversed-phase technique (in which the stationary phase is the lipophilic rather than the hydrophilic one), but the conventional mode has dominated because of the availability of suitable supports. Column packings specifically designed for reversed-phase chromatography are now becoming more widely available and promise to be extensively applied to the ion-pair technique.

A rational approach to ion-pair partition chromatography requires some basic knowledge of the principles of liquid–liquid partition chromatography. A thorough treatment of this subject is available in other textbooks.[33,34] The following brief description, extending the principles developed in the previous section, will introduce the subject.

The partition of a solute on a chromatographic column can be expressed in terms of the capacity factor, k', which is a measure of the relative retention time:

$$k' = \frac{(t_R - t_0)}{t_0} \tag{8}$$

where t_R and t_0 are the retention times of a solute and of an unretained component respectively. In liquid–liquid chromatography the partition ratio, D_Q, the bed-volume of the stationary phase, V_s, and the retention-volume of the mobile phase, V_m, will determine the magnitude of k'.

$$k' = \frac{V_s}{V_m D_Q} \tag{9}$$

In ion-pair chromatography k' may be correlated with the extraction constant and with the concentration of the counter-ion [see Eqn (6)].

$$k' = \frac{V_s}{V_m E_{QX} C_X} \tag{10}$$

If the phase volume ratio V_s/V_m is known and the extraction constant has been determined, the retention time can be calculated from Eqns (8) and (10). Good agreement between predicted and observed values of t_R has been demonstrated in several cases.[1,2]

The separation factor α, i.e. the ratio between the capacity factors of two different migrating solutes, is used as a measure of the selectivity of a chromatographic system. In ion-pair chromatography, α is given as the ratio of the appropriate extraction constants:

$$\alpha = \frac{k'_2}{k'_1} = \frac{E_{QX1}}{E_{QX2}} \tag{11}$$

The separation of two components on a chromatographic column is usually expressed by the resolution, R_s, which is composed of the separation factor of the phase-system and of the efficiency of the column:

$$R_s = \frac{(\alpha - 1)\sqrt{N_{eff}}}{4\alpha} \tag{12}$$

where N_{eff} is correlated with the number of theoretical plates, N, in the following way:

$$N_{eff} = \left[\frac{k'}{1+k'}\right]^2 N \tag{13}$$

For a complete separation of two sample components, an R_s value of 1.5 is sufficient, provided that they are present in comparable amounts. A sufficiently high R_s value can be achieved either by a high α or by a high N_{eff}, or by a combination of both, but improved resolution is more readily obtained by an increase in α, because increases in N_{eff} will count only in proportion to its square root. Liquid–liquid chromatographic systems lend themselves to the exploitation of selectivity to achieve good resolution.

It is convenient in liquid chromatography to choose conditions designed to give capacity factors within the range 2–10. Interference from relatively poorly retained impurities usually rules out capacity factors below 2, and such low values also correspond to low values of N_{eff} as seen from Eqn (13). On the other hand, values of k' above 10 are avoided since the retention times will be inconveniently long. Suitable values of the capacity factor are obtained in ion-pair chromatography by selecting the appropriate counter-ion and its concentration. A major part is also played by the mobile phase. Counter-ions with extraction constants in the range 0.1–1 are usually chosen, since k' values of about the right size are then obtained with a high concentration of the counter-ion in the aqueous phase. This is advantageous since such phase-systems give a high sample capacity, as discussed by Eksborg and Schill.[1]

Ion-pair partition chromatography has also been done on thin layers, the technique being developed by Gröningsson and co-workers.[5,6] The R_f value is given by:

$$R_f = \left(1 + \frac{V_s}{V_m E_{QX} C_Q}\right)^{-1} \tag{14}$$

Appropriate R_f values are obtained by selecting phase-systems of the same kind as in column chromatography. There is a linear relationship between the R_f value and the counter-ion concentration in the aqueous stationary phase, as obtained by transformation of Eqn (14):

$$\frac{R_f}{(1-R_f)} = \frac{E_{QX} C_Q V_m}{V_s} \tag{15}$$

A plot of $R_f/(1-R_f)$ versus C_Q gives a straight line through the origin. This is illustrated for tetrabutylammonium ion-pairs of two sulfonates in Fig. 7. Clearly the migration rate of an ion may easily be manipulated by varying the concentration of the counter-ion.

Thin layer chromatography is well suited to rapid screening of partition properties. Unlike column chromatography, strongly retained solutes are readily detectable.[6] The two techniques are thus complementary.

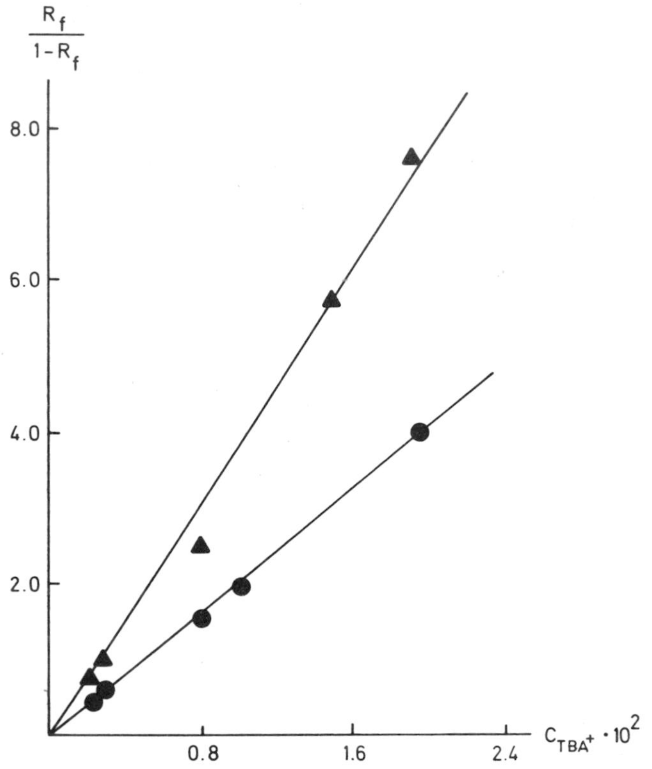

Fig. 7 Sulfonic acids as ion-pairs with tetrabutylammonium (TBA).[5] Stationary phase—tetrabutylammonium hydroxide; mobile phase—chloroform. ● = Toluene-4-sulfonic acid; ▲ = Naphthalene-2-sulfonic acid.

3.2 Practical considerations

3.2.1 COLUMN CHROMATOGRAPHY

In liquid–liquid chromatography the two liquid phases have to be equilibrated before use by mixing the mobile and stationary phases for several hours at the temperature to be used for chromatography. The phases are then separated by filtering the organic phase through glass wool or by centrifugation. The final equilibration of the mobile phase is made by passing it through a large pre-column of the same composition as the separation column.

Different methods have been used to prepare separation columns. Porous cellulose, diatomaceous earth or silica gel of medium particle size have usually been used to support the stationary phase.[1,2,14] The support is mixed with 25–40% of aqueous stationary phase, the mixture is suspended in the mobile phase, and the whole is homogenized for about a minute. The column is filled by adding small portions of the slurry and tamping with a rod at constant pressure, taking care to keep the packing covered with mobile phase. The pre-column is packed in the same way.[1]

Small accurately graded particles with average diameters in the region of 5–20μ have found wide use in liquid chromatography in the last few years, because columns with high efficiency can be obtained. Such microparticles are in most cases packed into the columns by a balanced density slurry technique.[3] The particles are dried at 150 °C for several hours, cooled, and then stored in a desiccator. A suspension of the particles in a solvent mixture consisting of tetrabromoethane:CCl_4:dioxane (4:3:3 by volume) is prepared by homogenization in an ultrasonic bath for a few minutes. The slurry is then transferred to a stainless-steel container or wide-bore tubing and forced into the column at the greatest possible flow-rate. Due to the small size of the particles there is a high pressure-drop, and stainless-steel columns of 3–4.5 mm i.d. and 10–30 cm length are commonly used. Pre-packed columns are becoming increasingly available from commercial sources. The silica gel particles are coated with stationary phase *in situ.*[3] Stationary aqueous phase and acetone (3:1, 20–40 ml) is passed through the column at 1–2 ml min^{-1}. When the particles are coated, 100–200 ml of hexane saturated with stationary phase are passed through with an initial flow-rate of 0.5–1 ml min^{-1} in order to remove acetone and surplus stationary phase from between the particles. Pre-columns are prepared from less expensive grades of silica gel mixed 2:1 with stationary phase and dry-packed. Equilibrated mobile phase (50–100 ml) is passed through the pre-column at 5 ml min^{-1}, the separation column is then attached, and the system is equilibrated by recycling for several hours.

In liquid–liquid partition chromatography, long-term stability of the column system requires equilibration of the two liquid phases and thermostatic control. A constant temperature is important for a stable base-line, particularly where there is high mutual solubility of the two liquid phases and where the counter-ion has a pronounced detector-response.[1]

Solutes should be injected as ion-pairs with the counter-ion which is used in the aqueous phase in order to preserve the stability of the system. However, for practical reasons it is often desirable to inject acids and bases in the uncharged forms. This is quite possible if the amount of sample is kept low, and the counter-ion concentration in the column is kept high.

In analytical methods it sometimes happens that a counter-ion which is more hydrophobic than that in the stationary phase has to be used in order to make the initial ion-pair extraction quantitative. Disturbance of the base-line can then be prevented if the more hydrophobic counter-ion is first removed by a displacement technique.[12,35] The sample should be dissolved in the mobile phase, otherwise peak deformation and variable retention times can be expected. If the sample solvent has a much higher extracting ability than the mobile phase it should be removed and replaced by the mobile phase.

Reversed-phase columns have been prepared in a similar way, and the same precautions have to be taken in equilibration, thermostatic control and sample composition.[36–38]

Gradient elution, enabling the elution of sample components with widely differing partition properties, is not very common in liquid–liquid chromatography, because of the difficulties encountered in equilibration. The use of stepwise and linear gradients has only been reported in a few papers.[2,36]

3.2.2 THIN-LAYER CHROMATOGRAPHY

Cellulose has generally been used as stationary phase in most of the applications of ion-pair partition to TLC.[5,39] Two different techniques have been developed for applying the stationary phase containing the counter-ion. In *slurry impregnation* the thin layer is prepared from a slurry of cellulose powder in a diluted solution of the stationary phase.[5] The layer is then dried on a hot-plate or in a stream of warm air until the right amount of stationary phase is left. In the other technique, conventionally prepared or commercially available thin-layer plates of cellulose are sprayed with the stationary phase.[6] This is more convenient, and the extent of the impregnation is quite reproducible. Reversed-phase thin-layer chromatography of ion-pairs has been done using a dry layer of acetylated cellulose first developed with the lipophilic stationary phase in a tank.[40] The plate was then dried until the right amount of stationary phase was left.

When the ion-pair TLC plate is to be used samples are applied to the layer before the final evaporation of the excess of stationary phase in slurry impregnation, or before the spraying with stationary phase in the other method.[5,6] The plates can be developed in sandwich chambers in conventional TLC, while in reversed-phase systems tanks give better results.[40]

3.3 Selection of phase-systems for analytical applications

The selection of phase-systems for ion-pair chromatography is based on extraction constants and other equilibrium constants. When calculating the separation conditions it is most important that the partition ratio should be independent of the concentration of the sample components, i.e. side-reactions that are dependent upon sample concentration have to be avoided or suppressed.[1] Since literature data on ion-pair extraction systems are rather limited, in most cases it is first necessary to determine extraction constants and constants for side-reactions. This can be very tedious if side-reactions are involved, and the simplified method of Lagerström can be recommended.[41] The partition ratio of the sample compounds is determined by batch extraction with the counter-ion present in the aqueous phase at the concentration intended for use in the chromatographic system. The sample concentration is varied within limits that include the intended chromatographic conditions. A system is suitable for chromatographic use with the sample concentration range over which the partition ratio remains constant.

3.3.1 STATIONARY PHASE

The stationary phase is selected first, since final adjustments of the capacity factor can be made by modifying the properties of the mobile phase. Regarding an aqueous stationary phase, the nature and concentration of the counter-ion and the pH are the crucial variables. If one or both of the ion-pair components are weak protolytes, a pH is chosen at which ionic forms dominate in the aqueous phase and at which partition of uncharged forms to the organic phase can be disregarded. When the counter-ion is the anion of a strong acid, a low pH is used, as for example in the separation of amines and other ammonium compounds. Biogenic amines can be separated on a column with a stationary phase containing perchloric acid as shown in Fig. 8. For hydrophobic tricyclic

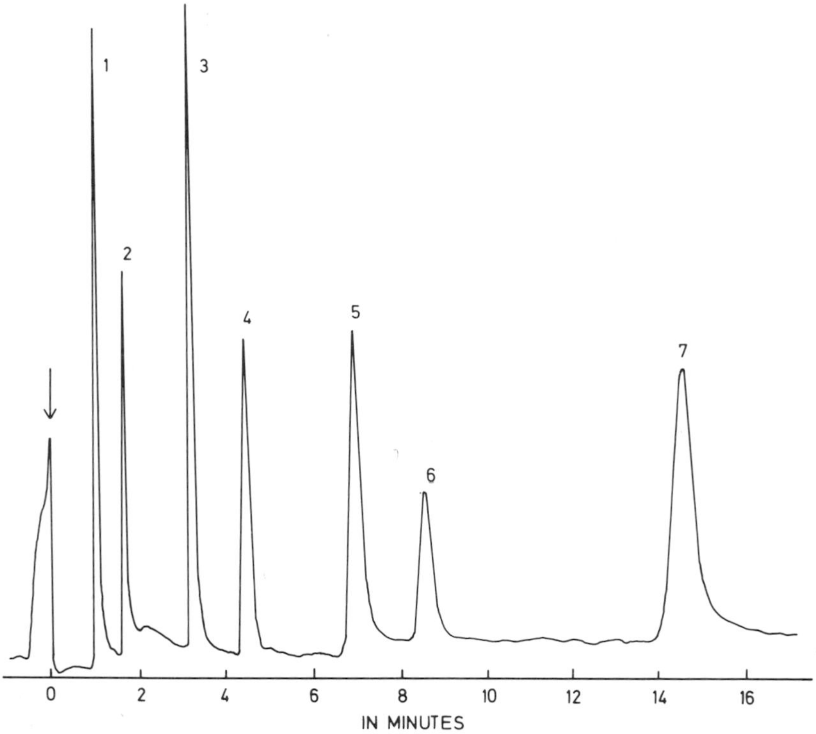

Fig. 8 Separation of biogenic amines. Support—LiChrospher SI 100 (mean diam. 10 μm) stationary phase—0.2 M $HClO_4$ + 0.8 M $NaClO_4$; mobile phase—butanol : methylene chloride (30 : 70). 1, toluene; 2, phenethylamine; 3, tyramine; 4, 3-methoxytyramine; 5, dopamine; 6, metanephrine; 7, adrenaline.

amines hydrochloric[29] and perchloric acid[42] have been used. Figure 9 shows the separation of three amines related to imipramine, extracted from plasma and chromatographed with methanesulfonic acid solution as stationary phase.[43]

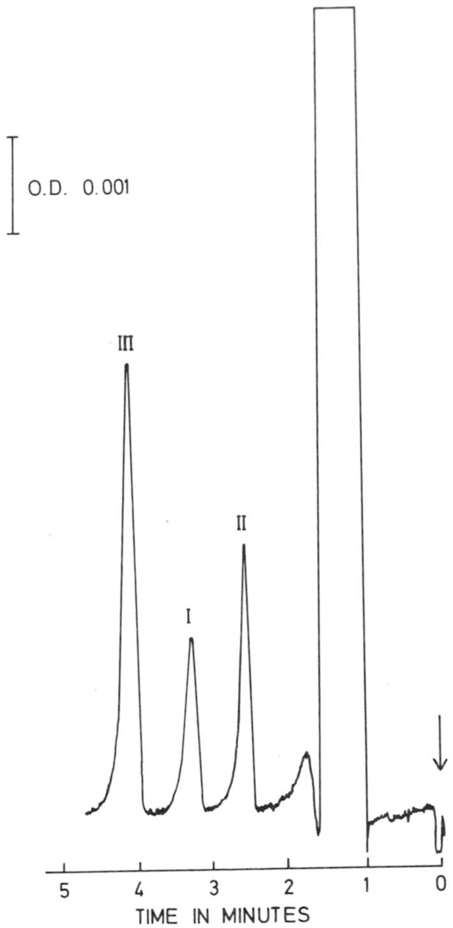

Fig. 9 Separation of chloroimipramine (I), its desmethylderivative (II) and desipramine (III) from plasma extract.[43] Stationary phase—0.1 M methanesulfonic acid; mobile phase—butanol : methylene chloride : hexane (10 : 45 : 45); sample—extract from 1 ml of plasma spiked with chloroimipramine (50 ng), desmethylchloroimipramine (50 ng) and desipramine (100 ng).

In most cases the pH of the stationary phase is maintained by buffers. The nature of the buffers can have a considerable influence on the blank extraction of the counter-ion. Thus much lower blanks were obtained with phosphate than with carbonate in extractions of anions with *N,N*-dimethylprotriptyline.[44]

High solubility of the counter-ion in the aqueous stationary phase is of great importance for sample capacity, as mentioned earlier.[1] Inorganic anions and sulfonates of low or moderate molecular weight can be used in 0.1–1 M aqueous solutions. Among the cationic counter-ions, symmetrical quaternary ammonium compounds have corresponding properties, and have been used in the separation of sulfonates and sulfonamides on thin layers[5,6] and in columns.[45] Another example is given in Fig. 10, where four carboxylates are separated as ion-pairs with tetrabutylammonium.[3]

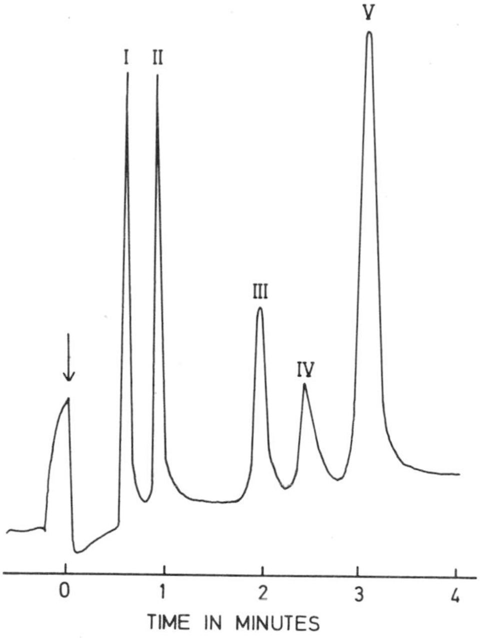

Fig. 10 Separation of carboxylic acids.[3] Support—silica gel (mean diam. 6 μm); stationary phase—0.1 M tetrabutylammonium HSO_4 (pH 7.5); mobile phase—butanol : CH_2Cl_2 : hexane (20 : 30 : 50); I, toluene; II, indole-3-acetic acid; III, homovanillic acid; IV, vanilmandelic acid; V, 5-hydroxyindole-3-acetic acid.

3.3.2 SENSITIVITY

The counter-ion is intended to give the sample components appropriate migration rates on the column, but it also offers the possibility of increasing the response to the chromatographic detector.

Photometric detectors have been widely used for some time because of their high stability and sensitivity. In ion-pair chromatography a high detector response can be obtained even for compounds that on their own do not absorb strongly, by selecting a counter-ion of high molar absorbance at the measured wavelength. Figure 11 shows how the detector response for anionic compounds is increased by this technique. Three carboxylic acids are separated as ion-pairs on a column with quaternized protriptyline as counter-ion in the stationary

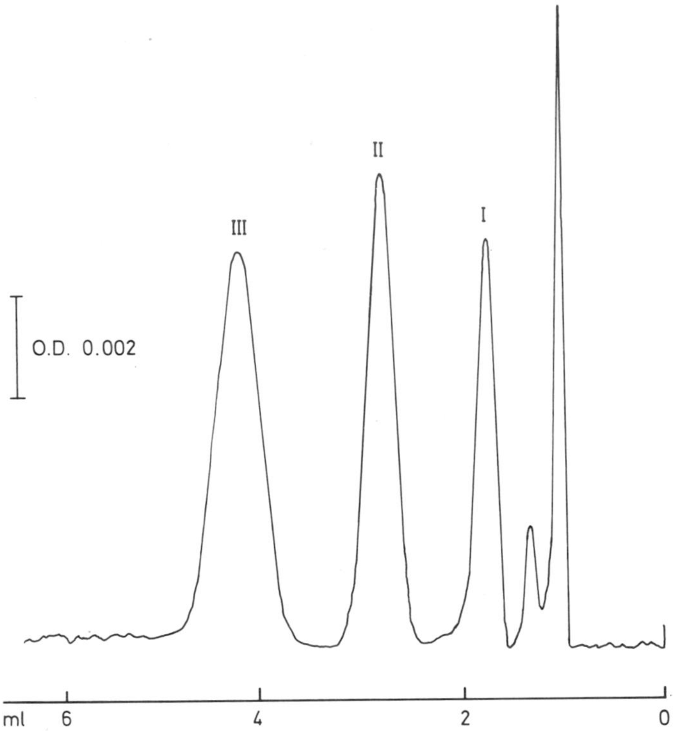

Fig. 11 Separation of benzilic acid (I), phenylbutyric acid (II) and salicylic acid (III).[46] Support—cellulose (37–74 μm); stationary phase—0.036 M *N,N*-dimethylprotriptyline (pH 9.0); mobile phase—cyclohexane : $CHCl_3$: 1-pentanol (15 : 4 : 1); sample—I, 0.7 nmol; II, 1.1 nmol; III, 1.4 nmol.

phase.[46] The ion-pairs have a molar absorbance of $10^{3.5}$ at 254 nm and $10^{4.1}$ at 293 nm, permitting the determination of less than 25 ng of each.

Picrate, with a molar absorbance of 10^4 at 254 nm, has been used as counter-ion in the separation of cationic compounds such as quaternary ammonium ions. Non-absorbing acetyl choline and choline extracted from biological samples have been isolated by liquid chromatography on picrate columns.[12,13] In these studies the chromatographic detector was not used for quantitative evaluation, and sensitivity was only moderate. The high sensitivity that can be achieved by the combination of a strongly absorbing counter-ion like picrate with a modern photometric detector is demonstrated in Fig. 12, showing a chromatographic peak corresponding to 10 ng of acetyl choline.[1] A bioanalytical method in the same concentration range has recently been developed.[47] Picrate is not very useful as a counter-ion for the separation of the cations given by amines, since its partition into chloroform, methylene chloride and similar solvents as free picric acid is too high at pH values below 6. A lower pH is, however, often necessary in the ion-pair separation of amines if

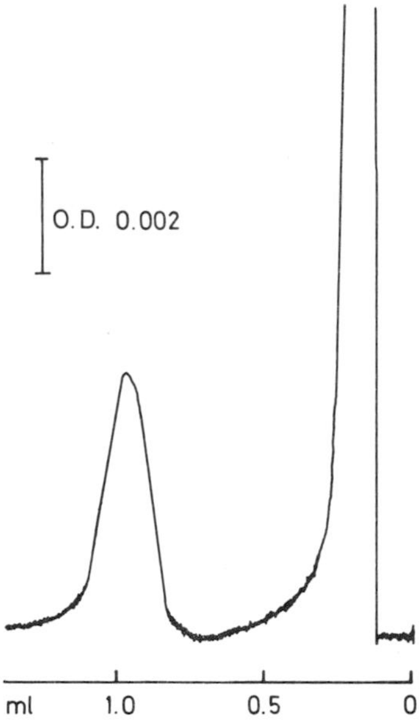

Fig. 12 Picrate ion-pair of acetyl choline. Support—cellulose; mobile phase—chloroform: pentanol (19 : 1); stationary phase—0.06 M picrate in buffer pH 6.5; sample—acetyl choline 10 ng as dipicrylamine ion-pair; displacer—tetrabutylammonium-picrate 1 nmol.

base-partition is to be avoided. In such cases sulfonic acids of high absorbance are better suited as anion-forming agents. An example is given in Fig. 13, where alkylamines are separated as naphthalene-2-sulfonate ion-pairs with a molar absorbance of $10^{3.5}$ at 254 nm permitting the detection of 15 ng of the non-absorbent cations.[14]

The technique can also be modified by making the transformation to a highly absorbing ion-pair in a separate column connected at the end of the separation column. Eksborg separated alkylammonium ions as chloride ion-pairs, and exchanged chloride for naphthalene-2-sulfonate in a column fitted just before the detector.[48]

A limited number of counter-ions with high absorbances have so far found use in ion-pair partition chromatography. To exploit the possibilities of this particular technique still further, counter ions will have to be tailored to obtain high solubility, suitable protolytic and ion-pair extraction properties and high optical absorbance. However, compromises will no doubt be necessary, since for example, high absorbance is often accompanied by low solubility.

3.3.3 MOBILE PHASE

In ion-pair extraction of ammonium compounds, pure methylene chloride and chloroform have been extensively used. A high selectivity is in many cases combined with good extraction ability and a relative absence of side-reactions.

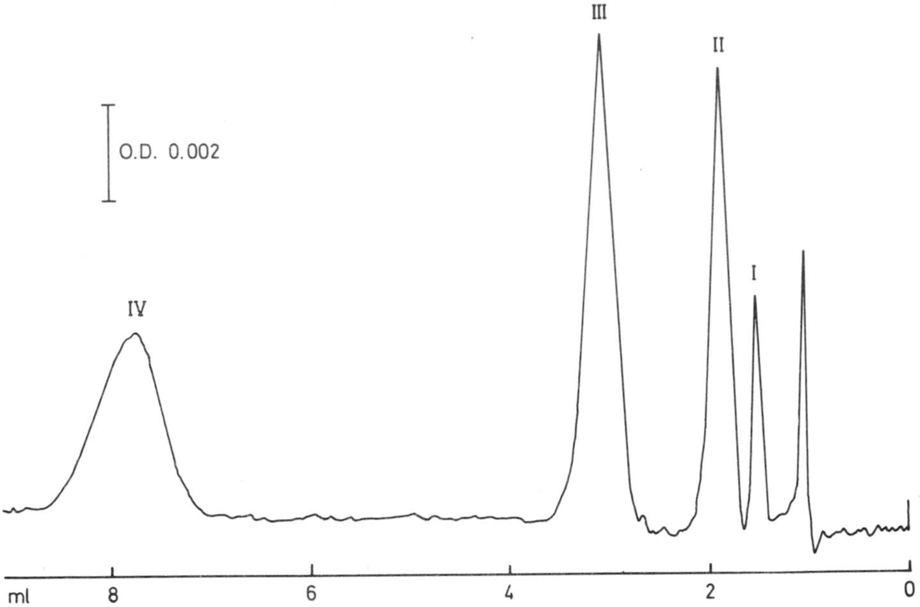

Fig. 13 Separation of primary aliphatic amines.[14] Support—cellulose (37–74 μm); stationary phase—0.1 M naphthalene-2-sulfonate; mobile phase—chloroform: pentanol (9:1) I, Hexylamine; II, Pentylamine; III, Butylamine; IV, Propylamine.

In liquid–liquid chromatography, small additions of a lipophilic alcohol have been found to be helpful, since the chromatographic behaviour of the ion-pairs is improved, as shown by improved peak symmetry and better agreement between observed and calculated retention parameters.[1,14] The addition of alcohol is usually limited to 5–10%, in order to maintain the high selectivity of the liquid–liquid system (see Fig. 4). The addition of alcohol to alkyl chloride solvents results in a decrease in capacity factors, while an increase can be obtained by substituting hexane for some of the alkyl chloride. When the composition of the mobile phase is made up of three organic solvents of different properties, a very versatile separation system is created. The same ion-pair column can then be used for sample components with a wide range of partition properties. Chloride as counter-ion in the stationary phase has accordingly been used both in separation of the hydrophobic tricyclic amines[29] and also for the hydrophilic amino acids.[32] In the first study dilution with nine parts of cyclohexane is used as the organic mobile phase,[29] and in the latter, pure pentanol was used.[32] An increased content of alcohol or other polar solvent in the mobile phase decreases the selectivity, but also rapidly increases the extraction of the counter-ion as ion-pair with the ions in the buffer solution.[30,44] A high blank extraction is particularly damaging to systems with counter-ions of high detector response, and the alcohol content must then be kept to a minimum, in order to maintain a stable base-line.

3.3.4 SELECTIVITY

High separation factors are often obtained in ion-pair partition chromatography, and such systems have found wide use in the chromatographic isolation of drugs and related compounds from complex biological matrices. An illustration of the kind of selectivity that can be achieved is given in Fig. 13, which shows the separation of homologous primary amines. The separation factor between two successive homologues is about three, which agrees with the results of batch extractions.[16] Small structural differences between two components are often enough to give complete separation. Dihydroquinidine is present as an impurity in quinidine samples, having an ethyl instead of a vinyl group. In spite of this small difference, a separation factor of 1.4 is obtained when separated as perchlorate ion-pairs with methylene chloride : butanol : hexane (6: 1: 3) as mobile phase (Fig. 14).

When the differences in structure comprise polar substituents, particularly good selectivity can be expected. Three positional isomers of monohydroxylated alprenolol have been separated as ion-pairs with perchlorate (Fig. 15). The separation factor is about 4 between the 6-hydroxy and 5-hydroxy compounds, and 1.5 between the 5- and 4-isomers. Intramolecular hydrogen bonding is a probable explanation for the higher hydrophobicity of the 6-hydroxy isomer.

In drug metabolism, aromatic hydroxylation and dealkylation of alkylamino groups are two major routes, both of which lead to high separation factors in ion-pair partition chromatography. With cyclohexane : 1-pentanol as the mobile phase, the separation factor is about 5 between the amine, alprenolol, and the de-isopropyl derivative, and the corresponding value is about 50 in relation to the 4-hydroxy derivative.[2] A chromatographic separation of these three compounds is shown in Fig. 16.

Strongly hydrophilic compounds require ion-pair systems of high solvating ability. For acetyl choline and choline, a hydrophobic counter-ion, picrate, together with chloroform or methylene chloride as mobile phase, is a highly selective separation system.[12,13] Extracting ability of the same sort of magnitude can be obtained with less hydrophobic counter-ions if the organic phase has a high content of strongly hydrogen-bonding agents such as butanol or tributyl phosphate. This kind of system offers poor selectivity, but compounds with large structural differences are in many cases still separated if the chromatographic columns are efficient. Figure 8 shows the separation of some biogenic amines derived from phenylethylamine, as ion-pairs with perchlorate in a mixture of methylene chloride and butanol as the mobile phase. A similar phase-system has been used for the separation of tryptamine, tryptophan and serotonin, as can be seen in Fig. 17, where tyramine and dopamine are also included.[3]

Halogen substitution into aromatic rings also gives large separation factors.[49] This is illustrated by the separation of tyrosine from its mono- and

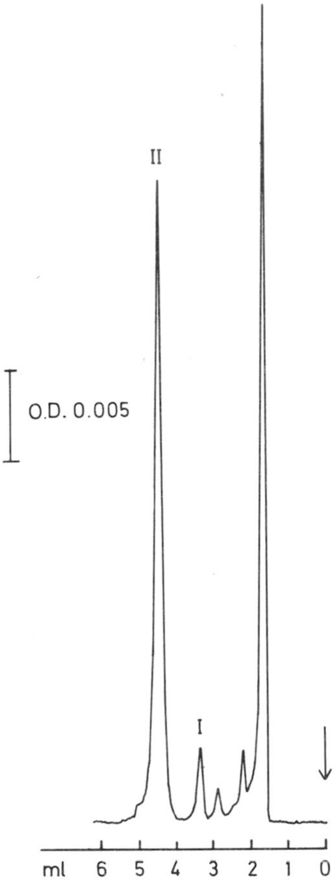

Fig. 14.

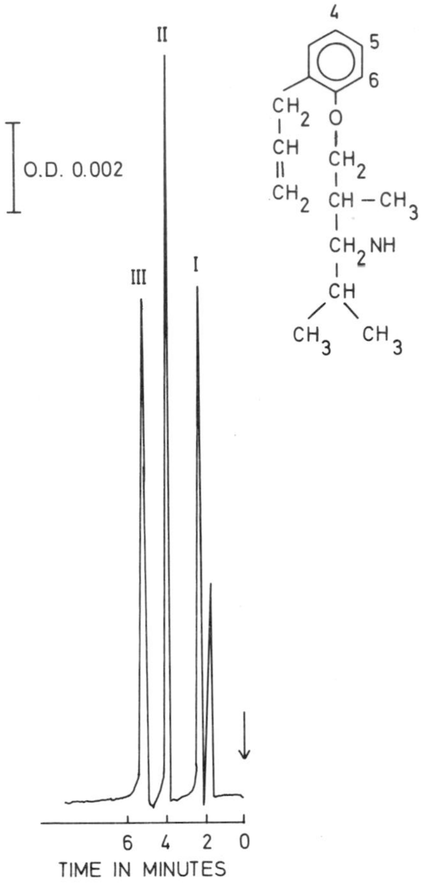

Fig. 15.

Fig. 14 Separation of dihydroquinidine (I) and quinidine (II) from plasma extract. Support—LiChrosorb SI 100 (mean diam. 10 μm); stationary phase—0.2 M $HClO_4$ + 0.8 M $NaClO_4$; mobile phase—butanol : methylene chloride : hexane (10 : 60 : 30); sample—extract from authentic sample containing 400 ng of quinidine in 1 ml of plasma.

Fig. 15 Separation of positional isomer of hydroxyalprenolol as perchlorate ion-pairs. Support—Partisil 10 (mean diam. 10 μm); stationary phase—0.2 M $HClO_4$ + 0.8 M $NaClO_4$; mobile phase—1,2-dichloroethane:1-butanol (95:5); I, 6-hydroxyalprenolol; II, 5-hydroxyalprenolol; III, 4-hydroxyalprenolol.

diiodo derivatives in Fig. 18. The amino acids are present mainly as cations in the aqueous phase under the conditions used, and are migrating as perchlorate ion-pairs.[3]

There are a few examples of reversed-phase ion-pair partition chromatography. The technique is attractive, since in most applications in the analytical

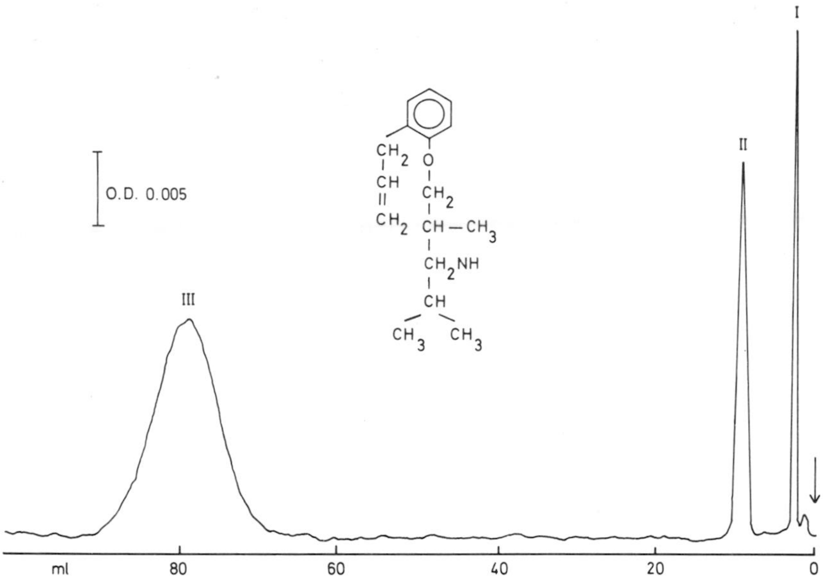

Fig. 16 Separation of alprenolol (I), de-isopropylalprenolol (II) and 4-hydroxyalprenolol (III).[2] Support—cellulose (37–74 μm); stationary phase—0.1 M $HClO_4 + 0.9$ M $NaClO_4$; mobile phase—cyclohexane: 1-pentanol (93:7).

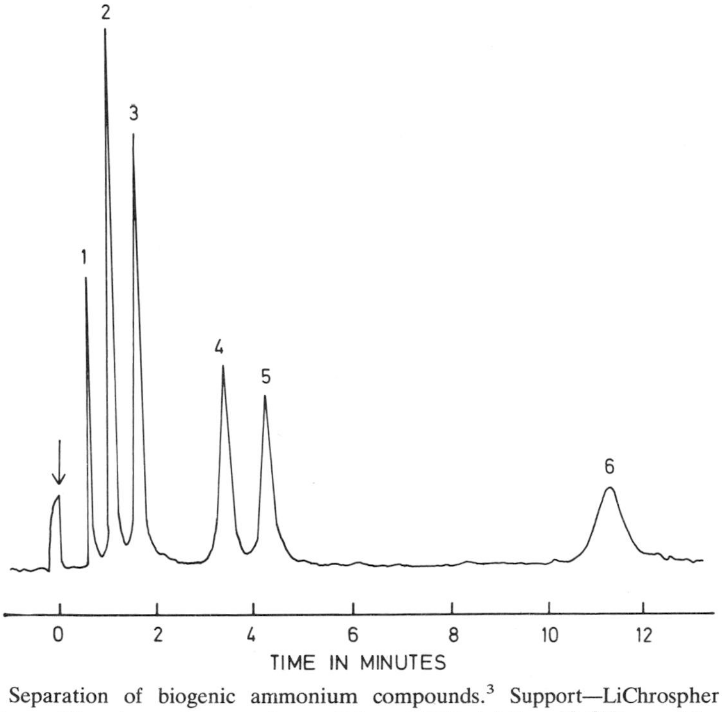

Fig. 17 Separation of biogenic ammonium compounds.[3] Support—LiChrospher SI 100 (mean diam. 10 μm), stationary phase—0.2 M $HClO_4 + 0.8$ M $NaClO_4$ mobile phase—Butanol: methylene chloride (20:80); 1, toluene; 2, tryptamine; 3, tryptophan; 4, tyramine; 5, serotonin; 6, dopamine.

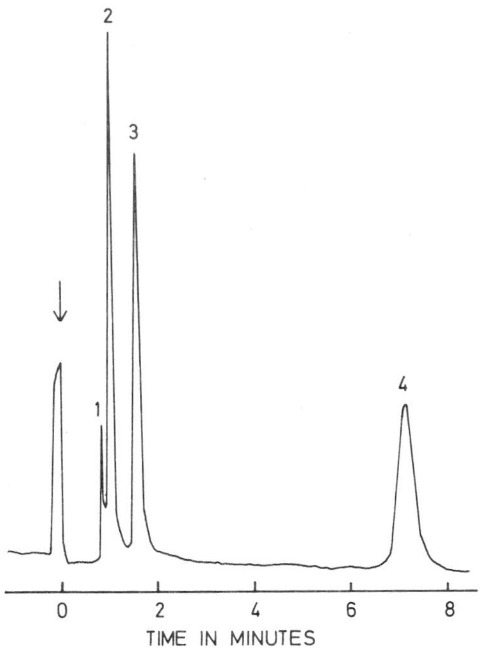

Fig. 18 Separation of tyrosine and its iododerivatives.[45] Support—LiChrospher SI 100 (mean diam. 10 μm); stationary phase—0.2 M $HClO_4 + 0.8$ M $NaClO_4$; mobile phase—butanol : methylene chloride (30 : 70); 1, toluene; 2, diiodotyrosine; 3, monoiodotyrosine; 4, tyrosine.

field the samples are in aqueous solution to start with. The reversed-phase systems are particularly useful in the separation of hydrophilic compounds which are difficult to extract into an organic phase. As discussed above, the chromatographic systems are similar to the conventional ones, except that an increased hydrophobicity and concentration of the counter-ion in the mobile phase give an increased retention and higher capacity factors.

Lipophilic alcohols have been used as stationary phases in studies by Wahlund *et al.* Amines of different kinds were separated with inorganic anions as counter-ions in the mobile phase.[37] In a recent investigation, organic acids were chromatographed with good separation efficiency as ion-pairs with tetrabutylammonium.[36] A separation of five benzoic acids and toluene sulfonic acid is shown in Fig. 19. Another reversed-phase ion-pair system for the separation of organic acids was proposed by Kraak,[38] who used trioctylamine as the stationary phase and aqueous perchloric acid solutions as eluents. An application of ion-pair and adduct formation was also given. A solution of bis-(ethylhexyl)phosphoric acid in chloroform was used as a stationary phase, and phenylethylamine derivatives were eluted with aqueous buffer solutions.[46]

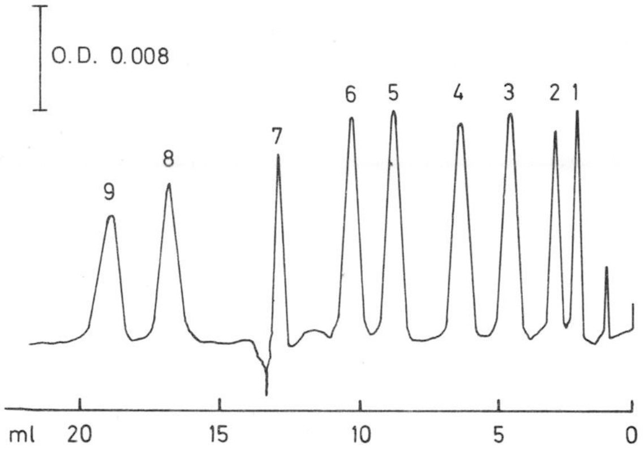

Fig. 19 Reversed phase ion-pair partition chromatography with gradient elution.[36] Support—LiChrosorb SI 60 silanized (mean diam. 30 μm); stationary phase—1-pentanol; mobile phosphate buffer pH 7.4 with a linear tetrabutylammonium gradient from 0.1 M to zero in the first 5 ml (0.4 ml min^{-1}). 1, 4-aminobenzoic acid; 2, 3-aminobenzoic acid; 3, 4-hydroxybenzoic acid; 4, 3-hydroxybenzoic acid; 5, benzenesulfonic acid; 6, benzoic acid; 7, toluene-4-sulfonic acid; 8, 2,4-dimethylbenzenesulfonic acid; 9, 2-hydroxybenzoic acid.

4 CONCLUSION

The examples presented give some indication of the potential of ion-pair extraction both for isolation and applied as ion-pair partition chromatography. Since isolation, particularly from complex biological matrices, is often a difficult part of any analytical method, some effort spent on mastering the principles of ion-pair extraction is likely to be a good investment. Although these methods have been applied to a limited extent as yet, it is clear that ion-pair partition chromatography will play an increasing role in liquid–liquid partition chromatography using the high-performance liquid chromatographs now becoming widely available. The versatility of such systems means that many different separations can be achieved with only a limited number of columns.

APPENDIX

Detailed theoretical treatment of ion-pair equilibria

Equation (16) expresses the fundamental equilibrium of ion-pair extraction, but in many cases the ions and/or the extracted ion-pair can participate in other equilibria which have to be taken into account in the calculations necessary to arrive at the optimum extraction conditions.

$$Q^+_{aq} + X^-_{aq} \rightleftharpoons QX_{org} \qquad (16)$$

The equilibrium of Eqn (16) is characterized by the extraction constant E_{QX}, which is defined by Eqn (17).

$$E_{QX} = \frac{[QX]_{org}}{[Q^+][X^-]} \tag{17}$$

In cases where the situation is affected by other equilibria, a conditional extraction constant E^*_{QX}, can be used to express the overall equilibrium; this is shown in Eqn (18).

$$E^*_{QX} = \frac{c'_{QXorg}}{c'_Q c'_X} \tag{18}$$

In Eqn (18) c'_{QXorg} is the concentration of Q^+ extracted into the organic phase *only* as an ion pair with X^-. The expressions c'_Q and c'_X represent the remaining concentration that has not been extracted as an ion pair; this includes any Q^+ and X^- involved in side equilibria. The conditional extraction constant, E^*_{QX}, can be related to the stoichiometric constant E_{QX} by introduction of α-coefficients[20] as shown in Eqn (19).

$$E^*_{QX} = \frac{E_{QX}\alpha_{QX}}{\alpha_Q \alpha_X} \tag{19}$$

The α-coefficients are the ratio between the primed and stoichiometric concentrations, as shown in Eqn (20).

$$\alpha_{QX} = \frac{c'_{QXorg}}{[QX]_{org}} \tag{20}$$

When the ion-pair components are weak protolytes, they can be subjected to protolysis in the aqueous phase and also be transferred to the organic phase in unchanged form. For example, considering the anion X^- which is the anion of a weak acid and as such is partly associated in the aqueous phase, equilibrium (21) holds.

$$X^- + H^+ \underset{K'_{HX}}{\rightleftharpoons} HX \overset{k_{d(HX)}}{\rightleftharpoons} HX_{org} \tag{21}$$

In these equilibria, K'_{HX} is the apparent acid dissociation constant and $k_{d(HX)}$ is the partition coefficient between the two phases. These equilibria will affect α_X only and not α_Q or α_{QX}. Consequently Eqn (19) reduces to Eqn (22) because the ratio of α_Q to α_{QX} is equal to unity.

$$E^*_{QX} = \frac{E_{QX}}{\alpha_X} \tag{22}$$

The α-coefficient α_X is thus the total concentration of non-ion-pair X divided by $[X^-]$ as shown in Eqn (23).

$$\alpha_X = \frac{[X^-] + [HX] + [HX]_{org}}{[X^-]} \tag{23}$$

The following manipulations give an expression for α_X:

from (23)

$$\alpha_X = 1 + \frac{[HX]}{[X^-]} + \frac{[HX]_{org}}{[X^-]} \tag{24}$$

but by definition

$$K'_{HX} = \frac{[H^+][X^-]}{[HX]} \tag{25}$$

so eliminating $[X^-]$ from (24) and (25):

$$\alpha_X = 1 + \frac{[H^+]}{K'_{HX}} + \frac{[HX]_{org}[H^+]}{[HX]K'_{HX}} \tag{26}$$

but,

$$k_{d(HX)} = \frac{[HX]_{org}}{[HX]} \quad \text{by definition} \tag{27}$$

so eliminating $[HX]_{org}$ and $[HX]$ from (26) and (27):

$$\alpha_X = 1 + \frac{[H^+]}{K'_{HX}}(1 + k_{d(HX)}) \tag{28}$$

Expression (28) is true provided that the aqueous and organic phase volumes are equal.

From Eqn (28) it may be seen that the conditional extraction constant E^*_{QX}, which is dependent upon α_X is also dependent upon hydrogen ion concentration (pH) and this will affect the partition ratio of Q^+. The partition ratio of Q^+ is called D_Q and defined as in Eqn (29).

$$D_Q = \frac{c'_{QX,org}}{c'_Q} \tag{29}$$

Eliminating c'_{QXorg} and c'_Q from Eqn (18) and rearranging:

$$D_Q = E^*_{QX} c'_x \tag{30}$$

The expression c'_X is itself related to α_X and so D_Q is dependent upon α_X and the pH of the system.

All processes which increase α_X and/or α_Q will decrease the value of E^*_{QX} and therefore D_Q; this in turn reduces the extraction efficiency. Protolysis, association in the aqueous phase, polymerization of Q^+ and micelle formation are such processes.[17,21] Similarly other equilibria in the organic phase will tend to increase α_{QX}, thus increasing c'_{QX} and thence D_Q from Eqn (29); this will improve the extraction efficiency. Dimerization and dissociation of the ion-pair in the organic phase are examples of such processes.

REFERENCES

1. S. Eksborg and G. Schill, *Anal. Chem.* **45**, 2092 (1973).
2. K. O. Borg, M. Gabrielsson and T.-E. Jönsson, *Acta Pharm. Suec.* **11**, 313 (1974).
3. B.-A. Persson and B. Karger, *J. Chromatogr. Sci.* **12**, 521 (1974).
4. K. O. Borg and A. Mikaelsson, *Acta Pharm. Suec.* **7**, 673 (1970).
5. K. Gröningsson and G. Schill, *Acta Pharm. Suec.* **6**, 447 (1969).
6. K. Gröningsson, H. Westerlind and R. Modin, *Acta Pharm. Suec.* **12**, 97 (1975).
7. G. Schill, *Acta Pharm. Suec.* **2**, 13 (1965).
8. G. Schill, *Anal. Chim. Acta* **21**, 341 (1959).
9. K. Gustavii and G. Schill, *Acta Pharm. Suec.* **3**, 241 (1966).
10. K. O. Borg and D. Westerlund, *Z. Anal. Chem.* **252**, 275 (1970).
11. D. Westerlund and K. O. Borg, *Anal. Chim. Acta* **67**, 89 (1973).
12. S. Eksborg and B.-A. Persson, *Acta Pharm. Suec.* **8**, 205 (1971).
13. S. Eksborg and B.-A. Persson, *Acta Pharm. Suec.* **8**, 605 (1971).
14. S. Eksborg, *Acta Pharm. Suec.* **12**, 19 (1975).
15. P.-O. Lagerström, *Acta Pharm. Suec.* **12**, 215 (1975).
16. K. Gustavii, *Acta Pharm. Suec.* **4**, 233 (1967).
17. R. Modin and G. Schill, *Acta Pharm. Suec.* **4**, 301 (1967).
18. R. Modin and A. Tilly, *Acta Pharm. Suec.* **5**, 311 (1968).
19. R. Modin and M. Schröder-Nielsen, *Acta Pharm. Suec.* **8**, 573 (1971).
20. A. Ringbom, *Complexation in Analytical Chemistry*, Wiley, New York, 1963.
21. G. Schill, *Acta Pharm. Suec.* **2**, 109 (1965).
22. R. Modin, B.-A. Persson and G. Schill, *Proceedings of the International Solvent Extraction Conference ISEC 71*, Vol. II, Society of Chemical Industry, London 1971, p. 1211.
23. P.-O. Lagerström, K. O. Borg and D. Westerlund, *Acta Pharm. Suec.* **9**, 53 (1972).
24. R. Modin, *Acta Pharm. Suec.* **8**, 509 (1971).
25. R. Modin and M. Johansson, *Acta Pharm. Suec.* **8**, 561 (1971).
26. R. Modin and G. Schill, to be published.
27. R. Modin and S. Bäck, *Acta Pharm. Suec.* **8**, 585 (1971).
28. T. Higuchi, A. Michaelis, T. Tan and A. Hurwitz, *Anal. Chem.* **39**, 974 (1967).
29. B.-A. Persson, *Acta Pharm. Suec.* **5**, 343 (1968).
30. K. O. Borg, *Acta Pharm. Suec.* **6**, 425 (1969).
31. S. S. Davis, *J. Pharm. Pharmacol.* **25**, 1 (1973).
32. B.-A. Persson, *Acta Pharm. Suec.* **8**, 193 (1971).
33. J. J. Kirkland (Ed.), *Modern Practice of Liquid Chromatography*, Wiley-Interscience, New York, 1971.
34. L. R. Snyder and J. J. Kirkland, *Introduction to Modern Liquid Chormatography*, Wiley–Interscience, New York, 1974.
35. S. Eksborg and G. Schill, *Acta Pharm. Suec.* **12**, 1 (1975).
36. K. G. Wahlund, *J. Chromatogr.* **115**, 411 (1975).
37. K. G. Wahlund and K. Gröningsson, *Acta Pharm. Suec.* **7**, 615 (1970).
38. J. C. Kraak and J. F. K. Huber, *J. Chromatogr.* **102**, 333 (1974).
39. K. Gröningsson and M. Weimers, *Acta Pharm. Suec.* **12**, 97 (1975).
40. K. Gröningsson, *Acta Pharm. Suec.* **7**, 635 (1970).
41. P.-O. Lagerström and A. Theodorsen, *Acta Pharm. Suec.* **12**, 429 (1975).
42. J. H. Knox and J. Jurand, *J. Chromatogr.* **83**, 405 (1973).
43. P.-O. Lagerström, I. Carlsson and B.-A. Persson, *Acta Pharm. Suec.* **13**, 157 (1976).
44. A. Tilly, *Acta Pharm. Suec.* **10**, 111 (1973).
45. B. Karger, S. Su, S. Marchese and B.-A. Persson, *J. Chromatogr. Sci.* **12**, 678 (1974).
46. S. Eksborg, P.-O. Lagerström, R. Modin and G. Schill, *J. Chromatogr.* **83**, 99 (1973).

47. B. Ulin, K. Gustavii and B.-A. Persson, *J. Pharm. Pharmacol.* **28**, 672 (1976).
48. S. Eksborg, *Acta Pharm. Suec.* **12**, 243 (1975).
49. B.-A. Persson, *Acta Pharm. Suec.* **5**, 335 (1968).

Index

Numbers in bold print indicate pages on which the topic is discussed in depth, or on which experimental details are given. Where a chapter states that a derivative has been prepared, gives no details but cites the original paper, this index gives the page number followed by the number of the page on which the full reference is given and then in parentheses the number of that reference. This will enable the reader to proceed directly from the index to the original reference.

In order to keep the length of this necessarily full index within reasonable bounds while still retaining the maximum amount of useful information, some compromises have been made with certain indexing practices. No references to gas chromatography as such are included as these occur on the majority of pages. Similarly there is no cross-referencing under the heading 'Derivatization of—' of compounds for which derivatives have been prepared, although derivatization is mentioned under the individual compound names. Where several sub-headings of an entry refer to the same page or pages these are grouped together rather than quoted separately e.g.

3,4-Dimethoxyphenylisopropylamine, deriv, detect, sep, 258

The abbreviations used in this index, although largely self-explanatory, are listed below. There is a full list of abbreviations for compounds and for processes on p. xv at the front of the book. The applications of compounds as catalysts, oxidising agents, resolving agents etc. are not included under the individual compound names but these names are listed under the general application headings 'Catalysts' etc. Where a variety of names is used to refer to a chemical or subject then detailed indexing is only included for one of the names, but all of the others are cross-referenced to it.

Abbreviations used in this Index

cmpd	Compound	prot	Protection of
config	Configuration, assignment of	purif	Purification of
deriv	Derivatization of	pyrol	Pyrolysis of
detect	Detection of	react	Reactivity of
determ	Determination of	reduct	Reduction of
diff	Differentiation of	regen	Regeneration of
elut	Elution order	remov	Removal of (generally) excess of
extr	Extraction of	resol	Resolution of
ident	Identification of	sep	Separation of
interf	Interference from	subtr	Subtraction of
opt	Optical purity of		
prep	Preparation of		

C

D

E

G

H

I

K

N

O

Q

R

S

U

Y

Z